STUDENT'S SOLUTIONS MANUAL

MULTIVARIABLE

WILLIAM ARDIS
Collin County Community College

THOMAS' CALCULUS

TWELFTH EDITION

BASED ON THE ORIGINAL WORK BY

George B. Thomas, Jr.
Massachusetts Institute of Technology

AS REVISED BY

Maurice D. Weir
Naval Postgraduate School

Joel Hass
University of California, Davis

Addison-Wesley
is an imprint of

The author and publisher of this book have used their best efforts in preparing this book. These efforts include the development, research, and testing of the theories and programs to determine their effectiveness. The author and publisher make no warranty of any kind, expressed or implied, with regard to these programs or the documentation contained in this book. The author and publisher shall not be liable in any event for incidental or consequential damages in connection with, or arising out of, the furnishing, performance, or use of these programs.

Reproduced by Addison-Wesley from electronic files supplied by the author.

Publishing as Pearson Addison-Wesley, 75 Arlington Street, Boston, MA 02116.

ISBN-13: 978-0-321-60071-4
ISBN-10: 0-321-60071-1

1 2 3 4 5 6 BB 14 13 12 11 10

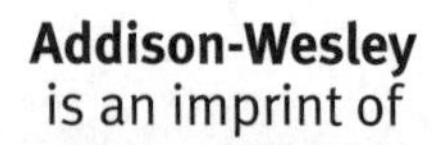

www.pearsonhighered.com

PREFACE TO THE STUDENT

The Student's Solutions Manual contains the solutions to all of the odd-numbered exercise in the 12th Edition of THOMAS' CALCULUS by Maurice Weir and Joel Hass, excluding the Computer Algebra System (CAS) exercises. We have worked each solution to ensure that it

- conforms exactly to the methods, procedures and steps presented in the text
- is mathematically correct
- includes all of the steps necessary so you can follow the logical argument and algebra
- includes a graph or figure whenever called for by the exercise, or if needed to help with the explanation
- is formatted in an appropriate style to aid in its understanding

How to use a solution's manual

- solve the assigned problem yourself
- if you get stuck along the way, refer to the solution in the manual as an aid but continue to solve the problem on your own
- if you cannot continue, reread the textbook section, or work through that section in the Student Study Guide, or consult your instructor
- if your answer is correct by your solution procedure seems to differ from the one in the manual, and you are unsure your method is correct, consult your instructor
- if your answer is incorrect and you cannot find your error, consult your instructor

For more information about other resources available with Thomas' Calculus, visit http://pearsonhighered.com.

TABLE OF CONTENTS

14 Partial Derivatives 393

15 Multiple Integrals 435

16 Integration in Vector Fields 463

CHAPTER 10 INFINITE SEQUENCES AND SERIES

10.1 SEQUENCES

1. $a_1 = \frac{1-1}{1^2} = 0,\ a_2 = \frac{1-2}{2^2} = -\frac{1}{4},\ a_3 = \frac{1-3}{3^2} = -\frac{2}{9},\ a_4 = \frac{1-4}{4^2} = -\frac{3}{16}$

3. $a_1 = \frac{(-1)^2}{2-1} = 1,\ a_2 = \frac{(-1)^3}{4-1} = -\frac{1}{3},\ a_3 = \frac{(-1)^4}{6-1} = \frac{1}{5},\ a_4 = \frac{(-1)^5}{8-1} = -\frac{1}{7}$

5. $a_1 = \frac{2}{2^2} = \frac{1}{2},\ a_2 = \frac{2^2}{2^3} = \frac{1}{2},\ a_3 = \frac{2^3}{2^4} = \frac{1}{2},\ a_4 = \frac{2^4}{2^5} = \frac{1}{2}$

7. $a_1 = 1,\ a_2 = 1 + \frac{1}{2} = \frac{3}{2},\ a_3 = \frac{3}{2} + \frac{1}{2^2} = \frac{7}{4},\ a_4 = \frac{7}{4} + \frac{1}{2^3} = \frac{15}{8},\ a_5 = \frac{15}{8} + \frac{1}{2^4} = \frac{31}{16},\ a_6 = \frac{63}{32},$
$a_7 = \frac{127}{64},\ a_8 = \frac{255}{128},\ a_9 = \frac{511}{256},\ a_{10} = \frac{1023}{512}$

9. $a_1 = 2,\ a_2 = \frac{(-1)^2(2)}{2} = 1,\ a_3 = \frac{(-1)^3(1)}{2} = -\frac{1}{2},\ a_4 = \frac{(-1)^4\left(-\frac{1}{2}\right)}{2} = -\frac{1}{4},\ a_5 = \frac{(-1)^5\left(-\frac{1}{4}\right)}{2} = \frac{1}{8},$
$a_6 = \frac{1}{16},\ a_7 = -\frac{1}{32},\ a_8 = -\frac{1}{64},\ a_9 = \frac{1}{128},\ a_{10} = \frac{1}{256}$

11. $a_1 = 1,\ a_2 = 1,\ a_3 = 1 + 1 = 2,\ a_4 = 2 + 1 = 3,\ a_5 = 3 + 2 = 5,\ a_6 = 8,\ a_7 = 13,\ a_8 = 21,\ a_9 = 34,\ a_{10} = 55$

13. $a_n = (-1)^{n+1},\ n = 1, 2, \ldots$

15. $a_n = (-1)^{n+1} n^2,\ n = 1, 2, \ldots$

17. $a_n = \frac{2^{n-1}}{3(n+2)},\ n = 1, 2, \ldots$

19. $a_n = n^2 - 1,\ n = 1, 2, \ldots$

21. $a_n = 4n - 3,\ n = 1, 2, \ldots$

23. $a_n = \frac{3n+2}{n!},\ n = 1, 2, \ldots$

25. $a_n = \frac{1+(-1)^{n+1}}{2},\ n = 1, 2, \ldots$

27. $\lim_{n\to\infty} 2 + (0.1)^n = 2 \Rightarrow$ converges (Theorem 5, #4)

29. $\lim_{n\to\infty} \frac{1-2n}{1+2n} = \lim_{n\to\infty} \frac{\left(\frac{1}{n}\right)-2}{\left(\frac{1}{n}\right)+2} = \lim_{n\to\infty} \frac{-2}{2} = -1 \Rightarrow$ converges

31. $\lim_{n\to\infty} \frac{1-5n^4}{n^4+8n^3} = \lim_{n\to\infty} \frac{\left(\frac{1}{n^4}\right)-5}{1+\left(\frac{8}{n}\right)} = -5 \Rightarrow$ converges

33. $\lim_{n\to\infty} \frac{n^2-2n+1}{n-1} = \lim_{n\to\infty} \frac{(n-1)(n-1)}{n-1} = \lim_{n\to\infty} (n-1) = \infty \Rightarrow$ diverges

35. $\lim_{n\to\infty} (1 + (-1)^n)$ does not exist $\Rightarrow$ diverges

37. $\lim_{n\to\infty} \left(\frac{n+1}{2n}\right)\left(1 - \frac{1}{n}\right) = \lim_{n\to\infty} \left(\frac{1}{2} + \frac{1}{2n}\right)\left(1 - \frac{1}{n}\right) = \frac{1}{2} \Rightarrow$ converges

39. $\lim_{n\to\infty} \frac{(-1)^{n+1}}{2n-1} = 0 \Rightarrow$ converges

41. $\lim_{n\to\infty} \sqrt{\frac{2n}{n+1}} = \sqrt{\lim_{n\to\infty} \frac{2n}{n+1}} = \sqrt{\lim_{n\to\infty} \left(\frac{2}{1+\frac{1}{n}}\right)} = \sqrt{2} \Rightarrow$ converges

43. $\lim_{n \to \infty} \sin\left(\frac{\pi}{2} + \frac{1}{n}\right) = \sin\left(\lim_{n \to \infty} \left(\frac{\pi}{2} + \frac{1}{n}\right)\right) = \sin\frac{\pi}{2} = 1 \Rightarrow$ converges

45. $\lim_{n \to \infty} \frac{\sin n}{n} = 0$ because $-\frac{1}{n} \le \frac{\sin n}{n} \le \frac{1}{n} \Rightarrow$ converges by the Sandwich Theorem for sequences

47. $\lim_{n \to \infty} \frac{n}{2^n} = \lim_{n \to \infty} \frac{1}{2^n \ln 2} = 0 \Rightarrow$ converges (using l'Hôpital's rule)

49. $\lim_{n \to \infty} \frac{\ln(n+1)}{\sqrt{n}} = \lim_{n \to \infty} \frac{\left(\frac{1}{n+1}\right)}{\left(\frac{1}{2\sqrt{n}}\right)} = \lim_{n \to \infty} \frac{2\sqrt{n}}{n+1} = \lim_{n \to \infty} \frac{\left(\frac{2}{\sqrt{n}}\right)}{1+\left(\frac{1}{n}\right)} = 0 \Rightarrow$ converges

51. $\lim_{n \to \infty} 8^{1/n} = 1 \Rightarrow$ converges (Theorem 5, #3)

53. $\lim_{n \to \infty} \left(1 + \frac{7}{n}\right)^n = e^7 \Rightarrow$ converges (Theorem 5, #5)

55. $\lim_{n \to \infty} \sqrt[n]{10n} = \lim_{n \to \infty} 10^{1/n} \cdot n^{1/n} = 1 \cdot 1 = 1 \Rightarrow$ converges (Theorem 5, #3 and #2)

57. $\lim_{n \to \infty} \left(\frac{3}{n}\right)^{1/n} = \frac{\lim_{n\to\infty} 3^{1/n}}{\lim_{n\to\infty} n^{1/n}} = \frac{1}{1} = 1 \Rightarrow$ converges (Theorem 5, #3 and #2)

59. $\lim_{n \to \infty} \frac{\ln n}{n^{1/n}} = \frac{\lim_{n\to\infty} \ln n}{\lim_{n\to\infty} n^{1/n}} = \frac{\infty}{1} = \infty \Rightarrow$ diverges (Theorem 5, #2)

61. $\lim_{n \to \infty} \sqrt[n]{4^n n} = \lim_{n \to \infty} 4\sqrt[n]{n} = 4 \cdot 1 = 4 \Rightarrow$ converges (Theorem 5, #2)

63. $\lim_{n \to \infty} \frac{n!}{n^n} = \lim_{n \to \infty} \frac{1\cdot 2\cdot 3\cdots(n-1)(n)}{n\cdot n\cdot n\cdots n\cdot n} \le \lim_{n \to \infty} \left(\frac{1}{n}\right) = 0$ and $\frac{n!}{n^n} \ge 0 \Rightarrow \lim_{n \to \infty} \frac{n!}{n^n} = 0 \Rightarrow$ converges

65. $\lim_{n \to \infty} \frac{n!}{10^{6n}} = \lim_{n \to \infty} \frac{1}{\left(\frac{(10^6)^n}{n!}\right)} = \infty \Rightarrow$ diverges (Theorem 5, #6)

67. $\lim_{n \to \infty} \left(\frac{1}{n}\right)^{1/(\ln n)} = \lim_{n \to \infty} \exp\left(\frac{1}{\ln n} \ln\left(\frac{1}{n}\right)\right) = \lim_{n \to \infty} \exp\left(\frac{\ln 1 - \ln n}{\ln n}\right) = e^{-1} \Rightarrow$ converges

69. $\lim_{n \to \infty} \left(\frac{3n+1}{3n-1}\right)^n = \lim_{n \to \infty} \exp\left(n \ln\left(\frac{3n+1}{3n-1}\right)\right) = \lim_{n \to \infty} \exp\left(\frac{\ln(3n+1) - \ln(3n-1)}{\frac{1}{n}}\right)$

$= \lim_{n \to \infty} \exp\left(\frac{\frac{3}{3n+1} - \frac{3}{3n-1}}{\left(-\frac{1}{n^2}\right)}\right) = \lim_{n \to \infty} \exp\left(\frac{6n^2}{(3n+1)(3n-1)}\right) = \exp\left(\frac{6}{9}\right) = e^{2/3} \Rightarrow$ converges

71. $\lim_{n \to \infty} \left(\frac{x^n}{2n+1}\right)^{1/n} = \lim_{n \to \infty} x\left(\frac{1}{2n+1}\right)^{1/n} = x \lim_{n \to \infty} \exp\left(\frac{1}{n} \ln\left(\frac{1}{2n+1}\right)\right) = x \lim_{n \to \infty} \exp\left(\frac{-\ln(2n+1)}{n}\right)$

$= x \lim_{n \to \infty} \exp\left(\frac{-2}{2n+1}\right) = xe^0 = x,\ x > 0 \Rightarrow$ converges

73. $\lim_{n \to \infty} \frac{3^n \cdot 6^n}{2^{-n} \cdot n!} = \lim_{n \to \infty} \frac{36^n}{n!} = 0 \Rightarrow$ converges (Theorem 5, #6)

75. $\lim_{n \to \infty} \tanh n = \lim_{n \to \infty} \frac{e^n - e^{-n}}{e^n + e^{-n}} = \lim_{n \to \infty} \frac{e^{2n} - 1}{e^{2n} + 1} = \lim_{n \to \infty} \frac{2e^{2n}}{2e^{2n}} = \lim_{n \to \infty} 1 = 1 \Rightarrow$ converges

77. $\lim_{n \to \infty} \frac{n^2 \sin\left(\frac{1}{n}\right)}{2n-1} = \lim_{n \to \infty} \frac{\sin\left(\frac{1}{n}\right)}{\left(\frac{2}{n} - \frac{1}{n^2}\right)} = \lim_{n \to \infty} \frac{-\left(\cos\left(\frac{1}{n}\right)\right)\left(\frac{1}{n^2}\right)}{\left(-\frac{2}{n^2} + \frac{2}{n^3}\right)} = \lim_{n \to \infty} \frac{-\cos\left(\frac{1}{n}\right)}{-2 + \left(\frac{2}{n}\right)} = \frac{1}{2} \Rightarrow$ converges

79. $\lim_{n\to\infty} \sqrt{n}\sin\left(\frac{1}{\sqrt{n}}\right) = \lim_{n\to\infty} \frac{\sin\left(\frac{1}{\sqrt{n}}\right)}{\frac{1}{\sqrt{n}}} = \lim_{n\to\infty} \frac{\cos\left(\frac{1}{\sqrt{n}}\right)\left(-\frac{1}{2n^{3/2}}\right)}{-\frac{1}{2n^{3/2}}} = \lim_{n\to\infty} \cos\left(\frac{1}{\sqrt{n}}\right) = \cos 0 = 1 \Rightarrow$ converges

81. $\lim_{n\to\infty} \tan^{-1} n = \frac{\pi}{2} \Rightarrow$ converges

83. $\lim_{n\to\infty} \left(\frac{1}{3}\right)^n + \frac{1}{\sqrt{2^n}} = \lim_{n\to\infty} \left(\left(\frac{1}{3}\right)^n + \left(\frac{1}{\sqrt{2}}\right)^n\right) = 0 \Rightarrow$ converges (Theorem 5, #4)

85. $\lim_{n\to\infty} \frac{(\ln n)^{200}}{n} = \lim_{n\to\infty} \frac{200(\ln n)^{199}}{n} = \lim_{n\to\infty} \frac{200\cdot 199(\ln n)^{198}}{n} = \dots = \lim_{n\to\infty} \frac{200!}{n} = 0 \Rightarrow$ converges

87. $\lim_{n\to\infty} \left(n - \sqrt{n^2-n}\right) = \lim_{n\to\infty} \left(n - \sqrt{n^2-n}\right)\left(\frac{n+\sqrt{n^2-n}}{n+\sqrt{n^2-n}}\right) = \lim_{n\to\infty} \frac{n}{n+\sqrt{n^2-n}} = \lim_{n\to\infty} \frac{1}{1+\sqrt{1-\frac{1}{n}}}$
$= \frac{1}{2} \Rightarrow$ converges

89. $\lim_{n\to\infty} \frac{1}{n}\int_1^n \frac{1}{x}\,dx = \lim_{n\to\infty} \frac{\ln n}{n} = \lim_{n\to\infty} \frac{1}{n} = 0 \Rightarrow$ converges (Theorem 5, #1)

91. Since a_n converges $\Rightarrow \lim_{n\to\infty} a_n = L \Rightarrow \lim_{n\to\infty} a_{n+1} = \lim_{n\to\infty} \frac{72}{1+a_n} \Rightarrow L = \frac{72}{1+L} \Rightarrow L(1+L) = 72 \Rightarrow L^2 + L - 72 = 0$
$\Rightarrow L = -9$ or $L = 8$; since $a_n > 0$ for $n \ge 1 \Rightarrow L = 8$

93. Since a_n converges $\Rightarrow \lim_{n\to\infty} a_n = L \Rightarrow \lim_{n\to\infty} a_{n+1} = \lim_{n\to\infty} \sqrt{8+2a_n} \Rightarrow L = \sqrt{8+2L} \Rightarrow L^2 - 2L - 8 = 0 \Rightarrow L = -2$ or $L = 4$; since $a_n > 0$ for $n \ge 3 \Rightarrow L = 4$

95. Since a_n converges $\Rightarrow \lim_{n\to\infty} a_n = L \Rightarrow \lim_{n\to\infty} a_{n+1} = \lim_{n\to\infty} \sqrt{5a_n} \Rightarrow L = \sqrt{5L} \Rightarrow L^2 - 5L = 0 \Rightarrow L = 0$ or $L = 5$; since $a_n > 0$ for $n \ge 1 \Rightarrow L = 5$

97. $a_{n+1} = 2 + \frac{1}{a_n}$, $n \ge 1$, $a_1 = 2$. Since a_n converges $\Rightarrow \lim_{n\to\infty} a_n = L \Rightarrow \lim_{n\to\infty} a_{n+1} = \lim_{n\to\infty}\left(2 + \frac{1}{a_n}\right) \Rightarrow L = 2 + \frac{1}{L}$
$\Rightarrow L^2 - 2L - 1 = 0 \Rightarrow L = 1 \pm \sqrt{2}$; since $a_n > 0$ for $n \ge 1 \Rightarrow L = 1 + \sqrt{2}$

99. $1, 1, 2, 4, 8, 16, 32, \dots = 1, 2^0, 2^1, 2^2, 2^3, 2^4, 2^5, \dots \Rightarrow x_1 = 1$ and $x_n = 2^{n-2}$ for $n \ge 2$

101. (a) $f(x) = x^2 - 2$; the sequence converges to $1.414213562 \approx \sqrt{2}$
(b) $f(x) = \tan(x) - 1$; the sequence converges to $0.7853981635 \approx \frac{\pi}{4}$
(c) $f(x) = e^x$; the sequence $1, 0, -1, -2, -3, -4, -5, \dots$ diverges

103. (a) If $a = 2n + 1$, then $b = \left\lfloor \frac{a^2}{2} \right\rfloor = \left\lfloor \frac{4n^2+4n+1}{2} \right\rfloor = \left\lfloor 2n^2 + 2n + \frac{1}{2} \right\rfloor = 2n^2 + 2n$, $c = \left\lceil \frac{a^2}{2} \right\rceil = \left\lceil 2n^2 + 2n + \frac{1}{2} \right\rceil$
$= 2n^2 + 2n + 1$ and $a^2 + b^2 = (2n+1)^2 + (2n^2+2n)^2 = 4n^2 + 4n + 1 + 4n^4 + 8n^3 + 4n^2$
$= 4n^4 + 8n^3 + 8n^2 + 4n + 1 = (2n^2 + 2n + 1)^2 = c^2$.
(b) $\lim_{a\to\infty} \frac{\left\lfloor \frac{a^2}{2} \right\rfloor}{\left\lceil \frac{a^2}{2} \right\rceil} = \lim_{a\to\infty} \frac{2n^2+2n}{2n^2+2n+1} = 1$ or $\lim_{a\to\infty} \frac{\left\lfloor \frac{a^2}{2} \right\rfloor}{\left\lceil \frac{a^2}{2} \right\rceil} = \lim_{a\to\infty} \sin\theta = \lim_{\theta\to\pi/2} \sin\theta = 1$

105. (a) $\lim_{n\to\infty} \frac{\ln n}{n^c} = \lim_{n\to\infty} \frac{\left(\frac{1}{n}\right)}{cn^{c-1}} = \lim_{n\to\infty} \frac{1}{cn^c} = 0$
(b) For all $\epsilon > 0$, there exists an $N = e^{-(\ln\epsilon)/c}$ such that $n > e^{-(\ln\epsilon)/c} \Rightarrow \ln n > -\frac{\ln\epsilon}{c} \Rightarrow \ln n^c > \ln\left(\frac{1}{\epsilon}\right)$
$\Rightarrow n^c > \frac{1}{\epsilon} \Rightarrow \frac{1}{n^c} < \epsilon \Rightarrow \left|\frac{1}{n^c} - 0\right| < \epsilon \Rightarrow \lim_{n\to\infty} \frac{1}{n^c} = 0$

107. $\lim_{n\to\infty} n^{1/n} = \lim_{n\to\infty} \exp\left(\frac{1}{n}\ln n\right) = \lim_{n\to\infty} \exp\left(\frac{1}{n}\right) = e^0 = 1$

109. Assume the hypotheses of the theorem and let ϵ be a positive number. For all ϵ there exists a N_1 such that when $n > N_1$ then $|a_n - L| < \epsilon \Rightarrow -\epsilon < a_n - L < \epsilon \Rightarrow L - \epsilon < a_n$, and there exists a N_2 such that when $n > N_2$ then $|c_n - L| < \epsilon \Rightarrow -\epsilon < c_n - L < \epsilon \Rightarrow c_n < L + \epsilon$. If $n > \max\{N_1, N_2\}$, then $L - \epsilon < a_n \le b_n \le c_n < L + \epsilon \Rightarrow |b_n - L| < \epsilon \Rightarrow \lim_{n\to\infty} b_n = L$.

111. $a_{n+1} \ge a_n \Rightarrow \frac{3(n+1)+1}{(n+1)+1} > \frac{3n+1}{n+1} \Rightarrow \frac{3n+4}{n+2} > \frac{3n+1}{n+1} \Rightarrow 3n^2 + 3n + 4n + 4 > 3n^2 + 6n + n + 2$ $\Rightarrow 4 > 2$; the steps are reversible so the sequence is nondecreasing; $\frac{3n+1}{n+1} < 3 \Rightarrow 3n + 1 < 3n + 3$ $\Rightarrow 1 < 3$; the steps are reversible so the sequence is bounded above by 3

113. $a_{n+1} \le a_n \Rightarrow \frac{2^{n+1}3^{n+1}}{(n+1)!} \le \frac{2^n 3^n}{n!} \Rightarrow \frac{2^{n+1}3^{n+1}}{2^n 3^n} \le \frac{(n+1)!}{n!} \Rightarrow 2 \cdot 3 \le n + 1$ which is true for $n \ge 5$; the steps are reversible so the sequence is decreasing after a_5, but it is not nondecreasing for all its terms; $a_1 = 6$, $a_2 = 18$, $a_3 = 36$, $a_4 = 54$, $a_5 = \frac{324}{5} = 64.8 \Rightarrow$ the sequence is bounded from above by 64.8

115. $a_n = 1 - \frac{1}{n}$ converges because $\frac{1}{n} \to 0$ by Example 1; also it is a nondecreasing sequence bounded above by 1

117. $a_n = \frac{2^n - 1}{2^n} = 1 - \frac{1}{2^n}$ and $0 < \frac{1}{2^n} < \frac{1}{n}$; since $\frac{1}{n} \to 0$ (by Example 1) $\Rightarrow \frac{1}{2^n} \to 0$, the sequence converges; also it is a nondecreasing sequence bounded above by 1

119. $a_n = ((-1)^n + 1)\left(\frac{n+1}{n}\right)$ diverges because $a_n = 0$ for n odd, while for n even $a_n = 2\left(1 + \frac{1}{n}\right)$ converges to 2; it diverges by definition of divergence

121. $a_n \ge a_{n+1} \Leftrightarrow \frac{1+\sqrt{2n}}{\sqrt{n}} \ge \frac{1+\sqrt{2(n+1)}}{\sqrt{n+1}} \Leftrightarrow \sqrt{n+1} + \sqrt{2n^2+2n} \ge \sqrt{n} + \sqrt{2n^2+2n} \Leftrightarrow \sqrt{n+1} \ge \sqrt{n}$ and $\frac{1+\sqrt{2n}}{\sqrt{n}} \ge \sqrt{2}$; thus the sequence is nonincreasing and bounded below by $\sqrt{2} \Rightarrow$ it converges

123. $\frac{4^{n+1}+3^n}{4^n} = 4 + \left(\frac{3}{4}\right)^n$ so $a_n \ge a_{n+1} \Leftrightarrow 4 + \left(\frac{3}{4}\right)^n \ge 4 + \left(\frac{3}{4}\right)^{n+1} \Leftrightarrow \left(\frac{3}{4}\right)^n \ge \left(\frac{3}{4}\right)^{n+1} \Leftrightarrow 1 \ge \frac{3}{4}$ and $4 + \left(\frac{3}{4}\right)^n \ge 4$; thus the sequence is nonincreasing and bounded below by 4 $\Rightarrow$ it converges

125. Let $0 < M < 1$ and let N be an integer greater than $\frac{M}{1-M}$. Then $n > N \Rightarrow n > \frac{M}{1-M} \Rightarrow n - nM > M$ $\Rightarrow n > M + nM \Rightarrow n > M(n+1) \Rightarrow \frac{n}{n+1} > M$.

127. The sequence $a_n = 1 + \frac{(-1)^n}{2}$ is the sequence $\frac{1}{2}, \frac{3}{2}, \frac{1}{2}, \frac{3}{2}, \ldots$. This sequence is bounded above by $\frac{3}{2}$, but it clearly does not converge, by definition of convergence.

129. Given an $\epsilon > 0$, by definition of convergence there corresponds an N such that for all $n > N$, $|L_1 - a_n| < \epsilon$ and $|L_2 - a_n| < \epsilon$. Now $|L_2 - L_1| = |L_2 - a_n + a_n - L_1| \le |L_2 - a_n| + |a_n - L_1| < \epsilon + \epsilon = 2\epsilon$. $|L_2 - L_1| < 2\epsilon$ says that the difference between two fixed values is smaller than any positive number 2ϵ. The only nonnegative number smaller than every positive number is 0, so $|L_1 - L_2| = 0$ or $L_1 = L_2$.

131. $a_{2k} \to L \Leftrightarrow$ given an $\epsilon > 0$ there corresponds an N_1 such that $[2k > N_1 \Rightarrow |a_{2k} - L| < \epsilon]$. Similarly, $a_{2k+1} \to L \Leftrightarrow [2k + 1 > N_2 \Rightarrow |a_{2k+1} - L| < \epsilon]$. Let $N = \max\{N_1, N_2\}$. Then $n > N \Rightarrow |a_n - L| < \epsilon$ whether n is even or odd, and hence $a_n \to L$.

133. (a) $f(x) = x^2 - a \Rightarrow f'(x) = 2x \Rightarrow x_{n+1} = x_n - \frac{x_n^2 - a}{2x_n} \Rightarrow x_{n+1} = \frac{2x_n^2 - (x_n^2 - a)}{2x_n} = \frac{x_n^2 + a}{2x_n} = \frac{\left(x_n + \frac{a}{x_n}\right)}{2}$

(b) $x_1 = 2$, $x_2 = 1.75$, $x_3 = 1.732142857$, $x_4 = 1.73205081$, $x_5 = 1.732050808$; we are finding the positive number where $x^2 - 3 = 0$; that is, where $x^2 = 3$, $x > 0$, or where $x = \sqrt{3}$.

10.2 INFINITE SERIES

1. $s_n = \frac{a(1-r^n)}{(1-r)} = \frac{2\left(1-\left(\frac{1}{3}\right)^n\right)}{1-\left(\frac{1}{3}\right)} \Rightarrow \lim_{n\to\infty} s_n = \frac{2}{1-\left(\frac{1}{3}\right)} = 3$

3. $s_n = \frac{a(1-r^n)}{(1-r)} = \frac{1-\left(-\frac{1}{2}\right)^n}{1-\left(-\frac{1}{2}\right)} \Rightarrow \lim_{n\to\infty} s_n = \frac{1}{\left(\frac{3}{2}\right)} = \frac{2}{3}$

5. $\frac{1}{(n+1)(n+2)} = \frac{1}{n+1} - \frac{1}{n+2} \Rightarrow s_n = \left(\frac{1}{2}-\frac{1}{3}\right) + \left(\frac{1}{3}-\frac{1}{4}\right) + \ldots + \left(\frac{1}{n+1}-\frac{1}{n+2}\right) = \frac{1}{2} - \frac{1}{n+2} \Rightarrow \lim_{n\to\infty} s_n = \frac{1}{2}$

7. $1 - \frac{1}{4} + \frac{1}{16} - \frac{1}{64} + \ldots$, the sum of this geometric series is $\frac{1}{1-\left(-\frac{1}{4}\right)} = \frac{1}{1+\left(\frac{1}{4}\right)} = \frac{4}{5}$

9. $\frac{7}{4} + \frac{7}{16} + \frac{7}{64} + \ldots$, the sum of this geometric series is $\frac{\left(\frac{7}{4}\right)}{1-\left(\frac{1}{4}\right)} = \frac{7}{3}$

11. $(5+1) + \left(\frac{5}{2}+\frac{1}{3}\right) + \left(\frac{5}{4}+\frac{1}{9}\right) + \left(\frac{5}{8}+\frac{1}{27}\right) + \ldots$, is the sum of two geometric series; the sum is $\frac{5}{1-\left(\frac{1}{2}\right)} + \frac{1}{1-\left(\frac{1}{3}\right)} = 10 + \frac{3}{2} = \frac{23}{2}$

13. $(1+1) + \left(\frac{1}{2}-\frac{1}{5}\right) + \left(\frac{1}{4}+\frac{1}{25}\right) + \left(\frac{1}{8}-\frac{1}{125}\right) + \ldots$, is the sum of two geometric series; the sum is $\frac{1}{1-\left(\frac{1}{2}\right)} + \frac{1}{1+\left(\frac{1}{5}\right)} = 2 + \frac{5}{6} = \frac{17}{6}$

15. Series is geometric with $r = \frac{2}{5} \Rightarrow \left|\frac{2}{5}\right| < 1 \Rightarrow$ Converges to $\frac{1}{1-\frac{2}{5}} = \frac{5}{3}$

17. Series is geometric with $r = \frac{1}{8} \Rightarrow \left|\frac{1}{8}\right| < 1 \Rightarrow$ Converges to $\frac{\frac{1}{8}}{1-\frac{1}{8}} = \frac{1}{7}$

19. $0.\overline{23} = \sum_{n=0}^{\infty} \frac{23}{100}\left(\frac{1}{10^2}\right)^n = \frac{\left(\frac{23}{100}\right)}{1-\left(\frac{1}{100}\right)} = \frac{23}{99}$

21. $0.\overline{7} = \sum_{n=0}^{\infty} \frac{7}{10}\left(\frac{1}{10}\right)^n = \frac{\left(\frac{7}{10}\right)}{1-\left(\frac{1}{10}\right)} = \frac{7}{9}$

23. $0.0\overline{6} = \sum_{n=0}^{\infty} \left(\frac{1}{10}\right)\left(\frac{6}{10}\right)\left(\frac{1}{10}\right)^n = \frac{\left(\frac{6}{100}\right)}{1-\left(\frac{1}{10}\right)} = \frac{6}{90} = \frac{1}{15}$

25. $1.24\overline{123} = \frac{124}{100} + \sum_{n=0}^{\infty} \frac{123}{10^5}\left(\frac{1}{10^3}\right)^n = \frac{124}{100} + \frac{\left(\frac{123}{10^5}\right)}{1-\left(\frac{1}{10^3}\right)} = \frac{124}{100} + \frac{123}{10^5-10^2} = \frac{124}{100} + \frac{123}{99{,}900} = \frac{123{,}999}{99{,}900} = \frac{41{,}333}{33{,}300}$

27. $\lim_{n\to\infty} \frac{n}{n+10} = \lim_{n\to\infty} \frac{1}{1} = 1 \neq 0 \Rightarrow$ diverges

29. $\lim_{n\to\infty} \frac{1}{n+4} = 0 \Rightarrow$ test inconclusive

31. $\lim_{n\to\infty} \cos\frac{1}{n} = \cos 0 = 1 \neq 0 \Rightarrow$ diverges

33. $\lim_{n\to\infty} \ln\frac{1}{n} = -\infty \neq 0 \Rightarrow$ diverges

35. $s_k = \left(1-\frac{1}{2}\right) + \left(\frac{1}{2}-\frac{1}{3}\right) + \left(\frac{1}{3}-\frac{1}{4}\right) + \ldots + \left(\frac{1}{k-1}-\frac{1}{k}\right) + \left(\frac{1}{k}-\frac{1}{k+1}\right) = 1 - \frac{1}{k+1} \Rightarrow \lim_{k\to\infty} s_k$ $= \lim_{k\to\infty}\left(1-\frac{1}{k+1}\right) = 1$, series converges to 1

37. $s_k = \left(\ln\sqrt{2} - \ln\sqrt{1}\right) + \left(\ln\sqrt{3} - \ln\sqrt{2}\right) + \left(\ln\sqrt{4} - \ln\sqrt{3}\right) + \ldots + \left(\ln\sqrt{k} - \ln\sqrt{k-1}\right) + \left(\ln\sqrt{k+1} - \ln\sqrt{k}\right)$
$= \ln\sqrt{k+1} - \ln\sqrt{1} = \ln\sqrt{k+1} \Rightarrow \lim_{k\to\infty} s_k = \lim_{k\to\infty} \ln\sqrt{k+1} = \infty$; series diverges

39. $s_k = \left(\cos^{-1}\left(\frac{1}{2}\right) - \cos^{-1}\left(\frac{1}{3}\right)\right) + \left(\cos^{-1}\left(\frac{1}{3}\right) - \cos^{-1}\left(\frac{1}{4}\right)\right) + \left(\cos^{-1}\left(\frac{1}{4}\right) - \cos^{-1}\left(\frac{1}{5}\right)\right) + \ldots$
$+ \left(\cos^{-1}\left(\frac{1}{k}\right) - \cos^{-1}\left(\frac{1}{k+1}\right)\right) + \left(\cos^{-1}\left(\frac{1}{k+1}\right) - \cos^{-1}\left(\frac{1}{k+2}\right)\right) = \frac{\pi}{3} - \cos^{-1}\left(\frac{1}{k+2}\right)$
$\Rightarrow \lim_{k\to\infty} s_k = \lim_{k\to\infty} \left[\frac{\pi}{3} - \cos^{-1}\left(\frac{1}{k+2}\right)\right] = \frac{\pi}{3} - \frac{\pi}{2} = \frac{\pi}{6}$, series converges to $\frac{\pi}{6}$

41. $\frac{4}{(4n-3)(4n+1)} = \frac{1}{4n-3} - \frac{1}{4n+1} \Rightarrow s_k = \left(1 - \frac{1}{5}\right) + \left(\frac{1}{5} - \frac{1}{9}\right) + \left(\frac{1}{9} - \frac{1}{13}\right) + \ldots + \left(\frac{1}{4k-7} - \frac{1}{4k-3}\right)$
$+ \left(\frac{1}{4k-3} - \frac{1}{4k+1}\right) = 1 - \frac{1}{4k+1} \Rightarrow \lim_{k\to\infty} s_k = \lim_{k\to\infty} \left(1 - \frac{1}{4k+1}\right) = 1$

43. $\frac{40n}{(2n-1)^2(2n+1)^2} = \frac{A}{(2n-1)} + \frac{B}{(2n-1)^2} + \frac{C}{(2n+1)} + \frac{D}{(2n+1)^2} = \frac{A(2n-1)(2n+1)^2 + B(2n+1)^2 + C(2n+1)(2n-1)^2 + D(2n-1)^2}{(2n-1)^2(2n+1)^2}$
$\Rightarrow A(2n-1)(2n+1)^2 + B(2n+1)^2 + C(2n+1)(2n-1)^2 + D(2n-1)^2 = 40n$
$\Rightarrow A\left(8n^3 + 4n^2 - 2n - 1\right) + B\left(4n^2 + 4n + 1\right) + C\left(8n^3 - 4n^2 - 2n + 1\right) = D\left(4n^2 - 4n + 1\right) = 40n$
$\Rightarrow (8A + 8C)n^3 + (4A + 4B - 4C + 4D)n^2 + (-2A + 4B - 2C - 4D)n + (-A + B + C + D) = 40n$

$$\Rightarrow \begin{cases} 8A + 8C = 0 \\ 4A + 4B - 4C + 4D = 0 \\ -2A + 4B - 2C - 4D = 40 \\ -A + B + C + D = 0 \end{cases} \Rightarrow \begin{cases} 8A + 8C = 0 \\ A + B - C + D = 0 \\ -A + 2B - C - 2D = 20 \\ -A + B + C + D = 0 \end{cases} \Rightarrow \begin{cases} B + D = 0 \\ 2B - 2D = 20 \end{cases} \Rightarrow 4B = 20 \Rightarrow B = 5$$

and $D = -5 \Rightarrow \begin{cases} A + C = 0 \\ -A + 5 + C - 5 = 0 \end{cases} \Rightarrow C = 0$ and $A = 0$. Hence, $\sum_{n=1}^{k} \left[\frac{40n}{(2n-1)^2(2n+1)^2}\right]$
$= 5\sum_{n=1}^{k} \left[\frac{1}{(2n-1)^2} - \frac{1}{(2n+1)^2}\right] = 5\left(\frac{1}{1} - \frac{1}{9} + \frac{1}{9} - \frac{1}{25} + \frac{1}{25} - \ldots - \frac{1}{(2(k-1)+1)^2} + \frac{1}{(2k-1)^2} - \frac{1}{(2k+1)^2}\right)$
$= 5\left(1 - \frac{1}{(2k+1)^2}\right) \Rightarrow$ the sum is $\lim_{n\to\infty} 5\left(1 - \frac{1}{(2k+1)^2}\right) = 5$

45. $s_k = \left(1 - \frac{1}{\sqrt{2}}\right) + \left(\frac{1}{\sqrt{2}} - \frac{1}{\sqrt{3}}\right) + \left(\frac{1}{\sqrt{3}} - \frac{1}{\sqrt{4}}\right) + \ldots + \left(\frac{1}{\sqrt{k-1}} + \frac{1}{\sqrt{k}}\right) + \left(\frac{1}{\sqrt{k}} - \frac{1}{\sqrt{k+1}}\right) = 1 - \frac{1}{\sqrt{k+1}}$
$\Rightarrow \lim_{k\to\infty} s_k = \lim_{k\to\infty} \left(1 - \frac{1}{\sqrt{k+1}}\right) = 1$

47. $s_k = \left(\frac{1}{\ln 3} - \frac{1}{\ln 2}\right) + \left(\frac{1}{\ln 4} - \frac{1}{\ln 3}\right) + \left(\frac{1}{\ln 5} - \frac{1}{\ln 4}\right) + \ldots + \left(\frac{1}{\ln(k+1)} - \frac{1}{\ln k}\right) + \left(\frac{1}{\ln(k+2)} - \frac{1}{\ln(k+1)}\right)$
$= -\frac{1}{\ln 2} + \frac{1}{\ln(k+2)} \Rightarrow \lim_{k\to\infty} s_k = -\frac{1}{\ln 2}$

49. convergent geometric series with sum $\frac{1}{1-\left(\frac{1}{\sqrt{2}}\right)} = \frac{\sqrt{2}}{\sqrt{2}-1} = 2 + \sqrt{2}$

51. convergent geometric series with sum $\frac{\left(\frac{3}{2}\right)}{1-\left(-\frac{1}{2}\right)} = 1$

53. $\lim_{n\to\infty} \cos(n\pi) = \lim_{n\to\infty} (-1)^n \neq 0 \Rightarrow$ diverges

55. convergent geometric series with sum $\frac{1}{1-\left(\frac{1}{e^2}\right)} = \frac{e^2}{e^2-1}$

57. convergent geometric series with sum $\frac{2}{1-\left(\frac{1}{10}\right)} - 2 = \frac{20}{9} - \frac{18}{9} = \frac{2}{9}$

59. difference of two geometric series with sum $\frac{1}{1-\left(\frac{2}{3}\right)} - \frac{1}{1-\left(\frac{1}{3}\right)} = 3 - \frac{3}{2} = \frac{3}{2}$

61. $\lim_{n \to \infty} \frac{n!}{1000^n} = \infty \neq 0 \Rightarrow$ diverges

63. $\sum_{n=1}^{\infty} \frac{2^n + 3^n}{4^n} = \sum_{n=1}^{\infty} \frac{2^n}{4^n} + \sum_{n=1}^{\infty} \frac{3^n}{4^n} = \sum_{n=1}^{\infty} \left(\frac{1}{2}\right)^n + \sum_{n=1}^{\infty} \left(\frac{3}{4}\right)^n$; both $= \sum_{n=1}^{\infty} \left(\frac{1}{2}\right)^n$ and $\sum_{n=1}^{\infty} \left(\frac{3}{4}\right)^n$ are geometric series, and both converge since $r = \frac{1}{2} \Rightarrow \left|\frac{1}{2}\right| < 1$ and $r = \frac{3}{4} \Rightarrow \left|\frac{3}{4}\right| < 1$, respectivley $\Rightarrow \sum_{n=1}^{\infty} \left(\frac{1}{2}\right)^n = \frac{\frac{1}{2}}{1-\frac{1}{2}} = 1$ and $\sum_{n=1}^{\infty} \left(\frac{3}{4}\right)^n = \frac{\frac{3}{4}}{1-\frac{3}{4}} = 3 \Rightarrow$ $\sum_{n=1}^{\infty} \frac{2^n + 3^n}{4^n} = 1 + 3 = 4$ by Theorem 8, part (1)

65. $\sum_{n=1}^{\infty} \ln\left(\frac{n}{n+1}\right) = \sum_{n=1}^{\infty} [\ln(n) - \ln(n+1)] \Rightarrow s_k = [\ln(1) - \ln(2)] + [\ln(2) - \ln(3)] + [\ln(3) - \ln(4)] + \ldots$ $+ [\ln(k-1) - \ln(k)] + [\ln(k) - \ln(k+1)] = -\ln(k+1) \Rightarrow \lim_{k \to \infty} s_k = -\infty, \Rightarrow$ diverges

67. convergent geometric series with sum $\frac{1}{1-\left(\frac{e}{\pi}\right)} = \frac{\pi}{\pi - e}$

69. $\sum_{n=0}^{\infty} (-1)^n x^n = \sum_{n=0}^{\infty} (-x)^n$; $a = 1$, $r = -x$; converges to $\frac{1}{1-(-x)} = \frac{1}{1+x}$ for $|x| < 1$

71. $a = 3$, $r = \frac{x-1}{2}$; converges to $\frac{3}{1-\left(\frac{x-1}{2}\right)} = \frac{6}{3-x}$ for $-1 < \frac{x-1}{2} < 1$ or $-1 < x < 3$

73. $a = 1$, $r = 2x$; converges to $\frac{1}{1-2x}$ for $|2x| < 1$ or $|x| < \frac{1}{2}$

75. $a = 1$, $r = -(x+1)^n$; converges to $\frac{1}{1+(x+1)} = \frac{1}{2+x}$ for $|x+1| < 1$ or $-2 < x < 0$

77. $a = 1$, $r = \sin x$; converges to $\frac{1}{1-\sin x}$ for $x \neq (2k+1)\frac{\pi}{2}$, k an integer

79. (a) $\sum_{n=-2}^{\infty} \frac{1}{(n+4)(n+5)}$ (b) $\sum_{n=0}^{\infty} \frac{1}{(n+2)(n+3)}$ (c) $\sum_{n=5}^{\infty} \frac{1}{(n-3)(n-2)}$

81. (a) one example is $\frac{1}{2} + \frac{1}{4} + \frac{1}{8} + \frac{1}{16} + \ldots = \frac{\left(\frac{1}{2}\right)}{1-\left(\frac{1}{2}\right)} = 1$

(b) one example is $-\frac{3}{2} - \frac{3}{4} - \frac{3}{8} - \frac{3}{16} - \ldots = \frac{\left(-\frac{3}{2}\right)}{1-\left(\frac{1}{2}\right)} = -3$

(c) one example is $1 - \frac{1}{2} - \frac{1}{4} - \frac{1}{8} - \frac{1}{16} - \ldots = 1 - \frac{\left(\frac{1}{2}\right)}{1-\left(\frac{1}{2}\right)} = 0.$

83. Let $a_n = b_n = \left(\frac{1}{2}\right)^n$. Then $\sum_{n=1}^{\infty} a_n = \sum_{n=1}^{\infty} b_n = \sum_{n=1}^{\infty} \left(\frac{1}{2}\right)^n = 1$, while $\sum_{n=1}^{\infty} \left(\frac{a_n}{b_n}\right) = \sum_{n=1}^{\infty} (1)$ diverges.

85. Let $a_n = \left(\frac{1}{4}\right)^n$ and $b_n = \left(\frac{1}{2}\right)^n$. Then $A = \sum_{n=1}^{\infty} a_n = \frac{1}{3}$, $B = \sum_{n=1}^{\infty} b_n = 1$ and $\sum_{n=1}^{\infty} \left(\frac{a_n}{b_n}\right) = \sum_{n=1}^{\infty} \left(\frac{1}{2}\right)^n = 1 \neq \frac{A}{B}$.

87. Since the sum of a finite number of terms is finite, adding or subtracting a finite number of terms from a series that diverges does not change the divergence of the series.

89. (a) $\frac{2}{1-r} = 5 \Rightarrow \frac{2}{5} = 1 - r \Rightarrow r = \frac{3}{5}$; $2 + 2\left(\frac{3}{5}\right) + 2\left(\frac{3}{5}\right)^2 + \ldots$

(b) $\frac{\left(\frac{13}{2}\right)}{1-r} = 5 \Rightarrow \frac{13}{10} = 1 - r \Rightarrow r = -\frac{3}{10}$; $\frac{13}{2} - \frac{13}{2}\left(\frac{3}{10}\right) + \frac{13}{2}\left(\frac{3}{10}\right)^2 - \frac{13}{2}\left(\frac{3}{10}\right)^3 + \ldots$

91. $s_n = 1 + 2r + r^2 + 2r^3 + r^4 + 2r^5 + \ldots + r^{2n} + 2r^{2n+1}, n = 0, 1, \ldots$
$\Rightarrow s_n = (1 + r^2 + r^4 + \ldots + r^{2n}) + (2r + 2r^3 + 2r^5 + \ldots + 2r^{2n+1}) \Rightarrow \lim_{n\to\infty} s_n = \frac{1}{1-r^2} + \frac{2r}{1-r^2}$
$= \frac{1+2r}{1-r^2}$, if $|r^2| < 1$ or $|r| < 1$

93. area $= 2^2 + \left(\sqrt{2}\right)^2 + (1)^2 + \left(\frac{1}{\sqrt{2}}\right)^2 + \ldots = 4 + 2 + 1 + \frac{1}{2} + \ldots = \frac{4}{1-\frac{1}{2}} = 8\ \text{m}^2$

10.3 THE INTEGRAL TEST

1. $f(x) = \frac{1}{x^2}$ is positive, continuous, and decreasing for $x \ge 1$; $\int_1^\infty \frac{1}{x^2}\,dx = \lim_{b\to\infty} \int_1^b \frac{1}{x^2}\,dx = \lim_{b\to\infty} \left[-\frac{1}{x}\right]_1^b$
$= \lim_{b\to\infty} \left(-\frac{1}{b} + 1\right) = 1 \Rightarrow \int_1^\infty \frac{1}{x^2}\,dx$ converges $\Rightarrow \sum_{n=1}^\infty \frac{1}{n^2}$ converges

3. $f(x) = \frac{1}{x^2+4}$ is positive, continuous, and decreasing for $x \ge 1$; $\int_1^\infty \frac{1}{x^2+4}\,dx = \lim_{b\to\infty} \int_1^b \frac{1}{x^2+4}\,dx = \lim_{b\to\infty} \left[\frac{1}{2}\tan^{-1}\frac{x}{2}\right]_1^b$
$= \lim_{b\to\infty} \left(\frac{1}{2}\tan^{-1}\frac{b}{2} - \frac{1}{2}\tan^{-1}\frac{1}{2}\right) = \frac{\pi}{4} - \frac{1}{2}\tan^{-1}\frac{1}{2} \Rightarrow \int_1^\infty \frac{1}{x^2+4}\,dx$ converges $\Rightarrow \sum_{n=1}^\infty \frac{1}{n^2+4}$ converges

5. $f(x) = e^{-2x}$ is positive, continuous, and decreasing for $x \ge 1$; $\int_1^\infty e^{-2x}\,dx = \lim_{b\to\infty} \int_1^b e^{-2x}\,dx = \lim_{b\to\infty} \left[-\frac{1}{2}e^{-2x}\right]_1^b$
$= \lim_{b\to\infty} \left(-\frac{1}{2e^{2b}} + \frac{1}{2e^2}\right) = \frac{1}{2e^2} \Rightarrow \int_1^\infty e^{-2x}\,dx$ converges $\Rightarrow \sum_{n=1}^\infty e^{-2n}$ converges

7. $f(x) = \frac{x}{x^2+4}$ is positive and continuous for $x \ge 1$, $f'(x) = \frac{4-x^2}{(x^2+4)^2} < 0$ for $x > 2$, thus f is decreasing for $x \ge 3$;
$\int_3^\infty \frac{x}{x^2+4}\,dx = \lim_{b\to\infty} \int_3^b \frac{x}{x^2+4}\,dx = \lim_{b\to\infty} \left[\frac{1}{2}\ln(x^2+4)\right]_3^b = \lim_{b\to\infty} \left(\frac{1}{2}\ln(b^2+4) - \frac{1}{2}\ln(13)\right) = \infty \Rightarrow \int_3^\infty \frac{x}{x^2+4}\,dx$
diverges $\Rightarrow \sum_{n=3}^\infty \frac{n}{n^2+4}$ diverges $\Rightarrow \sum_{n=1}^\infty \frac{n}{n^2+4} = \frac{1}{5} + \frac{2}{8} + \sum_{n=3}^\infty \frac{n}{n^2+4}$ diverges

9. $f(x) = \frac{x^2}{e^{x/3}}$ is positive and continuous for $x \ge 1$, $f'(x) = \frac{-x(x-6)}{3e^{x/3}} < 0$ for $x > 6$, thus f is decreasing for $x \ge 7$;
$\int_7^\infty \frac{x^2}{e^{x/3}}\,dx = \lim_{b\to\infty} \int_7^b \frac{x^2}{e^{x/3}}\,dx = \lim_{b\to\infty} \left[-\frac{3x^2}{e^{x/3}} - \frac{18x}{e^{x/3}} - \frac{54}{e^{x/3}}\right]_7^b = \lim_{b\to\infty} \left(\frac{-3b^2-18b-54}{e^{b/3}} + \frac{327}{e^{7/3}}\right) =$
$= \lim_{b\to\infty} \left(\frac{3(-6b-18)}{e^{b/3}}\right) + \frac{327}{e^{7/3}} = \lim_{b\to\infty} \left(\frac{-54}{e^{b/3}}\right) + \frac{327}{e^{7/3}} = \frac{327}{e^{7/3}} \Rightarrow \int_7^\infty \frac{x^2}{e^{x/3}}\,dx$ converges $\Rightarrow \sum_{n=7}^\infty \frac{n^2}{e^{n/3}}$ converges
$\Rightarrow \sum_{n=1}^\infty \frac{n^2}{e^{n/3}} = \frac{1}{e^{1/3}} + \frac{4}{e^{2/3}} + \frac{9}{e^1} + \frac{16}{e^{4/3}} + \frac{25}{e^{5/3}} + \frac{36}{e^2} + \sum_{n=7}^\infty \frac{n^2}{e^{n/3}}$ converges

11. converges; a geometric series with $r = \frac{1}{10} < 1$

13. diverges; by the nth-Term Test for Divergence, $\lim_{n\to\infty} \frac{n}{n+1} = 1 \ne 0$

15. diverges; $\sum_{n=1}^\infty \frac{3}{\sqrt{n}} = 3\sum_{n=1}^\infty \frac{1}{\sqrt{n}}$, which is a divergent p-series $\left(p = \frac{1}{2}\right)$

17. converges; a geometric series with $r = \frac{1}{8} < 1$

19. diverges by the Integral Test: $\int_2^n \frac{\ln x}{x}\,dx = \frac{1}{2}\left(\ln^2 n - \ln 2\right) \Rightarrow \int_2^\infty \frac{\ln x}{x}\,dx \to \infty$

21. converges; a geometric series with $r = \frac{2}{3} < 1$

23. diverges; $\sum_{n=0}^{\infty} \frac{-2}{n+1} = -2 \sum_{n=0}^{\infty} \frac{1}{n+1}$, which diverges by the Integral Test

25. diverges; $\lim_{n \to \infty} a_n = \lim_{n \to \infty} \frac{2^n}{n+1} = \lim_{n \to \infty} \frac{2^n \ln 2}{1} = \infty \neq 0$

27. diverges; $\lim_{n \to \infty} \frac{\sqrt{n}}{\ln n} = \lim_{n \to \infty} \frac{\left(\frac{1}{2\sqrt{n}}\right)}{\left(\frac{1}{n}\right)} = \lim_{n \to \infty} \frac{\sqrt{n}}{2} = \infty \neq 0$

29. diverges; a geometric series with $r = \frac{1}{\ln 2} \approx 1.44 > 1$

31. converges by the Integral Test: $\int_3^{\infty} \frac{\left(\frac{1}{x}\right)}{(\ln x)\sqrt{(\ln x)^2 - 1}}\, dx$; $\left[\begin{array}{l} u = \ln x \\ du = \frac{1}{x}\, dx \end{array}\right] \to \int_{\ln 3}^{\infty} \frac{1}{u\sqrt{u^2 - 1}}\, du$
$= \lim_{b \to \infty} \left[\sec^{-1} |u|\right]_{\ln 3}^{b} = \lim_{b \to \infty} \left[\sec^{-1} b - \sec^{-1} (\ln 3)\right] = \lim_{b \to \infty} \left[\cos^{-1}\left(\frac{1}{b}\right) - \sec^{-1} (\ln 3)\right]$
$= \cos^{-1}(0) - \sec^{-1}(\ln 3) = \frac{\pi}{2} - \sec^{-1}(\ln 3) \approx 1.1439$

33. diverges by the nth-Term Test for divergence; $\lim_{n \to \infty} n \sin\left(\frac{1}{n}\right) = \lim_{n \to \infty} \frac{\sin\left(\frac{1}{n}\right)}{\left(\frac{1}{n}\right)} = \lim_{x \to 0} \frac{\sin x}{x} = 1 \neq 0$

35. converges by the Integral Test: $\int_1^{\infty} \frac{e^x}{1+e^{2x}}\, dx$; $\left[\begin{array}{l} u = e^x \\ du = e^x\, dx \end{array}\right] \to \int_e^{\infty} \frac{1}{1+u^2}\, du = \lim_{n \to \infty} \left[\tan^{-1} u\right]_e^b$
$= \lim_{b \to \infty} \left(\tan^{-1} b - \tan^{-1} e\right) = \frac{\pi}{2} - \tan^{-1} e \approx 0.35$

37. converges by the Integral Test: $\int_1^{\infty} \frac{8 \tan^{-1} x}{1+x^2}\, dx$; $\left[\begin{array}{l} u = \tan^{-1} x \\ du = \frac{dx}{1+x^2} \end{array}\right] \to \int_{\pi/4}^{\pi/2} 8u\, du = \left[4u^2\right]_{\pi/4}^{\pi/2} = 4\left(\frac{\pi^2}{4} - \frac{\pi^2}{16}\right) = \frac{3\pi^2}{4}$

39. converges by the Integral Test: $\int_1^{\infty} \operatorname{sech} x\, dx = 2 \lim_{b \to \infty} \int_1^b \frac{e^x}{1+(e^x)^2}\, dx = 2 \lim_{b \to \infty} \left[\tan^{-1} e^x\right]_1^b$
$= 2 \lim_{b \to \infty} \left(\tan^{-1} e^b - \tan^{-1} e\right) = \pi - 2 \tan^{-1} e \approx 0.71$

41. $\int_1^{\infty} \left(\frac{a}{x+2} - \frac{1}{x+4}\right) dx = \lim_{b \to \infty} \left[a \ln |x+2| - \ln |x+4|\right]_1^b = \lim_{b \to \infty} \ln \frac{(b+2)^a}{b+4} - \ln\left(\frac{3^a}{5}\right)$;
$\lim_{b \to \infty} \frac{(b+2)^a}{b+4} = a \lim_{b \to \infty} (b+2)^{a-1} = \begin{cases} \infty, & a > 1 \\ 1, & a = 1 \end{cases} \Rightarrow$ the series converges to $\ln\left(\frac{5}{3}\right)$ if $a = 1$ and diverges to ∞ if $a > 1$. If $a < 1$, the terms of the series eventually become negative and the Integral Test does not apply. From that point on, however, the series behaves like a negative multiple of the harmonic series, and so it diverges.

43. (a)

(b) There are $(13)(365)(24)(60)(60)\left(10^9\right)$ seconds in 13 billion years; by part (a) $s_n \leq 1 + \ln n$ where $n = (13)(365)(24)(60)(60)\left(10^9\right) \Rightarrow s_n \leq 1 + \ln\left((13)(365)(24)(60)(60)\left(10^9\right)\right)$
$= 1 + \ln(13) + \ln(365) + \ln(24) + 2\ln(60) + 9\ln(10) \approx 41.55$

45. Yes. If $\sum_{n=1}^{\infty} a_n$ is a divergent series of positive numbers, then $\left(\frac{1}{2}\right)\sum_{n=1}^{\infty} a_n = \sum_{n=1}^{\infty}\left(\frac{a_n}{2}\right)$ also diverges and $\frac{a_n}{2} < a_n$.

There is no "smallest" divergent series of positive numbers: for any divergent series $\sum_{n=1}^{\infty} a_n$ of positive numbers $\sum_{n=1}^{\infty}\left(\frac{a_n}{2}\right)$ has smaller terms and still diverges.

47. (a) Both integrals can represent the area under the curve $f(x) = \frac{1}{\sqrt{x+1}}$, and the sum s_{50} can be considered an approximation of either integral using rectangles with $\Delta x = 1$. The sum $s_{50} = \sum_{n=1}^{50} \frac{1}{\sqrt{n+1}}$ is an overestimate of the integral $\int_1^{51} \frac{1}{\sqrt{x+1}}dx$. The sum s_{50} represents a left-hand sum (that is, the we are choosing the left-hand endpoint of each subinterval for c_i) and because f is a decreasing function, the value of f is a maximum at the left-hand endpoint of each sub interval. The area of each rectangle overestimates the true area, thus $\int_1^{51} \frac{1}{\sqrt{x+1}}dx < \sum_{n=1}^{50} \frac{1}{\sqrt{n+1}}$. In a similar manner, s_{50} underestimates the integral $\int_0^{50} \frac{1}{\sqrt{x+1}}dx$. In this case, the sum s_{50} represents a right-hand sum and because f is a decreasing function, the value of f is aminimum at the right-hand endpoint of each subinterval. The area of each rectangle underestimates the true area, thus $\sum_{n=1}^{50} \frac{1}{\sqrt{n+1}} < \int_0^{50} \frac{1}{\sqrt{x+1}}dx$. Evaluating the integrals we find $\int_1^{51} \frac{1}{\sqrt{x+1}}dx$ $= \left[2\sqrt{x+1}\right]_1^{51} = 2\sqrt{52} - 2\sqrt{2} \approx 11.6$ and $\int_0^{50} \frac{1}{\sqrt{x+1}}dx = \left[2\sqrt{x+1}\right]_0^{50} = 2\sqrt{51} - 2\sqrt{1} \approx 12.3$. Thus, $11.6 < \sum_{n=1}^{50} \frac{1}{\sqrt{n+1}} < 12.3$.

(b) $s_n > 1000 \Rightarrow \int_1^{n+1} \frac{1}{\sqrt{x+1}}dx = \left[2\sqrt{x+1}\right]_1^{n+1} = 2\sqrt{n+1} - 2\sqrt{2} > 1000 \Rightarrow n > \left(500 + 2\sqrt{2}\right)^2 - \approx 251414.2$ $\Rightarrow n \geq 251415$.

49. We want $S - s_n < 0.01 \Rightarrow \int_n^{\infty} \frac{1}{x^3}dx < 0.01 \Rightarrow \int_n^{\infty} \frac{1}{x^3}dx = \lim_{b\to\infty} \int_n^b \frac{1}{x^3}dx = \lim_{b\to\infty}\left[-\frac{1}{2x^2}\right]_n^b = \lim_{b\to\infty}\left(-\frac{1}{2b^2} + \frac{1}{2n^2}\right)$ $= \frac{1}{2n^2} < 0.01 \Rightarrow n > \sqrt{50} \approx 7.071 \Rightarrow n \geq 8 \Rightarrow S \approx s_8 = \sum_{n=1}^{8} \frac{1}{n^3} \approx 1.195$

51. $S - s_n < 0.00001 \Rightarrow \int_n^{\infty} \frac{1}{x^{1.1}}dx < 0.00001 \Rightarrow \int_n^{\infty} \frac{1}{x^{1.1}}dx = \lim_{b\to\infty} \int_n^b \frac{1}{x^{1.1}}dx = \lim_{b\to\infty}\left[-\frac{10}{x^{0.1}}\right]_n^b = \lim_{b\to\infty}\left(-\frac{10}{b^{0.1}} + \frac{10}{n^{0.1}}\right)$ $= \frac{10}{n^{0.1}} < 0.00001 \Rightarrow n > 1000000^{10} \Rightarrow n > 10^{60}$

53. Let $A_n = \sum_{k=1}^{n} a_k$ and $B_n = \sum_{k=1}^{n} 2^k a_{(2^k)}$, where $\{a_k\}$ is a nonincreasing sequence of positive terms converging to 0. Note that $\{A_n\}$ and $\{B_n\}$ are nondecreasing sequences of positive terms. Now,

$B_n = 2a_2 + 4a_4 + 8a_8 + \ldots + 2^n a_{(2^n)} = 2a_2 + (2a_4 + 2a_4) + (2a_8 + 2a_8 + 2a_8 + 2a_8) + \ldots$

$+ \underbrace{\left(2a_{(2^n)} + 2a_{(2^n)} + \ldots + 2a_{(2^n)}\right)}_{2^{n-1}\text{ terms}} \leq 2a_1 + 2a_2 + (2a_3 + 2a_4) + (2a_5 + 2a_6 + 2a_7 + 2a_8) + \ldots$

$+ \left(2a_{(2^{n-1})} + 2a_{(2^{n-1}+1)} + \ldots + 2a_{(2^n)}\right) = 2A_{(2^n)} \leq 2\sum_{k=1}^{\infty} a_k$. Therefore if $\sum a_k$ converges, then $\{B_n\}$ is bounded above $\Rightarrow \sum 2^k a_{(2^k)}$ converges. Conversely,

$A_n = a_1 + (a_2 + a_3) + (a_4 + a_5 + a_6 + a_7) + \ldots + a_n < a_1 + 2a_2 + 4a_4 + \ldots + 2^n a_{(2^n)} = a_1 + B_n < a_1 + \sum_{k=1}^{\infty} 2^k a_{(2^k)}$.

Therefore, if $\sum_{k=1}^{\infty} 2^k a_{(2^k)}$ converges, then $\{A_n\}$ is bounded above and hence converges.

55. (a) $\int_2^{\infty} \frac{dx}{x(\ln x)^p}$; $\begin{bmatrix} u = \ln x \\ du = \frac{dx}{x} \end{bmatrix} \to \int_{\ln 2}^{\infty} u^{-p}\,du = \lim_{b \to \infty} \left[\frac{u^{-p+1}}{-p+1}\right]_{\ln 2}^{b} = \lim_{b \to \infty} \left(\frac{1}{1-p}\right)\left[b^{-p+1} - (\ln 2)^{-p+1}\right]$

$= \begin{cases} \frac{1}{p-1}(\ln 2)^{-p+1}, & p > 1 \\ \infty, & p < 1 \end{cases}$ $\Rightarrow$ the improper integral converges if $p > 1$ and diverges if $p < 1$.

For $p = 1$: $\int_2^{\infty} \frac{dx}{x \ln x} = \lim_{b \to \infty} [\ln(\ln x)]_2^b = \lim_{b \to \infty} [\ln(\ln b) - \ln(\ln 2)] = \infty$, so the improper integral diverges if $p = 1$.

(b) Since the series and the integral converge or diverge together, $\sum_{n=2}^{\infty} \frac{1}{n(\ln n)^p}$ converges if and only if $p > 1$.

57. (a) From Fig. 10.11(a) in the text with $f(x) = \frac{1}{x}$ and $a_k = \frac{1}{k}$, we have $\int_1^{n+1} \frac{1}{x}\,dx \le 1 + \frac{1}{2} + \frac{1}{3} + \ldots + \frac{1}{n}$ $\le 1 + \int_1^n f(x)\,dx \Rightarrow \ln(n+1) \le 1 + \frac{1}{2} + \frac{1}{3} + \ldots + \frac{1}{n} \le 1 + \ln n \Rightarrow 0 \le \ln(n+1) - \ln n$ $\le \left(1 + \frac{1}{2} + \frac{1}{3} + \ldots + \frac{1}{n}\right) - \ln n \le 1$. Therefore the sequence $\left\{\left(1 + \frac{1}{2} + \frac{1}{3} + \ldots + \frac{1}{n}\right) - \ln n\right\}$ is bounded above by 1 and below by 0.

(b) From the graph in Fig. 10.11(b) with $f(x) = \frac{1}{x}$, $\frac{1}{n+1} < \int_n^{n+1} \frac{1}{x}\,dx = \ln(n+1) - \ln n$

$\Rightarrow 0 > \frac{1}{n+1} - [\ln(n+1) - \ln n] = \left(1 + \frac{1}{2} + \frac{1}{3} + \ldots + \frac{1}{n+1} - \ln(n+1)\right) - \left(1 + \frac{1}{2} + \frac{1}{3} + \ldots + \frac{1}{n} - \ln n\right)$.

If we define $a_n = 1 + \frac{1}{2} = \frac{1}{3} + \frac{1}{n} - \ln n$, then $0 > a_{n+1} - a_n \Rightarrow a_{n+1} < a_n \Rightarrow \{a_n\}$ is a decreasing sequence of nonnegative terms.

59. (a) $s_{10} = \sum_{n=1}^{10} \frac{1}{n^3} = 1.97531986$; $\int_{11}^{\infty} \frac{1}{x^3}\,dx = \lim_{b \to \infty} \int_{11}^{b} x^{-3}\,dx = \lim_{b \to \infty} \left[-\frac{x^{-2}}{2}\right]_{11}^{b} = \lim_{b \to \infty} \left(-\frac{1}{2b^2} + \frac{1}{242}\right) = \frac{1}{242}$ and

$\int_{10}^{\infty} \frac{1}{x^3}\,dx = \lim_{b \to \infty} \int_{10}^{b} x^{-3}\,dx = \lim_{b \to \infty} \left[-\frac{x^{-2}}{2}\right]_{10}^{b} = \lim_{b \to \infty} \left(-\frac{1}{2b^2} + \frac{1}{200}\right) = \frac{1}{200}$

$\Rightarrow 1.97531986 + \frac{1}{242} < s < 1.97531986 + \frac{1}{200} \Rightarrow 1.20166 < s < 1.20253$

(b) $s = \sum_{n=1}^{\infty} \frac{1}{n^3} \approx \frac{1.20166 + 1.20253}{2} = 1.202095$; error $\le \frac{1.20253 - 1.20166}{2} = 0.000435$

10.4 COMPARISON TESTS

1. Compare with $\sum_{n=1}^{\infty} \frac{1}{n^2}$, which is a convergent p-series, since $p = 2 > 1$. Both series have nonnegative terms for $n \ge 1$. For $n \ge 1$, we have $n^2 \le n^2 + 30 \Rightarrow \frac{1}{n^2} \ge \frac{1}{n^2+30}$. Then by Comparison Test, $\sum_{n=1}^{\infty} \frac{1}{n^2+30}$ converges.

3. Compare with $\sum_{n=2}^{\infty} \frac{1}{\sqrt{n}}$, which is a divergent p-series, since $p = \frac{1}{2} \le 1$. Both series have nonnegative terms for $n \ge 2$. For $n \ge 2$, we have $\sqrt{n} - 1 \le \sqrt{n} \Rightarrow \frac{1}{\sqrt{n}-1} \ge \frac{1}{\sqrt{n}}$. Then by Comparison Test, $\sum_{n=2}^{\infty} \frac{1}{\sqrt{n}-1}$ diverges.

5. Compare with $\sum_{n=1}^{\infty} \frac{1}{n^{3/2}}$, which is a convergent p-series, since $p = \frac{3}{2} > 1$. Both series have nonnegative terms for $n \ge 1$. For $n \ge 1$, we have $0 \le \cos^2 n \le 1 \Rightarrow \frac{\cos^2 n}{n^{3/2}} \le \frac{1}{n^{3/2}}$. Then by Comparison Test, $\sum_{n=1}^{\infty} \frac{\cos^2 n}{n^{3/2}}$ converges.

7. Compare with $\sum_{n=1}^{\infty} \frac{\sqrt{5}}{n^{3/2}}$. The series $\sum_{n=1}^{\infty} \frac{1}{n^{3/2}}$ is a convergent p-series, since $p = \frac{3}{2} > 1$, and the series $\sum_{n=1}^{\infty} \frac{\sqrt{5}}{n^{3/2}}$ $= \sqrt{5} \sum_{n=1}^{\infty} \frac{1}{n^{3/2}}$ converges by Theorem 8 part 3. Both series have nonnegative terms for $n \ge 1$. For $n \ge 1$, we have

$n^3 \le n^4 \Rightarrow 4n^3 \le 4n^4 \Rightarrow n^4 + 4n^3 \le n^4 + 4n^4 = 5n^4 \Rightarrow n^4 + 4n^3 \le 5n^4 + 20 = 5(n^4 + 4) \Rightarrow \frac{n^4+4n^3}{n^4+4} \le 5.$

$\Rightarrow \frac{n^3(n+4)}{n^4+4} \le 5 \Rightarrow \frac{n+4}{n^4+4} \le \frac{5}{n^3} \Rightarrow \sqrt{\frac{n+4}{n^4+4}} \le \sqrt{\frac{5}{n^3}} = \frac{\sqrt{5}}{n^{3/2}}$ Then by Comparison Test, $\sum_{n=1}^{\infty} \sqrt{\frac{n+4}{n^4+4}}$ converges.

9. Compare with $\sum_{n=1}^{\infty} \frac{1}{n^2}$, which is a convergent p-series, since $p = 2 > 1$. Both series have positive terms for $n \ge 1$. $\lim_{n\to\infty} \frac{a_n}{b_n}$ $= \lim_{n\to\infty} \frac{\frac{n-2}{n^3-n^2+3}}{1/n^2} = \lim_{n\to\infty} \frac{n^3-2n^2}{n^3-n^2+3} = \lim_{n\to\infty} \frac{3n^2-4n}{3n^2-2n} = \lim_{n\to\infty} \frac{6n-4}{6n-2} = \lim_{n\to\infty} \frac{6}{6} = 1 > 0$. Then by Limit Comparison Test, $\sum_{n=1}^{\infty} \frac{n-2}{n^3-n^2+3}$ converges.

11. Compare with $\sum_{n=2}^{\infty} \frac{1}{n}$, which is a divergent p-series, since $p = 1 \le 1$. Both series have positive terms for $n \ge 2$. $\lim_{n\to\infty} \frac{a_n}{b_n}$ $= \lim_{n\to\infty} \frac{\frac{n(n+1)}{(n^2+1)(n-1)}}{1/n} = \lim_{n\to\infty} \frac{n^3+n^2}{n^3-n^2+n-1} = \lim_{n\to\infty} \frac{3n^2+2n}{3n^2-2n+1} = \lim_{n\to\infty} \frac{6n+2}{6n-2} = \lim_{n\to\infty} \frac{6}{6} = 1 > 0$. Then by Limit Comparison Test, $\sum_{n=2}^{\infty} \frac{n(n+1)}{(n^2+1)(n-1)}$ diverges.

13. Compare with $\sum_{n=1}^{\infty} \frac{1}{\sqrt{n}}$, which is a divergent p-series, since $p = \frac{1}{2} \le 1$. Both series have positive terms for $n \ge 1$. $\lim_{n\to\infty} \frac{a_n}{b_n}$ $= \lim_{n\to\infty} \frac{\frac{5^n}{\sqrt{n}\cdot 4^n}}{1/\sqrt{n}} = \lim_{n\to\infty} \frac{5^n}{4^n} = \lim_{n\to\infty} \left(\frac{5}{4}\right)^n = \infty$. Then by Limit Comparison Test, $\sum_{n=1}^{\infty} \frac{5^n}{\sqrt{n}\cdot 4^n}$ diverges.

15. Compare with $\sum_{n=2}^{\infty} \frac{1}{n}$, which is a divergent p-series, since $p = 1 \le 1$. Both series have positive terms for $n \ge 2$. $\lim_{n\to\infty} \frac{a_n}{b_n}$ $= \lim_{n\to\infty} \frac{\frac{1}{\ln n}}{1/n} = \lim_{n\to\infty} \frac{n}{\ln n} = \lim_{n\to\infty} \frac{1}{1/n} = \lim_{n\to\infty} n = \infty$. Then by Limit Comparison Test, $\sum_{n=2}^{\infty} \frac{1}{\ln n}$ diverges.

17. diverges by the Limit Comparison Test (part 1) when compared with $\sum_{n=1}^{\infty} \frac{1}{\sqrt{n}}$, a divergent p-series:

$$\lim_{n\to\infty} \frac{\left(\frac{1}{2\sqrt{n}+\sqrt[3]{n}}\right)}{\left(\frac{1}{\sqrt{n}}\right)} = \lim_{n\to\infty} \frac{\sqrt{n}}{2\sqrt{n}+\sqrt[3]{n}} = \lim_{n\to\infty} \left(\frac{1}{2+n^{-1/6}}\right) = \frac{1}{2}$$

19. converges by the Direct Comparison Test; $\frac{\sin^2 n}{2^n} \le \frac{1}{2^n}$, which is the nth term of a convergent geometric series

21. diverges since $\lim_{n\to\infty} \frac{2n}{3n-1} = \frac{2}{3} \ne 0$

23. converges by the Limit Comparison Test (part 1) with $\frac{1}{n^2}$, the nth term of a convergent p-series:

$$\lim_{n\to\infty} \frac{\left(\frac{10n+1}{n(n+1)(n+2)}\right)}{\left(\frac{1}{n^2}\right)} = \lim_{n\to\infty} \frac{10n^2+n}{n^2+3n+2} = \lim_{n\to\infty} \frac{20n+1}{2n+3} = \lim_{n\to\infty} \frac{20}{2} = 10$$

25. converges by the Direct Comparison Test; $\left(\frac{n}{3n+1}\right)^n < \left(\frac{n}{3n}\right)^n = \left(\frac{1}{3}\right)^n$, the nth term of a convergent geometric series

27. diverges by the Direct Comparison Test; $n > \ln n \Rightarrow \ln n > \ln \ln n \Rightarrow \frac{1}{n} < \frac{1}{\ln n} < \frac{1}{\ln(\ln n)}$ and $\sum_{n=3}^{\infty} \frac{1}{n}$ diverges

29. diverges by the Limit Comparison Test (part 3) with $\frac{1}{n}$, the nth term of the divergent harmonic series:

$$\lim_{n\to\infty} \frac{\left[\frac{1}{\sqrt{n}\ln n}\right]}{\left(\frac{1}{n}\right)} = \lim_{n\to\infty} \frac{\sqrt{n}}{\ln n} = \lim_{n\to\infty} \frac{\left(\frac{1}{2\sqrt{n}}\right)}{\left(\frac{1}{n}\right)} = \lim_{n\to\infty} \frac{\sqrt{n}}{2} = \infty$$

31. diverges by the Limit Comparison Test (part 3) with $\frac{1}{n}$, the nth term of the divergent harmonic series:

$$\lim_{n\to\infty} \frac{\left(\frac{1}{1+\ln n}\right)}{\left(\frac{1}{n}\right)} = \lim_{n\to\infty} \frac{n}{1+\ln n} = \lim_{n\to\infty} \frac{1}{\left(\frac{1}{n}\right)} = \lim_{n\to\infty} n = \infty$$

33. converges by the Direct Comparison Test with $\frac{1}{n^{3/2}}$, the nth term of a convergent p-series: $n^2 - 1 > n$ for $n \ge 2 \Rightarrow n^2(n^2 - 1) > n^3 \Rightarrow n\sqrt{n^2 - 1} > n^{3/2} \Rightarrow \frac{1}{n^{3/2}} > \frac{1}{n\sqrt{n^2 - 1}}$ or use Limit Comparison Test with $\frac{1}{n^2}$.

35. converges because $\sum_{n=1}^{\infty} \frac{1-n}{n2^n} = \sum_{n=1}^{\infty} \frac{1}{n2^n} + \sum_{n=1}^{\infty} \frac{-1}{2^n}$ which is the sum of two convergent series: $\sum_{n=1}^{\infty} \frac{1}{n2^n}$ converges by the Direct Comparison Test since $\frac{1}{n2^n} < \frac{1}{2^n}$, and $\sum_{n=1}^{\infty} \frac{-1}{2^n}$ is a convergent geometric series

37. converges by the Direct Comparison Test: $\frac{1}{3^{n-1}+1} < \frac{1}{3^{n-1}}$, which is the nth term of a convergent geometric series

39. converges by Limit Comparison Test: compare with $\sum_{n=1}^{\infty} \left(\frac{1}{5}\right)^n$, which is a convergent geometric series with $|r| = \frac{1}{5} < 1$,

$$\lim_{n\to\infty} \frac{\left(\frac{n+1}{n^2+3n} \cdot \frac{1}{5^n}\right)}{(1/5)^n} = \lim_{n\to\infty} \frac{n+1}{n^2+3n} = \lim_{n\to\infty} \frac{1}{2n+3} = 0.$$

41. diverges by Limit Comparison Test: compare with $\sum_{n=1}^{\infty} \frac{1}{n}$, which is a divergent p-series, $\lim_{n\to\infty} \frac{\left(\frac{2^n - n}{n \cdot 2^n}\right)}{1/n} = \lim_{n\to\infty} \frac{2^n - n}{2^n}$ $= \lim_{n\to\infty} \frac{2^n \ln 2 - 1}{2^n \ln 2} = \lim_{n\to\infty} \frac{2^n (\ln 2)^2}{2^n (\ln 2)^2} = 1 > 0.$

43. converges by Comparison Test with $\sum_{n=2}^{\infty} \frac{1}{n(n-1)}$ which converges since $\sum_{n=2}^{\infty} \frac{1}{n(n-1)} = \sum_{n=2}^{\infty} \left[\frac{1}{n-1} - \frac{1}{n}\right]$, and $s_k = \left(1 - \frac{1}{2}\right) + \left(\frac{1}{2} - \frac{1}{3}\right) + \ldots + \left(\frac{1}{k-2} - \frac{1}{k-1}\right) + \left(\frac{1}{k-1} - \frac{1}{k}\right) = 1 - \frac{1}{k} \Rightarrow \lim_{k\to\infty} s_k = 1$; for $n \ge 2$, $(n-2)! \ge 1$ $\Rightarrow n(n-1)(n-2)! \ge n(n-1) \Rightarrow n! \ge n(n-1) \Rightarrow \frac{1}{n!} \le \frac{1}{n(n-1)}$

45. diverges by the Limit Comparison Test (part 1) with $\frac{1}{n}$, the nth term of the divergent harmonic series:

$$\lim_{n\to\infty} \frac{\left(\sin \frac{1}{n}\right)}{\left(\frac{1}{n}\right)} = \lim_{x\to 0} \frac{\sin x}{x} = 1$$

47. converges by the Direct Comparison Test: $\frac{\tan^{-1} n}{n^{1.1}} < \frac{\frac{\pi}{2}}{n^{1.1}}$ and $\sum_{n=1}^{\infty} \frac{\frac{\pi}{2}}{n^{1.1}} = \frac{\pi}{2} \sum_{n=1}^{\infty} \frac{1}{n^{1.1}}$ is the product of a convergent p-series and a nonzero constant

49. converges by the Limit Comparison Test (part 1) with $\frac{1}{n^2}$: $\lim_{n\to\infty} \frac{\left(\frac{\coth n}{n^2}\right)}{\left(\frac{1}{n^2}\right)} = \lim_{n\to\infty} \coth n = \lim_{n\to\infty} \frac{e^n + e^{-n}}{e^n - e^{-n}}$ $= \lim_{n\to\infty} \frac{1+e^{-2n}}{1-e^{-2n}} = 1$

51. diverges by the Limit Comparison Test (part 1) with $\frac{1}{n}$: $\lim_{n\to\infty} \frac{\left(\frac{1}{n\sqrt[n]{n}}\right)}{\left(\frac{1}{n}\right)} = \lim_{n\to\infty} \frac{1}{\sqrt[n]{n}} = 1.$

53. $\frac{1}{1+2+3+\ldots+n} = \frac{1}{\left(\frac{n(n+1)}{2}\right)} = \frac{2}{n(n+1)}$. The series converges by the Limit Comparison Test (part 1) with $\frac{1}{n^2}$:

$$\lim_{n\to\infty} \frac{\left(\frac{2}{n(n+1)}\right)}{\left(\frac{1}{n^2}\right)} = \lim_{n\to\infty} \frac{2n^2}{n^2+n} = \lim_{n\to\infty} \frac{4n}{2n+1} = \lim_{n\to\infty} \frac{4}{2} = 2.$$

55. (a) If $\lim_{n\to\infty} \frac{a_n}{b_n} = 0$, then there exists an integer N such that for all $n > N$, $\left|\frac{a_n}{b_n} - 0\right| < 1 \Rightarrow -1 < \frac{a_n}{b_n} < 1$ $\Rightarrow a_n < b_n$. Thus, if $\sum b_n$ converges, then $\sum a_n$ converges by the Direct Comparison Test.

(b) If $\lim_{n\to\infty} \frac{a_n}{b_n} = \infty$, then there exists an integer N such that for all $n > N$, $\frac{a_n}{b_n} > 1 \Rightarrow a_n > b_n$. Thus, if $\sum b_n$ diverges, then $\sum a_n$ diverges by the Direct Comparison Test.

57. $\lim_{n\to\infty} \frac{a_n}{b_n} = \infty \Rightarrow$ there exists an integer N such that for all $n > N$, $\frac{a_n}{b_n} > 1 \Rightarrow a_n > b_n$. If $\sum a_n$ converges, then $\sum b_n$ converges by the Direct Comparison Test

59. Since $a_n > 0$ and $\lim_{n\to\infty} a_n = \infty \neq 0$, by n^{th} term test for divergence, $\sum a_n$ diverges.

61. Let $-\infty < q < \infty$ and $p > 1$. If $q = 0$, then $\sum_{n=2}^{\infty} \frac{(\ln n)^q}{n^p} = \sum_{n=2}^{\infty} \frac{1}{n^p}$, which is a convergent p-series. If $q \neq 0$, compare with $\sum_{n=2}^{\infty} \frac{1}{n^r}$ where $1 < r < p$, then $\lim_{n\to\infty} \frac{\frac{(\ln n)^q}{n^p}}{1/n^r} = \lim_{n\to\infty} \frac{(\ln n)^q}{n^{p-r}}$, and $p - r > 0$. If $q < 0 \Rightarrow -q > 0$ and $\lim_{n\to\infty} \frac{(\ln n)^q}{n^{p-r}}$ $= \lim_{n\to\infty} \frac{1}{(\ln n)^{-q} n^{p-r}} = 0$. If $q > 0$, $\lim_{n\to\infty} \frac{(\ln n)^q}{n^{p-r}} = \lim_{n\to\infty} \frac{q(\ln n)^{q-1}\left(\frac{1}{n}\right)}{(p-r)n^{p-r-1}} = \lim_{n\to\infty} \frac{q(\ln n)^{q-1}}{(p-r)n^{p-r}}$. If $q - 1 \le 0 \Rightarrow 1 - q \ge 0$ and $\lim_{n\to\infty} \frac{q(\ln n)^{q-1}}{(p-r)n^{p-r}} = \lim_{n\to\infty} \frac{q}{(p-r)n^{p-r}(\ln n)^{1-q}} = 0$, otherwise, we apply L'Hopital's Rule again. $\lim_{n\to\infty} \frac{q(q-1)(\ln n)^{q-2}\left(\frac{1}{n}\right)}{(p-r)^2 n^{p-r-1}}$ $= \lim_{n\to\infty} \frac{q(q-1)(\ln n)^{q-2}}{(p-r)^2 n^{p-r}}$. If $q - 2 \le 0 \Rightarrow 2 - q \ge 0$ and $\lim_{n\to\infty} \frac{q(q-1)(\ln n)^{q-2}}{(p-r)^2 n^{p-r}} = \lim_{n\to\infty} \frac{q(q-1)}{(p-r)^2 n^{p-r}(\ln n)^{2-q}} = 0$; otherwise, we apply L'Hopital's Rule again. Since q is finite, there is a positive integer k such that $q - k \le 0 \Rightarrow k - q \ge 0$. Thus, after k applications of L'Hopital's Rule we obtain $\lim_{n\to\infty} \frac{q(q-1)\cdots(q-k+1)(\ln n)^{q-k}}{(p-r)^k n^{p-r}} = \lim_{n\to\infty} \frac{q(q-1)\cdots(q-k+1)}{(p-r)^k n^{p-r}(\ln n)^{k-q}} = 0$. Since the limit is 0 in every case, by Limit Comparison Test, the series $\sum_{n=1}^{\infty} \frac{(\ln n)^q}{n^p}$ converges.

63. Converges by Exercise 61 with $q = 3$ and $p = 4$.

65. Converges by Exercise 61 with $q = 1000$ and $p = 1.001$.

67. Converges by Exercise 61 with $q = -3$ and $p = 1.1$.

10.5 THE RATIO AND ROOT TESTS

1. $\frac{2^n}{n!} > 0$ for all $n \ge 1$; $\lim_{n\to\infty} \left(\frac{\frac{2^{n+1}}{(n+1)!}}{\frac{2^n}{n!}}\right) = \lim_{n\to\infty} \left(\frac{2^n \cdot 2}{(n+1)\cdot n!} \cdot \frac{n!}{2^n}\right) = \lim_{n\to\infty} \left(\frac{2}{n+1}\right) = 0 < 1 \Rightarrow \sum_{n=1}^{\infty} \frac{2^n}{n!}$ converges

3. $\frac{(n-1)!}{(n+1)^2} > 0$ for all $n \ge 1$; $\lim_{n\to\infty} \left(\frac{\frac{((n+1)-1)!}{((n+1)+1)^2}}{\frac{(n-1)!}{(n+1)^2}}\right) = \lim_{n\to\infty} \left(\frac{n\cdot(n-1)!}{(n+2)^2} \cdot \frac{(n+1)^2}{(n-1)!}\right) = \lim_{n\to\infty} \left(\frac{n^3+2n^2+n}{n^2+4n+4}\right) = \lim_{n\to\infty} \left(\frac{3n^2+4n+1}{2n+4}\right)$ $= \lim_{n\to\infty} \left(\frac{6n+4}{2}\right) = \infty > 1 \Rightarrow \sum_{n=1}^{\infty} \frac{(n-1)!}{(n+1)^2}$ diverges

5. $\frac{n^4}{4^n} > 0$ for all $n \geq 1$; $\lim_{n\to\infty} \left(\frac{\frac{(n+1)^4}{4^{n+1}}}{\frac{n^4}{4^n}} \right) = \lim_{n\to\infty} \left(\frac{(n+1)^4}{4^n \cdot 4} \cdot \frac{4^n}{n^4} \right) = \lim_{n\to\infty} \left(\frac{n^4+4n^3+6n^2+4n+1}{4n^4} \right)$

$= \lim_{n\to\infty} \left(\frac{1}{4} + \frac{1}{n} + \frac{3}{2n^2} + \frac{1}{n^3} + \frac{1}{4n^4} \right) = \frac{1}{4} < 1 \Rightarrow \sum_{n=1}^{\infty} \frac{n^4}{4^n}$ converges

7. $\frac{n^2(n+2)!}{n!3^{2n}} > 0$ for all $n \geq 1$; $\lim_{n\to\infty} \left(\frac{\frac{(n+1)^2((n+1)+2)!}{(n+1)!3^{2(n+1)}}}{\frac{n^2(n+2)!}{n!3^{2n}}} \right) = \lim_{n\to\infty} \left(\frac{(n+1)^2(n+3)(n+2)!}{(n+1)\cdot n!3^{2n}\cdot 3^2} \cdot \frac{n!3^{2n}}{n^2(n+2)!} \right) = \lim_{n\to\infty} \left(\frac{n^3+5n^2+7n+3}{9n^3+9n^2} \right)$

$= \lim_{n\to\infty} \left(\frac{3n^2+15n+7}{27n^2+18n} \right) = \lim_{n\to\infty} \left(\frac{6n+15}{54n+18} \right) = \lim_{n\to\infty} \left(\frac{6}{54} \right) = \frac{1}{9} < 1 \Rightarrow \sum_{n=1}^{\infty} \frac{n^2(n+2)!}{n!3^{2n}}$ converges

9. $\frac{7}{(2n+5)^n} \geq 0$ for all $n \geq 1$; $\lim_{n\to\infty} \sqrt[n]{\frac{7}{(2n+5)^n}} = \lim_{n\to\infty} \left(\frac{\sqrt[n]{7}}{2n+5} \right) = 0 < 1 \Rightarrow \sum_{n=1}^{\infty} \frac{7}{(2n+5)^n}$ converges

11. $\left(\frac{4n+3}{3n-5} \right)^n \geq 0$ for all $n \geq 2$; $\lim_{n\to\infty} \sqrt[n]{\left(\frac{4n+3}{3n-5} \right)^n} = \lim_{n\to\infty} \left(\frac{4n+3}{3n-5} \right) = \lim_{n\to\infty} \left(\frac{4}{3} \right) = \frac{4}{3} > 1 \Rightarrow \sum_{n=1}^{\infty} \left(\frac{4n+3}{3n-5} \right)^n$ diverges

13. $\frac{8}{\left(3+\frac{1}{n}\right)^{2n}} \geq 0$ for all $n \geq 1$; $\lim_{n\to\infty} \sqrt[n]{\frac{8}{\left(3+\frac{1}{n}\right)^{2n}}} = \lim_{n\to\infty} \left(\frac{\sqrt[n]{8}}{\left(3+\frac{1}{n}\right)^2} \right) = \frac{1}{9} < 1 \Rightarrow \sum_{n=1}^{\infty} \frac{8}{\left(3+\frac{1}{n}\right)^{2n}}$ converges

15. $\left(1-\frac{1}{n}\right)^{n^2} \geq 0$ for all $n \geq 1$; $\lim_{n\to\infty} \sqrt[n]{\left(1-\frac{1}{n}\right)^{n^2}} = \lim_{n\to\infty} \left(1-\frac{1}{n}\right)^n = e^{-1} < 1 \Rightarrow \sum_{n=1}^{\infty} \left(1-\frac{1}{n}\right)^{n^2}$ converges

17. converges by the Ratio Test: $\lim_{n\to\infty} \frac{a_{n+1}}{a_n} = \lim_{n\to\infty} \frac{\left[\frac{(n+1)\sqrt{2}}{2^{n+1}}\right]}{\left[\frac{n\sqrt{2}}{2^n}\right]} = \lim_{n\to\infty} \frac{(n+1)^{\sqrt{2}}}{2^{n+1}} \cdot \frac{2^n}{n^{\sqrt{2}}} = \lim_{n\to\infty} \left(1+\frac{1}{n}\right)^{\sqrt{2}} \left(\frac{1}{2}\right) = \frac{1}{2} < 1$

19. diverges by the Ratio Test: $\lim_{n\to\infty} \frac{a_{n+1}}{a_n} = \lim_{n\to\infty} \frac{\left(\frac{(n+1)!}{e^{n+1}}\right)}{\left(\frac{n!}{e^n}\right)} = \lim_{n\to\infty} \frac{(n+1)!}{e^{n+1}} \cdot \frac{e^n}{n!} = \lim_{n\to\infty} \frac{n+1}{e} = \infty$

21. converges by the Ratio Test: $\lim_{n\to\infty} \frac{a_{n+1}}{a_n} = \lim_{n\to\infty} \frac{\left(\frac{(n+1)^{10}}{10^{n+1}}\right)}{\left(\frac{n^{10}}{10^n}\right)} = \lim_{n\to\infty} \frac{(n+1)^{10}}{10^{n+1}} \cdot \frac{10^n}{n^{10}} = \lim_{n\to\infty} \left(1+\frac{1}{n}\right)^{10} \left(\frac{1}{10}\right) = \frac{1}{10} < 1$

23. converges by the Direct Comparison Test: $\frac{2+(-1)^n}{(1.25)^n} = \left(\frac{4}{5}\right)^n [2+(-1)^n] \leq \left(\frac{4}{5}\right)^n (3)$ which is the nth term of a convergent geometric series

25. diverges; $\lim_{n\to\infty} a_n = \lim_{n\to\infty} \left(1-\frac{3}{n}\right)^n = \lim_{n\to\infty} \left(1+\frac{-3}{n}\right)^n = e^{-3} \approx 0.05 \neq 0$

27. converges by the Direct Comparison Test: $\frac{\ln n}{n^3} < \frac{n}{n^3} = \frac{1}{n^2}$ for $n \geq 2$, the nth term of a convergent p-series.

29. diverges by the Direct Comparison Test: $\frac{1}{n} - \frac{1}{n^2} = \frac{n-1}{n^2} > \frac{1}{2}\left(\frac{1}{n}\right)$ for $n > 2$ or by the Limit Comparison Test (part 1) with $\frac{1}{n}$.

31. diverges by the Direct Comparison Test: $\frac{\ln n}{n} > \frac{1}{n}$ for $n \geq 3$

33. converges by the Ratio Test: $\lim_{n\to\infty} \frac{a_{n+1}}{a_n} = \lim_{n\to\infty} \frac{(n+2)(n+3)}{(n+1)!} \cdot \frac{n!}{(n+1)(n+2)} = 0 < 1$

35. converges by the Ratio Test: $\lim_{n\to\infty} \frac{a_{n+1}}{a_n} = \lim_{n\to\infty} \frac{(n+4)!}{3!(n+1)!3^{n+1}} \cdot \frac{3!n!3^n}{(n+3)!} = \lim_{n\to\infty} \frac{n+4}{3(n+1)} = \frac{1}{3} < 1$

37. converges by the Ratio Test: $\lim_{n\to\infty} \frac{a_{n+1}}{a_n} = \lim_{n\to\infty} \frac{(n+1)!}{(2n+3)!} \cdot \frac{(2n+1)!}{n!} = \lim_{n\to\infty} \frac{n+1}{(2n+3)(2n+2)} = 0 < 1$

39. converges by the Root Test: $\lim_{n\to\infty} \sqrt[n]{a_n} = \lim_{n\to\infty} \sqrt[n]{\frac{n}{(\ln n)^n}} = \lim_{n\to\infty} \frac{\sqrt[n]{n}}{\ln n} = \lim_{n\to\infty} \frac{1}{\ln n} = 0 < 1$

41. converges by the Direct Comparison Test: $\frac{n! \ln n}{n(n+2)!} = \frac{\ln n}{n(n+1)(n+2)} < \frac{n}{n(n+1)(n+2)} = \frac{1}{(n+1)(n+2)} < \frac{1}{n^2}$ which is the nth-term of a convergent p-series

43. converges by the Ratio Test: $\lim_{n\to\infty} \frac{a_{n+1}}{a_n} = \lim_{n\to\infty} \frac{[(n+1)!]^2}{[2(n+1)]!} \cdot \frac{(2n)!}{[n!]^2} = \lim_{n\to\infty} \frac{(n+1)^2}{(2n+2)(2n+1)} = \lim_{n\to\infty} \frac{n^2+2n+1}{4n^2+6n+2} = \frac{1}{4} < 1$

45. converges by the Ratio Test: $\lim_{n\to\infty} \frac{a_{n+1}}{a_n} = \lim_{n\to\infty} \frac{\left(\frac{1+\sin n}{n}\right) a_n}{a_n} = 0 < 1$

47. diverges by the Ratio Test: $\lim_{n\to\infty} \frac{a_{n+1}}{a_n} = \lim_{n\to\infty} \frac{\left(\frac{3n-1}{2n+5}\right) a_n}{a_n} = \lim_{n\to\infty} \frac{3n-1}{2n+5} = \frac{3}{2} > 1$

49. converges by the Ratio Test: $\lim_{n\to\infty} \frac{a_{n+1}}{a_n} = \lim_{n\to\infty} \frac{\left(\frac{2}{n}\right) a_n}{a_n} = \lim_{n\to\infty} \frac{2}{n} = 0 < 1$

51. converges by the Ratio Test: $\lim_{n\to\infty} \frac{a_{n+1}}{a_n} = \lim_{n\to\infty} \frac{\left(\frac{1+\ln n}{n}\right) a_n}{a_n} = \lim_{n\to\infty} \frac{1+\ln n}{n} = \lim_{n\to\infty} \frac{1}{n} = 0 < 1$

53. diverges by the nth-Term Test: $a_1 = \frac{1}{3}$, $a_2 = \sqrt[2]{\frac{1}{3}}$, $a_3 = \sqrt[3]{\sqrt[2]{\frac{1}{3}}} = \sqrt[6]{\frac{1}{3}}$, $a_4 = \sqrt[4]{\sqrt[3]{\sqrt[2]{\frac{1}{3}}}} = \sqrt[4!]{\frac{1}{3}}, \ldots,$

$a_n = \sqrt[n!]{\frac{1}{3}} \Rightarrow \lim_{n\to\infty} a_n = 1$ because $\left\{\sqrt[n!]{\frac{1}{3}}\right\}$ is a subsequence of $\left\{\sqrt[n]{\frac{1}{3}}\right\}$ whose limit is 1 by Table 8.1

55. converges by the Ratio Test: $\lim_{n\to\infty} \frac{a_{n+1}}{a_n} = \lim_{n\to\infty} \frac{2^{n+1}(n+1)!\,(n+1)!}{(2n+2)!} \cdot \frac{(2n)!}{2^n n!\, n!} = \lim_{n\to\infty} \frac{2(n+1)(n+1)}{(2n+2)(2n+1)}$
$= \lim_{n\to\infty} \frac{n+1}{2n+1} = \frac{1}{2} < 1$

57. diverges by the Root Test: $\lim_{n\to\infty} \sqrt[n]{a_n} \equiv \lim_{n\to\infty} \sqrt[n]{\frac{(n!)^n}{(n^n)^2}} = \lim_{n\to\infty} \frac{n!}{n^2} = \infty > 1$

59. converges by the Root Test: $\lim_{n\to\infty} \sqrt[n]{a_n} = \lim_{n\to\infty} \sqrt[n]{\frac{n^n}{2^{n^2}}} = \lim_{n\to\infty} \frac{n}{2^n} = \lim_{n\to\infty} \frac{1}{2^n \ln 2} = 0 < 1$

61. converges by the Ratio Test: $\lim_{n\to\infty} \frac{a_{n+1}}{a_n} = \lim_{n\to\infty} \frac{1\cdot 3\cdot \cdots \cdot(2n-1)(2n+1)}{4^{n+1}2^{n+1}(n+1)!} \cdot \frac{4^n\, 2^n\, n!}{1\cdot 3\cdot \cdots \cdot(2n-1)} = \lim_{n\to\infty} \frac{2n+1}{(4\cdot 2)(n+1)} = \frac{1}{4} < 1$

63. Ratio: $\lim_{n\to\infty} \frac{a_{n+1}}{a_n} = \lim_{n\to\infty} \frac{1}{(n+1)^p} \cdot \frac{n^p}{1} = \lim_{n\to\infty} \left(\frac{n}{n+1}\right)^p = 1^p = 1 \Rightarrow$ no conclusion

Root: $\lim_{n\to\infty} \sqrt[n]{a_n} = \lim_{n\to\infty} \sqrt[n]{\frac{1}{n^p}} = \lim_{n\to\infty} \frac{1}{(\sqrt[n]{n})^p} = \frac{1}{(1)^p} = 1 \Rightarrow$ no conclusion

65. $a_n \le \frac{n}{2^n}$ for every n and the series $\sum_{n=1}^{\infty} \frac{n}{2^n}$ converges by the Ratio Test since $\lim_{n\to\infty} \frac{(n+1)}{2^{n+1}} \cdot \frac{2^n}{n} = \frac{1}{2} < 1$

$\Rightarrow \sum_{n=1}^{\infty} a_n$ converges by the Direct Comparison Test

10.6 ALTERNATING SERIES, ABSOLUTE AND CONDITIONAL CONVERGENCE

1. converges by the Alternating Convergence Test since: $u_n = \frac{1}{\sqrt{n}} > 0$ for all $n \geq 1$; $n \geq 1 \Rightarrow n+1 \geq n \Rightarrow \sqrt{n+1} \geq \sqrt{n}$ $\Rightarrow \frac{1}{\sqrt{n+1}} \leq \frac{1}{\sqrt{n}} \Rightarrow u_{n+1} \leq u_n$; $\lim_{n\to\infty} u_n = \lim_{n\to\infty} \frac{1}{\sqrt{n}} = 0$.

3. converges $\Rightarrow$ converges by Alternating Series Test since: $u_n = \frac{1}{n3^n} > 0$ for all $n \geq 1$; $n \geq 1 \Rightarrow n+1 \geq n \Rightarrow 3^{n+1} \geq 3^n$ $\Rightarrow (n+1)3^{n+1} \geq n\,3^n \Rightarrow \frac{1}{(n+1)3^{n+1}} \leq \frac{1}{n\,3^n} \Rightarrow u_{n+1} \leq u_n$; $\lim_{n\to\infty} u_n = \lim_{n\to\infty} \frac{1}{n\,3^n} = 0$.

5. converges $\Rightarrow$ converges by Alternating Series Test since: $u_n = \frac{n}{n^2+1} > 0$ for all $n \geq 1$; $n \geq 1 \Rightarrow 2n^2 + 2n \geq n^2 + n + 1$ $\Rightarrow n^3 + 2n^2 + 2n \geq n^3 + n^2 + n + 1 \Rightarrow n(n^2 + 2n + 2) \geq n^3 + n^2 + n + 1 \Rightarrow n\left((n+1)^2 + 1\right) \geq (n^2+1)(n+1)$ $\Rightarrow \frac{n}{n^2+1} \geq \frac{n+1}{(n+1)^2+1} \Rightarrow u_{n+1} \leq u_n$; $\lim_{n\to\infty} u_n = \lim_{n\to\infty} \frac{n}{n^2+1} = 0$.

7. diverges $\Rightarrow$ diverges by n^{th} Term Test for Divergence since: $\lim_{n\to\infty} \frac{2^n}{n^2} = \infty \Rightarrow \lim_{n\to\infty} (-1)^{n+1}\frac{2^n}{n^2}$ = does not exist

9. diverges by the nth-Term Test since for $n > 10 \Rightarrow \frac{n}{10} > 1 \Rightarrow \lim_{n\to\infty} \left(\frac{n}{10}\right)^n \neq 0 \Rightarrow \sum_{n=1}^{\infty} (-1)^{n+1}\left(\frac{n}{10}\right)^n$ diverges

11. converges by the Alternating Series Test since $f(x) = \frac{\ln x}{x} \Rightarrow f'(x) = \frac{1-\ln x}{x^2} < 0$ when $x > e \Rightarrow f(x)$ is decreasing $\Rightarrow u_n \geq u_{n+1}$; also $u_n \geq 0$ for $n \geq 1$ and $\lim_{n\to\infty} u_n = \lim_{n\to\infty} \frac{\ln n}{n} = \lim_{n\to\infty} \frac{\left(\frac{1}{n}\right)}{1} = 0$

13. converges by the Alternating Series Test since $f(x) = \frac{\sqrt{x}+1}{x+1} \Rightarrow f'(x) = \frac{1-x-2\sqrt{x}}{2\sqrt{x}(x+1)^2} < 0 \Rightarrow f(x)$ is decreasing $\Rightarrow u_n \geq u_{n+1}$; also $u_n \geq 0$ for $n \geq 1$ and $\lim_{n\to\infty} u_n = \lim_{n\to\infty} \frac{\sqrt{n}+1}{n+1} = 0$

15. converges absolutely since $\sum_{n=1}^{\infty} |a_n| = \sum_{n=1}^{\infty} \left(\frac{1}{10}\right)^n$ a convergent geometric series

17. converges conditionally since $\frac{1}{\sqrt{n}} > \frac{1}{\sqrt{n+1}} > 0$ and $\lim_{n\to\infty} \frac{1}{\sqrt{n}} = 0 \Rightarrow$ convergence; but $\sum_{n=1}^{\infty} |a_n| = \sum_{n=1}^{\infty} \frac{1}{n^{1/2}}$ is a divergent p-series

19. converges absolutely since $\sum_{n=1}^{\infty} |a_n| = \sum_{n=1}^{\infty} \frac{n}{n^3+1}$ and $\frac{n}{n^3+1} < \frac{1}{n^2}$ which is the nth-term of a converging p-series

21. converges conditionally since $\frac{1}{n+3} > \frac{1}{(n+1)+3} > 0$ and $\lim_{n\to\infty} \frac{1}{n+3} = 0 \Rightarrow$ convergence; but $\sum_{n=1}^{\infty} |a_n|$ $= \sum_{n=1}^{\infty} \frac{1}{n+3}$ diverges because $\frac{1}{n+3} \geq \frac{1}{4n}$ and $\sum_{n=1}^{\infty} \frac{1}{n}$ is a divergent series

23. diverges by the nth-Term Test since $\lim_{n\to\infty} \frac{3+n}{5+n} = 1 \neq 0$

25. converges conditionally since $f(x) = \frac{1}{x^2} + \frac{1}{x} \Rightarrow f'(x) = -\left(\frac{2}{x^3} + \frac{1}{x^2}\right) < 0 \Rightarrow f(x)$ is decreasing and hence $u_n > u_{n+1} > 0$ for $n \geq 1$ and $\lim_{n\to\infty} \left(\frac{1}{n^2} + \frac{1}{n}\right) = 0 \Rightarrow$ convergence; but $\sum_{n=1}^{\infty} |a_n| = \sum_{n=1}^{\infty} \frac{1+n}{n^2}$ $= \sum_{n=1}^{\infty} \frac{1}{n^2} + \sum_{n=1}^{\infty} \frac{1}{n}$ is the sum of a convergent and divergent series, and hence diverges

27. converges absolutely by the Ratio Test: $\lim_{n\to\infty}\left(\frac{u_{n+1}}{u_n}\right) = \lim_{n\to\infty}\left[\frac{(n+1)^2\left(\frac{2}{3}\right)^{n+1}}{n^2\left(\frac{2}{3}\right)^n}\right] = \frac{2}{3} < 1$

29. converges absolutely by the Integral Test since $\int_1^{\infty}(\tan^{-1}x)\left(\frac{1}{1+x^2}\right)dx = \lim_{b\to\infty}\left[\frac{(\tan^{-1}x)^2}{2}\right]_1^b$
$= \lim_{b\to\infty}\left[(\tan^{-1}b)^2 - (\tan^{-1}1)^2\right] = \frac{1}{2}\left[\left(\frac{\pi}{2}\right)^2 - \left(\frac{\pi}{4}\right)^2\right] = \frac{3\pi^2}{32}$

31. diverges by the nth-Term Test since $\lim_{n\to\infty}\frac{n}{n+1} = 1 \neq 0$

33. converges absolutely by the Ratio Test: $\lim_{n\to\infty}\left(\frac{u_{n+1}}{u_n}\right) = \lim_{n\to\infty}\frac{(100)^{n+1}}{(n+1)!}\cdot\frac{n!}{(100)^n} = \lim_{n\to\infty}\frac{100}{n+1} = 0 < 1$

35. converges absolutely since $\sum_{n=1}^{\infty}|a_n| = \sum_{n=1}^{\infty}\left|\frac{(-1)^n}{n\sqrt{n}}\right| = \sum_{n=1}^{\infty}\frac{1}{n^{3/2}}$ is a convergent p-series

37. converges absolutely by the Root Test: $\lim_{n\to\infty}\sqrt[n]{|a_n|} = \lim_{n\to\infty}\left(\frac{(n+1)^n}{(2n)^n}\right)^{1/n} = \lim_{n\to\infty}\frac{n+1}{2n} = \frac{1}{2} < 1$

39. diverges by the nth-Term Test since $\lim_{n\to\infty}|a_n| = \lim_{n\to\infty}\frac{(2n)!}{2^n n!\,n} = \lim_{n\to\infty}\frac{(n+1)(n+2)\cdots(2n)}{2^n n}$
$= \lim_{n\to\infty}\frac{(n+1)(n+2)\cdots(n+(n-1))}{2^{n-1}} > \lim_{n\to\infty}\left(\frac{n+1}{2}\right)^{n-1} = \infty \neq 0$

41. converges conditionally since $\frac{\sqrt{n+1}-\sqrt{n}}{1}\cdot\frac{\sqrt{n+1}+\sqrt{n}}{\sqrt{n+1}+\sqrt{n}} = \frac{1}{\sqrt{n+1}+\sqrt{n}}$ and $\left\{\frac{1}{\sqrt{n+1}+\sqrt{n}}\right\}$ is a decreasing sequence of positive terms which converges to 0 $\Rightarrow \sum_{n=1}^{\infty}\frac{(-1)^n}{\sqrt{n+1}+\sqrt{n}}$ converges; but $\sum_{n=1}^{\infty}|a_n| = \sum_{n=1}^{\infty}\frac{1}{\sqrt{n+1}+\sqrt{n}}$ diverges by the Limit Comparison Test (part 1) with $\frac{1}{\sqrt{n}}$; a divergent p-series:
$\lim_{n\to\infty}\left(\frac{\frac{1}{\sqrt{n+1}+\sqrt{n}}}{\frac{1}{\sqrt{n}}}\right) = \lim_{n\to\infty}\frac{\sqrt{n}}{\sqrt{n+1}+\sqrt{n}} = \lim_{n\to\infty}\frac{1}{\sqrt{1+\frac{1}{n}}+1} = \frac{1}{2}$

43. diverges by the nth-Term Test since $\lim_{n\to\infty}\left(\sqrt{n+\sqrt{n}}-\sqrt{n}\right) = \lim_{n\to\infty}\left[\left(\sqrt{n+\sqrt{n}}-\sqrt{n}\right)\left(\frac{\sqrt{n+\sqrt{n}}+\sqrt{n}}{\sqrt{n+\sqrt{n}}+\sqrt{n}}\right)\right]$
$= \lim_{n\to\infty}\frac{\sqrt{n}}{\sqrt{n+\sqrt{n}}+\sqrt{n}} = \lim_{n\to\infty}\frac{1}{\sqrt{1+\frac{1}{\sqrt{n}}}+1} = \frac{1}{2} \neq 0$

45. converges absolutely by the Direct Comparison Test since $\operatorname{sech}(n) = \frac{2}{e^n+e^{-n}} = \frac{2e^n}{e^{2n}+1} < \frac{2e^n}{e^{2n}} = \frac{2}{e^n}$ which is the nth term of a convergent geometric series

47. $\frac{1}{4}-\frac{1}{6}+\frac{1}{8}-\frac{1}{10}+\frac{1}{12}-\frac{1}{14}+\ldots = \sum_{n=1}^{\infty}\frac{(-1)^{n+1}}{2(n+1)}$; converges by Alternating Series Test since: $u_n = \frac{1}{2(n+1)} > 0$ for all $n \geq 1$;
$n+2 \geq n+1 \Rightarrow 2(n+2) \geq 2(n+1) \Rightarrow \frac{1}{2((n+1)+1)} \leq \frac{1}{2(n+1)} \Rightarrow u_{n+1} \leq u_n$; $\lim_{n\to\infty}u_n = \lim_{n\to\infty}\frac{1}{2(n+1)} = 0.$

49. $|\text{error}| < \left|(-1)^6\left(\frac{1}{5}\right)\right| = 0.2$

51. $|\text{error}| < \left|(-1)^6\frac{(0.01)^5}{5}\right| = 2\times 10^{-11}$

53. $|\text{error}| < 0.001 \Rightarrow u_{n+1} < 0.001 \Rightarrow \frac{1}{(n+1)^2+3} < 0.001 \Rightarrow (n+1)^2+3 > 1000 \Rightarrow n > -1+\sqrt{997} \approx 30.5753 \Rightarrow n \geq 31$

55. $|\text{error}| < 0.001 \Rightarrow u_{n+1} < 0.001 \Rightarrow \frac{1}{\left((n+1)+3\sqrt{n+1}\right)^3} < 0.001 \Rightarrow \left((n+1)+3\sqrt{n+1}\right)^3 > 1000$
$\Rightarrow \left(\sqrt{n+1}\right)^2 + 3\sqrt{n+1} - 10 > 0 \Rightarrow \sqrt{n+1} = -\frac{3+\sqrt{9+40}}{2} = 2 \Rightarrow n = 3 \Rightarrow n \ge 4$

57. $\frac{1}{(2n)!} < \frac{5}{10^6} \Rightarrow (2n)! > \frac{10^6}{5} = 200{,}000 \Rightarrow n \ge 5 \Rightarrow 1 - \frac{1}{2!} + \frac{1}{4!} - \frac{1}{6!} + \frac{1}{8!} \approx 0.54030$

59. (a) $a_n \ge a_{n+1}$ fails since $\frac{1}{3} < \frac{1}{2}$

(b) Since $\sum_{n=1}^{\infty} |a_n| = \sum_{n=1}^{\infty} \left[\left(\frac{1}{3}\right)^n + \left(\frac{1}{2}\right)^n\right] = \sum_{n=1}^{\infty} \left(\frac{1}{3}\right)^n + \sum_{n=1}^{\infty} \left(\frac{1}{2}\right)^n$ is the sum of two absolutely convergent series, we can rearrange the terms of the original series to find its sum:
$\left(\frac{1}{3} + \frac{1}{9} + \frac{1}{27} + \ldots\right) - \left(\frac{1}{2} + \frac{1}{4} + \frac{1}{8} + \ldots\right) = \frac{\left(\frac{1}{3}\right)}{1-\left(\frac{1}{3}\right)} - \frac{\left(\frac{1}{2}\right)}{1-\left(\frac{1}{2}\right)} = \frac{1}{2} - 1 = -\frac{1}{2}$

61. The unused terms are $\sum_{j=n+1}^{\infty} (-1)^{j+1} a_j = (-1)^{n+1}(a_{n+1} - a_{n+2}) + (-1)^{n+3}(a_{n+3} - a_{n+4}) + \ldots$
$= (-1)^{n+1}\left[(a_{n+1} - a_{n+2}) + (a_{n+3} - a_{n+4}) + \ldots\right]$. Each grouped term is positive, so the remainder has the same sign as $(-1)^{n+1}$, which is the sign of the first unused term.

63. Theorem 16 states that $\sum_{n=1}^{\infty} |a_n|$ converges $\Rightarrow \sum_{n=1}^{\infty} a_n$ converges. But this is equivalent to $\sum_{n=1}^{\infty} a_n$ diverges $\Rightarrow \sum_{n=1}^{\infty} |a_n|$ diverges

65. (a) $\sum_{n=1}^{\infty} |a_n + b_n|$ converges by the Direct Comparison Test since $|a_n + b_n| \le |a_n| + |b_n|$ and hence $\sum_{n=1}^{\infty} (a_n + b_n)$ converges absolutely

(b) $\sum_{n=1}^{\infty} |b_n|$ converges $\Rightarrow \sum_{n=1}^{\infty} -b_n$ converges absolutely; since $\sum_{n=1}^{\infty} a_n$ converges absolutely and $\sum_{n=1}^{\infty} -b_n$ converges absolutely, we have $\sum_{n=1}^{\infty} [a_n + (-b_n)] = \sum_{n=1}^{\infty} (a_n - b_n)$ converges absolutely by part (a)

(c) $\sum_{n=1}^{\infty} |a_n|$ converges $\Rightarrow |k| \sum_{n=1}^{\infty} |a_n| = \sum_{n=1}^{\infty} |ka_n|$ converges $\Rightarrow \sum_{n=1}^{\infty} ka_n$ converges absolutely

67. $s_1 = -\frac{1}{2}$, $s_2 = -\frac{1}{2} + 1 = \frac{1}{2}$,
$s_3 = -\frac{1}{2} + 1 - \frac{1}{4} - \frac{1}{6} - \frac{1}{8} - \frac{1}{10} - \frac{1}{12} - \frac{1}{14} - \frac{1}{16} - \frac{1}{18} - \frac{1}{20} - \frac{1}{22} \approx -0.5099$,
$s_4 = s_3 + \frac{1}{3} \approx -0.1766$,
$s_5 = s_4 - \frac{1}{24} - \frac{1}{26} - \frac{1}{28} - \frac{1}{30} - \frac{1}{32} - \frac{1}{34} - \frac{1}{36} - \frac{1}{38} - \frac{1}{40} - \frac{1}{42} - \frac{1}{44} \approx -0.512$,
$s_6 = s_5 + \frac{1}{5} \approx -0.312$,
$s_7 = s_6 - \frac{1}{46} - \frac{1}{48} - \frac{1}{50} - \frac{1}{52} - \frac{1}{54} - \frac{1}{56} - \frac{1}{58} - \frac{1}{60} - \frac{1}{62} - \frac{1}{64} - \frac{1}{66} \approx -0.51106$

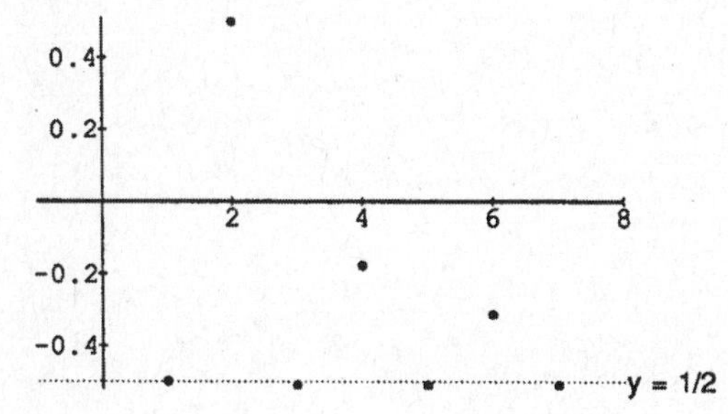

10.7 POWER SERIES

1. $\lim_{n\to\infty}\left|\frac{u_{n+1}}{u_n}\right| < 1 \Rightarrow \lim_{n\to\infty}\left|\frac{x^{n+1}}{x^n}\right| < 1 \Rightarrow |x| < 1 \Rightarrow -1 < x < 1$; when $x = -1$ we have $\sum_{n=1}^{\infty} (-1)^n$, a divergent series; when $x = 1$ we have $\sum_{n=1}^{\infty} 1$, a divergent series

   (a) the radius is 1; the interval of convergence is $-1 < x < 1$
   (b) the interval of absolute convergence is $-1 < x < 1$
   (c) there are no values for which the series converges conditionally

3. $\lim_{n\to\infty}\left|\frac{u_{n+1}}{u_n}\right| < 1 \Rightarrow \lim_{n\to\infty}\left|\frac{(4x+1)^{n+1}}{(4x+1)^n}\right| < 1 \Rightarrow |4x+1| < 1 \Rightarrow -1 < 4x+1 < 1 \Rightarrow -\frac{1}{2} < x < 0$; when $x = -\frac{1}{2}$ we have $\sum_{n=1}^{\infty}(-1)^n(-1)^n = \sum_{n=1}^{\infty}(-1)^{2n} = \sum_{n=1}^{\infty} 1^n$, a divergent series; when $x = 0$ we have $\sum_{n=1}^{\infty}(-1)^n(1)^n = \sum_{n=1}^{\infty}(-1)^n$, a divergent series

   (a) the radius is $\frac{1}{4}$; the interval of convergence is $-\frac{1}{2} < x < 0$
   (b) the interval of absolute convergence is $-\frac{1}{2} < x < 0$
   (c) there are no values for which the series converges conditionally

5. $\lim_{n\to\infty}\left|\frac{u_{n+1}}{u_n}\right| < 1 \Rightarrow \lim_{n\to\infty}\left|\frac{(x-2)^{n+1}}{10^{n+1}} \cdot \frac{10^n}{(x-2)^n}\right| < 1 \Rightarrow \frac{|x-2|}{10} < 1 \Rightarrow |x-2| < 10 \Rightarrow -10 < x-2 < 10$ $\Rightarrow -8 < x < 12$; when $x = -8$ we have $\sum_{n=1}^{\infty}(-1)^n$, a divergent series; when $x = 12$ we have $\sum_{n=1}^{\infty} 1$, a divergent series

   (a) the radius is 10; the interval of convergence is $-8 < x < 12$
   (b) the interval of absolute convergence is $-8 < x < 12$
   (c) there are no values for which the series converges conditionally

7. $\lim_{n\to\infty}\left|\frac{u_{n+1}}{u_n}\right| < 1 \Rightarrow \lim_{n\to\infty}\left|\frac{(n+1)x^{n+1}}{(n+3)} \cdot \frac{(n+2)}{nx^n}\right| < 1 \Rightarrow |x| \lim_{n\to\infty}\frac{(n+1)(n+2)}{(n+3)(n)} < 1 \Rightarrow |x| < 1$ $\Rightarrow -1 < x < 1$; when $x = -1$ we have $\sum_{n=1}^{\infty}(-1)^n \frac{n}{n+2}$, a divergent series by the nth-term Test; when $x = 1$ we have $\sum_{n=1}^{\infty}\frac{n}{n+2}$, a divergent series

   (a) the radius is 1; the interval of convergence is $-1 < x < 1$
   (b) the interval of absolute convergence is $-1 < x < 1$
   (c) there are no values for which the series converges conditionally

9. $\lim_{n\to\infty}\left|\frac{u_{n+1}}{u_n}\right| < 1 \Rightarrow \lim_{n\to\infty}\left|\frac{x^{n+1}}{(n+1)\sqrt{n+1}\,3^{n+1}} \cdot \frac{n\sqrt{n}\,3^n}{x^n}\right| < 1 \Rightarrow \frac{|x|}{3}\left(\lim_{n\to\infty}\frac{n}{n+1}\right)\left(\sqrt{\lim_{n\to\infty}\frac{n}{n+1}}\right) < 1$ $\Rightarrow \frac{|x|}{3}(1)(1) < 1 \Rightarrow |x| < 3 \Rightarrow -3 < x < 3$; when $x = -3$ we have $\sum_{n=1}^{\infty}\frac{(-1)^n}{n^{3/2}}$, an absolutely convergent series; when $x = 3$ we have $\sum_{n=1}^{\infty}\frac{1}{n^{3/2}}$, a convergent p-series

   (a) the radius is 3; the interval of convergence is $-3 \le x \le 3$
   (b) the interval of absolute convergence is $-3 \le x \le 3$
   (c) there are no values for which the series converges conditionally

11. $\lim_{n\to\infty}\left|\frac{u_{n+1}}{u_n}\right| < 1 \Rightarrow \lim_{n\to\infty}\left|\frac{x^{n+1}}{(n+1)!} \cdot \frac{n!}{x^n}\right| < 1 \Rightarrow |x| \lim_{n\to\infty}\left(\frac{1}{n+1}\right) < 1$ for all x

    (a) the radius is ∞; the series converges for all x
    (b) the series converges absolutely for all x
    (c) there are no values for which the series converges conditionally

13. $\lim_{n\to\infty}\left|\frac{u_{n+1}}{u_n}\right| < 1 \Rightarrow \lim_{n\to\infty}\left|\frac{4^{n+1}x^{2n+2}}{n+1}\cdot\frac{n}{4^n x^{2n}}\right| < 1 \Rightarrow x^2\lim_{n\to\infty}\left(\frac{4n}{n+1}\right) = 4x^2 < 1 \Rightarrow x^2 < \frac{1}{4}$

$\Rightarrow -\frac{1}{2} < x < \frac{1}{2}$; when $x = -\frac{1}{2}$ we have $\sum_{n=1}^{\infty}\frac{4^n}{n}\left(-\frac{1}{2}\right)^{2n} = \sum_{n=1}^{\infty}\frac{1}{n}$, a divergent p-series; when $x = \frac{1}{2}$ we have

$\sum_{n=1}^{\infty}\frac{4^n}{n}\left(\frac{1}{2}\right)^{2n} = \sum_{n=1}^{\infty}\frac{1}{n}$, a divergent p-series

(a) the radius is $\frac{1}{2}$; the interval of convergence is $-\frac{1}{2} < x < \frac{1}{2}$

(b) the interval of absolute convergence is $-\frac{1}{2} < x < \frac{1}{2}$

(c) there are no values for which the series converges conditionally

15. $\lim_{n\to\infty}\left|\frac{u_{n+1}}{u_n}\right| < 1 \Rightarrow \lim_{n\to\infty}\left|\frac{x^{n+1}}{\sqrt{(n+1)^2+3}}\cdot\frac{\sqrt{n^2+3}}{x^n}\right| < 1 \Rightarrow |x|\sqrt{\lim_{n\to\infty}\frac{n^2+3}{n^2+2n+4}} < 1 \Rightarrow |x| < 1$

$\Rightarrow -1 < x < 1$; when $x = -1$ we have $\sum_{n=1}^{\infty}\frac{(-1)^n}{\sqrt{n^2+3}}$, a conditionally convergent series; when $x = 1$ we have

$\sum_{n=1}^{\infty}\frac{1}{\sqrt{n^2+3}}$, a divergent series

(a) the radius is 1; the interval of convergence is $-1 \le x < 1$

(b) the interval of absolute convergence is $-1 < x < 1$

(c) the series converges conditionally at $x = -1$

17. $\lim_{n\to\infty}\left|\frac{u_{n+1}}{u_n}\right| < 1 \Rightarrow \lim_{n\to\infty}\left|\frac{(n+1)(x+3)^{n+1}}{5^{n+1}}\cdot\frac{5^n}{n(x+3)^n}\right| < 1 \Rightarrow \frac{|x+3|}{5}\lim_{n\to\infty}\left(\frac{n+1}{n}\right) < 1 \Rightarrow \frac{|x+3|}{5} < 1$

$\Rightarrow |x+3| < 5 \Rightarrow -5 < x+3 < 5 \Rightarrow -8 < x < 2$; when $x = -8$ we have $\sum_{n=1}^{\infty}\frac{n(-5)^n}{5^n} = \sum_{n=1}^{\infty}(-1)^n n$, a divergent

series; when $x = 2$ we have $\sum_{n=1}^{\infty}\frac{n5^n}{5^n} = \sum_{n=1}^{\infty} n$, a divergent series

(a) the radius is 5; the interval of convergence is $-8 < x < 2$

(b) the interval of absolute convergence is $-8 < x < 2$

(c) there are no values for which the series converges conditionally

19. $\lim_{n\to\infty}\left|\frac{u_{n+1}}{u_n}\right| < 1 \Rightarrow \lim_{n\to\infty}\left|\frac{\sqrt{n+1}\,x^{n+1}}{3^{n+1}}\cdot\frac{3^n}{\sqrt{n}\,x^n}\right| < 1 \Rightarrow \frac{|x|}{3}\sqrt{\lim_{n\to\infty}\left(\frac{n+1}{n}\right)} < 1 \Rightarrow \frac{|x|}{3} < 1 \Rightarrow |x| < 3$

$\Rightarrow -3 < x < 3$; when $x = -3$ we have $\sum_{n=1}^{\infty}(-1)^n\sqrt{n}$, a divergent series; when $x = 3$ we have $\sum_{n=1}^{\infty}\sqrt{n}$, a divergent series

(a) the radius is 3; the interval of convergence is $-3 < x < 3$

(b) the interval of absolute convergence is $-3 < x < 3$

(c) there are no values for which the series converges conditionally

21. First, rewrite the series as $\sum_{n=1}^{\infty}(2+(-1)^n)(x+1)^{n-1} = \sum_{n=1}^{\infty}2(x+1)^{n-1} + \sum_{n=1}^{\infty}(-1)^n(x+1)^{n-1}$. For the series

$\sum_{n=1}^{\infty}2(x+1)^{n-1}$: $\lim_{n\to\infty}\left|\frac{u_{n+1}}{u_n}\right| < 1 \Rightarrow \lim_{n\to\infty}\left|\frac{2(x+1)^n}{2(x+1)^{n-1}}\right| < 1 \Rightarrow |x+1|\lim_{n\to\infty}1 = |x+1| < 1 \Rightarrow -2 < x < 0$; For the

series $\sum_{n=1}^{\infty}(-1)^n(x+1)^{n-1}$: $\lim_{n\to\infty}\left|\frac{u_{n+1}}{u_n}\right| < 1 \Rightarrow \lim_{n\to\infty}\left|\frac{(-1)^{n+1}(x+1)^n}{(-1)^n(x+1)^{n-1}}\right| < 1 \Rightarrow |x+1|\lim_{n\to\infty}1 = |x+1| < 1$

$\Rightarrow -2 < x < 0$; when $x = -2$ we have $\sum_{n=1}^{\infty}(2+(-1)^n)(-1)^{n-1}$, a divergent series; when $x = 0$ we have

$\sum_{n=1}^{\infty}(2+(-1)^n)$, a divergent series

(a) the radius is 1; the interval of convergence is $-2 < x < 0$

(b) the interval of absolute convergence is $-2 < x < 0$

(c) there are no values for which the series converges conditionally

23. $\lim_{n\to\infty} \left|\frac{u_{n+1}}{u_n}\right| < 1 \Rightarrow \lim_{n\to\infty} \left|\frac{\left(1+\frac{1}{n+1}\right)^{n+1} x^{n+1}}{\left(1+\frac{1}{n}\right)^{n} x^{n}}\right| < 1 \Rightarrow |x| \left(\frac{\lim_{t\to\infty}\left(1+\frac{1}{t}\right)^{t}}{\lim_{n\to\infty}\left(1+\frac{1}{n}\right)^{n}}\right) < 1 \Rightarrow |x|\left(\frac{e}{e}\right) < 1 \Rightarrow |x| < 1$

$\Rightarrow -1 < x < 1$; when $x = -1$ we have $\sum_{n=1}^{\infty} (-1)^n \left(1+\frac{1}{n}\right)^n$, a divergent series by the nth-Term Test since $\lim_{n\to\infty} \left(1+\frac{1}{n}\right)^n = e \neq 0$; when $x = 1$ we have $\sum_{n=1}^{\infty} \left(1+\frac{1}{n}\right)^n$, a divergent series

(a) the radius is 1; the interval of convergence is $-1 < x < 1$

(b) the interval of absolute convergence is $-1 < x < 1$

(c) there are no values for which the series converges conditionally

25. $\lim_{n\to\infty} \left|\frac{u_{n+1}}{u_n}\right| < 1 \Rightarrow \lim_{n\to\infty} \left|\frac{(n+1)^{n+1}x^{n+1}}{n^n x^n}\right| < 1 \Rightarrow |x| \left(\lim_{n\to\infty} \left(1+\frac{1}{n}\right)^n\right)\left(\lim_{n\to\infty} (n+1)\right) < 1$

$\Rightarrow e|x| \lim_{n\to\infty} (n+1) < 1 \Rightarrow$ only $x = 0$ satisfies this inequality

(a) the radius is 0; the series converges only for $x = 0$

(b) the series converges absolutely only for $x = 0$

(c) there are no values for which the series converges conditionally

27. $\lim_{n\to\infty} \left|\frac{u_{n+1}}{u_n}\right| < 1 \Rightarrow \lim_{n\to\infty} \left|\frac{(x+2)^{n+1}}{(n+1)\,2^{n+1}} \cdot \frac{n2^n}{(x+2)^n}\right| < 1 \Rightarrow \frac{|x+2|}{2} \lim_{n\to\infty} \left(\frac{n}{n+1}\right) < 1 \Rightarrow \frac{|x+2|}{2} < 1 \Rightarrow |x+2| < 2$

$\Rightarrow -2 < x+2 < 2 \Rightarrow -4 < x < 0$; when $x = -4$ we have $\sum_{n=1}^{\infty} \frac{-1}{n}$, a divergent series; when $x = 0$ we have $\sum_{n=1}^{\infty} \frac{(-1)^{n+1}}{n}$, the alternating harmonic series which converges conditionally

(a) the radius is 2; the interval of convergence is $-4 < x \le 0$

(b) the interval of absolute convergence is $-4 < x < 0$

(c) the series converges conditionally at $x = 0$

29. $\lim_{n\to\infty} \left|\frac{u_{n+1}}{u_n}\right| < 1 \Rightarrow \lim_{n\to\infty} \left|\frac{x^{n+1}}{(n+1)(\ln(n+1))^2} \cdot \frac{n(\ln n)^2}{x^n}\right| < 1 \Rightarrow |x| \left(\lim_{n\to\infty} \frac{n}{n+1}\right)\left(\lim_{n\to\infty} \frac{\ln n}{\ln(n+1)}\right)^2 < 1$

$\Rightarrow |x|(1)\left(\lim_{n\to\infty} \frac{\left(\frac{1}{n}\right)}{\left(\frac{1}{n+1}\right)}\right)^2 < 1 \Rightarrow |x|\left(\lim_{n\to\infty} \frac{n+1}{n}\right)^2 < 1 \Rightarrow |x| < 1 \Rightarrow -1 < x < 1$; when $x = -1$ we have $\sum_{n=1}^{\infty} \frac{(-1)^n}{n(\ln n)^2}$ which converges absolutely; when $x = 1$ we have $\sum_{n=1}^{\infty} \frac{1}{n(\ln n)^2}$ which converges

(a) the radius is 1; the interval of convergence is $-1 \le x \le 1$

(b) the interval of absolute convergence is $-1 \le x \le 1$

(c) there are no values for which the series converges conditionally

31. $\lim_{n\to\infty} \left|\frac{u_{n+1}}{u_n}\right| < 1 \Rightarrow \lim_{n\to\infty} \left|\frac{(4x-5)^{2n+3}}{(n+1)^{3/2}} \cdot \frac{n^{3/2}}{(4x-5)^{2n+1}}\right| < 1 \Rightarrow (4x-5)^2 \left(\lim_{n\to\infty} \frac{n}{n+1}\right)^{3/2} < 1 \Rightarrow (4x-5)^2 < 1$

$\Rightarrow |4x-5| < 1 \Rightarrow -1 < 4x-5 < 1 \Rightarrow 1 < x < \frac{3}{2}$; when $x = 1$ we have $\sum_{n=1}^{\infty} \frac{(-1)^{2n+1}}{n^{3/2}} = \sum_{n=1}^{\infty} \frac{-1}{n^{3/2}}$ which is absolutely convergent; when $x = \frac{3}{2}$ we have $\sum_{n=1}^{\infty} \frac{(1)^{2n+1}}{n^{3/2}}$, a convergent p-series

(a) the radius is $\frac{1}{4}$; the interval of convergence is $1 \le x \le \frac{3}{2}$

(b) the interval of absolute convergence is $1 \le x \le \frac{3}{2}$

(c) there are no values for which the series converges conditionally

33. $\lim_{n\to\infty} \left|\frac{u_{n+1}}{u_n}\right| < 1 \Rightarrow \lim_{n\to\infty} \left|\frac{x^{n+1}}{2\cdot4\cdot6\cdots(2n)(2(n+1))} \cdot \frac{2\cdot4\cdot6\cdots(2n)}{x^n}\right| < 1 \Rightarrow |x| \lim_{n\to\infty} \left(\frac{1}{2n+2}\right) < 1$ for all x

(a) the radius is ∞; the series converges for all x

(b) the series converges absolutely for all x

(c) there are no values for which the series converges conditionally

35. For the series $\sum_{n=1}^{\infty} \frac{1+2+\cdots+n}{1^2+2^2+\cdots+n^2} x^n$, recall $1+2+\cdots+n = \frac{n(n+1)}{2}$ and $1^2+2^2+\cdots+n^2 = \frac{n(n+1)(2n+1)}{6}$ so that we can rewrite the series as $\sum_{n=1}^{\infty} \left(\frac{\frac{n(n+1)}{2}}{\frac{n(n+1)(2n+1)}{6}} \right) x^n = \sum_{n=1}^{\infty} \left(\frac{3}{2n+1}\right) x^n$; then $\lim_{n\to\infty} \left|\frac{u_{n+1}}{u_n}\right| < 1 \Rightarrow \lim_{n\to\infty} \left| \frac{3x^{n+1}}{(2(n+1)+1)} \cdot \frac{(2n+1)}{3x^n} \right| < 1$ $\Rightarrow |x| \lim_{n\to\infty} \left|\frac{(2n+1)}{(2n+3)}\right| < 1 \Rightarrow |x| < 1 \Rightarrow -1 < x < 1$; when $x = -1$ we have $\sum_{n=1}^{\infty} \left(\frac{3}{2n+1}\right)(-1)^n$, a conditionally convergent series; when $x = 1$ we have $\sum_{n=1}^{\infty} \left(\frac{3}{2n+1}\right)$, a divergent series.

(a) the radius is 1; the interval of convergence is $-1 \le x < 1$
(b) the interval of absolute convergence is $-1 < x < 1$
(c) the series converges conditionally at $x = -1$

37. $\lim_{n\to\infty} \left|\frac{u_{n+1}}{u_n}\right| < 1 \Rightarrow \lim_{n\to\infty} \left| \frac{(n+1)!x^{n+1}}{3\cdot6\cdot9\cdots(3n)(3(n+1))} \cdot \frac{3\cdot6\cdot9\cdots(3n)}{n!\,x^n} \right| < 1 \Rightarrow |x| \lim_{n\to\infty} \left|\frac{(n+1)}{3(n+1)}\right| < 1 \Rightarrow \frac{|x|}{3} < 1 \Rightarrow |x| < 3 \Rightarrow R = 3$

39. $\lim_{n\to\infty} \left|\frac{u_{n+1}}{u_n}\right| < 1 \Rightarrow \lim_{n\to\infty} \left| \frac{((n+1)!)^2 x^{n+1}}{2^{n+1}(2(n+1))!} \cdot \frac{2^n(2n)!}{(n!)^2 x^n} \right| < 1 \Rightarrow |x| \lim_{n\to\infty} \left| \frac{(n+1)^2}{2(2n+2)(2n+1)} \right| < 1 \Rightarrow \frac{|x|}{8} < 1 \Rightarrow |x| < 8 \Rightarrow R = 8$

41. $\lim_{n\to\infty} \left|\frac{u_{n+1}}{u_n}\right| < 1 \Rightarrow \lim_{n\to\infty} \left| \frac{3^{n+1} x^{n+1}}{3^n x^n} \right| < 1 \Rightarrow |x| \lim_{n\to\infty} 3 < 1 \Rightarrow |x| < \frac{1}{3} \Rightarrow -\frac{1}{3} < x < \frac{1}{3}$; at $x = -\frac{1}{3}$ we have $\sum_{n=0}^{\infty} 3^n \left(-\frac{1}{3}\right)^n = \sum_{n=0}^{\infty} (-1)^n$, which diverges; at $x = \frac{1}{3}$ we have $\sum_{n=0}^{\infty} 3^n \left(\frac{1}{3}\right)^n = \sum_{n=0}^{\infty} 1$, which diverges. The series $\sum_{n=0}^{\infty} 3^n x^n$ $= \sum_{n=0}^{\infty} (3x)^n$ is a convergent geometric series when $-\frac{1}{3} < x < \frac{1}{3}$ and the sum is $\frac{1}{1-3x}$.

43. $\lim_{n\to\infty} \left|\frac{u_{n+1}}{u_n}\right| < 1 \Rightarrow \lim_{n\to\infty} \left| \frac{(x-1)^{2n+2}}{4^{n+1}} \cdot \frac{4^n}{(x-1)^{2n}} \right| < 1 \Rightarrow \frac{(x-1)^2}{4} \lim_{n\to\infty} |1| < 1 \Rightarrow (x-1)^2 < 4 \Rightarrow |x-1| < 2$ $\Rightarrow -2 < x-1 < 2 \Rightarrow -1 < x < 3$; at $x = -1$ we have $\sum_{n=0}^{\infty} \frac{(-2)^{2n}}{4^n} = \sum_{n=0}^{\infty} \frac{4^n}{4^n} = \sum_{n=0}^{\infty} 1$, which diverges; at $x = 3$ we have $\sum_{n=0}^{\infty} \frac{2^{2n}}{4^n} = \sum_{n=0}^{\infty} \frac{4^n}{4^n} = \sum_{n=0}^{\infty} 1$, a divergent series; the interval of convergence is $-1 < x < 3$; the series $\sum_{n=0}^{\infty} \frac{(x-1)^{2n}}{4^n} = \sum_{n=0}^{\infty} \left(\left(\frac{x-1}{2}\right)^2 \right)^n$ is a convergent geometric series when $-1 < x < 3$ and the sum is

$$\frac{1}{1-\left(\frac{x-1}{2}\right)^2} = \frac{1}{\left[\frac{4-(x-1)^2}{4}\right]} = \frac{4}{4-x^2+2x-1} = \frac{4}{3+2x-x^2}$$

45. $\lim_{n\to\infty} \left|\frac{u_{n+1}}{u_n}\right| < 1 \Rightarrow \lim_{n\to\infty} \left| \frac{(\sqrt{x}-2)^{n+1}}{2^{n+1}} \cdot \frac{2^n}{(\sqrt{x}-2)^n} \right| < 1 \Rightarrow |\sqrt{x}-2| < 2 \Rightarrow -2 < \sqrt{x}-2 < 2 \Rightarrow 0 < \sqrt{x} < 4$ $\Rightarrow 0 < x < 16$; when $x = 0$ we have $\sum_{n=0}^{\infty} (-1)^n$, a divergent series; when $x = 16$ we have $\sum_{n=0}^{\infty} (1)^n$, a divergent series; the interval of convergence is $0 < x < 16$; the series $\sum_{n=0}^{\infty} \left(\frac{\sqrt{x}-2}{2}\right)^n$ is a convergent geometric series when $0 < x < 16$ and its sum is $\frac{1}{1-\left(\frac{\sqrt{x}-2}{2}\right)} = \frac{1}{\left(\frac{2-\sqrt{x}+2}{2}\right)} = \frac{2}{4-\sqrt{x}}$

47. $\lim_{n\to\infty} \left|\frac{u_{n+1}}{u_n}\right| < 1 \Rightarrow \lim_{n\to\infty} \left| \left(\frac{x^2+1}{3}\right)^{n+1} \cdot \left(\frac{3}{x^2+1}\right)^n \right| < 1 \Rightarrow \frac{(x^2+1)}{3} \lim_{n\to\infty} |1| < 1 \Rightarrow \frac{x^2+1}{3} < 1 \Rightarrow x^2 < 2$ $\Rightarrow |x| < \sqrt{2} \Rightarrow -\sqrt{2} < x < \sqrt{2}$; at $x = \pm\sqrt{2}$ we have $\sum_{n=0}^{\infty} (1)^n$ which diverges; the interval of convergence is $-\sqrt{2} < x < \sqrt{2}$; the series $\sum_{n=0}^{\infty} \left(\frac{x^2+1}{3}\right)^n$ is a convergent geometric series when $-\sqrt{2} < x < \sqrt{2}$ and its sum is $\frac{1}{1-\left(\frac{x^2+1}{3}\right)} = \frac{1}{\left(\frac{3-x^2-1}{3}\right)} = \frac{3}{2-x^2}$

49. $\lim_{n \to \infty} \left| \frac{(x-3)^{n+1}}{2^{n+1}} \cdot \frac{2^n}{(x-3)^n} \right| < 1 \Rightarrow |x-3| < 2 \Rightarrow 1 < x < 5$; when $x = 1$ we have $\sum_{n=1}^{\infty} (1)^n$ which diverges; when $x = 5$ we have $\sum_{n=1}^{\infty} (-1)^n$ which also diverges; the interval of convergence is $1 < x < 5$; the sum of this convergent geometric series is $\frac{1}{1+\left(\frac{x-3}{2}\right)} = \frac{2}{x-1}$. If $f(x) = 1 - \frac{1}{2}(x-3) + \frac{1}{4}(x-3)^2 + \ldots + \left(-\frac{1}{2}\right)^n (x-3)^n + \ldots$ $= \frac{2}{x-1}$ then $f'(x) = -\frac{1}{2} + \frac{1}{2}(x-3) + \ldots + \left(-\frac{1}{2}\right)^n n(x-3)^{n-1} + \ldots$ is convergent when $1 < x < 5$, and diverges when $x = 1$ or 5. The sum for $f'(x)$ is $\frac{-2}{(x-1)^2}$, the derivative of $\frac{2}{x-1}$.

51. (a) Differentiate the series for sin x to get $\cos x = 1 - \frac{3x^2}{3!} + \frac{5x^4}{5!} - \frac{7x^6}{7!} + \frac{9x^8}{9!} - \frac{11x^{10}}{11!} + \ldots$ $= 1 - \frac{x^2}{2!} + \frac{x^4}{4!} - \frac{x^6}{6!} + \frac{x^8}{8!} - \frac{x^{10}}{10!} + \ldots$. The series converges for all values of x since $\lim_{n \to \infty} \left| \frac{x^{2n+2}}{(2n+2)!} \cdot \frac{(2n)!}{x^{2n}} \right| = x^2 \lim_{n \to \infty} \left(\frac{1}{(2n+1)(2n+2)} \right) = 0 < 1$ for all x.

(b) $\sin 2x = 2x - \frac{2^3x^3}{3!} + \frac{2^5x^5}{5!} - \frac{2^7x^7}{7!} + \frac{2^9x^9}{9!} - \frac{2^{11}x^{11}}{11!} + \ldots = 2x - \frac{8x^3}{3!} + \frac{32x^5}{5!} - \frac{128x^7}{7!} + \frac{512x^9}{9!} - \frac{2048x^{11}}{11!} + \ldots$

(c) $2 \sin x \cos x = 2\left[(0 \cdot 1) + (0 \cdot 0 + 1 \cdot 1)x + \left(0 \cdot \frac{-1}{2} + 1 \cdot 0 + 0 \cdot 1\right)x^2 + \left(0 \cdot 0 - 1 \cdot \frac{1}{2} + 0 \cdot 0 - 1 \cdot \frac{1}{3!}\right)x^3 \right.$ $+ \left(0 \cdot \frac{1}{4!} + 1 \cdot 0 - 0 \cdot \frac{1}{2} - 0 \cdot \frac{1}{3!} + 0 \cdot 1\right)x^4 + \left(0 \cdot 0 + 1 \cdot \frac{1}{4!} + 0 \cdot 0 + \frac{1}{2} \cdot \frac{1}{3!} + 0 \cdot 0 + 1 \cdot \frac{1}{5!}\right)x^5$ $\left. + \left(0 \cdot \frac{1}{6!} + 1 \cdot 0 + 0 \cdot \frac{1}{4!} + 0 \cdot \frac{1}{3!} + 0 \cdot \frac{1}{2} + 0 \cdot \frac{1}{5!} + 0 \cdot 1\right)x^6 + \ldots\right] = 2\left[x - \frac{4x^3}{3!} + \frac{16x^5}{5!} - \ldots\right]$ $= 2x - \frac{2^3x^3}{3!} + \frac{2^5x^5}{5!} - \frac{2^7x^7}{7!} + \frac{2^9x^9}{9!} - \frac{2^{11}x^{11}}{11!} + \ldots$

53. (a) $\ln|\sec x| + C = \int \tan x\, dx = \int \left(x + \frac{x^3}{3} + \frac{2x^5}{15} + \frac{17x^7}{315} + \frac{62x^9}{2835} + \ldots\right) dx$ $= \frac{x^2}{2} + \frac{x^4}{12} + \frac{x^6}{45} + \frac{17x^8}{2520} + \frac{31x^{10}}{14{,}175} + \ldots + C$; $x = 0 \Rightarrow C = 0 \Rightarrow \ln|\sec x| = \frac{x^2}{2} + \frac{x^4}{12} + \frac{x^6}{45} + \frac{17x^8}{2520} + \frac{31x^{10}}{14{,}175} + \ldots$, converges when $-\frac{\pi}{2} < x < \frac{\pi}{2}$

(b) $\sec^2 x = \frac{d(\tan x)}{dx} = \frac{d}{dx}\left(x + \frac{x^3}{3} + \frac{2x^5}{15} + \frac{17x^7}{315} + \frac{62x^9}{2835} + \ldots\right) = 1 + x^2 + \frac{2x^4}{3} + \frac{17x^6}{45} + \frac{62x^8}{315} + \ldots$, converges when $-\frac{\pi}{2} < x < \frac{\pi}{2}$

(c) $\sec^2 x = (\sec x)(\sec x) = \left(1 + \frac{x^2}{2} + \frac{5x^4}{24} + \frac{61x^6}{720} + \ldots\right)\left(1 + \frac{x^2}{2} + \frac{5x^4}{24} + \frac{61x^6}{720} + \ldots\right)$ $= 1 + \left(\frac{1}{2} + \frac{1}{2}\right)x^2 + \left(\frac{5}{24} + \frac{1}{4} + \frac{5}{24}\right)x^4 + \left(\frac{61}{720} + \frac{5}{48} + \frac{5}{48} + \frac{61}{720}\right)x^6 + \ldots$ $= 1 + x^2 + \frac{2x^4}{3} + \frac{17x^6}{45} + \frac{62x^8}{315} + \ldots, -\frac{\pi}{2} < x < \frac{\pi}{2}$

55. (a) If $f(x) = \sum_{n=0}^{\infty} a_n x^n$, then $f^{(k)}(x) = \sum_{n=k}^{\infty} n(n-1)(n-2)\cdots(n-(k-1))\, a_n x^{n-k}$ and $f^{(k)}(0) = k!a_k$ $\Rightarrow a_k = \frac{f^{(k)}(0)}{k!}$; likewise if $f(x) = \sum_{n=0}^{\infty} b_n x^n$, then $b_k = \frac{f^{(k)}(0)}{k!} \Rightarrow a_k = b_k$ for every nonnegative integer k

(b) If $f(x) = \sum_{n=0}^{\infty} a_n x^n = 0$ for all x, then $f^{(k)}(x) = 0$ for all x $\Rightarrow$ from part (a) that $a_k = 0$ for every nonnegative integer k

10.8 TAYLOR AND MACLAURIN SERIES

1. $f(x) = e^{2x}$, $f'(x) = 2e^{2x}$, $f''(x) = 4e^{2x}$, $f'''(x) = 8e^{2x}$; $f(0) = e^{2(0)} = 1$, $f'(0) = 2$, $f''(0) = 4$, $f'''(0) = 8 \Rightarrow P_0(x) = 1$, $P_1(x) = 1 + 2x$, $P_2(x) = 1 + x + 2x^2$, $P_3(x) = 1 + x + 2x^2 + \frac{4}{3}x^3$

3. $f(x) = \ln x$, $f'(x) = \frac{1}{x}$, $f''(x) = -\frac{1}{x^2}$, $f'''(x) = \frac{2}{x^3}$; $f(1) = \ln 1 = 0$, $f'(1) = 1$, $f''(1) = -1$, $f'''(1) = 2 \Rightarrow P_0(x) = 0$, $P_1(x) = (x-1)$, $P_2(x) = (x-1) - \frac{1}{2}(x-1)^2$, $P_3(x) = (x-1) - \frac{1}{2}(x-1)^2 + \frac{1}{3}(x-1)^3$

5. $f(x) = \frac{1}{x} = x^{-1}$, $f'(x) = -x^{-2}$, $f''(x) = 2x^{-3}$, $f'''(x) = -6x^{-4}$; $f(2) = \frac{1}{2}$, $f'(2) = -\frac{1}{4}$, $f''(2) = \frac{1}{4}$, $f'''(x) = -\frac{3}{8}$
$\Rightarrow P_0(x) = \frac{1}{2}$, $P_1(x) = \frac{1}{2} - \frac{1}{4}(x-2)$, $P_2(x) = \frac{1}{2} - \frac{1}{4}(x-2) + \frac{1}{8}(x-2)^2$,
$P_3(x) = \frac{1}{2} - \frac{1}{4}(x-2) + \frac{1}{8}(x-2)^2 - \frac{1}{16}(x-2)^3$

7. $f(x) = \sin x$, $f'(x) = \cos x$, $f''(x) = -\sin x$, $f'''(x) = -\cos x$; $f\left(\frac{\pi}{4}\right) = \sin\frac{\pi}{4} = \frac{\sqrt{2}}{2}$, $f'\left(\frac{\pi}{4}\right) = \cos\frac{\pi}{4} = \frac{\sqrt{2}}{2}$,
$f''\left(\frac{\pi}{4}\right) = -\sin\frac{\pi}{4} = -\frac{\sqrt{2}}{2}$, $f'''\left(\frac{\pi}{4}\right) = -\cos\frac{\pi}{4} = -\frac{\sqrt{2}}{2} \Rightarrow P_0 = \frac{\sqrt{2}}{2}$, $P_1(x) = \frac{\sqrt{2}}{2} + \frac{\sqrt{2}}{2}\left(x - \frac{\pi}{4}\right)$,
$P_2(x) = \frac{\sqrt{2}}{2} + \frac{\sqrt{2}}{2}\left(x - \frac{\pi}{4}\right) - \frac{\sqrt{2}}{4}\left(x - \frac{\pi}{4}\right)^2$, $P_3(x) = \frac{\sqrt{2}}{2} + \frac{\sqrt{2}}{2}\left(x - \frac{\pi}{4}\right) - \frac{\sqrt{2}}{4}\left(x - \frac{\pi}{4}\right)^2 - \frac{\sqrt{2}}{12}\left(x - \frac{\pi}{4}\right)^3$

9. $f(x) = \sqrt{x} = x^{1/2}$, $f'(x) = \left(\frac{1}{2}\right)x^{-1/2}$, $f''(x) = \left(-\frac{1}{4}\right)x^{-3/2}$, $f'''(x) = \left(\frac{3}{8}\right)x^{-5/2}$; $f(4) = \sqrt{4} = 2$,
$f'(4) = \left(\frac{1}{2}\right)4^{-1/2} = \frac{1}{4}$, $f''(4) = \left(-\frac{1}{4}\right)4^{-3/2} = -\frac{1}{32}$, $f'''(4) = \left(\frac{3}{8}\right)4^{-5/2} = \frac{3}{256} \Rightarrow P_0(x) = 2$, $P_1(x) = 2 + \frac{1}{4}(x-4)$,
$P_2(x) = 2 + \frac{1}{4}(x-4) - \frac{1}{64}(x-4)^2$, $P_3(x) = 2 + \frac{1}{4}(x-4) - \frac{1}{64}(x-4)^2 + \frac{1}{512}(x-4)^3$

11. $f(x) = e^{-x}$, $f'(x) = -e^{-x}$, $f''(x) = e^{-x}$, $f'''(x) = -e^{-x} \Rightarrow \ldots f^{(k)}(x) = (-1)^k e^{-x}$; $f(0) = e^{-(0)} = 1$, $f'(0) = -1$,
$f''(0) = 1$, $f'''(0) = -1, \ldots, f^{(k)}(0) = (-1)^k \Rightarrow e^{-x} = 1 - x + \frac{1}{2}x^2 - \frac{1}{6}x^3 + \ldots = \sum_{n=0}^{\infty} \frac{(-1)^n}{n!}x^n$

13. $f(x) = (1+x)^{-1} \Rightarrow f'(x) = -(1+x)^{-2}$, $f''(x) = 2(1+x)^{-3}$, $f'''(x) = -3!(1+x)^{-4} \Rightarrow \ldots f^{(k)}(x)$
$= (-1)^k k!(1+x)^{-k-1}$; $f(0) = 1$, $f'(0) = -1$, $f''(0) = 2$, $f'''(0) = -3!, \ldots, f^{(k)}(0) = (-1)^k k!$
$\Rightarrow 1 - x + x^2 - x^3 + \ldots = \sum_{n=0}^{\infty}(-x)^n = \sum_{n=0}^{\infty}(-1)^n x^n$

15. $\sin x = \sum_{n=0}^{\infty} \frac{(-1)^n x^{2n+1}}{(2n+1)!} \Rightarrow \sin 3x = \sum_{n=0}^{\infty} \frac{(-1)^n (3x)^{2n+1}}{(2n+1)!} = \sum_{n=0}^{\infty} \frac{(-1)^n 3^{2n+1} x^{2n+1}}{(2n+1)!} = 3x - \frac{3^3x^3}{3!} + \frac{3^5x^5}{5!} - \ldots$

17. $7\cos(-x) = 7\cos x = 7\sum_{n=0}^{\infty} \frac{(-1)^n x^{2n}}{(2n)!} = 7 - \frac{7x^2}{2!} + \frac{7x^4}{4!} - \frac{7x^6}{6!} + \ldots$, since the cosine is an even function

19. $\cosh x = \frac{e^x + e^{-x}}{2} = \frac{1}{2}\left[\left(1 + x^2 + \frac{x^2}{2!} + \frac{x^3}{3!} + \frac{x^4}{4!} + \ldots\right) + \left(1 - x + \frac{x^2}{2!} - \frac{x^3}{3!} + \frac{x^4}{4!} - \ldots\right)\right] = 1 + \frac{x^2}{2!} + \frac{x^4}{4!} + \frac{x^6}{6!} + \ldots$
$= \sum_{n=0}^{\infty} \frac{x^{2n}}{(2n)!}$

21. $f(x) = x^4 - 2x^3 - 5x + 4 \Rightarrow f'(x) = 4x^3 - 6x^2 - 5$, $f''(x) = 12x^2 - 12x$, $f'''(x) = 24x - 12$, $f^{(4)}(x) = 24$
$\Rightarrow f^{(n)}(x) = 0$ if $n \ge 5$; $f(0) = 4$, $f'(0) = -5$, $f''(0) = 0$, $f'''(0) = -12$, $f^{(4)}(0) = 24$, $f^{(n)}(0) = 0$ if $n \ge 5$
$\Rightarrow x^4 - 2x^3 - 5x + 4 = 4 - 5x - \frac{12}{3!}x^3 + \frac{24}{4!}x^4 = x^4 - 2x^3 - 5x + 4$

23. $f(x) = x^3 - 2x + 4 \Rightarrow f'(x) = 3x^2 - 2$, $f''(x) = 6x$, $f'''(x) = 6 \Rightarrow f^{(n)}(x) = 0$ if $n \ge 4$; $f(2) = 8$, $f'(2) = 10$,
$f''(2) = 12$, $f'''(2) = 6$, $f^{(n)}(2) = 0$ if $n \ge 4 \Rightarrow x^3 - 2x + 4 = 8 + 10(x-2) + \frac{12}{2!}(x-2)^2 + \frac{6}{3!}(x-2)^3$
$= 8 + 10(x-2) + 6(x-2)^2 + (x-2)^3$

25. $f(x) = x^4 + x^2 + 1 \Rightarrow f'(x) = 4x^3 + 2x$, $f''(x) = 12x^2 + 2$, $f'''(x) = 24x$, $f^{(4)}(x) = 24$, $f^{(n)}(x) = 0$ if $n \ge 5$;
$f(-2) = 21$, $f'(-2) = -36$, $f''(-2) = 50$, $f'''(-2) = -48$, $f^{(4)}(-2) = 24$, $f^{(n)}(-2) = 0$ if $n \ge 5 \Rightarrow x^4 + x^2 + 1$
$= 21 - 36(x+2) + \frac{50}{2!}(x+2)^2 - \frac{48}{3!}(x+2)^3 + \frac{24}{4!}(x+2)^4 = 21 - 36(x+2) + 25(x+2)^2 - 8(x+2)^3 + (x+2)^4$

27. $f(x) = x^{-2} \Rightarrow f'(x) = -2x^{-3}, f''(x) = 3!\,x^{-4}, f'''(x) = -4!\,x^{-5} \Rightarrow f^{(n)}(x) = (-1)^n(n+1)!\,x^{-n-2};$
$f(1) = 1, f'(1) = -2, f''(1) = 3!, f'''(1) = -4!, f^{(n)}(1) = (-1)^n(n+1)! \Rightarrow \frac{1}{x^2}$
$= 1 - 2(x-1) + 3(x-1)^2 - 4(x-1)^3 + \ldots = \sum_{n=0}^{\infty} (-1)^n(n+1)(x-1)^n$

29. $f(x) = e^x \Rightarrow f'(x) = e^x, f''(x) = e^x \Rightarrow f^{(n)}(x) = e^x; f(2) = e^2, f'(2) = e^2, \ldots f^{(n)}(2) = e^2$
$\Rightarrow e^x = e^2 + e^2(x-2) + \frac{e^2}{2}(x-2)^2 + \frac{e^3}{3!}(x-2)^3 + \ldots = \sum_{n=0}^{\infty} \frac{e^2}{n!}(x-2)^n$

31. $f(x) = \cos\left(2x + \frac{\pi}{2}\right), f'(x) = -2\sin\left(2x + \frac{\pi}{2}\right), f''(x) = -4\cos\left(2x + \frac{\pi}{2}\right), f'''(x) = 8\sin\left(2x + \frac{\pi}{2}\right),$
$f^{(4)}(x) = 2^4\cos\left(2x + \frac{\pi}{2}\right), f^{(5)}(x) = -2^5\sin\left(2x + \frac{\pi}{2}\right), \ldots; f\left(\frac{\pi}{4}\right) = -1, f'\left(\frac{\pi}{4}\right) = 0, f''\left(\frac{\pi}{4}\right) = 4, f'''\left(\frac{\pi}{4}\right) = 0, f^{(4)}\left(\frac{\pi}{4}\right) = 2^4,$
$f^{(5)}\left(\frac{\pi}{4}\right) = 0, \ldots, f^{(2n)}\left(\frac{\pi}{4}\right) = (-1)^n 2^{2n} \Rightarrow \cos\left(2x + \frac{\pi}{2}\right) = -1 + 2\left(x - \frac{\pi}{4}\right)^2 - \frac{2}{3}\left(x - \frac{\pi}{4}\right)^4 + \ldots$
$= \sum_{n=0}^{\infty} \frac{(-1)^n 2^{2n}}{(2n)!}\left(x - \frac{\pi}{4}\right)^{2n}$

33. The Maclaurin series generated by $\cos x$ is $\sum_{n=0}^{\infty} \frac{(-1)^n}{(2n)!}x^{2n}$ which converges on $(-\infty, \infty)$ and the Maclaurin series generated by $\frac{2}{1-x}$ is $2\sum_{n=0}^{\infty} x^n$ which converges on $(-1, 1)$. Thus the Maclaurin series generated by $f(x) = \cos x - \frac{2}{1-x}$ is given by $\sum_{n=0}^{\infty} \frac{(-1)^n}{(2n)!}x^{2n} - 2\sum_{n=0}^{\infty} x^n = -1 - 2x - \frac{5}{2}x^2 - \ldots.$ which converges on the intersection of $(-\infty, \infty)$ and $(-1, 1)$, so the interval of convergence is $(-1, 1)$.

35. The Maclaurin series generated by $\sin x$ is $\sum_{n=0}^{\infty} \frac{(-1)^n}{(2n+1)!}x^{2n+1}$ which converges on $(-\infty, \infty)$ and the Maclaurin series generated by $\ln(1+x)$ is $\sum_{n=1}^{\infty} \frac{(-1)^{n-1}}{n}x^n$ which converges on $(-1, 1)$. Thus the Maclaurin series genereated by $f(x) = \sin x \cdot \ln(1+x)$ is given by $\left(\sum_{n=0}^{\infty} \frac{(-1)^n}{(2n+1)!}x^{2n+1}\right)\left(\sum_{n=1}^{\infty} \frac{(-1)^{n-1}}{n}x^n\right) = x^2 - \frac{1}{2}x^3 + \frac{1}{6}x^4 - \ldots.$ which converges on the intersection of $(-\infty, \infty)$ and $(-1, 1)$, so the interval of convergence is $(-1, 1)$.

37. If $e^x = \sum_{n=0}^{\infty} \frac{f^{(n)}(a)}{n!}(x-a)^n$ and $f(x) = e^x$, we have $f^{(n)}(a) = e^a$ f or all $n = 0, 1, 2, 3, \ldots$
$\Rightarrow e^x = e^a\left[\frac{(x-a)^0}{0!} + \frac{(x-a)^1}{1!} + \frac{(x-a)^2}{2!} + \ldots\right] = e^a\left[1 + (x-a) + \frac{(x-a)^2}{2!} + \ldots\right]$ at $x = a$

39. $f(x) = f(a) + f'(a)(x-a) + \frac{f''(a)}{2}(x-a)^2 + \frac{f'''(a)}{3!}(x-a)^3 + \ldots \Rightarrow f'(x)$
$= f'(a) + f''(a)(x-a) + \frac{f'''(a)}{3!}3(x-a)^2 + \ldots \Rightarrow f''(x) = f''(a) + f'''(a)(x-a) + \frac{f^{(4)}(a)}{4!}4 \cdot 3(x-a)^2 + \ldots$
$\Rightarrow f^{(n)}(x) = f^{(n)}(a) + f^{(n+1)}(a)(x-a) + \frac{f^{(n+2)}(a)}{2}(x-a)^2 + \ldots$
$\Rightarrow f(a) = f(a) + 0, f'(a) = f'(a) + 0, \ldots, f^{(n)}(a) = f^{(n)}(a) + 0$

41. $f(x) = \ln(\cos x) \Rightarrow f'(x) = -\tan x$ and $f''(x) = -\sec^2 x; f(0) = 0, f'(0) = 0, f''(0) = -1 \Rightarrow L(x) = 0$ and $Q(x) = -\frac{x^2}{2}$

43. $f(x) = (1-x^2)^{-1/2} \Rightarrow f'(x) = x(1-x^2)^{-3/2}$ and $f''(x) = (1-x^2)^{-3/2} + 3x^2(1-x^2)^{-5/2}; f(0) = 1, f'(0) = 0,$
$f''(0) = 1 \Rightarrow L(x) = 1$ and $Q(x) = 1 + \frac{x^2}{2}$

45. $f(x) = \sin x \Rightarrow f'(x) = \cos x$ and $f''(x) = -\sin x; f(0) = 0, f'(0) = 1, f''(0) = 0 \Rightarrow L(x) = x$ and $Q(x) = x$

10.9 CONVERGENCE OF TAYLOR SERIES

1. $e^x = 1 + x + \frac{x^2}{2!} + \ldots = \sum_{n=0}^{\infty} \frac{x^n}{n!} \Rightarrow e^{-5x} = 1 + (-5x) + \frac{(-5x)^2}{2!} + \ldots = 1 - 5x + \frac{5^2x^2}{2!} - \frac{5^3x^3}{3!} + \ldots = \sum_{n=0}^{\infty} \frac{(-1)^n 5^n x^n}{n!}$

3. $\sin x = x - \frac{x^3}{3!} + \frac{x^5}{5!} - \ldots = \sum_{n=0}^{\infty} \frac{(-1)^n x^{2n+1}}{(2n+1)!} \Rightarrow 5\sin(-x) = 5\left[(-x) - \frac{(-x)^3}{3!} + \frac{(-x)^5}{5!} - \ldots\right] = \sum_{n=0}^{\infty} \frac{5(-1)^{n+1} x^{2n+1}}{(2n+1)!}$

5. $\cos x = \sum_{n=0}^{\infty} \frac{(-1)^n x^{2n}}{(2n)!} \Rightarrow \cos 5x^2 = \sum_{n=0}^{\infty} \frac{(-1)^n \left(5x^2\right)^{2n}}{(2n)!} = \sum_{n=0}^{\infty} \frac{(-1)^n 5^{2n} x^{4n}}{(2n)!} = 1 - \frac{25x^4}{2!} + \frac{625x^8}{4!} - \frac{15625x^{12}}{6!} + \ldots$

7. $\ln(1+x) = \sum_{n=1}^{\infty} \frac{(-1)^{n-1} x^n}{n} \Rightarrow \ln(1+x^2) = \sum_{n=1}^{\infty} \frac{(-1)^{n-1} (x^2)^n}{n} = \sum_{n=1}^{\infty} \frac{(-1)^{n-1} x^{2n}}{n} = x^2 - \frac{x^4}{2} + \frac{x^6}{3} - \frac{x^8}{4} + \ldots$

9. $\frac{1}{1+x} = \sum_{n=0}^{\infty} (-1)^n x^n \Rightarrow \frac{1}{1+\frac{3}{4}x^3} = \sum_{n=0}^{\infty} (-1)^n \left(\frac{3}{4}x^3\right)^n = \sum_{n=0}^{\infty} (-1)^n \left(\frac{3}{4}\right)^n x^{3n} = 1 - \frac{3}{4}x^3 + \frac{9}{16}x^6 - \frac{27}{64}x^9 + \ldots$

11. $e^x = \sum_{n=0}^{\infty} \frac{x^n}{n!} \Rightarrow xe^x = x\left(\sum_{n=0}^{\infty} \frac{x^n}{n!}\right) = \sum_{n=0}^{\infty} \frac{x^{n+1}}{n!} = x + x^2 + \frac{x^3}{2!} + \frac{x^4}{3!} + \frac{x^5}{4!} + \ldots$

13. $\cos x = \sum_{n=0}^{\infty} \frac{(-1)^n x^{2n}}{(2n)!} \Rightarrow \frac{x^2}{2} - 1 + \cos x = \frac{x^2}{2} - 1 + \sum_{n=0}^{\infty} \frac{(-1)^n x^{2n}}{(2n)!} = \frac{x^2}{2} - 1 + 1 - \frac{x^2}{2} + \frac{x^4}{4!} - \frac{x^6}{6!} + \frac{x^8}{8!} - \frac{x^{10}}{10!} + \ldots$
$= \frac{x^4}{4!} - \frac{x^6}{6!} + \frac{x^8}{8!} - \frac{x^{10}}{10!} + \ldots = \sum_{n=2}^{\infty} \frac{(-1)^n x^{2n}}{(2n)!}$

15. $\cos x = \sum_{n=0}^{\infty} \frac{(-1)^n x^{2n}}{(2n)!} \Rightarrow x\cos \pi x = x\sum_{n=0}^{\infty} \frac{(-1)^n (\pi x)^{2n}}{(2n)!} = \sum_{n=0}^{\infty} \frac{(-1)^n \pi^{2n} x^{2n+1}}{(2n)!} = x - \frac{\pi^2 x^3}{2!} + \frac{\pi^4 x^5}{4!} - \frac{\pi^6 x^7}{6!} + \ldots$

17. $\cos^2 x = \frac{1}{2} + \frac{\cos 2x}{2} = \frac{1}{2} + \frac{1}{2}\sum_{n=0}^{\infty} \frac{(-1)^n (2x)^{2n}}{(2n)!} = \frac{1}{2} + \frac{1}{2}\left[1 - \frac{(2x)^2}{2!} + \frac{(2x)^4}{4!} - \frac{(2x)^6}{6!} + \frac{(2x)^8}{8!} - \ldots\right]$
$= 1 - \frac{(2x)^2}{2\cdot 2!} + \frac{(2x)^4}{2\cdot 4!} - \frac{(2x)^6}{2\cdot 6!} + \frac{(2x)^8}{2\cdot 8!} - \ldots = 1 + \sum_{n=1}^{\infty} \frac{(-1)^n (2x)^{2n}}{2\cdot(2n)!} = 1 + \sum_{n=1}^{\infty} \frac{(-1)^n 2^{2n-1} x^{2n}}{(2n)!}$

19. $\frac{x^2}{1-2x} = x^2\left(\frac{1}{1-2x}\right) = x^2 \sum_{n=0}^{\infty} (2x)^n = \sum_{n=0}^{\infty} 2^n x^{n+2} = x^2 + 2x^3 + 2^2x^4 + 2^3x^5 + \ldots$

21. $\frac{1}{1-x} = \sum_{n=0}^{\infty} x^n = 1 + x + x^2 + x^3 + \ldots \Rightarrow \frac{d}{dx}\left(\frac{1}{1-x}\right) = \frac{1}{(1-x)^2} = 1 + 2x + 3x^2 + \ldots = \sum_{n=1}^{\infty} nx^{n-1} = \sum_{n=0}^{\infty} (n+1)x^n$

23. $\tan^{-1}x = x - \frac{1}{3}x^3 + \frac{1}{5}x^5 - \frac{1}{7}x^7 + \ldots \Rightarrow x\tan^{-1}x^2 = x\left(x^2 - \frac{1}{3}(x^2)^3 + \frac{1}{5}(x^2)^5 - \frac{1}{7}(x^2)^7 + \ldots\right)$
$= x^3 - \frac{1}{3}x^7 + \frac{1}{5}x^{11} - \frac{1}{7}x^{15} + \ldots = \sum_{n=1}^{\infty} \frac{(-1)^n x^{4n-1}}{2n-1}$

25. $e^x = 1 + x + \frac{x^2}{2!} + \frac{x^3}{3!} + \ldots$ and $\frac{1}{1+x} = 1 - x + x^2 - x^3 + \ldots \Rightarrow e^x + \frac{1}{1+x}$
$= \left(1 + x + \frac{x^2}{2!} + \frac{x^3}{3!} + \ldots\right) + (1 - x + x^2 - x^3 + \ldots) = 2 + \frac{3}{2}x^2 - \frac{5}{6}x^3 + \frac{25}{24}x^4 + \ldots = \sum_{n=0}^{\infty}\left(\frac{1}{n!} + (-1)^n\right)x^n$

27. $\ln(1+x) = x - \frac{1}{2}x^2 + \frac{1}{3}x^3 - \frac{1}{4}x^4 + \ldots \Rightarrow \frac{x}{3}\ln(1+x^2) = \frac{x}{3}\left(x^2 - \frac{1}{2}(x^2)^2 + \frac{1}{3}(x^2)^3 - \frac{1}{4}(x^2)^4 + \ldots\right)$
$= \frac{1}{3}x^3 - \frac{1}{6}x^5 + \frac{1}{9}x^7 - \frac{1}{12}x^9 + \ldots = \sum_{n=1}^{\infty} \frac{(-1)^{n-1}}{3n} x^{2n+1}$

29. $e^x = 1 + x + \frac{x^2}{2!} + \frac{x^3}{3!} + \ldots$ and $\sin x = x - \frac{x^3}{3!} + \frac{x^5}{5!} - \frac{x^7}{7!} + \ldots \Rightarrow e^x \cdot \sin x$
$= \left(1 + x + \frac{x^2}{2!} + \frac{x^3}{3!} + \ldots\right)\left(x - \frac{x^3}{3!} + \frac{x^5}{5!} - \frac{x^7}{7!} + \ldots\right) = x + x^2 + \frac{1}{3}x^3 - \frac{1}{30}x^5 - \ldots.$

31. $\tan^{-1}x = x - \frac{1}{3}x^3 + \frac{1}{5}x^5 - \frac{1}{7}x^7 + \ldots \Rightarrow (\tan^{-1}x)^2 = (\tan^{-1}x)(\tan^{-1}x)$
$= (x - \frac{1}{3}x^3 + \frac{1}{5}x^5 - \frac{1}{7}x^7 + \ldots)(x - \frac{1}{3}x^3 + \frac{1}{5}x^5 - \frac{1}{7}x^7 + \ldots) = x^2 - \frac{2}{3}x^4 - \frac{23}{45}x^6 - \frac{44}{105}x^8 + \ldots.$

33. $\sin x = x - \frac{x^3}{3!} + \frac{x^5}{5!} - \frac{x^7}{7!} + \ldots$ and $e^x = 1 + x + \frac{x^2}{2!} + \frac{x^3}{3!} + \ldots$
$\Rightarrow e^{\sin x} = 1 + \left(x - \frac{x^3}{3!} + \frac{x^5}{5!} - \frac{x^7}{7!} + \ldots\right) + \frac{1}{2}\left(x - \frac{x^3}{3!} + \frac{x^5}{5!} - \frac{x^7}{7!} + \ldots\right)^2 + \frac{1}{6}\left(x - \frac{x^3}{3!} + \frac{x^5}{5!} - \frac{x^7}{7!} + \ldots\right)^3 + \ldots$
$= 1 + x + \frac{1}{2}x^2 - \frac{1}{8}x^4 + \ldots.$

35. Since $n = 3$, then $f^{(4)}(x) = \sin x$, $|f^{(4)}(x)| \le M$ on $[0, 0.1] \Rightarrow |\sin x| \le 1$ on $[0, 0.1] \Rightarrow M = 1$. Then $|R_3(0.1)| \le 1\frac{|0.1-0|^4}{4!}$
$= 4.2 \times 10^{-6} \Rightarrow$ error $\le 4.2 \times 10^{-6}$

37. By the Alternating Series Estimation Theorem, the error is less than $\frac{|x|^5}{5!} \Rightarrow |x|^5 < (5!)(5 \times 10^{-4}) \Rightarrow |x|^5 < 600 \times 10^{-4}$
$\Rightarrow |x| < \sqrt[5]{6 \times 10^{-2}} \approx 0.56968$

39. If $\sin x = x$ and $|x| < 10^{-3}$, then the error is less than $\frac{(10^{-3})^3}{3!} \approx 1.67 \times 10^{-10}$, by Alternating Series Estimation Theorem; The Alternating Series Estimation Theorem says $R_2(x)$ has the same sign as $-\frac{x^3}{3!}$. Moreover, $x < \sin x$
$\Rightarrow 0 < \sin x - x = R_2(x) \Rightarrow x < 0 \Rightarrow -10^{-3} < x < 0.$

41. $|R_2(x)| = \left|\frac{e^c x^3}{3!}\right| < \frac{3^{(0.1)}(0.1)^3}{3!} < 1.87 \times 10^{-4}$, where c is between 0 and x

43. $\sin^2 x = \left(\frac{1-\cos 2x}{2}\right) = \frac{1}{2} - \frac{1}{2}\cos 2x = \frac{1}{2} - \frac{1}{2}\left(1 - \frac{(2x)^2}{2!} + \frac{(2x)^4}{4!} - \frac{(2x)^6}{6!} + \ldots\right) = \frac{2x^2}{2!} - \frac{2^3x^4}{4!} + \frac{2^5x^6}{6!} - \ldots$
$\Rightarrow \frac{d}{dx}(\sin^2 x) = \frac{d}{dx}\left(\frac{2x^2}{2!} - \frac{2^3x^4}{4!} + \frac{2^5x^6}{6!} - \ldots\right) = 2x - \frac{(2x)^3}{3!} + \frac{(2x)^5}{5!} - \frac{(2x)^7}{7!} + \ldots \Rightarrow 2\sin x\cos x$
$= 2x - \frac{(2x)^3}{3!} + \frac{(2x)^5}{5!} - \frac{(2x)^7}{7!} + \ldots = \sin 2x$, which checks

45. A special case of Taylor's Theorem is $f(b) = f(a) + f'(c)(b-a)$, where c is between a and b $\Rightarrow f(b) - f(a) = f'(c)(b-a)$, the Mean Value Theorem.

47. (a) $f'' \le 0$, $f'(a) = 0$ and $x = a$ interior to the interval I $\Rightarrow f(x) - f(a) = \frac{f''(c_2)}{2}(x-a)^2 \le 0$ throughout I
$\Rightarrow f(x) \le f(a)$ throughout I $\Rightarrow$ f has a local maximum at $x = a$
(b) similar reasoning gives $f(x) - f(a) = \frac{f''(c_2)}{2}(x-a)^2 \ge 0$ throughout I $\Rightarrow f(x) \ge f(a)$ throughout I $\Rightarrow$ f has a local minimum at $x = a$

49. (a) $f(x) = (1+x)^k \Rightarrow f'(x) = k(1+x)^{k-1} \Rightarrow f''(x) = k(k-1)(1+x)^{k-2}$; $f(0) = 1$, $f'(0) = k$, and $f''(0) = k(k-1)$
$\Rightarrow Q(x) = 1 + kx + \frac{k(k-1)}{2}x^2$
(b) $|R_2(x)| = \left|\frac{3\cdot 2\cdot 1}{3!}x^3\right| < \frac{1}{100} \Rightarrow |x^3| < \frac{1}{100} \Rightarrow 0 < x < \frac{1}{100^{1/3}}$ or $0 < x < .21544$

51. If $f(x) = \sum_{n=0}^{\infty} a_n x^n$, then $f^{(k)}(x) = \sum_{n=k}^{\infty} n(n-1)(n-2)\cdots(n-k+1)a_n x^{n-k}$ and $f^{(k)}(0) = k!\, a_k$ $\Rightarrow a_k = \frac{f^{(k)}(0)}{k!}$ for k a nonnegative integer. Therefore, the coefficients of f(x) are identical with the corresponding coefficients in the Maclaurin series of f(x) and the statement follows.

10.10 THE BINOMIAL SERIES

1. $(1+x)^{1/2} = 1 + \frac{1}{2}x + \frac{\left(\frac{1}{2}\right)\left(-\frac{1}{2}\right)x^2}{2!} + \frac{\left(\frac{1}{2}\right)\left(-\frac{1}{2}\right)\left(-\frac{3}{2}\right)x^3}{3!} + \ldots = 1 + \frac{1}{2}x - \frac{1}{8}x^2 + \frac{1}{16}x^3 - \ldots$

3. $(1-x)^{-1/2} = 1 - \frac{1}{2}(-x) + \frac{\left(-\frac{1}{2}\right)\left(-\frac{3}{2}\right)(-x)^2}{2!} + \frac{\left(-\frac{1}{2}\right)\left(-\frac{3}{2}\right)\left(-\frac{5}{2}\right)(-x)^3}{3!} + \ldots = 1 + \frac{1}{2}x + \frac{3}{8}x^2 + \frac{5}{16}x^3 + \ldots$

5. $\left(1+\frac{x}{2}\right)^{-2} = 1 - 2\left(\frac{x}{2}\right) + \frac{(-2)(-3)\left(\frac{x}{2}\right)^2}{2!} + \frac{(-2)(-3)(-4)\left(\frac{x}{2}\right)^3}{3!} + \ldots = 1 - x + \frac{3}{4}x^2 - \frac{1}{2}x^3$

7. $\left(1+x^3\right)^{-1/2} = 1 - \frac{1}{2}x^3 + \frac{\left(-\frac{1}{2}\right)\left(-\frac{3}{2}\right)\left(x^3\right)^2}{2!} + \frac{\left(-\frac{1}{2}\right)\left(-\frac{3}{2}\right)\left(-\frac{5}{2}\right)\left(x^3\right)^3}{3!} + \ldots = 1 - \frac{1}{2}x^3 + \frac{3}{8}x^6 - \frac{5}{16}x^9 + \ldots$

9. $\left(1+\frac{1}{x}\right)^{1/2} = 1 + \frac{1}{2}\left(\frac{1}{x}\right) + \frac{\left(\frac{1}{2}\right)\left(-\frac{1}{2}\right)\left(\frac{1}{x}\right)^2}{2!} + \frac{\left(\frac{1}{2}\right)\left(-\frac{1}{2}\right)\left(-\frac{3}{2}\right)\left(\frac{1}{x}\right)^3}{3!} + \ldots = 1 + \frac{1}{2x} - \frac{1}{8x^2} + \frac{1}{16x^3} + \ldots$

11. $(1+x)^4 = 1 + 4x + \frac{(4)(3)x^2}{2!} + \frac{(4)(3)(2)x^3}{3!} + \frac{(4)(3)(2)x^4}{4!} = 1 + 4x + 6x^2 + 4x^3 + x^4$

13. $(1-2x)^3 = 1 + 3(-2x) + \frac{(3)(2)(-2x)^2}{2!} + \frac{(3)(2)(1)(-2x)^3}{3!} = 1 - 6x + 12x^2 - 8x^3$

15. $\int_0^{0.2} \sin x^2\, dx = \int_0^{0.2} \left(x^2 - \frac{x^6}{3!} + \frac{x^{10}}{5!} - \ldots\right) dx = \left[\frac{x^3}{3} - \frac{x^7}{7\cdot 3!} + \ldots\right]_0^{0.2} \approx \left[\frac{x^3}{3}\right]_0^{0.2} \approx 0.00267$ with error $|E| \le \frac{(.2)^7}{7\cdot 3!} \approx 0.0000003$

17. $\int_0^{0.1} \frac{1}{\sqrt{1+x^4}}\, dx = \int_0^{0.1} \left(1 - \frac{x^4}{2} + \frac{3x^8}{8} - \ldots\right) dx = \left[x - \frac{x^5}{10} + \ldots\right]_0^{0.1} \approx [x]_0^{0.1} \approx 0.1$ with error $|E| \le \frac{(0.1)^5}{10} = 0.000001$

19. $\int_0^{0.1} \frac{\sin x}{x}\, dx = \int_0^{0.1} \left(1 - \frac{x^2}{3!} + \frac{x^4}{5!} - \frac{x^6}{7!} + \ldots\right) dx = \left[x - \frac{x^3}{3\cdot 3!} + \frac{x^5}{5\cdot 5!} - \frac{x^7}{7\cdot 7!} + \ldots\right]_0^{0.1} \approx \left[x - \frac{x^3}{3\cdot 3!} + \frac{x^5}{5\cdot 5!}\right]_0^{0.1}$ ≈ 0.0999444611, $|E| \le \frac{(0.1)^7}{7\cdot 7!} \approx 2.8 \times 10^{-12}$

21. $\left(1+x^4\right)^{1/2} = (1)^{1/2} + \frac{\left(\frac{1}{2}\right)}{1}(1)^{-1/2}\left(x^4\right) + \frac{\left(\frac{1}{2}\right)\left(-\frac{1}{2}\right)}{2!}(1)^{-3/2}\left(x^4\right)^2 + \frac{\left(\frac{1}{2}\right)\left(-\frac{1}{2}\right)\left(-\frac{3}{2}\right)}{3!}(1)^{-5/2}\left(x^4\right)^3$ $+ \frac{\left(\frac{1}{2}\right)\left(-\frac{1}{2}\right)\left(-\frac{3}{2}\right)\left(-\frac{5}{2}\right)}{4!}(1)^{-7/2}\left(x^4\right)^4 + \ldots = 1 + \frac{x^4}{2} - \frac{x^8}{8} + \frac{x^{12}}{16} - \frac{5x^{16}}{128} + \ldots$ $\Rightarrow \int_0^{0.1} \left(1 + \frac{x^4}{2} - \frac{x^8}{8} + \frac{x^{12}}{16} - \frac{5x^{16}}{128} + \ldots\right) dx \approx \left[x + \frac{x^5}{10}\right]_0^{0.1} \approx 0.100001$, $|E| \le \frac{(0.1)^9}{72} \approx 1.39 \times 10^{-11}$

23. $\int_0^1 \cos t^2\, dt = \int_0^1 \left(1 - \frac{t^4}{2} + \frac{t^8}{4!} - \frac{t^{12}}{6!} + \ldots\right) dt = \left[t - \frac{t^5}{10} + \frac{t^9}{9\cdot 4!} - \frac{t^{13}}{13\cdot 6!} + \ldots\right]_0^1 \Rightarrow |\text{error}| < \frac{1}{13\cdot 6!} \approx .00011$

25. $F(x) = \int_0^x \left(t^2 - \frac{t^6}{3!} + \frac{t^{10}}{5!} - \frac{t^{14}}{7!} + \ldots\right) dt = \left[\frac{t^3}{3} - \frac{t^7}{7\cdot 3!} + \frac{t^{11}}{11\cdot 5!} - \frac{t^{15}}{15\cdot 7!} + \ldots\right]_0^x \approx \frac{x^3}{3} - \frac{x^7}{7\cdot 3!} + \frac{x^{11}}{11\cdot 5!}$ $\Rightarrow |\text{error}| < \frac{1}{15\cdot 7!} \approx 0.000013$

27. (a) $F(x) = \int_0^x \left(t - \frac{t^3}{3} + \frac{t^5}{5} - \frac{t^7}{7} + \ldots\right) dt = \left[\frac{t^2}{2} - \frac{t^4}{12} + \frac{t^6}{30} - \ldots\right]_0^x \approx \frac{x^2}{2} - \frac{x^4}{12} \Rightarrow |\text{error}| < \frac{(0.5)^6}{30} \approx .00052$

(b) $|\text{error}| < \frac{1}{33 \cdot 34} \approx .00089$ when $F(x) \approx \frac{x^2}{2} - \frac{x^4}{3 \cdot 4} + \frac{x^6}{5 \cdot 6} - \frac{x^8}{7 \cdot 8} + \ldots + (-1)^{15} \frac{x^{32}}{31 \cdot 32}$

29. $\frac{1}{x^2}\left(e^x - (1+x)\right) = \frac{1}{x^2}\left(\left(1 + x + \frac{x^2}{2} + \frac{x^3}{3!} + \ldots\right) - 1 - x\right) = \frac{1}{2} + \frac{x}{3!} + \frac{x^2}{4!} + \ldots \Rightarrow \lim_{x \to 0} \frac{e^x - (1+x)}{x^2}$
$= \lim_{x \to 0} \left(\frac{1}{2} + \frac{x}{3!} + \frac{x^2}{4!} + \ldots\right) = \frac{1}{2}$

31. $\frac{1}{t^4}\left(1 - \cos t - \frac{t^2}{2}\right) = \frac{1}{t^4}\left[1 - \frac{t^2}{2} - \left(1 - \frac{t^2}{2} + \frac{t^4}{4!} - \frac{t^6}{6!} + \ldots\right)\right] = -\frac{1}{4!} + \frac{t^2}{6!} - \frac{t^4}{8!} + \ldots \Rightarrow \lim_{t \to 0} \frac{1 - \cos t - \left(\frac{t^2}{2}\right)}{t^4}$
$= \lim_{t \to 0} \left(-\frac{1}{4!} + \frac{t^2}{6!} - \frac{t^4}{8!} + \ldots\right) = -\frac{1}{24}$

33. $\frac{1}{y^3}\left(y - \tan^{-1} y\right) = \frac{1}{y^3}\left[y - \left(y - \frac{y^3}{3} + \frac{y^5}{5} - \ldots\right)\right] = \frac{1}{3} - \frac{y^2}{5} + \frac{y^4}{7} - \ldots \Rightarrow \lim_{y \to 0} \frac{y - \tan^{-1} y}{y^3} = \lim_{y \to 0} \left(\frac{1}{3} - \frac{y^2}{5} + \frac{y^4}{7} - \ldots\right)$
$= \frac{1}{3}$

35. $x^2\left(-1 + e^{-1/x^2}\right) = x^2\left(-1 + 1 - \frac{1}{x^2} + \frac{1}{2x^4} - \frac{1}{6x^6} + \ldots\right) = -1 + \frac{1}{2x^2} - \frac{1}{6x^4} + \ldots \Rightarrow \lim_{x \to \infty} x^2\left(e^{-1/x^2} - 1\right)$
$= \lim_{x \to \infty} \left(-1 + \frac{1}{2x^2} - \frac{1}{6x^4} + \ldots\right) = -1$

37. $\frac{\ln(1+x^2)}{1 - \cos x} = \frac{\left(x^2 - \frac{x^4}{2} + \frac{x^6}{3} - \ldots\right)}{1 - \left(1 - \frac{x^2}{2!} + \frac{x^4}{4!} - \ldots\right)} = \frac{\left(1 - \frac{x^2}{2} + \frac{x^4}{3} - \ldots\right)}{\left(\frac{1}{2!} - \frac{x^2}{4!} + \ldots\right)} \Rightarrow \lim_{x \to 0} \frac{\ln(1+x^2)}{1 - \cos x} = \lim_{x \to 0} \frac{\left(1 - \frac{x^2}{2} + \frac{x^4}{3} - \ldots\right)}{\left(\frac{1}{2!} - \frac{x^2}{4!} + \ldots\right)} = 2! = 2$

39. $\sin 3x^2 = 3x^2 - \frac{9}{2}x^6 + \frac{81}{40}x^{10} - \ldots$ and $1 - \cos 2x = 2x^2 - \frac{2}{3}x^4 + \frac{4}{45}x^6 - \ldots \Rightarrow \lim_{x \to 0} \frac{\sin 3x^2}{1 - \cos 2x}$
$= \lim_{x \to 0} \frac{3x^2 - \frac{9}{2}x^6 + \frac{81}{40}x^{10} - \ldots}{2x^2 - \frac{2}{3}x^4 + \frac{4}{45}x^6 - \ldots} = \lim_{x \to 0} \frac{3 - \frac{9}{2}x^4 + \frac{81}{40}x^8 - \ldots}{2 - \frac{2}{3}x^2 + \frac{4}{45}x^4 - \ldots} = \frac{3}{2}$

41. $1 + 1 + \frac{1}{2!} + \frac{1}{3!} + \frac{1}{4!} + \ldots = e^1 = e$

43. $1 - \frac{3^2}{4^2 2!} + \frac{3^4}{4^4 4!} - \frac{3^6}{4^6 6!} + \ldots = 1 - \frac{1}{2!}\left(\frac{3}{4}\right)^2 + \frac{1}{4!}\left(\frac{3}{4}\right)^4 - \frac{1}{6!}\left(\frac{3}{4}\right)^6 + \ldots = \cos\left(\frac{3}{4}\right)$

45. $\frac{\pi}{3} - \frac{\pi^3}{3^3 3!} + \frac{\pi^5}{3^5 5!} - \frac{\pi^7}{3^7 7!} + \ldots = \frac{\pi}{3} - \frac{1}{3!}\left(\frac{\pi}{3}\right)^3 + \frac{1}{5!}\left(\frac{\pi}{3}\right)^5 - \frac{1}{7!}\left(\frac{\pi}{3}\right)^7 + \ldots = \sin\left(\frac{\pi}{3}\right) = \frac{\sqrt{3}}{2}$

47. $x^3 + x^4 + x^5 + x^6 + \ldots = x^3\left(1 + x + x^2 + x^3 + \ldots\right) = x^3\left(\frac{1}{1-x}\right) = \frac{x^3}{1-x}$

49. $x^3 - x^5 + x^7 - x^9 + \ldots = x^3\left(1 - x^2 + (x^2)^2 - (x^2)^3 + \ldots\right) = x^3\left(\frac{1}{1+x^2}\right) = \frac{x^3}{1+x^2}$

51. $-1 + 2x - 3x^2 + 4x^3 - 5x^4 + \ldots = \frac{d}{dx}\left(1 - x + x^2 - x^3 + x^4 - x^5 + \ldots\right) = \frac{d}{dx}\left(\frac{1}{1+x}\right) = \frac{-1}{(1+x)^2}$

53. $\ln\left(\frac{1+x}{1-x}\right) = \ln(1+x) - \ln(1-x) = \left(x - \frac{x^2}{2} + \frac{x^3}{3} - \frac{x^4}{4} + \ldots\right) - \left(-x - \frac{x^2}{2} - \frac{x^3}{3} - \frac{x^4}{4} - \ldots\right) = 2\left(x + \frac{x^3}{3} + \frac{x^5}{5} + \ldots\right)$

55. $\tan^{-1} x = x - \frac{x^3}{3} + \frac{x^5}{5} - \frac{x^7}{7} + \frac{x^9}{9} - \ldots + \frac{(-1)^{n-1}x^{2n-1}}{2n-1} + \ldots \Rightarrow |\text{error}| = \left|\frac{(-1)^{n-1}x^{2n-1}}{2n-1}\right| = \frac{1}{2n-1}$ when $x = 1$;
$\frac{1}{2n-1} < \frac{1}{10^3} \Rightarrow n > \frac{1001}{2} = 500.5 \Rightarrow$ the first term not used is the $501^{\text{st}} \Rightarrow$ we must use 500 terms

57. $\tan^{-1} x = x - \frac{x^3}{3} + \frac{x^5}{5} - \frac{x^7}{7} + \frac{x^9}{9} - \ldots + \frac{(-1)^{n-1}x^{2n-1}}{2n-1} + \ldots$ and when the series representing $48\tan^{-1}\left(\frac{1}{18}\right)$ has an error less than $\frac{1}{3} \cdot 10^{-6}$, then the series representing the sum $48\tan^{-1}\left(\frac{1}{18}\right) + 32\tan^{-1}\left(\frac{1}{57}\right) - 20\tan^{-1}\left(\frac{1}{239}\right)$ also has an error of magnitude less than 10^{-6}; thus $|\text{error}| = 48\frac{\left(\frac{1}{18}\right)^{2n-1}}{2n-1} < \frac{1}{3\cdot 10^6} \Rightarrow n \geq 4$ using a calculator $\Rightarrow$ 4 terms

59. (a) $(1-x^2)^{-1/2} \approx 1 + \frac{x^2}{2} + \frac{3x^4}{8} + \frac{5x^6}{16} \Rightarrow \sin^{-1} x \approx x + \frac{x^3}{6} + \frac{3x^5}{40} + \frac{5x^7}{112}$; Using the Ratio Test:
$\lim_{n\to\infty}\left|\frac{1\cdot3\cdot5\cdots(2n-1)(2n+1)x^{2n+3}}{2\cdot4\cdot6\cdots(2n)(2n+2)(2n+3)} \cdot \frac{2\cdot4\cdot6\cdots(2n)(2n+1)}{1\cdot3\cdot5\cdots(2n-1)x^{2n+1}}\right| < 1 \Rightarrow x^2 \lim_{n\to\infty}\left|\frac{(2n+1)(2n+1)}{(2n+2)(2n+3)}\right| < 1$
$\Rightarrow |x| < 1 \Rightarrow$ the radius of convergence is 1. See Exercise 69.

(b) $\frac{d}{dx}(\cos^{-1} x) = -(1-x^2)^{-1/2} \Rightarrow \cos^{-1} x = \frac{\pi}{2} - \sin^{-1} x \approx \frac{\pi}{2} - \left(x + \frac{x^3}{6} + \frac{3x^5}{40} + \frac{5x^7}{112}\right) \approx \frac{\pi}{2} - x - \frac{x^3}{6} - \frac{3x^5}{40} - \frac{5x^7}{112}$

61. $\frac{-1}{1+x} = -\frac{1}{1-(-x)} = -1 + x - x^2 + x^3 - \ldots \Rightarrow \frac{d}{dx}\left(\frac{-1}{1+x}\right) = \frac{1}{1+x^2} = \frac{d}{dx}(-1 + x - x^2 + x^3 - \ldots)$
$= 1 - 2x + 3x^2 - 4x^3 + \ldots$

63. Wallis' formula gives the approximation $\pi \approx 4\left[\frac{2\cdot4\cdot4\cdot6\cdot6\cdot8\cdots(2n-2)\cdot(2n)}{3\cdot3\cdot5\cdot5\cdot7\cdot7\cdots(2n-1)\cdot(2n-1)}\right]$ to produce the table

n	$\sim \pi$
10	3.221088998
20	3.181104886
30	3.167880758
80	3.151425420
90	3.150331383
93	3.150049112
94	3.149959030
95	3.149870848
100	3.149456425

At n = 1929 we obtain the first approximation accurate to 3 decimals: 3.141999845. At n = 30,000 we still do not obtain accuracy to 4 decimals: 3.141617732, so the convergence to π is very slow. Here is a Maple CAS procedure to produce these approximations:

```
pie :=
  proc(n)
  local i,j;
    a(2) := evalf(8/9);
    for i from 3 to n do a(i) := evalf(2*(2*i-2)*i/(2*i-1)^2*a(i-1)) od;
    [[j,4*a(j)] $ (j = n-5 .. n)]
  end
```

65. $(1-x^2)^{-1/2} = (1+(-x^2))^{-1/2} = (1)^{-1/2} + \left(-\frac{1}{2}\right)(1)^{-3/2}(-x^2) + \frac{\left(-\frac{1}{2}\right)\left(-\frac{3}{2}\right)(1)^{-5/2}(-x^2)^2}{2!}$
$+ \frac{\left(-\frac{1}{2}\right)\left(-\frac{3}{2}\right)\left(-\frac{5}{2}\right)(1)^{-7/2}(-x^2)^3}{3!} + \ldots = 1 + \frac{x^2}{2} + \frac{1\cdot3x^4}{2^2\cdot2!} + \frac{1\cdot3\cdot5x^6}{2^3\cdot3!} + \ldots = 1 + \sum_{n=1}^{\infty} \frac{1\cdot3\cdot5\cdots(2n-1)x^{2n}}{2^n\cdot n!}$
$\Rightarrow \sin^{-1} x = \int_0^x (1-t^2)^{-1/2}\, dt = \int_0^x \left(1 + \sum_{n=1}^{\infty} \frac{1\cdot3\cdot5\cdots(2n-1)x^{2n}}{2^n\cdot n!}\right) dt = x + \sum_{n=1}^{\infty} \frac{1\cdot3\cdot5\cdots(2n-1)x^{2n+1}}{2\cdot4\cdots(2n)(2n+1)}$,
where $|x| < 1$

67. (a) $e^{-i\pi} = \cos(-\pi) + i\sin(-\pi) = -1 + i(0) = -1$

(b) $e^{i\pi/4} = \cos\left(\frac{\pi}{4}\right) + i\sin\left(\frac{\pi}{4}\right) = \frac{1}{\sqrt{2}} + \frac{i}{\sqrt{2}} = \left(\frac{1}{\sqrt{2}}\right)(1+i)$

(c) $e^{-i\pi/2} = \cos\left(-\frac{\pi}{2}\right) + i\sin\left(-\frac{\pi}{2}\right) = 0 + i(-1) = -i$

69. $e^x = 1 + x + \frac{x^2}{2!} + \frac{x^3}{3!} + \frac{x^4}{4!} + \ldots \Rightarrow e^{i\theta} = 1 + i\theta + \frac{(i\theta)^2}{2!} + \frac{(i\theta)^3}{3!} + \frac{(i\theta)^4}{4!} + \ldots$ and
$e^{-i\theta} = 1 - i\theta + \frac{(-i\theta)^2}{2!} + \frac{(-i\theta)^3}{3!} + \frac{(-i\theta)^4}{4!} + \ldots = 1 - i\theta + \frac{(i\theta)^2}{2!} - \frac{(i\theta)^3}{3!} + \frac{(i\theta)^4}{4!} - \ldots$
$\Rightarrow \frac{e^{i\theta}+e^{-i\theta}}{2} = \frac{\left(1+i\theta+\frac{(i\theta)^2}{2!}+\frac{(i\theta)^3}{3!}+\frac{(i\theta)^4}{4!}+\ldots\right)+\left(1-i\theta+\frac{(i\theta)^2}{2!}-\frac{(i\theta)^3}{3!}+\frac{(i\theta)^4}{4!}-\ldots\right)}{2}$
$= 1 - \frac{\theta^2}{2!} + \frac{\theta^4}{4!} - \frac{\theta^6}{6!} + \ldots = \cos\theta;$
$\frac{e^{i\theta}-e^{-i\theta}}{2i} = \frac{\left(1+i\theta+\frac{(i\theta)^2}{2!}+\frac{(i\theta)^3}{3!}+\frac{(i\theta)^4}{4!}+\ldots\right)-\left(1-i\theta+\frac{(i\theta)^2}{2!}-\frac{(i\theta)^3}{3!}+\frac{(i\theta)^4}{4!}-\ldots\right)}{2i}$
$= \theta - \frac{\theta^3}{3!} + \frac{\theta^5}{5!} - \frac{\theta^7}{7!} + \ldots = \sin\theta$

71. $e^x \sin x = \left(1 + x + \frac{x^2}{2!} + \frac{x^3}{3!} + \frac{x^4}{4!} + \ldots\right)\left(x - \frac{x^3}{3!} + \frac{x^5}{5!} - \frac{x^7}{7!} + \ldots\right)$
$= (1)x + (1)x^2 + \left(-\frac{1}{6} + \frac{1}{2}\right)x^3 + \left(-\frac{1}{6} + \frac{1}{6}\right)x^4 + \left(\frac{1}{120} - \frac{1}{12} + \frac{1}{24}\right)x^5 + \ldots = x + x^2 + \frac{1}{3}x^3 - \frac{1}{30}x^5 + \ldots;$
$e^x \cdot e^{ix} = e^{(1+i)x} = e^x(\cos x + i\sin x) = e^x\cos x + i(e^x\sin x) \Rightarrow e^x\sin x$ is the series of the imaginary part of $e^{(1+i)x}$ which we calculate next; $e^{(1+i)x} = \sum_{n=0}^{\infty} \frac{(x+ix)^n}{n!} = 1 + (x+ix) + \frac{(x+ix)^2}{2!} + \frac{(x+ix)^3}{3!} + \frac{(x+ix)^4}{4!} + \ldots$
$= 1 + x + ix + \frac{1}{2!}(2ix^2) + \frac{1}{3!}(2ix^3 - 2x^3) + \frac{1}{4!}(-4x^4) + \frac{1}{5!}(-4x^5 - 4ix^5) + \frac{1}{6!}(-8ix^6) + \ldots \Rightarrow$ the imaginary part of $e^{(1+i)x}$ is $x + \frac{2}{2!}x^2 + \frac{2}{3!}x^3 - \frac{4}{5!}x^5 - \frac{8}{6!}x^6 + \ldots = x + x^2 + \frac{1}{3}x^3 - \frac{1}{30}x^5 - \frac{1}{90}x^6 + \ldots$ in agreement with our product calculation. The series for $e^x \sin x$ converges for all values of x.

73. (a) $e^{i\theta_1}e^{i\theta_2} = (\cos\theta_1 + i\sin\theta_1)(\cos\theta_2 + i\sin\theta_2) = (\cos\theta_1\cos\theta_2 - \sin\theta_1\sin\theta_2) + i(\sin\theta_1\cos\theta_2 + \sin\theta_2\cos\theta_1)$
$= \cos(\theta_1 + \theta_2) + i\sin(\theta_1 + \theta_2) = e^{i(\theta_1+\theta_2)}$
(b) $e^{-i\theta} = \cos(-\theta) + i\sin(-\theta) = \cos\theta - i\sin\theta = (\cos\theta - i\sin\theta)\left(\frac{\cos\theta + i\sin\theta}{\cos\theta + i\sin\theta}\right) = \frac{1}{\cos\theta + i\sin\theta} = \frac{1}{e^{i\theta}}$

CHAPTER 10 PRACTICE EXERCISES

1. converges to 1, since $\lim_{n\to\infty} a_n = \lim_{n\to\infty}\left(1 + \frac{(-1)^n}{n}\right) = 1$

3. converges to -1, since $\lim_{n\to\infty} a_n = \lim_{n\to\infty}\left(\frac{1-2^n}{2^n}\right) = \lim_{n\to\infty}\left(\frac{1}{2^n} - 1\right) = -1$

5. diverges, since $\left\{\sin\frac{n\pi}{2}\right\} = \{0, 1, 0, -1, 0, 1, \ldots\}$

7. converges to 0, since $\lim_{n\to\infty} a_n = \lim_{n\to\infty}\frac{\ln n^2}{n} = 2\lim_{n\to\infty}\frac{\left(\frac{1}{n}\right)}{1} = 0$

9. converges to 1, since $\lim_{n\to\infty} a_n = \lim_{n\to\infty}\left(\frac{n+\ln n}{n}\right) = \lim_{n\to\infty}\frac{1+\left(\frac{1}{n}\right)}{1} = 1$

11. converges to e^{-5}, since $\lim_{n\to\infty} a_n = \lim_{n\to\infty}\left(\frac{n-5}{n}\right)^n = \lim_{n\to\infty}\left(1 + \frac{(-5)}{n}\right)^n = e^{-5}$ by Theorem 5

13. converges to 3, since $\lim_{n\to\infty} a_n = \lim_{n\to\infty}\left(\frac{3^n}{n}\right)^{1/n} = \lim_{n\to\infty}\frac{3}{n^{1/n}} = \frac{3}{1} = 3$ by Theorem 5

15. converges to ln 2, since $\lim_{n\to\infty} a_n = \lim_{n\to\infty} n\left(2^{1/n} - 1\right) = \lim_{n\to\infty}\frac{2^{1/n}-1}{\left(\frac{1}{n}\right)} = \lim_{n\to\infty}\frac{\left[\frac{\left(-2^{1/n}\ln 2\right)}{n^2}\right]}{\left(\frac{-1}{n^2}\right)} = \lim_{n\to\infty} 2^{1/n}\ln 2$
$= 2^0 \cdot \ln 2 = \ln 2$

17. diverges, since $\lim_{n\to\infty} a_n = \lim_{n\to\infty} \frac{(n+1)!}{n!} = \lim_{n\to\infty} (n+1) = \infty$

19. $\frac{1}{(2n-3)(2n-1)} = \frac{\left(\frac{1}{2}\right)}{2n-3} - \frac{\left(\frac{1}{2}\right)}{2n-1} \Rightarrow s_n = \left[\frac{\left(\frac{1}{2}\right)}{3} - \frac{\left(\frac{1}{2}\right)}{5}\right] + \left[\frac{\left(\frac{1}{2}\right)}{5} - \frac{\left(\frac{1}{2}\right)}{7}\right] + \ldots + \left[\frac{\left(\frac{1}{2}\right)}{2n-3} - \frac{\left(\frac{1}{2}\right)}{2n-1}\right] = \frac{\left(\frac{1}{2}\right)}{3} - \frac{\left(\frac{1}{2}\right)}{2n-1}$

$\Rightarrow \lim_{n\to\infty} s_n = \lim_{n\to\infty} \left[\frac{1}{6} - \frac{\left(\frac{1}{2}\right)}{2n-1}\right] = \frac{1}{6}$

21. $\frac{9}{(3n-1)(3n+2)} = \frac{3}{3n-1} - \frac{3}{3n+2} \Rightarrow s_n = \left(\frac{3}{2} - \frac{3}{5}\right) + \left(\frac{3}{5} - \frac{3}{8}\right) + \left(\frac{3}{8} - \frac{3}{11}\right) + \ldots + \left(\frac{3}{3n-1} - \frac{3}{3n+2}\right)$
$= \frac{3}{2} - \frac{3}{3n+2} \Rightarrow \lim_{n\to\infty} s_n = \lim_{n\to\infty} \left(\frac{3}{2} - \frac{3}{3n+2}\right) = \frac{3}{2}$

23. $\sum_{n=0}^{\infty} e^{-n} = \sum_{n=0}^{\infty} \frac{1}{e^n}$, a convergent geometric series with $r = \frac{1}{e}$ and $a = 1 \Rightarrow$ the sum is $\frac{1}{1-\left(\frac{1}{e}\right)} = \frac{e}{e-1}$

25. diverges, a p-series with $p = \frac{1}{2}$

27. Since $f(x) = \frac{1}{x^{1/2}} \Rightarrow f'(x) = -\frac{1}{2x^{3/2}} < 0 \Rightarrow f(x)$ is decreasing $\Rightarrow a_{n+1} < a_n$, and $\lim_{n\to\infty} a_n = \lim_{n\to\infty} \frac{1}{\sqrt{n}} = 0$, the series $\sum_{n=1}^{\infty} \frac{(-1)^n}{\sqrt{n}}$ converges by the Alternating Series Test. Since $\sum_{n=1}^{\infty} \frac{1}{\sqrt{n}}$ diverges, the given series converges conditionally.

29. The given series does not converge absolutely by the Direct Comparison Test since $\frac{1}{\ln(n+1)} > \frac{1}{n+1}$, which is the nth term of a divergent series. Since $f(x) = \frac{1}{\ln(x+1)} \Rightarrow f'(x) = -\frac{1}{(\ln(x+1))^2(x+1)} < 0 \Rightarrow f(x)$ is decreasing $\Rightarrow a_{n+1} < a_n$, and $\lim_{n\to\infty} a_n = \lim_{n\to\infty} \frac{1}{\ln(n+1)} = 0$, the given series converges conditionally by the Alternating Series Test.

31. converges absolutely by the Direct Comparison Test since $\frac{\ln n}{n^3} < \frac{n}{n^3} = \frac{1}{n^2}$, the nth term of a convergent p-series

33. $\lim_{n\to\infty} \frac{\left(\frac{1}{n\sqrt{n^2+1}}\right)}{\left(\frac{1}{n^2}\right)} = \sqrt{\lim_{n\to\infty} \frac{n^2}{n^2+1}} = \sqrt{1} = 1 \Rightarrow$ converges absolutely by the Limit Comparison Test

35. converges absolutely by the Ratio Test since $\lim_{n\to\infty} \left[\frac{n+2}{(n+1)!} \cdot \frac{n!}{n+1}\right] = \lim_{n\to\infty} \frac{n+2}{(n+1)^2} = 0 < 1$

37. converges absolutely by the Ratio Test since $\lim_{n\to\infty} \left[\frac{3^{n+1}}{(n+1)!} \cdot \frac{n!}{3^n}\right] = \lim_{n\to\infty} \frac{3}{n+1} = 0 < 1$

39. converges absolutely by the Limit Comparison Test since $\lim_{n\to\infty} \frac{\left(\frac{1}{n^{3/2}}\right)}{\left(\frac{1}{\sqrt{n(n+1)(n+2)}}\right)} = \sqrt{\lim_{n\to\infty} \frac{n(n+1)(n+2)}{n^3}} = 1$

41. $\lim_{n\to\infty} \left|\frac{u_{n+1}}{u_n}\right| < 1 \Rightarrow \lim_{n\to\infty} \left|\frac{(x+4)^{n+1}}{(n+1)3^{n+1}} \cdot \frac{n3^n}{(x+4)^n}\right| < 1 \Rightarrow \frac{|x+4|}{3} \lim_{n\to\infty} \left(\frac{n}{n+1}\right) < 1 \Rightarrow \frac{|x+4|}{3} < 1$
$\Rightarrow |x+4| < 3 \Rightarrow -3 < x+4 < 3 \Rightarrow -7 < x < -1$; at $x = -7$ we have $\sum_{n=1}^{\infty} \frac{(-1)^n 3^n}{n3^n} = \sum_{n=1}^{\infty} \frac{(-1)^n}{n}$, the alternating harmonic series, which converges conditionally; at $x = -1$ we have $\sum_{n=1}^{\infty} \frac{3^n}{n3^n} = \sum_{n=1}^{\infty} \frac{1}{n}$, the divergent harmonic series

(a) the radius is 3; the interval of convergence is $-7 \le x < -1$
(b) the interval of absolute convergence is $-7 < x < -1$
(c) the series converges conditionally at $x = -7$

43. $\lim_{n\to\infty}\left|\frac{u_{n+1}}{u_n}\right| < 1 \Rightarrow \lim_{n\to\infty}\left|\frac{(3x-1)^{n+1}}{(n+1)^2}\cdot\frac{n^2}{(3x-1)^n}\right| < 1 \Rightarrow |3x-1|\lim_{n\to\infty}\frac{n^2}{(n+1)^2} < 1 \Rightarrow |3x-1| < 1$

$\Rightarrow -1 < 3x-1 < 1 \Rightarrow 0 < 3x < 2 \Rightarrow 0 < x < \frac{2}{3}$; at $x = 0$ we have $\sum_{n=1}^{\infty}\frac{(-1)^{n-1}(-1)^n}{n^2} = \sum_{n=1}^{\infty}\frac{(-1)^{2n-1}}{n^2}$

$= -\sum_{n=1}^{\infty}\frac{1}{n^2}$, a nonzero constant multiple of a convergent p-series, which is absolutely convergent; at $x = \frac{2}{3}$ we have $\sum_{n=1}^{\infty}\frac{(-1)^{n-1}(1)^n}{n^2} = \sum_{n=1}^{\infty}\frac{(-1)^{n-1}}{n^2}$, which converges absolutely

(a) the radius is $\frac{1}{3}$; the interval of convergence is $0 \le x \le \frac{2}{3}$
(b) the interval of absolute convergence is $0 \le x \le \frac{2}{3}$
(c) there are no values for which the series converges conditionally

45. $\lim_{n\to\infty}\left|\frac{u_{n+1}}{u_n}\right| < 1 \Rightarrow \lim_{n\to\infty}\left|\frac{x^{n+1}}{(n+1)^{n+1}}\cdot\frac{n^n}{x^n}\right| < 1 \Rightarrow |x|\lim_{n\to\infty}\left|\left(\frac{n}{n+1}\right)^n\left(\frac{1}{n+1}\right)\right| < 1 \Rightarrow \frac{|x|}{e}\lim_{n\to\infty}\left(\frac{1}{n+1}\right) < 1$

$\Rightarrow \frac{|x|}{e}\cdot 0 < 1$, which holds for all x

(a) the radius is ∞; the series converges for all x
(b) the series converges absolutely for all x
(c) there are no values for which the series converges conditionally

47. $\lim_{n\to\infty}\left|\frac{u_{n+1}}{u_n}\right| < 1 \Rightarrow \lim_{n\to\infty}\left|\frac{(n+2)x^{2n+1}}{3^{n+1}}\cdot\frac{3^n}{(n+1)x^{2n-1}}\right| < 1 \Rightarrow \frac{x^2}{3}\lim_{n\to\infty}\left(\frac{n+2}{n+1}\right) < 1 \Rightarrow -\sqrt{3} < x < \sqrt{3}$;

the series $\sum_{n=1}^{\infty} -\frac{n+1}{\sqrt{3}}$ and $\sum_{n=1}^{\infty}\frac{n+1}{\sqrt{3}}$, obtained with $x = \pm\sqrt{3}$, both diverge

(a) the radius is $\sqrt{3}$; the interval of convergence is $-\sqrt{3} < x < \sqrt{3}$
(b) the interval of absolute convergence is $-\sqrt{3} < x < \sqrt{3}$
(c) there are no values for which the series converges conditionally

49. $\lim_{n\to\infty}\left|\frac{u_{n+1}}{u_n}\right| < 1 \Rightarrow \lim_{n\to\infty}\left|\frac{\operatorname{csch}(n+1)x^{n+1}}{\operatorname{csch}(n)x^n}\right| < 1 \Rightarrow |x|\lim_{n\to\infty}\left|\frac{\left(\frac{2}{e^{n+1}-e^{-n-1}}\right)}{\left(\frac{2}{e^n-e^{-n}}\right)}\right| < 1$

$\Rightarrow |x|\lim_{n\to\infty}\left|\frac{e^{-1}-e^{-2n-1}}{1-e^{-2n-2}}\right| < 1 \Rightarrow \frac{|x|}{e} < 1 \Rightarrow -e < x < e$; the series $\sum_{n=1}^{\infty}(\pm e)^n \operatorname{csch} n$, obtained with $x = \pm e$, both diverge since $\lim_{n\to\infty}(\pm e)^n \operatorname{csch} n \ne 0$

(a) the radius is e; the interval of convergence is $-e < x < e$
(b) the interval of absolute convergence is $-e < x < e$
(c) there are no values for which the series converges conditionally

51. The given series has the form $1 - x + x^2 - x^3 + \ldots + (-x)^n + \ldots = \frac{1}{1+x}$, where $x = \frac{1}{4}$; the sum is $\frac{1}{1+\left(\frac{1}{4}\right)} = \frac{4}{5}$

53. The given series has the form $x - \frac{x^3}{3!} + \frac{x^5}{5!} - \ldots + (-1)^n\frac{x^{2n+1}}{(2n+1)!} + \ldots = \sin x$, where $x = \pi$; the sum is $\sin\pi = 0$

55. The given series has the form $1 + x + \frac{x^2}{2!} + \frac{x^2}{3!} + \ldots + \frac{x^n}{n!} + \ldots = e^x$, where $x = \ln 2$; the sum is $e^{\ln(2)} = 2$

57. Consider $\frac{1}{1-2x}$ as the sum of a convergent geometric series with $a = 1$ and $r = 2x \Rightarrow \frac{1}{1-2x}$
$= 1 + (2x) + (2x)^2 + (2x)^3 + \ldots = \sum_{n=0}^{\infty}(2x)^n = \sum_{n=0}^{\infty}2^n x^n$ where $|2x| < 1 \Rightarrow |x| < \frac{1}{2}$

59. $\sin x = \sum_{n=0}^{\infty}\frac{(-1)^n x^{2n+1}}{(2n+1)!} \Rightarrow \sin\pi x = \sum_{n=0}^{\infty}\frac{(-1)^n(\pi x)^{2n+1}}{(2n+1)!} = \sum_{n=0}^{\infty}\frac{(-1)^n\pi^{2n+1}x^{2n+1}}{(2n+1)!}$

61. $\cos x = \sum_{n=0}^{\infty} \frac{(-1)^n x^{2n}}{(2n)!} \Rightarrow \cos\left(x^{5/3}\right) = \sum_{n=0}^{\infty} \frac{(-1)^n \left(x^{5/3}\right)^{2n}}{(2n)!} = \sum_{n=0}^{\infty} \frac{(-1)^n x^{10n/3}}{(2n)!}$

63. $e^x = \sum_{n=0}^{\infty} \frac{x^n}{n!} \Rightarrow e^{(\pi x/2)} = \sum_{n=0}^{\infty} \frac{\left(\frac{\pi x}{2}\right)^n}{n!} = \sum_{n=0}^{\infty} \frac{\pi^n x^n}{2^n n!}$

65. $f(x) = \sqrt{3+x^2} = (3+x^2)^{1/2} \Rightarrow f'(x) = x\,(3+x^2)^{-1/2} \Rightarrow f''(x) = -x^2\,(3+x^2)^{-3/2} + (3+x^2)^{-1/2}$
$\Rightarrow f'''(x) = 3x^3\,(3+x^2)^{-5/2} - 3x\,(3+x^2)^{-3/2}$; $f(-1) = 2$, $f'(-1) = -\frac{1}{2}$, $f''(-1) = -\frac{1}{8} + \frac{1}{2} = \frac{3}{8}$,
$f'''(-1) = -\frac{3}{32} + \frac{3}{8} = \frac{9}{32} \Rightarrow \sqrt{3+x^2} = 2 - \frac{(x+1)}{2\cdot 1!} + \frac{3(x+1)^2}{2^3\cdot 2!} + \frac{9(x+1)^3}{2^5\cdot 3!} + \ldots$

67. $f(x) = \frac{1}{x+1} = (x+1)^{-1} \Rightarrow f'(x) = -(x+1)^{-2} \Rightarrow f''(x) = 2(x+1)^{-3} \Rightarrow f'''(x) = -6(x+1)^{-4}$; $f(3) = \frac{1}{4}$,
$f'(3) = -\frac{1}{4^2}$, $f''(3) = \frac{2}{4^3}$, $f'''(2) = \frac{-6}{4^4} \Rightarrow \frac{1}{x+1} = \frac{1}{4} - \frac{1}{4^2}(x-3) + \frac{1}{4^3}(x-3)^2 - \frac{1}{4^4}(x-3)^3 + \ldots$

69. $\int_0^{1/2} \exp(-x^3)\,dx = \int_0^{1/2} \left(1 - x^3 + \frac{x^6}{2!} - \frac{x^9}{3!} + \frac{x^{12}}{4!} + \ldots\right) dx = \left[x - \frac{x^4}{4} + \frac{x^7}{7\cdot 2!} - \frac{x^{10}}{10\cdot 3!} + \frac{x^{13}}{13\cdot 4!} - \ldots\right]_0^{1/2}$
$\approx \frac{1}{2} - \frac{1}{2^4\cdot 4} + \frac{1}{2^7\cdot 7\cdot 2!} - \frac{1}{2^{10}\cdot 10\cdot 3!} + \frac{1}{2^{13}\cdot 13\cdot 4!} - \frac{1}{2^{16}\cdot 16\cdot 5!} \approx 0.484917143$

71. $\int_1^{1/2} \frac{\tan^{-1} x}{x}\,dx = \int_1^{1/2} \left(1 - \frac{x^2}{3} + \frac{x^4}{5} - \frac{x^6}{7} + \frac{x^8}{9} - \frac{x^{10}}{11} + \ldots\right) dx = \left[x - \frac{x^3}{9} + \frac{x^5}{25} - \frac{x^7}{49} + \frac{x^9}{81} - \frac{x^{11}}{121} + \ldots\right]_0^{1/2}$
$\approx \frac{1}{2} - \frac{1}{9\cdot 2^3} + \frac{1}{5^2\cdot 2^5} - \frac{1}{7^2\cdot 2^7} + \frac{1}{9^2\cdot 2^9} - \frac{1}{11^2\cdot 2^{11}} + \frac{1}{13^2\cdot 2^{13}} - \frac{1}{15^2\cdot 2^{15}} + \frac{1}{17^2\cdot 2^{17}} - \frac{1}{19^2\cdot 2^{19}} + \frac{1}{21^2\cdot 2^{21}} \approx 0.4872223583$

73. $\lim_{x\to 0} \frac{7\sin x}{e^{2x}-1} = \lim_{x\to 0} \frac{7\left(x - \frac{x^3}{3!} + \frac{x^5}{5!} - \ldots\right)}{\left(2x + \frac{2^2x^2}{2!} + \frac{2^3x^3}{3!} + \ldots\right)} = \lim_{x\to 0} \frac{7\left(1 - \frac{x^2}{3!} + \frac{x^4}{5!} - \ldots\right)}{\left(2 + \frac{2^2x}{2!} + \frac{2^3x^2}{3!} + \ldots\right)} = \frac{7}{2}$

75. $\lim_{t\to 0} \left(\frac{1}{2-2\cos t} - \frac{1}{t^2}\right) = \lim_{t\to 0} \frac{t^2 - 2 + 2\cos t}{2t^2(1-\cos t)} = \lim_{t\to 0} \frac{t^2 - 2 + 2\left(1 - \frac{t^2}{2} + \frac{t^4}{4!} - \ldots\right)}{2t^2\left(1 - 1 + \frac{t^2}{2} - \frac{t^4}{4!} + \ldots\right)} = \lim_{t\to 0} \frac{2\left(\frac{t^4}{4!} - \frac{t^6}{6!} + \ldots\right)}{\left(t^4 - \frac{2t^6}{4!} + \ldots\right)}$
$= \lim_{t\to 0} \frac{2\left(\frac{1}{4!} - \frac{t^2}{6!} + \ldots\right)}{\left(1 - \frac{2t^2}{4!} + \ldots\right)} = \frac{1}{12}$

77. $\lim_{z\to 0} \frac{1-\cos^2 z}{\ln(1-z) + \sin z} = \lim_{z\to 0} \frac{1 - \left(1 - z^2 + \frac{z^4}{3} - \ldots\right)}{\left(-z - \frac{z^2}{2} - \frac{z^3}{3} - \ldots\right) + \left(z - \frac{z^3}{3!} + \frac{z^5}{5!} - \ldots\right)} = \lim_{z\to 0} \frac{\left(z^2 - \frac{z^4}{3} + \ldots\right)}{\left(-\frac{z^2}{2} - \frac{2z^3}{3} - \frac{z^4}{4} - \ldots\right)}$
$= \lim_{z\to 0} \frac{\left(1 - \frac{z^2}{3} + \ldots\right)}{\left(-\frac{1}{2} - \frac{2z}{3} - \frac{z^2}{4} - \ldots\right)} = -2$

79. $\lim_{x\to 0} \left(\frac{\sin 3x}{x^3} + \frac{r}{x^2} + s\right) = \lim_{x\to 0} \left[\frac{\left(3x - \frac{(3x)^3}{6} + \frac{(3x)^5}{120} - \ldots\right)}{x^3} + \frac{r}{x^2} + s\right] = \lim_{x\to 0} \left(\frac{3}{x^2} - \frac{9}{2} + \frac{81x^2}{40} + \ldots + \frac{r}{x^2} + s\right) = 0$
$\Rightarrow \frac{r}{x^2} + \frac{3}{x^2} = 0$ and $s - \frac{9}{2} = 0 \Rightarrow r = -3$ and $s = \frac{9}{2}$

81. $\lim_{n\to\infty} \left|\frac{2\cdot 5\cdot 8\cdots(3n-1)(3n+2)x^{n+1}}{2\cdot 4\cdot 6\cdots(2n)(2n+2)} \cdot \frac{2\cdot 4\cdot 6\cdots(2n)}{2\cdot 5\cdot 8\cdots(3n-1)x^n}\right| < 1 \Rightarrow |x| \lim_{n\to\infty} \left|\frac{3n+2}{2n+2}\right| < 1 \Rightarrow |x| < \frac{2}{3}$
$\Rightarrow$ the radius of convergence is $\frac{2}{3}$

83. $\sum_{k=2}^{n} \ln\left(1 - \frac{1}{k^2}\right) = \sum_{k=2}^{n} \left[\ln\left(1 + \frac{1}{k}\right) + \ln\left(1 - \frac{1}{k}\right)\right] = \sum_{k=2}^{n} \left[\ln(k+1) - \ln k + \ln(k-1) - \ln k\right]$
$= [\ln 3 - \ln 2 + \ln 1 - \ln 2] + [\ln 4 - \ln 3 + \ln 2 - \ln 3] + [\ln 5 - \ln 4 + \ln 3 - \ln 4] + [\ln 6 - \ln 5 + \ln 4 - \ln 5]$

$+ \ldots + [\ln(n+1) - \ln n + \ln(n-1) - \ln n] = [\ln 1 - \ln 2] + [\ln(n+1) - \ln n]$ after cancellation

$\Rightarrow \sum_{k=2}^{n} \ln\left(1 - \frac{1}{k^2}\right) = \ln\left(\frac{n+1}{2n}\right) \Rightarrow \sum_{k=2}^{\infty} \ln\left(1 - \frac{1}{k^2}\right) = \lim_{n \to \infty} \ln\left(\frac{n+1}{2n}\right) = \ln \frac{1}{2}$ is the sum

85. (a) $\lim_{n \to \infty} \left| \frac{1 \cdot 4 \cdot 7 \cdots (3n-2)(3n+1)x^{3n+3}}{(3n+3)!} \cdot \frac{(3n)!}{1 \cdot 4 \cdot 7 \cdots (3n-2)x^{3n}} \right| < 1 \Rightarrow |x^3| \lim_{n \to \infty} \frac{(3n+1)}{(3n+1)(3n+2)(3n+3)}$

$= |x^3| \cdot 0 < 1 \Rightarrow$ the radius of convergence is ∞

(b) $y = 1 + \sum_{n=1}^{\infty} \frac{1 \cdot 4 \cdot 7 \cdots (3n-2)}{(3n)!} x^{3n} \Rightarrow \frac{dy}{dx} = \sum_{n=1}^{\infty} \frac{1 \cdot 4 \cdot 7 \cdots (3n-2)}{(3n-1)!} x^{3n-1}$

$\Rightarrow \frac{d^2y}{dx^2} = \sum_{n=1}^{\infty} \frac{1 \cdot 4 \cdot 7 \cdots (3n-2)}{(3n-2)!} x^{3n-2} = x + \sum_{n=2}^{\infty} \frac{1 \cdot 4 \cdot 7 \cdots (3n-5)}{(3n-3)!} x^{3n-2}$

$= x\left(1 + \sum_{n=1}^{\infty} \frac{1 \cdot 4 \cdot 7 \cdots (3n-2)}{(3n)!} x^{3n}\right) = xy + 0 \Rightarrow a = 1$ and $b = 0$

87. Yes, the series $\sum_{n=1}^{\infty} a_n b_n$ converges as we now show. Since $\sum_{n=1}^{\infty} a_n$ converges it follows that $a_n \to 0 \Rightarrow a_n < 1$ for $n >$ some index $N \Rightarrow a_n b_n < b_n$ for $n > N \Rightarrow \sum_{n=1}^{\infty} a_n b_n$ converges by the Direct Comparison Test with $\sum_{n=1}^{\infty} b_n$

89. $\sum_{n=1}^{\infty} (x_{n+1} - x_n) = \lim_{n \to \infty} \sum_{k=1}^{n} (x_{k+1} - x_k) = \lim_{n \to \infty} (x_{n+1} - x_1) = \lim_{n \to \infty} (x_{n+1}) - x_1 \Rightarrow$ both the series and sequence must either converge or diverge.

91. $\sum_{n=1}^{\infty} \frac{a_n}{n} = a_1 + \frac{a_2}{2} + \frac{a_3}{3} + \frac{a_4}{4} + \ldots \geq a_1 + \left(\frac{1}{2}\right) a_2 + \left(\frac{1}{3} + \frac{1}{4}\right) a_4 + \left(\frac{1}{5} + \frac{1}{6} + \frac{1}{7} + \frac{1}{8}\right) a_8$
$+ \left(\frac{1}{9} + \frac{1}{10} + \frac{1}{11} + \ldots + \frac{1}{16}\right) a_{16} + \ldots \geq \frac{1}{2}(a_2 + a_4 + a_8 + a_{16} + \ldots)$ which is a divergent series

CHAPTER 10 ADDITIONAL AND ADVANCED EXERCISES

1. converges since $\frac{1}{(3n-2)^{(2n+1)/2}} < \frac{1}{(3n-2)^{3/2}}$ and $\sum_{n=1}^{\infty} \frac{1}{(3n-2)^{3/2}}$ converges by the Limit Comparison Test:

$\lim_{n \to \infty} \frac{\left(\frac{1}{n^{3/2}}\right)}{\left(\frac{1}{(3n-2)^{3/2}}\right)} = \lim_{n \to \infty} \left(\frac{3n-2}{n}\right)^{3/2} = 3^{3/2}$

3. diverges by the nth-Term Test since $\lim_{n \to \infty} a_n = \lim_{n \to \infty} (-1)^n \tanh n = \lim_{b \to \infty} (-1)^n \left(\frac{1-e^{-2n}}{1+e^{-2n}}\right) = \lim_{n \to \infty} (-1)^n$ does not exist

5. converges by the Direct Comparison Test: $a_1 = 1 = \frac{12}{(1)(3)(2)^2}$, $a_2 = \frac{1 \cdot 2}{3 \cdot 4} = \frac{12}{(2)(4)(3)^2}$, $a_3 = \left(\frac{2 \cdot 3}{4 \cdot 5}\right)\left(\frac{1 \cdot 2}{3 \cdot 4}\right)$
$= \frac{12}{(3)(5)(4)^2}$, $a_4 = \left(\frac{3 \cdot 4}{5 \cdot 6}\right)\left(\frac{2 \cdot 3}{4 \cdot 5}\right)\left(\frac{1 \cdot 2}{3 \cdot 4}\right) = \frac{12}{(4)(6)(5)^2}, \ldots \Rightarrow 1 + \sum_{n=1}^{\infty} \frac{12}{(n+1)(n+3)(n+2)^2}$ represents the given series and $\frac{12}{(n+1)(n+3)(n+2)^2} < \frac{12}{n^4}$, which is the nth-term of a convergent p-series

7. diverges by the nth-Term Test since if $a_n \to L$ as $n \to \infty$, then $L = \frac{1}{1+L} \Rightarrow L^2 + L - 1 = 0 \Rightarrow L = \frac{-1 \pm \sqrt{5}}{2} \neq 0$

9. $f(x) = \cos x$ with $a = \frac{\pi}{3} \Rightarrow f\left(\frac{\pi}{3}\right) = 0.5$, $f'\left(\frac{\pi}{3}\right) = -\frac{\sqrt{3}}{2}$, $f''\left(\frac{\pi}{3}\right) = -0.5$, $f'''\left(\frac{\pi}{3}\right) = \frac{\sqrt{3}}{2}$, $f^{(4)}\left(\frac{\pi}{3}\right) = 0.5$;
$\cos x = \frac{1}{2} - \frac{\sqrt{3}}{2}\left(x - \frac{\pi}{3}\right) - \frac{1}{4}\left(x - \frac{\pi}{3}\right)^2 + \frac{\sqrt{3}}{12}\left(x - \frac{\pi}{3}\right)^3 + \ldots$

11. $e^x = 1 + x + \frac{x^2}{2!} + \frac{x^3}{3!} + \ldots$ with $a = 0$

13. $f(x) = \cos x$ with $a = 22\pi \Rightarrow f(22\pi) = 1, f'(22\pi) = 0, f''(22\pi) = -1, f'''(22\pi) = 0, f^{(4)}(22\pi) = 1,$ $f^{(5)}(22\pi) = 0, f^{(6)}(22\pi) = -1; \cos x = 1 - \frac{1}{2}(x - 22\pi)^2 + \frac{1}{4!}(x - 22\pi)^4 - \frac{1}{6!}(x - 22\pi)^6 + \ldots$

15. Yes, the sequence converges: $c_n = (a^n + b^n)^{1/n} \Rightarrow c_n = b\left(\left(\frac{a}{b}\right)^n + 1\right)^{1/n} \Rightarrow \lim_{n\to\infty} c_n = \ln b + \lim_{n\to\infty} \frac{\ln\left(\left(\frac{a}{b}\right)^n + 1\right)}{n}$
$= \ln b + \lim_{n\to\infty} \frac{\left(\frac{a}{b}\right)^n \ln\left(\frac{a}{b}\right)}{\left(\frac{a}{b}\right)^n + 1} = \ln b + \frac{0 \cdot \ln\left(\frac{a}{b}\right)}{0+1} = \ln b$ since $0 < a < b$. Thus, $\lim_{n\to\infty} c_n = e^{\ln b} = b$.

17. $s_n = \sum_{k=0}^{n-1} \int_k^{k+1} \frac{dx}{1+x^2} \Rightarrow s_n = \int_0^1 \frac{dx}{1+x^2} + \int_1^2 \frac{dx}{1+x^2} + \ldots + \int_{n-1}^n \frac{dx}{1+x^2} \Rightarrow s_n = \int_0^n \frac{dx}{1+x^2}$
$\Rightarrow \lim_{n\to\infty} s_n = \lim_{n\to\infty} (\tan^{-1} n - \tan^{-1} 0) = \frac{\pi}{2}$

19. (a) No, the limit does not appear to depend on the value of the constant a
(b) Yes, the limit depends on the value of b
(c) $s = \left(1 - \frac{\cos\left(\frac{a}{n}\right)}{n}\right)^n \Rightarrow \ln s = \frac{\ln\left(1 - \frac{\cos\left(\frac{a}{n}\right)}{n}\right)}{\left(\frac{1}{n}\right)} \Rightarrow \lim_{n\to\infty} \ln s = \frac{\left(\frac{1}{1 - \frac{\cos\left(\frac{a}{n}\right)}{n}}\right)\left(\frac{-\frac{a}{n}\sin\left(\frac{a}{n}\right) + \cos\left(\frac{a}{n}\right)}{n^2}\right)}{\left(-\frac{1}{n^2}\right)}$
$= \lim_{n\to\infty} \frac{\frac{a}{n}\sin\left(\frac{a}{n}\right) - \cos\left(\frac{a}{n}\right)}{1 - \frac{\cos\left(\frac{a}{n}\right)}{n}} = \frac{0-1}{1-0} = -1 \Rightarrow \lim_{n\to\infty} s = e^{-1} \approx 0.3678794412$; similarly,
$\lim_{n\to\infty} \left(1 - \frac{\cos\left(\frac{a}{n}\right)}{bn}\right)^n = e^{-1/b}$

21. $\lim_{n\to\infty} \left|\frac{u_{n+1}}{u_n}\right| < 1 \Rightarrow \lim_{n\to\infty} \left|\frac{b^{n+1}x^{n+1}}{\ln(n+1)} \cdot \frac{\ln n}{b^n x^n}\right| < 1 \Rightarrow |bx| < 1 \Rightarrow -\frac{1}{b} < x < \frac{1}{b} = 5 \Rightarrow b = \pm\frac{1}{5}$

23. $\lim_{x\to 0} \frac{\sin(ax) - \sin x - x}{x^3} = \lim_{x\to 0} \frac{\left(ax - \frac{a^3x^3}{3!} + \ldots\right) - \left(x - \frac{x^3}{3!} + \ldots\right) - x}{x^3} = \lim_{x\to 0} \left[\frac{a-2}{x^2} - \frac{a^3}{3!} + \frac{1}{3!} - \left(\frac{a^5}{5!} - \frac{1}{5!}\right)x^2 + \ldots\right]$
is finite if $a - 2 = 0 \Rightarrow a = 2$; $\lim_{x\to 0} \frac{\sin 2x - \sin x - x}{x^3} = -\frac{2^3}{3!} + \frac{1}{3!} = -\frac{7}{6}$

25. (a) $\frac{u_n}{u_{n+1}} = \frac{(n+1)^2}{n^2} = 1 + \frac{2}{n} + \frac{1}{n^2} \Rightarrow C = 2 > 1$ and $\sum_{n=1}^{\infty} \frac{1}{n^2}$ converges
(b) $\frac{u_n}{u_{n+1}} = \frac{n+1}{n} = 1 + \frac{1}{n} + \frac{0}{n^2} \Rightarrow C = 1 \le 1$ and $\sum_{n=1}^{\infty} \frac{1}{n}$ diverges

27. (a) $\sum_{n=1}^{\infty} a_n = L \Rightarrow a_n^2 \le a_n \sum_{n=1}^{\infty} a_n = a_n L \Rightarrow \sum_{n=1}^{\infty} a_n^2$ converges by the Direct Comparison Test
(b) converges by the Limit Comparison Test: $\lim_{n\to\infty} \frac{\left(\frac{a_n}{1-a_n}\right)}{a_n} = \lim_{n\to\infty} \frac{1}{1-a_n} = 1$ since $\sum_{n=1}^{\infty} a_n$ converges and therefore $\lim_{x\to\infty} a_n = 0$

29. $(1-x)^{-1} = 1 + \sum_{n=1}^{\infty} x^n$ where $|x| < 1 \Rightarrow \frac{1}{(1-x)^2} = \frac{d}{dx}(1-x)^{-1} = \sum_{n=1}^{\infty} nx^{n-1}$ and when $x = \frac{1}{2}$ we have
$4 = 1 + 2\left(\frac{1}{2}\right) + 3\left(\frac{1}{2}\right)^2 + 4\left(\frac{1}{2}\right)^3 + \ldots + n\left(\frac{1}{2}\right)^{n-1} + \ldots$

31. (a) $\frac{1}{(1-x)^2} = \frac{d}{dx}\left(\frac{1}{1-x}\right) = \frac{d}{dx}(1 + x + x^2 + x^3 + \ldots) = 1 + 2x + 3x^2 + 4x^3 + \ldots = \sum_{n=1}^{\infty} nx^{n-1}$
(b) from part (a) we have $\sum_{n=1}^{\infty} n\left(\frac{5}{6}\right)^{n-1}\left(\frac{1}{6}\right) = \left(\frac{1}{6}\right)\left[\frac{1}{1-\left(\frac{5}{6}\right)}\right]^2 = 6$
(c) from part (a) we have $\sum_{n=1}^{\infty} np^{n-1}q = \frac{q}{(1-p)^2} = \frac{q}{q^2} = \frac{1}{q}$

33. (a) $R_n = C_0e^{-kt_0} + C_0e^{-2kt_0} + \ldots + C_0e^{-nkt_0} = \frac{C_0e^{-kt_0}(1-e^{-nkt_0})}{1-e^{-kt_0}} \Rightarrow R = \lim_{n\to\infty} R_n = \frac{C_0e^{-kt_0}}{1-e^{-kt_0}} = \frac{C_0}{e^{kt_0}-1}$

(b) $R_n = \frac{e^{-1}(1-e^{-n})}{1-e^{-1}} \Rightarrow R_1 = e^{-1} \approx 0.36787944$ and $R_{10} = \frac{e^{-1}(1-e^{-10})}{1-e^{-1}} \approx 0.58195028$;
$R = \frac{1}{e-1} \approx 0.58197671$; $R - R_{10} \approx 0.00002643 \Rightarrow \frac{R-R_{10}}{R} < 0.0001$

(c) $R_n = \frac{e^{-.1}(1-e^{-.1n})}{1-e^{-.1}}$, $\frac{R}{2} = \frac{1}{2}\left(\frac{1}{e^{.1}-1}\right) \approx 4.7541659$; $R_n > \frac{R}{2} \Rightarrow \frac{1-e^{-.1n}}{e^{.1}-1} > \left(\frac{1}{2}\right)\left(\frac{1}{e^{.1}-1}\right)$
$\Rightarrow 1 - e^{-n/10} > \frac{1}{2} \Rightarrow e^{-n/10} < \frac{1}{2} \Rightarrow -\frac{n}{10} < \ln\left(\frac{1}{2}\right) \Rightarrow \frac{n}{10} > -\ln\left(\frac{1}{2}\right) \Rightarrow n > 6.93 \Rightarrow n = 7$

CHAPTER 11 PARAMETRIC EQUATIONS AND POLAR COORDINATES

11.1 PARAMETRIZATIONS OF PLANE CURVES

1. $x = 3t,\ y = 9t^2,\ -\infty < t < \infty \Rightarrow y = x^2$

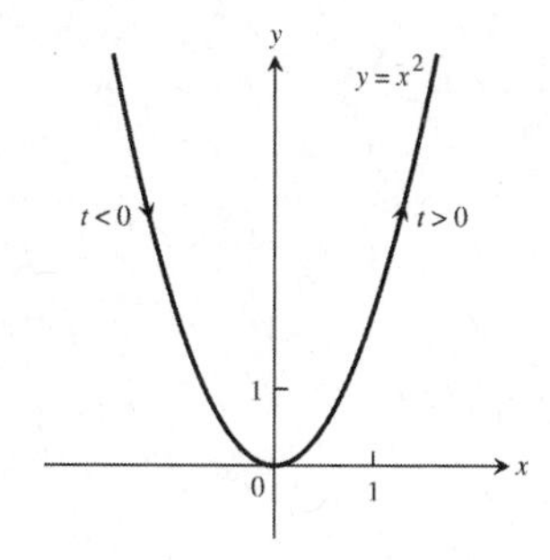

3. $x = 2t - 5,\ y = 4t - 7,\ -\infty < t < \infty$
$\Rightarrow x + 5 = 2t \Rightarrow 2(x + 5) = 4t$
$\Rightarrow y = 2(x + 5) - 7 \Rightarrow y = 2x + 3$

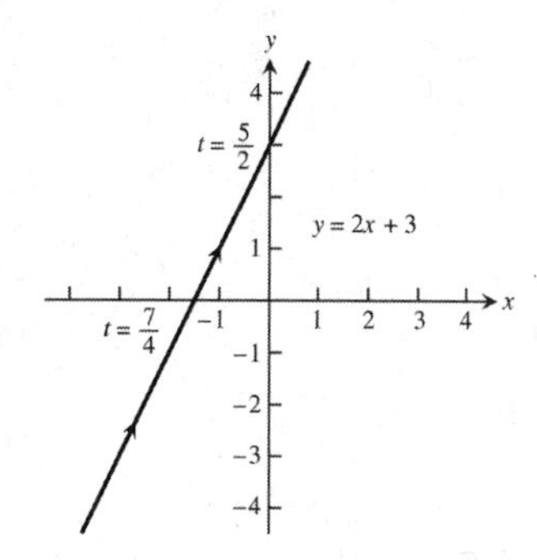

5. $x = \cos 2t,\ y = \sin 2t,\ 0 \le t \le \pi$
$\Rightarrow \cos^2 2t + \sin^2 2t = 1 \Rightarrow x^2 + y^2 = 1$

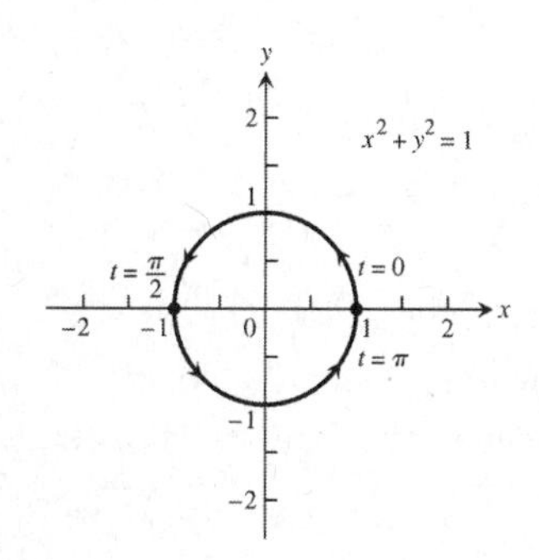

7. $x = 4 \cos t,\ y = 2 \sin t,\ 0 \le t \le 2\pi$
$\Rightarrow \frac{16 \cos^2 t}{16} + \frac{4 \sin^2 t}{4} = 1 \Rightarrow \frac{x^2}{16} + \frac{y^2}{4} = 1$

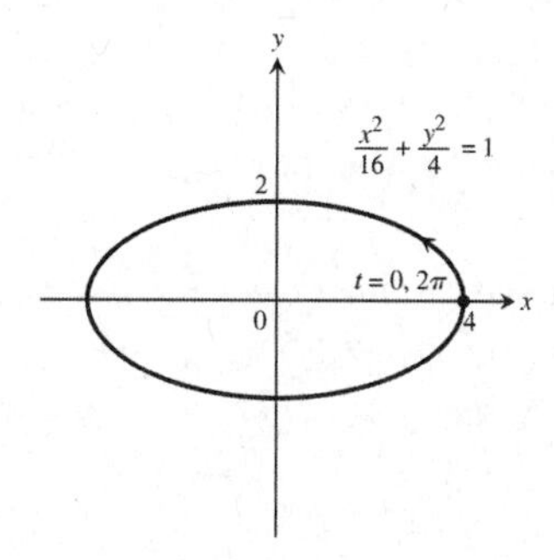

9. $x = \sin t,\ y = \cos 2t,\ -\frac{\pi}{2} \le t \le \frac{\pi}{2}$
$\Rightarrow y = \cos 2t = 1 - 2\sin^2 t \Rightarrow y = 1 - 2x^2$

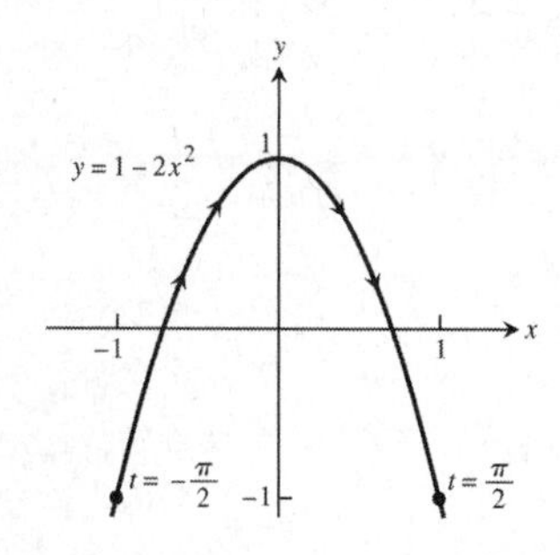

11. $x = t^2,\ y = t^6 - 2t^4,\ -\infty < t < \infty$
$\Rightarrow y = (t^2)^3 - 2(t^2)^2 \Rightarrow y = x^3 - 2x^2$

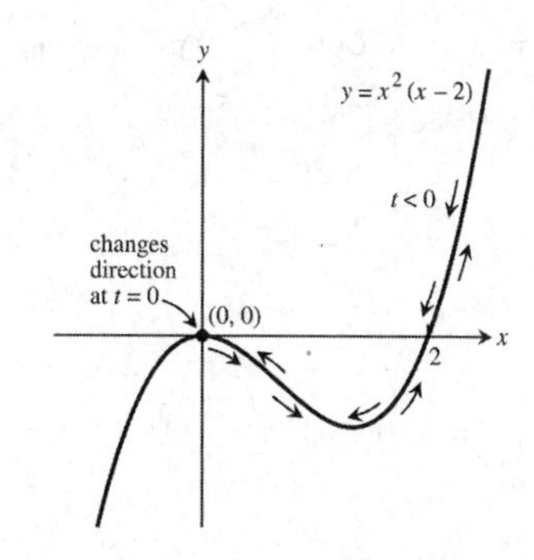

13. $x = t,\ y = \sqrt{1 - t^2},\ -1 \le t \le 0$
$\Rightarrow y = \sqrt{1 - x^2}$

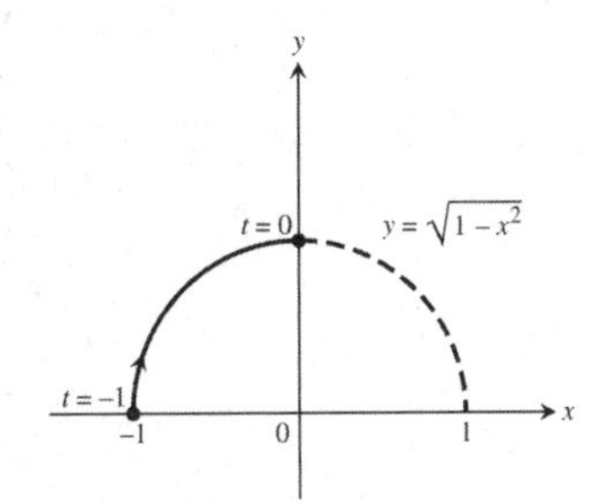

15. $x = \sec^2 t - 1,\ y = \tan t,\ -\frac{\pi}{2} < t < \frac{\pi}{2}$
$\Rightarrow \sec^2 t - 1 = \tan^2 t \Rightarrow x = y^2$

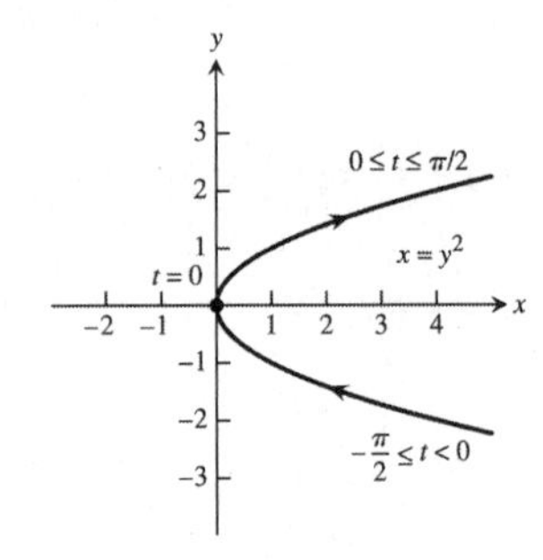

17. $x = -\cosh t,\ y = \sinh t,\ -\infty < 1 < \infty$
$\Rightarrow \cosh^2 t - \sinh^2 t = 1 \Rightarrow x^2 - y^2 = 1$

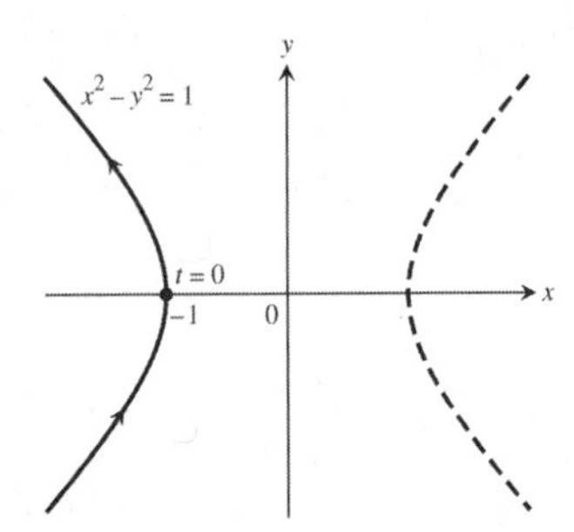

19. (a) $x = a\cos t,\ y = -a\sin t,\ 0 \le t \le 2\pi$
(b) $x = a\cos t,\ y = a\sin t,\ 0 \le t \le 2\pi$
(c) $x = a\cos t,\ y = -a\sin t,\ 0 \le t \le 4\pi$
(d) $x = a\cos t,\ y = a\sin t,\ 0 \le t \le 4\pi$

21. Using $(-1, -3)$ we create the parametric equations $x = -1 + at$ and $y = -3 + bt$, representing a line which goes through $(-1, -3)$ at $t = 0$. We determine a and b so that the line goes through $(4, 1)$ when $t = 1$. Since $4 = -1 + a \Rightarrow a = 5$. Since $1 = -3 + b \Rightarrow b = 4$. Therefore, one possible parameterization is $x = -1 + 5t$, $y = -3 + 4t,\ 0 \le t \le 1$.

23. The lower half of the parabola is given by $x = y^2 + 1$ for $y \le 0$. Substituting t for y, we obtain one possible parameterization $x = t^2 + 1,\ y = t,\ t \le 0$.

25. For simplicity, we assume that x and y are linear functions of t and that the point(x, y) starts at $(2, 3)$ for $t = 0$ and passes through $(-1, -1)$ at $t = 1$. Then $x = f(t)$, where $f(0) = 2$ and $f(1) = -1$.
Since slope $= \frac{\Delta x}{\Delta t} = \frac{-1-2}{1-0} = -3$, $x = f(t) = -3t + 2 = 2 - 3t$. Also, $y = g(t)$, where $g(0) = 3$ and $g(1) = -1$.
Since slope $= \frac{\Delta y}{\Delta t} = \frac{-1-3}{1-0} = -4$. $y = g(t) = -4t + 3 = 3 - 4t$.
One possible parameterization is: $x = 2 - 3t,\ y = 3 - 4t,\ t \ge 0$.

27. Since we only want the top half of a circle, $y \ge 0$, so let $x = 2\cos t,\ y = 2|\sin t|,\ 0 \le t \le 4\pi$

29. $x^2 + y^2 = a^2 \Rightarrow 2x + 2y\frac{dy}{dx} = 0 \Rightarrow \frac{dy}{dx} = -\frac{x}{y}$; let $t = \frac{dy}{dx} \Rightarrow -\frac{x}{y} = t \Rightarrow x = -yt$. Substitution yields $y^2t^2 + y^2 = a^2 \Rightarrow y = \frac{a}{\sqrt{1+t^2}}$ and $x = \frac{-at}{\sqrt{1+t}}$, $-\infty < t < \infty$

31. Drop a vertical line from the point (x, y) to the x-axis, then θ is an angle in a right triangle, and from trigonometry we know that $\tan\theta = \frac{y}{x} \Rightarrow y = x\tan\theta$. The equation of the line through $(0, 2)$ and $(4, 0)$ is given by $y = -\frac{1}{2}x + 2$. Thus $x\tan\theta = -\frac{1}{2}x + 2 \Rightarrow x = \frac{4}{2\tan\theta + 1}$ and $y = \frac{4\tan\theta}{2\tan\theta + 1}$ where $0 \le \theta < \frac{\pi}{2}$.

33. The equation of the circle is given by $(x-2)^2 + y^2 = 1$. Drop a vertical line from the point (x, y) on the circle to the x-axis, then θ is an angle in a right triangle. So that we can start at (1, 0) and rotate in a clockwise direction, let $x = 2 - \cos\theta,\ y = \sin\theta,\ 0 \le \theta \le 2\pi$.

35. Extend the vertical line through A to the x-axis and let C be the point of intersection. Then OC = AQ = x and $\tan t = \frac{2}{OC} = \frac{2}{x} \Rightarrow x = \frac{2}{\tan t} = 2\cot t$; $\sin t = \frac{2}{OA} \Rightarrow OA = \frac{2}{\sin t}$; and $(AB)(OA) = (AQ)^2 \Rightarrow AB\left(\frac{2}{\sin t}\right) = x^2$ $\Rightarrow AB\left(\frac{2}{\sin t}\right) = \left(\frac{2}{\tan t}\right)^2 \Rightarrow AB = \frac{2\sin t}{\tan^2 t}$. Next $y = 2 - AB\sin t \Rightarrow y = 2 - \left(\frac{2\sin t}{\tan^2 t}\right)\sin t =$ $2 - \frac{2\sin^2 t}{\tan^2 t} = 2 - 2\cos^2 t = 2\sin^2 t$. Therefore let $x = 2\cot t$ and $y = 2\sin^2 t$, $0 < t < \pi$.

37. Draw line AM in the figure and note that $\angle AMO$ is a right angle since it is an inscribed angle which spans the diameter of a circle. Then $AN^2 = MN^2 + AM^2$. Now, OA = a, $\frac{AN}{a} = \tan t$, and $\frac{AM}{a} = \sin t$. Next MN = OP
$\Rightarrow OP^2 = AN^2 - AM^2 = a^2\tan^2 t - a^2\sin^2 t$
$\Rightarrow OP = \sqrt{a^2\tan^2 t - a^2\sin^2 t}$
$= (a\sin t)\sqrt{\sec^2 t - 1} = \frac{a\sin^2 t}{\cos t}$. In triangle BPO,
$x = OP\sin t = \frac{a\sin^3 t}{\cos t} = a\sin^2 t\tan t$ and
$y = OP\cos t = a\sin^2 t \Rightarrow x = a\sin^2 t\tan t$ and $y = a\sin^2 t$.

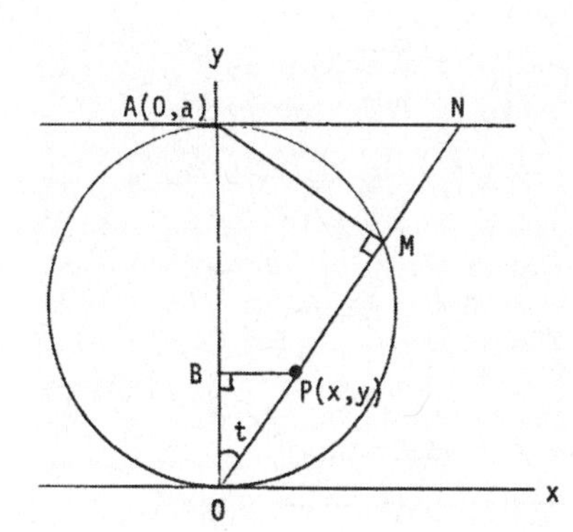

39. $D = \sqrt{(x-2)^2 + \left(y - \frac{1}{2}\right)^2} \Rightarrow D^2 = (x-2)^2 + \left(y-\frac{1}{2}\right)^2 = (t-2)^2 + \left(t^2 - \frac{1}{2}\right)^2 \Rightarrow D^2 = t^4 - 4t + \frac{17}{4}$ $\Rightarrow \frac{d(D^2)}{dt} = 4t^3 - 4 = 0 \Rightarrow t = 1$. The second derivative is always positive for $t \neq 0 \Rightarrow t = 1$ gives a local minimum for D^2 (and hence D) which is an absolute minimum since it is the only extremum $\Rightarrow$ the closest point on the parabola is (1, 1).

41. (a)

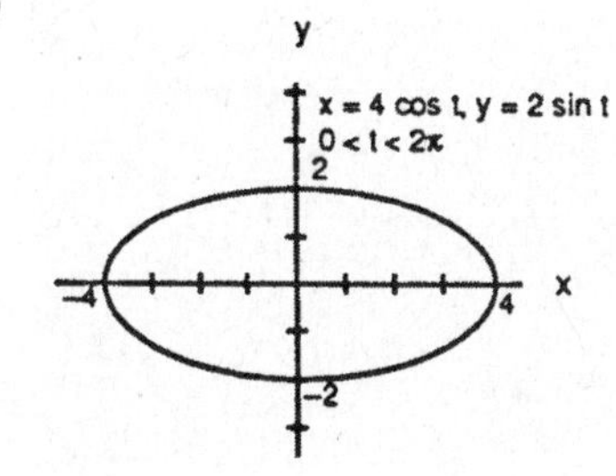

(b)

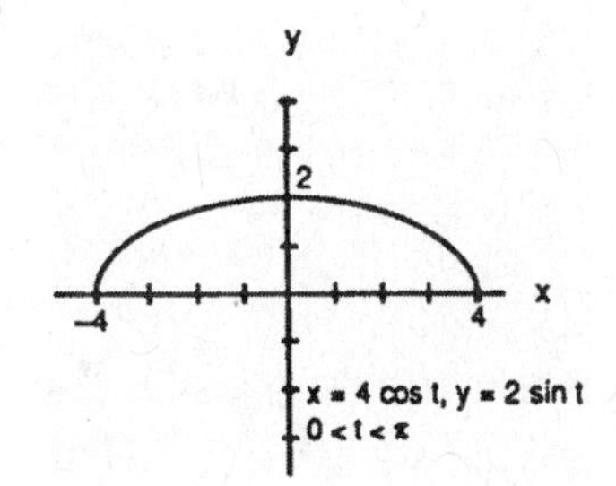

(c)

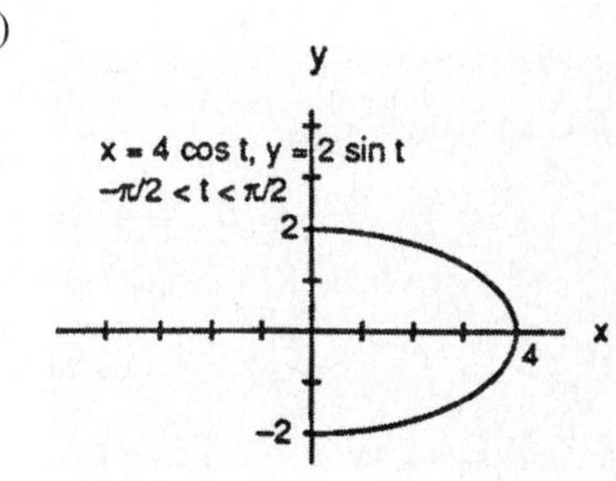

43.

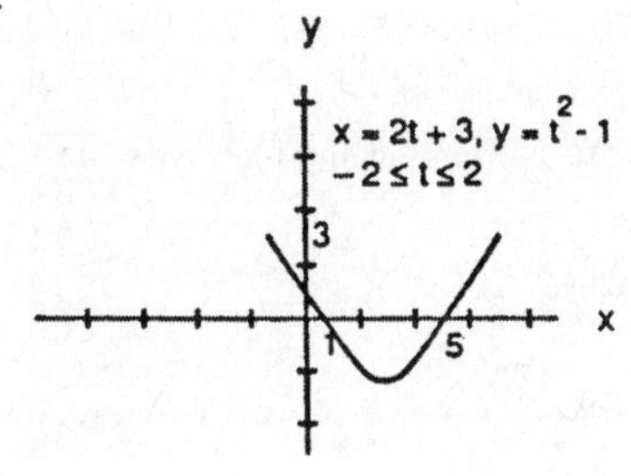

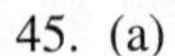

45. (a)

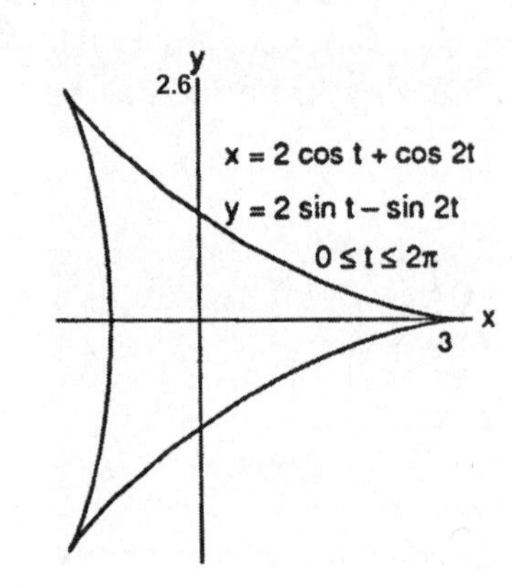

(b)

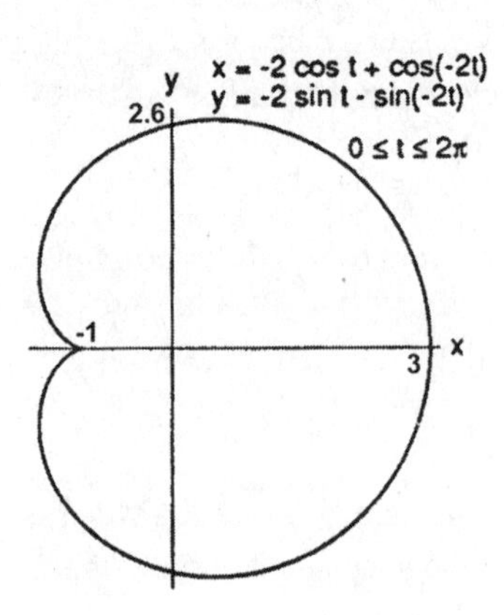

47. (a)

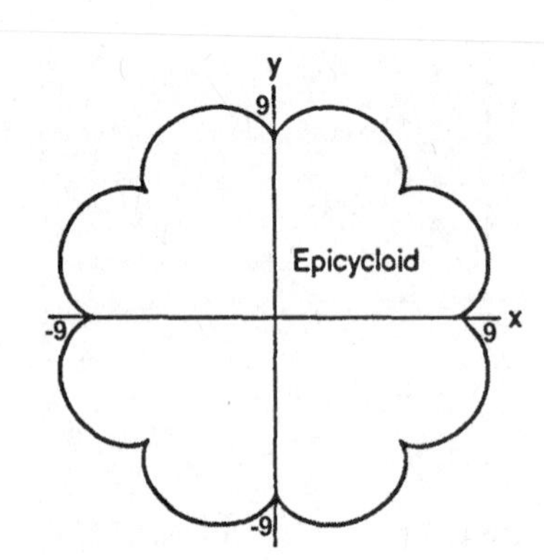

(b)

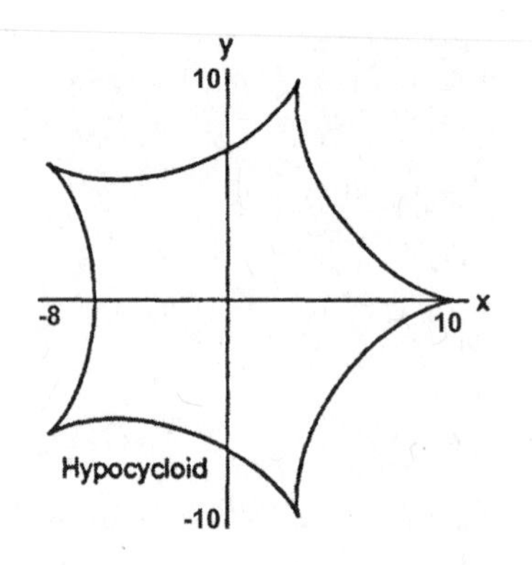

(c)

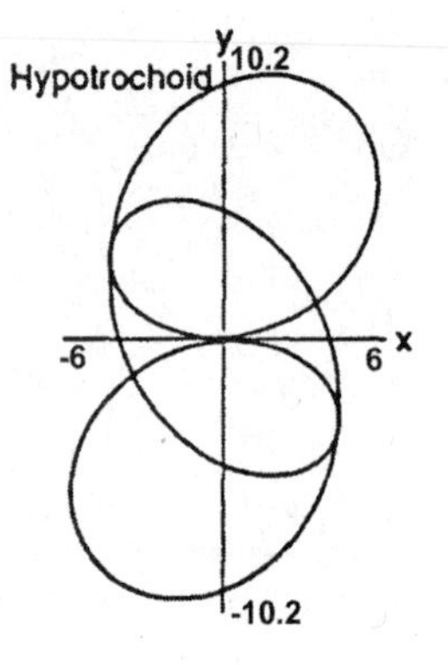

11.2 CALCULUS WITH PARAMETRIC CURVES

1. $t = \frac{\pi}{4} \Rightarrow x = 2\cos\frac{\pi}{4} = \sqrt{2}$, $y = 2\sin\frac{\pi}{4} = \sqrt{2}$; $\frac{dx}{dt} = -2\sin t$, $\frac{dy}{dt} = 2\cos t \Rightarrow \frac{dy}{dx} = \frac{dy/dt}{dx/dt} = \frac{2\cos t}{-2\sin t} = -\cot t$ $\Rightarrow \left.\frac{dy}{dx}\right|_{t=\frac{\pi}{4}} = -\cot\frac{\pi}{4} = -1$; tangent line is $y - \sqrt{2} = -1\left(x - \sqrt{2}\right)$ or $y = -x + 2\sqrt{2}$; $\frac{dy'}{dt} = \csc^2 t$ $\Rightarrow \frac{d^2y}{dx^2} = \frac{dy'/dt}{dx/dt} = \frac{\csc^2 t}{-2\sin t} = -\frac{1}{2\sin^3 t} \Rightarrow \left.\frac{d^2y}{dx^2}\right|_{t=\frac{\pi}{4}} = -\sqrt{2}$

3. $t = \frac{\pi}{4} \Rightarrow x = 4\sin\frac{\pi}{4} = 2\sqrt{2}$, $y = 2\cos\frac{\pi}{4} = \sqrt{2}$; $\frac{dx}{dt} = 4\cos t$, $\frac{dy}{dt} = -2\sin t \Rightarrow \frac{dy}{dx} = \frac{dy/dt}{dx/dt} = \frac{-2\sin t}{4\cos t}$ $= -\frac{1}{2}\tan t \Rightarrow \left.\frac{dy}{dx}\right|_{t=\frac{\pi}{4}} = -\frac{1}{2}\tan\frac{\pi}{4} = -\frac{1}{2}$; tangent line is $y - \sqrt{2} = -\frac{1}{2}\left(x - 2\sqrt{2}\right)$ or $y = -\frac{1}{2}x + 2\sqrt{2}$; $\frac{dy'}{dt} = -\frac{1}{2}\sec^2 t \Rightarrow \frac{d^2y}{dx^2} = \frac{dy'/dt}{dx/dt} = \frac{-\frac{1}{2}\sec^2 t}{4\cos t} = -\frac{1}{8\cos^3 t} \Rightarrow \left.\frac{d^2y}{dx^2}\right|_{t=\frac{\pi}{4}} = -\frac{\sqrt{2}}{4}$

5. $t = \frac{1}{4} \Rightarrow x = \frac{1}{4}$, $y = \frac{1}{2}$; $\frac{dx}{dt} = 1$, $\frac{dy}{dt} = \frac{1}{2\sqrt{t}} \Rightarrow \frac{dy}{dx} = \frac{dy/dt}{dx/dt} = \frac{1}{2\sqrt{t}} \Rightarrow \left.\frac{dy}{dx}\right|_{t=\frac{1}{4}} = \frac{1}{2\sqrt{\frac{1}{4}}} = 1$; tangent line is $y - \frac{1}{2} = 1\cdot\left(x - \frac{1}{4}\right)$ or $y = x + \frac{1}{4}$; $\frac{dy'}{dt} = -\frac{1}{4}t^{-3/2} \Rightarrow \frac{d^2y}{dx^2} = \frac{dy'/dt}{dx/dt} = -\frac{1}{4}t^{-3/2} \Rightarrow \left.\frac{d^2y}{dx^2}\right|_{t=\frac{1}{4}} = -2$

7. $t = \frac{\pi}{6} \Rightarrow x = \sec\frac{\pi}{6} = \frac{2}{\sqrt{3}}$, $y = \tan\frac{\pi}{6} = \frac{1}{\sqrt{3}}$; $\frac{dx}{dt} = \sec t\tan t$, $\frac{dy}{dt} = \sec^2 t \Rightarrow \frac{dy}{dx} = \frac{dy/dt}{dx/dt}$ $= \frac{\sec^2 t}{\sec t\tan t} = \csc t \Rightarrow \left.\frac{dy}{dx}\right|_{t=\frac{\pi}{6}} = \csc\frac{\pi}{6} = 2$; tangent line is $y - \frac{1}{\sqrt{3}} = 2\left(x - \frac{2}{\sqrt{3}}\right)$ or $y = 2x - \sqrt{3}$; $\frac{dy'}{dt} = -\csc t\cot t \Rightarrow \frac{d^2y}{dx^2} = \frac{dy'/dt}{dx/dt} = \frac{-\csc t\cot t}{\sec t\tan t} = -\cot^3 t \Rightarrow \left.\frac{d^2y}{dx^2}\right|_{t=\frac{\pi}{6}} = -3\sqrt{3}$

9. $t = -1 \Rightarrow x = 5$, $y = 1$; $\frac{dx}{dt} = 4t$, $\frac{dy}{dt} = 4t^3 \Rightarrow \frac{dy}{dx} = \frac{dy/dt}{dx/dt} = \frac{4t^3}{4t} = t^2 \Rightarrow \left.\frac{dy}{dx}\right|_{t=-1} = (-1)^2 = 1$; tangent line is $y - 1 = 1\cdot(x - 5)$ or $y = x - 4$; $\frac{dy'}{dt} = 2t \Rightarrow \frac{d^2y}{dx^2} = \frac{dy'/dt}{dx/dt} = \frac{2t}{4t} = \frac{1}{2} \Rightarrow \left.\frac{d^2y}{dx^2}\right|_{t=-1} = \frac{1}{2}$

11. $t = \frac{\pi}{3} \Rightarrow x = \frac{\pi}{3} - \sin\frac{\pi}{3} = \frac{\pi}{3} - \frac{\sqrt{3}}{2}$, $y = 1 - \cos\frac{\pi}{3} = 1 - \frac{1}{2} = \frac{1}{2}$; $\frac{dx}{dt} = 1 - \cos t$, $\frac{dy}{dt} = \sin t \Rightarrow \frac{dy}{dx} = \frac{dy/dt}{dx/dt}$

$= \frac{\sin t}{1-\cos t} \Rightarrow \left.\frac{dy}{dx}\right|_{t=\frac{\pi}{3}} = \frac{\sin\left(\frac{\pi}{3}\right)}{1-\cos\left(\frac{\pi}{3}\right)} = \frac{\left(\frac{\sqrt{3}}{2}\right)}{\left(\frac{1}{2}\right)} = \sqrt{3}$; tangent line is $y - \frac{1}{2} = \sqrt{3}\left(x - \frac{\pi}{3} + \frac{\sqrt{3}}{2}\right)$

$\Rightarrow y = \sqrt{3}x - \frac{\pi\sqrt{3}}{3} + 2$; $\frac{dy'}{dt} = \frac{(1-\cos t)(\cos t) - (\sin t)(\sin t)}{(1-\cos t)^2} = \frac{-1}{1-\cos t} \Rightarrow \frac{d^2y}{dx^2} = \frac{dy'/dt}{dx/dt} = \frac{\left(\frac{-1}{1-\cos t}\right)}{1-\cos t}$

$= \frac{-1}{(1-\cos t)^2} \Rightarrow \left.\frac{d^2y}{dx^2}\right|_{t=\frac{\pi}{3}} = -4$

13. $t = 2 \Rightarrow x = \frac{1}{2+1} = \frac{1}{3}$, $y = \frac{2}{2-1} = 2$; $\frac{dx}{dt} = \frac{-1}{(t+1)^2}$, $\frac{dy}{dt} = \frac{-1}{(t-1)^2} \Rightarrow \frac{dy}{dx} = \frac{(t+1)^2}{(t-1)^2} \Rightarrow \left.\frac{dy}{dx}\right|_{t=2} = \frac{(2+1)^2}{(2-1)^2} = 9$;

tangent line is $y = 9x - 1$; $\frac{dy'}{dt} = -\frac{4(t+1)}{(t-1)^3} \Rightarrow \frac{d^2y}{dx^2} = \frac{4(t+1)^3}{(t-1)^3} \Rightarrow \left.\frac{d^2y}{dx^2}\right|_{t=2} = \frac{4(2+1)^3}{(2-1)^3} = 108$

15. $x^3 + 2t^2 = 9 \Rightarrow 3x^2\frac{dx}{dt} + 4t = 0 \Rightarrow 3x^2\frac{dx}{dt} = -4t \Rightarrow \frac{dx}{dt} = \frac{-4t}{3x^2}$;

$2y^3 - 3t^2 = 4 \Rightarrow 6y^2\frac{dy}{dt} - 6t = 0 \Rightarrow \frac{dy}{dt} = \frac{6t}{6y^2} = \frac{t}{y^2}$; thus $\frac{dy}{dx} = \frac{dy/dt}{dx/dt} = \frac{\left(\frac{t}{y^2}\right)}{\left(\frac{-4t}{3x^2}\right)} = \frac{t(3x^2)}{y^2(-4t)} = \frac{3x^2}{-4y^2}$; $t = 2$

$\Rightarrow x^3 + 2(2)^2 = 9 \Rightarrow x^3 + 8 = 9 \Rightarrow x^3 = 1 \Rightarrow x = 1$; $t = 2 \Rightarrow 2y^3 - 3(2)^2 = 4$

$\Rightarrow 2y^3 = 16 \Rightarrow y^3 = 8 \Rightarrow y = 2$; therefore $\left.\frac{dy}{dx}\right|_{t=2} = \frac{3(1)^2}{-4(2)^2} = -\frac{3}{16}$

17. $x + 2x^{3/2} = t^2 + t \Rightarrow \frac{dx}{dt} + 3x^{1/2}\frac{dx}{dt} = 2t + 1 \Rightarrow \left(1 + 3x^{1/2}\right)\frac{dx}{dt} = 2t + 1 \Rightarrow \frac{dx}{dt} = \frac{2t+1}{1+3x^{1/2}}$; $y\sqrt{t+1} + 2t\sqrt{y} = 4$

$\Rightarrow \frac{dy}{dt}\sqrt{t+1} + y\left(\frac{1}{2}\right)(t+1)^{-1/2} + 2\sqrt{y} + 2t\left(\frac{1}{2}y^{-1/2}\right)\frac{dy}{dt} = 0 \Rightarrow \frac{dy}{dt}\sqrt{t+1} + \frac{y}{2\sqrt{t+1}} + 2\sqrt{y} + \left(\frac{t}{\sqrt{y}}\right)\frac{dy}{dt} = 0$

$\Rightarrow \left(\sqrt{t+1} + \frac{t}{\sqrt{y}}\right)\frac{dy}{dt} = \frac{-y}{2\sqrt{t+1}} - 2\sqrt{y} \Rightarrow \frac{dy}{dt} = \frac{\left(\frac{-y}{2\sqrt{t+1}} - 2\sqrt{y}\right)}{\left(\sqrt{t+1} + \frac{t}{\sqrt{y}}\right)} = \frac{-y\sqrt{y} - 4y\sqrt{t+1}}{2\sqrt{y}(t+1) + 2t\sqrt{t+1}}$; thus

$\frac{dy}{dx} = \frac{dy/dt}{dx/dt} = \frac{\left(\frac{-y\sqrt{y} - 4y\sqrt{t+1}}{2\sqrt{y}(t+1) + 2t\sqrt{t+1}}\right)}{\left(\frac{2t+1}{1+3x^{1/2}}\right)}$; $t = 0 \Rightarrow x + 2x^{3/2} = 0 \Rightarrow x\left(1 + 2x^{1/2}\right) = 0 \Rightarrow x = 0$; $t = 0$

$\Rightarrow y\sqrt{0+1} + 2(0)\sqrt{y} = 4 \Rightarrow y = 4$; therefore $\left.\frac{dy}{dx}\right|_{t=0} = \frac{\left(\frac{-4\sqrt{4} - 4(4)\sqrt{0+1}}{2\sqrt{4}(0+1) + 2(0)\sqrt{0+1}}\right)}{\left(\frac{2(0)+1}{1+3(0)^{1/2}}\right)} = -6$

19. $x = t^3 + t$, $y + 2t^3 = 2x + t^2 \Rightarrow \frac{dx}{dt} = 3t^2 + 1$, $\frac{dy}{dt} + 6t^2 = 2\frac{dx}{dt} + 2t \Rightarrow \frac{dy}{dt} = 2(3t^2 + 1) + 2t - 6t^2 = 2t + 2$

$\Rightarrow \frac{dy}{dx} = \frac{2t+2}{3t^2+1} \Rightarrow \left.\frac{dy}{dx}\right|_{t=1} = \frac{2(1)+2}{3(1)^2+1} = 1$

21. $A = \int_0^{2\pi} y\,dx = \int_0^{2\pi} a(1-\cos t)a(1-\cos t)dt = a^2\int_0^{2\pi}(1-\cos t)^2dt = a^2\int_0^{2\pi}(1 - 2\cos t + \cos^2 t)dt$

$= a^2\int_0^{2\pi}\left(1 - 2\cos t + \frac{1+\cos 2t}{2}\right)dt = a^2\int_0^{2\pi}\left(\frac{3}{2} - 2\cos t + \frac{1}{2}\cos 2t\right)dt = a^2\left[\frac{3}{2}t - 2\sin t + \frac{1}{4}\sin 2t\right]_0^{2\pi}$

$= a^2(3\pi - 0 + 0) - 0 = 3\pi a^2$

23. $A = 2\int_\pi^0 y\,dx = 2\int_\pi^0 (b\sin t)(-a\sin t)dt = 2ab\int_0^\pi \sin^2 t\,dt = 2ab\int_0^\pi \frac{1-\cos 2t}{2}\,dt = ab\int_0^\pi (1 - \cos 2t)\,dt$

$= ab\left[t - \frac{1}{2}\sin 2t\right]_0^\pi = ab((\pi - 0) - 0) = \pi ab$

25. $\frac{dx}{dt} = -\sin t$ and $\frac{dy}{dt} = 1 + \cos t \Rightarrow \sqrt{\left(\frac{dx}{dt}\right)^2 + \left(\frac{dy}{dt}\right)^2} = \sqrt{(-\sin t)^2 + (1+\cos t)^2} = \sqrt{2 + 2\cos t}$

$\Rightarrow \text{Length} = \int_0^\pi \sqrt{2 + 2\cos t}\,dt = \sqrt{2}\int_0^\pi \sqrt{\left(\frac{1-\cos t}{1-\cos t}\right)(1+\cos t)}\,dt = \sqrt{2}\int_0^\pi \sqrt{\frac{\sin^2 t}{1-\cos t}}\,dt$

$= \sqrt{2} \int_0^{\pi} \frac{\sin t}{\sqrt{1-\cos t}}\, dt$ (since $\sin t \ge 0$ on $[0, \pi]$); $[u = 1 - \cos t \Rightarrow du = \sin t\, dt;\ t = 0 \Rightarrow u = 0,$

$t = \pi \Rightarrow u = 2] \rightarrow \sqrt{2} \int_0^2 u^{-1/2}\, du = \sqrt{2}\left[2u^{1/2}\right]_0^2 = 4$

27. $\frac{dx}{dt} = t$ and $\frac{dy}{dt} = (2t+1)^{1/2} \Rightarrow \sqrt{\left(\frac{dx}{dt}\right)^2 + \left(\frac{dy}{dt}\right)^2} = \sqrt{t^2 + (2t+1)} = \sqrt{(t+1)^2} = |t+1| = t+1$ since $0 \le t \le 4$

$\Rightarrow \text{Length} = \int_0^4 (t+1)\, dt = \left[\frac{t^2}{2} + t\right]_0^4 = (8+4) = 12$

29. $\frac{dx}{dt} = 8t \cos t$ and $\frac{dy}{dt} = 8t \sin t \Rightarrow \sqrt{\left(\frac{dx}{dt}\right)^2 + \left(\frac{dy}{dt}\right)^2} = \sqrt{(8t\cos t)^2 + (8t \sin t)^2} = \sqrt{64t^2 \cos^2 t + 64t^2 \sin^2 t}$

$= |8t| = 8t$ since $0 \le t \le \frac{\pi}{2} \Rightarrow \text{Length} = \int_0^{\pi/2} 8t\, dt = \left[4t^2\right]_0^{\pi/2} = \pi^2$

31. $\frac{dx}{dt} = -\sin t$ and $\frac{dy}{dt} = \cos t \Rightarrow \sqrt{\left(\frac{dx}{dt}\right)^2 + \left(\frac{dy}{dt}\right)^2} = \sqrt{(-\sin t)^2 + (\cos t)^2} = 1 \Rightarrow \text{Area} = \int 2\pi y\, ds$

$= \int_0^{2\pi} 2\pi(2 + \sin t)(1)dt = 2\pi\left[2t - \cos t\right]_0^{2\pi} = 2\pi[(4\pi - 1) - (0 - 1)] = 8\pi^2$

33. $\frac{dx}{dt} = 1$ and $\frac{dy}{dt} = t + \sqrt{2} \Rightarrow \sqrt{\left(\frac{dx}{dt}\right)^2 + \left(\frac{dy}{dt}\right)^2} = \sqrt{1^2 + \left(t + \sqrt{2}\right)^2} = \sqrt{t^2 + 2\sqrt{2}t + 3} \Rightarrow \text{Area} = \int 2\pi x\, ds$

$= \int_{-\sqrt{2}}^{\sqrt{2}} 2\pi\left(t + \sqrt{2}\right)\sqrt{t^2 + 2\sqrt{2}t + 3}\, dt;\ \left[u = t^2 + 2\sqrt{2}t + 3 \Rightarrow du = \left(2t + 2\sqrt{2}\right) dt;\ t = -\sqrt{2} \Rightarrow u = 1,\right.$

$\left. t = \sqrt{2} \Rightarrow u = 9\right] \rightarrow \int_1^9 \pi\sqrt{u}\, du = \left[\frac{2}{3}\pi u^{3/2}\right]_1^9 = \frac{2\pi}{3}(27 - 1) = \frac{52\pi}{3}$

35. $\frac{dx}{dt} = 2$ and $\frac{dy}{dt} = 1 \Rightarrow \sqrt{\left(\frac{dx}{dt}\right)^2 + \left(\frac{dy}{dt}\right)^2} = \sqrt{2^2 + 1^2} = \sqrt{5} \Rightarrow \text{Area} = \int 2\pi y\, ds = \int_0^1 2\pi(t+1)\sqrt{5}\, dt$

$= 2\pi\sqrt{5}\left[\frac{t^2}{2} + t\right]_0^1 = 3\pi\sqrt{5}$. Check: slant height is $\sqrt{5} \Rightarrow$ Area is $\pi(1+2)\sqrt{5} = 3\pi\sqrt{5}$.

37. Let the density be $\delta = 1$. Then $x = \cos t + t \sin t \Rightarrow \frac{dx}{dt} = t \cos t$, and $y = \sin t - t \cos t \Rightarrow \frac{dy}{dt} = t \sin t$

$\Rightarrow dm = 1 \cdot ds = \sqrt{\left(\frac{dx}{dt}\right)^2 + \left(\frac{dy}{dt}\right)^2}\, dt = \sqrt{(t\cos t)^2 + (t \sin t)^2} = |t|\, dt = t\, dt$ since $0 \le t \le \frac{\pi}{2}$. The curve's mass is

$M = \int dm = \int_0^{\pi/2} t\, dt = \frac{\pi^2}{8}$. Also $M_x = \int \tilde{y}\, dm = \int_0^{\pi/2} (\sin t - t\cos t)\, t\, dt = \int_0^{\pi/2} t \sin t\, dt - \int_0^{\pi/2} t^2 \cos t\, dt$

$= \left[\sin t - t\cos t\right]_0^{\pi/2} - \left[t^2 \sin t - 2\sin t + 2t \cos t\right]_0^{\pi/2} = 3 - \frac{\pi^2}{4}$, where we integrated by parts. Therefore,

$\overline{y} = \frac{M_x}{M} = \frac{\left(3 - \frac{\pi^2}{4}\right)}{\left(\frac{\pi^2}{8}\right)} = \frac{24}{\pi^2} - 2$. Next, $M_y = \int \tilde{x}\, dm = \int_0^{\pi/2} (\cos t + t \sin t)\, t\, dt = \int_0^{\pi/2} t\cos t\, dt + \int_0^{\pi/2} t^2 \sin t\, dt$

$= \left[\cos t + t \sin t\right]_0^{\pi/2} + \left[-t^2 \cos t + 2\cos t + 2t \sin t\right]_0^{\pi/2} = \frac{3\pi}{2} - 3$, again integrating by parts. Hence

$\overline{x} = \frac{M_y}{M} = \frac{\left(\frac{3\pi}{2} - 3\right)}{\left(\frac{\pi^2}{8}\right)} = \frac{12}{\pi} - \frac{24}{\pi^2}$. Therefore $(\overline{x}, \overline{y}) = \left(\frac{12}{\pi} - \frac{24}{\pi^2}, \frac{24}{\pi^2} - 2\right)$.

39. Let the density be $\delta = 1$. Then $x = \cos t \Rightarrow \frac{dx}{dt} = -\sin t$, and $y = t + \sin t \Rightarrow \frac{dy}{dt} = 1 + \cos t$

$\Rightarrow dm = 1 \cdot ds = \sqrt{\left(\frac{dx}{dt}\right)^2 + \left(\frac{dy}{dt}\right)^2}\, dt = \sqrt{(-\sin t)^2 + (1 + \cos t)^2}\, dt = \sqrt{2 + 2\cos t}\, dt$. The curve's mass

is $M = \int dm = \int_0^{\pi} \sqrt{2 + 2\cos t}\, dt = \sqrt{2}\int_0^{\pi} \sqrt{1 + \cos t}\, dt = \sqrt{2}\int_0^{\pi} \sqrt{2\cos^2\left(\frac{t}{2}\right)}\, dt = 2\int_0^{\pi} \left|\cos\left(\frac{t}{2}\right)\right| dt$

$= 2\int_0^{\pi} \cos\left(\frac{t}{2}\right) dt$ (since $0 \le t \le \pi \Rightarrow 0 \le \frac{t}{2} \le \frac{\pi}{2}$) $= 2\left[2\sin\left(\frac{t}{2}\right)\right]_0^{\pi} = 4$. Also $M_x = \int \tilde{y}\, dm$

$= \int_0^{\pi} (t + \sin t)\left(2\cos\frac{t}{2}\right) dt = \int_0^{\pi} 2t\cos\left(\frac{t}{2}\right) dt + \int_0^{\pi} 2\sin t \cos\left(\frac{t}{2}\right) dt$

$= 2\left[4\cos\left(\frac{t}{2}\right) + 2t\sin\left(\frac{t}{2}\right)\right]_0^{\pi} + 2\left[-\frac{1}{3}\cos\left(\frac{3}{2}t\right) - \cos\left(\frac{1}{2}t\right)\right]_0^{\pi} = 4\pi - \frac{16}{3} \Rightarrow \bar{y} = \frac{M_x}{M} = \frac{\left(4\pi - \frac{16}{3}\right)}{4} = \pi - \frac{4}{3}$.

Next $M_y = \int \tilde{x}\, dm = \int_0^{\pi} (\cos t)\left(2\cos\frac{t}{2}\right) dt = \int_0^{\pi} \cos t \cos\left(\frac{t}{2}\right) dt = 2\left[\sin\left(\frac{t}{2}\right) + \frac{\sin\left(\frac{3}{2}t\right)}{3}\right]_0^{\pi} = 2 - \frac{2}{3}$

$= \frac{4}{3} \Rightarrow \bar{x} = \frac{M_y}{M} = \frac{\left(\frac{4}{3}\right)}{4} = \frac{1}{3}$. Therefore $(\bar{x}, \bar{y}) = \left(\frac{1}{3}, \pi - \frac{4}{3}\right)$.

41. (a) $\frac{dx}{dt} = -2\sin 2t$ and $\frac{dy}{dt} = 2\cos 2t \Rightarrow \sqrt{\left(\frac{dx}{dt}\right)^2 + \left(\frac{dy}{dt}\right)^2} = \sqrt{(-2\sin 2t)^2 + (2\cos 2t)^2} = 2$

$\Rightarrow \text{Length} = \int_0^{\pi/2} 2\, dt = [2t]_0^{\pi/2} = \pi$

(b) $\frac{dx}{dt} = \pi\cos\pi t$ and $\frac{dy}{dt} = -\pi\sin\pi t \Rightarrow \sqrt{\left(\frac{dx}{dt}\right)^2 + \left(\frac{dy}{dt}\right)^2} = \sqrt{(\pi\cos\pi t)^2 + (-\pi\sin\pi t)^2} = \pi$

$\Rightarrow \text{Length} = \int_{-1/2}^{1/2} \pi\, dt = [\pi t]_{-1/2}^{1/2} = \pi$

43. $x = (1 + 2\sin\theta)\cos\theta,\ y = (1 + 2\sin\theta)\sin\theta \Rightarrow \frac{dx}{d\theta} = 2\cos^2\theta - \sin\theta(1 + 2\sin\theta),\ \frac{dy}{d\theta} = 2\cos\theta\sin\theta + \cos\theta(1 + 2\sin\theta)$

$\Rightarrow \frac{dy}{dx} = \frac{2\cos\theta\sin\theta + \cos\theta(1+2\sin\theta)}{2\cos^2\theta - \sin\theta(1+2\sin\theta)} = \frac{4\cos\theta\sin\theta + \cos\theta}{2\cos^2\theta - 2\sin^2\theta - \sin\theta} = \frac{2\sin 2\theta + \cos\theta}{2\cos 2\theta - \sin\theta}$

(a) $x = (1 + 2\sin(0))\cos(0) = 1,\ y = (1 + 2\sin(0))\sin(0) = 0;\ \left.\frac{dy}{dx}\right|_{\theta=0} = \frac{2\sin(2(0)) + \cos(0)}{2\cos(2(0)) - \sin(0)} = \frac{0+1}{2-0} = \frac{1}{2}$

(b) $x = \left(1 + 2\sin\left(\frac{\pi}{2}\right)\right)\cos\left(\frac{\pi}{2}\right) = 0,\ y = \left(1 + 2\sin\left(\frac{\pi}{2}\right)\right)\sin\left(\frac{\pi}{2}\right) = 3;\ \left.\frac{dy}{dx}\right|_{\theta=\pi/2} = \frac{2\sin\left(2\left(\frac{\pi}{2}\right)\right) + \cos\left(\frac{\pi}{2}\right)}{2\cos\left(2\left(\frac{\pi}{2}\right)\right) - \sin\left(\frac{\pi}{2}\right)} = \frac{0+0}{-2-1} = 0$

(c) $x = \left(1 + 2\sin\left(\frac{4\pi}{3}\right)\right)\cos\left(\frac{4\pi}{3}\right) = \frac{\sqrt{3}-1}{2},\ y = \left(1 + 2\sin\left(\frac{4\pi}{3}\right)\right)\sin\left(\frac{4\pi}{3}\right) = \frac{3-\sqrt{3}}{2};\ \left.\frac{dy}{dx}\right|_{\theta=4\pi/3} = \frac{2\sin\left(2\left(\frac{4\pi}{3}\right)\right) + \cos\left(\frac{4\pi}{3}\right)}{2\cos\left(2\left(\frac{4\pi}{3}\right)\right) - \sin\left(\frac{4\pi}{3}\right)}$

$= \frac{\sqrt{3} - \frac{1}{2}}{-1 + \frac{\sqrt{3}}{2}} = \frac{2\sqrt{3}-1}{\sqrt{3}-2} = -\left(4 + 3\sqrt{3}\right)$

45. $\frac{dx}{dt} = \cos t$ and $\frac{dy}{dt} = 2\cos 2t \Rightarrow \frac{dy}{dx} = \frac{dy/dt}{dx/dt} = \frac{2\cos 2t}{\cos t} = \frac{2(2\cos^2 t - 1)}{\cos t}$; then $\frac{dy}{dx} = 0 \Rightarrow \frac{2(2\cos^2 t - 1)}{\cos t} = 0$

$\Rightarrow 2\cos^2 t - 1 = 0 \Rightarrow \cos t = \pm\frac{1}{\sqrt{2}} \Rightarrow t = \frac{\pi}{4}, \frac{3\pi}{4}, \frac{5\pi}{4}, \frac{7\pi}{4}$. In the 1st quadrant: $t = \frac{\pi}{4} \Rightarrow x = \sin\frac{\pi}{4} = \frac{\sqrt{2}}{2}$ and

$y = \sin 2\left(\frac{\pi}{4}\right) = 1 \Rightarrow \left(\frac{\sqrt{2}}{2}, 1\right)$ is the point where the tangent line is horizontal. At the origin: $x = 0$ and $y = 0$

$\Rightarrow \sin t = 0 \Rightarrow t = 0$ or $t = \pi$ and $\sin 2t = 0 \Rightarrow t = 0, \frac{\pi}{2}, \pi, \frac{3\pi}{2}$; thus $t = 0$ and $t = \pi$ give the tangent lines at the origin. Tangents at origin: $\left.\frac{dy}{dx}\right|_{t=0} = 2 \Rightarrow y = 2x$ and $\left.\frac{dy}{dx}\right|_{t=\pi} = -2 \Rightarrow y = -2x$

47. (a) $x = a(t - \sin t),\ y = a(1 - \cos t),\ 0 \le t \le 2\pi \Rightarrow \frac{dx}{dt} = a(1 - \cos t),\ \frac{dy}{dt} = a\sin t \Rightarrow$ Length

$= \int_0^{2\pi} \sqrt{(a(1-\cos t))^2 + (a\sin t)^2}\, dt = \int_0^{2\pi} \sqrt{a^2 - 2a^2\cos t + a^2\cos^2 t + a^2\sin^2 t}\, dt$

$= a\sqrt{2}\int_0^{2\pi} \sqrt{1 - \cos t}\, dt = a\sqrt{2}\int_0^{2\pi} \sqrt{2\sin^2\left(\frac{t}{2}\right)}\, dt = 2a\int_0^{2\pi} \sin\left(\frac{t}{2}\right) dt = \left[-4a\cos\left(\frac{t}{2}\right)\right]_0^{2\pi}$

$= -4a\cos\pi + 4a\cos(0) = 8a$

(b) $a = 1 \Rightarrow x = t - \sin t,\ y = 1 - \cos t,\ 0 \le t \le 2\pi \Rightarrow \frac{dx}{dt} = 1 - \cos t,\ \frac{dy}{dt} = \sin t \Rightarrow$ Surface area $=$

$= \int_0^{2\pi} 2\pi(1 - \cos t)\sqrt{(1-\cos t)^2 + (\sin t)^2}\, dt = \int_0^{2\pi} 2\pi(1 - \cos t)\sqrt{1 - 2\cos t + \cos^2 t + \sin^2 t}\, dt$

$= 2\pi\int_0^{2\pi} (1 - \cos t)\sqrt{2 - 2\cos t}\, dt = 2\sqrt{2}\pi\int_0^{2\pi} (1 - \cos t)^{3/2}\, dt = 2\sqrt{2}\pi\int_0^{2\pi} \left(1 - \cos\left(2 \cdot \frac{t}{2}\right)\right)^{3/2} dt$

$= 2\sqrt{2}\pi\int_0^{2\pi} \left(2\sin^2\left(\frac{t}{2}\right)\right)^{3/2} dt = 8\pi\int_0^{2\pi} \sin^3\left(\frac{t}{2}\right) dt$

$$\left[u = \frac{t}{2} \Rightarrow du = \frac{1}{2}dt \Rightarrow dt = 2\, du;\ t = 0 \Rightarrow u = 0,\ t = 2\pi \Rightarrow u = \pi\right]$$

$= 16\pi\int_0^{\pi} \sin^3 u\, du = 16\pi\int_0^{\pi} \sin^2 u \sin u\, du = 16\pi\int_0^{\pi} (1 - \cos^2 u)\sin u\, du = 16\pi\int_0^{\pi} \sin u\, du - 16\pi\int_0^{\pi} \cos^2 u \sin u\, du$

$= \left[-16\pi\cos u + \frac{16\pi}{3}\cos^3 u\right]_0^{\pi} = \left(16\pi - \frac{16\pi}{3}\right) - \left(-16\pi + \frac{16\pi}{3}\right) = \frac{64\pi}{3}$

11.3 POLAR COORDINATES

1. a, e; b, g; c, h; d, f

3. (a) $\left(2, \frac{\pi}{2} + 2n\pi\right)$ and $\left(-2, \frac{\pi}{2} + (2n+1)\pi\right)$, n an integer
   (b) $(2, 2n\pi)$ and $(-2, (2n+1)\pi)$, n an integer
   (c) $\left(2, \frac{3\pi}{2} + 2n\pi\right)$ and $\left(-2, \frac{3\pi}{2} + (2n+1)\pi\right)$, n an integer
   (d) $(2, (2n+1)\pi)$ and $(-2, 2n\pi)$, n an integer

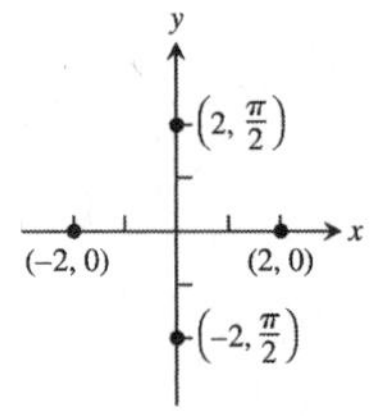

5. (a) $x = r\cos\theta = 3\cos 0 = 3$, $y = r\sin\theta = 3\sin 0 = 0 \Rightarrow$ Cartesian coordinates are $(3, 0)$
   (b) $x = r\cos\theta = -3\cos 0 = -3$, $y = r\sin\theta = -3\sin 0 = 0 \Rightarrow$ Cartesian coordinates are $(-3, 0)$
   (c) $x = r\cos\theta = 2\cos\frac{2\pi}{3} = -1$, $y = r\sin\theta = 2\sin\frac{2\pi}{3} = \sqrt{3} \Rightarrow$ Cartesian coordinates are $\left(-1, \sqrt{3}\right)$
   (d) $x = r\cos\theta = 2\cos\frac{7\pi}{3} = 1$, $y = r\sin\theta = 2\sin\frac{7\pi}{3} = \sqrt{3} \Rightarrow$ Cartesian coordinates are $\left(1, \sqrt{3}\right)$
   (e) $x = r\cos\theta = -3\cos\pi = 3$, $y = r\sin\theta = -3\sin\pi = 0 \Rightarrow$ Cartesian coordinates are $(3, 0)$
   (f) $x = r\cos\theta = 2\cos\frac{\pi}{3} = 1$, $y = r\sin\theta = 2\sin\frac{\pi}{3} = \sqrt{3} \Rightarrow$ Cartesian coordinates are $\left(1, \sqrt{3}\right)$
   (g) $x = r\cos\theta = -3\cos 2\pi = -3$, $y = r\sin\theta = -3\sin 2\pi = 0 \Rightarrow$ Cartesian coordinates are $(-3, 0)$
   (h) $x = r\cos\theta = -2\cos\left(-\frac{\pi}{3}\right) = -1$, $y = r\sin\theta = -2\sin\left(-\frac{\pi}{3}\right) = \sqrt{3} \Rightarrow$ Cartesian coordinates are $\left(-1, \sqrt{3}\right)$

7. (a) $(1, 1) \Rightarrow r = \sqrt{1^2 + 1^2} = \sqrt{2}$, $\sin\theta = \frac{1}{\sqrt{2}}$ and $\cos\theta = \frac{1}{\sqrt{2}} \Rightarrow \theta = \frac{\pi}{4} \Rightarrow$ Polar coordinates are $\left(\sqrt{2}, \frac{\pi}{4}\right)$
   (b) $(-3, 0) \Rightarrow r = \sqrt{(-3)^2 + 0^2} = 3$, $\sin\theta = 0$ and $\cos\theta = -1 \Rightarrow \theta = \pi \Rightarrow$ Polar coordinates are $(3, \pi)$
   (c) $\left(\sqrt{3}, -1\right) \Rightarrow r = \sqrt{\left(\sqrt{3}\right)^2 + (-1)^2} = 2$, $\sin\theta = -\frac{1}{2}$ and $\cos\theta = \frac{\sqrt{3}}{2} \Rightarrow \theta = \frac{11\pi}{6} \Rightarrow$ Polar coordinates are $\left(2, \frac{11\pi}{6}\right)$
   (d) $(-3, 4) \Rightarrow r = \sqrt{(-3)^2 + 4^2} = 5$, $\sin\theta = \frac{4}{5}$ and $\cos\theta = -\frac{3}{5} \Rightarrow \theta = \pi - \arctan\left(\frac{4}{3}\right) \Rightarrow$ Polar coordinates are $\left(5, \pi - \arctan\left(\frac{4}{3}\right)\right)$

9. (a) $(3, 3) \Rightarrow r = -\sqrt{3^2 + 3^2} = -3\sqrt{2}$, $\sin\theta = -\frac{1}{\sqrt{2}}$ and $\cos\theta = -\frac{1}{\sqrt{2}} \Rightarrow \theta = \frac{5\pi}{4} \Rightarrow$ Polar coordinates are $\left(-3\sqrt{2}, \frac{5\pi}{4}\right)$
   (b) $(-1, 0) \Rightarrow r = -\sqrt{(-1)^2 + 0^2} = -1$, $\sin\theta = 0$ and $\cos\theta = 1 \Rightarrow \theta = 0 \Rightarrow$ Polar coordinates are $(-1, 0)$
   (c) $\left(-1, \sqrt{3}\right) \Rightarrow r = -\sqrt{(-1)^2 + \left(\sqrt{3}\right)^2} = -2$, $\sin\theta = -\frac{\sqrt{3}}{2}$ and $\cos\theta = \frac{1}{2} \Rightarrow \theta = \frac{5\pi}{3} \Rightarrow$ Polar coordinates are $\left(-2, \frac{5\pi}{3}\right)$
   (d) $(4, -3) \Rightarrow r = -\sqrt{4^2 + (-3)^2} = -5$, $\sin\theta = \frac{3}{5}$ and $\cos\theta = -\frac{4}{5} \Rightarrow \theta = \pi - \arctan\left(\frac{3}{4}\right) \Rightarrow$ Polar coordinates are $\left(-5, \pi - \arctan\left(\frac{4}{3}\right)\right)$

11.

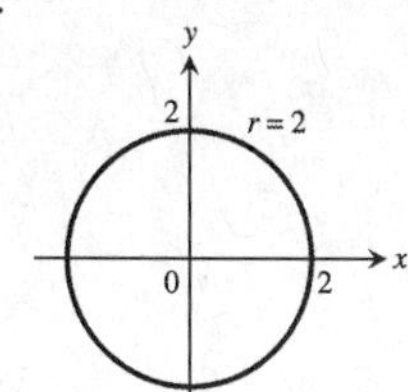

13.

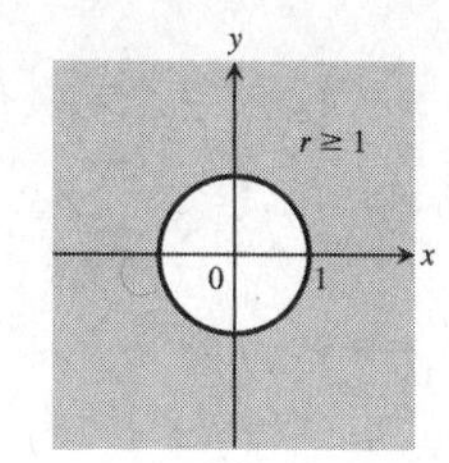

15.

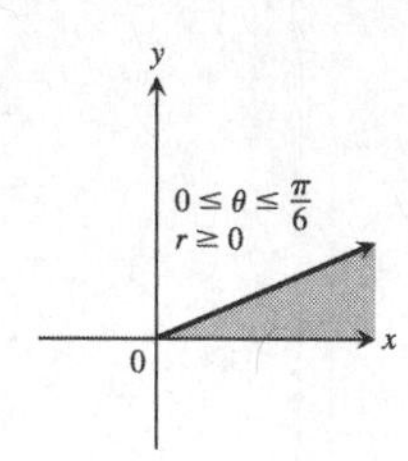

17.

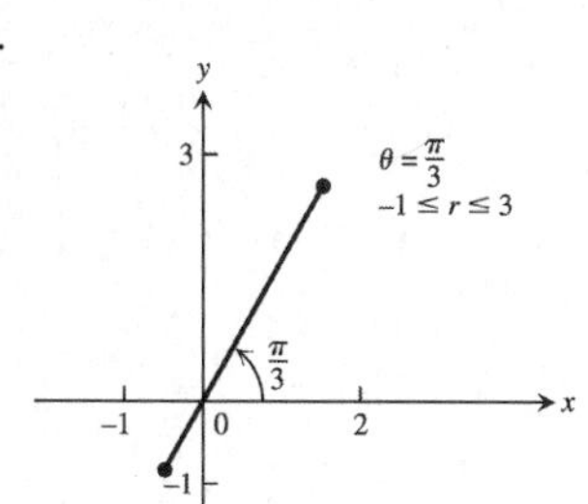

19.

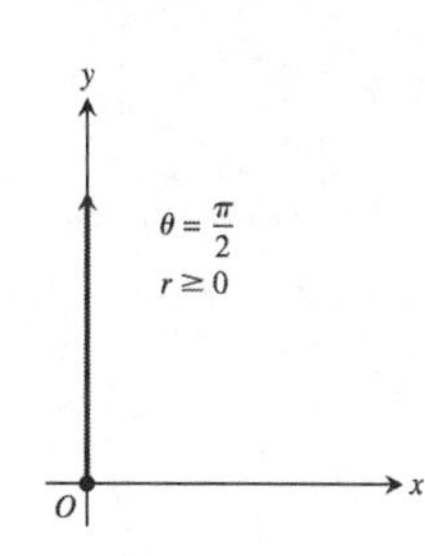

21.

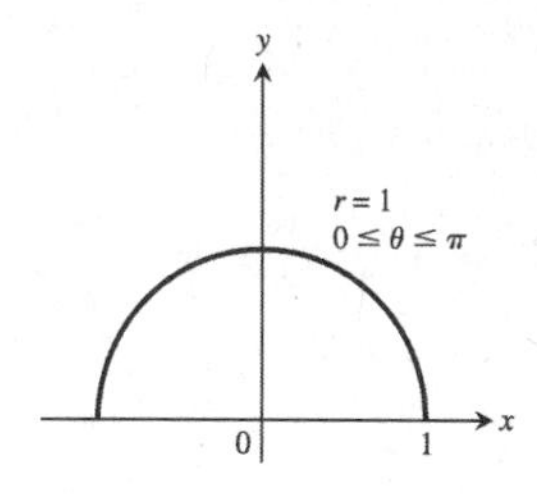

23.

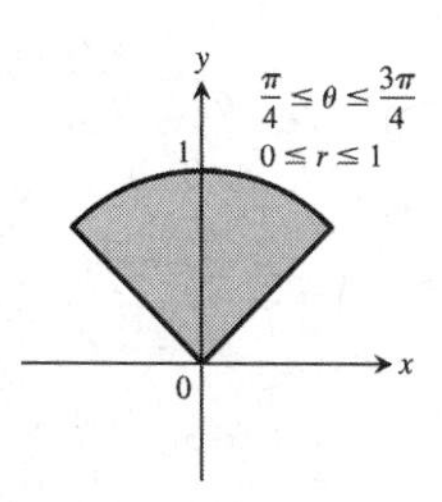

25.

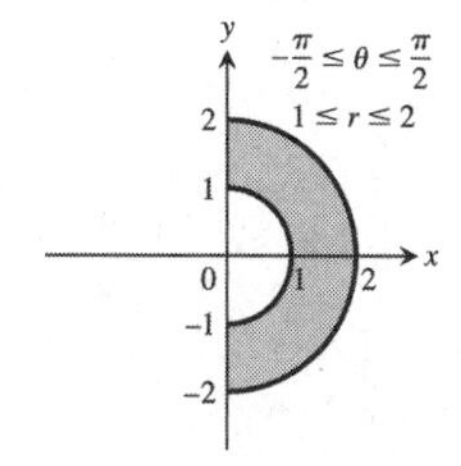

27. $r\cos\theta = 2 \Rightarrow x = 2$, vertical line through $(2, 0)$

29. $r\sin\theta = 0 \Rightarrow y = 0$, the x-axis

31. $r = 4\csc\theta \Rightarrow r = \frac{4}{\sin\theta} \Rightarrow r\sin\theta = 4 \Rightarrow y = 4$, a horizontal line through $(0, 4)$

33. $r\cos\theta + r\sin\theta = 1 \Rightarrow x + y = 1$, line with slope $m = -1$ and intercept $b = 1$

35. $r^2 = 1 \Rightarrow x^2 + y^2 = 1$, circle with center $C = (0, 0)$ and radius 1

37. $r = \frac{5}{\sin\theta - 2\cos\theta} \Rightarrow r\sin\theta - 2r\cos\theta = 5 \Rightarrow y - 2x = 5$, line with slope $m = 2$ and intercept $b = 5$

39. $r = \cot\theta\csc\theta = \left(\frac{\cos\theta}{\sin\theta}\right)\left(\frac{1}{\sin\theta}\right) \Rightarrow r\sin^2\theta = \cos\theta \Rightarrow r^2\sin^2\theta = r\cos\theta \Rightarrow y^2 = x$, parabola with vertex $(0, 0)$ which opens to the right

41. $r = (\csc\theta)\, e^{r\cos\theta} \Rightarrow r\sin\theta = e^{r\cos\theta} \Rightarrow y = e^x$, graph of the natural exponential function

43. $r^2 + 2r^2\cos\theta\sin\theta = 1 \Rightarrow x^2 + y^2 + 2xy = 1 \Rightarrow x^2 + 2xy + y^2 = 1 \Rightarrow (x + y)^2 = 1 \Rightarrow x + y = \pm 1$, two parallel straight lines of slope -1 and y-intercepts $b = \pm 1$

45. $r^2 = -4r\cos\theta \Rightarrow x^2 + y^2 = -4x \Rightarrow x^2 + 4x + y^2 = 0 \Rightarrow x^2 + 4x + 4 + y^2 = 4 \Rightarrow (x + 2)^2 + y^2 = 4$, a circle with center $C(-2, 0)$ and radius 2

47. $r = 8\sin\theta \Rightarrow r^2 = 8r\sin\theta \Rightarrow x^2 + y^2 = 8y \Rightarrow x^2 + y^2 - 8y = 0 \Rightarrow x^2 + y^2 - 8y + 16 = 16 \Rightarrow x^2 + (y - 4)^2 = 16$, a circle with center $C(0, 4)$ and radius 4

49. $r = 2\cos\theta + 2\sin\theta \Rightarrow r^2 = 2r\cos\theta + 2r\sin\theta \Rightarrow x^2 + y^2 = 2x + 2y \Rightarrow x^2 - 2x + y^2 - 2y = 0$
$\Rightarrow (x-1)^2 + (y-1)^2 = 2$, a circle with center C(1, 1) and radius $\sqrt{2}$

51. $r\sin\left(\theta + \frac{\pi}{6}\right) = 2 \Rightarrow r\left(\sin\theta\cos\frac{\pi}{6} + \cos\theta\sin\frac{\pi}{6}\right) = 2 \Rightarrow \frac{\sqrt{3}}{2}r\sin\theta + \frac{1}{2}r\cos\theta = 2 \Rightarrow \frac{\sqrt{3}}{2}y + \frac{1}{2}x = 2$
$\Rightarrow \sqrt{3}y + x = 4$, line with slope $m = -\frac{1}{\sqrt{3}}$ and intercept $b = \frac{4}{\sqrt{3}}$

53. $x = 7 \Rightarrow r\cos\theta = 7$

55. $x = y \Rightarrow r\cos\theta = r\sin\theta \Rightarrow \theta = \frac{\pi}{4}$

57. $x^2 + y^2 = 4 \Rightarrow r^2 = 4 \Rightarrow r = 2$ or $r = -2$

59. $\frac{x^2}{9} + \frac{y^2}{4} = 1 \Rightarrow 4x^2 + 9y^2 = 36 \Rightarrow 4r^2\cos^2\theta + 9r^2\sin^2\theta = 36$

61. $y^2 = 4x \Rightarrow r^2\sin^2\theta = 4r\cos\theta \Rightarrow r\sin^2\theta = 4\cos\theta$

63. $x^2 + (y-2)^2 = 4 \Rightarrow x^2 + y^2 - 4y + 4 = 4 \Rightarrow x^2 + y^2 = 4y \Rightarrow r^2 = 4r\sin\theta \Rightarrow r = 4\sin\theta$

65. $(x-3)^2 + (y+1)^2 = 4 \Rightarrow x^2 - 6x + 9 + y^2 + 2y + 1 = 4 \Rightarrow x^2 + y^2 = 6x - 2y - 6 \Rightarrow r^2 = 6r\cos\theta - 2r\sin\theta - 6$

67. $(0, \theta)$ where θ is any angle

11.4 GRAPHING IN POLAR COORDINATES

1. $1 + \cos(-\theta) = 1 + \cos\theta = r \Rightarrow$ symmetric about the x-axis; $1 + \cos(-\theta) \neq -r$ and $1 + \cos(\pi - \theta)$ $= 1 - \cos\theta \neq r \Rightarrow$ not symmetric about the y-axis; therefore not symmetric about the origin

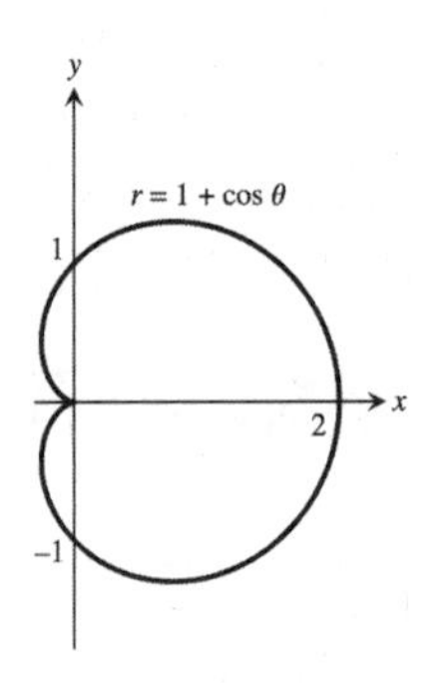

3. $1 - \sin(-\theta) = 1 + \sin\theta \neq r$ and $1 - \sin(\pi - \theta)$ $= 1 - \sin\theta \neq -r \Rightarrow$ not symmetric about the x-axis; $1 - \sin(\pi - \theta) = 1 - \sin\theta = r \Rightarrow$ symmetric about the y-axis; therefore not symmetric about the origin

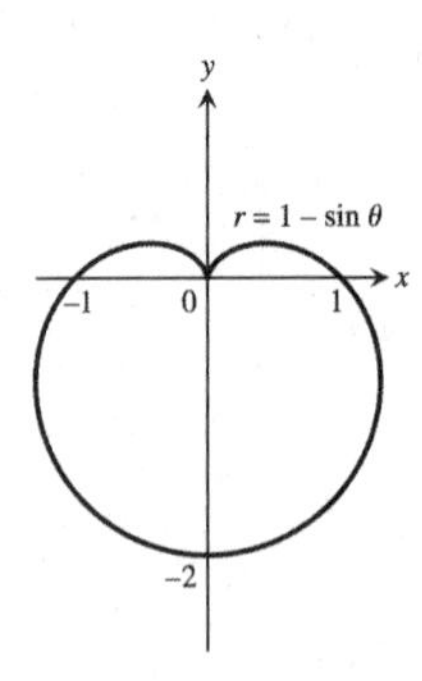

5. $2 + \sin(-\theta) = 2 - \sin\theta \neq r$ and $2 + \sin(\pi - \theta)$
$= 2 + \sin\theta \neq -r \Rightarrow$ not symmetric about the x-axis;
$2 + \sin(\pi - \theta) = 2 + \sin\theta = r \Rightarrow$ symmetric about the y-axis; therefore not symmetric about the origin

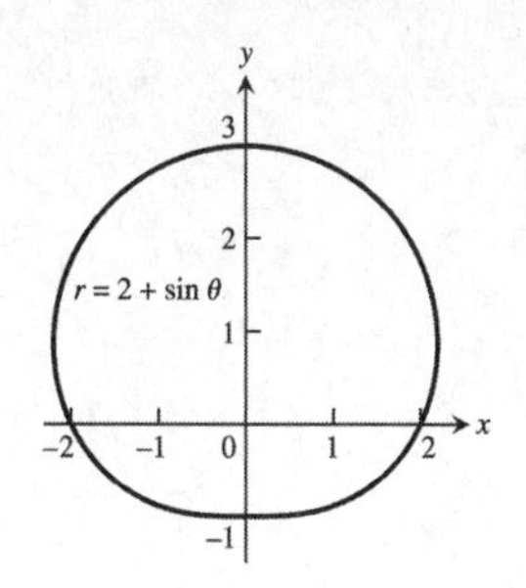

7. $\sin\left(-\frac{\theta}{2}\right) = -\sin\left(\frac{\theta}{2}\right) = -r \Rightarrow$ symmetric about the y-axis;
$\sin\left(\frac{2\pi-\theta}{2}\right) = \sin\left(\frac{\theta}{2}\right)$, so the graph is symmetric about the x-axis, and hence the origin.

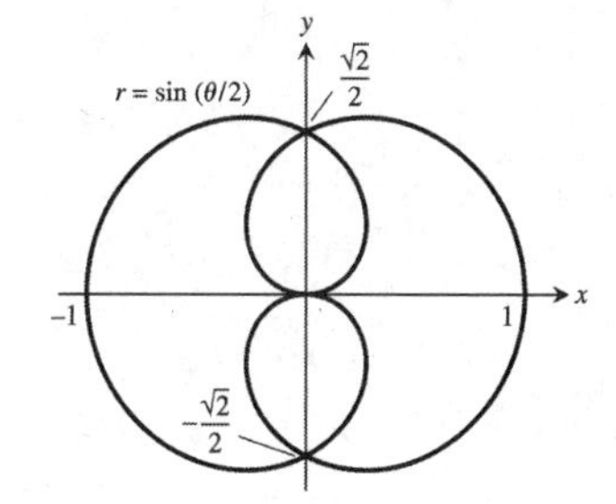

9. $\cos(-\theta) = \cos\theta = r^2 \Rightarrow (r, -\theta)$ and $(-r, -\theta)$ are on the graph when (r, θ) is on the graph $\Rightarrow$ symmetric about the x-axis and the y-axis; therefore symmetric about the origin

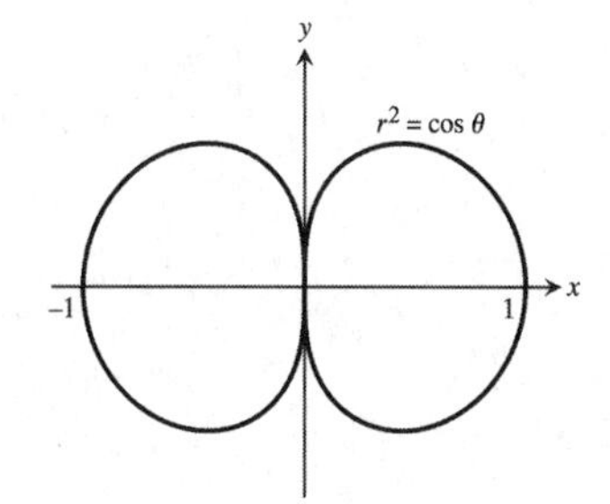

11. $-\sin(\pi - \theta) = -\sin\theta = r^2 \Rightarrow (r, \pi - \theta)$ and $(-r, \pi - \theta)$ are on the graph when (r, θ) is on the graph $\Rightarrow$ symmetric about the y-axis and the x-axis; therefore symmetric about the origin

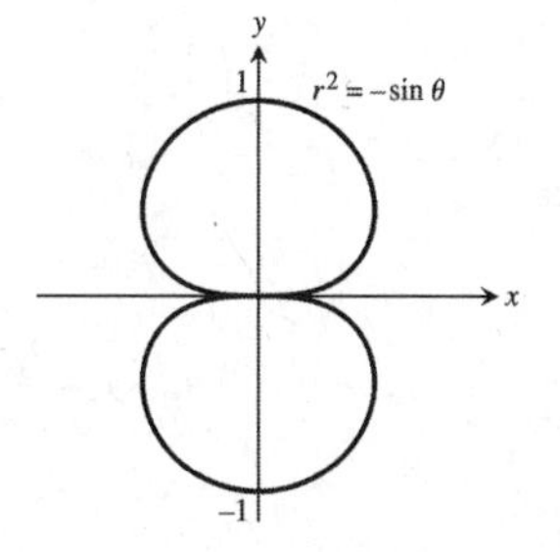

13. Since $(\pm r, -\theta)$ are on the graph when (r, θ) is on the graph $\left((\pm r)^2 = 4\cos 2(-\theta) \Rightarrow r^2 = 4\cos 2\theta\right)$, the graph is symmetric about the x-axis and the y-axis $\Rightarrow$ the graph is symmetric about the origin

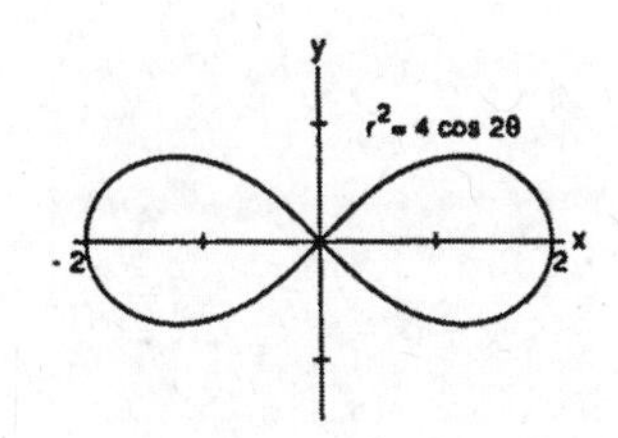

15. Since (r, θ) on the graph $\Rightarrow (-r, \theta)$ is on the graph $\left((\pm r)^2 = -\sin 2\theta \Rightarrow r^2 = -\sin 2\theta\right)$, the graph is symmetric about the origin. But $-\sin 2(-\theta) = -(-\sin 2\theta)$
$\sin 2\theta \neq r^2$ and $-\sin 2(\pi - \theta) = -\sin(2\pi - 2\theta)$
$= -\sin(-2\theta) = -(-\sin 2\theta) = \sin 2\theta \neq r^2 \Rightarrow$ the graph is not symmetric about the x-axis; therefore the graph is not symmetric about the y-axis

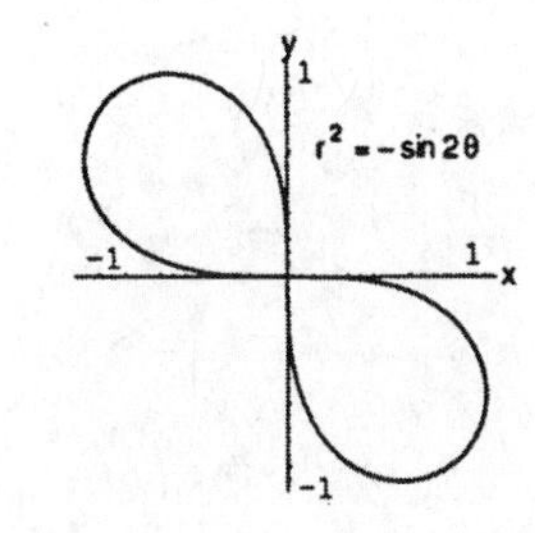

17. $\theta = \frac{\pi}{2} \Rightarrow r = -1 \Rightarrow \left(-1, \frac{\pi}{2}\right)$, and $\theta = -\frac{\pi}{2} \Rightarrow r = -1$
$\Rightarrow \left(-1, -\frac{\pi}{2}\right)$; $r' = \frac{dr}{d\theta} = -\sin\theta$; Slope $= \frac{r'\sin\theta + r\cos\theta}{r'\cos\theta - r\sin\theta}$
$= \frac{-\sin^2\theta + r\cos\theta}{-\sin\theta\cos\theta - r\sin\theta} \Rightarrow$ Slope at $\left(-1, \frac{\pi}{2}\right)$ is
$\frac{-\sin^2\left(\frac{\pi}{2}\right) + (-1)\cos\frac{\pi}{2}}{-\sin\frac{\pi}{2}\cos\frac{\pi}{2} - (-1)\sin\frac{\pi}{2}} = -1$; Slope at $\left(-1, -\frac{\pi}{2}\right)$ is
$\frac{-\sin^2\left(-\frac{\pi}{2}\right) + (-1)\cos\left(-\frac{\pi}{2}\right)}{-\sin\left(-\frac{\pi}{2}\right)\cos\left(-\frac{\pi}{2}\right) - (-1)\sin\left(-\frac{\pi}{2}\right)} = 1$

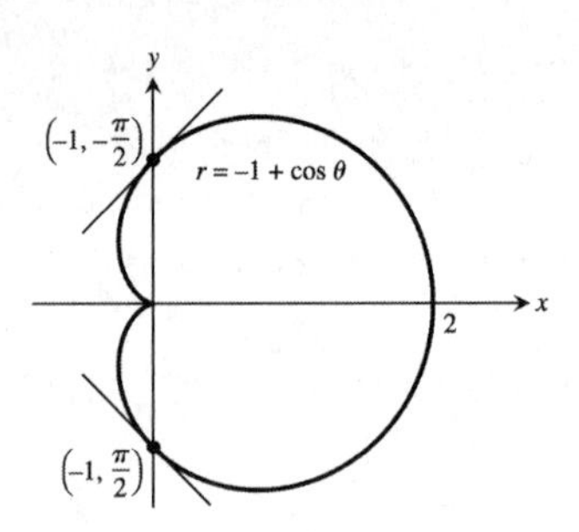

19. $\theta = \frac{\pi}{4} \Rightarrow r = 1 \Rightarrow \left(1, \frac{\pi}{4}\right)$; $\theta = -\frac{\pi}{4} \Rightarrow r = -1$
$\Rightarrow \left(-1, -\frac{\pi}{4}\right)$; $\theta = \frac{3\pi}{4} \Rightarrow r = -1 \Rightarrow \left(-1, \frac{3\pi}{4}\right)$;
$\theta = -\frac{3\pi}{4} \Rightarrow r = 1 \Rightarrow \left(1, -\frac{3\pi}{4}\right)$;
$r' = \frac{dr}{d\theta} = 2\cos 2\theta$;
Slope $= \frac{r'\sin\theta + r\cos\theta}{r'\cos\theta - r\sin\theta} = \frac{2\cos 2\theta\sin\theta + r\cos\theta}{2\cos 2\theta\cos\theta - r\sin\theta}$
$\Rightarrow$ Slope at $\left(1, \frac{\pi}{4}\right)$ is $\frac{2\cos\left(\frac{\pi}{2}\right)\sin\left(\frac{\pi}{4}\right) + (1)\cos\left(\frac{\pi}{4}\right)}{2\cos\left(\frac{\pi}{2}\right)\cos\left(\frac{\pi}{4}\right) - (1)\sin\left(\frac{\pi}{4}\right)} = -1$;

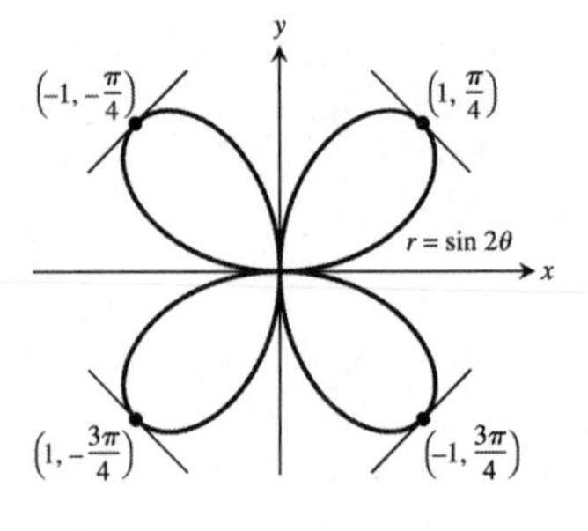

Slope at $\left(-1, -\frac{\pi}{4}\right)$ is $\frac{2\cos\left(-\frac{\pi}{2}\right)\sin\left(-\frac{\pi}{4}\right) + (-1)\cos\left(-\frac{\pi}{4}\right)}{2\cos\left(-\frac{\pi}{2}\right)\cos\left(-\frac{\pi}{4}\right) - (-1)\sin\left(-\frac{\pi}{4}\right)} = 1$;

Slope at $\left(-1, \frac{3\pi}{4}\right)$ is $\frac{2\cos\left(\frac{3\pi}{2}\right)\sin\left(\frac{3\pi}{4}\right) + (-1)\cos\left(\frac{3\pi}{4}\right)}{2\cos\left(\frac{3\pi}{2}\right)\cos\left(\frac{3\pi}{4}\right) - (-1)\sin\left(\frac{3\pi}{4}\right)} = 1$;

Slope at $\left(1, -\frac{3\pi}{4}\right)$ is $\frac{2\cos\left(-\frac{3\pi}{2}\right)\sin\left(-\frac{3\pi}{4}\right) + (1)\cos\left(-\frac{3\pi}{4}\right)}{2\cos\left(-\frac{3\pi}{2}\right)\cos\left(-\frac{3\pi}{4}\right) - (1)\sin\left(-\frac{3\pi}{4}\right)} = -1$

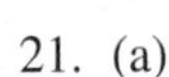

21. (a)

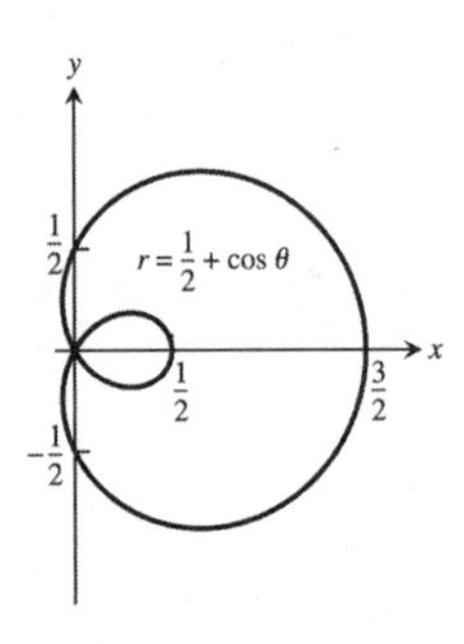

(b)

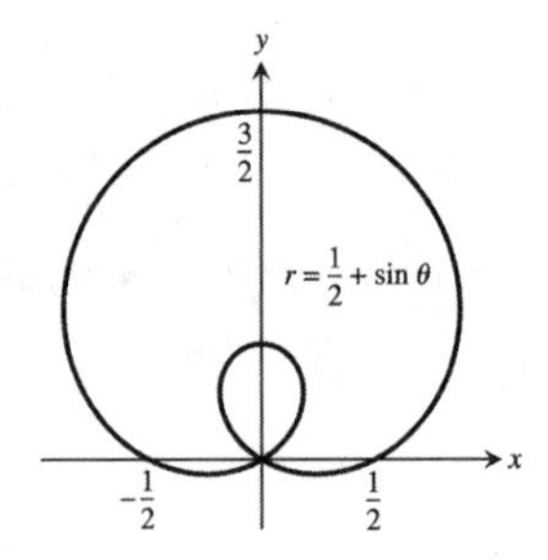

23. (a)

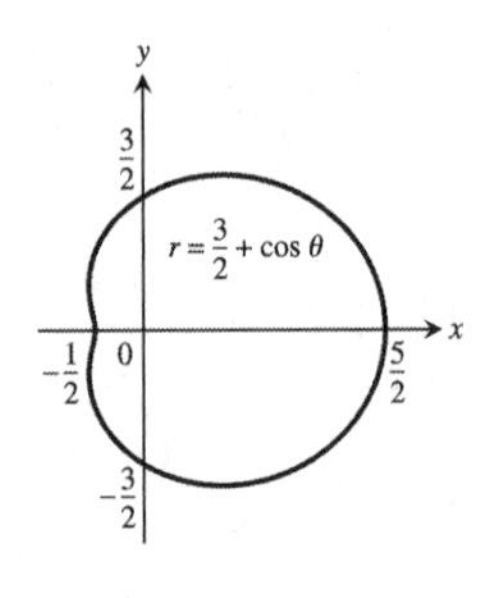

(b)

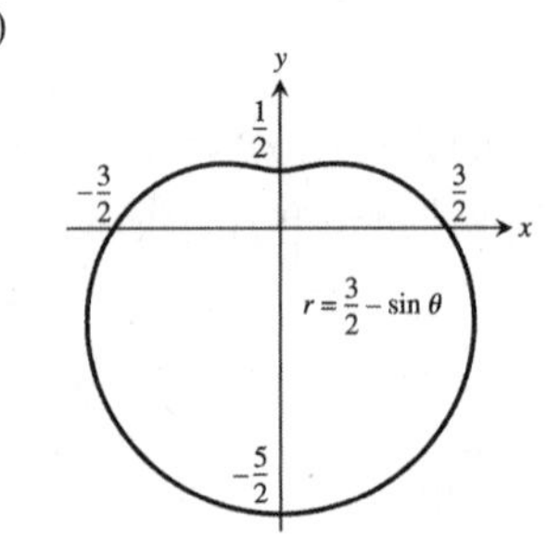

25.

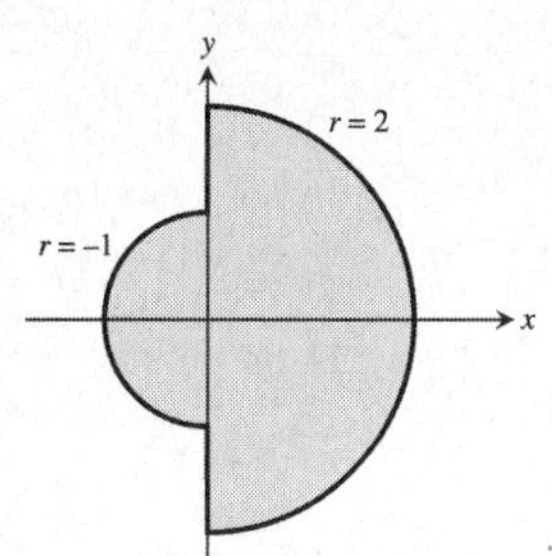

27.

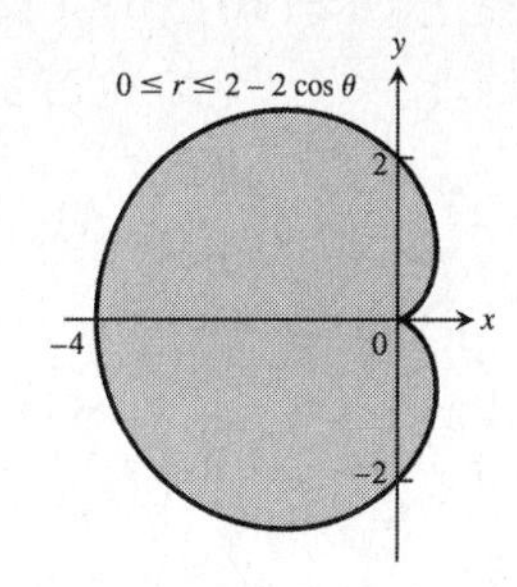

29. Note that (r, θ) and $(-r, \theta + \pi)$ describe the same point in the plane. Then $r = 1 - \cos\theta \Leftrightarrow -1 - \cos(\theta + \pi)$ $= -1 - (\cos\theta \cos\pi - \sin\theta \sin\pi) = -1 + \cos\theta = -(1 - \cos\theta) = -r$; therefore (r, θ) is on the graph of $r = 1 - \cos\theta \Leftrightarrow (-r, \theta + \pi)$ is on the graph of $r = -1 - \cos\theta \Rightarrow$ the answer is (a).

r = 1 - cos θ

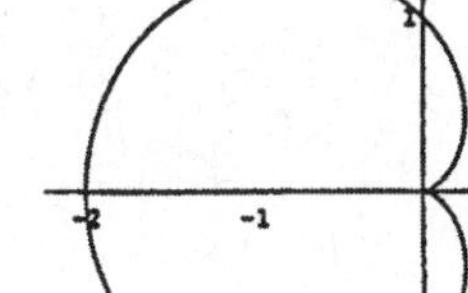

r = -1 - cos θ

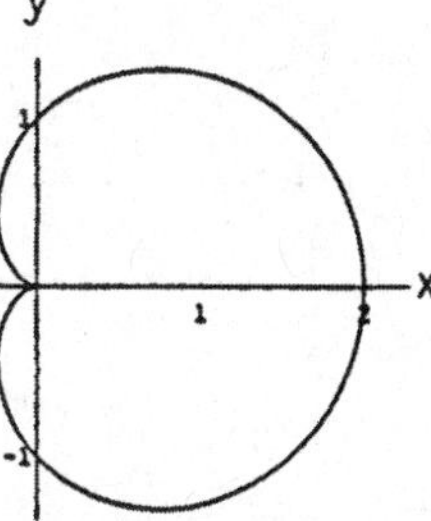

r = 1 + cos θ

31.

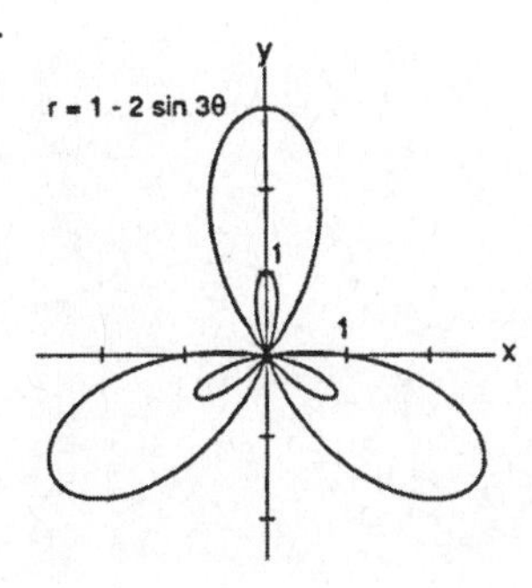

33. (a)

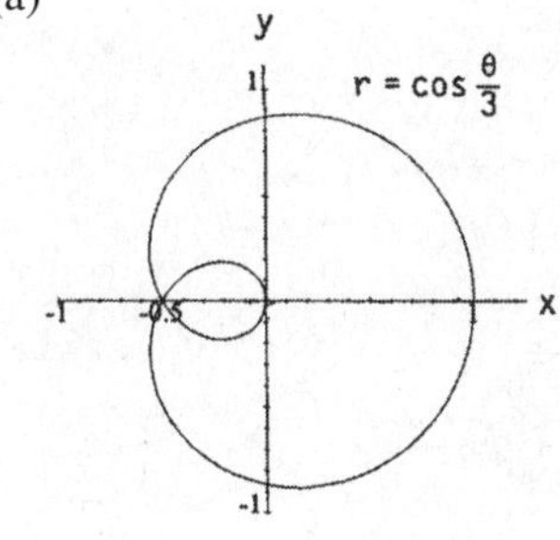

(b)

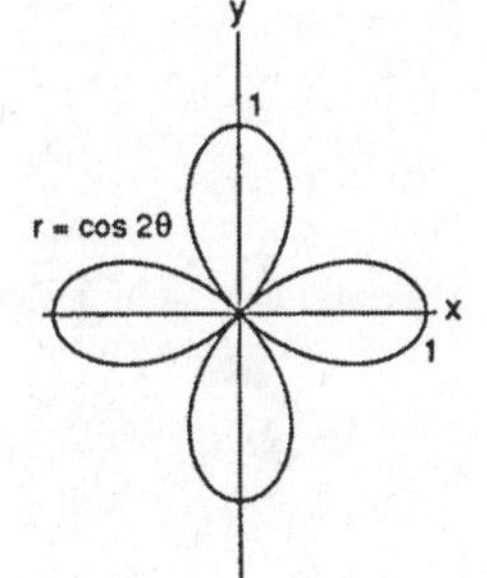

(c)

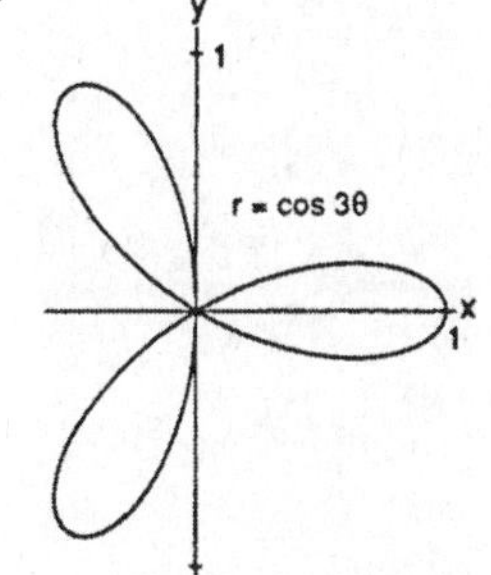

(d)

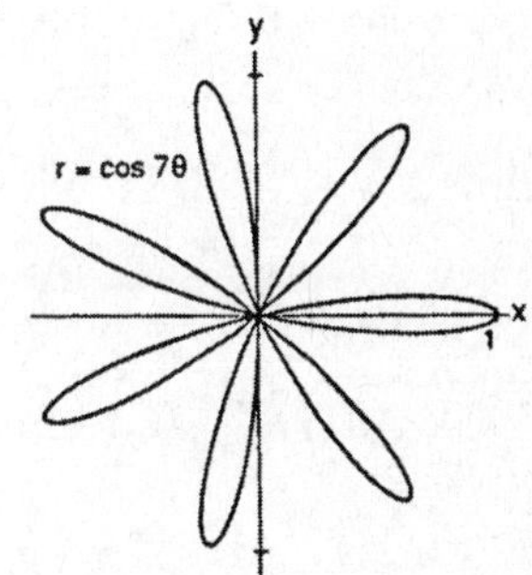

11.5 AREA AND LENGTHS IN POLAR COORDINATES

1. $A = \int_0^{\pi} \frac{1}{2}\theta^2\, d\theta = \left[\frac{1}{6}\theta^3\right]_0^{\pi} = \frac{\pi^3}{6}$

3. $A = \int_0^{2\pi} \frac{1}{2}(4 + 2\cos\theta)^2\, d\theta = \int_0^{2\pi} \frac{1}{2}(16 + 16\cos\theta + 4\cos^2\theta)\, d\theta = \int_0^{2\pi} \left[8 + 8\cos\theta + 2\left(\frac{1+\cos 2\theta}{2}\right)\right] d\theta$
$= \int_0^{2\pi} (9 + 8\cos\theta + \cos 2\theta)\, d\theta = \left[9\theta + 8\sin\theta + \frac{1}{2}\sin 2\theta\right]_0^{2\pi} = 18\pi$

5. $A = 2\int_0^{\pi/4} \frac{1}{2}\cos^2 2\theta\, d\theta = \int_0^{\pi/4} \frac{1+\cos 4\theta}{2}\, d\theta = \frac{1}{2}\left[\theta + \frac{\sin 4\theta}{4}\right]_0^{\pi/4} = \frac{\pi}{8}$

7. $A = \int_0^{\pi/2} \frac{1}{2}(4\sin 2\theta)\, d\theta = \int_0^{\pi/2} 2\sin 2\theta\, d\theta = [-\cos 2\theta]_0^{\pi/2} = 2$

9. $r = 2\cos\theta$ and $r = 2\sin\theta \Rightarrow 2\cos\theta = 2\sin\theta$
$\Rightarrow \cos\theta = \sin\theta \Rightarrow \theta = \frac{\pi}{4}$; therefore
$A = 2\int_0^{\pi/4} \frac{1}{2}(2\sin\theta)^2\, d\theta = \int_0^{\pi/4} 4\sin^2\theta\, d\theta$
$= \int_0^{\pi/4} 4\left(\frac{1-\cos 2\theta}{2}\right) d\theta = \int_0^{\pi/4} (2 - 2\cos 2\theta)\, d\theta$
$= [2\theta - \sin 2\theta]_0^{\pi/4} = \frac{\pi}{2} - 1$

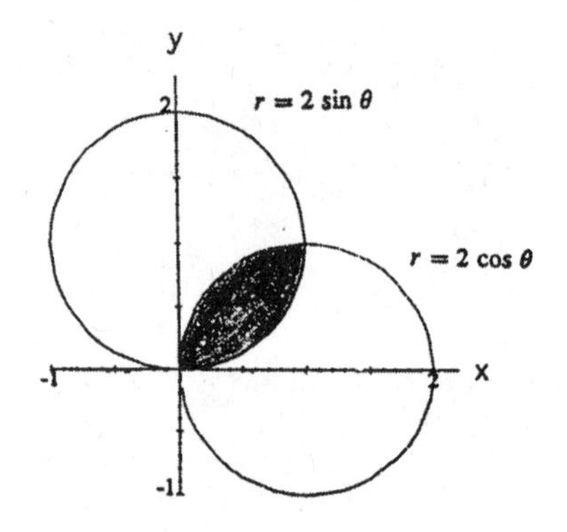

11. $r = 2$ and $r = 2(1 - \cos\theta) \Rightarrow 2 = 2(1 - \cos\theta)$
$\Rightarrow \cos\theta = 0 \Rightarrow \theta = \pm\frac{\pi}{2}$; therefore
$A = 2\int_0^{\pi/2} \frac{1}{2}[2(1 - \cos\theta)]^2\, d\theta + \frac{1}{2}\text{area of the circle}$
$= \int_0^{\pi/2} 4(1 - 2\cos\theta + \cos^2\theta)\, d\theta + \left(\frac{1}{2}\pi\right)(2)^2$
$= \int_0^{\pi/2} 4\left(1 - 2\cos\theta + \frac{1+\cos 2\theta}{2}\right) d\theta + 2\pi$
$= \int_0^{\pi/2} (4 - 8\cos\theta + 2 + 2\cos 2\theta)\, d\theta + 2\pi$
$= [6\theta - 8\sin\theta + \sin 2\theta]_0^{\pi/2} + 2\pi = 5\pi - 8$

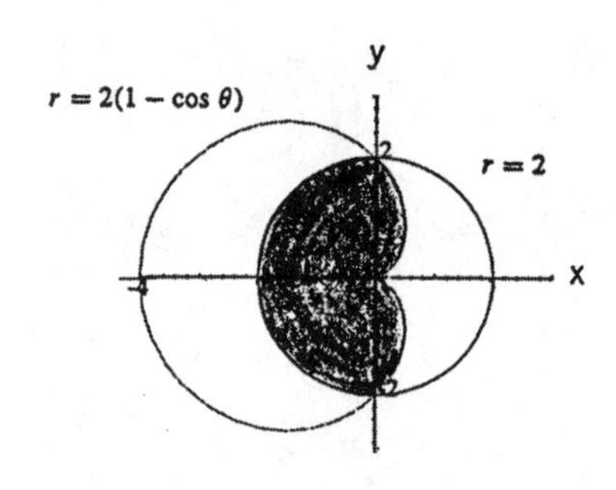

13. $r = \sqrt{3}$ and $r^2 = 6\cos 2\theta \Rightarrow 3 = 6\cos 2\theta \Rightarrow \cos 2\theta = \frac{1}{2}$
$\Rightarrow \theta = \frac{\pi}{6}$ (in the 1st quadrant); we use symmetry of the graph to find the area, so
$A = 4\int_0^{\pi/6} \left[\frac{1}{2}(6\cos 2\theta) - \frac{1}{2}\left(\sqrt{3}\right)^2\right] d\theta$
$= 2\int_0^{\pi/6} (6\cos 2\theta - 3)\, d\theta = 2[3\sin 2\theta - 3\theta]_0^{\pi/6}$
$= 3\sqrt{3} - \pi$

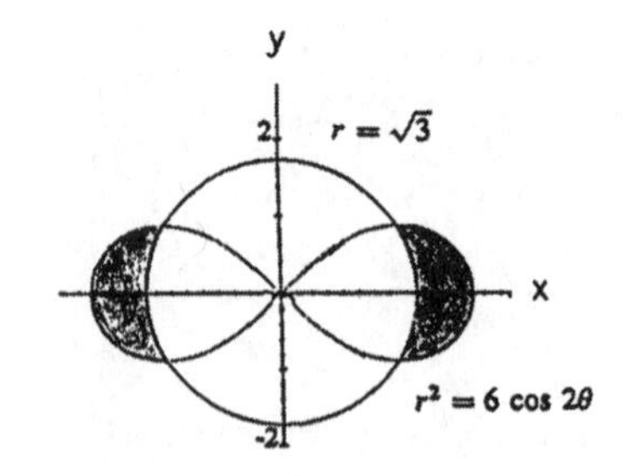

15. $r = 1$ and $r = -2\cos\theta \Rightarrow 1 = -2\cos\theta \Rightarrow \cos\theta = -\frac{1}{2}$
$\Rightarrow \theta = \frac{2\pi}{3}$ in quadrant II; therefore
$A = 2\int_{2\pi/3}^{\pi} \frac{1}{2}\left[(-2\cos\theta)^2 - 1^2\right] d\theta = \int_{2\pi/3}^{\pi} (4\cos^2\theta - 1)\, d\theta$
$= \int_{2\pi/3}^{\pi} [2(1 + \cos 2\theta) - 1]\, d\theta = \int_{2\pi/3}^{\pi} (1 + 2\cos 2\theta)\, d\theta$
$= [\theta + \sin 2\theta]_{2\pi/3}^{\pi} = \frac{\pi}{3} + \frac{\sqrt{3}}{2}$

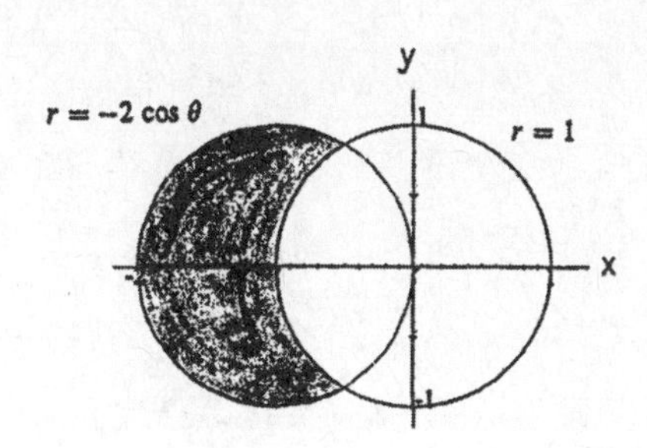

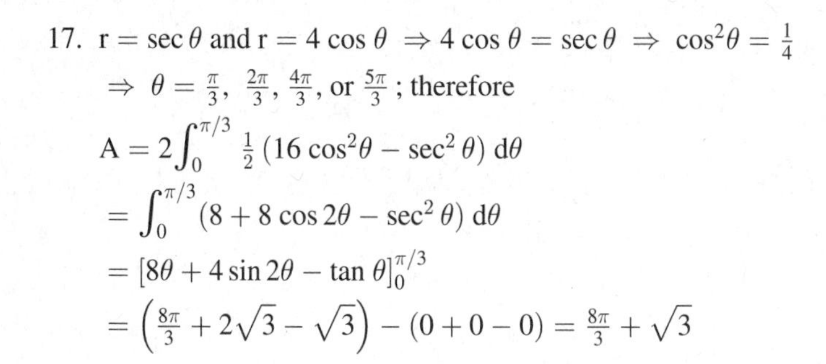

17. $r = \sec\theta$ and $r = 4\cos\theta \Rightarrow 4\cos\theta = \sec\theta \Rightarrow \cos^2\theta = \frac{1}{4}$
$\Rightarrow \theta = \frac{\pi}{3}, \frac{2\pi}{3}, \frac{4\pi}{3}$, or $\frac{5\pi}{3}$; therefore
$A = 2\int_0^{\pi/3} \frac{1}{2}\left(16\cos^2\theta - \sec^2\theta\right) d\theta$
$= \int_0^{\pi/3} \left(8 + 8\cos 2\theta - \sec^2\theta\right) d\theta$
$= [8\theta + 4\sin 2\theta - \tan\theta]_0^{\pi/3}$
$= \left(\frac{8\pi}{3} + 2\sqrt{3} - \sqrt{3}\right) - (0 + 0 - 0) = \frac{8\pi}{3} + \sqrt{3}$

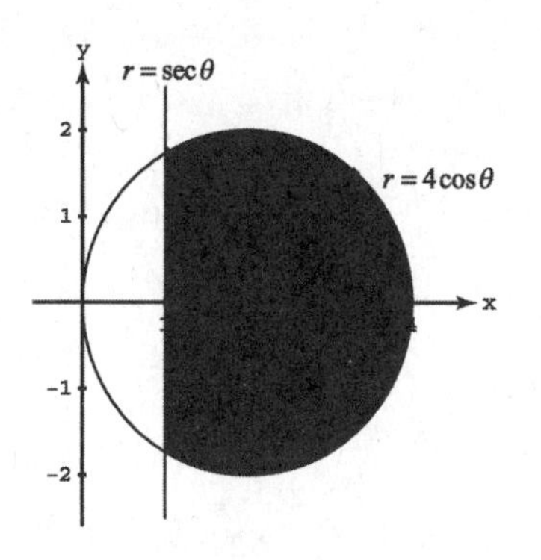

19. (a) $r = \tan\theta$ and $r = \left(\frac{\sqrt{2}}{2}\right)\csc\theta \Rightarrow \tan\theta = \left(\frac{\sqrt{2}}{2}\right)\csc\theta$
$\Rightarrow \sin^2\theta = \left(\frac{\sqrt{2}}{2}\right)\cos\theta \Rightarrow 1 - \cos^2\theta = \left(\frac{\sqrt{2}}{2}\right)\cos\theta$
$\Rightarrow \cos^2\theta + \left(\frac{\sqrt{2}}{2}\right)\cos\theta - 1 = 0 \Rightarrow \cos\theta = -\sqrt{2}$ or
$\frac{\sqrt{2}}{2}$ (use the quadratic formula) $\Rightarrow \theta = \frac{\pi}{4}$ (the solution in the first quadrant); therefore the area of R_1 is

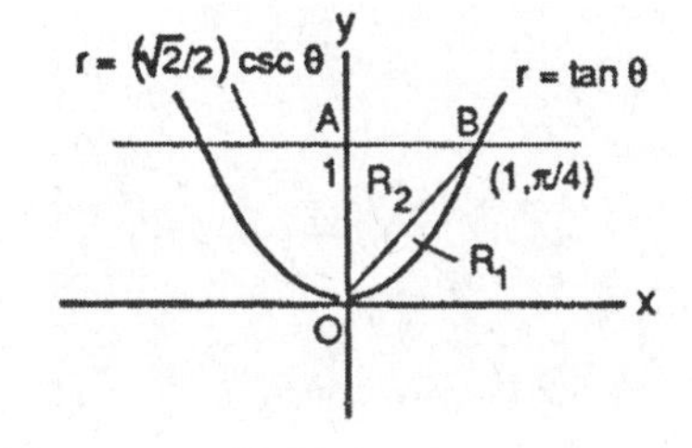

$A_1 = \int_0^{\pi/4} \frac{1}{2}\tan^2\theta\, d\theta = \frac{1}{2}\int_0^{\pi/4} (\sec^2\theta - 1)\, d\theta = \frac{1}{2}[\tan\theta - \theta]_0^{\pi/4} = \frac{1}{2}\left(\tan\frac{\pi}{4} - \frac{\pi}{4}\right) = \frac{1}{2} - \frac{\pi}{8}$; $AO = \left(\frac{\sqrt{2}}{2}\right)\csc\frac{\pi}{2}$
$= \frac{\sqrt{2}}{2}$ and $OB = \left(\frac{\sqrt{2}}{2}\right)\csc\frac{\pi}{4} = 1 \Rightarrow AB = \sqrt{1^2 - \left(\frac{\sqrt{2}}{2}\right)^2} = \frac{\sqrt{2}}{2} \Rightarrow$ the area of R_2 is $A_2 = \frac{1}{2}\left(\frac{\sqrt{2}}{2}\right)\left(\frac{\sqrt{2}}{2}\right) = \frac{1}{4}$;
therefore the area of the region shaded in the text is $2\left(\frac{1}{2} - \frac{\pi}{8} + \frac{1}{4}\right) = \frac{3}{2} - \frac{\pi}{4}$. Note: The area must be found this way since no common interval generates the region. For example, the interval $0 \le \theta \le \frac{\pi}{4}$ generates the arc OB of $r = \tan\theta$ but does not generate the segment AB of the liner $= \frac{\sqrt{2}}{2}\csc\theta$. Instead the interval generates the half-line from B to $+\infty$ on the line $r = \frac{\sqrt{2}}{2}\csc\theta$.

(b) $\lim_{\theta \to \pi/2^-} \tan\theta = \infty$ and the line $x = 1$ is $r = \sec\theta$ in polar coordinates; then $\lim_{\theta \to \pi/2^-} (\tan\theta - \sec\theta)$
$= \lim_{\theta \to \pi/2^-} \left(\frac{\sin\theta}{\cos\theta} - \frac{1}{\cos\theta}\right) = \lim_{\theta \to \pi/2^-} \left(\frac{\sin\theta - 1}{\cos\theta}\right) = \lim_{\theta \to \pi/2^-} \left(\frac{\cos\theta}{-\sin\theta}\right) = 0 \Rightarrow r = \tan\theta$ approaches

$r = \sec\theta$ as $\theta \to \frac{\pi^-}{2} \Rightarrow r = \sec\theta$ (or $x = 1$) is a vertical asymptote of $r = \tan\theta$. Similarly, $r = -\sec\theta$ (or $x = -1$) is a vertical asymptote of $r = \tan\theta$.

21. $r = \theta^2, 0 \le \theta \le \sqrt{5} \Rightarrow \frac{dr}{d\theta} = 2\theta$; therefore Length $= \int_0^{\sqrt{5}} \sqrt{(\theta^2)^2 + (2\theta)^2}\, d\theta = \int_0^{\sqrt{5}} \sqrt{\theta^4 + 4\theta^2}\, d\theta$
$= \int_0^{\sqrt{5}} |\theta| \sqrt{\theta^2 + 4}\, d\theta =$ (since $\theta \ge 0$) $\int_0^{\sqrt{5}} \theta\sqrt{\theta^2 + 4}\, d\theta$; $\left[u = \theta^2 + 4 \Rightarrow \frac{1}{2}\, du = \theta\, d\theta;\ \theta = 0 \Rightarrow u = 4,\right.$
$\left.\theta = \sqrt{5} \Rightarrow u = 9\right] \to \int_4^9 \frac{1}{2}\sqrt{u}\, du = \frac{1}{2}\left[\frac{2}{3}u^{3/2}\right]_4^9 = \frac{19}{3}$

23. $r = 1 + \cos\theta \Rightarrow \frac{dr}{d\theta} = -\sin\theta$; therefore Length $= \int_0^{2\pi} \sqrt{(1+\cos\theta)^2 + (-\sin\theta)^2}\, d\theta$
$= 2\int_0^{\pi} \sqrt{2+2\cos\theta}\, d\theta = 2\int_0^{\pi} \sqrt{\frac{4(1+\cos\theta)}{2}}\, d\theta = 4\int_0^{\pi} \sqrt{\frac{1+\cos\theta}{2}}\, d\theta = 4\int_0^{\pi} \cos\left(\frac{\theta}{2}\right) d\theta = 4\left[2\sin\frac{\theta}{2}\right]_0^{\pi} = 8$

25. $r = \frac{6}{1+\cos\theta},\ 0 \le \theta \le \frac{\pi}{2} \Rightarrow \frac{dr}{d\theta} = \frac{6\sin\theta}{(1+\cos\theta)^2}$; therefore Length $= \int_0^{\pi/2} \sqrt{\left(\frac{6}{1+\cos\theta}\right)^2 + \left(\frac{6\sin\theta}{(1+\cos\theta)^2}\right)^2}\, d\theta$
$= \int_0^{\pi/2} \sqrt{\frac{36}{(1+\cos\theta)^2} + \frac{36\sin^2\theta}{(1+\cos\theta)^4}}\, d\theta = 6\int_0^{\pi/2} \left|\frac{1}{1+\cos\theta}\right| \sqrt{1 + \frac{\sin^2\theta}{(1+\cos\theta)^2}}\, d\theta$
$= \left(\text{since } \frac{1}{1+\cos\theta} > 0 \text{ on } 0 \le \theta \le \frac{\pi}{2}\right)\ 6\int_0^{\pi/2} \left(\frac{1}{1+\cos\theta}\right) \sqrt{\frac{1+2\cos\theta+\cos^2\theta+\sin^2\theta}{(1+\cos\theta)^2}}\, d\theta$
$= 6\int_0^{\pi/2} \left(\frac{1}{1+\cos\theta}\right) \sqrt{\frac{2+2\cos\theta}{(1+\cos\theta)^2}}\, d\theta = 6\sqrt{2}\int_0^{\pi/2} \frac{d\theta}{(1+\cos\theta)^{3/2}} = 6\sqrt{2}\int_0^{\pi/2} \frac{d\theta}{\left(2\cos^2\frac{\theta}{2}\right)^{3/2}} = 3\int_0^{\pi/2} \left|\sec^3\frac{\theta}{2}\right| d\theta$
$= 3\int_0^{\pi/2} \sec^3\frac{\theta}{2}\, d\theta = 6\int_0^{\pi/4} \sec^3 u\, du = \text{(use tables) } 6\left(\left[\frac{\sec u \tan u}{2}\right]_0^{\pi/4} + \frac{1}{2}\int_0^{\pi/4} \sec u\, du\right)$
$= 6\left(\frac{1}{\sqrt{2}} + \left[\frac{1}{2}\ln|\sec u + \tan u|\right]_0^{\pi/4}\right) = 3\left[\sqrt{2} + \ln\left(1+\sqrt{2}\right)\right]$

27. $r = \cos^3\frac{\theta}{3} \Rightarrow \frac{dr}{d\theta} = -\sin\frac{\theta}{3}\cos^2\frac{\theta}{3}$; therefore Length $= \int_0^{\pi/4} \sqrt{\left(\cos^3\frac{\theta}{3}\right)^2 + \left(-\sin\frac{\theta}{3}\cos^2\frac{\theta}{3}\right)^2}\, d\theta$
$= \int_0^{\pi/4} \sqrt{\cos^6\left(\frac{\theta}{3}\right) + \sin^2\left(\frac{\theta}{3}\right)\cos^4\left(\frac{\theta}{3}\right)}\, d\theta = \int_0^{\pi/4} \left(\cos^2\frac{\theta}{3}\right) \sqrt{\cos^2\left(\frac{\theta}{3}\right) + \sin^2\left(\frac{\theta}{3}\right)}\, d\theta = \int_0^{\pi/4} \cos^2\left(\frac{\theta}{3}\right) d\theta$
$= \int_0^{\pi/4} \frac{1+\cos\left(\frac{2\theta}{3}\right)}{2}\, d\theta = \frac{1}{2}\left[\theta + \frac{3}{2}\sin\frac{2\theta}{3}\right]_0^{\pi/4} = \frac{\pi}{8} + \frac{3}{8}$

29. Let $r = f(\theta)$. Then $x = f(\theta)\cos\theta \Rightarrow \frac{dx}{d\theta} = f'(\theta)\cos\theta - f(\theta)\sin\theta \Rightarrow \left(\frac{dx}{d\theta}\right)^2 = [f'(\theta)\cos\theta - f(\theta)\sin\theta]^2$
$= [f'(\theta)]^2\cos^2\theta - 2f'(\theta)f(\theta)\sin\theta\cos\theta + [f(\theta)]^2\sin^2\theta$; $y = f(\theta)\sin\theta \Rightarrow \frac{dy}{d\theta} = f'(\theta)\sin\theta + f(\theta)\cos\theta$
$\Rightarrow \left(\frac{dy}{d\theta}\right)^2 = [f'(\theta)\sin\theta + f(\theta)\cos\theta]^2 = [f'(\theta)]^2\sin^2\theta + 2f'(\theta)f(\theta)\sin\theta\cos\theta + [f(\theta)]^2\cos^2\theta$. Therefore
$\left(\frac{dx}{d\theta}\right)^2 + \left(\frac{dy}{d\theta}\right)^2 = [f'(\theta)]^2(\cos^2\theta + \sin^2\theta) + [f(\theta)]^2(\cos^2\theta + \sin^2\theta) = [f'(\theta)]^2 + [f(\theta)]^2 = r^2 + \left(\frac{dr}{d\theta}\right)^2$.
Thus, $L = \int_\alpha^\beta \sqrt{\left(\frac{dx}{d\theta}\right)^2 + \left(\frac{dy}{d\theta}\right)^2}\, d\theta = \int_\alpha^\beta \sqrt{r^2 + \left(\frac{dr}{d\theta}\right)^2}\, d\theta.$

31. (a) $r_{av} = \frac{1}{2\pi - 0}\int_0^{2\pi} a(1-\cos\theta)\, d\theta = \frac{a}{2\pi}[\theta - \sin\theta]_0^{2\pi} = a$
(b) $r_{av} = \frac{1}{2\pi - 0}\int_0^{2\pi} a\, d\theta = \frac{1}{2\pi}[a\theta]_0^{2\pi} = a$
(c) $r_{av} = \frac{1}{\left(\frac{\pi}{2}\right) - \left(-\frac{\pi}{2}\right)}\int_{-\pi/2}^{\pi/2} a\cos\theta\, d\theta = \frac{1}{\pi}[a\sin\theta]_{-\pi/2}^{\pi/2} = \frac{2a}{\pi}$

11.6 CONIC SECTIONS

1. $x = \frac{y^2}{8} \Rightarrow 4p = 8 \Rightarrow p = 2$; focus is $(2, 0)$, directrix is $x = -2$

3. $y = -\frac{x^2}{6} \Rightarrow 4p = 6 \Rightarrow p = \frac{3}{2}$; focus is $\left(0, -\frac{3}{2}\right)$, directrix is $y = \frac{3}{2}$

5. $\frac{x^2}{4} - \frac{y^2}{9} = 1 \Rightarrow c = \sqrt{4+9} = \sqrt{13} \Rightarrow$ foci are $\left(\pm\sqrt{13}, 0\right)$; vertices are $(\pm 2, 0)$; asymptotes are $y = \pm\frac{3}{2}x$

7. $\frac{x^2}{2} + y^2 = 1 \Rightarrow c = \sqrt{2-1} = 1 \Rightarrow$ foci are $(\pm 1, 0)$; vertices are $\left(\pm\sqrt{2}, 0\right)$

9. $y^2 = 12x \Rightarrow x = \frac{y^2}{12} \Rightarrow 4p = 12 \Rightarrow p = 3;$
focus is $(3, 0)$, directrix is $x = -3$

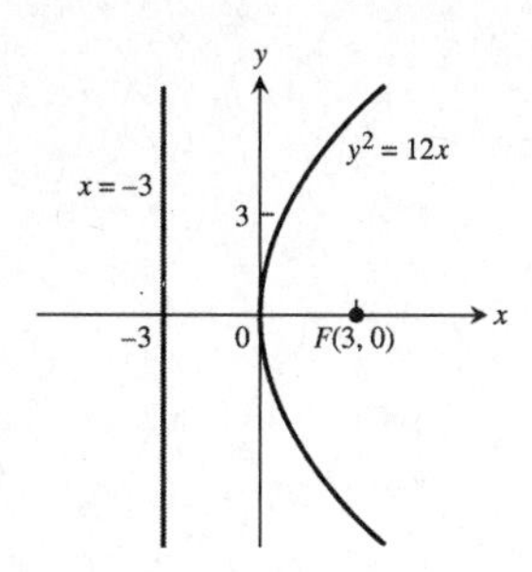

11. $x^2 = -8y \Rightarrow y = \frac{x^2}{-8} \Rightarrow 4p = 8 \Rightarrow p = 2;$
focus is $(0, -2)$, directrix is $y = 2$

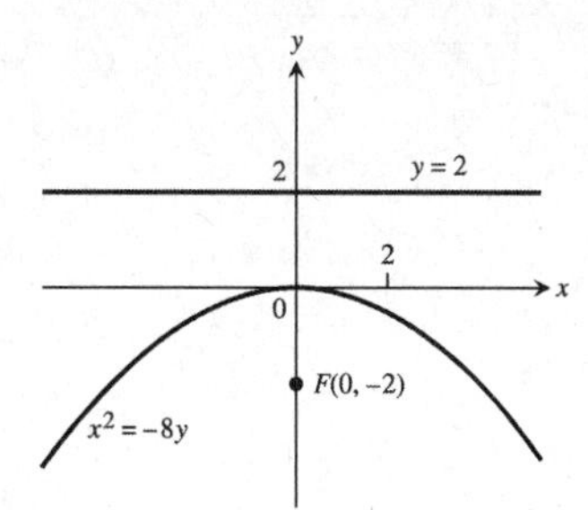

13. $y = 4x^2 \Rightarrow y = \frac{x^2}{\left(\frac{1}{4}\right)} \Rightarrow 4p = \frac{1}{4} \Rightarrow p = \frac{1}{16};$
focus is $\left(0, \frac{1}{16}\right)$, directrix is $y = -\frac{1}{16}$

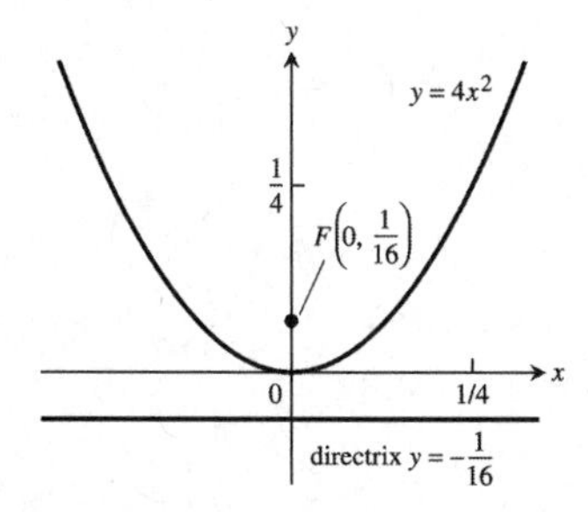

15. $x = -3y^2 \Rightarrow x = -\frac{y^2}{\left(\frac{1}{3}\right)} \Rightarrow 4p = \frac{1}{3} \Rightarrow p = \frac{1}{12};$
focus is $\left(-\frac{1}{12}, 0\right)$, directrix is $x = \frac{1}{12}$

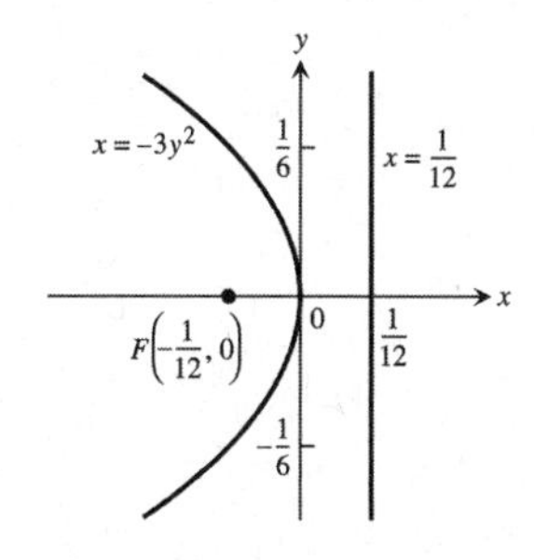

17. $16x^2 + 25y^2 = 400 \Rightarrow \frac{x^2}{25} + \frac{y^2}{16} = 1$
$\Rightarrow c = \sqrt{a^2 - b^2} = \sqrt{25 - 16} = 3$

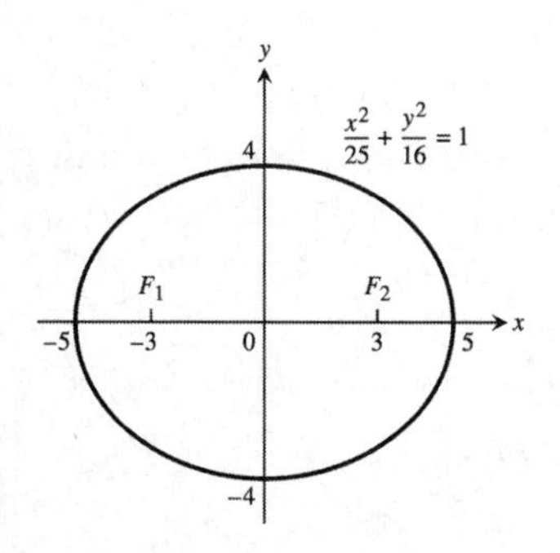

19. $2x^2 + y^2 = 2 \Rightarrow x^2 + \frac{y^2}{2} = 1$
$\Rightarrow c = \sqrt{a^2 - b^2} = \sqrt{2 - 1} = 1$

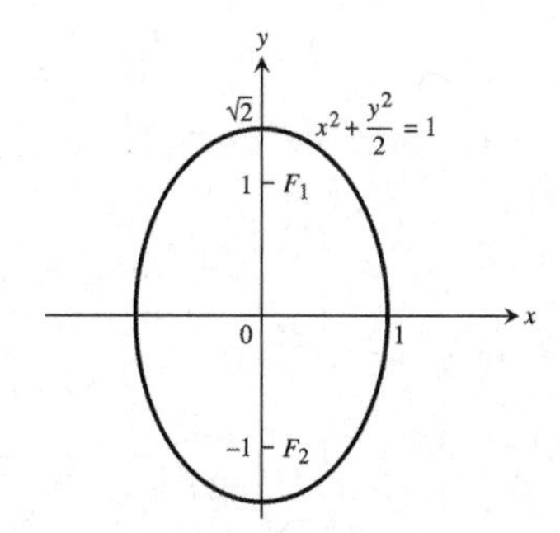

21. $3x^2 + 2y^2 = 6 \Rightarrow \frac{x^2}{2} + \frac{y^2}{3} = 1$
$\Rightarrow c = \sqrt{a^2 - b^2} = \sqrt{3 - 2} = 1$

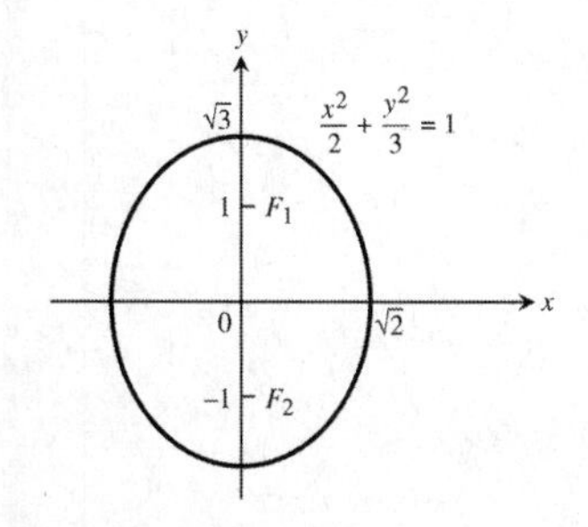

23. $6x^2 + 9y^2 = 54 \Rightarrow \frac{x^2}{9} + \frac{y^2}{6} = 1$
$\Rightarrow c = \sqrt{a^2 - b^2} = \sqrt{9 - 6} = \sqrt{3}$

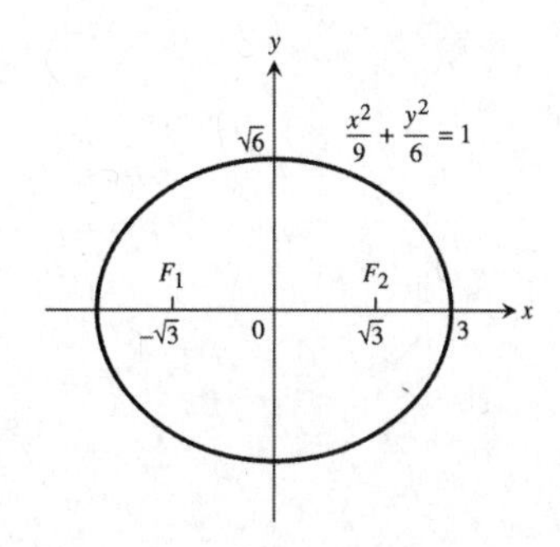

25. Foci: $\left(\pm\sqrt{2}, 0\right)$, Vertices: $(\pm 2, 0) \Rightarrow a = 2, c = \sqrt{2} \Rightarrow b^2 = a^2 - c^2 = 4 - \left(\sqrt{2}\right)^2 = 2 \Rightarrow \frac{x^2}{4} + \frac{y^2}{2} = 1$

27. $x^2 - y^2 = 1 \Rightarrow c = \sqrt{a^2 + b^2} = \sqrt{1 + 1} = \sqrt{2}$; asymptotes are $y = \pm x$

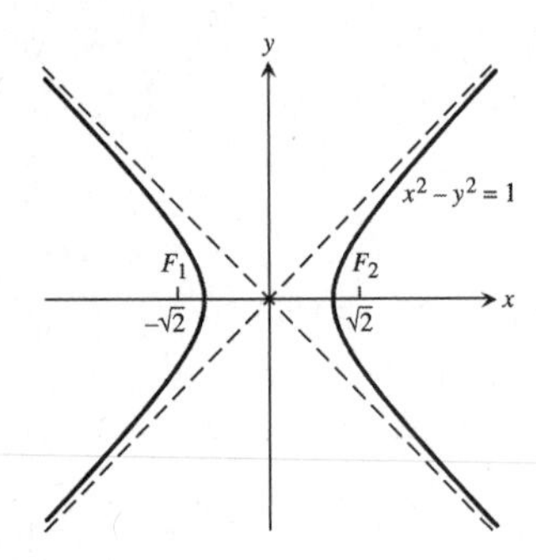

29. $y^2 - x^2 = 8 \Rightarrow \frac{y^2}{8} - \frac{x^2}{8} = 1 \Rightarrow c = \sqrt{a^2 + b^2} = \sqrt{8 + 8} = 4$; asymptotes are $y = \pm x$

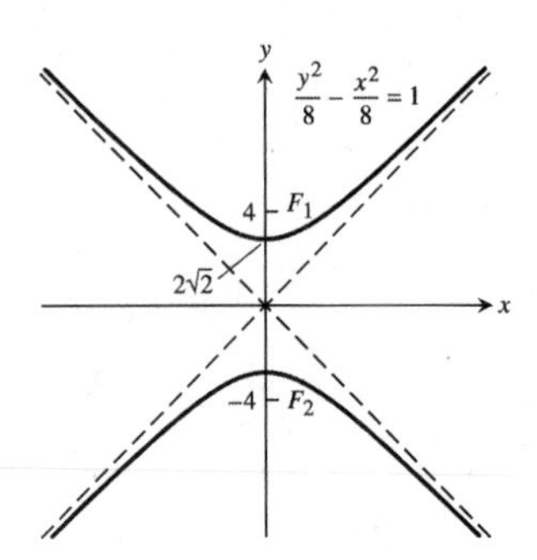

31. $8x^2 - 2y^2 = 16 \Rightarrow \frac{x^2}{2} - \frac{y^2}{8} = 1 \Rightarrow c = \sqrt{a^2 + b^2} = \sqrt{2 + 8} = \sqrt{10}$; asymptotes are $y = \pm 2x$

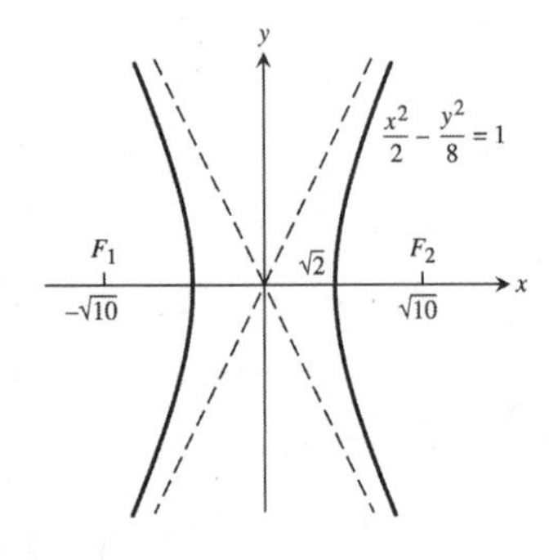

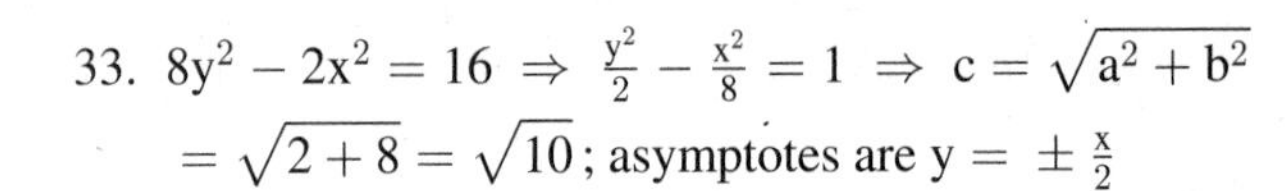
33. $8y^2 - 2x^2 = 16 \Rightarrow \frac{y^2}{2} - \frac{x^2}{8} = 1 \Rightarrow c = \sqrt{a^2 + b^2} = \sqrt{2 + 8} = \sqrt{10}$; asymptotes are $y = \pm \frac{x}{2}$

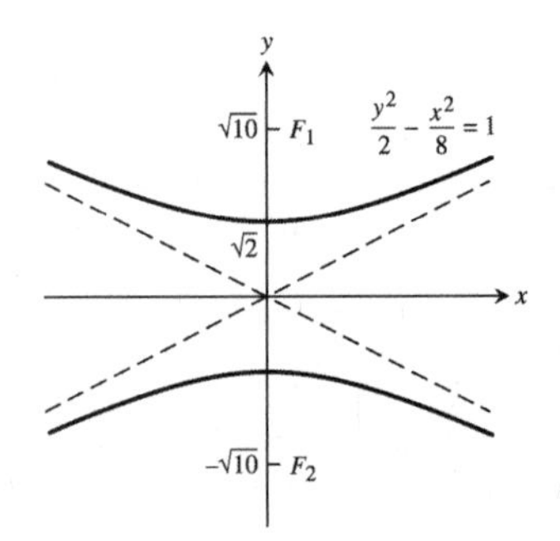

35. Foci: $\left(0, \pm\sqrt{2}\right)$, Asymptotes: $y = \pm x \Rightarrow c = \sqrt{2}$ and $\frac{a}{b} = 1 \Rightarrow a = b \Rightarrow c^2 = a^2 + b^2 = 2a^2 \Rightarrow 2 = 2a^2 \Rightarrow a = 1 \Rightarrow b = 1 \Rightarrow y^2 - x^2 = 1$

37. Vertices: $(\pm 3, 0)$, Asymptotes: $y = \pm \frac{4}{3}x \Rightarrow a = 3$ and $\frac{b}{a} = \frac{4}{3} \Rightarrow b = \frac{4}{3}(3) = 4 \Rightarrow \frac{x^2}{9} - \frac{y^2}{16} = 1$

39. (a) $y^2 = 8x \Rightarrow 4p = 8 \Rightarrow p = 2 \Rightarrow$ directrix is $x = -2$, focus is $(2, 0)$, and vertex is $(0, 0)$; therefore the new directrix is $x = -1$, the new focus is $(3, -2)$, and the new vertex is $(1, -2)$

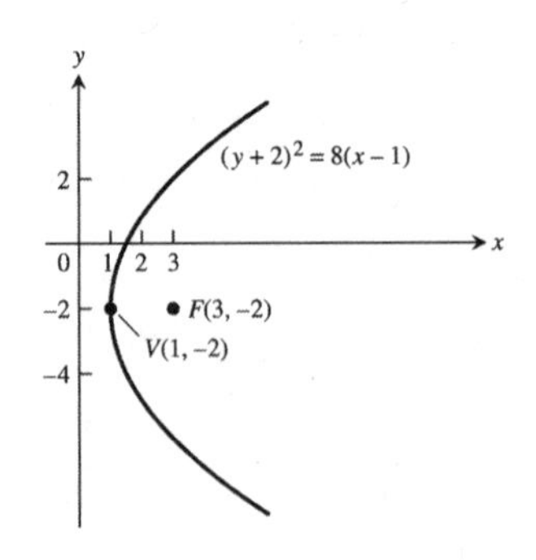

41. (a) $\frac{x^2}{16} + \frac{y^2}{9} = 1 \Rightarrow$ center is $(0,0)$, vertices are $(-4,0)$ and $(4,0)$; $c = \sqrt{a^2 - b^2} = \sqrt{7} \Rightarrow$ foci are $\left(\sqrt{7},0\right)$ and $\left(-\sqrt{7},0\right)$; therefore the new center is $(4,3)$, the new vertices are $(0,3)$ and $(8,3)$, and the new foci are $\left(4 \pm \sqrt{7},3\right)$

(b)

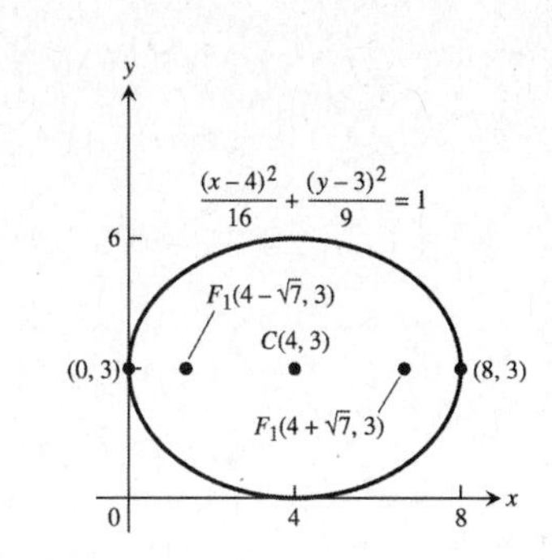

43. (a) $\frac{x^2}{16} - \frac{y^2}{9} = 1 \Rightarrow$ center is $(0,0)$, vertices are $(-4,0)$ and $(4,0)$, and the asymptotes are $\frac{x}{4} = \pm\frac{y}{3}$ or $y = \pm\frac{3x}{4}$; $c = \sqrt{a^2 + b^2} = \sqrt{25} = 5 \Rightarrow$ foci are $(-5,0)$ and $(5,0)$; therefore the new center is $(2,0)$, the new vertices are $(-2,0)$ and $(6,0)$, the new foci are $(-3,0)$ and $(7,0)$, and the new asymptotes are $y = \pm\frac{3(x-2)}{4}$

(b)

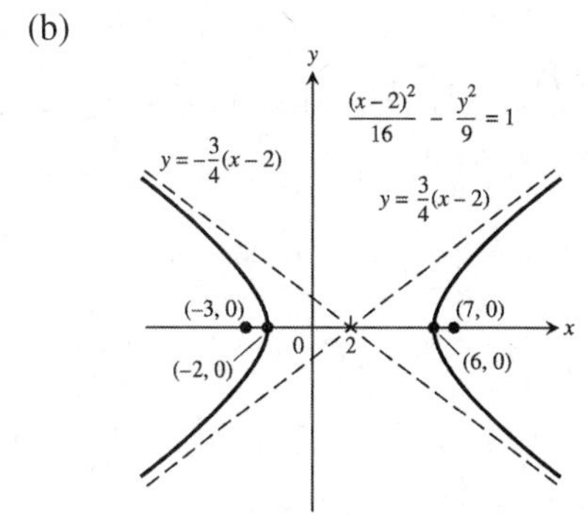

45. $y^2 = 4x \Rightarrow 4p = 4 \Rightarrow p = 1 \Rightarrow$ focus is $(1,0)$, directrix is $x = -1$, and vertex is $(0,0)$; therefore the new vertex is $(-2,-3)$, the new focus is $(-1,-3)$, and the new directrix is $x = -3$; the new equation is $(y+3)^2 = 4(x+2)$

47. $x^2 = 8y \Rightarrow 4p = 8 \Rightarrow p = 2 \Rightarrow$ focus is $(0,2)$, directrix is $y = -2$, and vertex is $(0,0)$; therefore the new vertex is $(1,-7)$, the new focus is $(1,-5)$, and the new directrix is $y = -9$; the new equation is $(x-1)^2 = 8(y+7)$

49. $\frac{x^2}{6} + \frac{y^2}{9} = 1 \Rightarrow$ center is $(0,0)$, vertices are $(0,3)$ and $(0,-3)$; $c = \sqrt{a^2 - b^2} = \sqrt{9-6} = \sqrt{3} \Rightarrow$ foci are $\left(0,\sqrt{3}\right)$ and $\left(0,-\sqrt{3}\right)$; therefore the new center is $(-2,-1)$, the new vertices are $(-2,2)$ and $(-2,-4)$, and the new foci are $\left(-2,-1 \pm \sqrt{3}\right)$; the new equation is $\frac{(x+2)^2}{6} + \frac{(y+1)^2}{9} = 1$

51. $\frac{x^2}{3} + \frac{y^2}{2} = 1 \Rightarrow$ center is $(0,0)$, vertices are $\left(\sqrt{3},0\right)$ and $\left(-\sqrt{3},0\right)$; $c = \sqrt{a^2 - b^2} = \sqrt{3-2} = 1 \Rightarrow$ foci are $(-1,0)$ and $(1,0)$; therefore the new center is $(2,3)$, the new vertices are $\left(2 \pm \sqrt{3},3\right)$, and the new foci are $(1,3)$ and $(3,3)$; the new equation is $\frac{(x-2)^2}{3} + \frac{(y-3)^2}{2} = 1$

53. $\frac{x^2}{4} - \frac{y^2}{5} = 1 \Rightarrow$ center is $(0,0)$, vertices are $(2,0)$ and $(-2,0)$; $c = \sqrt{a^2 + b^2} = \sqrt{4+5} = 3 \Rightarrow$ foci are $(3,0)$ and $(-3,0)$; the asymptotes are $\pm\frac{x}{2} = \frac{y}{\sqrt{5}} \Rightarrow y = \pm\frac{\sqrt{5}x}{2}$; therefore the new center is $(2,2)$, the new vertices are $(4,2)$ and $(0,2)$, and the new foci are $(5,2)$ and $(-1,2)$; the new asymptotes are $y - 2 = \pm\frac{\sqrt{5}(x-2)}{2}$; the new equation is $\frac{(x-2)^2}{4} - \frac{(y-2)^2}{5} = 1$

55. $y^2 - x^2 = 1 \Rightarrow$ center is $(0,0)$, vertices are $(0,1)$ and $(0,-1)$; $c = \sqrt{a^2 + b^2} = \sqrt{1+1} = \sqrt{2} \Rightarrow$ foci are $\left(0, \pm\sqrt{2}\right)$; the asymptotes are $y = \pm x$; therefore the new center is $(-1,-1)$, the new vertices are $(-1,0)$ and

$(-1,-2)$, and the new foci are $\left(-1,-1 \pm \sqrt{2}\right)$; the new asymptotes are $y+1 = \pm(x+1)$; the new equation is $(y+1)^2 - (x+1)^2 = 1$

57. $x^2 + 4x + y^2 = 12 \Rightarrow x^2 + 4x + 4 + y^2 = 12 + 4 \Rightarrow (x+2)^2 + y^2 = 16$; this is a circle: center at $C(-2,0)$, $a = 4$

59. $x^2 + 2x + 4y - 3 = 0 \Rightarrow x^2 + 2x + 1 = -4y + 3 + 1 \Rightarrow (x+1)^2 = -4(y-1)$; this is a parabola: $V(-1,1)$, $F(-1,0)$

61. $x^2 + 5y^2 + 4x = 1 \Rightarrow x^2 + 4x + 4 + 5y^2 = 5 \Rightarrow (x+2)^2 + 5y^2 = 5 \Rightarrow \frac{(x+2)^2}{5} + y^2 = 1$; this is an ellipse: the center is $(-2,0)$, the vertices are $\left(-2 \pm \sqrt{5}, 0\right)$; $c = \sqrt{a^2 - b^2} = \sqrt{5-1} = 2 \Rightarrow$ the foci are $(-4,0)$ and $(0,0)$

63. $x^2 + 2y^2 - 2x - 4y = -1 \Rightarrow x^2 - 2x + 1 + 2(y^2 - 2y + 1) = 2 \Rightarrow (x-1)^2 + 2(y-1)^2 = 2$ $\Rightarrow \frac{(x-1)^2}{2} + (y-1)^2 = 1$; this is an ellipse: the center is $(1,1)$, the vertices are $\left(1 \pm \sqrt{2}, 1\right)$; $c = \sqrt{a^2 - b^2} = \sqrt{2-1} = 1 \Rightarrow$ the foci are $(2,1)$ and $(0,1)$

65. $x^2 - y^2 - 2x + 4y = 4 \Rightarrow x^2 - 2x + 1 - (y^2 - 4y + 4) = 1 \Rightarrow (x-1)^2 - (y-2)^2 = 1$; this is a hyperbola: the center is $(1,2)$, the vertices are $(2,2)$ and $(0,2)$; $c = \sqrt{a^2 + b^2} = \sqrt{1+1} = \sqrt{2} \Rightarrow$ the foci are $\left(1 \pm \sqrt{2}, 2\right)$; the asymptotes are $y - 2 = \pm(x-1)$

67. $2x^2 - y^2 + 6y = 3 \Rightarrow 2x^2 - (y^2 - 6y + 9) = -6 \Rightarrow \frac{(y-3)^2}{6} - \frac{x^2}{3} = 1$; this is a hyperbola: the center is $(0,3)$, the vertices are $\left(0, 3 \pm \sqrt{6}\right)$; $c = \sqrt{a^2 + b^2} = \sqrt{6+3} = 3 \Rightarrow$ the foci are $(0,6)$ and $(0,0)$; the asymptotes are $\frac{y-3}{\sqrt{6}} = \pm \frac{x}{\sqrt{3}} \Rightarrow y = \pm\sqrt{2}x + 3$

69. (a) $y^2 = kx \Rightarrow x = \frac{y^2}{k}$; the volume of the solid formed by revolving R_1 about the y-axis is $V_1 = \int_0^{\sqrt{kx}} \pi \left(\frac{y^2}{k}\right)^2 dy$ $= \frac{\pi}{k^2} \int_0^{\sqrt{kx}} y^4\, dy = \frac{\pi x^2 \sqrt{kx}}{5}$; the volume of the right circular cylinder formed by revolving PQ about the y-axis is $V_2 = \pi x^2 \sqrt{kx} \Rightarrow$ the volume of the solid formed by revolving R_2 about the y-axis is $V_3 = V_2 - V_1 = \frac{4\pi x^2 \sqrt{kx}}{5}$. Therefore we can see the ratio of V_3 to V_1 is 4:1.

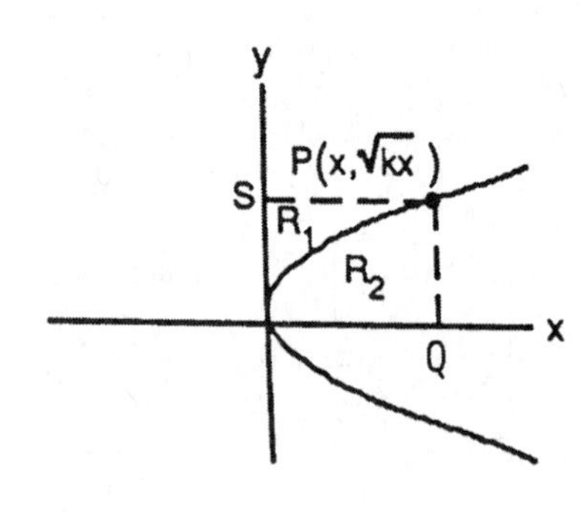

(b) The volume of the solid formed by revolving R_2 about the x-axis is $V_1 = \int_0^x \pi\left(\sqrt{kt}\right)^2 dt = \pi k \int_0^x t\, dt$ $= \frac{\pi k x^2}{2}$. The volume of the right circular cylinder formed by revolving PS about the x-axis is $V_2 = \pi\left(\sqrt{kx}\right)^2 x = \pi k x^2 \Rightarrow$ the volume of the solid formed by revolving R_1 about the x-axis is $V_3 = V_2 - V_1 = \pi k x^2 - \frac{\pi k x^2}{2} = \frac{\pi k x^2}{2}$. Therefore the ratio of V_3 to V_1 is 1:1.

71. $x^2 = 4py$ and $y = p \Rightarrow x^2 = 4p^2 \Rightarrow x = \pm 2p$. Therefore the line $y = p$ cuts the parabola at points $(-2p, p)$ and $(2p, p)$, and these points are $\sqrt{[2p - (-2p)]^2 + (p-p)^2} = 4p$ units apart.

73. Let $y = \sqrt{1 - \frac{x^2}{4}}$ on the interval $0 \le x \le 2$. The area of the inscribed rectangle is given by $A(x) = 2x\left(2\sqrt{1 - \frac{x^2}{4}}\right) = 4x\sqrt{1 - \frac{x^2}{4}}$ (since the length is 2x and the height is 2y)

$\Rightarrow A'(x) = 4\sqrt{1-\frac{x^2}{4}} - \frac{x^2}{\sqrt{1-\frac{x^2}{4}}}$. Thus $A'(x) = 0 \Rightarrow 4\sqrt{1-\frac{x^2}{4}} - \frac{x^2}{\sqrt{1-\frac{x^2}{4}}} = 0 \Rightarrow 4\left(1-\frac{x^2}{4}\right) - x^2 = 0 \Rightarrow x^2 = 2$

$\Rightarrow x = \sqrt{2}$ (only the positive square root lies in the interval). Since $A(0) = A(2) = 0$ we have that $A\left(\sqrt{2}\right) = 4$ is the maximum area when the length is $2\sqrt{2}$ and the height is $\sqrt{2}$.

75. $9x^2 - 4y^2 = 36 \Rightarrow y^2 = \frac{9x^2-36}{4} \Rightarrow y = \pm\frac{3}{2}\sqrt{x^2-4}$ on the interval $2 \le x \le 4 \Rightarrow V = \int_2^4 \pi\left(\frac{3}{2}\sqrt{x^2-4}\right)^2 dx$

$= \frac{9\pi}{4}\int_2^4 (x^2-4)\,dx = \frac{9\pi}{4}\left[\frac{x^3}{3} - 4x\right]_2^4 = \frac{9\pi}{4}\left[\left(\frac{64}{3}-16\right) - \left(\frac{8}{3}-8\right)\right] = \frac{9\pi}{4}\left(\frac{56}{3}-8\right) = \frac{3\pi}{4}(56-24) = 24\pi$

77. $(x-2)^2 + (y-1)^2 = 5 \Rightarrow 2(x-2) + 2(y-1)\frac{dy}{dx} = 0 \Rightarrow \frac{dy}{dx} = -\frac{x-2}{y-1}$; $y = 0 \Rightarrow (x-2)^2 + (0-1)^2 = 5$

$\Rightarrow (x-2)^2 = 4 \Rightarrow x = 4$ or $x = 0 \Rightarrow$ the circle crosses the x-axis at $(4,0)$ and $(0,0)$; $x = 0$

$\Rightarrow (0-2)^2 + (y-1)^2 = 5 \Rightarrow (y-1)^2 = 1 \Rightarrow y = 2$ or $y = 0 \Rightarrow$ the circle crosses the y-axis at $(0,2)$ and $(0,0)$.

At $(4,0)$: $\frac{dy}{dx} = -\frac{4-2}{0-1} = 2 \Rightarrow$ the tangent line is $y = 2(x-4)$ or $y = 2x-8$

At $(0,0)$: $\frac{dy}{dx} = -\frac{0-2}{0-1} = -2 \Rightarrow$ the tangent line is $y = -2x$

At $(0,2)$: $\frac{dy}{dx} = -\frac{0-2}{2-1} = 2 \Rightarrow$ the tangent line is $y - 2 = 2x$ or $y = 2x+2$

79. Let $y = \sqrt{16 - \frac{16}{9}x^2}$ on the interval $-3 \le x \le 3$. Since the plate is symmetric about the y-axis, $\overline{x} = 0$. For a vertical strip: $(\widetilde{x}, \widetilde{y}) = \left(x, \frac{\sqrt{16-\frac{16}{9}x^2}}{2}\right)$, length $= \sqrt{16-\frac{16}{9}x^2}$, width $= dx \Rightarrow$ area $= dA = \sqrt{16-\frac{16}{9}x^2}\,dx$

$\Rightarrow$ mass $= dm = \delta\,dA = \delta\sqrt{16-\frac{16}{9}x^2}\,dx$. Moment of the strip about the x-axis:

$\widetilde{y}\,dm = \frac{\sqrt{16-\frac{16}{9}x^2}}{2}\left(\delta\sqrt{16-\frac{16}{9}x^2}\right)dx = \delta\left(8-\frac{8}{9}x^2\right)dx$ so the moment of the plate about the x-axis is

$M_x = \int \widetilde{y}\,dm = \int_{-3}^{3}\delta\left(8-\frac{8}{9}x^2\right)dx = \delta\left[8x - \frac{8}{27}x^3\right]_{-3}^{3} = 32\delta$; also the mass of the plate is

$M = \int_{-3}^{3}\delta\sqrt{16-\frac{16}{9}x^2}\,dx = \int_{-3}^{3}4\delta\sqrt{1-\left(\frac{1}{3}x\right)^2}\,dx = 4\delta\int_{-1}^{1}3\sqrt{1-u^2}\,du$ where $u = \frac{x}{3} \Rightarrow 3\,du = dx$; $x = -3$

$\Rightarrow u = -1$ and $x = 3 \Rightarrow u = 1$. Hence, $4\delta\int_{-1}^{1}3\sqrt{1-u^2}\,du = 12\delta\int_{-1}^{1}\sqrt{1-u^2}\,du$

$= 12\delta\left[\frac{1}{2}\left(u\sqrt{1-u^2} + \sin^{-1}u\right)\right]_{-1}^{1} = 6\pi\delta \Rightarrow \overline{y} = \frac{M_x}{M} = \frac{32\delta}{6\pi\delta} = \frac{16}{3\pi}$. Therefore the center of mass is $\left(0, \frac{16}{3\pi}\right)$.

81. (a) $\tan\beta = m_L \Rightarrow \tan\beta = f'(x_0)$ where $f(x) = \sqrt{4px}$;

$f'(x) = \frac{1}{2}(4px)^{-1/2}(4p) = \frac{2p}{\sqrt{4px}} \Rightarrow f'(x_0) = \frac{2p}{\sqrt{4px_0}}$

$= \frac{2p}{y_0} \Rightarrow \tan\beta = \frac{2p}{y_0}$.

(b) $\tan\phi = m_{FP} = \frac{y_0 - 0}{x_0 - p} = \frac{y_0}{x_0 - p}$

(c) $\tan\alpha = \frac{\tan\phi - \tan\beta}{1 + \tan\phi\tan\beta} = \frac{\left(\frac{y_0}{x_0-p} - \frac{2p}{y_0}\right)}{1 + \left(\frac{y_0}{x_0-p}\right)\left(\frac{2p}{y_0}\right)}$

$= \frac{y_0^2 - 2p(x_0-p)}{y_0(x_0-p+2p)} = \frac{4px_0 - 2px_0 + 2p^2}{y_0(x_0+p)} = \frac{2p(x_0+p)}{y_0(x_0+p)} = \frac{2p}{y_0}$

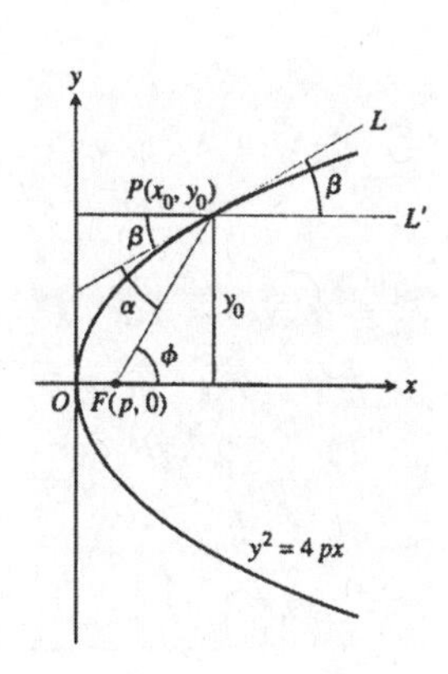

11.7 CONICS IN POLAR COORDINATES

1. $16x^2 + 25y^2 = 400 \Rightarrow \frac{x^2}{25} + \frac{y^2}{16} = 1 \Rightarrow c = \sqrt{a^2 - b^2}$
$= \sqrt{25 - 16} = 3 \Rightarrow e = \frac{c}{a} = \frac{3}{5}$; $F(\pm 3, 0)$;
directrices are $x = 0 \pm \frac{a}{e} = \pm \frac{5}{\left(\frac{3}{5}\right)} = \pm \frac{25}{3}$

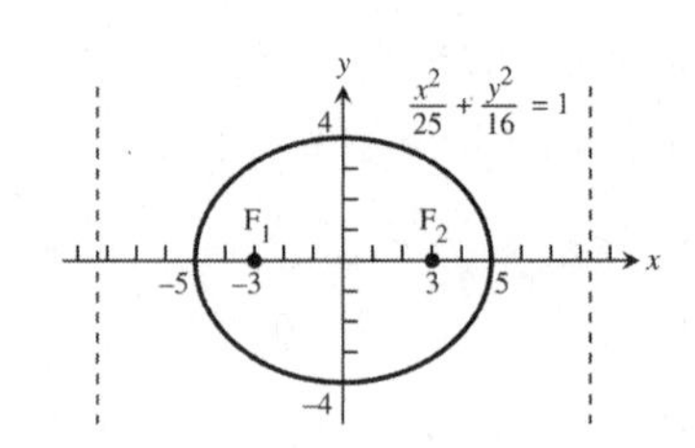

3. $2x^2 + y^2 = 2 \Rightarrow x^2 + \frac{y^2}{2} = 1 \Rightarrow c = \sqrt{a^2 - b^2}$
$= \sqrt{2 - 1} = 1 \Rightarrow e = \frac{c}{a} = \frac{1}{\sqrt{2}}$; $F(0, \pm 1)$;
directrices are $y = 0 \pm \frac{a}{e} = \pm \frac{\sqrt{2}}{\left(\frac{1}{\sqrt{2}}\right)} = \pm 2$

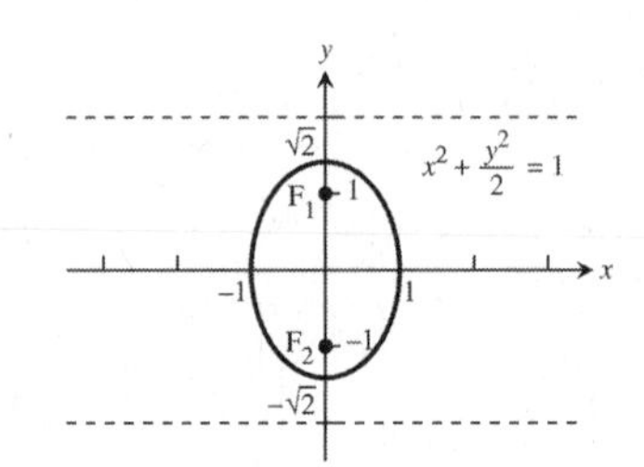

5. $3x^2 + 2y^2 = 6 \Rightarrow \frac{x^2}{2} + \frac{y^2}{3} = 1 \Rightarrow c = \sqrt{a^2 - b^2}$
$= \sqrt{3 - 2} = 1 \Rightarrow e = \frac{c}{a} = \frac{1}{\sqrt{3}}$; $F(0, \pm 1)$;
directrices are $y = 0 \pm \frac{a}{e} = \pm \frac{\sqrt{3}}{\left(\frac{1}{\sqrt{3}}\right)} = \pm 3$

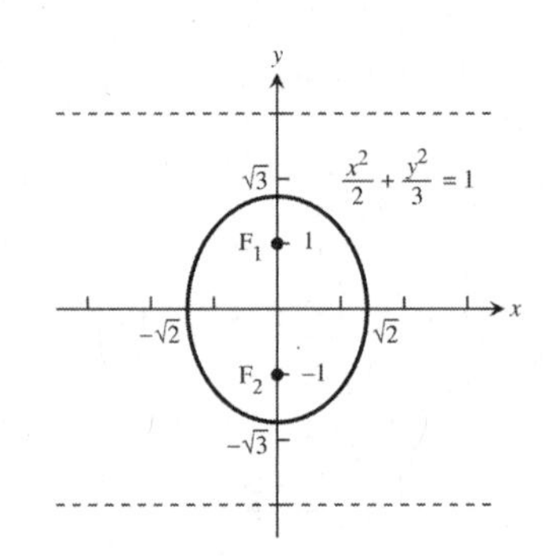

7. $6x^2 + 9y^2 = 54 \Rightarrow \frac{x^2}{9} + \frac{y^2}{6} = 1 \Rightarrow c = \sqrt{a^2 - b^2}$
$= \sqrt{9 - 6} = \sqrt{3} \Rightarrow e = \frac{c}{a} = \frac{\sqrt{3}}{3}$; $F\left(\pm\sqrt{3}, 0\right)$;
directrices are $x = 0 \pm \frac{a}{e} = \pm \frac{3}{\left(\frac{\sqrt{3}}{3}\right)} = \pm 3\sqrt{3}$

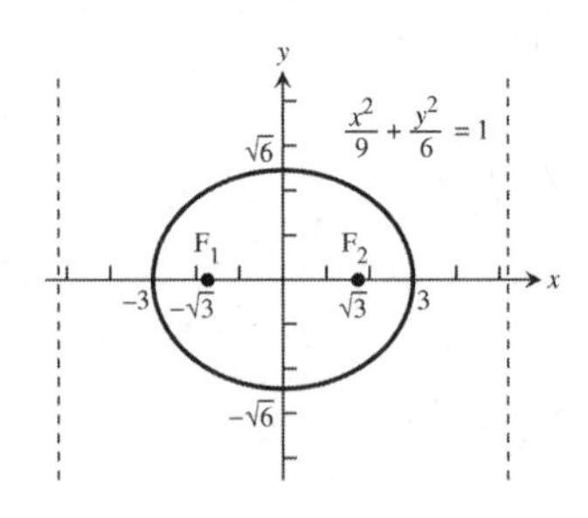

9. Foci: $(0, \pm 3)$, $e = 0.5 \Rightarrow c = 3$ and $a = \frac{c}{e} = \frac{3}{0.5} = 6 \Rightarrow b^2 = 36 - 9 = 27 \Rightarrow \frac{x^2}{27} + \frac{y^2}{36} = 1$

11. Vertices: $(0, \pm 70)$, $e = 0.1 \Rightarrow a = 70$ and $c = ae = 70(0.1) = 7 \Rightarrow b^2 = 4900 - 49 = 4851 \Rightarrow \frac{x^2}{4851} + \frac{y^2}{4900} = 1$

13. Focus: $\left(\sqrt{5}, 0\right)$, Directrix: $x = \frac{9}{\sqrt{5}} \Rightarrow c = ae = \sqrt{5}$ and $\frac{a}{e} = \frac{9}{\sqrt{5}} \Rightarrow \frac{ae}{e^2} = \frac{9}{\sqrt{5}} \Rightarrow \frac{\sqrt{5}}{e^2} = \frac{9}{\sqrt{5}} \Rightarrow e^2 = \frac{5}{9}$
$\Rightarrow e = \frac{\sqrt{5}}{3}$. Then $PF = \frac{\sqrt{5}}{3} PD \Rightarrow \sqrt{\left(x - \sqrt{5}\right)^2 + (y - 0)^2} = \frac{\sqrt{5}}{3}\left|x - \frac{9}{\sqrt{5}}\right| \Rightarrow \left(x - \sqrt{5}\right)^2 + y^2 = \frac{5}{9}\left(x - \frac{9}{\sqrt{5}}\right)^2$
$\Rightarrow x^2 - 2\sqrt{5}x + 5 + y^2 = \frac{5}{9}\left(x^2 - \frac{18}{\sqrt{5}}x + \frac{81}{5}\right) \Rightarrow \frac{4}{9}x^2 + y^2 = 4 \Rightarrow \frac{x^2}{9} + \frac{y^2}{4} = 1$

15. Focus: $(-4, 0)$, Directrix: $x = -16 \Rightarrow c = ae = 4$ and $\frac{a}{e} = 16 \Rightarrow \frac{ae}{e^2} = 16 \Rightarrow \frac{4}{e^2} = 16 \Rightarrow e^2 = \frac{1}{4} \Rightarrow e = \frac{1}{2}$. Then
$PF = \frac{1}{2} PD \Rightarrow \sqrt{(x + 4)^2 + (y - 0)^2} = \frac{1}{2}|x + 16| \Rightarrow (x + 4)^2 + y^2 = \frac{1}{4}(x + 16)^2 \Rightarrow x^2 + 8x + 16 + y^2$

$= \frac{1}{4}(x^2 + 32x + 256) \Rightarrow \frac{3}{4}x^2 + y^2 = 48 \Rightarrow \frac{x^2}{64} + \frac{y^2}{48} = 1$

17. $x^2 - y^2 = 1 \Rightarrow c = \sqrt{a^2 + b^2} = \sqrt{1+1} = \sqrt{2} \Rightarrow e = \frac{c}{a}$
$= \frac{\sqrt{2}}{1} = \sqrt{2}$; asymptotes are $y = \pm x$; $F\left(\pm\sqrt{2}, 0\right)$;
directrices are $x = 0 \pm \frac{a}{e} = \pm \frac{1}{\sqrt{2}}$

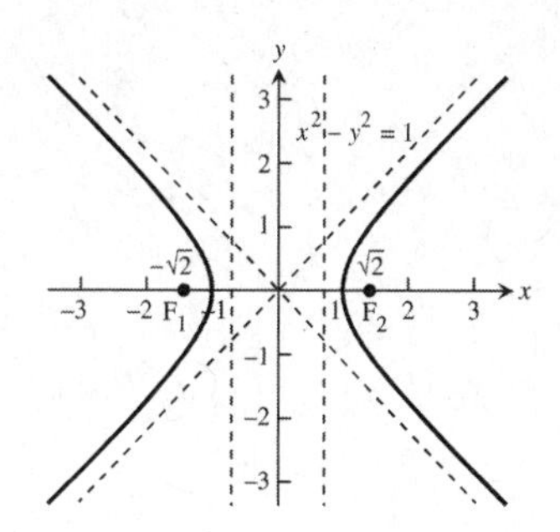

19. $y^2 - x^2 = 8 \Rightarrow \frac{y^2}{8} - \frac{x^2}{8} = 1 \Rightarrow c = \sqrt{a^2 + b^2}$
$= \sqrt{8+8} = 4 \Rightarrow e = \frac{c}{a} = \frac{4}{\sqrt{8}} = \sqrt{2}$; asymptotes are
$y = \pm x$; $F(0, \pm 4)$; directrices are $y = 0 \pm \frac{a}{e}$
$= \pm \frac{\sqrt{8}}{\sqrt{2}} = \pm 2$

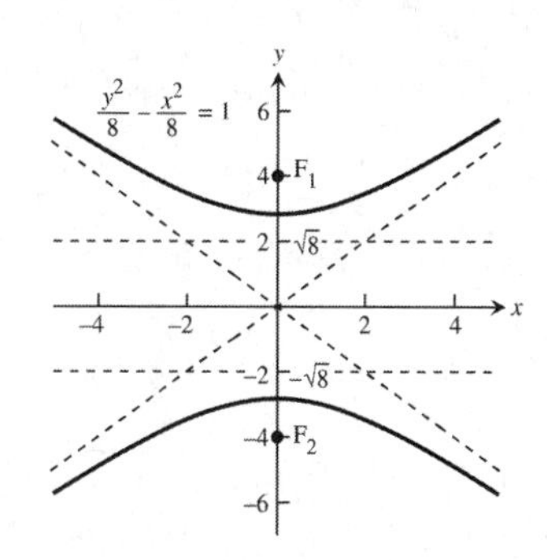

21. $8x^2 - 2y^2 = 16 \Rightarrow \frac{x^2}{2} - \frac{y^2}{8} = 1 \Rightarrow c = \sqrt{a^2 + b^2}$
$= \sqrt{2+8} = \sqrt{10} \Rightarrow e = \frac{c}{a} = \frac{\sqrt{10}}{\sqrt{2}} = \sqrt{5}$; asymptotes
are $y = \pm 2x$; $F\left(\pm\sqrt{10}, 0\right)$; directrices are $x = 0 \pm \frac{a}{e}$
$= \pm \frac{\sqrt{2}}{\sqrt{5}} = \pm \frac{2}{\sqrt{10}}$

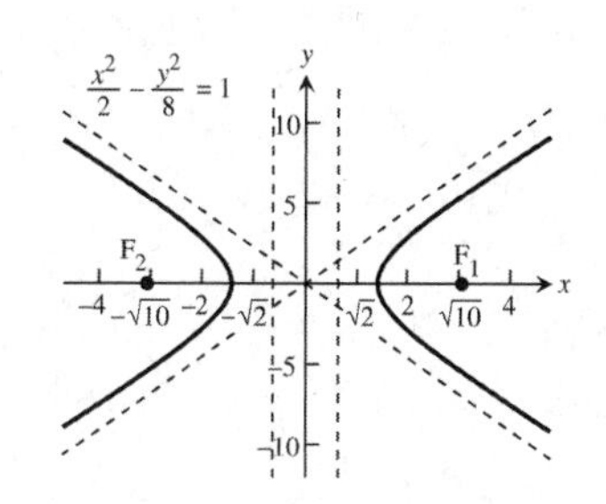

23. $8y^2 - 2x^2 = 16 \Rightarrow \frac{y^2}{2} - \frac{x^2}{8} = 1 \Rightarrow c = \sqrt{a^2 + b^2}$
$= \sqrt{2+8} = \sqrt{10} \Rightarrow e = \frac{c}{a} = \frac{\sqrt{10}}{\sqrt{2}} = \sqrt{5}$; asymptotes
are $y = \pm \frac{x}{2}$; $F\left(0, \pm\sqrt{10}\right)$; directrices are $y = 0 \pm \frac{a}{e}$
$= \pm \frac{\sqrt{2}}{\sqrt{5}} = \pm \frac{2}{\sqrt{10}}$

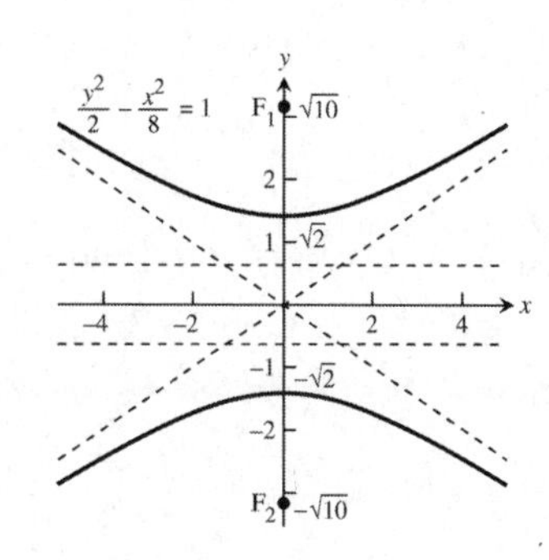

25. Vertices $(0, \pm 1)$ and $e = 3 \Rightarrow a = 1$ and $e = \frac{c}{a} = 3 \Rightarrow c = 3a = 3 \Rightarrow b^2 = c^2 - a^2 = 9 - 1 = 8 \Rightarrow y^2 - \frac{x^2}{8} = 1$

27. Foci $(\pm 3, 0)$ and $e = 3 \Rightarrow c = 3$ and $e = \frac{c}{a} = 3 \Rightarrow c = 3a \Rightarrow a = 1 \Rightarrow b^2 = c^2 - a^2 = 9 - 1 = 8 \Rightarrow x^2 - \frac{y^2}{8} = 1$

29. $e = 1, x = 2 \Rightarrow k = 2 \Rightarrow r = \frac{2(1)}{1+(1)\cos\theta} = \frac{2}{1+\cos\theta}$

31. $e = 5, y = -6 \Rightarrow k = 6 \Rightarrow r = \frac{6(5)}{1 - 5\sin\theta} = \frac{30}{1-5\sin\theta}$

33. $e = \frac{1}{2}, x = 1 \Rightarrow k = 1 \Rightarrow r = \frac{\left(\frac{1}{2}\right)(1)}{1 + \left(\frac{1}{2}\right)\cos\theta} = \frac{1}{2 + \cos\theta}$

35. $e = \frac{1}{5}, y = -10 \Rightarrow k = 10 \Rightarrow r = \frac{\left(\frac{1}{5}\right)(10)}{1 - \left(\frac{1}{5}\right)\sin\theta} = \frac{10}{5 - \sin\theta}$

37. $r = \frac{1}{1 + \cos\theta} \Rightarrow e = 1, k = 1 \Rightarrow x = 1$

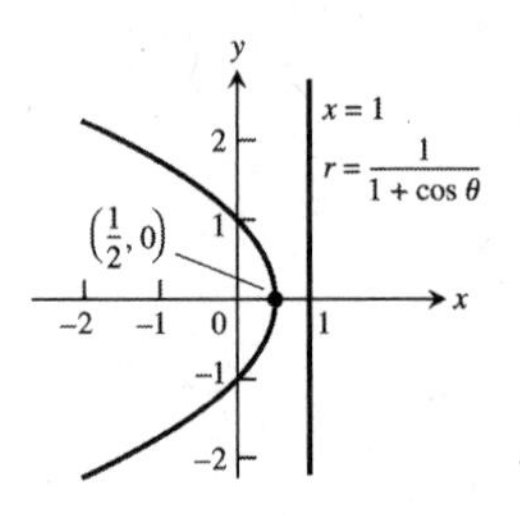

39. $r = \frac{25}{10 - 5\cos\theta} \Rightarrow r = \frac{\left(\frac{25}{10}\right)}{1 - \left(\frac{5}{10}\right)\cos\theta} = \frac{\left(\frac{5}{2}\right)}{1 - \left(\frac{1}{2}\right)\cos\theta}$
$\Rightarrow e = \frac{1}{2}, k = 5 \Rightarrow x = -5; a\left(1 - e^2\right) = ke$
$\Rightarrow a\left[1 - \left(\frac{1}{2}\right)^2\right] = \frac{5}{2} \Rightarrow \frac{3}{4}a = \frac{5}{2} \Rightarrow a = \frac{10}{3} \Rightarrow ea = \frac{5}{3}$

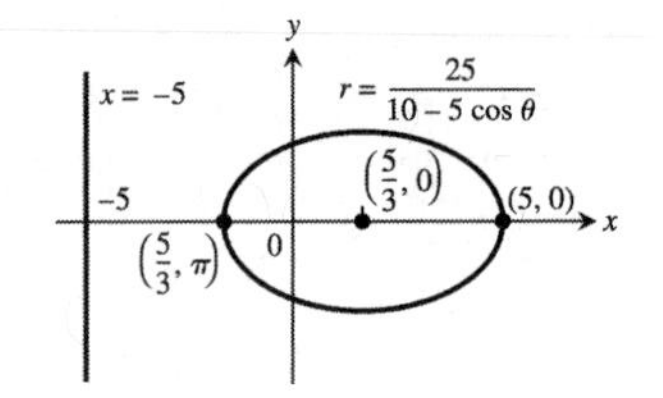

41. $r = \frac{400}{16 + 8\sin\theta} \Rightarrow r = \frac{\left(\frac{400}{16}\right)}{1 + \left(\frac{8}{16}\right)\sin\theta} \Rightarrow r = \frac{25}{1 + \left(\frac{1}{2}\right)\sin\theta}$
$e = \frac{1}{2}, k = 50 \Rightarrow y = 50; a\left(1 - e^2\right) = ke$
$\Rightarrow a\left[1 - \left(\frac{1}{2}\right)^2\right] = 25 \Rightarrow \frac{3}{4}a = 25 \Rightarrow a = \frac{100}{3}$
$\Rightarrow ea = \frac{50}{3}$

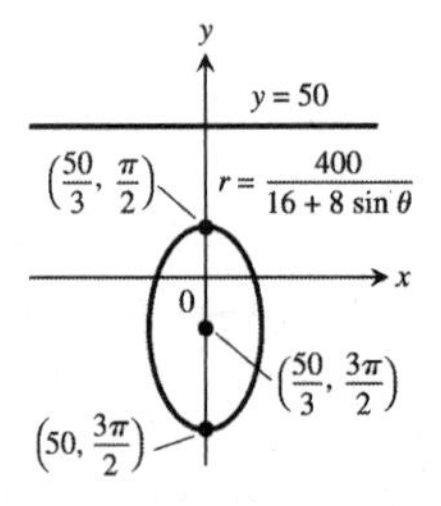

43. $r = \frac{8}{2 - 2\sin\theta} \Rightarrow r = \frac{4}{1 - \sin\theta} \Rightarrow e = 1,$
$k = 4 \Rightarrow y = -4$

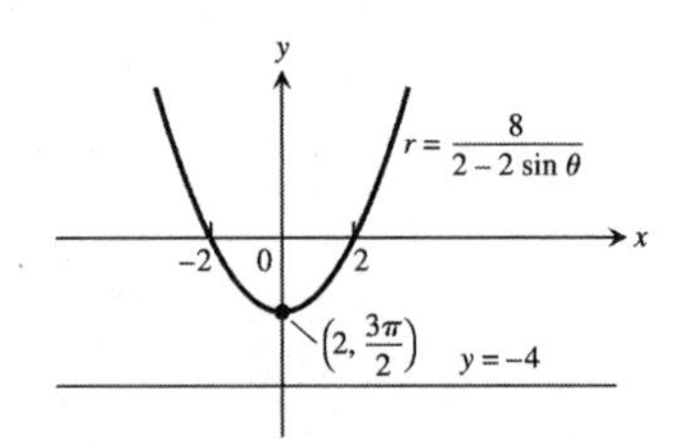

45. $r\cos\left(\theta - \frac{\pi}{4}\right) = \sqrt{2} \Rightarrow r\left(\cos\theta\cos\frac{\pi}{4} + \sin\theta\sin\frac{\pi}{4}\right)$
$= \sqrt{2} \Rightarrow \frac{1}{\sqrt{2}}r\cos\theta + \frac{1}{\sqrt{2}}r\sin\theta = \sqrt{2} \Rightarrow \frac{1}{\sqrt{2}}x + \frac{1}{\sqrt{2}}y$
$= \sqrt{2} \Rightarrow x + y = 2 \Rightarrow y = 2 - x$

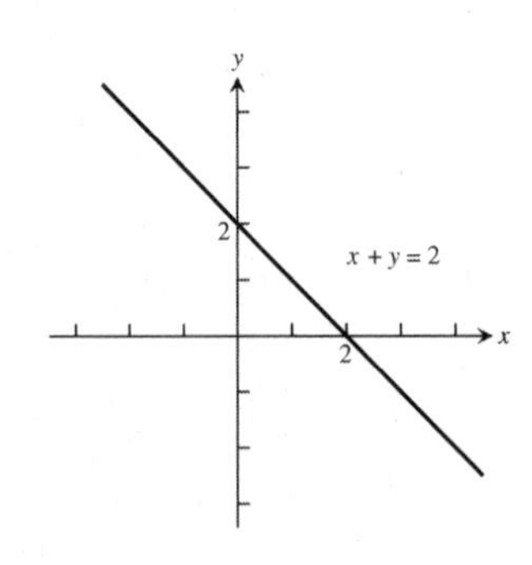

47. $r\cos\left(\theta - \frac{2\pi}{3}\right) = 3 \Rightarrow r\left(\cos\theta\cos\frac{2\pi}{3} + \sin\theta\sin\frac{2\pi}{3}\right) = 3$
$\Rightarrow -\frac{1}{2}r\cos\theta + \frac{\sqrt{3}}{2}r\sin\theta = 3 \Rightarrow -\frac{1}{2}x + \frac{\sqrt{3}}{2}y = 3$
$\Rightarrow -x + \sqrt{3}y = 6 \Rightarrow y = \frac{\sqrt{3}}{3}x + 2\sqrt{3}$

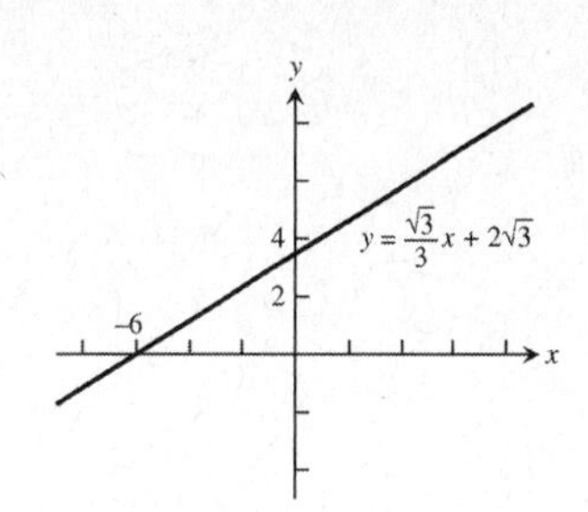

49. $\sqrt{2}x + \sqrt{2}y = 6 \Rightarrow \sqrt{2}r\cos\theta + \sqrt{2}r\sin\theta = 6 \Rightarrow r\left(\frac{\sqrt{2}}{2}\cos\theta + \frac{\sqrt{2}}{2}\sin\theta\right) = 3 \Rightarrow r\left(\cos\frac{\pi}{4}\cos\theta + \sin\frac{\pi}{4}\sin\theta\right)$
$= 3 \Rightarrow r\cos\left(\theta - \frac{\pi}{4}\right) = 3$

51. $y = -5 \Rightarrow r\sin\theta = -5 \Rightarrow -r\sin\theta = 5 \Rightarrow r\sin(-\theta) = 5 \Rightarrow r\cos\left(\frac{\pi}{2} - (-\theta)\right) = 5 \Rightarrow r\cos\left(\theta + \frac{\pi}{2}\right) = 5$

53.

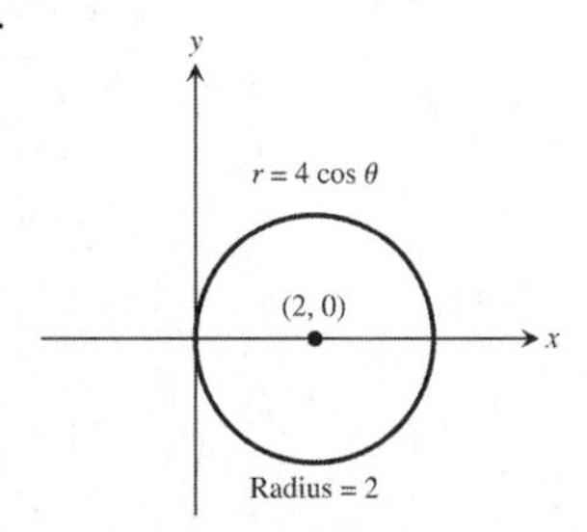

55.

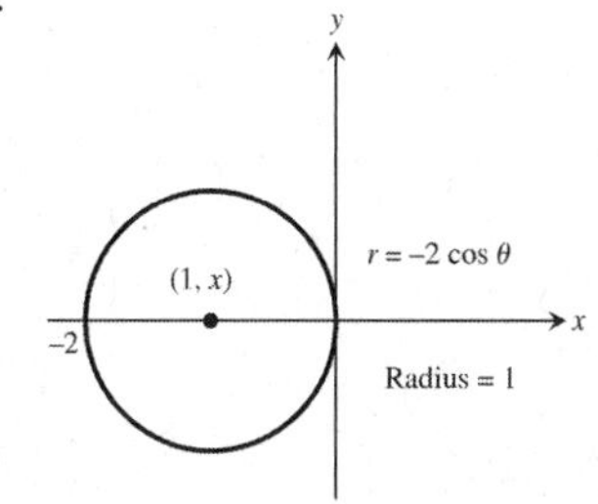

57. $(x-6)^2 + y^2 = 36 \Rightarrow C = (6, 0), a = 6$
$\Rightarrow r = 12\cos\theta$ is the polar equation

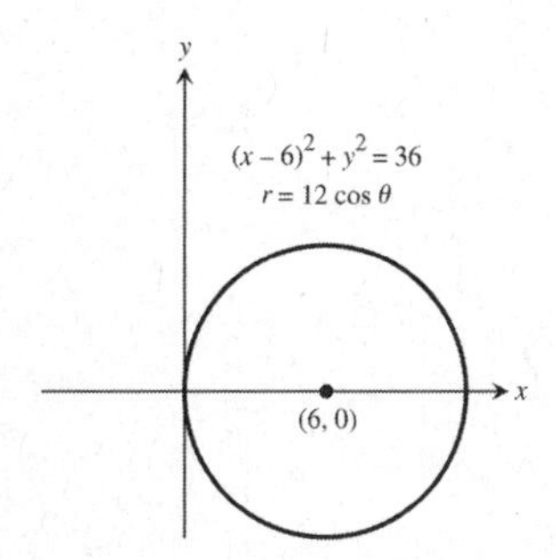

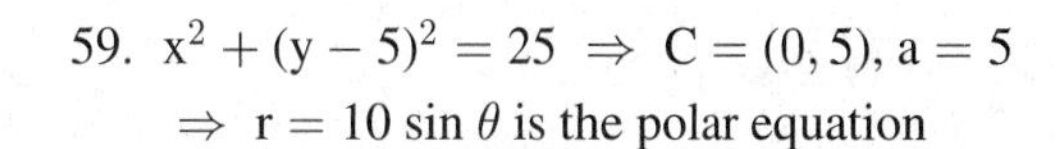

59. $x^2 + (y-5)^2 = 25 \Rightarrow C = (0, 5), a = 5$
$\Rightarrow r = 10\sin\theta$ is the polar equation

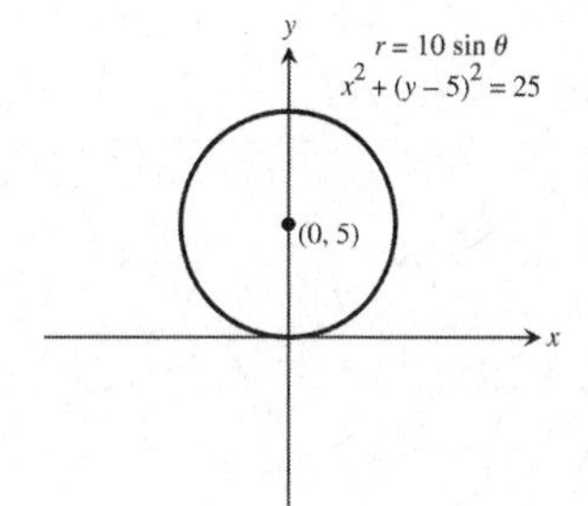

61. $x^2 + 2x + y^2 = 0 \Rightarrow (x+1)^2 + y^2 = 1$
$\Rightarrow C = (-1, 0), a = 1 \Rightarrow r = -2\cos\theta$ is the polar equation

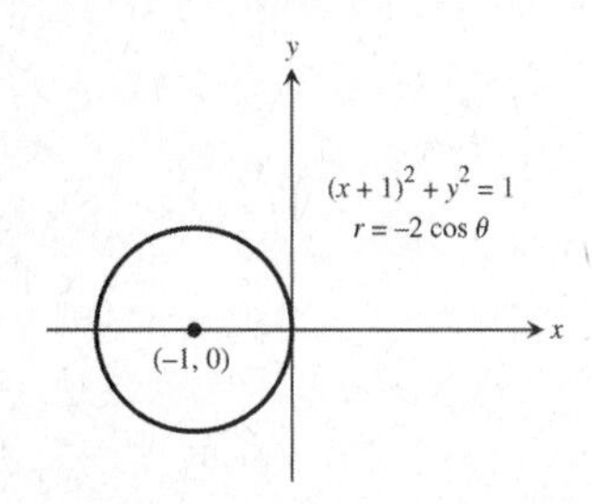

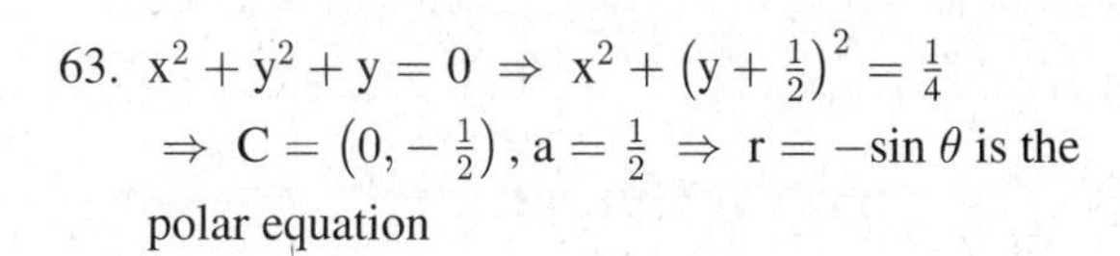

63. $x^2 + y^2 + y = 0 \Rightarrow x^2 + \left(y + \frac{1}{2}\right)^2 = \frac{1}{4}$
$\Rightarrow C = \left(0, -\frac{1}{2}\right), a = \frac{1}{2} \Rightarrow r = -\sin\theta$ is the polar equation

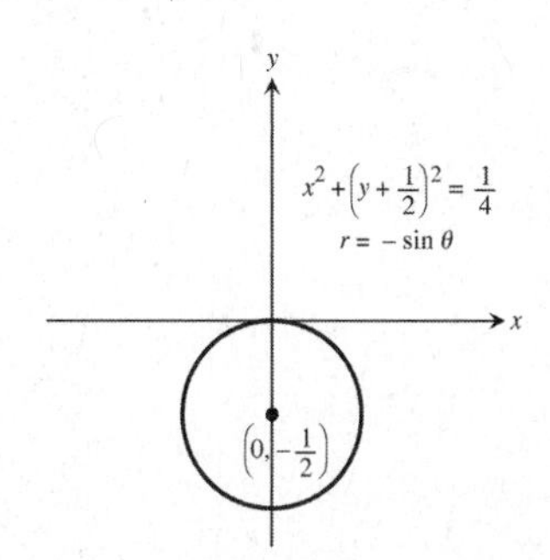

65.

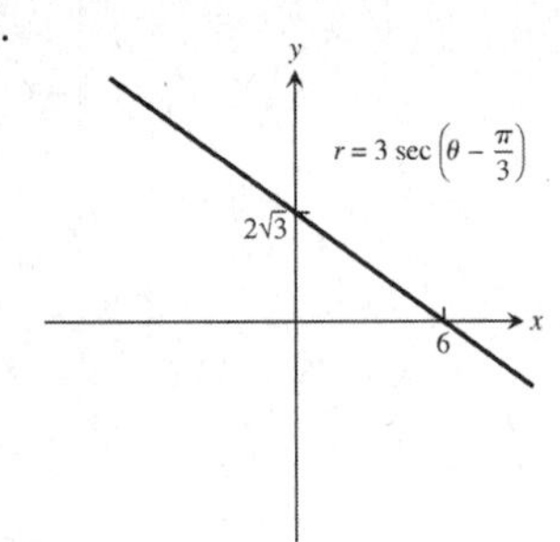

67.

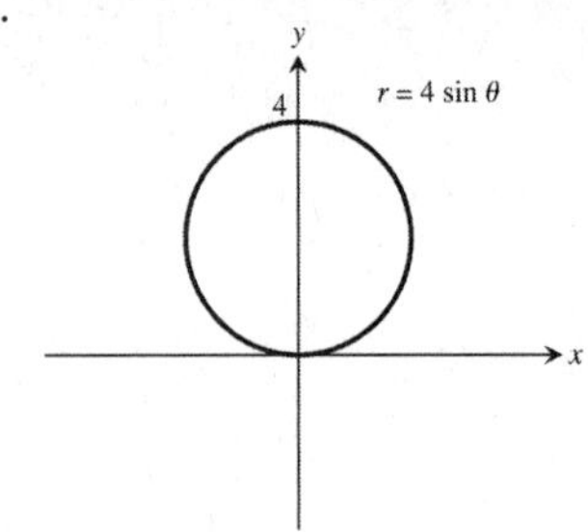

69.

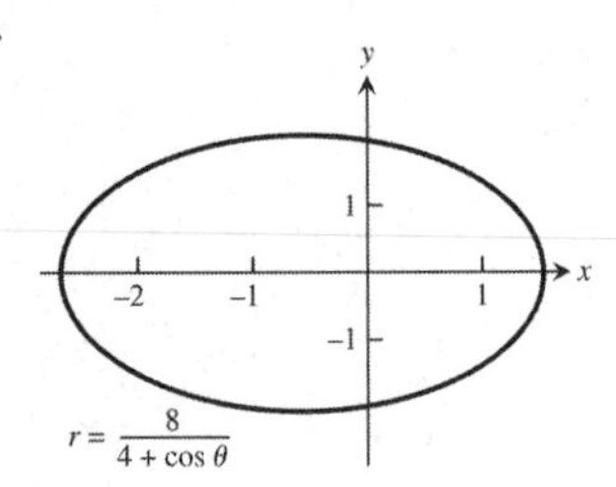

71.

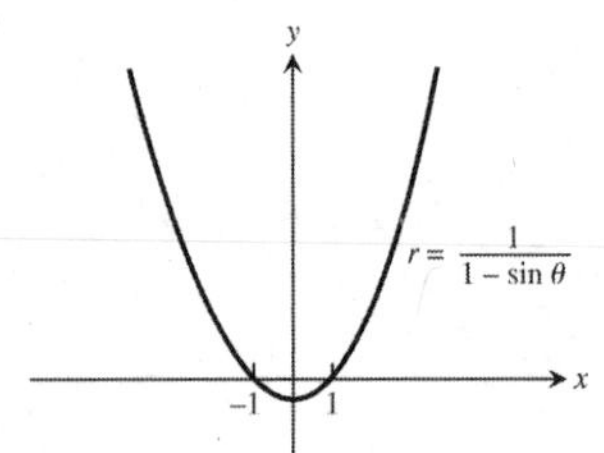

73.

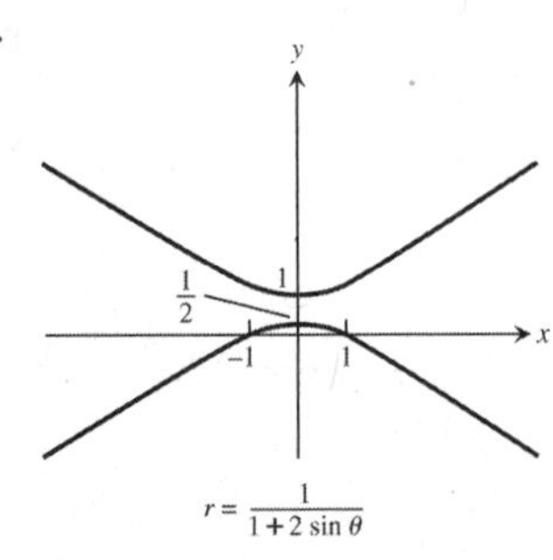

75. (a) Perihelion $= a - ae = a(1-e)$, Aphelion $= ea + a = a(1+e)$

(b)

Planet	Perihelion	Aphelion
Mercury	0.3075 AU	0.4667 AU
Venus	0.7184 AU	0.7282 AU
Earth	0.9833 AU	1.0167 AU
Mars	1.3817 AU	1.6663 AU
Jupiter	4.9512 AU	5.4548 AU
Saturn	9.0210 AU	10.0570 AU
Uranus	18.2977 AU	20.0623 AU
Neptune	29.8135 AU	30.3065 AU

CHAPTER 11 PRACTICE EXERCISES

1. $x = \frac{t}{2}$ and $y = t + 1 \Rightarrow 2x = t \Rightarrow y = 2x + 1$

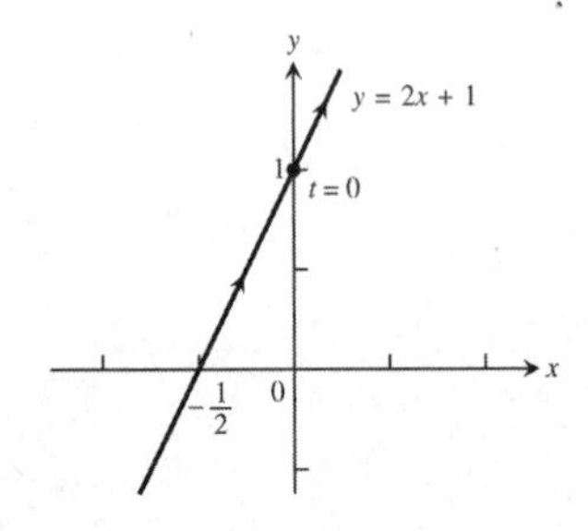

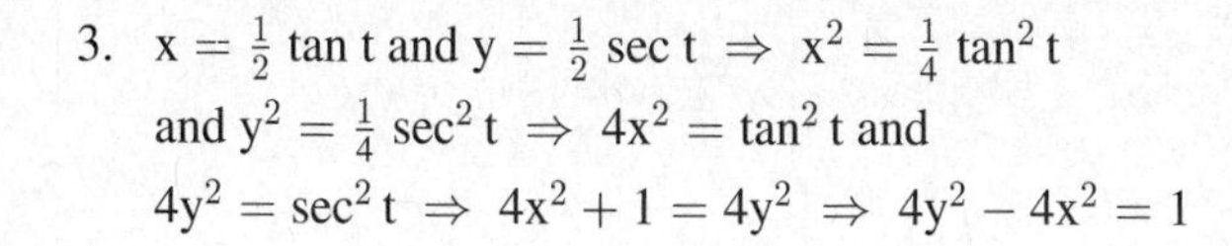

3. $x = \frac{1}{2}\tan t$ and $y = \frac{1}{2}\sec t \Rightarrow x^2 = \frac{1}{4}\tan^2 t$ and $y^2 = \frac{1}{4}\sec^2 t \Rightarrow 4x^2 = \tan^2 t$ and $4y^2 = \sec^2 t \Rightarrow 4x^2 + 1 = 4y^2 \Rightarrow 4y^2 - 4x^2 = 1$

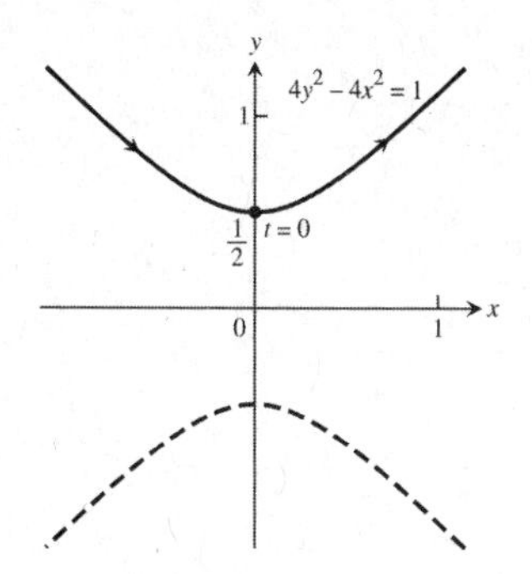

5. $x = -\cos t$ and $y = \cos^2 t \Rightarrow y = (-x)^2 = x^2$

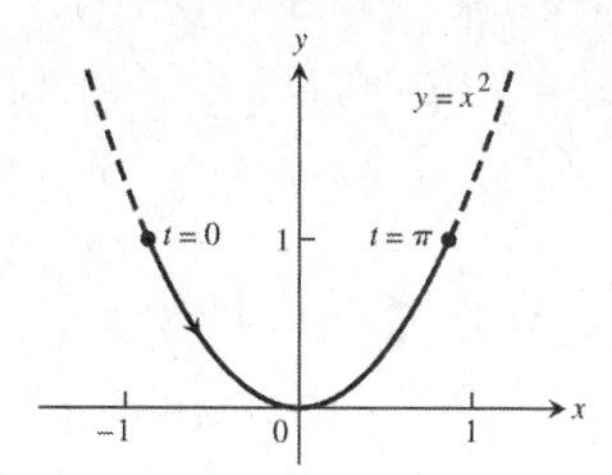

7. $16x^2 + 9y^2 = 144 \Rightarrow \frac{x^2}{9} + \frac{y^2}{16} = 1 \Rightarrow a = 3$ and $b = 4 \Rightarrow x = 3\cos t$ and $y = 4\sin t,\ 0 \le t \le 2\pi$

9. $x = \frac{1}{2}\tan t,\ y = \frac{1}{2}\sec t \Rightarrow \frac{dy}{dx} = \frac{dy/dt}{dx/dt} = \frac{\frac{1}{2}\sec t \tan t}{\frac{1}{2}\sec^2 t} = \frac{\tan t}{\sec t} = \sin t \Rightarrow \left.\frac{dy}{dx}\right|_{t=\pi/3} = \sin\frac{\pi}{3} = \frac{\sqrt{3}}{2};\ t = \frac{\pi}{3}$
$\Rightarrow x = \frac{1}{2}\tan\frac{\pi}{3} = \frac{\sqrt{3}}{2}$ and $y = \frac{1}{2}\sec\frac{\pi}{3} = 1 \Rightarrow y = \frac{\sqrt{3}}{2}x + \frac{1}{4};\ \frac{d^2y}{dx^2} = \frac{dy'/dt}{dx/dt} = \frac{\cos t}{\frac{1}{2}\sec^2 t} = 2\cos^3 t \Rightarrow \left.\frac{d^2y}{dx^2}\right|_{t=\pi/3}$
$= 2\cos^3\left(\frac{\pi}{3}\right) = \frac{1}{4}$

11. (a) $x = 4t^2,\ y = t^3 - 1 \Rightarrow t = \pm\frac{\sqrt{x}}{2} \Rightarrow y = \left(\pm\frac{\sqrt{x}}{2}\right)^3 - 1 = \pm\frac{x^{3/2}}{8} - 1$

(b) $x = \cos t,\ y = \tan t \Rightarrow \sec t = \frac{1}{x} \Rightarrow \tan^2 t + 1 = \sec^2 t \Rightarrow y^2 = \frac{1}{x^2} - 1 = \frac{1-x^2}{x^2} \Rightarrow y = \pm\frac{\sqrt{1-x^2}}{x}$

13. $y = x^{1/2} - \frac{x^{3/2}}{3} \Rightarrow \frac{dy}{dx} = \frac{1}{2}x^{-1/2} - \frac{1}{2}x^{1/2} \Rightarrow \left(\frac{dy}{dx}\right)^2 = \frac{1}{4}\left(\frac{1}{x} - 2 + x\right) \Rightarrow L = \int_1^4 \sqrt{1 + \frac{1}{4}\left(\frac{1}{x} - 2 + x\right)}\,dx$
$\Rightarrow L = \int_1^4 \sqrt{\frac{1}{4}\left(\frac{1}{x} + 2 + x\right)}\,dx = \int_1^4 \sqrt{\frac{1}{4}\left(x^{-1/2} + x^{1/2}\right)^2}\,dx = \int_1^4 \frac{1}{2}\left(x^{-1/2} + x^{1/2}\right)dx = \frac{1}{2}\left[2x^{1/2} + \frac{2}{3}x^{3/2}\right]_1^4$
$= \frac{1}{2}\left[\left(4 + \frac{2}{3}\cdot 8\right) - \left(2 + \frac{2}{3}\right)\right] = \frac{1}{2}\left(2 + \frac{14}{3}\right) = \frac{10}{3}$

15. $y = \frac{5}{12}x^{6/5} - \frac{5}{8}x^{4/5} \Rightarrow \frac{dy}{dx} = \frac{1}{2}x^{1/5} - \frac{1}{2}x^{-1/5} \Rightarrow \left(\frac{dy}{dx}\right)^2 = \frac{1}{4}\left(x^{2/5} - 2 + x^{-2/5}\right)$
$\Rightarrow L = \int_1^{32} \sqrt{1 + \frac{1}{4}\left(x^{2/5} - 2 + x^{-2/5}\right)}\,dx \Rightarrow L = \int_1^{32} \sqrt{\frac{1}{4}\left(x^{2/5} + 2 + x^{-2/5}\right)}\,dx = \int_1^{32} \sqrt{\frac{1}{4}\left(x^{1/5} + x^{-1/5}\right)^2}\,dx$
$= \int_1^{32} \frac{1}{2}\left(x^{1/5} + x^{-1/5}\right)dx = \frac{1}{2}\left[\frac{5}{6}x^{6/5} + \frac{5}{4}x^{4/5}\right]_1^{32} = \frac{1}{2}\left[\left(\frac{5}{6}\cdot 2^6 + \frac{5}{4}\cdot 2^4\right) - \left(\frac{5}{6} + \frac{5}{4}\right)\right] = \frac{1}{2}\left(\frac{315}{6} + \frac{75}{4}\right)$
$= \frac{1}{48}(1260 + 450) = \frac{1710}{48} = \frac{285}{8}$

17. $\frac{dx}{dt} = -5\sin t + 5\sin 5t$ and $\frac{dy}{dt} = 5\cos t - 5\cos 5t \Rightarrow \sqrt{\left(\frac{dx}{dt}\right)^2 + \left(\frac{dy}{dt}\right)^2}$
$= \sqrt{(-5\sin t + 5\sin 5t)^2 + (5\cos t - 5\cos 5t)^2}$
$= 5\sqrt{\sin^2 5t - 2\sin t\sin 5t + \sin^2 t + \cos^2 t - 2\cos t\cos 5t + \cos^2 5t} = 5\sqrt{2 - 2(\sin t\sin 5t + \cos t\cos 5t)}$
$= 5\sqrt{2(1 - \cos 4t)} = 5\sqrt{4\left(\frac{1}{2}\right)(1 - \cos 4t)} = 10\sqrt{\sin^2 2t} = 10|\sin 2t| = 10\sin 2t$ (since $0 \le t \le \frac{\pi}{2}$)
$\Rightarrow \text{Length} = \int_0^{\pi/2} 10\sin 2t\,dt = [-5\cos 2t]_0^{\pi/2} = (-5)(-1) - (-5)(1) = 10$

19. $\frac{dx}{d\theta} = -3 \sin \theta$ and $\frac{dy}{d\theta} = 3 \cos \theta \Rightarrow \sqrt{\left(\frac{dx}{d\theta}\right)^2 + \left(\frac{dy}{d\theta}\right)^2} = \sqrt{(-3 \sin \theta)^2 + (3 \cos \theta)^2} = \sqrt{3(\sin^2 \theta + \cos^2 \theta)} = 3$
$\Rightarrow$ Length $= \int_0^{3\pi/2} 3\, d\theta = 3\int_0^{3\pi/2} d\theta = 3\left(\frac{3\pi}{2} - 0\right) = \frac{9\pi}{2}$

21. $x = \frac{t^2}{2}$ and $y = 2t,\ 0 \le t \le \sqrt{5} \Rightarrow \frac{dx}{dt} = t$ and $\frac{dy}{dt} = 2 \Rightarrow$ Surface Area $= \int_0^{\sqrt{5}} 2\pi(2t)\sqrt{t^2+4}\, dt = \int_4^9 2\pi u^{1/2}\, du$
$= 2\pi\left[\frac{2}{3} u^{3/2}\right]_4^9 = \frac{76\pi}{3}$, where $u = t^2 + 4 \Rightarrow du = 2t\, dt;\ t = 0 \Rightarrow u = 4,\ t = \sqrt{5} \Rightarrow u = 9$

23. $r \cos\left(\theta + \frac{\pi}{3}\right) = 2\sqrt{3} \Rightarrow r\left(\cos \theta \cos \frac{\pi}{3} - \sin \theta \sin \frac{\pi}{3}\right)$
$= 2\sqrt{3} \Rightarrow \frac{1}{2} r \cos \theta - \frac{\sqrt{3}}{2} r \sin \theta = 2\sqrt{3}$
$\Rightarrow r \cos \theta - \sqrt{3} r \sin \theta = 4\sqrt{3} \Rightarrow x - \sqrt{3} y = 4\sqrt{3}$
$\Rightarrow y = \frac{\sqrt{3}}{3} x - 4$

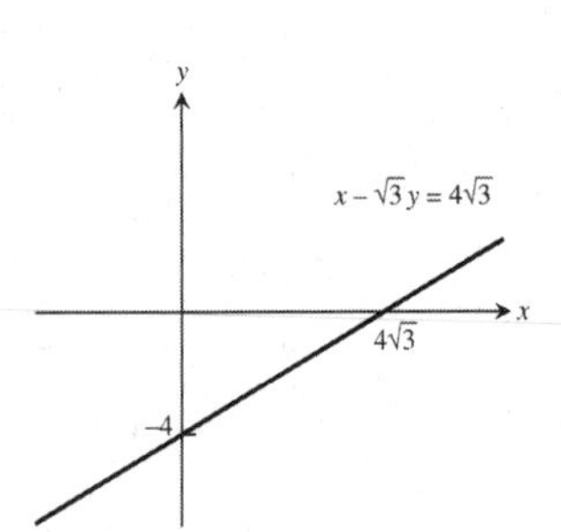

25. $r = 2 \sec \theta \Rightarrow r = \frac{2}{\cos \theta} \Rightarrow r \cos \theta = 2 \Rightarrow x = 2$

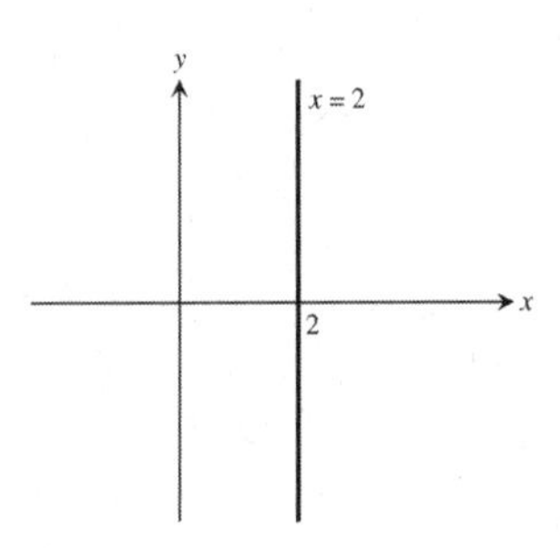

27. $r = -\frac{3}{2} \csc \theta \Rightarrow r \sin \theta = -\frac{3}{2} \Rightarrow y = -\frac{3}{2}$

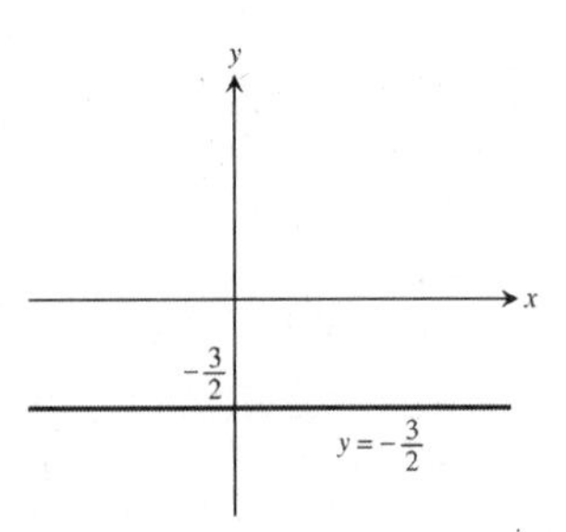

29. $r = -4 \sin \theta \Rightarrow r^2 = -4r \sin \theta \Rightarrow x^2 + y^2 + 4y = 0$
$\Rightarrow x^2 + (y+2)^2 = 4$; circle with center $(0, -2)$ and radius 2.

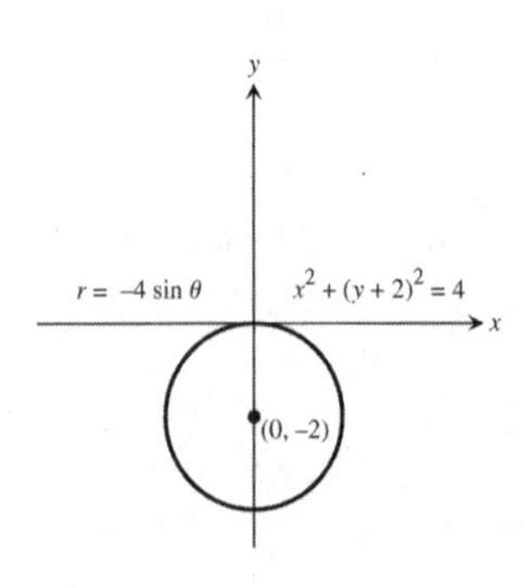

31. $r = 2\sqrt{2}\cos\theta \Rightarrow r^2 = 2\sqrt{2}\,r\cos\theta$
$\Rightarrow x^2 + y^2 - 2\sqrt{2}\,x = 0 \Rightarrow \left(x - \sqrt{2}\right)^2 + y^2 = 2;$
circle with center $\left(\sqrt{2}, 0\right)$ and radius $\sqrt{2}$

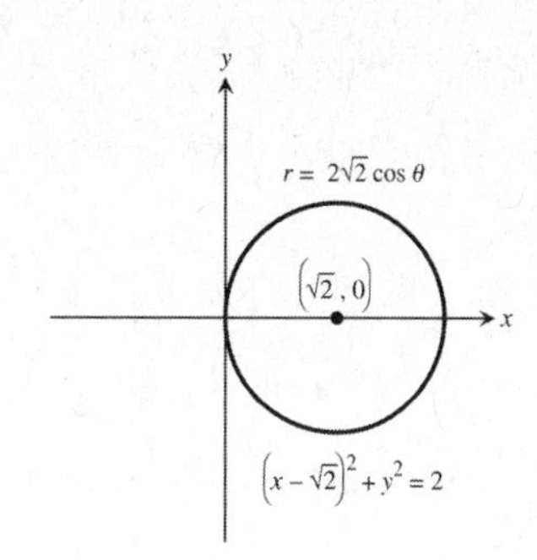

33. $x^2 + y^2 + 5y = 0 \Rightarrow x^2 + \left(y + \frac{5}{2}\right)^2 = \frac{25}{4} \Rightarrow C = \left(0, -\frac{5}{2}\right)$
and $a = \frac{5}{2}$; $r^2 + 5r\sin\theta = 0 \Rightarrow r = -5\sin\theta$

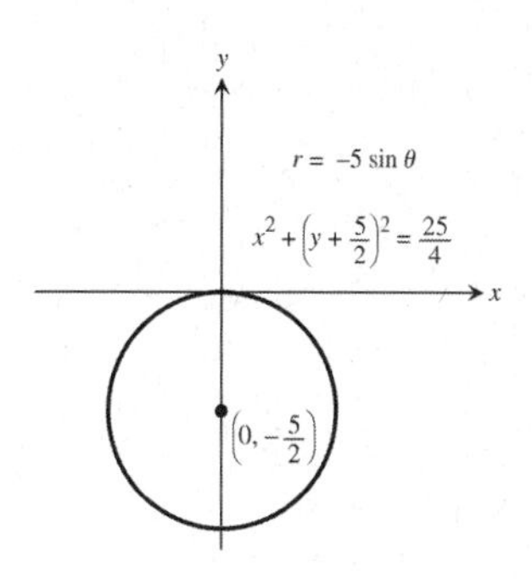

35. $x^2 + y^2 - 3x = 0 \Rightarrow \left(x - \frac{3}{2}\right)^2 + y^2 = \frac{9}{4} \Rightarrow C = \left(\frac{3}{2}, 0\right)$
and $a = \frac{3}{2}$; $r^2 - 3r\cos\theta = 0 \Rightarrow r = 3\cos\theta$

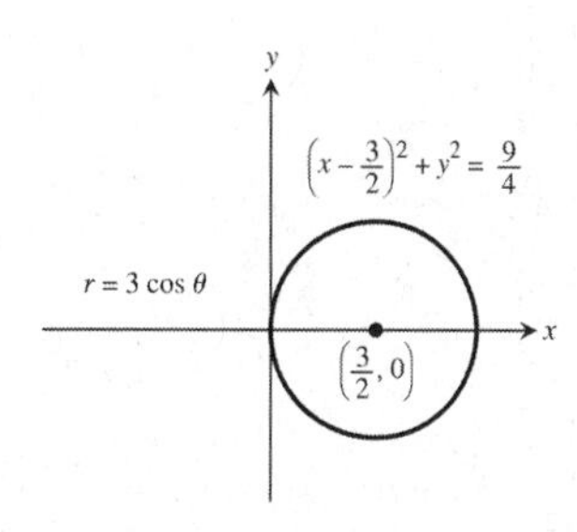

37.

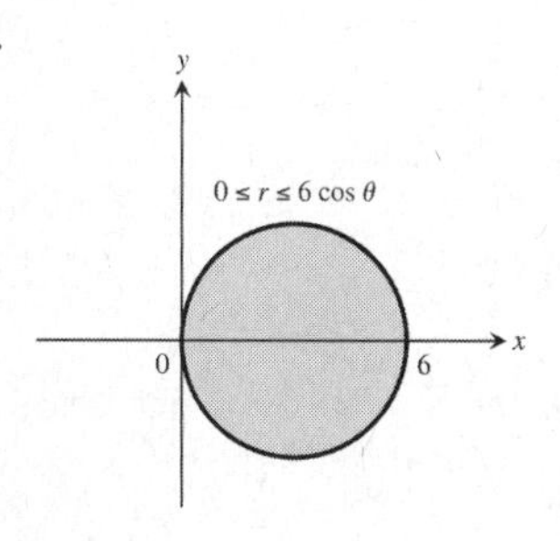

39. d

41. l

43. k

45. i

47. $A = 2\int_0^{\pi} \frac{1}{2} r^2\, d\theta = \int_0^{\pi} (2 - \cos\theta)^2\, d\theta = \int_0^{\pi} (4 - 4\cos\theta + \cos^2\theta)\, d\theta = \int_0^{\pi} \left(4 - 4\cos\theta + \frac{1 + \cos 2\theta}{2}\right) d\theta$
$= \int_0^{\pi} \left(\frac{9}{2} - 4\cos\theta + \frac{\cos 2\theta}{2}\right) d\theta = \left[\frac{9}{2}\theta - 4\sin\theta + \frac{\sin 2\theta}{4}\right]_0^{\pi} = \frac{9}{2}\pi$

49. $r = 1 + \cos 2\theta$ and $r = 1 \Rightarrow 1 = 1 + \cos 2\theta \Rightarrow 0 = \cos 2\theta \Rightarrow 2\theta = \frac{\pi}{2} \Rightarrow \theta = \frac{\pi}{4}$; therefore
$A = 4\int_0^{\pi/4} \frac{1}{2}\left[(1 + \cos 2\theta)^2 - 1^2\right] d\theta = 2\int_0^{\pi/4} (1 + 2\cos 2\theta + \cos^2 2\theta - 1)\, d\theta$
$= 2\int_0^{\pi/4} \left(2\cos 2\theta + \frac{1}{2} + \frac{\cos 4\theta}{2}\right) d\theta = 2\left[\sin 2\theta + \frac{1}{2}\theta + \frac{\sin 4\theta}{8}\right]_0^{\pi/4} = 2\left(1 + \frac{\pi}{8} + 0\right) = 2 + \frac{\pi}{4}$

51. $r = -1 + \cos\theta \Rightarrow \frac{dr}{d\theta} = -\sin\theta$; Length $= \int_0^{2\pi} \sqrt{(-1+\cos\theta)^2 + (-\sin\theta)^2}\, d\theta = \int_0^{2\pi} \sqrt{2 - 2\cos\theta}\, d\theta$
$= \int_0^{2\pi} \sqrt{\frac{4(1-\cos\theta)}{2}}\, d\theta = \int_0^{2\pi} 2\sin\frac{\theta}{2}\, d\theta = \left[-4\cos\frac{\theta}{2}\right]_0^{2\pi} = (-4)(-1) - (-4)(1) = 8$

53. $r = 8\sin^3\left(\frac{\theta}{3}\right), 0 \le \theta \le \frac{\pi}{4} \Rightarrow \frac{dr}{d\theta} = 8\sin^2\left(\frac{\theta}{3}\right)\cos\left(\frac{\theta}{3}\right); r^2 + \left(\frac{dr}{d\theta}\right)^2 = \left[8\sin^3\left(\frac{\theta}{3}\right)\right]^2 + \left[8\sin^2\left(\frac{\theta}{3}\right)\cos\left(\frac{\theta}{3}\right)\right]^2$
$= 64\sin^4\left(\frac{\theta}{3}\right) \Rightarrow L = \int_0^{\pi/4} \sqrt{64\sin^4\left(\frac{\theta}{3}\right)}\, d\theta = \int_0^{\pi/4} 8\sin^2\left(\frac{\theta}{3}\right) d\theta = \int_0^{\pi/4} 8\left[\frac{1-\cos\left(\frac{2\theta}{3}\right)}{2}\right] d\theta$
$= \int_0^{\pi/4} \left[4 - 4\cos\left(\frac{2\theta}{3}\right)\right] d\theta = \left[4\theta - 6\sin\left(\frac{2\theta}{3}\right)\right]_0^{\pi/4} = 4\left(\frac{\pi}{4}\right) - 6\sin\left(\frac{\pi}{6}\right) - 0 = \pi - 3$

55. $x^2 = -4y \Rightarrow y = -\frac{x^2}{4} \Rightarrow 4p = 4 \Rightarrow p = 1$;
therefore Focus is $(0, -1)$, Directrix is $y = 1$

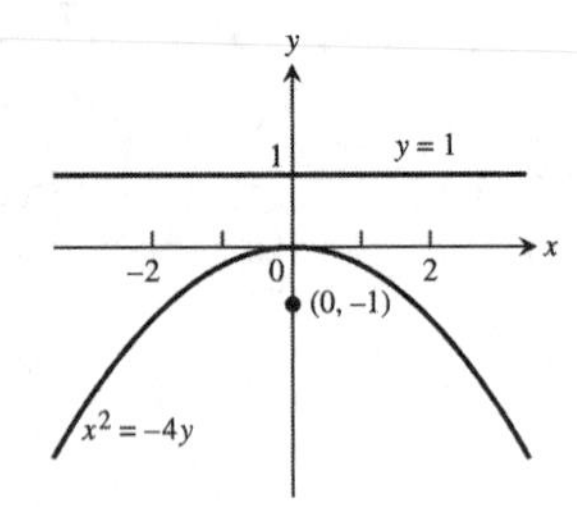

57. $y^2 = 3x \Rightarrow x = \frac{y^2}{3} \Rightarrow 4p = 3 \Rightarrow p = \frac{3}{4}$;
therefore Focus is $\left(\frac{3}{4}, 0\right)$, Directrix is $x = -\frac{3}{4}$

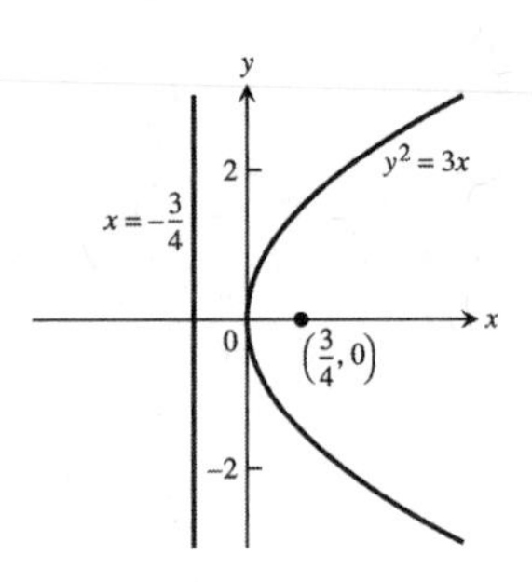

59. $16x^2 + 7y^2 = 112 \Rightarrow \frac{x^2}{7} + \frac{y^2}{16} = 1$
$\Rightarrow c^2 = 16 - 7 = 9 \Rightarrow c = 3; e = \frac{c}{a} = \frac{3}{4}$

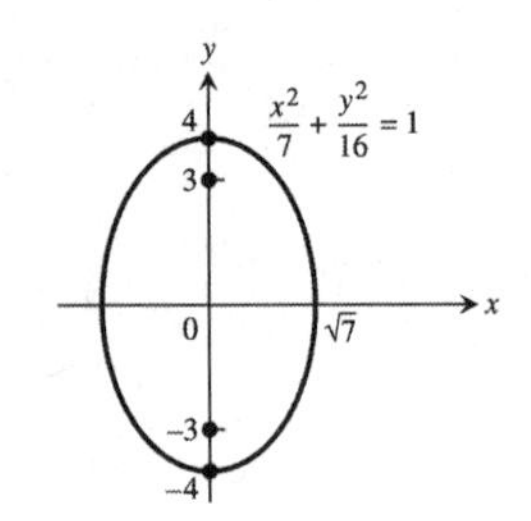

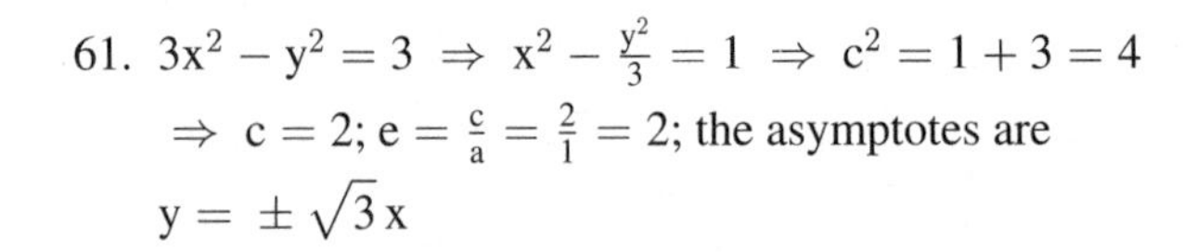

61. $3x^2 - y^2 = 3 \Rightarrow x^2 - \frac{y^2}{3} = 1 \Rightarrow c^2 = 1 + 3 = 4$
$\Rightarrow c = 2; e = \frac{c}{a} = \frac{2}{1} = 2$; the asymptotes are
$y = \pm\sqrt{3}x$

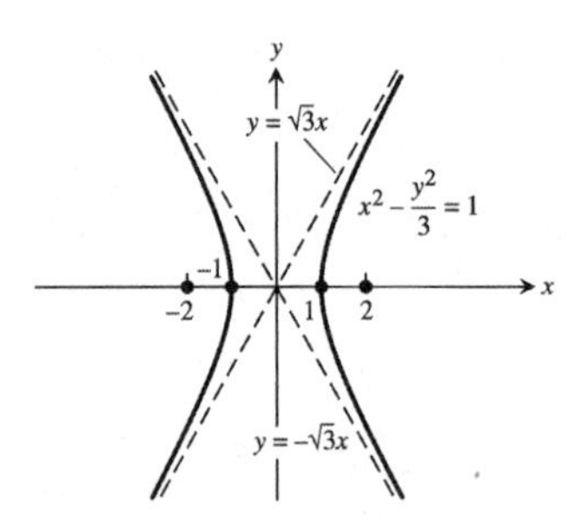

63. $x^2 = -12y \Rightarrow -\frac{x^2}{12} = y \Rightarrow 4p = 12 \Rightarrow p = 3 \Rightarrow$ focus is $(0, -3)$, directrix is $y = 3$, vertex is $(0, 0)$; therefore new vertex is $(2, 3)$, new focus is $(2, 0)$, new directrix is $y = 6$, and the new equation is $(x-2)^2 = -12(y-3)$

65. $\frac{x^2}{9} + \frac{y^2}{25} = 1 \Rightarrow a = 5$ and $b = 3 \Rightarrow c = \sqrt{25 - 9} = 4 \Rightarrow$ foci are $(0, \pm 4)$, vertices are $(0, \pm 5)$, center is $(0, 0)$; therefore the new center is $(-3, -5)$, new foci are $(-3, -1)$ and $(-3, -9)$, new vertices are $(-3, -10)$ and $(-3, 0)$, and the new equation is $\frac{(x+3)^2}{9} + \frac{(y+5)^2}{25} = 1$

67. $\frac{y^2}{8} - \frac{x^2}{2} = 1 \Rightarrow a = 2\sqrt{2}$ and $b = \sqrt{2} \Rightarrow c = \sqrt{8+2} = \sqrt{10} \Rightarrow$ foci are $\left(0, \pm\sqrt{10}\right)$, vertices are $\left(0, \pm 2\sqrt{2}\right)$, center is $(0, 0)$, and the asymptotes are $y = \pm 2x$; therefore the new center is $\left(2, 2\sqrt{2}\right)$, new foci are $\left(2, 2\sqrt{2} \pm \sqrt{10}\right)$, new vertices are $\left(2, 4\sqrt{2}\right)$ and $(2, 0)$, the new asymptotes are $y = 2x - 4 + 2\sqrt{2}$ and $y = -2x + 4 + 2\sqrt{2}$; the new equation is $\frac{\left(y - 2\sqrt{2}\right)^2}{8} - \frac{(x-2)^2}{2} = 1$

69. $x^2 - 4x - 4y^2 = 0 \Rightarrow x^2 - 4x + 4 - 4y^2 = 4 \Rightarrow (x-2)^2 - 4y^2 = 4 \Rightarrow \frac{(x-2)^2}{4} - y^2 = 1$, a hyperbola; $a = 2$ and $b = 1 \Rightarrow c = \sqrt{1+4} = \sqrt{5}$; the center is $(2, 0)$, the vertices are $(0, 0)$ and $(4, 0)$; the foci are $\left(2 \pm \sqrt{5}, 0\right)$ and the asymptotes are $y = \pm \frac{x-2}{2}$

71. $y^2 - 2y + 16x = -49 \Rightarrow y^2 - 2y + 1 = -16x - 48 \Rightarrow (y-1)^2 = -16(x+3)$, a parabola; the vertex is $(-3, 1)$; $4p = 16 \Rightarrow p = 4 \Rightarrow$ the focus is $(-7, 1)$ and the directrix is $x = 1$

73. $9x^2 + 16y^2 + 54x - 64y = -1 \Rightarrow 9(x^2 + 6x) + 16(y^2 - 4y) = -1 \Rightarrow 9(x^2 + 6x + 9) + 16(y^2 - 4y + 4) = 144$ $\Rightarrow 9(x+3)^2 + 16(y-2)^2 = 144 \Rightarrow \frac{(x+3)^2}{16} + \frac{(y-2)^2}{9} = 1$, an ellipse; the center is $(-3, 2)$; $a = 4$ and $b = 3$ $\Rightarrow c = \sqrt{16-9} = \sqrt{7}$; the foci are $\left(-3 \pm \sqrt{7}, 2\right)$; the vertices are $(1, 2)$ and $(-7, 2)$

75. $x^2 + y^2 - 2x - 2y = 0 \Rightarrow x^2 - 2x + 1 + y^2 - 2y + 1 = 2 \Rightarrow (x-1)^2 + (y-1)^2 = 2$, a circle with center $(1, 1)$ and radius $= \sqrt{2}$

77. $r = \frac{2}{1+\cos\theta} \Rightarrow e = 1 \Rightarrow$ parabola with vertex at $(1, 0)$

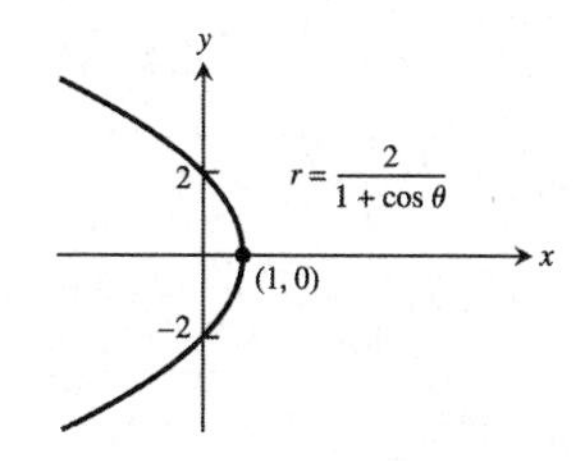

79. $r = \frac{6}{1-2\cos\theta} \Rightarrow e = 2 \Rightarrow$ hyperbola; $ke = 6 \Rightarrow 2k = 6$ $\Rightarrow k = 3 \Rightarrow$ vertices are $(2, \pi)$ and $(6, \pi)$

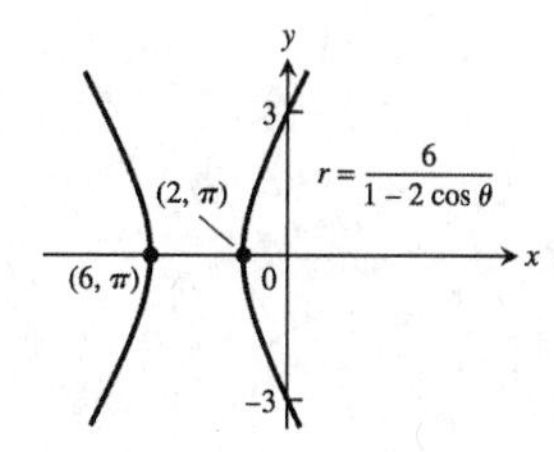

81. $e = 2$ and $r\cos\theta = 2 \Rightarrow x = 2$ is directrix $\Rightarrow k = 2$; the conic is a hyperbola; $r = \frac{ke}{1+e\cos\theta} \Rightarrow r = \frac{(2)(2)}{1+2\cos\theta}$ $\Rightarrow r = \frac{4}{1+2\cos\theta}$

83. $e = \frac{1}{2}$ and $r\sin\theta = 2 \Rightarrow y = 2$ is directrix $\Rightarrow k = 2$; the conic is an ellipse; $r = \frac{ke}{1+e\sin\theta} \Rightarrow r = \frac{(2)\left(\frac{1}{2}\right)}{1+\left(\frac{1}{2}\right)\sin\theta}$ $\Rightarrow r = \frac{2}{2+\sin\theta}$

85. (a) Around the x-axis: $9x^2 + 4y^2 = 36 \Rightarrow y^2 = 9 - \frac{9}{4}x^2 \Rightarrow y = \pm\sqrt{9 - \frac{9}{4}x^2}$ and we use the positive root:

$$V = 2\int_0^2 \pi\left(\sqrt{9 - \frac{9}{4}x^2}\right)^2 dx = 2\int_0^2 \pi\left(9 - \frac{9}{4}x^2\right) dx = 2\pi\left[9x - \frac{3}{4}x^3\right]_0^2 = 24\pi$$

(b) Around the y-axis: $9x^2 + 4y^2 = 36 \Rightarrow x^2 = 4 - \frac{4}{9}y^2 \Rightarrow x = \pm\sqrt{4 - \frac{4}{9}y^2}$ and we use the positive root:

$$V = 2\int_0^3 \pi\left(\sqrt{4 - \frac{4}{9}y^2}\right)^2 dy = 2\int_0^3 \pi\left(4 - \frac{4}{9}y^2\right) dy = 2\pi\left[4y - \frac{4}{27}y^3\right]_0^3 = 16\pi$$

87. (a) $r = \frac{k}{1+e\cos\theta} \Rightarrow r + er\cos\theta = k \Rightarrow \sqrt{x^2+y^2} + ex = k \Rightarrow \sqrt{x^2+y^2} = k - ex \Rightarrow x^2 + y^2$ $= k^2 - 2kex + e^2x^2 \Rightarrow x^2 - e^2x^2 + y^2 + 2kex - k^2 = 0 \Rightarrow (1 - e^2)x^2 + y^2 + 2kex - k^2 = 0$

(b) $e = 0 \Rightarrow x^2 + y^2 - k^2 = 0 \Rightarrow x^2 + y^2 = k^2 \Rightarrow$ circle;
$0 < e < 1 \Rightarrow e^2 < 1 \Rightarrow e^2 - 1 < 0 \Rightarrow B^2 - 4AC = 0^2 - 4(1 - e^2)(1) = 4(e^2 - 1) < 0 \Rightarrow$ ellipse;
$e = 1 \Rightarrow B^2 - 4AC = 0^2 - 4(0)(1) = 0 \Rightarrow$ parabola;
$e > 1 \Rightarrow e^2 > 1 \Rightarrow B^2 - 4AC = 0^2 - 4(1 - e^2)(1) = 4e^2 - 4 > 0 \Rightarrow$ hyperbola

CHAPTER 11 ADDITIONAL AND ADVANCED EXERCISES

1. Directrix $x = 3$ and focus $(4, 0) \Rightarrow$ vertex is $\left(\frac{7}{2}, 0\right)$
$\Rightarrow p = \frac{1}{2} \Rightarrow$ the equation is $x - \frac{7}{2} = \frac{y^2}{2}$

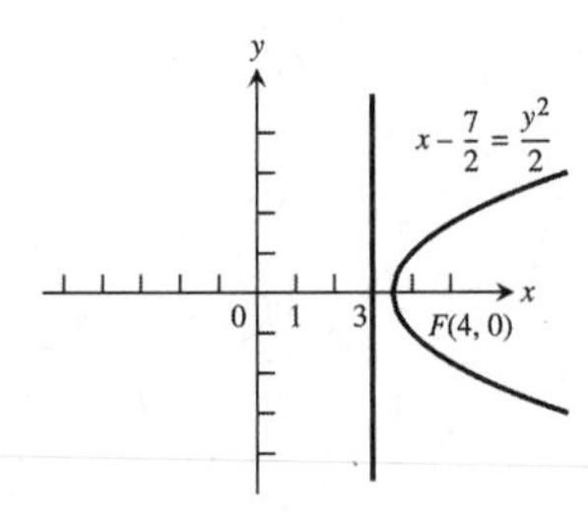

3. $x^2 = 4y \Rightarrow$ vertex is $(0, 0)$ and $p = 1 \Rightarrow$ focus is $(0, 1)$; thus the distance from $P(x, y)$ to the vertex is $\sqrt{x^2 + y^2}$ and the distance from P to the focus is $\sqrt{x^2 + (y - 1)^2} \Rightarrow \sqrt{x^2 + y^2} = 2\sqrt{x^2 + (y - 1)^2}$
$\Rightarrow x^2 + y^2 = 4\left[x^2 + (y - 1)^2\right] \Rightarrow x^2 + y^2 = 4x^2 + 4y^2 - 8y + 4 \Rightarrow 3x^2 + 3y^2 - 8y + 4 = 0$, which is a circle

5. Vertices are $(0, \pm 2) \Rightarrow a = 2$; $e = \frac{c}{a} \Rightarrow 0.5 = \frac{c}{2} \Rightarrow c = 1 \Rightarrow$ foci are $(0, \pm 1)$

7. Let the center of the hyperbola be $(0, y)$.
(a) Directrix $y = -1$, focus $(0, -7)$ and $e = 2 \Rightarrow c - \frac{a}{e} = 6 \Rightarrow \frac{a}{e} = c - 6 \Rightarrow a = 2c - 12$. Also $c = ae = 2a$
$\Rightarrow a = 2(2a) - 12 \Rightarrow a = 4 \Rightarrow c = 8$; $y - (-1) = \frac{a}{e} = \frac{4}{2} = 2 \Rightarrow y = 1 \Rightarrow$ the center is $(0, 1)$; $c^2 = a^2 + b^2$
$\Rightarrow b^2 = c^2 - a^2 = 64 - 16 = 48$; therefore the equation is $\frac{(y-1)^2}{16} - \frac{x^2}{48} = 1$
(b) $e = 5 \Rightarrow c - \frac{a}{e} = 6 \Rightarrow \frac{a}{e} = c - 6 \Rightarrow a = 5c - 30$. Also, $c = ae = 5a \Rightarrow a = 5(5a) - 30 \Rightarrow 24a = 30 \Rightarrow a = \frac{5}{4}$
$\Rightarrow c = \frac{25}{4}$; $y - (-1) = \frac{a}{e} = \frac{\left(\frac{5}{4}\right)}{5} = \frac{1}{4} \Rightarrow y = -\frac{3}{4} \Rightarrow$ the center is $\left(0, -\frac{3}{4}\right)$; $c^2 = a^2 + b^2 \Rightarrow b^2 = c^2 - a^2$
$= \frac{625}{16} - \frac{25}{16} = \frac{75}{2}$; therefore the equation is $\frac{\left(y + \frac{3}{4}\right)^2}{\left(\frac{25}{16}\right)} - \frac{x^2}{\left(\frac{75}{2}\right)} = 1$ or $\frac{16\left(y + \frac{3}{4}\right)^2}{25} - \frac{2x^2}{75} = 1$

9. $b^2x^2 + a^2y^2 = a^2b^2 \Rightarrow \frac{dy}{dx} = -\frac{b^2x}{a^2y}$; at (x_1, y_1) the tangent line is $y - y_1 = \left(-\frac{b^2x_1}{a^2y_1}\right)(x - x_1)$
$\Rightarrow a^2yy_1 + b^2xx_1 = b^2x_1^2 + a^2y_1^2 = a^2b^2 \Rightarrow b^2xx_1 + a^2yy_1 - a^2b^2 = 0$

11.

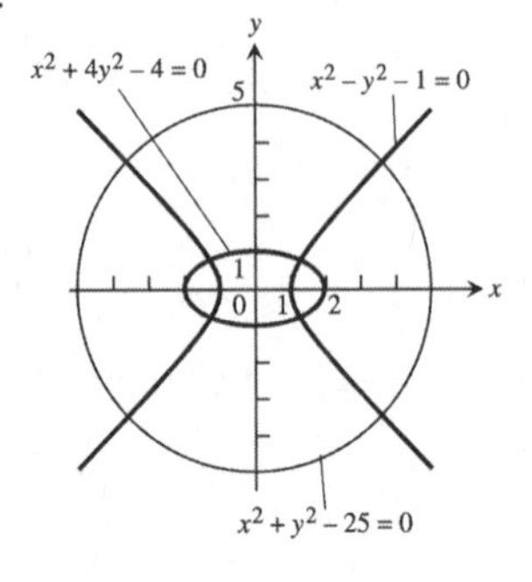

13.

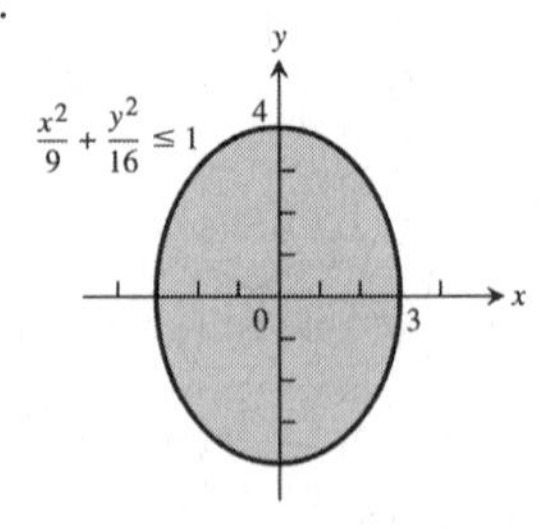

15. $(9x^2 + 4y^2 - 36)(4x^2 + 9y^2 - 16) \le 0$
$\Rightarrow 9x^2 + 4y^2 - 36 \le 0$ and $4x^2 + 9y^2 - 16 \ge 0$
or $9x^2 + 4y^2 - 36 \ge 0$ and $4x^2 + 9y^2 - 16 \le 0$

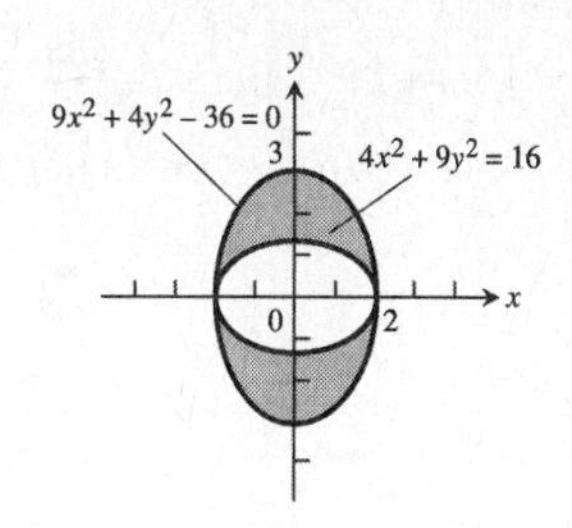

17. (a) $x = e^{2t}\cos t$ and $y = e^{2t}\sin t \Rightarrow x^2 + y^2 = e^{4t}\cos^2 t + e^{4t}\sin^2 t = e^{4t}$. Also $\frac{y}{x} = \frac{e^{2t}\sin t}{e^{2t}\cos t} = \tan t$
$\Rightarrow t = \tan^{-1}\left(\frac{y}{x}\right) \Rightarrow x^2 + y^2 = e^{4\tan^{-1}(y/x)}$ is the Cartesian equation. Since $r^2 = x^2 + y^2$ and
$\theta = \tan^{-1}\left(\frac{y}{x}\right)$, the polar equation is $r^2 = e^{4\theta}$ or $r = e^{2\theta}$ for $r > 0$

(b) $ds^2 = r^2\,d\theta^2 + dr^2$; $r = e^{2\theta} \Rightarrow dr = 2e^{2\theta}\,d\theta$
$\Rightarrow ds^2 = r^2\,d\theta^2 + \left(2e^{2\theta}\,d\theta\right)^2 = \left(e^{2\theta}\right)^2 d\theta^2 + 4e^{4\theta}\,d\theta^2$
$= 5e^{4\theta}\,d\theta^2 \Rightarrow ds = \sqrt{5}\,e^{2\theta}\,d\theta \Rightarrow L = \int_0^{2\pi} \sqrt{5}\,e^{2\theta}\,d\theta$
$= \left[\frac{\sqrt{5}\,e^{2\theta}}{2}\right]_0^{2\pi} = \frac{\sqrt{5}}{2}\left(e^{4\pi} - 1\right)$

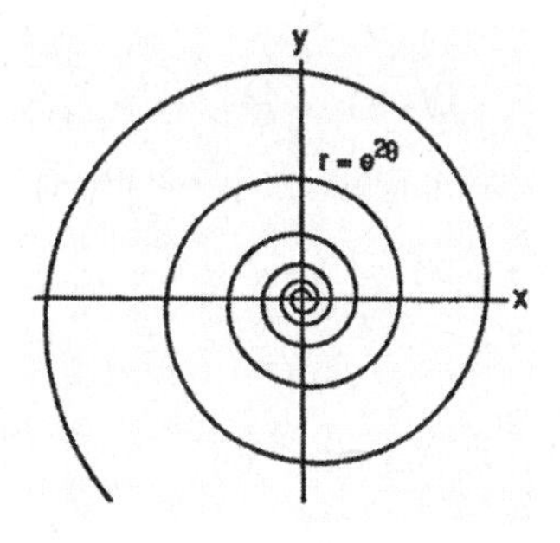

19. $e = 2$ and $r\cos\theta = 2 \Rightarrow x = 2$ is the directrix $\Rightarrow k = 2$; the conic is a hyperbola with $r = \frac{ke}{1 + e\cos\theta}$
$\Rightarrow r = \frac{(2)(2)}{1 + 2\cos\theta} = \frac{4}{1 + 2\cos\theta}$

21. $e = \frac{1}{2}$ and $r\sin\theta = 2 \Rightarrow y = 2$ is the directrix $\Rightarrow k = 2$; the conic is an ellipse with $r = \frac{ke}{1 + e\sin\theta}$
$\Rightarrow r = \frac{2\left(\frac{1}{2}\right)}{1 + \left(\frac{1}{2}\right)\sin\theta} = \frac{2}{2 + \sin\theta}$

23. Arc PF = Arc AF since each is the distance rolled;
$\angle PCF = \frac{\text{Arc PF}}{b} \Rightarrow \text{Arc PF} = b(\angle PCF)$; $\theta = \frac{\text{Arc AF}}{a}$
$\Rightarrow \text{Arc AF} = a\theta \Rightarrow a\theta = b(\angle PCF) \Rightarrow \angle PCF = \left(\frac{a}{b}\right)\theta$;
$\angle OCB = \frac{\pi}{2} - \theta$ and $\angle OCB = \angle PCF - \angle PCE$
$= \angle PCF - \left(\frac{\pi}{2} - \alpha\right) = \left(\frac{a}{b}\right)\theta - \left(\frac{\pi}{2} - \alpha\right) \Rightarrow \frac{\pi}{2} - \theta$
$= \left(\frac{a}{b}\right)\theta - \left(\frac{\pi}{2} - \alpha\right) \Rightarrow \frac{\pi}{2} - \theta = \left(\frac{a}{b}\right)\theta - \frac{\pi}{2} + \alpha$
$\Rightarrow \alpha = \pi - \theta - \left(\frac{a}{b}\right)\theta \Rightarrow \alpha = \pi - \left(\frac{a+b}{b}\right)\theta$.

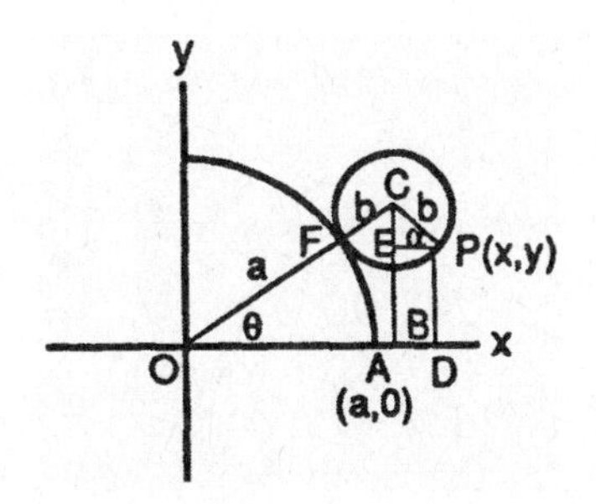

Now $x = OB + BD = OB + EP = (a + b)\cos\theta + b\cos\alpha = (a + b)\cos\theta + b\cos\left(\pi - \left(\frac{a+b}{b}\right)\theta\right)$
$= (a + b)\cos\theta + b\cos\pi\cos\left(\left(\frac{a+b}{b}\right)\theta\right) + b\sin\pi\sin\left(\left(\frac{a+b}{b}\right)\theta\right) = (a + b)\cos\theta - b\cos\left(\left(\frac{a+b}{b}\right)\theta\right)$ and
$y = PD = CB - CE = (a + b)\sin\theta - b\sin\alpha = (a + b)\sin\theta - b\sin\left(\left(\frac{a+b}{b}\right)\theta\right)$
$= (a + b)\sin\theta - b\sin\pi\cos\left(\left(\frac{a+b}{b}\right)\theta\right) + b\cos\pi\sin\left(\left(\frac{a+b}{b}\right)\theta\right) = (a + b)\sin\theta - b\sin\left(\left(\frac{a+b}{b}\right)\theta\right)$;
therefore $x = (a + b)\cos\theta - b\cos\left(\left(\frac{a+b}{b}\right)\theta\right)$ and $y = (a + b)\sin\theta - b\sin\left(\left(\frac{a+b}{b}\right)\theta\right)$

25. $\beta = \psi_2 - \psi_1 \Rightarrow \tan\beta = \tan(\psi_2 - \psi_1) = \frac{\tan\psi_2 - \tan\psi_1}{1 + \tan\psi_2 \tan\psi_1}$; the curves will be orthogonal when $\tan\beta$ is undefined, or when $\tan\psi_2 = \frac{-1}{\tan\psi_1} \Rightarrow \frac{r}{g'(\theta)} = \frac{-1}{\left[\frac{r}{f'(\theta)}\right]}$

$\Rightarrow r^2 = -f'(\theta)\, g'(\theta)$

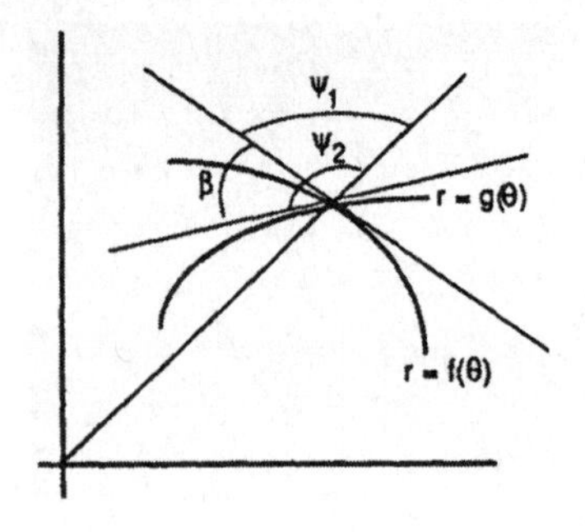

27. $r = 2a\sin 3\theta \Rightarrow \frac{dr}{d\theta} = 6a\cos 3\theta \Rightarrow \tan\psi = \frac{r}{\left(\frac{dr}{d\theta}\right)} = \frac{2a\sin 3\theta}{6a\cos 3\theta} = \frac{1}{3}\tan 3\theta$; when $\theta = \frac{\pi}{6}$, $\tan\psi = \frac{1}{3}\tan\frac{\pi}{2} \Rightarrow \psi = \frac{\pi}{2}$

29. $\tan\psi_1 = \frac{\sqrt{3}\cos\theta}{-\sqrt{3}\sin\theta} = -\cot\theta$ is $-\frac{1}{\sqrt{3}}$ at $\theta = \frac{\pi}{3}$; $\tan\psi_2 = \frac{\sin\theta}{\cos\theta} = \tan\theta$ is $\sqrt{3}$ at $\theta = \frac{\pi}{3}$; since the product of these slopes is -1, the tangents are perpendicular.

CHAPTER 12 VECTORS AND THE GEOMETRY OF SPACE

12.1 THREE-DIMENSIONAL COORDINATE SYSTEMS

1. The line through the point $(2, 3, 0)$ parallel to the z-axis

3. The x-axis

5. The circle $x^2 + y^2 = 4$ in the xy-plane

7. The circle $x^2 + z^2 = 4$ in the xz-plane

9. The circle $y^2 + z^2 = 1$ in the yz-plane

11. The circle $x^2 + y^2 = 16$ in the xy-plane

13. The ellipse formed by the intersection of the cylinder $x^2 + y^2 = 4$ and the plane $z = y$.

15. The parabola $y = x^2$ in the the xy-plane.

17. (a) The first quadrant of the xy-plane (b) The fourth quadrant of the xy-plane

19. (a) The solid ball of radius 1 centered at the origin
    (b) The exterior of the sphere of radius 1 centered at the origin

21. (a) The solid enclosed between the sphere of radius 1 and radius 2 centered at the origin
    (b) The solid upper hemisphere of radius 1 centered at the origin

23. (a) The region on or inside the parabola $y = x^2$ in the xy-plane and all points above this region.
    (b) The region on or to the left of the parabola $x = y^2$ in the xy-plane and all points above it that are 2 units or less away from the xy-plane.

25. (a) $x = 3$ (b) $y = -1$ (c) $z = -2$

27. (a) $z = 1$ (b) $x = 3$ (c) $y = -1$

29. (a) $x^2 + (y - 2)^2 = 4,\ z = 0$ (b) $(y - 2)^2 + z^2 = 4,\ x = 0$ (c) $x^2 + z^2 = 4,\ y = 2$

31. (a) $y = 3,\ z = -1$ (b) $x = 1,\ z = -1$ (c) $x = 1,\ y = 3$

33. $x^2 + y^2 + z^2 = 25,\ z = 3 \Rightarrow x^2 + y^2 = 16$ in the plane $z = 3$

35. $0 \le z \le 1$

37. $z \le 0$

39. (a) $(x - 1)^2 + (y - 1)^2 + (z - 1)^2 < 1$ (b) $(x - 1)^2 + (y - 1)^2 + (z - 1)^2 > 1$

41. $|P_1P_2| = \sqrt{(3-1)^2 + (3-1)^2 + (0-1)^2} = \sqrt{9} = 3$

43. $|P_1P_2| = \sqrt{(4-1)^2 + (-2-4)^2 + (7-5)^2} = \sqrt{49} = 7$

45. $|P_1P_2| = \sqrt{(2-0)^2 + (-2-0)^2 + (-2-0)^2} = \sqrt{3 \cdot 4} = 2\sqrt{3}$

47. center $(-2, 0, 2)$, radius $2\sqrt{2}$

49. center $\left(\sqrt{2}, \sqrt{2}, -\sqrt{2}\right)$, radius $\sqrt{2}$

51. $(x-1)^2 + (y-2)^2 + (z-3)^2 = 14$

53. $(x+1)^2 + \left(y - \frac{1}{2}\right)^2 + \left(z + \frac{2}{3}\right)^2 = \frac{16}{81}$

55. $x^2 + y^2 + z^2 + 4x - 4z = 0 \Rightarrow (x^2 + 4x + 4) + y^2 + (z^2 - 4z + 4) = 4 + 4$
$\Rightarrow (x+2)^2 + (y-0)^2 + (z-2)^2 = \left(\sqrt{8}\right)^2 \Rightarrow$ the center is at $(-2, 0, 2)$ and the radius is $\sqrt{8}$

57. $2x^2 + 2y^2 + 2z^2 + x + y + z = 9 \Rightarrow x^2 + \frac{1}{2}x + y^2 + \frac{1}{2}y + z^2 + \frac{1}{2}z = \frac{9}{2}$
$\Rightarrow \left(x^2 + \frac{1}{2}x + \frac{1}{16}\right) + \left(y^2 + \frac{1}{2}y + \frac{1}{16}\right) + \left(z^2 + \frac{1}{2}z + \frac{1}{16}\right) = \frac{9}{2} + \frac{3}{16} \Rightarrow \left(x + \frac{1}{4}\right)^2 + \left(y + \frac{1}{4}\right)^2 + \left(z + \frac{1}{4}\right)^2 = \left(\frac{5\sqrt{3}}{4}\right)^2$
$\Rightarrow$ the center is at $\left(-\frac{1}{4}, -\frac{1}{4}, -\frac{1}{4}\right)$ and the radius is $\frac{5\sqrt{3}}{4}$

59. (a) the distance between (x, y, z) and $(x, 0, 0)$ is $\sqrt{y^2 + z^2}$
(b) the distance between (x, y, z) and $(0, y, 0)$ is $\sqrt{x^2 + z^2}$
(c) the distance between (x, y, z) and $(0, 0, z)$ is $\sqrt{x^2 + y^2}$

61. $|AB| = \sqrt{(1-(-1))^2 + (-1-2)^2 + (3-1)^2} = \sqrt{4+9+4} = \sqrt{17}$
$|BC| = \sqrt{(3-1)^2 + (4-(-1))^2 + (5-3)^2} = \sqrt{4+25+4} = \sqrt{33}$
$|CA| = \sqrt{(-1-3)^2 + (2-4)^2 + (1-5)^2} = \sqrt{16+4+16} = \sqrt{36} = 6$
Thus the perimeter of triangle ABC is $\sqrt{17} + \sqrt{33} + 6$.

63. $\sqrt{(x-x)^2 + (y-(-1))^2 + (z-z)^2} = \sqrt{(x-x)^2 + (y-3)^2 + (z-z)^2} \Rightarrow (y+1)^2 = (y-3)^2 \Rightarrow 2y + 1 = -6y + 9$
$\Rightarrow y = 1$

65. (a) Since the entire sphere is below the xy-plane, the point on the sphere closest to the xy-plane is the point at the top of the sphere, which occurs when $x = 0$ and $y = 3 \Rightarrow 0^2 + (3-3)^2 + (z+5)^2 = 4 \Rightarrow z = -5 \pm 2 \Rightarrow z = -3$
$\Rightarrow (0, 3, -3)$.
(b) Both the center $(0, 3, -5)$ and the the point $(0, 7, -5)$ lie in the plane $z = -5$, so the point on the sphere closest to $(0, 7, -5)$ should also be in the same plane. In fact it should lie on the line segment between $(0, 3, -5)$ and $(0, 7, -5)$, thus the point occurs when $x = 0$ and $z = -5 \Rightarrow 0^2 + (y-3)^2 + (-5+5)^2 = 4 \Rightarrow y = 3 \pm 2 \Rightarrow y = 5$
$\Rightarrow (0, 5, -5)$.

12.2 VECTORS

1. (a) $\langle 3(3), 3(-2) \rangle = \langle 9, -6 \rangle$
(b) $\sqrt{9^2 + (-6)^2} = \sqrt{117} = 3\sqrt{13}$

3. (a) $\langle 3 + (-2), -2 + 5 \rangle = \langle 1, 3 \rangle$
(b) $\sqrt{1^2 + 3^2} = \sqrt{10}$

5. (a) $2\mathbf{u} = \langle 2(3), 2(-2)\rangle = \langle 6, -4\rangle$

$3\mathbf{v} = \langle 3(-2), 3(5)\rangle = \langle -6, 15\rangle$

$2\mathbf{u} - 3\mathbf{v} = \langle 6 - (-6), -4 - 15\rangle = \langle 12, -19\rangle$

(b) $\sqrt{12^2 + (-19)^2} = \sqrt{505}$

7. (a) $\frac{3}{5}\mathbf{u} = \left\langle \frac{3}{5}(3), \frac{3}{5}(-2)\right\rangle = \left\langle \frac{9}{5}, -\frac{6}{5}\right\rangle$

$\frac{4}{5}\mathbf{v} = \left\langle \frac{4}{5}(-2), \frac{4}{5}(5)\right\rangle = \left\langle -\frac{8}{5}, 4\right\rangle$

$\frac{3}{5}\mathbf{u} + \frac{4}{5}\mathbf{v} = \left\langle \frac{9}{5} + \left(-\frac{8}{5}\right), -\frac{6}{5} + 4\right\rangle = \left\langle \frac{1}{5}, \frac{14}{5}\right\rangle$

(b) $\sqrt{\left(\frac{1}{5}\right)^2 + \left(\frac{14}{5}\right)^2} = \frac{\sqrt{197}}{5}$

9. $\langle 2 - 1, -1 - 3\rangle = \langle 1, -4\rangle$

11. $\langle 0 - 2, 0 - 3\rangle = \langle -2, -3\rangle$

13. $\left\langle \cos\frac{2\pi}{3}, \sin\frac{2\pi}{3}\right\rangle = \left\langle -\frac{1}{2}, \frac{\sqrt{3}}{2}\right\rangle$

15. This is the unit vector which makes an angle of $120° + 90° = 210°$ with the positive x-axis;
$\langle \cos 210°, \sin 210°\rangle = \left\langle -\frac{\sqrt{3}}{2}, -\frac{1}{2}\right\rangle$

17. $\overrightarrow{P_1P_2} = (2 - 5)\mathbf{i} + (9 - 7)\mathbf{j} + (-2 - (-1))\mathbf{k} = -3\mathbf{i} + 2\mathbf{j} - \mathbf{k}$

19. $\overrightarrow{AB} = (-10 - (-7))\mathbf{i} + (8 - (-8))\mathbf{j} + (1 - 1)\mathbf{k} = -3\mathbf{i} + 16\mathbf{j}$

21. $5\mathbf{u} - \mathbf{v} = 5\langle 1, 1, -1\rangle - \langle 2, 0, 3\rangle = \langle 5, 5, -5\rangle - \langle 2, 0, 3\rangle = \langle 5 - 2, 5 - 0, -5 - 3\rangle = \langle 3, 5, -8\rangle = 3\mathbf{i} + 5\mathbf{j} - 8\mathbf{k}$

23. The vector $\mathbf{v}$ is horizontal and 1 in. long. The vectors $\mathbf{u}$ and $\mathbf{w}$ are $\frac{11}{16}$ in. long. $\mathbf{w}$ is vertical and $\mathbf{u}$ makes a 45° angle with the horizontal. All vectors must be drawn to scale.

(a)

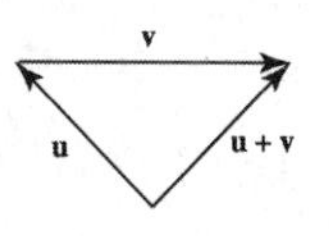

(b)

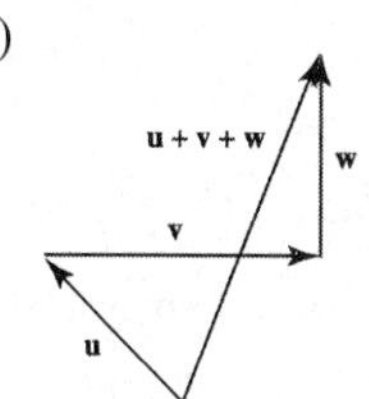

(c)

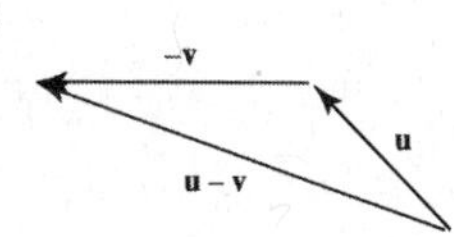

(d)

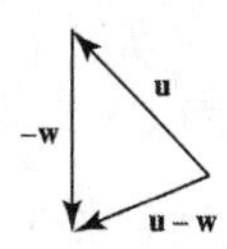

25. length $= |2\mathbf{i} + \mathbf{j} - 2\mathbf{k}| = \sqrt{2^2 + 1^2 + (-2)^2} = 3$, the direction is $\frac{2}{3}\mathbf{i} + \frac{1}{3}\mathbf{j} - \frac{2}{3}\mathbf{k} \Rightarrow 2\mathbf{i} + \mathbf{j} - 2\mathbf{k} = 3\left(\frac{2}{3}\mathbf{i} + \frac{1}{3}\mathbf{j} - \frac{2}{3}\mathbf{k}\right)$

27. length $= |5\mathbf{k}| = \sqrt{25} = 5$, the direction is $\mathbf{k} \Rightarrow 5\mathbf{k} = 5(\mathbf{k})$

29. length $= \left|\frac{1}{\sqrt{6}}\mathbf{i} - \frac{1}{\sqrt{6}}\mathbf{j} - \frac{1}{\sqrt{6}}\mathbf{k}\right| = \sqrt{3\left(\frac{1}{\sqrt{6}}\right)^2} = \sqrt{\frac{1}{2}}$, the direction is $\frac{1}{\sqrt{3}}\mathbf{i} - \frac{1}{\sqrt{3}}\mathbf{j} - \frac{1}{\sqrt{3}}\mathbf{k}$
$\Rightarrow \frac{1}{\sqrt{6}}\mathbf{i} - \frac{1}{\sqrt{6}}\mathbf{j} - \frac{1}{\sqrt{6}}\mathbf{k} = \sqrt{\frac{1}{2}}\left(\frac{1}{\sqrt{3}}\mathbf{i} - \frac{1}{\sqrt{3}}\mathbf{j} - \frac{1}{\sqrt{3}}\mathbf{k}\right)$

31. (a) $2\mathbf{i}$ (b) $-\sqrt{3}\mathbf{k}$ (c) $\frac{3}{10}\mathbf{j} + \frac{2}{5}\mathbf{k}$ (d) $6\mathbf{i} - 2\mathbf{j} + 3\mathbf{k}$

33. $|\mathbf{v}| = \sqrt{12^2 + 5^2} = \sqrt{169} = 13$; $\frac{\mathbf{v}}{|\mathbf{v}|} = \frac{1}{13}\mathbf{v} = \frac{1}{13}(12\mathbf{i} - 5\mathbf{k}) \Rightarrow$ the desired vector is $\frac{7}{13}(12\mathbf{i} - 5\mathbf{k})$

35. (a) $3\mathbf{i}+4\mathbf{j}-5\mathbf{k}=5\sqrt{2}\left(\frac{3}{5\sqrt{2}}\mathbf{i}+\frac{4}{5\sqrt{2}}\mathbf{j}-\frac{1}{\sqrt{2}}\mathbf{k}\right) \Rightarrow$ the direction is $\frac{3}{5\sqrt{2}}\mathbf{i}+\frac{4}{5\sqrt{2}}\mathbf{j}-\frac{1}{\sqrt{2}}\mathbf{k}$
(b) the midpoint is $\left(\frac{1}{2}, 3, \frac{5}{2}\right)$

37. (a) $-\mathbf{i}-\mathbf{j}-\mathbf{k}=\sqrt{3}\left(-\frac{1}{\sqrt{3}}\mathbf{i}-\frac{1}{\sqrt{3}}\mathbf{j}-\frac{1}{\sqrt{3}}\mathbf{k}\right) \Rightarrow$ the direction is $-\frac{1}{\sqrt{3}}\mathbf{i}-\frac{1}{\sqrt{3}}\mathbf{j}-\frac{1}{\sqrt{3}}\mathbf{k}$
(b) the midpoint is $\left(\frac{5}{2}, \frac{7}{2}, \frac{9}{2}\right)$

39. $\overrightarrow{AB}=(5-a)\mathbf{i}+(1-b)\mathbf{j}+(3-c)\mathbf{k}=\mathbf{i}+4\mathbf{j}-2\mathbf{k} \Rightarrow 5-a=1, 1-b=4$, and $3-c=-2 \Rightarrow a=4, b=-3$, and $c=5 \Rightarrow$ A is the point $(4,-3,5)$

41. $2\mathbf{i}+\mathbf{j}=a(\mathbf{i}+\mathbf{j})+b(\mathbf{i}-\mathbf{j})=(a+b)\mathbf{i}+(a-b)\mathbf{j} \Rightarrow a+b=2$ and $a-b=1 \Rightarrow 2a=3 \Rightarrow a=\frac{3}{2}$ and $b=a-1=\frac{1}{2}$

43. $25°$ west of north is $90°+25°=115°$ north of east. $800\langle\cos 115°, \sin 115°\rangle \approx \langle -338.095, 725.046\rangle$

45. $\mathbf{F}_1=\left\langle -|\mathbf{F}_1|\cos 30°, |\mathbf{F}_1|\sin 30°\right\rangle=\left\langle -\frac{\sqrt{3}}{2}|\mathbf{F}_1|, \frac{1}{2}|\mathbf{F}_1|\right\rangle$, $\mathbf{F}_2=\left\langle |\mathbf{F}_2|\cos 45°, |\mathbf{F}_2|\sin 45°\right\rangle=\left\langle \frac{1}{\sqrt{2}}|\mathbf{F}_2|, \frac{1}{\sqrt{2}}|\mathbf{F}_2|\right\rangle$, and $\mathbf{w}=\langle 0,-100\rangle$. Since $\mathbf{F}_1+\mathbf{F}_2=\langle 0, 100\rangle \Rightarrow \left\langle -\frac{\sqrt{3}}{2}|\mathbf{F}_1|+\frac{1}{\sqrt{2}}|\mathbf{F}_2|, \frac{1}{2}|\mathbf{F}_1|+\frac{1}{\sqrt{2}}|\mathbf{F}_2|\right\rangle=\langle 0, 100\rangle$
$\Rightarrow -\frac{\sqrt{3}}{2}|\mathbf{F}_1|+\frac{1}{\sqrt{2}}|\mathbf{F}_2|=0$ and $\frac{1}{2}|\mathbf{F}_1|+\frac{1}{\sqrt{2}}|\mathbf{F}_2|=100$. Solving the first equation for $|\mathbf{F}_2|$ results in: $|\mathbf{F}_2|=\frac{\sqrt{6}}{2}|\mathbf{F}_1|$.
Substituting this result into the second equation gives us: $\frac{1}{2}|\mathbf{F}_1|+\frac{1}{\sqrt{2}}\left(\frac{\sqrt{6}}{2}|\mathbf{F}_1|\right)=100 \Rightarrow |\mathbf{F}_1|=\frac{200}{1+\sqrt{3}} \approx 73.205$ N
$\Rightarrow |\mathbf{F}_2|=\frac{100\sqrt{6}}{1+\sqrt{3}} \approx 89.658$ N $\Rightarrow \mathbf{F}_1 \approx \langle -63.397, 36.603\rangle$ and $\mathbf{F}_2 = \approx \langle 63.397, 63.397\rangle$

47. $\mathbf{F}_1=\left\langle -|\mathbf{F}_1|\cos 40°, |\mathbf{F}_1|\sin 40°\right\rangle$, $\mathbf{F}_2=\langle 100\cos 35°, 100\sin 35°\rangle$, and $\mathbf{w}=\langle 0,-w\rangle$. Since $\mathbf{F}_1+\mathbf{F}_2=\langle 0, w\rangle$
$\Rightarrow \left\langle -|\mathbf{F}_1|\cos 40°+100\cos 35°, |\mathbf{F}_1|\sin 40°+100\sin 35°\right\rangle=\langle 0, w\rangle \Rightarrow -|\mathbf{F}_1|\cos 40°+100\cos 35°=0$ and $|\mathbf{F}_1|\sin 40°+100\sin 35°=w$. Solving the first equation for $|\mathbf{F}_1|$ results in: $|\mathbf{F}_1|=\frac{100\cos 35°}{\cos 40°} \approx 106.933$ N. Substituting this result into the second equation gives us: $w \approx 126.093$ N.

49. (a) The tree is located at the tip of the vector $\overrightarrow{OP}=(5\cos 60°)\mathbf{i}+(5\sin 60°)\mathbf{j}=\frac{5}{2}\mathbf{i}+\frac{5\sqrt{3}}{2}\mathbf{j} \Rightarrow P=\left(\frac{5}{2}, \frac{5\sqrt{3}}{2}\right)$
(b) The telephone pole is located at the point Q, which is the tip of the vector $\overrightarrow{OP}+\overrightarrow{PQ}$
$=\left(\frac{5}{2}\mathbf{i}+\frac{5\sqrt{3}}{2}\mathbf{j}\right)+(10\cos 315°)\mathbf{i}+(10\sin 315°)\mathbf{j}=\left(\frac{5}{2}+\frac{10\sqrt{2}}{2}\right)\mathbf{i}+\left(\frac{5\sqrt{3}}{2}-\frac{10\sqrt{2}}{2}\right)\mathbf{j}$
$\Rightarrow Q=\left(\frac{5+10\sqrt{2}}{2}, \frac{5\sqrt{3}-10\sqrt{2}}{2}\right)$

51. (a) the midpoint of AB is $M\left(\frac{5}{2}, \frac{5}{2}, 0\right)$ and $\overrightarrow{CM}=\left(\frac{5}{2}-1\right)\mathbf{i}+\left(\frac{5}{2}-1\right)\mathbf{j}+(0-3)\mathbf{k}=\frac{3}{2}\mathbf{i}+\frac{3}{2}\mathbf{j}-3\mathbf{k}$
(b) the desired vector is $\left(\frac{2}{3}\right)\overrightarrow{CM}=\frac{2}{3}\left(\frac{3}{2}\mathbf{i}+\frac{3}{2}\mathbf{j}-3\mathbf{k}\right)=\mathbf{i}+\mathbf{j}-2\mathbf{k}$
(c) the vector whose sum is the vector from the origin to C and the result of part (b) will terminate at the center of mass $\Rightarrow$ the terminal point of $(\mathbf{i}+\mathbf{j}+3\mathbf{k})+(\mathbf{i}+\mathbf{j}-2\mathbf{k})=2\mathbf{i}+2\mathbf{j}+\mathbf{k}$ is the point $(2,2,1)$, which is the location of the center of mass

53. Without loss of generality we identify the vertices of the quadrilateral such that $A(0,0,0)$, $B(x_b,0,0)$, $C(x_c,y_c,0)$ and $D(x_d,y_d,z_d) \Rightarrow$ the midpoint of AB is $M_{AB}\left(\frac{x_b}{2},0,0\right)$, the midpoint of BC is $M_{BC}\left(\frac{x_b+x_c}{2}, \frac{y_c}{2}, 0\right)$, the midpoint of CD is $M_{CD}\left(\frac{x_c+x_d}{2}, \frac{y_c+y_d}{2}, \frac{z_d}{2}\right)$ and the midpoint of AD is

$M_{AD}\left(\frac{x_d}{2}, \frac{y_d}{2}, \frac{z_d}{2}\right) \Rightarrow$ the midpoint of $M_{AB}M_{CD}$ is $\left(\frac{\frac{x_b}{2}+\frac{x_c+x_d}{2}}{2}, \frac{y_c+y_d}{4}, \frac{z_d}{4}\right)$ which is the same as the midpoint of $M_{AD}M_{BC} = \left(\frac{\frac{x_b+x_c}{2}+\frac{x_d}{2}}{2}, \frac{y_c+y_d}{4}, \frac{z_d}{4}\right)$.

55. Without loss of generality we can coordinatize the vertices of the triangle such that A(0, 0), B(b, 0) and $C(x_c, y_c) \Rightarrow$ a is located at $\left(\frac{b+x_c}{2}, \frac{y_c}{2}\right)$, b is at $\left(\frac{x_c}{2}, \frac{y_c}{2}\right)$ and c is at $\left(\frac{b}{2}, 0\right)$. Therefore, $\overrightarrow{Aa} = \left(\frac{b}{2}+\frac{x_c}{2}\right)\mathbf{i} + \left(\frac{y_c}{2}\right)\mathbf{j}$, $\overrightarrow{Bb} = \left(\frac{x_c}{2} - b\right)\mathbf{i} + \left(\frac{y_c}{2}\right)\mathbf{j}$, and $\overrightarrow{Cc} = \left(\frac{b}{2} - x_c\right)\mathbf{i} + (-y_c)\mathbf{j} \Rightarrow \overrightarrow{Aa} + \overrightarrow{Bb} + \overrightarrow{Cc} = \mathbf{0}$.

12.3 THE DOT PRODUCT

NOTE: In Exercises 1-8 below we calculate $\text{proj}_\mathbf{v}\,\mathbf{u}$ as the vector $\left(\frac{|\mathbf{u}|\cos\theta}{|\mathbf{v}|}\right)\mathbf{v}$, so the scalar multiplier of $\mathbf{v}$ is the number in column 5 divided by the number in column 2.

	$\mathbf{v}\cdot\mathbf{u}$	$\lvert\mathbf{v}\rvert$	$\lvert\mathbf{u}\rvert$	$\cos\theta$	$\lvert\mathbf{u}\rvert\cos\theta$	$\text{proj}_\mathbf{v}\,\mathbf{u}$
1.	-25	5	5	-1	-5	$-2\mathbf{i}+4\mathbf{j}-\sqrt{5}\mathbf{k}$
3.	25	15	5	$\frac{1}{3}$	$\frac{5}{3}$	$\frac{1}{9}(10\mathbf{i}+11\mathbf{j}-2\mathbf{k})$
5.	2	$\sqrt{34}$	$\sqrt{3}$	$\frac{2}{\sqrt{3}\sqrt{34}}$	$\frac{2}{\sqrt{34}}$	$\frac{1}{17}(5\mathbf{j}-3\mathbf{k})$
7.	$10+\sqrt{17}$	$\sqrt{26}$	$\sqrt{21}$	$\frac{10+\sqrt{17}}{\sqrt{546}}$	$\frac{10+\sqrt{17}}{\sqrt{26}}$	$\frac{10+\sqrt{17}}{26}(5\mathbf{i}+\mathbf{j})$

9. $\theta = \cos^{-1}\left(\frac{\mathbf{u}\cdot\mathbf{v}}{|\mathbf{u}|\,|\mathbf{v}|}\right) = \cos^{-1}\left(\frac{(2)(1)+(1)(2)+(0)(-1)}{\sqrt{2^2+1^2+0^2}\sqrt{1^2+2^2+(-1)^2}}\right) = \cos^{-1}\left(\frac{4}{\sqrt{5}\sqrt{6}}\right) = \cos^{-1}\left(\frac{4}{\sqrt{30}}\right) \approx 0.75$ rad

11. $\theta = \cos^{-1}\left(\frac{\mathbf{u}\cdot\mathbf{v}}{|\mathbf{u}|\,|\mathbf{v}|}\right) = \cos^{-1}\left(\frac{(\sqrt{3})(\sqrt{3})+(-7)(1)+(0)(-2)}{\sqrt{(\sqrt{3})^2+(-7)^2+0^2}\sqrt{(\sqrt{3})^2+(1)^2+(-2)^2}}\right) = \cos^{-1}\left(\frac{3-7}{\sqrt{52}\sqrt{8}}\right)$
$= \cos^{-1}\left(\frac{-1}{\sqrt{26}}\right) \approx 1.77$ rad

13. $\overrightarrow{AB} = \langle 3, 1\rangle$, $\overrightarrow{BC} = \langle -1, -3\rangle$, and $\overrightarrow{AC} = \langle 2, -2\rangle$. $\overrightarrow{BA} = \langle -3, -1\rangle$, $\overrightarrow{CB} = \langle 1, 3\rangle$, $\overrightarrow{CA} = \langle -2, 2\rangle$.
$\left|\overrightarrow{AB}\right| = \left|\overrightarrow{BA}\right| = \sqrt{10}$, $\left|\overrightarrow{BC}\right| = \left|\overrightarrow{CB}\right| = \sqrt{10}$, $\left|\overrightarrow{AC}\right| = \left|\overrightarrow{CA}\right| = 2\sqrt{2}$,

Angle at A $= \cos^{-1}\left(\frac{\overrightarrow{AB}\cdot\overrightarrow{AC}}{\left|\overrightarrow{AB}\right|\left|\overrightarrow{AC}\right|}\right) = \cos^{-1}\left(\frac{3(2)+1(-2)}{(\sqrt{10})(2\sqrt{2})}\right) = \cos^{-1}\left(\frac{1}{\sqrt{5}}\right) \approx 63.435°$

Angle at B $= \cos^{-1}\left(\frac{\overrightarrow{BC}\cdot\overrightarrow{BA}}{\left|\overrightarrow{BC}\right|\left|\overrightarrow{BA}\right|}\right) = \cos^{-1}\left(\frac{(-1)(-3)+(-3)(-1)}{(\sqrt{10})(\sqrt{10})}\right) = \cos^{-1}\left(\frac{3}{5}\right) \approx 53.130°$, and

Angle at C $= \cos^{-1}\left(\frac{\overrightarrow{CB}\cdot\overrightarrow{CA}}{\left|\overrightarrow{CB}\right|\left|\overrightarrow{CA}\right|}\right) = \cos^{-1}\left(\frac{1(-2)+3(2)}{(\sqrt{10})(2\sqrt{2})}\right) = \cos^{-1}\left(\frac{1}{\sqrt{5}}\right) \approx 63.435°$

15. (a) $\cos\alpha = \frac{\mathbf{i}\cdot\mathbf{v}}{|\mathbf{i}|\,|\mathbf{v}|} = \frac{a}{|\mathbf{v}|}$, $\cos\beta = \frac{\mathbf{j}\cdot\mathbf{v}}{|\mathbf{j}|\,|\mathbf{v}|} = \frac{b}{|\mathbf{v}|}$, $\cos\gamma = \frac{\mathbf{k}\cdot\mathbf{v}}{|\mathbf{k}|\,|\mathbf{v}|} = \frac{c}{|\mathbf{v}|}$ and
$\cos^2\alpha + \cos^2\beta + \cos^2\gamma = \left(\frac{a}{|\mathbf{v}|}\right)^2 + \left(\frac{b}{|\mathbf{v}|}\right)^2 + \left(\frac{c}{|\mathbf{v}|}\right)^2 = \frac{a^2+b^2+c^2}{|\mathbf{v}|\,|\mathbf{v}|} = \frac{|\mathbf{v}|\,|\mathbf{v}|}{|\mathbf{v}|\,|\mathbf{v}|} = 1$

(b) $|\mathbf{v}| = 1 \Rightarrow \cos\alpha = \frac{a}{|\mathbf{v}|} = a$, $\cos\beta = \frac{b}{|\mathbf{v}|} = b$ and $\cos\gamma = \frac{c}{|\mathbf{v}|} = c$ are the direction cosines of $\mathbf{v}$

17. The sum of two vectors of equal length is *always* orthogonal to their difference, as we can see from the equation $(\mathbf{v}_1 + \mathbf{v}_2) \cdot (\mathbf{v}_1 - \mathbf{v}_2) = \mathbf{v}_1 \cdot \mathbf{v}_1 + \mathbf{v}_2 \cdot \mathbf{v}_1 - \mathbf{v}_1 \cdot \mathbf{v}_2 - \mathbf{v}_2 \cdot \mathbf{v}_2 = |\mathbf{v}_1|^2 - |\mathbf{v}_2|^2 = 0$

19. Let $\mathbf{u}$ and $\mathbf{v}$ be the sides of a rhombus $\Rightarrow$ the diagonals are $\mathbf{d}_1 = \mathbf{u} + \mathbf{v}$ and $\mathbf{d}_2 = -\mathbf{u} + \mathbf{v}$ $\Rightarrow \mathbf{d}_1 \cdot \mathbf{d}_2 = (\mathbf{u} + \mathbf{v}) \cdot (-\mathbf{u} + \mathbf{v}) = -\mathbf{u} \cdot \mathbf{u} + \mathbf{u} \cdot \mathbf{v} - \mathbf{v} \cdot \mathbf{u} + \mathbf{v} \cdot \mathbf{v} = |\mathbf{v}|^2 - |\mathbf{u}|^2 = 0$ because $|\mathbf{u}| = |\mathbf{v}|$, since a rhombus has equal sides.

21. Clearly the diagonals of a rectangle are equal in length. What is not as obvious is the statement that equal diagonals happen only in a rectangle. We show this is true by letting the adjacent sides of a parallelogram be the vectors $(v_1\mathbf{i} + v_2\mathbf{j})$ and $(u_1\mathbf{i} + u_2\mathbf{j})$. The equal diagonals of the parallelogram are $\mathbf{d}_1 = (v_1\mathbf{i} + v_2\mathbf{j}) + (u_1\mathbf{i} + u_2\mathbf{j})$ and $\mathbf{d}_2 = (v_1\mathbf{i} + v_2\mathbf{j}) - (u_1\mathbf{i} + u_2\mathbf{j})$. Hence $|\mathbf{d}_1| = |\mathbf{d}_2| = |(v_1\mathbf{i} + v_2\mathbf{j}) + (u_1\mathbf{i} + u_2\mathbf{j})| = |(v_1\mathbf{i} + v_2\mathbf{j}) - (u_1\mathbf{i} + u_2\mathbf{j})|$ $\Rightarrow |(v_1 + u_1)\mathbf{i} + (v_2 + u_2)\mathbf{j}| = |(v_1 - u_1)\mathbf{i} + (v_2 - u_2)\mathbf{j}| \Rightarrow \sqrt{(v_1 + u_1)^2 + (v_2 + u_2)^2} = \sqrt{(v_1 - u_1)^2 + (v_2 - u_2)^2}$ $\Rightarrow v_1^2 + 2v_1u_1 + u_1^2 + v_2^2 + 2v_2u_2 + u_2^2 = v_1^2 - 2v_1u_1 + u_1^2 + v_2^2 - 2v_2u_2 + u_2^2 \Rightarrow 2(v_1u_1 + v_2u_2)$ $= -2(v_1u_1 + v_2u_2) \Rightarrow v_1u_1 + v_2u_2 = 0 \Rightarrow (v_1\mathbf{i} + v_2\mathbf{j}) \cdot (u_1\mathbf{i} + u_2\mathbf{j}) = 0 \Rightarrow$ the vectors $(v_1\mathbf{i} + v_2\mathbf{j})$ and $(u_1\mathbf{i} + u_2\mathbf{j})$ are perpendicular and the parallelogram must be a rectangle.

23. horizontal component: $1200 \cos(8°) \approx 1188$ ft/s; vertical component: $1200 \sin(8°) \approx 167$ ft/s

25. (a) Since $|\cos\theta| \le 1$, we have $|\mathbf{u} \cdot \mathbf{v}| = |\mathbf{u}|\,|\mathbf{v}|\,|\cos\theta| \le |\mathbf{u}|\,|\mathbf{v}|\,(1) = |\mathbf{u}|\,|\mathbf{v}|$.
(b) We have equality precisely when $|\cos\theta| = 1$ or when one or both of $\mathbf{u}$ and $\mathbf{v}$ is $\mathbf{0}$. In the case of nonzero vectors, we have equality when $\theta = 0$ or π, i.e., when the vectors are parallel.

27. $\mathbf{v} \cdot \mathbf{u}_1 = (a\mathbf{u}_1 + b\mathbf{u}_2) \cdot \mathbf{u}_1 = a\mathbf{u}_1 \cdot \mathbf{u}_1 + b\mathbf{u}_2 \cdot \mathbf{u}_1 = a|\mathbf{u}_1|^2 + b(\mathbf{u}_2 \cdot \mathbf{u}_1) = a(1)^2 + b(0) = a$

29. $\text{proj}_{\mathbf{v}}\,\mathbf{u} = \frac{\mathbf{u}\cdot\mathbf{v}}{|\mathbf{v}|^2}\mathbf{v} \Rightarrow \left(\mathbf{u} - \frac{\mathbf{u}\cdot\mathbf{v}}{|\mathbf{v}|^2}\mathbf{v}\right) \cdot \left(\frac{\mathbf{u}\cdot\mathbf{v}}{|\mathbf{v}|^2}\mathbf{v}\right) = \mathbf{u} \cdot \left(\frac{\mathbf{u}\cdot\mathbf{v}}{|\mathbf{v}|^2}\mathbf{v}\right) - \left(\frac{\mathbf{u}\cdot\mathbf{v}}{|\mathbf{v}|^2}\mathbf{v}\right) \cdot \left(\frac{\mathbf{u}\cdot\mathbf{v}}{|\mathbf{v}|^2}\mathbf{v}\right) = \left(\frac{\mathbf{u}\cdot\mathbf{v}}{|\mathbf{v}|^2}\right)(\mathbf{u} \cdot \mathbf{v}) - \left(\frac{\mathbf{u}\cdot\mathbf{v}}{|\mathbf{v}|^2}\right)^2 (\mathbf{v} \cdot \mathbf{v})$ $= \frac{(\mathbf{u}\cdot\mathbf{v})^2}{|\mathbf{v}|^2} - \frac{(\mathbf{u}\cdot\mathbf{v})^2}{|\mathbf{v}|^4}|\mathbf{v}|^2 = 0$

31. $P(x_1, y_1) = P\left(x_1, \frac{c}{b} - \frac{a}{b}x_1\right)$ and $Q(x_2, y_2) = Q\left(x_2, \frac{c}{b} - \frac{a}{b}x_2\right)$ are any two points P and Q on the line with $b \ne 0$ $\Rightarrow \overrightarrow{PQ} = (x_2 - x_1)\mathbf{i} + \frac{a}{b}(x_1 - x_2)\mathbf{j} \Rightarrow \overrightarrow{PQ} \cdot \mathbf{v} = \left[(x_2 - x_1)\mathbf{i} + \frac{a}{b}(x_1 - x_2)\mathbf{j}\right] \cdot (a\mathbf{i} + b\mathbf{j}) = a(x_2 - x_1) + b\left(\frac{a}{b}\right)(x_1 - x_2)$ $= 0 \Rightarrow \mathbf{v}$ is perpendicular to $\overrightarrow{PQ}$ for $b \ne 0$. If $b = 0$, then $\mathbf{v} = a\mathbf{i}$ is perpendicular to the vertical line $ax = c$. Alternatively, the slope of $\mathbf{v}$ is $\frac{b}{a}$ and the slope of the line $ax + by = c$ is $-\frac{a}{b}$, so the slopes are negative reciprocals $\Rightarrow$ the vector $\mathbf{v}$ and the line are perpendicular.

33. $\mathbf{v} = \mathbf{i} + 2\mathbf{j}$ is perpendicular to the line $x + 2y = c$; $P(2, 1)$ on the line $\Rightarrow 2 + 2 = c \Rightarrow x + 2y = 4$

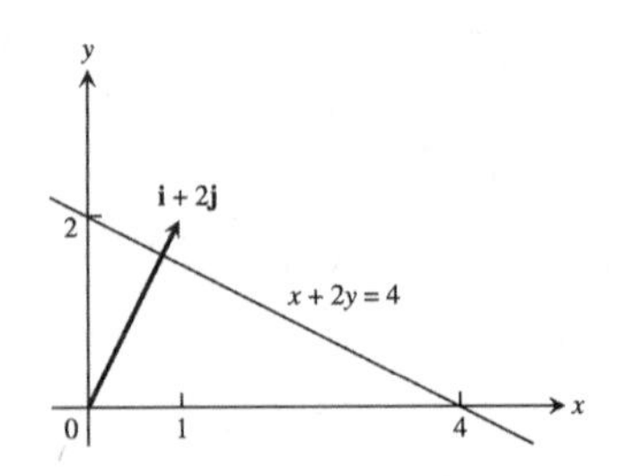

35. $\mathbf{v} = -2\mathbf{i} + \mathbf{j}$ is perpendicular to the line $-2x + y = c$;
$P(-2, -7)$ on the line $\Rightarrow (-2)(-2) - 7 = c$
$\Rightarrow -2x + y = -3$

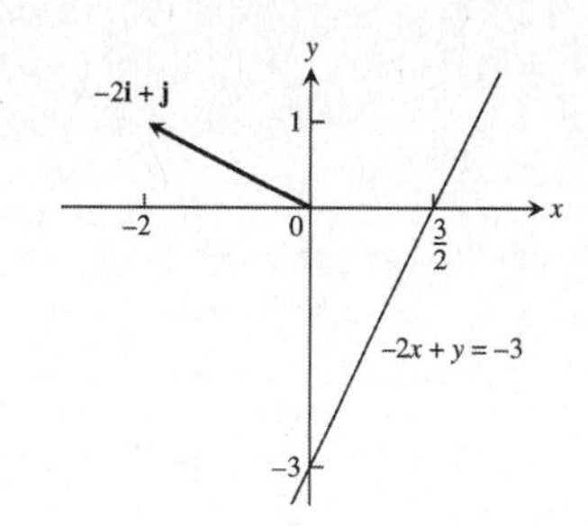

37. $\mathbf{v} = \mathbf{i} - \mathbf{j}$ is parallel to the line $-x - y = c$;
$P(-2, 1)$ on the line $\Rightarrow -(-2) - 1 = c \Rightarrow -x - y = 1$
or $x + y = -1$.

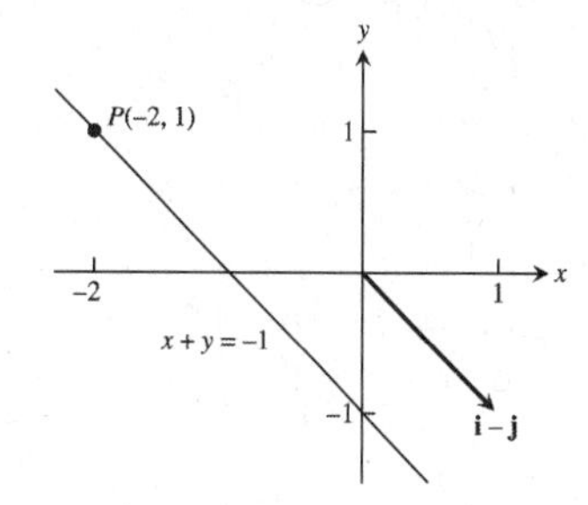

39. $\mathbf{v} = -\mathbf{i} - 2\mathbf{j}$ is parallel to the line $-2x + y = c$;
$P(1, 2)$ on the line $\Rightarrow -2(1) + 2 = c \Rightarrow -2x - y = 0$
or $2x - y = 0$.

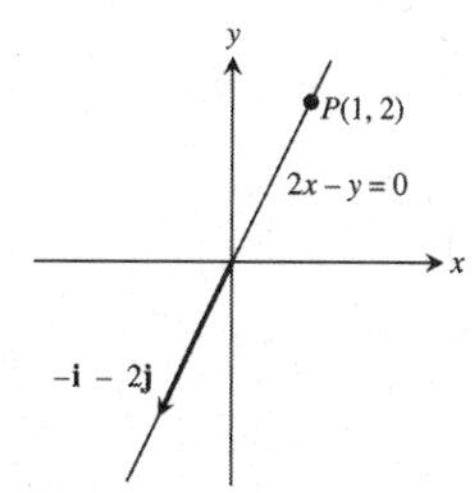

41. $P(0, 0)$, $Q(1, 1)$ and $\mathbf{F} = 5\mathbf{j} \Rightarrow \overrightarrow{PQ} = \mathbf{i} + \mathbf{j}$ and $\mathbf{W} = \mathbf{F} \cdot \overrightarrow{PQ} = (5\mathbf{j}) \cdot (\mathbf{i} + \mathbf{j}) = 5 \text{ N} \cdot \text{m} = 5 \text{ J}$

43. $\mathbf{W} = |\mathbf{F}| \left|\overrightarrow{PQ}\right| \cos\theta = (200)(20)(\cos 30°) = 2000\sqrt{3} = 3464.10 \text{ N} \cdot \text{m} = 3464.10 \text{ J}$

In Exercises 45-50 we use the fact that $\mathbf{n} = a\mathbf{i} + b\mathbf{j}$ is normal to the line $ax + by = c$.

45. $\mathbf{n}_1 = 3\mathbf{i} + \mathbf{j}$ and $\mathbf{n}_2 = 2\mathbf{i} - \mathbf{j} \Rightarrow \theta = \cos^{-1}\left(\frac{\mathbf{n}_1 \cdot \mathbf{n}_2}{|\mathbf{n}_1|\,|\mathbf{n}_2|}\right) = \cos^{-1}\left(\frac{6-1}{\sqrt{10}\sqrt{5}}\right) = \cos^{-1}\left(\frac{1}{\sqrt{2}}\right) = \frac{\pi}{4}$

47. $\mathbf{n}_1 = \sqrt{3}\mathbf{i} - \mathbf{j}$ and $\mathbf{n}_2 = \mathbf{i} - \sqrt{3}\mathbf{j} \Rightarrow \theta = \cos^{-1}\left(\frac{\mathbf{n}_1 \cdot \mathbf{n}_2}{|\mathbf{n}_1|\,|\mathbf{n}_2|}\right) = \cos^{-1}\left(\frac{\sqrt{3}+\sqrt{3}}{\sqrt{4}\sqrt{4}}\right) = \cos^{-1}\left(\frac{\sqrt{3}}{2}\right) = \frac{\pi}{6}$

49. $\mathbf{n}_1 = 3\mathbf{i} - 4\mathbf{j}$ and $\mathbf{n}_2 = \mathbf{i} - \mathbf{j} \Rightarrow \theta = \cos^{-1}\left(\frac{\mathbf{n}_1 \cdot \mathbf{n}_2}{|\mathbf{n}_1|\,|\mathbf{n}_2|}\right) = \cos^{-1}\left(\frac{3+4}{\sqrt{25}\sqrt{2}}\right) = \cos^{-1}\left(\frac{7}{5\sqrt{2}}\right) \approx 0.14 \text{ rad}$

12.4 THE CROSS PRODUCT

1. $\mathbf{u} \times \mathbf{v} = \begin{vmatrix} \mathbf{i} & \mathbf{j} & \mathbf{k} \\ 2 & -2 & -1 \\ 1 & 0 & -1 \end{vmatrix} = 3\left(\frac{2}{3}\mathbf{i} + \frac{1}{3}\mathbf{j} + \frac{2}{3}\mathbf{k}\right) \Rightarrow$ length $= 3$ and the direction is $\frac{2}{3}\mathbf{i} + \frac{1}{3}\mathbf{j} + \frac{2}{3}\mathbf{k}$;
$\mathbf{v} \times \mathbf{u} = -(\mathbf{u} \times \mathbf{v}) = -3\left(\frac{2}{3}\mathbf{i} + \frac{1}{3}\mathbf{j} + \frac{2}{3}\mathbf{k}\right) \Rightarrow$ length $= 3$ and the direction is $-\frac{2}{3}\mathbf{i} - \frac{1}{3}\mathbf{j} - \frac{2}{3}\mathbf{k}$

3. $\mathbf{u} \times \mathbf{v} = \begin{vmatrix} \mathbf{i} & \mathbf{j} & \mathbf{k} \\ 2 & -2 & 4 \\ -1 & 1 & -2 \end{vmatrix} = \mathbf{0} \Rightarrow$ length $= 0$ and has no direction

$\mathbf{v} \times \mathbf{u} = -(\mathbf{u} \times \mathbf{v}) = \mathbf{0} \Rightarrow$ length $= 0$ and has no direction

5. $\mathbf{u} \times \mathbf{v} = \begin{vmatrix} \mathbf{i} & \mathbf{j} & \mathbf{k} \\ 2 & 0 & 0 \\ 0 & -3 & 0 \end{vmatrix} = -6(\mathbf{k}) \Rightarrow$ length $= 6$ and the direction is $-\mathbf{k}$

$\mathbf{v} \times \mathbf{u} = -(\mathbf{u} \times \mathbf{v}) = 6(\mathbf{k}) \Rightarrow$ length $= 6$ and the direction is $\mathbf{k}$

7. $\mathbf{u} \times \mathbf{v} = \begin{vmatrix} \mathbf{i} & \mathbf{j} & \mathbf{k} \\ -8 & -2 & -4 \\ 2 & 2 & 1 \end{vmatrix} = 6\mathbf{i} - 12\mathbf{k} \Rightarrow$ length $= 6\sqrt{5}$ and the direction is $\frac{1}{\sqrt{5}}\mathbf{i} - \frac{2}{\sqrt{5}}\mathbf{k}$

$\mathbf{v} \times \mathbf{u} = -(\mathbf{u} \times \mathbf{v}) = -(6\mathbf{i} - 12\mathbf{k}) \Rightarrow$ length $= 6\sqrt{5}$ and the direction is $-\frac{1}{\sqrt{5}}\mathbf{i} + \frac{2}{\sqrt{5}}\mathbf{k}$

9. $\mathbf{u} \times \mathbf{v} = \begin{vmatrix} \mathbf{i} & \mathbf{j} & \mathbf{k} \\ 1 & 0 & 0 \\ 0 & 1 & 0 \end{vmatrix} = \mathbf{k}$

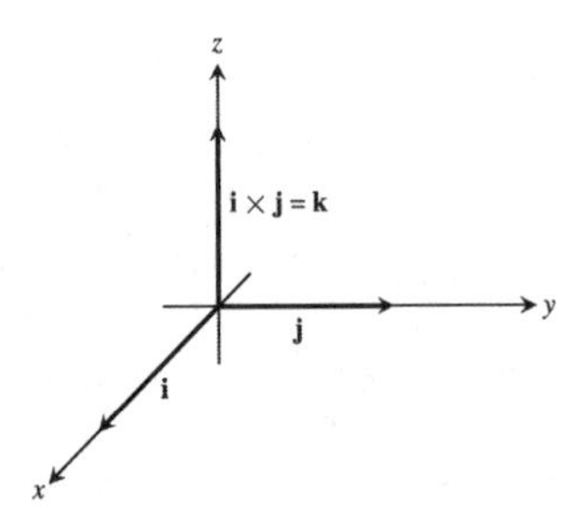

11. $\mathbf{u} \times \mathbf{v} = \begin{vmatrix} \mathbf{i} & \mathbf{j} & \mathbf{k} \\ 1 & 0 & -1 \\ 0 & 1 & 1 \end{vmatrix} = \mathbf{i} - \mathbf{j} + \mathbf{k}$

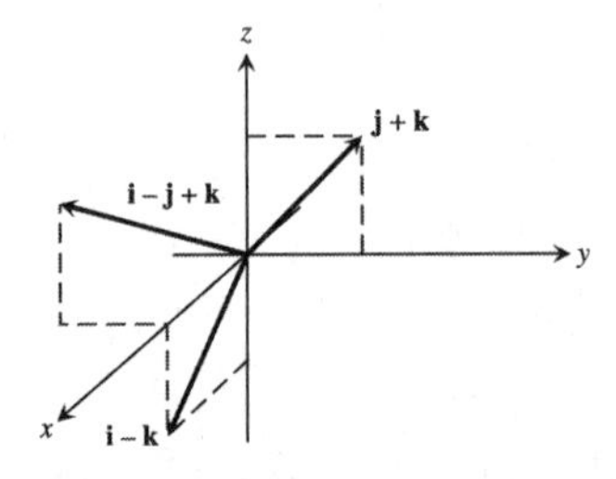

13. $\mathbf{u} \times \mathbf{v} = \begin{vmatrix} \mathbf{i} & \mathbf{j} & \mathbf{k} \\ 1 & 1 & 0 \\ 1 & -1 & 0 \end{vmatrix} = -2\mathbf{k}$

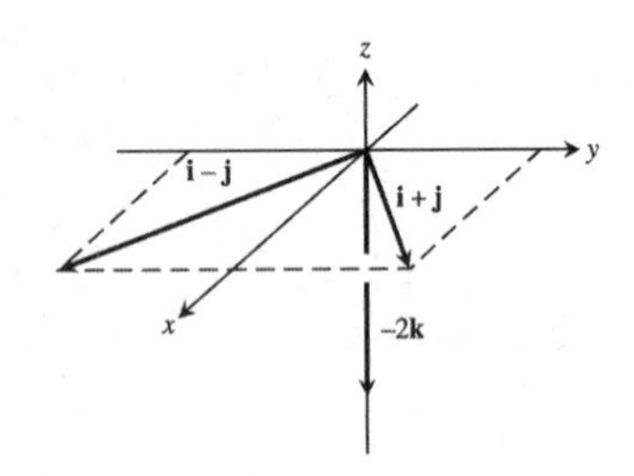

15. (a) $\overrightarrow{PQ} \times \overrightarrow{PR} = \begin{vmatrix} \mathbf{i} & \mathbf{j} & \mathbf{k} \\ 1 & 1 & -3 \\ -1 & 3 & -1 \end{vmatrix} = 8\mathbf{i} + 4\mathbf{j} + 4\mathbf{k} \Rightarrow$ Area $= \frac{1}{2}\left|\overrightarrow{PQ} \times \overrightarrow{PR}\right| = \frac{1}{2}\sqrt{64 + 16 + 16} = 2\sqrt{6}$

(b) $\mathbf{u} = \frac{\overrightarrow{PQ} \times \overrightarrow{PR}}{\left|\overrightarrow{PQ} \times \overrightarrow{PR}\right|} = \frac{1}{\sqrt{6}}(2\mathbf{i} + \mathbf{j} + \mathbf{k})$

17. (a) $\overrightarrow{PQ} \times \overrightarrow{PR} = \begin{vmatrix} \mathbf{i} & \mathbf{j} & \mathbf{k} \\ 1 & 1 & 1 \\ 1 & 1 & 0 \end{vmatrix} = -\mathbf{i} + \mathbf{j} \Rightarrow$ Area $= \frac{1}{2}\left|\overrightarrow{PQ} \times \overrightarrow{PR}\right| = \frac{1}{2}\sqrt{1+1} = \frac{\sqrt{2}}{2}$

(b) $\mathbf{u} = \frac{\overrightarrow{PQ} \times \overrightarrow{PR}}{\left|\overrightarrow{PQ} \times \overrightarrow{PR}\right|} = \frac{1}{\sqrt{2}}(-\mathbf{i} + \mathbf{j}) = -\frac{1}{\sqrt{2}}(\mathbf{i} - \mathbf{j})$

19. If $\mathbf{u} = a_1\mathbf{i} + a_2\mathbf{j} + a_3\mathbf{k}$, $\mathbf{v} = b_1\mathbf{i} + b_2\mathbf{j} + b_3\mathbf{k}$, and $\mathbf{w} = c_1\mathbf{i} + c_2\mathbf{j} + c_3\mathbf{k}$, then $(\mathbf{u} \times \mathbf{v}) \cdot \mathbf{w} = \begin{vmatrix} a_1 & a_2 & a_3 \\ b_1 & b_2 & b_3 \\ c_1 & c_2 & c_3 \end{vmatrix}$,

$(\mathbf{v} \times \mathbf{w}) \cdot \mathbf{u} = \begin{vmatrix} b_1 & b_2 & b_3 \\ c_1 & c_2 & c_3 \\ a_1 & a_2 & a_3 \end{vmatrix}$ and $(\mathbf{w} \times \mathbf{u}) \cdot \mathbf{v} = \begin{vmatrix} c_1 & c_2 & c_3 \\ a_1 & a_2 & a_3 \\ b_1 & b_2 & b_3 \end{vmatrix}$ which all have the same absolute value, since the interchanging of two rows in a determinant does not change its absolute value $\Rightarrow$ the volume is $|(\mathbf{u} \times \mathbf{v}) \cdot \mathbf{w}| = \text{abs} \begin{vmatrix} 2 & 0 & 0 \\ 0 & 2 & 0 \\ 0 & 0 & 2 \end{vmatrix} = 8$

21. $|(\mathbf{u} \times \mathbf{v}) \cdot \mathbf{w}| = \text{abs} \begin{vmatrix} 2 & 1 & 0 \\ 2 & -1 & 1 \\ 1 & 0 & 2 \end{vmatrix} = |-7| = 7$ (for details about verification, see Exercise 19)

23. (a) $\mathbf{u} \cdot \mathbf{v} = -6$, $\mathbf{u} \cdot \mathbf{w} = -81$, $\mathbf{v} \cdot \mathbf{w} = 18 \Rightarrow$ none are perpendicular

(b) $\mathbf{u} \times \mathbf{v} = \begin{vmatrix} \mathbf{i} & \mathbf{j} & \mathbf{k} \\ 5 & -1 & 1 \\ 0 & 1 & -5 \end{vmatrix} \neq \mathbf{0}$, $\mathbf{u} \times \mathbf{w} = \begin{vmatrix} \mathbf{i} & \mathbf{j} & \mathbf{k} \\ 5 & -1 & 1 \\ -15 & 3 & -3 \end{vmatrix} = \mathbf{0}$, $\mathbf{v} \times \mathbf{w} = \begin{vmatrix} \mathbf{i} & \mathbf{j} & \mathbf{k} \\ 0 & 1 & -5 \\ -15 & 3 & -3 \end{vmatrix} \neq \mathbf{0}$

$\Rightarrow$ **u** and **w** are parallel

25. $\left|\overrightarrow{PQ} \times \mathbf{F}\right| = \left|\overrightarrow{PQ}\right| |\mathbf{F}| \sin(60°) = \frac{2}{3} \cdot 30 \cdot \frac{\sqrt{3}}{2}$ ft · lb $= 10\sqrt{3}$ ft · lb

27. (a) true, $|\mathbf{u}| = \sqrt{a_1^2 + a_2^2 + a_3^2} = \sqrt{\mathbf{u} \cdot \mathbf{u}}$

(b) not always true, $\mathbf{u} \cdot \mathbf{u} = |\mathbf{u}|^2$

(c) true, $\mathbf{u} \times \mathbf{0} = \begin{vmatrix} \mathbf{i} & \mathbf{j} & \mathbf{k} \\ u_1 & u_2 & u_3 \\ 0 & 0 & 0 \end{vmatrix} = 0\mathbf{i} + 0\mathbf{j} + 0\mathbf{k} = \mathbf{0}$ and $\mathbf{0} \times \mathbf{u} = \begin{vmatrix} \mathbf{i} & \mathbf{j} & \mathbf{k} \\ 0 & 0 & 0 \\ u_1 & u_2 & u_3 \end{vmatrix} = 0\mathbf{i} + 0\mathbf{j} + 0\mathbf{k} = \mathbf{0}$

(d) true, $\mathbf{u} \times (-\mathbf{u}) = \begin{vmatrix} \mathbf{i} & \mathbf{j} & \mathbf{k} \\ u_1 & u_2 & u_3 \\ -u_1 & -u_2 & -u_3 \end{vmatrix} = (-u_2u_3 + u_2u_3)\mathbf{i} - (-u_1u_3 + u_1u_3)\mathbf{j} + (-u_1u_2 + u_1u_2)\mathbf{k} = \mathbf{0}$

(e) not always true, $\mathbf{i} \times \mathbf{j} = \mathbf{k} \neq -\mathbf{k} = \mathbf{j} \times \mathbf{i}$ for example

(f) true, distributive property of the cross product

(g) true, $(\mathbf{u} \times \mathbf{v}) \cdot \mathbf{v} = \mathbf{u} \cdot (\mathbf{v} \times \mathbf{v}) = \mathbf{u} \cdot \mathbf{0} = 0$

(h) true, the volume of a parallelpiped with **u**, **v**, and **w** along the three edges is the same whether the plane containing **u** and **v** or the plane containing **v** and **w** is used as the base plane, and the dot product is commutative.

29. (a) $\text{proj}_\mathbf{v}\, \mathbf{u} = \left(\frac{\mathbf{u} \cdot \mathbf{v}}{|\mathbf{v}||\mathbf{v}|}\right) \mathbf{v}$ (b) $(\mathbf{u} \times \mathbf{v})$ (c) $((\mathbf{u} \times \mathbf{v}) \times \mathbf{w})$ (d) $|(\mathbf{u} \times \mathbf{v}) \cdot \mathbf{w}|$

(e) $(\mathbf{u} \times \mathbf{v}) \times (\mathbf{u} \times \mathbf{w})$ (f) $|\mathbf{u}| \frac{\mathbf{v}}{|\mathbf{v}|}$

31. (a) yes, $\mathbf{u} \times \mathbf{v}$ and **w** are both vectors (b) no, **u** is a vector but $\mathbf{v} \cdot \mathbf{w}$ is a scalar

(c) yes, **u** and $\mathbf{u} \times \mathbf{w}$ are both vectors (d) no, **u** is a vector but $\mathbf{v} \cdot \mathbf{w}$ is a scalar

33. No, **v** need not equal **w**. For example, $\mathbf{i} + \mathbf{j} \neq -\mathbf{i} + \mathbf{j}$, but $\mathbf{i} \times (\mathbf{i} + \mathbf{j}) = \mathbf{i} \times \mathbf{i} + \mathbf{i} \times \mathbf{j} = \mathbf{0} + \mathbf{k} = \mathbf{k}$ and $\mathbf{i} \times (-\mathbf{i} + \mathbf{j}) = \mathbf{i} \times (-\mathbf{i}) + \mathbf{i} \times \mathbf{j} = \mathbf{0} + \mathbf{k} = \mathbf{k}$.

35. $\overrightarrow{AB} = -\mathbf{i} + \mathbf{j}$ and $\overrightarrow{AD} = -\mathbf{i} - \mathbf{j} \Rightarrow \overrightarrow{AB} \times \overrightarrow{AD} = \begin{vmatrix} \mathbf{i} & \mathbf{j} & \mathbf{k} \\ -1 & 1 & 0 \\ -1 & -1 & 0 \end{vmatrix} = 2\mathbf{k} \Rightarrow$ area $= \left|\overrightarrow{AB} \times \overrightarrow{AD}\right| = 2$

37. $\overrightarrow{AB} = 3\mathbf{i} - 2\mathbf{j}$ and $\overrightarrow{AD} = 5\mathbf{i} + \mathbf{j} \Rightarrow \overrightarrow{AB} \times \overrightarrow{AD} = \begin{vmatrix} \mathbf{i} & \mathbf{j} & \mathbf{k} \\ 3 & -2 & 0 \\ 5 & 1 & 0 \end{vmatrix} = 13\mathbf{k} \Rightarrow$ area $= \left|\overrightarrow{AB} \times \overrightarrow{AD}\right| = 13$

39. $\overrightarrow{AB} = 3\mathbf{i} + 2\mathbf{j} + 4\mathbf{k}$ and $\overrightarrow{DC} = 3\mathbf{i} + 2\mathbf{j} + 4\mathbf{k} \Rightarrow \overrightarrow{AB}$ is parallel to $\overrightarrow{DC}$; $\overrightarrow{BC} = 2\mathbf{i} - \mathbf{j}$ and $\overrightarrow{AD} = 2\mathbf{i} - \mathbf{j} \Rightarrow \overrightarrow{BC}$ is parallel to $\overrightarrow{AD}$. $\overrightarrow{AB} \times \overrightarrow{BC} = \begin{vmatrix} \mathbf{i} & \mathbf{j} & \mathbf{k} \\ 3 & 2 & 4 \\ 2 & -1 & 0 \end{vmatrix} = 4\mathbf{i} + 8\mathbf{j} - 7\mathbf{k} \Rightarrow$ area $= |\overrightarrow{AB} \times \overrightarrow{BC}| = \sqrt{129}$

41. $\overrightarrow{AB} = -2\mathbf{i} + 3\mathbf{j}$ and $\overrightarrow{AC} = 3\mathbf{i} + \mathbf{j} \Rightarrow \overrightarrow{AB} \times \overrightarrow{AC} = \begin{vmatrix} \mathbf{i} & \mathbf{j} & \mathbf{k} \\ -2 & 3 & 0 \\ 3 & 1 & 0 \end{vmatrix} = -11\mathbf{k} \Rightarrow$ area $= \frac{1}{2}\left|\overrightarrow{AB} \times \overrightarrow{AC}\right| = \frac{11}{2}$

43. $\overrightarrow{AB} = 6\mathbf{i} - 5\mathbf{j}$ and $\overrightarrow{AC} = 11\mathbf{i} - 5\mathbf{j} \Rightarrow \overrightarrow{AB} \times \overrightarrow{AC} = \begin{vmatrix} \mathbf{i} & \mathbf{j} & \mathbf{k} \\ 6 & -5 & 0 \\ 11 & -5 & 0 \end{vmatrix} = 25\mathbf{k} \Rightarrow$ area $= \frac{1}{2}\left|\overrightarrow{AB} \times \overrightarrow{AC}\right| = \frac{25}{2}$

45. $\overrightarrow{AB} = -\mathbf{i} + 2\mathbf{j}$ and $\overrightarrow{AC} = -\mathbf{i} - \mathbf{k} \Rightarrow \overrightarrow{AB} \times \overrightarrow{AC} = \begin{vmatrix} \mathbf{i} & \mathbf{j} & \mathbf{k} \\ -1 & 2 & 0 \\ -1 & 0 & -1 \end{vmatrix} = -2\mathbf{i} - \mathbf{j} + 2\mathbf{k} \Rightarrow$ area $= \frac{1}{2}\left|\overrightarrow{AB} \times \overrightarrow{AC}\right| = \frac{3}{2}$

47. $\overrightarrow{AB} = -\mathbf{i} + 2\mathbf{j}$ and $\overrightarrow{AC} = \mathbf{j} - 2\mathbf{k} \Rightarrow \overrightarrow{AB} \times \overrightarrow{AC} = \begin{vmatrix} \mathbf{i} & \mathbf{j} & \mathbf{k} \\ -1 & 2 & 0 \\ 0 & 1 & -2 \end{vmatrix} = -4\mathbf{i} - 2\mathbf{j} - \mathbf{k} \Rightarrow$ area $= \frac{1}{2}\left|\overrightarrow{AB} \times \overrightarrow{AC}\right| = \frac{\sqrt{21}}{2}$

49. If $\mathbf{A} = a_1\mathbf{i} + a_2\mathbf{j}$ and $\mathbf{B} = b_1\mathbf{i} + b_2\mathbf{j}$, then $\mathbf{A} \times \mathbf{B} = \begin{vmatrix} \mathbf{i} & \mathbf{j} & \mathbf{k} \\ a_1 & a_2 & 0 \\ b_1 & b_2 & 0 \end{vmatrix} = \begin{vmatrix} a_1 & a_2 \\ b_1 & b_2 \end{vmatrix}\mathbf{k}$ and the triangle's area is $\frac{1}{2}|\mathbf{A} \times \mathbf{B}| = \pm\frac{1}{2}\begin{vmatrix} a_1 & a_2 \\ b_1 & b_2 \end{vmatrix}$. The applicable sign is (+) if the acute angle from **A** to **B** runs counterclockwise in the xy-plane, and (−) if it runs clockwise, because the area must be a nonnegative number.

12.5 LINES AND PLANES IN SPACE

1. The direction $\mathbf{i} + \mathbf{j} + \mathbf{k}$ and $P(3, -4, -1) \Rightarrow x = 3 + t,\ y = -4 + t,\ z = -1 + t$

3. The direction $\overrightarrow{PQ} = 5\mathbf{i} + 5\mathbf{j} - 5\mathbf{k}$ and $P(-2, 0, 3) \Rightarrow x = -2 + 5t,\ y = 5t,\ z = 3 - 5t$

5. The direction $2\mathbf{j} + \mathbf{k}$ and $P(0, 0, 0) \Rightarrow x = 0,\ y = 2t,\ z = t$

7. The direction $\mathbf{k}$ and $P(1, 1, 1) \Rightarrow x = 1,\ y = 1,\ z = 1 + t$

9. The direction $\mathbf{i} + 2\mathbf{j} + 2\mathbf{k}$ and $P(0, -7, 0) \Rightarrow x = t,\ y = -7 + 2t,\ z = 2t$

11. The direction $\mathbf{i}$ and $P(0, 0, 0) \Rightarrow x = t,\ y = 0,\ z = 0$

13. The direction $\overrightarrow{PQ} = \mathbf{i} + \mathbf{j} + \frac{3}{2}\mathbf{k}$ and $P(0,0,0) \Rightarrow x = t$, $y = t$, $z = \frac{3}{2}t$, where $0 \le t \le 1$

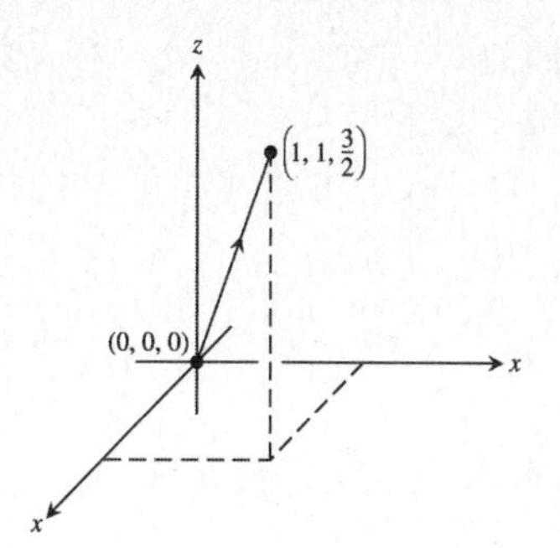

15. The direction $\overrightarrow{PQ} = \mathbf{j}$ and $P(1,1,0) \Rightarrow x = 1$, $y = 1 + t$, $z = 0$, where $-1 \le t \le 0$

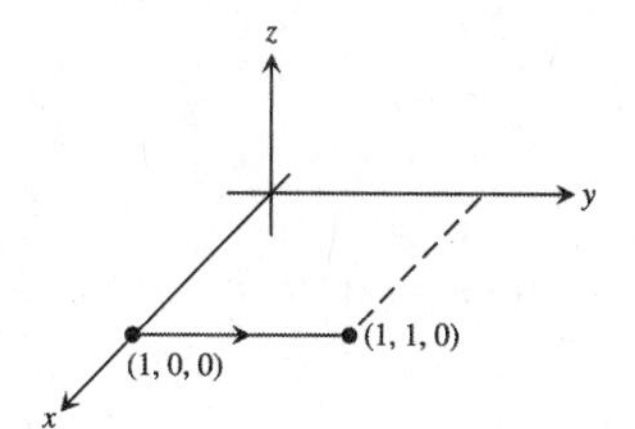

17. The direction $\overrightarrow{PQ} = -2\mathbf{j}$ and $P(0,1,1) \Rightarrow x = 0$, $y = 1 - 2t$, $z = 1$, where $0 \le t \le 1$

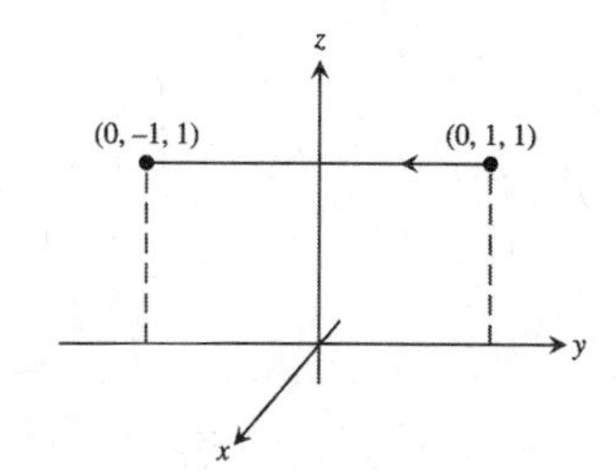

19. The direction $\overrightarrow{PQ} = -2\mathbf{i} + 2\mathbf{j} - 2\mathbf{k}$ and $P(2,0,2)$ $\Rightarrow x = 2 - 2t$, $y = 2t$, $z = 2 - 2t$, where $0 \le t \le 1$

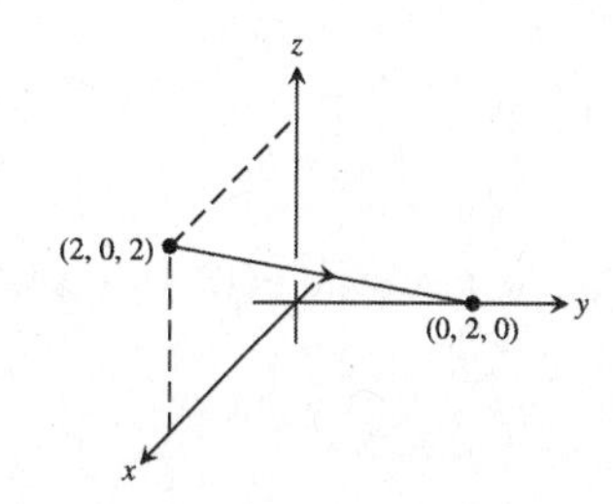

21. $3(x - 0) + (-2)(y - 2) + (-1)(z + 1) = 0 \Rightarrow 3x - 2y - z = -3$

23. $\overrightarrow{PQ} = \mathbf{i} - \mathbf{j} + 3\mathbf{k}$, $\overrightarrow{PS} = -\mathbf{i} - 3\mathbf{j} + 2\mathbf{k} \Rightarrow \overrightarrow{PQ} \times \overrightarrow{PS} = \begin{vmatrix} \mathbf{i} & \mathbf{j} & \mathbf{k} \\ 1 & -1 & 3 \\ -1 & -3 & 2 \end{vmatrix} = 7\mathbf{i} - 5\mathbf{j} - 4\mathbf{k}$ is normal to the plane $\Rightarrow 7(x - 2) + (-5)(y - 0) + (-4)(z - 2) = 0 \Rightarrow 7x - 5y - 4z = 6$

25. $\mathbf{n} = \mathbf{i} + 3\mathbf{j} + 4\mathbf{k}$, $P(2,4,5) \Rightarrow (1)(x - 2) + (3)(y - 4) + (4)(z - 5) = 0 \Rightarrow x + 3y + 4z = 34$

27. $\begin{cases} x = 2t + 1 = s + 2 \\ y = 3t + 2 = 2s + 4 \end{cases} \Rightarrow \begin{cases} 2t - s = 1 \\ 3t - 2s = 2 \end{cases} \Rightarrow \begin{cases} 4t - 2s = 2 \\ 3t - 2s = 2 \end{cases} \Rightarrow t = 0$ and $s = -1$; then $z = 4t + 3 = -4s - 1$ $\Rightarrow 4(0) + 3 = (-4)(-1) - 1$ is satisfied $\Rightarrow$ the lines intersect when $t = 0$ and $s = -1 \Rightarrow$ the point of intersection is $x = 1$, $y = 2$, and $z = 3$ or $P(1,2,3)$. A vector normal to the plane determined by these lines is $\mathbf{n}_1 \times \mathbf{n}_2 = \begin{vmatrix} \mathbf{i} & \mathbf{j} & \mathbf{k} \\ 2 & 3 & 4 \\ 1 & 2 & -4 \end{vmatrix} = -20\mathbf{i} + 12\mathbf{j} + \mathbf{k}$, where $\mathbf{n}_1$ and $\mathbf{n}_2$ are directions of the lines $\Rightarrow$ the plane containing the lines is represented by $(-20)(x - 1) + (12)(y - 2) + (1)(z - 3) = 0 \Rightarrow -20x + 12y + z = 7$.

29. The cross product of $\mathbf{i}+\mathbf{j}-\mathbf{k}$ and $-4\mathbf{i}+2\mathbf{j}-2\mathbf{k}$ has the same direction as the normal to the plane $\Rightarrow \mathbf{n} = \begin{vmatrix} \mathbf{i} & \mathbf{j} & \mathbf{k} \\ 1 & 1 & -1 \\ -4 & 2 & -2 \end{vmatrix} = 6\mathbf{j}+6\mathbf{k}$. Select a point on either line, such as $P(-1,2,1)$. Since the lines are given to intersect, the desired plane is $0(x+1)+6(y-2)+6(z-1)=0 \Rightarrow 6y+6z=18 \Rightarrow y+z=3$.

31. $\mathbf{n}_1 \times \mathbf{n}_2 = \begin{vmatrix} \mathbf{i} & \mathbf{j} & \mathbf{k} \\ 2 & 1 & -1 \\ 1 & 2 & 1 \end{vmatrix} = 3\mathbf{i}-3\mathbf{j}+3\mathbf{k}$ is a vector in the direction of the line of intersection of the planes $\Rightarrow 3(x-2)+(-3)(y-1)+3(z+1)=0 \Rightarrow 3x-3y+3z=0 \Rightarrow x-y+z=0$ is the desired plane containing $P_0(2,1,-1)$

33. $S(0,0,12)$, $P(0,0,0)$ and $\mathbf{v}=4\mathbf{i}-2\mathbf{j}+2\mathbf{k} \Rightarrow \overrightarrow{PS} \times \mathbf{v} = \begin{vmatrix} \mathbf{i} & \mathbf{j} & \mathbf{k} \\ 0 & 0 & 12 \\ 4 & -2 & 2 \end{vmatrix} = 24\mathbf{i}+48\mathbf{j} = 24(\mathbf{i}+2\mathbf{j})$

$\Rightarrow d = \frac{|\overrightarrow{PS}\times\mathbf{v}|}{|\mathbf{v}|} = \frac{24\sqrt{1+4}}{\sqrt{16+4+4}} = \frac{24\sqrt{5}}{\sqrt{24}} = \sqrt{5\cdot 24} = 2\sqrt{30}$ is the distance from S to the line

35. $S(2,1,3)$, $P(2,1,3)$ and $\mathbf{v}=2\mathbf{i}+6\mathbf{j} \Rightarrow \overrightarrow{PS} \times \mathbf{v} = \mathbf{0} \Rightarrow d = \frac{|\overrightarrow{PS}\times\mathbf{v}|}{|\mathbf{v}|} = \frac{0}{\sqrt{40}} = 0$ is the distance from S to the line (i.e., the point S lies on the line)

37. $S(3,-1,4)$, $P(4,3,-5)$ and $\mathbf{v}=-\mathbf{i}+2\mathbf{j}+3\mathbf{k} \Rightarrow \overrightarrow{PS} \times \mathbf{v} = \begin{vmatrix} \mathbf{i} & \mathbf{j} & \mathbf{k} \\ -1 & -4 & 9 \\ -1 & 2 & 3 \end{vmatrix} = -30\mathbf{i}-6\mathbf{j}-6\mathbf{k}$

$\Rightarrow d = \frac{|\overrightarrow{PS}\times\mathbf{v}|}{|\mathbf{v}|} = \frac{\sqrt{900+36+36}}{\sqrt{1+4+9}} = \frac{\sqrt{972}}{\sqrt{14}} = \frac{\sqrt{486}}{\sqrt{7}} = \frac{\sqrt{81\cdot 6}}{\sqrt{7}} = \frac{9\sqrt{42}}{7}$ is the distance from S to the line

39. $S(2,-3,4)$, $x+2y+2z=13$ and $P(13,0,0)$ is on the plane $\Rightarrow \overrightarrow{PS} = -11\mathbf{i}-3\mathbf{j}+4\mathbf{k}$ and $\mathbf{n}=\mathbf{i}+2\mathbf{j}+2\mathbf{k}$ $\Rightarrow d = \left|\overrightarrow{PS}\cdot\frac{\mathbf{n}}{|\mathbf{n}|}\right| = \left|\frac{-11-6+8}{\sqrt{1+4+4}}\right| = \left|\frac{-9}{\sqrt{9}}\right| = 3$

41. $S(0,1,1)$, $4y+3z=-12$ and $P(0,-3,0)$ is on the plane $\Rightarrow \overrightarrow{PS} = 4\mathbf{j}+\mathbf{k}$ and $\mathbf{n}=4\mathbf{j}+3\mathbf{k}$ $\Rightarrow d = \left|\overrightarrow{PS}\cdot\frac{\mathbf{n}}{|\mathbf{n}|}\right| = \left|\frac{16+3}{\sqrt{16+9}}\right| = \frac{19}{5}$

43. $S(0,-1,0)$, $2x+y+2z=4$ and $P(2,0,0)$ is on the plane $\Rightarrow \overrightarrow{PS} = -2\mathbf{i}-\mathbf{j}$ and $\mathbf{n}=2\mathbf{i}+\mathbf{j}+2\mathbf{k}$ $\Rightarrow d = \left|\overrightarrow{PS}\cdot\frac{\mathbf{n}}{|\mathbf{n}|}\right| = \left|\frac{-4-1+0}{\sqrt{4+1+4}}\right| = \frac{5}{3}$

45. The point $P(1,0,0)$ is on the first plane and $S(10,0,0)$ is a point on the second plane $\Rightarrow \overrightarrow{PS} = 9\mathbf{i}$, and $\mathbf{n}=\mathbf{i}+2\mathbf{j}+6\mathbf{k}$ is normal to the first plane $\Rightarrow$ the distance from S to the first plane is $d = \left|\overrightarrow{PS}\cdot\frac{\mathbf{n}}{|\mathbf{n}|}\right| = \left|\frac{9}{\sqrt{1+4+36}}\right| = \frac{9}{\sqrt{41}}$, which is also the distance between the planes.

47. $\mathbf{n}_1=\mathbf{i}+\mathbf{j}$ and $\mathbf{n}_2=2\mathbf{i}+\mathbf{j}-2\mathbf{k} \Rightarrow \theta = \cos^{-1}\left(\frac{\mathbf{n}_1\cdot\mathbf{n}_2}{|\mathbf{n}_1|\,|\mathbf{n}_2|}\right) = \cos^{-1}\left(\frac{2+1}{\sqrt{2}\sqrt{9}}\right) = \cos^{-1}\left(\frac{1}{\sqrt{2}}\right) = \frac{\pi}{4}$

49. $\mathbf{n}_1=2\mathbf{i}+2\mathbf{j}+2\mathbf{k}$ and $\mathbf{n}_2=2\mathbf{i}-2\mathbf{j}-\mathbf{k} \Rightarrow \theta = \cos^{-1}\left(\frac{\mathbf{n}_1\cdot\mathbf{n}_2}{|\mathbf{n}_1|\,|\mathbf{n}_2|}\right) = \cos^{-1}\left(\frac{4-4-2}{\sqrt{12}\sqrt{9}}\right) = \cos^{-1}\left(\frac{-1}{3\sqrt{3}}\right) \approx 1.76$ rad

51. $\mathbf{n}_1 = 2\mathbf{i} + 2\mathbf{j} - \mathbf{k}$ and $\mathbf{n}_2 = \mathbf{i} + 2\mathbf{j} + \mathbf{k} \Rightarrow \theta = \cos^{-1}\left(\frac{\mathbf{n}_1 \cdot \mathbf{n}_2}{|\mathbf{n}_1|\,|\mathbf{n}_2|}\right) = \cos^{-1}\left(\frac{2+4-1}{\sqrt{9}\sqrt{6}}\right) = \cos^{-1}\left(\frac{5}{3\sqrt{6}}\right) \approx 0.82$ rad

53. $2x - y + 3z = 6 \Rightarrow 2(1-t) - (3t) + 3(1+t) = 6 \Rightarrow -2t + 5 = 6 \Rightarrow t = -\frac{1}{2} \Rightarrow x = \frac{3}{2}, y = -\frac{3}{2}$ and $z = \frac{1}{2}$ $\Rightarrow \left(\frac{3}{2}, -\frac{3}{2}, \frac{1}{2}\right)$ is the point

55. $x + y + z = 2 \Rightarrow (1+2t) + (1+5t) + (3t) = 2 \Rightarrow 10t + 2 = 2 \Rightarrow t = 0 \Rightarrow x = 1, y = 1$ and $z = 0$ $\Rightarrow (1, 1, 0)$ is the point

57. $\mathbf{n}_1 = \mathbf{i} + \mathbf{j} + \mathbf{k}$ and $\mathbf{n}_2 = \mathbf{i} + \mathbf{j} \Rightarrow \mathbf{n}_1 \times \mathbf{n}_2 = \begin{vmatrix} \mathbf{i} & \mathbf{j} & \mathbf{k} \\ 1 & 1 & 1 \\ 1 & 1 & 0 \end{vmatrix} = -\mathbf{i} + \mathbf{j}$, the direction of the desired line; $(1, 1, -1)$ is on both planes $\Rightarrow$ the desired line is $x = 1 - t, y = 1 + t, z = -1$

59. $\mathbf{n}_1 = \mathbf{i} - 2\mathbf{j} + 4\mathbf{k}$ and $\mathbf{n}_2 = \mathbf{i} + \mathbf{j} - 2\mathbf{k} \Rightarrow \mathbf{n}_1 \times \mathbf{n}_2 = \begin{vmatrix} \mathbf{i} & \mathbf{j} & \mathbf{k} \\ 1 & -2 & 4 \\ 1 & 1 & -2 \end{vmatrix} = 6\mathbf{j} + 3\mathbf{k}$, the direction of the desired line; $(4, 3, 1)$ is on both planes $\Rightarrow$ the desired line is $x = 4, y = 3 + 6t, z = 1 + 3t$

61. <u>L1 & L2</u>: $x = 3 + 2t = 1 + 4s$ and $y = -1 + 4t = 1 + 2s \Rightarrow \begin{cases} 2t - 4s = -2 \\ 4t - 2s = 2 \end{cases} \Rightarrow \begin{cases} 2t - 4s = -2 \\ 2t - s = 1 \end{cases}$
$\Rightarrow -3s = -3 \Rightarrow s = 1$ and $t = 1 \Rightarrow$ on L1, $z = 1$ and on L2, $z = 1 \Rightarrow$ L1 and L2 intersect at $(5, 3, 1)$.
<u>L2 & L3</u>: The direction of L2 is $\frac{1}{6}(4\mathbf{i} + 2\mathbf{j} + 4\mathbf{k}) = \frac{1}{3}(2\mathbf{i} + \mathbf{j} + 2\mathbf{k})$ which is the same as the direction $\frac{1}{3}(2\mathbf{i} + \mathbf{j} + 2\mathbf{k})$ of L3; hence L2 and L3 are parallel.
<u>L1 & L3</u>: $x = 3 + 2t = 3 + 2r$ and $y = -1 + 4t = 2 + r \Rightarrow \begin{cases} 2t - 2r = 0 \\ 4t - r = 3 \end{cases} \Rightarrow \begin{cases} t - r = 0 \\ 4t - r = 3 \end{cases} \Rightarrow 3t = 3$
$\Rightarrow t = 1$ and $r = 1 \Rightarrow$ on L1, $z = 2$ while on L3, $z = 0 \Rightarrow$ L1 and L2 do not intersect. The direction of L1 is $\frac{1}{\sqrt{21}}(2\mathbf{i} + 4\mathbf{j} - \mathbf{k})$ while the direction of L3 is $\frac{1}{3}(2\mathbf{i} + \mathbf{j} + 2\mathbf{k})$ and neither is a multiple of the other; hence L1 and L3 are skew.

63. $x = 2 + 2t, y = -4 - t, z = 7 + 3t; x = -2 - t, y = -2 + \frac{1}{2}t, z = 1 - \frac{3}{2}t$

65. $x = 0 \Rightarrow t = -\frac{1}{2}, y = -\frac{1}{2}, z = -\frac{3}{2} \Rightarrow \left(0, -\frac{1}{2}, -\frac{3}{2}\right); y = 0 \Rightarrow t = -1, x = -1, z = -3 \Rightarrow (-1, 0, -3); z = 0$ $\Rightarrow t = 0, x = 1, y = -1 \Rightarrow (1, -1, 0)$

67. With substitution of the line into the plane we have $2(1 - 2t) + (2 + 5t) - (-3t) = 8 \Rightarrow 2 - 4t + 2 + 5t + 3t = 8$ $\Rightarrow 4t + 4 = 8 \Rightarrow t = 1 \Rightarrow$ the point $(-1, 7, -3)$ is contained in both the line and plane, so they are not parallel.

69. There are many possible answers. One is found as follows: eliminate t to get $t = x - 1 = 2 - y = \frac{z-3}{2}$ $\Rightarrow x - 1 = 2 - y$ and $2 - y = \frac{z-3}{2} \Rightarrow x + y = 3$ and $2y + z = 7$ are two such planes.

71. The points $(a, 0, 0)$, $(0, b, 0)$ and $(0, 0, c)$ are the x, y, and z intercepts of the plane. Since a, b, and c are all nonzero, the plane must intersect all three coordinate axes and cannot pass through the origin. Thus, $\frac{x}{a} + \frac{y}{b} + \frac{z}{c} = 1$ describes all planes <u>except</u> those through the origin or parallel to a coordinate axis.

73. (a) $\overrightarrow{EP} = c\overrightarrow{EP_1} \Rightarrow -x_0\mathbf{i} + y\mathbf{j} + z\mathbf{k} = c\left[(x_1 - x_0)\mathbf{i} + y_1\mathbf{j} + z_1\mathbf{k}\right] \Rightarrow -x_0 = c(x_1 - x_0), y = cy_1$ and $z = cz_1$, where c is a positive real number

(b) At $x_1 = 0 \Rightarrow c = 1 \Rightarrow y = y_1$ and $z = z_1$; at $x_1 = x_0 \Rightarrow x_0 = 0, y = 0, z = 0$; $\lim_{x_0 \to \infty} c = \lim_{x_0 \to \infty} \frac{-x_0}{x_1 - x_0}$ $= \lim_{x_0 \to \infty} \frac{-1}{-1} = 1 \Rightarrow c \to 1$ so that $y \to y_1$ and $z \to z_1$

12.6 CYLINDERS AND QUADRIC SURFACES

1. d, ellipsoid

3. a, cylinder

5. l, hyperbolic paraboloid

7. b, cylinder

9. k, hyperbolic paraboloid

11. h, cone

13. $x^2 + y^2 = 4$

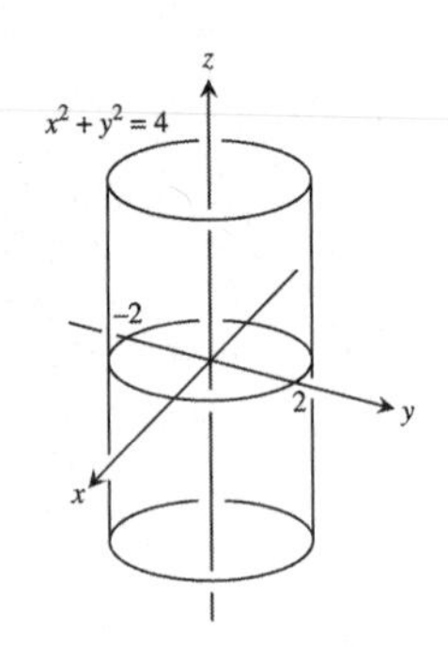

15. $x^2 + 4z^2 = 16$

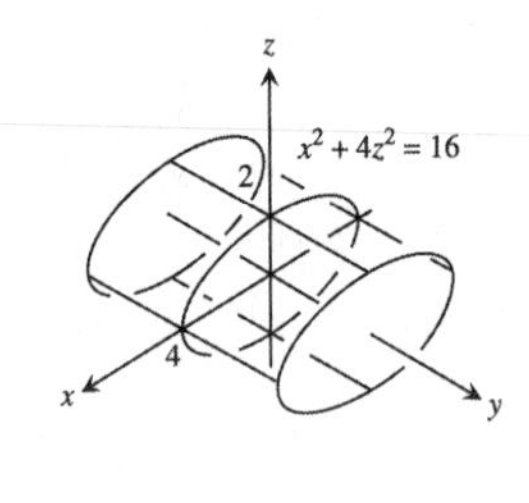

17. $9x^2 + y^2 + z^2 = 9$

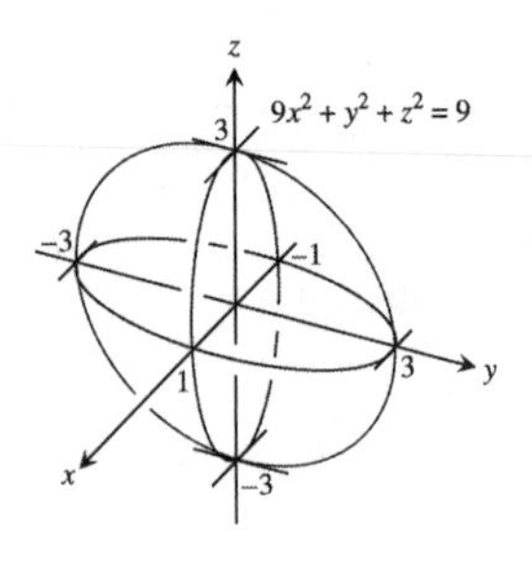

19. $4x^2 + 9y^2 + 4z^2 = 36$

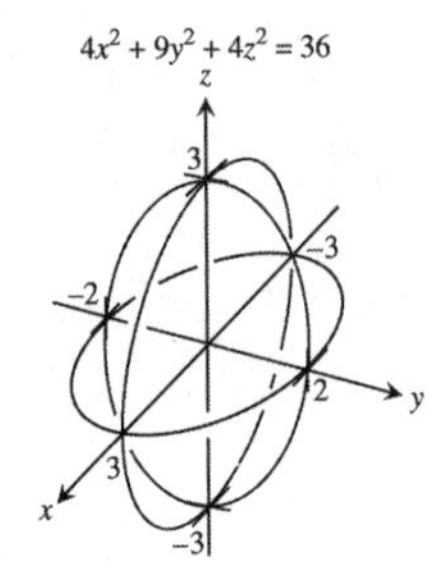

21. $x^2 + 4y^2 = z$

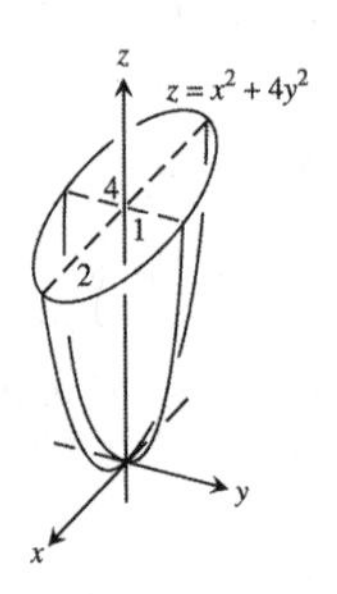

23. $x = 4 - 4y^2 - z^2$

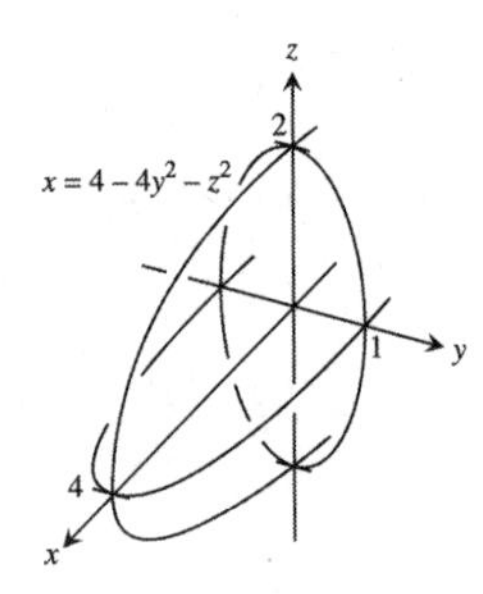

25. $x^2 + y^2 = z^2$

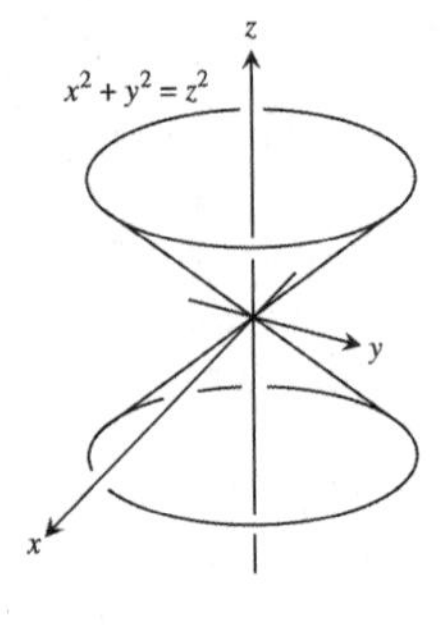

27. $x^2 + y^2 - z^2 = 1$

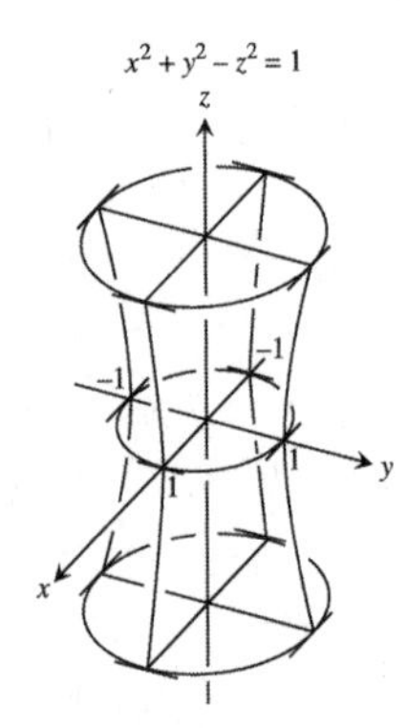

29. $z^2 - x^2 - y^2 = 1$

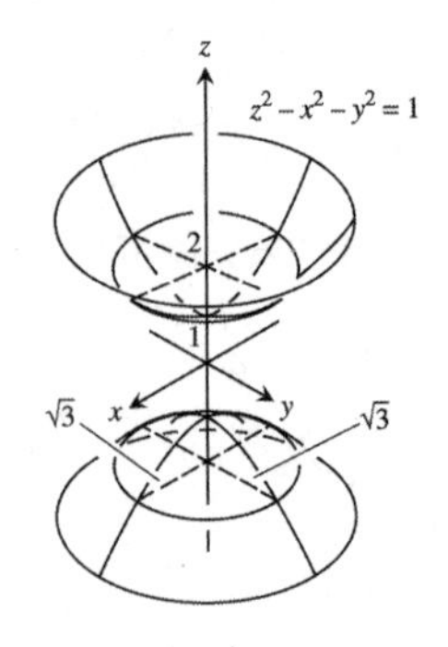

31. $y^2 - x^2 = z$

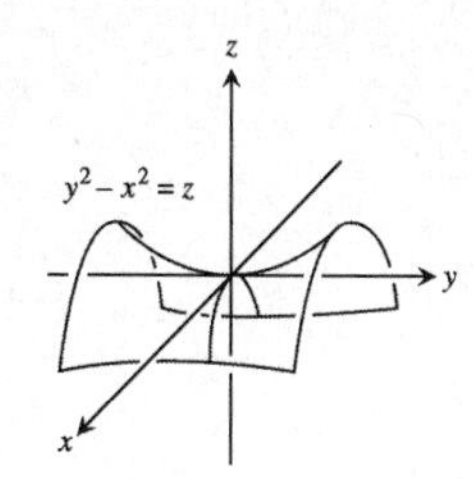

33. $z = 1 + y^2 - x^2$

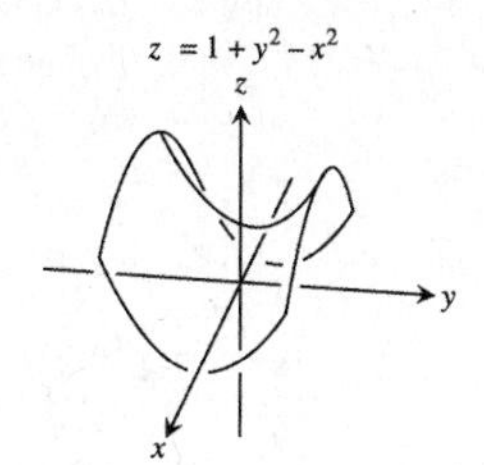

35. $y = -(x^2 + z^2)$

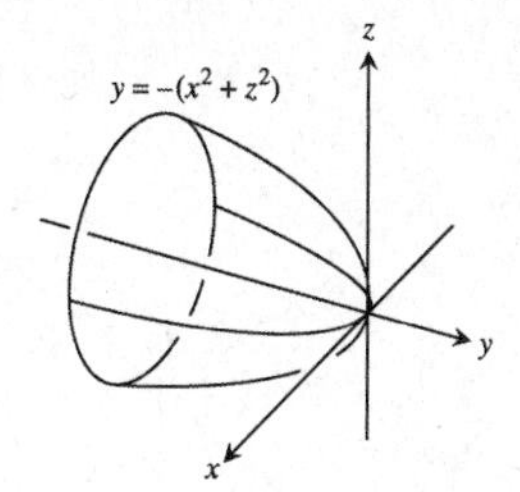

37. $x^2 + y^2 - z^2 = 4$

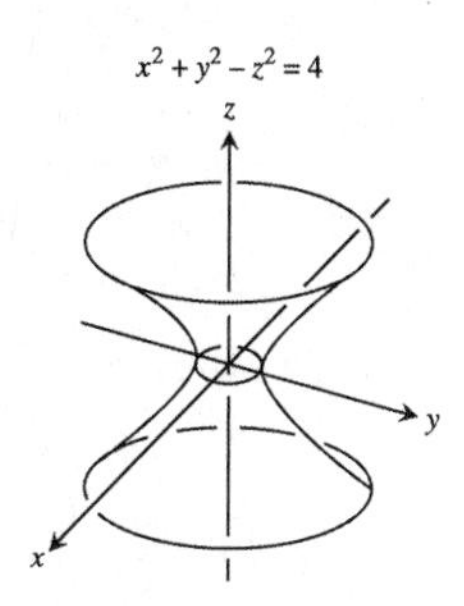

39. $x^2 + z^2 = 1$

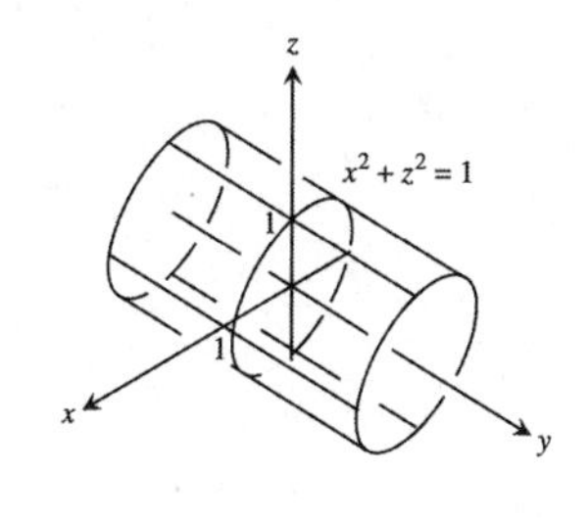

41. $z = -(x^2 + y^2)$

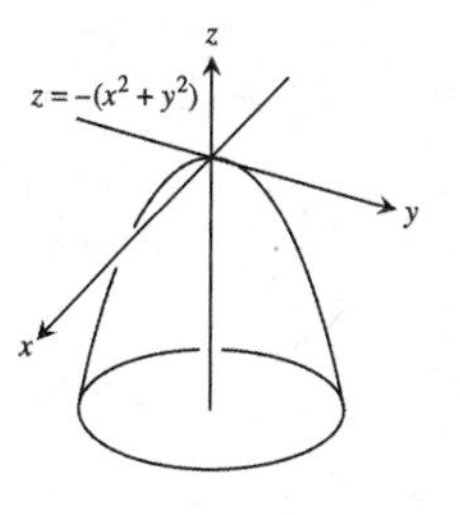

43. $4y^2 + z^2 - 4x^2 = 4$

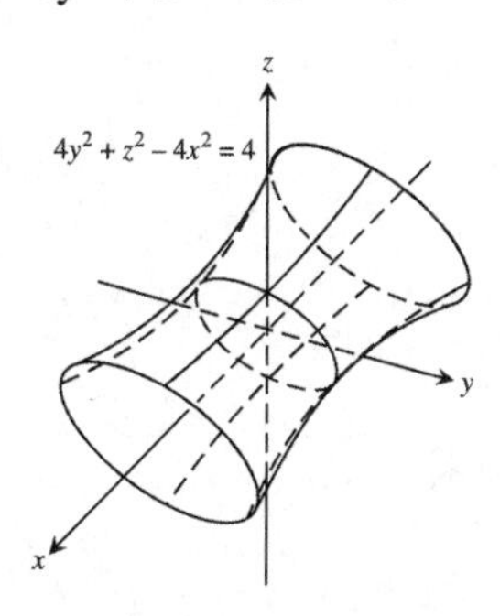

45. (a) If $x^2 + \frac{y^2}{4} + \frac{z^2}{9} = 1$ and $z = c$, then $x^2 + \frac{y^2}{4} = \frac{9-c^2}{9} \Rightarrow \frac{x^2}{\left(\frac{9-c^2}{9}\right)} + \frac{y^2}{\left[\frac{4(9-c^2)}{9}\right]} = 1 \Rightarrow A = ab\pi$
$= \pi\left(\frac{\sqrt{9-c^2}}{3}\right)\left(\frac{2\sqrt{9-c^2}}{3}\right) = \frac{2\pi(9-c^2)}{9}$

(b) From part (a), each slice has the area $\frac{2\pi(9-z^2)}{9}$, where $-3 \le z \le 3$. Thus $V = 2\int_0^3 \frac{2\pi}{9}(9 - z^2)\,dz$
$= \frac{4\pi}{9}\int_0^3 (9 - z^2)\,dz = \frac{4\pi}{9}\left[9z - \frac{z^3}{3}\right]_0^3 = \frac{4\pi}{9}(27 - 9) = 8\pi$

(c) $\frac{x^2}{a^2} + \frac{y^2}{b^2} + \frac{z^2}{c^2} = 1 \Rightarrow \frac{x^2}{\left[\frac{a^2(c^2-z^2)}{c^2}\right]} + \frac{y^2}{\left[\frac{b^2(c^2-z^2)}{c^2}\right]} = 1 \Rightarrow A = \pi\left(\frac{a\sqrt{c^2-z^2}}{c}\right)\left(\frac{b\sqrt{c^2-z^2}}{c}\right)$

$\Rightarrow V = 2\int_0^c \frac{\pi ab}{c^2}(c^2 - z^2)\,dz = \frac{2\pi ab}{c^2}\left[c^2 z - \frac{z^3}{3}\right]_0^c = \frac{2\pi ab}{c^2}\left(\frac{2}{3}c^3\right) = \frac{4\pi abc}{3}$. Note that if $r = a = b = c$, then $V = \frac{4\pi r^3}{3}$, which is the volume of a sphere.

47. We calculate the volume by the slicing method, taking slices parallel to the xy-plane. For fixed z, $\frac{x^2}{a^2} + \frac{y^2}{b^2} = \frac{z}{c}$ gives the ellipse $\frac{x^2}{\left(\frac{za^2}{c}\right)} + \frac{y^2}{\left(\frac{zb^2}{c}\right)} = 1$. The area of this ellipse is $\pi\left(a\sqrt{\frac{z}{c}}\right)\left(b\sqrt{\frac{z}{c}}\right) = \frac{\pi abz}{c}$ (see Exercise 45a). Hence the volume is given by $V = \int_0^h \frac{\pi abz}{c}\,dz = \left[\frac{\pi abz^2}{2c}\right]_0^h = \frac{\pi abh^2}{c}$. Now the area of the elliptic base when $z = h$ is $A = \frac{\pi abh}{c}$, as determined previously. Thus, $V = \frac{\pi abh^2}{c} = \frac{1}{2}\left(\frac{\pi abh}{c}\right)h = \frac{1}{2}$(base)(altitude), as claimed.

49. $z = y^2$

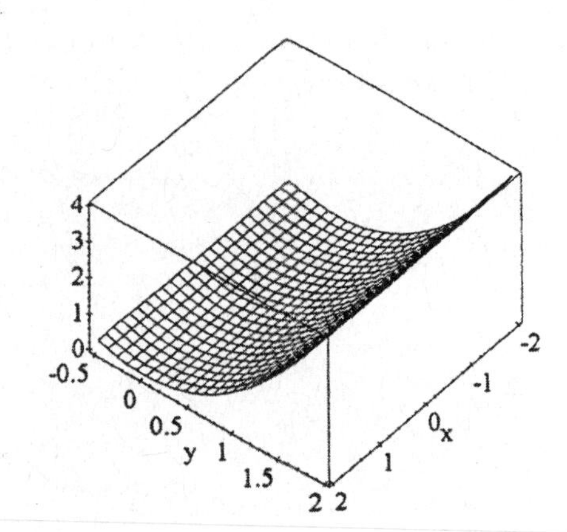

51. $z = x^2 + y^2$

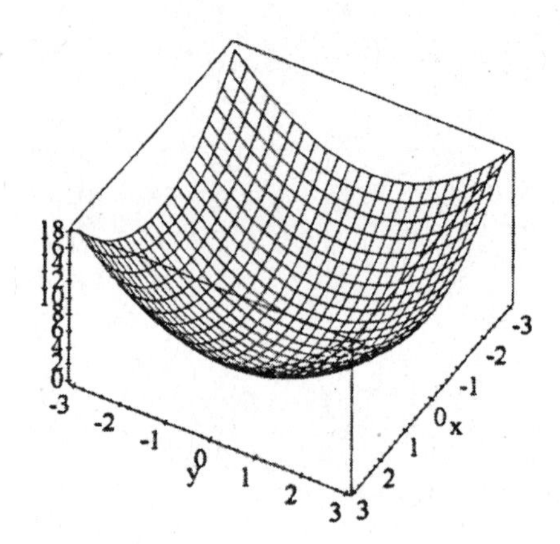

CHAPTER 12 PRACTICE EXERCISES

1. (a) $3\langle -3, 4\rangle - 4\langle 2, -5\rangle = \langle -9-8, 12+20\rangle = \langle -17, 32\rangle$
   (b) $\sqrt{17^2 + 32^2} = \sqrt{1313}$

3. (a) $\langle -2(-3), -2(4)\rangle = \langle 6, -8\rangle$
   (b) $\sqrt{6^2 + (-8)^2} = 10$

5. $\frac{\pi}{6}$ radians below the negative x-axis: $\left\langle -\frac{\sqrt{3}}{2}, -\frac{1}{2}\right\rangle$ [assuming counterclockwise].

7. $2\left(\frac{1}{\sqrt{4^2+1^2}}\right)(4\mathbf{i} - \mathbf{j}) = \left(\frac{8}{\sqrt{17}}\mathbf{i} - \frac{2}{\sqrt{17}}\mathbf{j}\right)$

9. length $= \left|\sqrt{2}\mathbf{i} + \sqrt{2}\mathbf{j}\right| = \sqrt{2+2} = 2$, $\sqrt{2}\mathbf{i} + \sqrt{2}\mathbf{j} = 2\left(\frac{1}{\sqrt{2}}\mathbf{i} + \frac{1}{\sqrt{2}}\mathbf{j}\right) \Rightarrow$ the direction is $\frac{1}{\sqrt{2}}\mathbf{i} + \frac{1}{\sqrt{2}}\mathbf{j}$

11. $t = \frac{\pi}{2} \Rightarrow \mathbf{v} = \left(-2\sin\frac{\pi}{2}\right)\mathbf{i} + \left(2\cos\frac{\pi}{2}\right)\mathbf{j} = -2\mathbf{i}$; length $= |-2\mathbf{i}| = \sqrt{4+0} = 2$; $-2\mathbf{i} = 2(-\mathbf{i}) \Rightarrow$ the direction is $-\mathbf{i}$

13. length $= |2\mathbf{i} - 3\mathbf{j} + 6\mathbf{k}| = \sqrt{4+9+36} = 7$, $2\mathbf{i} - 3\mathbf{j} + 6\mathbf{k} = 7\left(\frac{2}{7}\mathbf{i} - \frac{3}{7}\mathbf{j} + \frac{6}{7}\mathbf{k}\right) \Rightarrow$ the direction is $\frac{2}{7}\mathbf{i} - \frac{3}{7}\mathbf{j} + \frac{6}{7}\mathbf{k}$

15. $2\frac{\mathbf{v}}{|\mathbf{v}|} = 2 \cdot \frac{4\mathbf{i} - \mathbf{j} + 4\mathbf{k}}{\sqrt{4^2 + (-1)^2 + 4^2}} = 2 \cdot \frac{4\mathbf{i} - \mathbf{j} + 4\mathbf{k}}{\sqrt{33}} = \frac{8}{\sqrt{33}}\mathbf{i} - \frac{2}{\sqrt{33}}\mathbf{j} + \frac{8}{\sqrt{33}}\mathbf{k}$

17. $|\mathbf{v}| = \sqrt{1+1} = \sqrt{2}$, $|\mathbf{u}| = \sqrt{4+1+4} = 3$, $\mathbf{v}\cdot\mathbf{u} = 3$, $\mathbf{u}\cdot\mathbf{v} = 3$, $\mathbf{v}\times\mathbf{u} = \begin{vmatrix} \mathbf{i} & \mathbf{j} & \mathbf{k} \\ 1 & 1 & 0 \\ 2 & 1 & -2 \end{vmatrix} = -2\mathbf{i} + 2\mathbf{j} - \mathbf{k}$,

    $\mathbf{u}\times\mathbf{v} = -(\mathbf{v}\times\mathbf{u}) = 2\mathbf{i} - 2\mathbf{j} + \mathbf{k}$, $|\mathbf{v}\times\mathbf{u}| = \sqrt{4+4+1} = 3$, $\theta = \cos^{-1}\left(\frac{\mathbf{v}\cdot\mathbf{u}}{|\mathbf{v}|\,|\mathbf{u}|}\right) = \cos^{-1}\left(\frac{1}{\sqrt{2}}\right) = \frac{\pi}{4}$,

    $|\mathbf{u}|\cos\theta = \frac{3}{\sqrt{2}}$, $\text{proj}_{\mathbf{v}}\,\mathbf{u} = \left(\frac{\mathbf{v}\cdot\mathbf{u}}{|\mathbf{v}||\mathbf{v}|}\right)\mathbf{v} = \frac{3}{2}(\mathbf{i} + \mathbf{j})$

19. $\text{proj}_{\mathbf{v}}\mathbf{u} = \left(\frac{\mathbf{v}\cdot\mathbf{u}}{|\mathbf{v}||\mathbf{v}|}\right)\mathbf{v} = \frac{4}{3}(2\mathbf{i} + \mathbf{j} - \mathbf{k})$ where $\mathbf{v}\cdot\mathbf{u} = 8$ and $\mathbf{v}\cdot\mathbf{v} = 6$

21. $\mathbf{u}\times\mathbf{v}=\begin{vmatrix}\mathbf{i}&\mathbf{j}&\mathbf{k}\\1&0&0\\1&1&0\end{vmatrix}=\mathbf{k}$

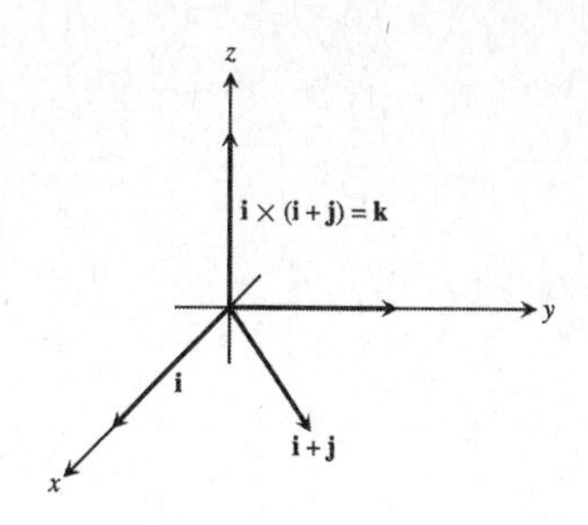

23. Let $\mathbf{v}=v_1\mathbf{i}+v_2\mathbf{j}+v_3\mathbf{k}$ and $\mathbf{w}=w_1\mathbf{i}+w_2\mathbf{j}+w_3\mathbf{k}$. Then $|\mathbf{v}-2\mathbf{w}|^2=|(v_1\mathbf{i}+v_2\mathbf{j}+v_3\mathbf{k})-2(w_1\mathbf{i}+w_2\mathbf{j}+w_3\mathbf{k})|^2$
$=|(v_1-2w_1)\mathbf{i}+(v_2-2w_2)\mathbf{j}+(v_3-2w_3)\mathbf{k}|^2=\left(\sqrt{(v_1-2w_1)^2+(v_2-2w_2)^2+(v_3-2w_3)^2}\right)^2$
$=(v_1^2+v_2^2+v_3^2)-4(v_1w_1+v_2w_2+v_3w_3)+4(w_1^2+w_2^2+w_3^2)=|\mathbf{v}|^2-4\mathbf{v}\cdot\mathbf{w}+4|\mathbf{w}|^2$
$=|\mathbf{v}|^2-4|\mathbf{v}||\mathbf{w}|\cos\theta+4|\mathbf{w}|^2=4-4(2)(3)\left(\cos\frac{\pi}{3}\right)+36=40-24\left(\frac{1}{2}\right)=40-12=28\Rightarrow|\mathbf{v}-2\mathbf{w}|=\sqrt{28}$
$=2\sqrt{7}$

25. (a) area $=|\mathbf{u}\times\mathbf{v}|=\text{abs}\begin{vmatrix}\mathbf{i}&\mathbf{j}&\mathbf{k}\\1&1&-1\\2&1&1\end{vmatrix}=|2\mathbf{i}-3\mathbf{j}-\mathbf{k}|=\sqrt{4+9+1}=\sqrt{14}$

(b) volume $=(\mathbf{u}\times\mathbf{v})\cdot\mathbf{w}=\begin{vmatrix}1&1&-1\\2&1&1\\-1&-2&3\end{vmatrix}=1(3+2)-1(6-(-1))-1(-4+1)=1$

27. The desired vector is $\mathbf{n}\times\mathbf{v}$ or $\mathbf{v}\times\mathbf{n}$ since $\mathbf{n}\times\mathbf{v}$ is perpendicular to both $\mathbf{n}$ and $\mathbf{v}$ and, therefore, also parallel to the plane.

29. The line L passes through the point $P(0,0,-1)$ parallel to $\mathbf{v}=-\mathbf{i}+\mathbf{j}+\mathbf{k}$. With $\overrightarrow{PS}=2\mathbf{i}+2\mathbf{j}+\mathbf{k}$ and
$\overrightarrow{PS}\times\mathbf{v}=\begin{vmatrix}\mathbf{i}&\mathbf{j}&\mathbf{k}\\2&2&1\\-1&1&1\end{vmatrix}=(2-1)\mathbf{i}-(2+1)\mathbf{j}+(2+2)\mathbf{k}=\mathbf{i}-3\mathbf{j}+4\mathbf{k}$, we find the distance
$d=\frac{|\overrightarrow{PS}\times\mathbf{v}|}{|\mathbf{v}|}=\frac{\sqrt{1+9+16}}{\sqrt{1+1+1}}=\frac{\sqrt{26}}{\sqrt{3}}=\frac{\sqrt{78}}{3}$.

31. Parametric equations for the line are $x=1-3t$, $y=2$, $z=3+7t$.

33. The point $P(4,0,0)$ lies on the plane $x-y=4$, and $\overrightarrow{PS}=(6-4)\mathbf{i}+0\mathbf{j}+(-6+0)\mathbf{k}=2\mathbf{i}-6\mathbf{k}$ with $\mathbf{n}=\mathbf{i}-\mathbf{j}$
$\Rightarrow d=\frac{|\mathbf{n}\cdot\overrightarrow{PS}|}{|\mathbf{n}|}=\left|\frac{2+0+0}{\sqrt{1+1+0}}\right|=\frac{2}{\sqrt{2}}=\sqrt{2}$.

35. $P(3,-2,1)$ and $\mathbf{n}=2\mathbf{i}+\mathbf{j}+\mathbf{k}\Rightarrow(2)(x-3)+(1)(y-(-2))+(1)(z-1)=0\Rightarrow 2x+y+z=5$

37. $P(1,-1,2)$, $Q(2,1,3)$ and $R(-1,2,-1)\Rightarrow\overrightarrow{PQ}=\mathbf{i}+2\mathbf{j}+\mathbf{k}$, $\overrightarrow{PR}=-2\mathbf{i}+3\mathbf{j}-3\mathbf{k}$ and $\overrightarrow{PQ}\times\overrightarrow{PR}$
$=\begin{vmatrix}\mathbf{i}&\mathbf{j}&\mathbf{k}\\1&2&1\\-2&3&-3\end{vmatrix}=-9\mathbf{i}+\mathbf{j}+7\mathbf{k}$ is normal to the plane $\Rightarrow(-9)(x-1)+(1)(y+1)+(7)(z-2)=0$
$\Rightarrow -9x+y+7z=4$

39. $\left(0,-\frac{1}{2},-\frac{3}{2}\right)$, since $t=-\frac{1}{2}$, $y=-\frac{1}{2}$ and $z=-\frac{3}{2}$ when $x=0$; $(-1,0,-3)$, since $t=-1$, $x=-1$ and $z=-3$ when $y=0$; $(1,-1,0)$, since $t=0$, $x=1$ and $y=-1$ when $z=0$

41. $\mathbf{n}_1 = \mathbf{i}$ and $\mathbf{n}_2 = \mathbf{i} + \mathbf{j} + \sqrt{2}\mathbf{k} \Rightarrow$ the desired angle is $\cos^{-1}\left(\frac{\mathbf{n}_1 \cdot \mathbf{n}_2}{|\mathbf{n}_1|\,|\mathbf{n}_2|}\right) = \cos^{-1}\left(\frac{1}{2}\right) = \frac{\pi}{3}$

43. The direction of the line is $\mathbf{n}_1 \times \mathbf{n}_2 = \begin{vmatrix} \mathbf{i} & \mathbf{j} & \mathbf{k} \\ 1 & 2 & 1 \\ 1 & -1 & 2 \end{vmatrix} = 5\mathbf{i} - \mathbf{j} - 3\mathbf{k}$. Since the point $(-5, 3, 0)$ is on both planes, the desired line is $x = -5 + 5t$, $y = 3 - t$, $z = -3t$.

45. (a) The corresponding normals are $\mathbf{n}_1 = 3\mathbf{i} + 6\mathbf{k}$ and $\mathbf{n}_2 = 2\mathbf{i} + 2\mathbf{j} - \mathbf{k}$ and since $\mathbf{n}_1 \cdot \mathbf{n}_2$ $= (3)(2) + (0)(2) + (6)(-1) = 6 + 0 - 6 = 0$, we have that the planes are orthogonal

(b) The line of intersection is parallel to $\mathbf{n}_1 \times \mathbf{n}_2 = \begin{vmatrix} \mathbf{i} & \mathbf{j} & \mathbf{k} \\ 3 & 0 & 6 \\ 2 & 2 & -1 \end{vmatrix} = -12\mathbf{i} + 15\mathbf{j} + 6\mathbf{k}$. Now to find a point in the intersection, solve $\begin{cases} 3x + 6z = 1 \\ 2x + 2y - z = 3 \end{cases} \Rightarrow \begin{cases} 3x + 6z = 1 \\ 12x + 12y - 6z = 18 \end{cases} \Rightarrow 15x + 12y = 19 \Rightarrow x = 0$ and $y = \frac{19}{12}$ $\Rightarrow \left(0, \frac{19}{12}, \frac{1}{6}\right)$ is a point on the line we seek. Therefore, the line is $x = -12t$, $y = \frac{19}{12} + 15t$ and $z = \frac{1}{6} + 6t$.

47. Yes; $\mathbf{v} \cdot \mathbf{n} = (2\mathbf{i} - 4\mathbf{j} + \mathbf{k}) \cdot (2\mathbf{i} + \mathbf{j} + 0\mathbf{k}) = 2 \cdot 2 - 4 \cdot 1 + 1 \cdot 0 = 0 \Rightarrow$ the vector is orthogonal to the plane's normal $\Rightarrow \mathbf{v}$ is parallel to the plane

49. A normal to the plane is $\mathbf{n} = \overrightarrow{AB} \times \overrightarrow{AC} = \begin{vmatrix} \mathbf{i} & \mathbf{j} & \mathbf{k} \\ 2 & 0 & -1 \\ 2 & -1 & 0 \end{vmatrix} = -\mathbf{i} - 2\mathbf{j} - 2\mathbf{k} \Rightarrow$ the distance is $d = \left|\frac{\overrightarrow{AP} \cdot \mathbf{n}}{\mathbf{n}}\right|$ $= \left|\frac{(\mathbf{i} + 4\mathbf{j}) \cdot (-\mathbf{i} - 2\mathbf{j} - 2\mathbf{k})}{\sqrt{1+4+4}}\right| = \left|\frac{-1-8+0}{3}\right| = 3$

51. $\mathbf{n} = 2\mathbf{i} - \mathbf{j} - \mathbf{k}$ is normal to the plane $\Rightarrow \mathbf{n} \times \mathbf{v} = \begin{vmatrix} \mathbf{i} & \mathbf{j} & \mathbf{k} \\ 2 & -1 & -1 \\ 1 & 1 & 1 \end{vmatrix} = 0\mathbf{i} - 3\mathbf{j} + 3\mathbf{k} = -3\mathbf{j} + 3\mathbf{k}$ is orthogonal to $\mathbf{v}$ and parallel to the plane

53. A vector parallel to the line of intersection is $\mathbf{v} = \mathbf{n}_1 \times \mathbf{n}_2 = \begin{vmatrix} \mathbf{i} & \mathbf{j} & \mathbf{k} \\ 1 & 2 & 1 \\ 1 & -1 & 2 \end{vmatrix} = 5\mathbf{i} - \mathbf{j} - 3\mathbf{k}$ $\Rightarrow |\mathbf{v}| = \sqrt{25 + 1 + 9} = \sqrt{35} \Rightarrow 2\left(\frac{\mathbf{v}}{|\mathbf{v}|}\right) = \frac{2}{\sqrt{35}}(5\mathbf{i} - \mathbf{j} - 3\mathbf{k})$ is the desired vector.

55. The line is represented by $x = 3 + 2t$, $y = 2 - t$, and $z = 1 + 2t$. It meets the plane $2x - y + 2z = -2$ when $2(3 + 2t) - (2 - t) + 2(1 + 2t) = -2 \Rightarrow t = -\frac{8}{9} \Rightarrow$ the point is $\left(\frac{11}{9}, \frac{26}{9}, -\frac{7}{9}\right)$.

57. The intersection occurs when $(3 + 2t) + 3(2t) - t = -4 \Rightarrow t = -1 \Rightarrow$ the point is $(1, -2, -1)$. The required line must be perpendicular to both the given line and to the normal, and hence is parallel to $\begin{vmatrix} \mathbf{i} & \mathbf{j} & \mathbf{k} \\ 2 & 2 & 1 \\ 1 & 3 & -1 \end{vmatrix}$ $= -5\mathbf{i} + 3\mathbf{j} + 4\mathbf{k} \Rightarrow$ the line is represented by $x = 1 - 5t$, $y = -2 + 3t$, and $z = -1 + 4t$.

59. The vector $\overrightarrow{AB} \times \overrightarrow{CD} = \begin{vmatrix} \mathbf{i} & \mathbf{j} & \mathbf{k} \\ 3 & -2 & 4 \\ \frac{26}{5} & 0 & -\frac{26}{5} \end{vmatrix} = \frac{26}{5}(2\mathbf{i} + 7\mathbf{j} + 2\mathbf{k})$ is normal to the plane and $A(-2, 0, -3)$ lies on the plane $\Rightarrow 2(x + 2) + 7(y - 0) + 2(z - (-3)) = 0 \Rightarrow 2x + 7y + 2z + 10 = 0$ is an equation of the plane.

61. The vector $\overrightarrow{PQ} \times \overrightarrow{PR} = \begin{vmatrix} \mathbf{i} & \mathbf{j} & \mathbf{k} \\ 2 & -1 & 3 \\ -3 & 0 & 1 \end{vmatrix} = -\mathbf{i} - 11\mathbf{j} - 3\mathbf{k}$ is normal to the plane.

(a) No, the plane is not orthogonal to $\overrightarrow{PQ} \times \overrightarrow{PR}$.

(b) No, these equations represent a line, not a plane.

(c) No, the plane $(x+2) + 11(y-1) - 3z = 0$ has normal $\mathbf{i} + 11\mathbf{j} - 3\mathbf{k}$ which is not parallel to $\overrightarrow{PQ} \times \overrightarrow{PR}$.

(d) No, this vector equation is equivalent to the equations $3y + 3z = 3$, $3x - 2z = -6$, and $3x + 2y = -4$ $\Rightarrow x = -\frac{4}{3} - \frac{2}{3}t$, $y = t$, $z = 1 - t$, which represents a line, not a plane.

(e) Yes, this is a plane containing the point $R(-2, 1, 0)$ with normal $\overrightarrow{PQ} \times \overrightarrow{PR}$.

63. $\overrightarrow{AB} = -2\mathbf{i} + \mathbf{j} + \mathbf{k}$, $\overrightarrow{CD} = \mathbf{i} + 4\mathbf{j} - \mathbf{k}$, and $\overrightarrow{AC} = 2\mathbf{i} + \mathbf{j} \Rightarrow \mathbf{n} = \begin{vmatrix} \mathbf{i} & \mathbf{j} & \mathbf{k} \\ -2 & 1 & 1 \\ 1 & 4 & -1 \end{vmatrix} = -5\mathbf{i} - \mathbf{j} - 9\mathbf{k} \Rightarrow$ the distance is

$d = \left| \frac{(2\mathbf{i}+\mathbf{j})\cdot(-5\mathbf{i}-\mathbf{j}-9\mathbf{k})}{\sqrt{25+1+81}} \right| = \frac{11}{\sqrt{107}}$

65. $x^2 + y^2 + z^2 = 4$

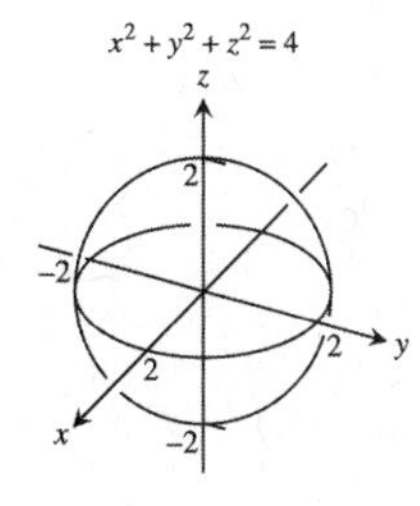

67. $4x^2 + 4y^2 + z^2 = 4$

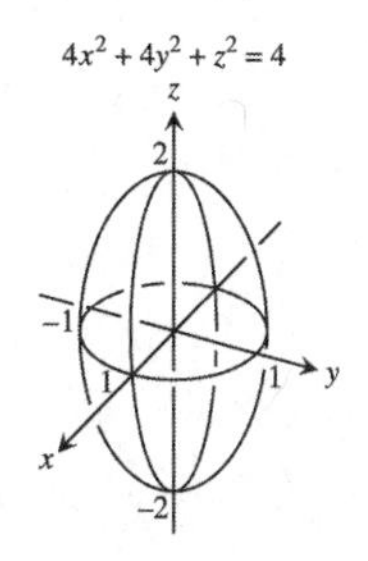

69. $z = -(x^2 + y^2)$

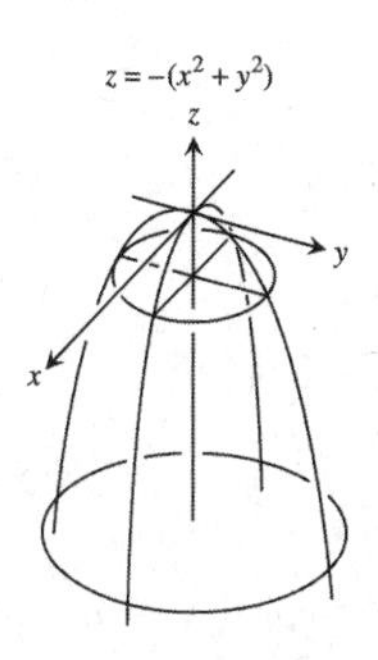

71. $x^2 + y^2 = z^2$

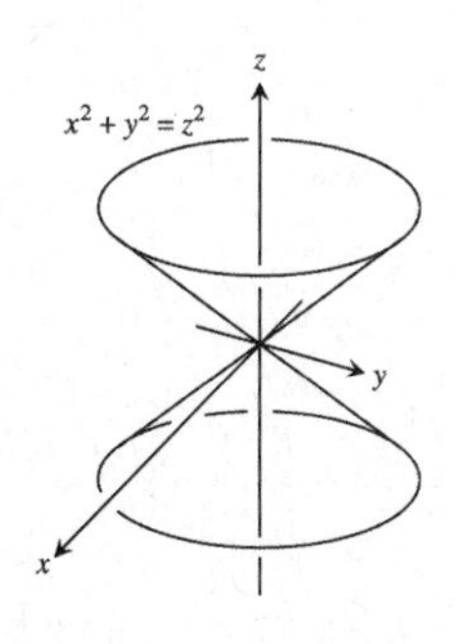

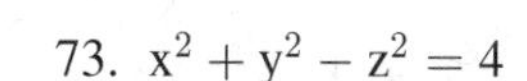
73. $x^2 + y^2 - z^2 = 4$

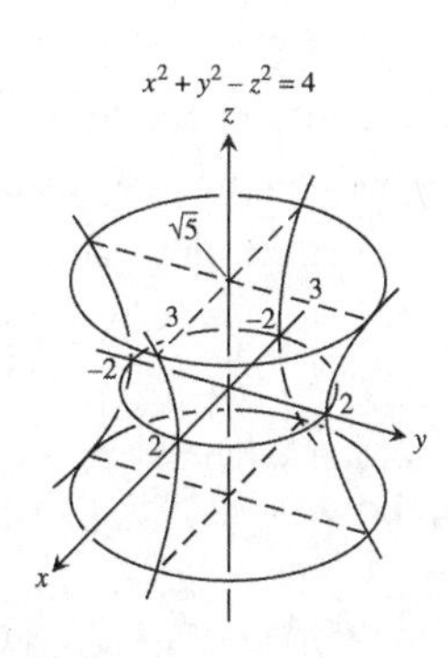

75. $y^2 - x^2 - z^2 = 1$

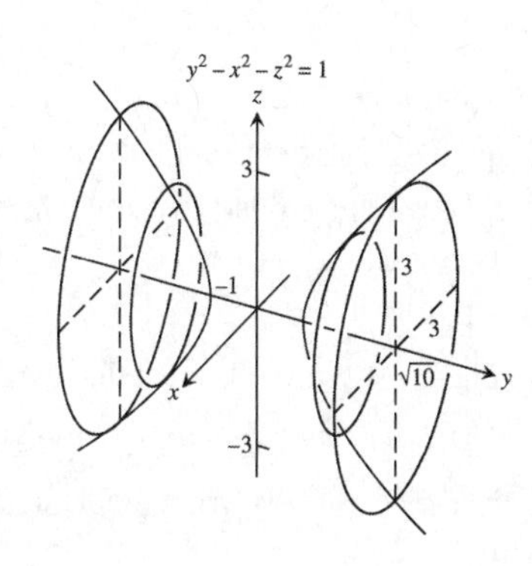

CHAPTER 12 ADDITIONAL AND ADVANCED EXERCISES

1. Information from ship A indicates the submarine is now on the line L_1: $x = 4 + 2t$, $y = 3t$, $z = -\frac{1}{3}t$; information from ship B indicates the submarine is now on the line L_2: $x = 18s$, $y = 5 - 6s$, $z = -s$. The current position of the sub is $\left(6, 3, -\frac{1}{3}\right)$ and occurs when the lines intersect at $t = 1$ and $s = \frac{1}{3}$. The straight line path of the submarine contains both points $P\left(2, -1, -\frac{1}{3}\right)$ and $Q\left(6, 3, -\frac{1}{3}\right)$; the line representing this path is L: $x = 2 + 4t$, $y = -1 + 4t$, $z = -\frac{1}{3}$. The submarine traveled the distance between P and Q in 4 minutes $\Rightarrow$ a speed of $\frac{|\overrightarrow{PQ}|}{4} = \frac{\sqrt{32}}{4} = \sqrt{2}$ thousand ft/min. In 20 minutes the submarine will move $20\sqrt{2}$ thousand ft from Q along the line L

$\Rightarrow 20\sqrt{2} = \sqrt{(2+4t-6)^2+(-1+4t-3)^2+0^2} \Rightarrow 800 = 16(t-1)^2 + 16(t-1)^2 = 32(t-1)^2 \Rightarrow (t-1)^2 = \frac{800}{32}$ $= 25 \Rightarrow t = 6 \Rightarrow$ the submarine will be located at $\left(26, 23, -\frac{1}{3}\right)$ in 20 minutes.

3. Torque $= \left|\overrightarrow{PQ} \times \mathbf{F}\right| \Rightarrow 15$ ft-lb $= \left|\overrightarrow{PQ}\right| |\mathbf{F}| \sin\frac{\pi}{2} = \frac{3}{4}$ ft $\cdot |\mathbf{F}| \Rightarrow |\mathbf{F}| = 20$ lb

5. (a) By the Law of Cosines we have $\cos\alpha = \frac{3^2+5^2-4^2}{2(3)(5)} = \frac{3}{5}$ and $\cos\beta = \frac{4^2+5^2-3^2}{2(4)(5)} = \frac{4}{5} \Rightarrow \sin\alpha = \frac{4}{5}$ and $\sin\beta = \frac{3}{5}$ $\Rightarrow \mathbf{F}_1 = \left\langle -|\mathbf{F}_1|\cos\alpha, |\mathbf{F}_1|\sin\alpha \right\rangle = \left\langle -\frac{3}{5}|\mathbf{F}_1|, \frac{4}{5}|\mathbf{F}_1| \right\rangle$, $\mathbf{F}_2 = \left\langle |\mathbf{F}_2|\cos\beta, |\mathbf{F}_2|\sin\beta \right\rangle = \left\langle \frac{4}{5}|\mathbf{F}_2|, \frac{3}{5}|\mathbf{F}_2| \right\rangle$, and $\mathbf{w} = \langle 0, -100\rangle$. Since $\mathbf{F}_1 + \mathbf{F}_2 = \langle 0, 100\rangle \Rightarrow \left\langle -\frac{3}{5}|\mathbf{F}_1| + \frac{4}{5}|\mathbf{F}_2|, \frac{4}{5}|\mathbf{F}_1| + \frac{3}{5}|\mathbf{F}_2| \right\rangle = \langle 0, 100\rangle \Rightarrow -\frac{3}{5}|\mathbf{F}_1| + \frac{4}{5}|\mathbf{F}_2| = 0$ and $\frac{4}{5}|\mathbf{F}_1| + \frac{3}{5}|\mathbf{F}_2| = 100$. Solving the first equation for $|\mathbf{F}_2|$ results in: $|\mathbf{F}_2| = \frac{3}{4}|\mathbf{F}_1|$. Substituting this result into the second equation gives us: $\frac{4}{5}|\mathbf{F}_1| + \frac{9}{20}|\mathbf{F}_1| = 100 \Rightarrow |\mathbf{F}_1| = 80$ lb. $\Rightarrow |\mathbf{F}_2| = 60$ lb. $\Rightarrow \mathbf{F}_1 = \langle -48, 64\rangle$ and $\mathbf{F}_2 = \langle 48, 36\rangle$, and $\alpha = \tan^{-1}\left(\frac{4}{3}\right)$ and $\beta = \tan^{-1}\left(\frac{3}{4}\right)$

(b) By the Law of Cosines we have $\cos\alpha = \frac{5^2+13^2-12^2}{2(5)(13)} = \frac{5}{13}$ and $\cos\beta = \frac{12^2+13^2-5^2}{2(12)(13)} = \frac{12}{13} \Rightarrow \sin\alpha = \frac{12}{13}$ and $\sin\beta = \frac{5}{13}$ $\Rightarrow \mathbf{F}_1 = \left\langle -|\mathbf{F}_1|\cos\alpha, |\mathbf{F}_1|\sin\alpha \right\rangle = \left\langle -\frac{5}{13}|\mathbf{F}_1|, \frac{12}{13}|\mathbf{F}_1| \right\rangle$, $\mathbf{F}_2 = \left\langle |\mathbf{F}_2|\cos\beta, |\mathbf{F}_2|\sin\beta \right\rangle = \left\langle \frac{12}{13}|\mathbf{F}_2|, \frac{5}{13}|\mathbf{F}_2| \right\rangle$, and $\mathbf{w} = \langle 0, -200\rangle$. Since $\mathbf{F}_1 + \mathbf{F}_2 = \langle 0, 200\rangle \Rightarrow \left\langle -\frac{5}{13}|\mathbf{F}_1| + \frac{12}{13}|\mathbf{F}_2|, \frac{12}{13}|\mathbf{F}_1| + \frac{5}{13}|\mathbf{F}_2| \right\rangle = \langle 0, 200\rangle$ $\Rightarrow -\frac{5}{13}|\mathbf{F}_1| + \frac{12}{13}|\mathbf{F}_2| = 0$ and $\frac{12}{13}|\mathbf{F}_1| + \frac{5}{13}|\mathbf{F}_2| = 200$. Solving the first equation for $|\mathbf{F}_2|$ results in: $|\mathbf{F}_2| = \frac{5}{12}|\mathbf{F}_1|$. Substituting this result into the second equation gives us: $\frac{12}{13}|\mathbf{F}_1| + \frac{25}{156}|\mathbf{F}_1| = 200 \Rightarrow |\mathbf{F}_1| = \frac{2400}{13} \approx 184.615$ lb. $\Rightarrow |\mathbf{F}_2| = \frac{1000}{13} \approx 76.923$ lb. $\Rightarrow \mathbf{F}_1 = \left\langle -\frac{12000}{1169}, \frac{28800}{1169} \right\rangle \approx \langle -71.006, 170.414\rangle$ and $\mathbf{F}_2 = \left\langle \frac{12000}{1169}, \frac{5000}{1169} \right\rangle$ $\approx \langle 71.006, 29.586\rangle$.

7. (a) If $P(x, y, z)$ is a point in the plane determined by the three points $P_1(x_1, y_1, z_1)$, $P_2(x_2, y_2, z_2)$ and $P_3(x_3, y_3, z_3)$, then the vectors $\overrightarrow{PP_1}$, $\overrightarrow{PP_2}$ and $\overrightarrow{PP_3}$ all lie in the plane. Thus $\overrightarrow{PP_1} \cdot (\overrightarrow{PP_2} \times \overrightarrow{PP_3}) = 0$

$$\Rightarrow \begin{vmatrix} x_1 - x & y_1 - y & z_1 - z \\ x_2 - x & y_2 - y & z_2 - z \\ x_3 - x & y_3 - y & z_3 - z \end{vmatrix} = 0$$ by the determinant formula for the triple scalar product in Section 12.4.

(b) Subtract row 1 from rows 2, 3, and 4 and evaluate the resulting determinant (which has the same value as the given determinant) by cofactor expansion about column 4. This expansion is exactly the determinant in part (a) so we have all points $P(x, y, z)$ in the plane determined by $P_1(x_1, y_1, z_1)$, $P_2(x_2, y_2, z_2)$, and $P_3(x_3, y_3, z_3)$.

9. (a) Place the tetrahedron so that A is at $(0, 0, 0)$, the point P is on the y-axis, and $\triangle ABC$ lies in the xy-plane. Since $\triangle ABC$ is an equilateral triangle, all the angles in the triangle are $60°$ and since AP bisects BC $\Rightarrow \triangle ABP$ is a $30°$-$60°$-$90°$ trinagle. Thus the coordinates of P are $\left(0, \sqrt{3}, 0\right)$, the coordinates of B are $\left(1, \sqrt{3}, 0\right)$, and the coordinates of C are $\left(-1, \sqrt{3}, 0\right)$. Let the coordinates of D be given by (a, b, c). Since all of the faces are equilateral trinagles $\Rightarrow$ all the angles in each of the triangles are $60° \Rightarrow \cos(\angle DAB) = \cos(60°) = \frac{\overrightarrow{AD}\cdot\overrightarrow{AB}}{|\overrightarrow{AD}||\overrightarrow{AB}|} = \frac{a+b\sqrt{3}}{(2)(2)} = \frac{1}{2}$ $\Rightarrow a + b\sqrt{3} = 2$ and $\cos(\angle DAC) = \cos(60°) = \frac{\overrightarrow{AD}\cdot\overrightarrow{AC}}{|\overrightarrow{AD}||\overrightarrow{AC}|} = \frac{-a+b\sqrt{3}}{(2)(2)} = \frac{1}{2} \Rightarrow -a + b\sqrt{3} = 2$. Add the two equations to obtain: $2b\sqrt{3} = 4 \Rightarrow b = \frac{2}{\sqrt{3}}$. Substituting this value for b in the first equation gives us: $a + \left(\frac{2}{\sqrt{3}}\right)\sqrt{3} = 2$ $\Rightarrow a = 0$. Since $|\overrightarrow{AD}| = \sqrt{a^2+b^2+c^2} = 2 \Rightarrow 0^2 + \left(\frac{2}{\sqrt{3}}\right)^2 + c^2 = 4 \Rightarrow c = \frac{2\sqrt{2}}{\sqrt{3}}$. Thus the coordinates of D are $\left(0, \frac{2}{\sqrt{3}}, \frac{2\sqrt{2}}{\sqrt{3}}\right)$. $\cos\theta = \cos(\angle DAP) = \frac{\overrightarrow{AD}\cdot\overrightarrow{AP}}{|\overrightarrow{AD}||\overrightarrow{AP}|} = \frac{2}{2\sqrt{3}} \Rightarrow \theta = \cos^{-1}\left(\frac{1}{\sqrt{3}}\right) \Rightarrow 57.74°$.

(b) Since $\triangle ABC$ lies in the xy-plane $\Rightarrow$ the normal to the face given by $\triangle ABC$ is $\mathbf{n}_1 = \mathbf{k}$. The face given by $\triangle BCD$ is an adjacent face. The vectors $\overrightarrow{DB} = \mathbf{i} + \frac{1}{\sqrt{3}}\mathbf{j} - \frac{2\sqrt{2}}{\sqrt{3}}\mathbf{k}$ and $\overrightarrow{DC} = -\mathbf{i} + \frac{1}{\sqrt{3}}\mathbf{j} - \frac{2\sqrt{2}}{\sqrt{3}}\mathbf{k}$ both lie in the plane containing $\triangle BCD$. The normal to this plane is given by $\mathbf{n}_2 = \begin{vmatrix} \mathbf{i} & \mathbf{j} & \mathbf{k} \\ 1 & \frac{1}{\sqrt{3}} & -\frac{2\sqrt{2}}{\sqrt{3}} \\ -1 & \frac{1}{\sqrt{3}} & -\frac{2\sqrt{2}}{\sqrt{3}} \end{vmatrix} = \frac{4\sqrt{2}}{\sqrt{3}}\mathbf{j} + \frac{2}{\sqrt{3}}\mathbf{k}$. The angle θ between two adjacent faces is given by $\cos\theta = \cos(\angle DAP) = \frac{\mathbf{n}_1 \cdot \mathbf{n}_2}{|\mathbf{n}_1||\mathbf{n}_2|} = \frac{2/\sqrt{3}}{(1)\left(6/\sqrt{3}\right)} \Rightarrow \theta = \cos^{-1}\left(\frac{1}{3}\right) \approx 70.53°$.

11. If Q(x, y) is a point on the line ax + by = c, then $\overrightarrow{P_1Q} = (x - x_1)\mathbf{i} + (y - y_1)\mathbf{j}$, and $\mathbf{n} = a\mathbf{i} + b\mathbf{j}$ is normal to the line. The distance is $\left|\text{proj}_{\mathbf{n}}\, \overrightarrow{P_1Q}\right| = \left|\frac{[(x - x_1)\mathbf{i} + (y - y_1)\mathbf{j}]\cdot(a\mathbf{i} + b\mathbf{j})}{\sqrt{a^2 + b^2}}\right| = \frac{|a(x - x_1) + b(y - y_1)|}{\sqrt{a^2 + b^2}}$
$= \frac{|ax_1 + by_1 - c|}{\sqrt{a^2 + b^2}}$, since c = ax + by.

13. (a) If (x_1, y_1, z_1) is on the plane $Ax + By + Cz = D_1$, then the distance d between the planes is $d = \frac{|Ax_1 + By_1 + Cz_1 - D_2|}{\sqrt{A^2 + B^2 + C^2}} = \frac{|D_1 - D_2|}{|A\mathbf{i} + B\mathbf{j} + C\mathbf{k}|}$, since $Ax_1 + By_1 + Cz_1 = D_1$, by Exercise 12(a).

(b) $d = \frac{|12 - 6|}{\sqrt{4 + 9 + 1}} = \frac{6}{\sqrt{14}}$

(c) $\frac{|2(3) + (-1)(2) + 2(-1) + 4|}{\sqrt{14}} = \frac{|2(3) + (-1)(2) + 2(-1) - D|}{\sqrt{14}} \Rightarrow D = 8$ or $-4 \Rightarrow$ the desired plane is $2x - y + 2x = 8$

(d) Choose the point (2, 0, 1) on the plane. Then $\frac{|3 - D|}{\sqrt{6}} = 5 \Rightarrow D = 3 \pm 5\sqrt{6} \Rightarrow$ the desired planes are $x - 2y + z = 3 + 5\sqrt{6}$ and $x - 2y + z = 3 - 5\sqrt{6}$.

15. $\mathbf{n} = \mathbf{i} + 2\mathbf{j} + 6\mathbf{k}$ is normal to the plane $x + 2y + 6z = 6$; $\mathbf{v} \times \mathbf{n} = \begin{vmatrix} \mathbf{i} & \mathbf{j} & \mathbf{k} \\ 1 & 1 & 1 \\ 1 & 2 & 6 \end{vmatrix} = 4\mathbf{i} - 5\mathbf{j} + \mathbf{k}$ is parallel to the plane and perpendicular to the plane of $\mathbf{v}$ and $\mathbf{n} \Rightarrow \mathbf{w} = \mathbf{n} \times (\mathbf{v} \times \mathbf{n}) = \begin{vmatrix} \mathbf{i} & \mathbf{j} & \mathbf{k} \\ 1 & 2 & 6 \\ 4 & -5 & 1 \end{vmatrix} = 32\mathbf{i} + 23\mathbf{j} - 13\mathbf{k}$ is a vector parallel to the plane $x + 2y + 6z = 6$ in the direction of the projection vector $\text{proj}_P\, \mathbf{v}$. Therefore, $\text{proj}_P\, \mathbf{v} = \text{proj}_{\mathbf{w}}\, \mathbf{v} = \left(\mathbf{v} \cdot \frac{\mathbf{w}}{|\mathbf{w}|}\right)\frac{\mathbf{w}}{|\mathbf{w}|} = \left(\frac{\mathbf{v}\cdot\mathbf{w}}{|\mathbf{w}|^2}\right)\mathbf{w} = \left(\frac{32 + 23 - 13}{32^2 + 23^2 + 13^2}\right)\mathbf{w} = \frac{42}{1722}\mathbf{w} = \frac{1}{41}\mathbf{w} = \frac{32}{41}\mathbf{i} + \frac{23}{41}\mathbf{j} - \frac{13}{41}\mathbf{k}$

17. (a) $\mathbf{u} \times \mathbf{v} = 2\mathbf{i} \times 2\mathbf{j} = 4\mathbf{k} \Rightarrow (\mathbf{u} \times \mathbf{v}) \times \mathbf{w} = \mathbf{0}$; $(\mathbf{u} \cdot \mathbf{w})\mathbf{v} - (\mathbf{v} \cdot \mathbf{w})\mathbf{u} = 0\mathbf{v} - 0\mathbf{u} = \mathbf{0}$; $\mathbf{v} \times \mathbf{w} = 4\mathbf{i} \Rightarrow \mathbf{u} \times (\mathbf{v} \times \mathbf{w}) = \mathbf{0}$; $(\mathbf{u} \cdot \mathbf{w})\mathbf{v} - (\mathbf{u} \cdot \mathbf{v})\mathbf{w} = 0\mathbf{v} - 0\mathbf{w} = \mathbf{0}$

(b) $\mathbf{u} \times \mathbf{v} = \begin{vmatrix} \mathbf{i} & \mathbf{j} & \mathbf{k} \\ 1 & -1 & 1 \\ 2 & 1 & -2 \end{vmatrix} = \mathbf{i} + 4\mathbf{j} + 3\mathbf{k} \Rightarrow (\mathbf{u} \times \mathbf{v}) \times \mathbf{w} = \begin{vmatrix} \mathbf{i} & \mathbf{j} & \mathbf{k} \\ 1 & 4 & 3 \\ -1 & 2 & -1 \end{vmatrix} = -10\mathbf{i} - 2\mathbf{j} + 6\mathbf{k}$;

$(\mathbf{u} \cdot \mathbf{w})\mathbf{v} - (\mathbf{v} \cdot \mathbf{w})\mathbf{u} = -4(2\mathbf{i} + \mathbf{j} - 2\mathbf{k}) - 2(\mathbf{i} - \mathbf{j} + \mathbf{k}) = -10\mathbf{i} - 2\mathbf{j} + 6\mathbf{k}$;

$\mathbf{v} \times \mathbf{w} = \begin{vmatrix} \mathbf{i} & \mathbf{j} & \mathbf{k} \\ 2 & 1 & -2 \\ -1 & 2 & -1 \end{vmatrix} = 3\mathbf{i} + 4\mathbf{j} + 5\mathbf{k} \Rightarrow \mathbf{u} \times (\mathbf{v} \times \mathbf{w}) = \begin{vmatrix} \mathbf{i} & \mathbf{j} & \mathbf{k} \\ 1 & -1 & 1 \\ 3 & 4 & 5 \end{vmatrix} = -9\mathbf{i} - 2\mathbf{j} + 7\mathbf{k}$;

$(\mathbf{u} \cdot \mathbf{w})\mathbf{v} - (\mathbf{u} \cdot \mathbf{v})\mathbf{w} = -4(2\mathbf{i} + \mathbf{j} - 2\mathbf{k}) - (-1)(-\mathbf{i} + 2\mathbf{j} - \mathbf{k}) = -9\mathbf{i} - 2\mathbf{j} + 7\mathbf{k}$

(c) $\mathbf{u} \times \mathbf{v} = \begin{vmatrix} \mathbf{i} & \mathbf{j} & \mathbf{k} \\ 2 & 1 & 0 \\ 2 & -1 & 1 \end{vmatrix} = \mathbf{i} - 2\mathbf{j} - 4\mathbf{k} \Rightarrow (\mathbf{u} \times \mathbf{v}) \times \mathbf{w} = \begin{vmatrix} \mathbf{i} & \mathbf{j} & \mathbf{k} \\ 1 & -2 & -4 \\ 1 & 0 & 2 \end{vmatrix} = -4\mathbf{i} - 6\mathbf{j} + 2\mathbf{k}$;

$(\mathbf{u} \cdot \mathbf{w})\mathbf{v} - (\mathbf{v} \cdot \mathbf{w})\mathbf{u} = 2(2\mathbf{i} - \mathbf{j} + \mathbf{k}) - 4(2\mathbf{i} + \mathbf{j}) = -4\mathbf{i} - 6\mathbf{j} + 2\mathbf{k}$;

$\mathbf{v} \times \mathbf{w} = \begin{vmatrix} \mathbf{i} & \mathbf{j} & \mathbf{k} \\ 2 & -1 & 1 \\ 1 & 0 & 2 \end{vmatrix} = -2\mathbf{i} - 3\mathbf{j} + \mathbf{k} \Rightarrow \mathbf{u} \times (\mathbf{v} \times \mathbf{w}) = \begin{vmatrix} \mathbf{i} & \mathbf{j} & \mathbf{k} \\ 2 & 1 & 0 \\ -2 & -3 & 1 \end{vmatrix} = \mathbf{i} - 2\mathbf{j} - 4\mathbf{k};$

$(\mathbf{u} \cdot \mathbf{w})\mathbf{v} - (\mathbf{u} \cdot \mathbf{v})\mathbf{w} = 2(2\mathbf{i} - \mathbf{j} + \mathbf{k}) - 3(\mathbf{i} + 2\mathbf{k}) = \mathbf{i} - 2\mathbf{j} - 4\mathbf{k}$

(d) $\mathbf{u} \times \mathbf{v} = \begin{vmatrix} \mathbf{i} & \mathbf{j} & \mathbf{k} \\ 1 & 1 & -2 \\ -1 & 0 & -1 \end{vmatrix} = -\mathbf{i} + 3\mathbf{j} + \mathbf{k} \Rightarrow (\mathbf{u} \times \mathbf{v}) \times \mathbf{w} = \begin{vmatrix} \mathbf{i} & \mathbf{j} & \mathbf{k} \\ -1 & 3 & 1 \\ 2 & 4 & -2 \end{vmatrix} = -10\mathbf{i} - 10\mathbf{k};$

$(\mathbf{u} \cdot \mathbf{w})\mathbf{v} - (\mathbf{v} \cdot \mathbf{w})\mathbf{u} = 10(-\mathbf{i} - \mathbf{k}) - 0(\mathbf{i} + \mathbf{j} - 2\mathbf{k}) = -10\mathbf{i} - 10\mathbf{k};$

$\mathbf{v} \times \mathbf{w} = \begin{vmatrix} \mathbf{i} & \mathbf{j} & \mathbf{k} \\ -1 & 0 & -1 \\ 2 & 4 & -2 \end{vmatrix} = 4\mathbf{i} - 4\mathbf{j} - 4\mathbf{k} \Rightarrow \mathbf{u} \times (\mathbf{v} \times \mathbf{w}) = \begin{vmatrix} \mathbf{i} & \mathbf{j} & \mathbf{k} \\ 1 & 1 & -2 \\ 4 & -4 & -4 \end{vmatrix} = -12\mathbf{i} - 4\mathbf{j} - 8\mathbf{k};$

$(\mathbf{u} \cdot \mathbf{w})\mathbf{v} - (\mathbf{u} \cdot \mathbf{v})\mathbf{w} = 10(-\mathbf{i} - \mathbf{k}) - 1(2\mathbf{i} + 4\mathbf{j} - 2\mathbf{k}) = -12\mathbf{i} - 4\mathbf{j} - 8\mathbf{k}$

19. The formula is always true; $\mathbf{u} \times [\mathbf{u} \times (\mathbf{u} \times \mathbf{v})] \cdot \mathbf{w} = \mathbf{u} \times [(\mathbf{u} \cdot \mathbf{v})\mathbf{u} - (\mathbf{u} \cdot \mathbf{u})\mathbf{v}] \cdot \mathbf{w}$
$= [(\mathbf{u} \cdot \mathbf{v})\mathbf{u} \times \mathbf{u} - (\mathbf{u} \cdot \mathbf{u})\mathbf{u} \times \mathbf{v}] \cdot \mathbf{w} = -|\mathbf{u}|^2 \mathbf{u} \times \mathbf{v} \cdot \mathbf{w} = -|\mathbf{u}|^2 \mathbf{u} \cdot \mathbf{v} \times \mathbf{w}$

21. If $\mathbf{u} = a\mathbf{i} + b\mathbf{j}$ and $\mathbf{v} = c\mathbf{i} + d\mathbf{j}$, then $\mathbf{u} \cdot \mathbf{v} = |\mathbf{u}|\,|\mathbf{v}| \cos\theta \Rightarrow ac + bd = \sqrt{a^2 + b^2}\,\sqrt{c^2 + d^2} \cos\theta$
$\Rightarrow (ac + bd)^2 = (a^2 + b^2)(c^2 + d^2) \cos^2\theta \Rightarrow (ac + bd)^2 \le (a^2 + b^2)(c^2 + d^2)$, since $\cos^2\theta \le 1$.

23. $|\mathbf{u} + \mathbf{v}|^2 = (\mathbf{u} + \mathbf{v}) \cdot (\mathbf{u} + \mathbf{v}) = \mathbf{u} \cdot \mathbf{u} + 2\mathbf{u} \cdot \mathbf{v} + \mathbf{v} \cdot \mathbf{v} \le |\mathbf{u}|^2 + 2\,|\mathbf{u}|\,|\mathbf{v}| + |\mathbf{v}|^2 = (|\mathbf{u}| + |\mathbf{v}|)^2 \Rightarrow |\mathbf{u} + \mathbf{v}| \le |\mathbf{u}| + |\mathbf{v}|$

25. $(|\mathbf{u}|\mathbf{v} + |\mathbf{v}|\mathbf{u}) \cdot (|\mathbf{v}|\mathbf{u} - |\mathbf{u}|\mathbf{v}) = |\mathbf{u}|\mathbf{v} \cdot |\mathbf{v}|\mathbf{u} + |\mathbf{v}|\mathbf{u} \cdot |\mathbf{v}|\mathbf{u} - |\mathbf{u}|\mathbf{v} \cdot |\mathbf{u}|\mathbf{v} - |\mathbf{v}|\mathbf{u} \cdot |\mathbf{u}|\mathbf{v}$
$= |\mathbf{v}|\mathbf{u} \cdot |\mathbf{u}|\mathbf{v} + |\mathbf{v}|^2 \mathbf{u} \cdot \mathbf{u} - |\mathbf{u}|^2 \mathbf{v} \cdot \mathbf{v} - |\mathbf{v}|\mathbf{u} \cdot |\mathbf{u}|\mathbf{v} = |\mathbf{v}|^2 |\mathbf{u}|^2 - |\mathbf{u}|^2 |\mathbf{v}|^2 = 0$

CHAPTER 13 VECTOR-VALUED FUNCTIONS AND MOTION IN SPACE

13.1 CURVES IN SPACE AND THEIR TANGENTS

1. $x = t + 1$ and $y = t^2 - 1 \Rightarrow y = (x-1)^2 - 1 = x^2 - 2x$; $\mathbf{v} = \frac{d\mathbf{r}}{dt} = \mathbf{i} + 2t\mathbf{j} \Rightarrow \mathbf{a} = \frac{d\mathbf{v}}{dt} = 2\mathbf{j} \Rightarrow \mathbf{v} = \mathbf{i} + 2\mathbf{j}$ and $\mathbf{a} = 2\mathbf{j}$ at $t = 1$

3. $x = e^t$ and $y = \frac{2}{9}e^{2t} \Rightarrow y = \frac{2}{9}x^2$; $\mathbf{v} = \frac{d\mathbf{r}}{dt} = e^t\mathbf{i} + \frac{4}{9}e^{2t}\mathbf{j} \Rightarrow \mathbf{a} = e^t\mathbf{i} + \frac{8}{9}e^{2t}\mathbf{j} \Rightarrow \mathbf{v} = 3\mathbf{i} + 4\mathbf{j}$ and $\mathbf{a} = 3\mathbf{i} + 8\mathbf{j}$ at $t = \ln 3$

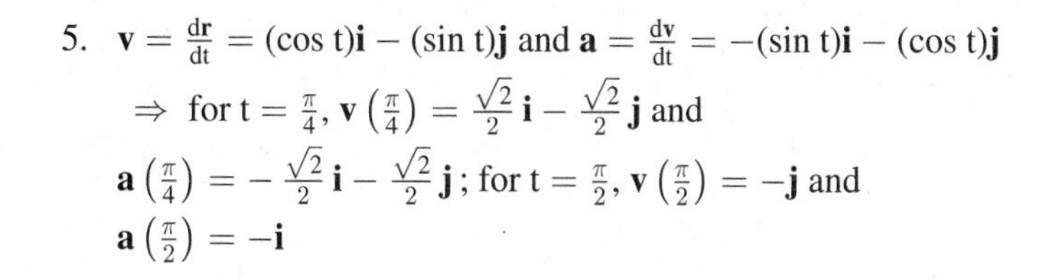

5. $\mathbf{v} = \frac{d\mathbf{r}}{dt} = (\cos t)\mathbf{i} - (\sin t)\mathbf{j}$ and $\mathbf{a} = \frac{d\mathbf{v}}{dt} = -(\sin t)\mathbf{i} - (\cos t)\mathbf{j}$
$\Rightarrow$ for $t = \frac{\pi}{4}$, $\mathbf{v}\left(\frac{\pi}{4}\right) = \frac{\sqrt{2}}{2}\mathbf{i} - \frac{\sqrt{2}}{2}\mathbf{j}$ and
$\mathbf{a}\left(\frac{\pi}{4}\right) = -\frac{\sqrt{2}}{2}\mathbf{i} - \frac{\sqrt{2}}{2}\mathbf{j}$; for $t = \frac{\pi}{2}$, $\mathbf{v}\left(\frac{\pi}{2}\right) = -\mathbf{j}$ and
$\mathbf{a}\left(\frac{\pi}{2}\right) = -\mathbf{i}$

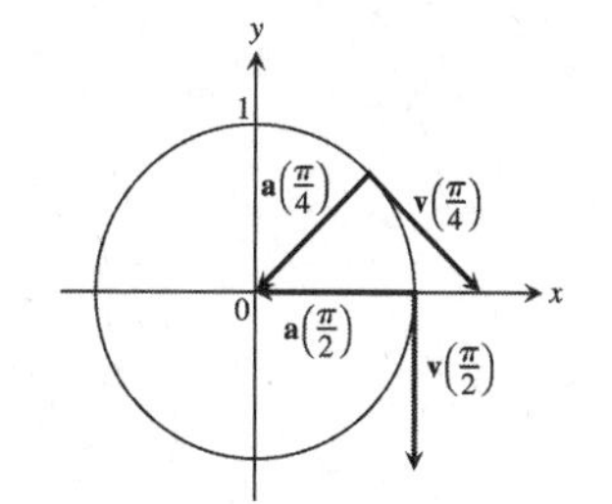

7. $\mathbf{v} = \frac{d\mathbf{r}}{dt} = (1 - \cos t)\mathbf{i} + (\sin t)\mathbf{j}$ and $\mathbf{a} = \frac{d\mathbf{v}}{dt}$
$= (\sin t)\mathbf{i} + (\cos t)\mathbf{j} \Rightarrow$ for $t = \pi$, $\mathbf{v}(\pi) = 2\mathbf{i}$ and $\mathbf{a}(\pi) = -\mathbf{j}$;
for $t = \frac{3\pi}{2}$, $\mathbf{v}\left(\frac{3\pi}{2}\right) = \mathbf{i} - \mathbf{j}$ and $\mathbf{a}\left(\frac{3\pi}{2}\right) = -\mathbf{i}$

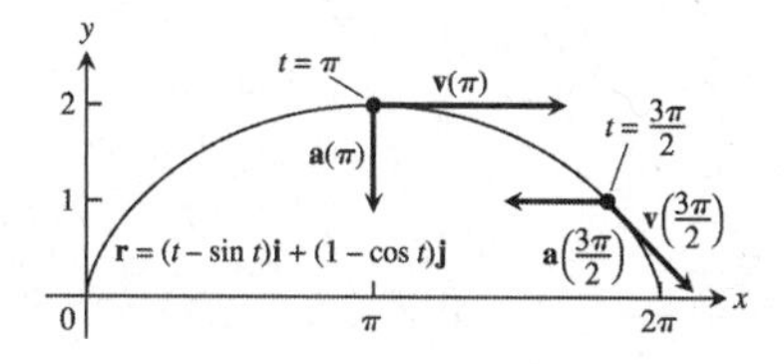

9. $\mathbf{r} = (t+1)\mathbf{i} + (t^2 - 1)\mathbf{j} + 2t\mathbf{k} \Rightarrow \mathbf{v} = \frac{d\mathbf{r}}{dt} = \mathbf{i} + 2t\mathbf{j} + 2\mathbf{k} \Rightarrow \mathbf{a} = \frac{d^2\mathbf{r}}{dt^2} = 2\mathbf{j}$; Speed: $|\mathbf{v}(1)| = \sqrt{1^2 + (2(1))^2 + 2^2} = 3$;
Direction: $\frac{\mathbf{v}(1)}{|\mathbf{v}(1)|} = \frac{\mathbf{i} + 2(1)\mathbf{j} + 2\mathbf{k}}{3} = \frac{1}{3}\mathbf{i} + \frac{2}{3}\mathbf{j} + \frac{2}{3}\mathbf{k} \Rightarrow \mathbf{v}(1) = 3\left(\frac{1}{3}\mathbf{i} + \frac{2}{3}\mathbf{j} + \frac{2}{3}\mathbf{k}\right)$

11. $\mathbf{r} = (2\cos t)\mathbf{i} + (3\sin t)\mathbf{j} + 4t\mathbf{k} \Rightarrow \mathbf{v} = \frac{d\mathbf{r}}{dt} = (-2\sin t)\mathbf{i} + (3\cos t)\mathbf{j} + 4\mathbf{k} \Rightarrow \mathbf{a} = \frac{d^2\mathbf{r}}{dt^2} = (-2\cos t)\mathbf{i} - (3\sin t)\mathbf{j}$;
Speed: $\left|\mathbf{v}\left(\frac{\pi}{2}\right)\right| = \sqrt{\left(-2\sin\frac{\pi}{2}\right)^2 + \left(3\cos\frac{\pi}{2}\right)^2 + 4^2} = 2\sqrt{5}$; Direction: $\frac{\mathbf{v}\left(\frac{\pi}{2}\right)}{\left|\mathbf{v}\left(\frac{\pi}{2}\right)\right|}$
$= \left(-\frac{2}{2\sqrt{5}}\sin\frac{\pi}{2}\right)\mathbf{i} + \left(\frac{3}{2\sqrt{5}}\cos\frac{\pi}{2}\right)\mathbf{j} + \frac{4}{2\sqrt{5}}\mathbf{k} = -\frac{1}{\sqrt{5}}\mathbf{i} + \frac{2}{\sqrt{5}}\mathbf{k} \Rightarrow \mathbf{v}\left(\frac{\pi}{2}\right) = 2\sqrt{5}\left(-\frac{1}{\sqrt{5}}\mathbf{i} + \frac{2}{\sqrt{5}}\mathbf{k}\right)$

13. $\mathbf{r} = (2\ln(t+1))\mathbf{i} + t^2\mathbf{j} + \frac{t^2}{2}\mathbf{k} \Rightarrow \mathbf{v} = \frac{d\mathbf{r}}{dt} = \left(\frac{2}{t+1}\right)\mathbf{i} + 2t\mathbf{j} + t\mathbf{k} \Rightarrow \mathbf{a} = \frac{d^2\mathbf{r}}{dt^2} = \left[\frac{-2}{(t+1)^2}\right]\mathbf{i} + 2\mathbf{j} + \mathbf{k}$;
Speed: $|\mathbf{v}(1)| = \sqrt{\left(\frac{2}{1+1}\right)^2 + (2(1))^2 + 1^2} = \sqrt{6}$; Direction: $\frac{\mathbf{v}(1)}{|\mathbf{v}(1)|} = \frac{\left(\frac{2}{1+1}\right)\mathbf{i} + 2(1)\mathbf{j} + (1)\mathbf{k}}{\sqrt{6}}$
$= \frac{1}{\sqrt{6}}\mathbf{i} + \frac{2}{\sqrt{6}}\mathbf{j} + \frac{1}{\sqrt{6}}\mathbf{k} \Rightarrow \mathbf{v}(1) = \sqrt{6}\left(\frac{1}{\sqrt{6}}\mathbf{i} + \frac{2}{\sqrt{6}}\mathbf{j} + \frac{1}{\sqrt{6}}\mathbf{k}\right)$

15. $\mathbf{v} = 3\mathbf{i} + \sqrt{3}\mathbf{j} + 2t\mathbf{k}$ and $\mathbf{a} = 2\mathbf{k} \Rightarrow \mathbf{v}(0) = 3\mathbf{i} + \sqrt{3}\mathbf{j}$ and $\mathbf{a}(0) = 2\mathbf{k} \Rightarrow |\mathbf{v}(0)| = \sqrt{3^2 + \left(\sqrt{3}\right)^2 + 0^2} = \sqrt{12}$ and
$|\mathbf{a}(0)| = \sqrt{2^2} = 2$; $\mathbf{v}(0) \cdot \mathbf{a}(0) = 0 \Rightarrow \cos\theta = 0 \Rightarrow \theta = \frac{\pi}{2}$

17. $\mathbf{v} = \left(\frac{2t}{t^2+1}\right)\mathbf{i} + \left(\frac{1}{t^2+1}\right)\mathbf{j} + t(t^2+1)^{-1/2}\mathbf{k}$ and $\mathbf{a} = \left[\frac{-2t^2+2}{(t^2+1)^2}\right]\mathbf{i} - \left[\frac{2t}{(t^2+1)^2}\right]\mathbf{j} + \left[\frac{1}{(t^2+1)^{3/2}}\right]\mathbf{k} \Rightarrow \mathbf{v}(0) = \mathbf{j}$ and
$\mathbf{a}(0) = 2\mathbf{i} + \mathbf{k} \Rightarrow |\mathbf{v}(0)| = 1$ and $|\mathbf{a}(0)| = \sqrt{2^2 + 1^2} = \sqrt{5}$; $\mathbf{v}(0) \cdot \mathbf{a}(0) = 0 \Rightarrow \cos\theta = 0 \Rightarrow \theta = \frac{\pi}{2}$

19. $\mathbf{r}(t) = (\sin t)\mathbf{i} + (t^2 - \cos t)\,\mathbf{j} + e^t\mathbf{k} \Rightarrow \mathbf{v}(t) = (\cos t)\mathbf{i} + (2t + \sin t)\mathbf{j} + e^t\mathbf{k}\,;\ t_0 = 0 \Rightarrow \mathbf{v}(t_0) = \mathbf{i} + \mathbf{k}$ and $\mathbf{r}(t_0) = P_0 = (0, -1, 1) \Rightarrow x = 0 + t = t,\ y = -1$, and $z = 1 + t$ are parametric equations of the tangent line

21. $\mathbf{r}(t) = (\ln t)\mathbf{i} + \frac{t-1}{t+2}\,\mathbf{j} + (t \ln t)\mathbf{k} \Rightarrow \mathbf{v}(t) = \frac{1}{t}\,\mathbf{i} + \frac{3}{(t+2)^2}\mathbf{j} + (\ln t + 1)\mathbf{k}\,;\ t_0 = 1 \Rightarrow \mathbf{v}(1) = \mathbf{i} + \frac{1}{3}\,\mathbf{j} + \mathbf{k}$ and $\mathbf{r}(t_0) = P_0 = (0, 0, 0) \Rightarrow x = 0 + t = t,\ y = 0 + \frac{1}{3}t = \frac{1}{3}t$, and $z = 0 + t = t$ are parametric equations of the tangent line

23. (a) $\mathbf{v}(t) = -(\sin t)\mathbf{i} + (\cos t)\mathbf{j} \Rightarrow \mathbf{a}(t) = -(\cos t)\mathbf{i} - (\sin t)\mathbf{j}\,;$
   (i) $|\mathbf{v}(t)| = \sqrt{(-\sin t)^2 + (\cos t)^2} = 1 \Rightarrow$ constant speed;
   (ii) $\mathbf{v} \cdot \mathbf{a} = (\sin t)(\cos t) - (\cos t)(\sin t) = 0 \Rightarrow$ yes, orthogonal;
   (iii) counterclockwise movement;
   (iv) yes, $\mathbf{r}(0) = \mathbf{i} + 0\mathbf{j}$

(b) $\mathbf{v}(t) = -(2 \sin 2t)\mathbf{i} + (2 \cos 2t)\mathbf{j} \Rightarrow \mathbf{a}(t) = -(4 \cos 2t)\mathbf{i} - (4 \sin 2t)\mathbf{j};$
   (i) $|\mathbf{v}(t)| = \sqrt{4 \sin^2 2t + 4 \cos^2 2t} = 2 \Rightarrow$ constant speed;
   (ii) $\mathbf{v} \cdot \mathbf{a} = 8 \sin 2t \cos 2t - 8 \cos 2t \sin 2t = 0 \Rightarrow$ yes, orthogonal;
   (iii) counterclockwise movement;
   (iv) yes, $\mathbf{r}(0) = \mathbf{i} + 0\mathbf{j}$

(c) $\mathbf{v}(t) = -\sin\left(t - \frac{\pi}{2}\right)\mathbf{i} + \cos\left(t - \frac{\pi}{2}\right)\mathbf{j} \Rightarrow \mathbf{a}(t) = -\cos\left(t - \frac{\pi}{2}\right)\mathbf{i} - \sin\left(t - \frac{\pi}{2}\right)\mathbf{j}\,;$
   (i) $|\mathbf{v}(t)| = \sqrt{\sin^2\left(t - \frac{\pi}{2}\right) + \cos^2\left(t - \frac{\pi}{2}\right)} = 1 \Rightarrow$ constant speed;
   (ii) $\mathbf{v} \cdot \mathbf{a} = \sin\left(t - \frac{\pi}{2}\right)\cos\left(t - \frac{\pi}{2}\right) - \cos\left(t - \frac{\pi}{2}\right)\sin\left(t - \frac{\pi}{2}\right) = 0 \Rightarrow$ yes, orthogonal;
   (iii) counterclockwise movement;
   (iv) no, $\mathbf{r}(0) = 0\mathbf{i} - \mathbf{j}$ instead of $\mathbf{i} + 0\mathbf{j}$

(d) $\mathbf{v}(t) = -(\sin t)\mathbf{i} - (\cos t)\mathbf{j} \Rightarrow \mathbf{a}(t) = -(\cos t)\mathbf{i} + (\sin t)\mathbf{j}\,;$
   (i) $|\mathbf{v}(t)| = \sqrt{(-\sin t)^2 + (-\cos t)^2} = 1 \Rightarrow$ constant speed;
   (ii) $\mathbf{v} \cdot \mathbf{a} = (\sin t)(\cos t) - (\cos t)(\sin t) = 0 \Rightarrow$ yes, orthogonal;
   (iii) clockwise movement;
   (iv) yes, $\mathbf{r}(0) = \mathbf{i} - 0\mathbf{j}$

(e) $\mathbf{v}(t) = -(2t \sin t)\mathbf{i} + (2t \cos t)\mathbf{j} \Rightarrow \mathbf{a}(t) = -(2 \sin t + 2t \cos t)\mathbf{i} + (2 \cos t - 2t \sin t)\mathbf{j}\,;$
   (i) $|\mathbf{v}(t)| = \sqrt{[-(2t \sin t)]^2 + (2t \cos t)^2} = \sqrt{4t^2(\sin^2 t + \cos^2 t)} = 2|t| = 2t,\ t \geq 0$
   $\Rightarrow$ variable speed;
   (ii) $\mathbf{v} \cdot \mathbf{a} = 4\,(t \sin^2 t + t^2 \sin t \cos t) + 4\,(t \cos^2 t - t^2 \cos t \sin t) = 4t \neq 0$ in general $\Rightarrow$ not orthogonal in general;
   (iii) counterclockwise movement;
   (iv) yes, $\mathbf{r}(0) = \mathbf{i} + 0\mathbf{j}$

25. The velocity vector is tangent to the graph of $y^2 = 2x$ at the point $(2, 2)$, has length 5, and a positive $\mathbf{i}$ component. Now, $y^2 = 2x \Rightarrow 2y\,\frac{dy}{dx} = 2 \Rightarrow \left.\frac{dy}{dx}\right|_{(2,2)} = \frac{2}{2 \cdot 2} = \frac{1}{2} \Rightarrow$ the tangent vector lies in the direction of the vector $\mathbf{i} + \frac{1}{2}\,\mathbf{j} \Rightarrow$ the velocity vector is $\mathbf{v} = \frac{5}{\sqrt{1 + \frac{1}{4}}}\left(\mathbf{i} + \frac{1}{2}\,\mathbf{j}\right) = \frac{5}{\left(\frac{\sqrt{5}}{2}\right)}\left(\mathbf{i} + \frac{1}{2}\,\mathbf{j}\right) = 2\sqrt{5}\mathbf{i} + \sqrt{5}\mathbf{j}$

27. $\frac{d}{dt}(\mathbf{r} \cdot \mathbf{r}) = \mathbf{r} \cdot \frac{d\mathbf{r}}{dt} + \frac{d\mathbf{r}}{dt} \cdot \mathbf{r} = 2\mathbf{r} \cdot \frac{d\mathbf{r}}{dt} = 2 \cdot 0 = 0 \Rightarrow \mathbf{r} \cdot \mathbf{r}$ is a constant $\Rightarrow |\mathbf{r}| = \sqrt{\mathbf{r} \cdot \mathbf{r}}$ is constant

29. (a) $\mathbf{u} = f(t)\mathbf{i} + g(t)\mathbf{j} + h(t)\mathbf{k} \Rightarrow c\mathbf{u} = cf(t)\mathbf{i} + cg(t)\mathbf{j} + ch(t)\mathbf{k} \Rightarrow \frac{d}{dt}(c\mathbf{u}) = c\,\frac{df}{dt}\,\mathbf{i} + c\,\frac{dg}{dt}\,\mathbf{j} + c\,\frac{dh}{dt}\,\mathbf{k}$
$= c\left(\frac{df}{dt}\,\mathbf{i} + \frac{dg}{dt}\,\mathbf{j} + \frac{dh}{dt}\,\mathbf{k}\right) = c\,\frac{d\mathbf{u}}{dt}$

(b) $f\mathbf{u} = ff(t)\mathbf{i} + fg(t)\mathbf{j} + fh(t)\mathbf{k} \Rightarrow \frac{d}{dt}(f\mathbf{u}) = \left[\frac{df}{dt}\,f(t) + f\,\frac{df}{dt}\right]\mathbf{i} + \left[\frac{df}{dt}\,g(t) + f\,\frac{dg}{dt}\right]\mathbf{j} + \left[\frac{df}{dt}\,h(t) + f\,\frac{dh}{dt}\right]\mathbf{k}$
$= \frac{df}{dt}\,[f(t)\mathbf{i} + g(t)\mathbf{j} + h(t)\mathbf{k}] + f\left[\frac{df}{dt}\,\mathbf{i} + \frac{dg}{dt}\,\mathbf{j} + \frac{dh}{dt}\,\mathbf{k}\right] = \frac{df}{dt}\,\mathbf{u} + f\,\frac{d\mathbf{u}}{dt}$

31. Suppose $\mathbf{r}$ is continuous at $t = t_0$. Then $\lim_{t \to t_0} \mathbf{r}(t) = \mathbf{r}(t_0) \Leftrightarrow \lim_{t \to t_0} [f(t)\mathbf{i} + g(t)\mathbf{j} + h(t)\mathbf{k}]$ $= f(t_0)\mathbf{i} + g(t_0)\mathbf{j} + h(t_0)\mathbf{k} \Leftrightarrow \lim_{t \to t_0} f(t) = f(t_0)$, $\lim_{t \to t_0} g(t) = g(t_0)$, and $\lim_{t \to t_0} h(t) = h(t_0) \Leftrightarrow$ f, g, and h are continuous at $t = t_0$.

33. $\mathbf{r}'(t_0)$ exists $\Rightarrow f'(t_0)\mathbf{i} + g'(t_0)\mathbf{j} + h'(t_0)\mathbf{k}$ exists $\Rightarrow f'(t_0), g'(t_0), h'(t_0)$ all exist $\Rightarrow$ f, g, and h are continuous at $t = t_0 \Rightarrow \mathbf{r}(t)$ is continuous at $t = t_0$

13.2 INTEGRALS OF VECTOR FUNCTIONS; PROJECTILE MOTION

1. $\int_0^1 [t^3\mathbf{i} + 7\mathbf{j} + (t+1)\mathbf{k}]\,dt = \left[\frac{t^4}{4}\right]_0^1 \mathbf{i} + [7t]_0^1 \mathbf{j} + \left[\frac{t^2}{2} + t\right]_0^1 \mathbf{k} = \frac{1}{4}\mathbf{i} + 7\mathbf{j} + \frac{3}{2}\mathbf{k}$

3. $\int_{-\pi/4}^{\pi/4} [(\sin t)\mathbf{i} + (1 + \cos t)\mathbf{j} + (\sec^2 t)\mathbf{k}]\,dt = [-\cos t]_{-\pi/4}^{\pi/4} \mathbf{i} + [t + \sin t]_{-\pi/4}^{\pi/4} \mathbf{j} + [\tan t]_{-\pi/4}^{\pi/4} \mathbf{k} = \left(\frac{\pi + 2\sqrt{2}}{2}\right)\mathbf{j} + 2\mathbf{k}$

5. $\int_1^4 \left(\frac{1}{t}\mathbf{i} + \frac{1}{5-t}\mathbf{j} + \frac{1}{2t}\mathbf{k}\right) dt = = [\ln t]_1^4 \mathbf{i} + [-\ln(5-t)]_1^4 \mathbf{j} + \left[\frac{1}{2}\ln t\right]_1^4 \mathbf{k} = (\ln 4)\mathbf{i} + (\ln 4)\mathbf{j} + (\ln 2)\mathbf{k}$

7. $\int_0^1 \left(te^{t^2}\mathbf{i} + e^{-t}\mathbf{j} + \mathbf{k}\right) dt = \left[\frac{1}{2}e^{t^2}\right]_0^1 \mathbf{i} - [e^{-t}]_0^1 \mathbf{j} + [t]_0^1 \mathbf{k} = \frac{e-1}{2}\mathbf{i} + \frac{e-1}{e}\mathbf{i} + \mathbf{k}$

9. $\int_0^{\pi/2} [(\cos t)\mathbf{i} - (\sin 2t)\mathbf{j} + (\sin^2 t)\mathbf{k}]\,dt = \int_0^{\pi/2} \left[(\cos t)\mathbf{i} - (\sin 2t)\mathbf{j} + \left(\frac{1}{2} - \frac{1}{2}\cos 2t\right)\mathbf{k}\right] dt =$ $= [\sin t]_0^{\pi/2} \mathbf{i} + \left[\frac{1}{2}\cos t\right]_0^{\pi/2} \mathbf{j} + \left[\frac{1}{2}t - \frac{1}{4}\sin 2t\right]_0^{\pi/2} \mathbf{k} = \mathbf{i} - \mathbf{j} + \frac{\pi}{4}\mathbf{k}$

11. $\mathbf{r} = \int (-t\mathbf{i} - t\mathbf{j} - t\mathbf{k})\,dt = -\frac{t^2}{2}\mathbf{i} - \frac{t^2}{2}\mathbf{j} - \frac{t^2}{2}\mathbf{k} + \mathbf{C}$; $\mathbf{r}(0) = 0\mathbf{i} - 0\mathbf{j} - 0\mathbf{k} + \mathbf{C} = \mathbf{i} + 2\mathbf{j} + 3\mathbf{k} \Rightarrow \mathbf{C} = \mathbf{i} + 2\mathbf{j} + 3\mathbf{k}$ $\Rightarrow \mathbf{r} = \left(-\frac{t^2}{2} + 1\right)\mathbf{i} + \left(-\frac{t^2}{2} + 2\right)\mathbf{j} + \left(-\frac{t^2}{2} + 3\right)\mathbf{k}$

13. $\mathbf{r} = \int \left[\left(\frac{3}{2}(t+1)^{1/2}\right)\mathbf{i} + e^{-t}\mathbf{j} + \left(\frac{1}{t+1}\right)\mathbf{k}\right] dt = (t+1)^{3/2}\mathbf{i} - e^{-t}\mathbf{j} + \ln(t+1)\mathbf{k} + \mathbf{C}$; $\mathbf{r}(0) = (0+1)^{3/2}\mathbf{i} - e^{-0}\mathbf{j} + \ln(0+1)\mathbf{k} + \mathbf{C} = \mathbf{k} \Rightarrow \mathbf{C} = -\mathbf{i} + \mathbf{j} + \mathbf{k}$ $\Rightarrow \mathbf{r} = \left[(t+1)^{3/2} - 1\right]\mathbf{i} + (1 - e^{-t})\mathbf{j} + [1 + \ln(t+1)]\mathbf{k}$

15. $\frac{d\mathbf{r}}{dt} = \int (-32\mathbf{k})\,dt = -32t\mathbf{k} + \mathbf{C}_1$; $\frac{d\mathbf{r}}{dt}(0) = 8\mathbf{i} + 8\mathbf{j} \Rightarrow -32(0)\mathbf{k} + \mathbf{C}_1 = 8\mathbf{i} + 8\mathbf{j} \Rightarrow \mathbf{C}_1 = 8\mathbf{i} + 8\mathbf{j}$ $\Rightarrow \frac{d\mathbf{r}}{dt} = 8\mathbf{i} + 8\mathbf{j} - 32t\mathbf{k}$; $\mathbf{r} = \int (8\mathbf{i} + 8\mathbf{j} - 32t\mathbf{k})\,dt = 8t\mathbf{i} + 8t\mathbf{j} - 16t^2\mathbf{k} + \mathbf{C}_2$; $\mathbf{r}(0) = 100\mathbf{k}$ $\Rightarrow 8(0)\mathbf{i} + 8(0)\mathbf{j} - 16(0)^2\mathbf{k} + \mathbf{C}_2 = 100\mathbf{k} \Rightarrow \mathbf{C}_2 = 100\mathbf{k} \Rightarrow \mathbf{r} = 8t\mathbf{i} + 8t\mathbf{j} + (100 - 16t^2)\mathbf{k}$

17. $\frac{d\mathbf{v}}{dt} = \mathbf{a} = 3\mathbf{i} - \mathbf{j} + \mathbf{k} \Rightarrow \mathbf{v}(t) = 3t\mathbf{i} - t\mathbf{j} + t\mathbf{k} + \mathbf{C}_1$; the particle travels in the direction of the vector $(4-1)\mathbf{i} + (1-2)\mathbf{j} + (4-3)\mathbf{k} = 3\mathbf{i} - \mathbf{j} + \mathbf{k}$ (since it travels in a straight line), and at time $t = 0$ it has speed $2 \Rightarrow \mathbf{v}(0) = \frac{2}{\sqrt{9+1+1}}(3\mathbf{i} - \mathbf{j} + \mathbf{k}) = \mathbf{C}_1 \Rightarrow \frac{d\mathbf{r}}{dt} = \mathbf{v}(t) = \left(3t + \frac{6}{\sqrt{11}}\right)\mathbf{i} - \left(t + \frac{2}{\sqrt{11}}\right)\mathbf{j} + \left(t + \frac{2}{\sqrt{11}}\right)\mathbf{k}$ $\Rightarrow \mathbf{r}(t) = \left(\frac{3}{2}t^2 + \frac{6}{\sqrt{11}}t\right)\mathbf{i} - \left(\frac{1}{2}t^2 + \frac{2}{\sqrt{11}}t\right)\mathbf{j} + \left(\frac{1}{2}t^2 + \frac{2}{\sqrt{11}}t\right)\mathbf{k} + \mathbf{C}_2$; $\mathbf{r}(0) = \mathbf{i} + 2\mathbf{j} + 3\mathbf{k} = \mathbf{C}_2$ $\Rightarrow \mathbf{r}(t) = \left(\frac{3}{2}t^2 + \frac{6}{\sqrt{11}}t + 1\right)\mathbf{i} - \left(\frac{1}{2}t^2 + \frac{2}{\sqrt{11}}t - 2\right)\mathbf{j} + \left(\frac{1}{2}t^2 + \frac{2}{\sqrt{11}}t + 3\right)\mathbf{k}$ $= \left(\frac{1}{2}t^2 + \frac{2}{\sqrt{11}}t\right)(3\mathbf{i} - \mathbf{j} + \mathbf{k}) + (\mathbf{i} + 2\mathbf{j} + 3\mathbf{k})$

19. $x = (v_0 \cos\alpha)t \Rightarrow (21\text{ km})\left(\frac{1000\text{ m}}{1\text{ km}}\right) = (840\text{ m/s})(\cos 60°)t \Rightarrow t = \frac{21{,}000\text{ m}}{(840\text{ m/s})(\cos 60°)} = 50$ seconds

21. (a) $t = \frac{2v_0 \sin\alpha}{g} = \frac{2(500 \text{ m/s})(\sin 45°)}{9.8 \text{ m/s}^2} \approx 72.2$ seconds; $R = \frac{v_0^2}{g} \sin 2\alpha = \frac{(500 \text{ m/s})^2}{9.8 \text{ m/s}^2} (\sin 90°) \approx 25{,}510.2$ m

(b) $x = (v_0 \cos\alpha)t \Rightarrow 5000 \text{ m} = (500 \text{ m/s})(\cos 45°)t \Rightarrow t = \frac{5000 \text{ m}}{(500 \text{ m/s})(\cos 45°)} \approx 14.14$ s; thus,

$y = (v_0 \sin\alpha)t - \frac{1}{2}gt^2 \Rightarrow y \approx (500 \text{ m/s})(\sin 45°)(14.14 \text{ s}) - \frac{1}{2}(9.8 \text{ m/s}^2)(14.14 \text{ s})^2 \approx 4020$ m

(c) $y_{max} = \frac{(v_0 \sin\alpha)^2}{2g} = \frac{((500 \text{ m/s})(\sin 45°))^2}{2(9.8 \text{ m/s}^2)} \approx 6378$ m

23. (a) $R = \frac{v_0^2}{g} \sin 2\alpha \Rightarrow 10 \text{ m} = \left(\frac{v_0^2}{9.8 \text{ m/s}^2}\right)(\sin 90°) \Rightarrow v_0^2 = 98 \text{ m}^2\text{s}^2 \Rightarrow v_0 \approx 9.9$ m/s;

(b) $6\text{m} \approx \frac{(9.9 \text{ m/s})^2}{9.8 \text{ m/s}^2}(\sin 2\alpha) \Rightarrow \sin 2\alpha \approx 0.59999 \Rightarrow 2\alpha \approx 36.87°$ or $143.12° \Rightarrow \alpha \approx 18.4°$ or $71.6°$

25. $R = \frac{v_0^2}{g} \sin 2\alpha \Rightarrow 16{,}000 \text{ m} = \frac{(400 \text{ m/s})^2}{9.8 \text{ m/s}^2} \sin 2\alpha \Rightarrow \sin 2\alpha = 0.98 \Rightarrow 2\alpha \approx 78.5°$ or $2\alpha \approx 101.5° \Rightarrow \alpha \approx 39.3°$ or $50.7°$

27. The projectile reaches its maximum height when its vertical component of velocity is zero $\Rightarrow \frac{dy}{dt} = v_0 \sin\alpha - gt = 0$ $\Rightarrow t = \frac{v_0 \sin\alpha}{g} \Rightarrow y_{max} = (v_0 \sin\alpha)\left(\frac{v_0 \sin\alpha}{g}\right) - \frac{1}{2}g\left(\frac{v_0 \sin\alpha}{g}\right)^2 = \frac{(v_0 \sin\alpha)^2}{g} - \frac{(v_0 \sin\alpha)^2}{2g} = \frac{(v_0 \sin\alpha)^2}{2g}$. To find the flight time we find the time when the projectile lands: $(v_0 \sin\alpha)t - \frac{1}{2}gt^2 = 0 \Rightarrow t\left(v_0 \sin\alpha - \frac{1}{2}gt\right) = 0 \Rightarrow t = 0$ or $t = \frac{2v_0 \sin\alpha}{g}$. $t = 0$ is the time when the projectile is fired, so $t = \frac{2v_0 \sin\alpha}{g}$ is the time when the projectile strikes the ground. The range is the value of the horizontal component when $t = \frac{2v_0 \sin\alpha}{g} \Rightarrow R = x = (v_0 \cos\alpha)\left(\frac{2v_0 \sin\alpha}{g}\right) = \frac{v_0^2}{g}(2 \sin\alpha \cos\alpha) = \frac{v_0^2}{g} \sin 2\alpha$. The range is largest when $\sin 2\alpha = 1 \Rightarrow \alpha = 45°$.

29. $\frac{d\mathbf{r}}{dt} = \int(-g\mathbf{j})\,dt = -gt\mathbf{j} + \mathbf{C}_1$ and $\frac{d\mathbf{r}}{dt}(0) = (v_0 \cos\alpha)\mathbf{i} + (v_0 \sin\alpha)\mathbf{j} \Rightarrow -g(0)\mathbf{j} + \mathbf{C}_1 = (v_0 \cos\alpha)\mathbf{i} + (v_0 \sin\alpha)\mathbf{j}$ $\Rightarrow \mathbf{C}_1 = (v_0 \cos\alpha)\mathbf{i} + (v_0 \sin\alpha)\mathbf{j} \Rightarrow \frac{d\mathbf{r}}{dt} = (v_0 \cos\alpha)\mathbf{i} + (v_0 \sin\alpha - gt)\mathbf{j}$; $\mathbf{r} = \int[(v_0 \cos\alpha)\mathbf{i} + (v_0 \sin\alpha - gt)\mathbf{j}]\,dt$ $= (v_0 t \cos\alpha)\mathbf{i} + \left(v_0 t \sin\alpha - \frac{1}{2}gt^2\right)\mathbf{j} + \mathbf{C}_2$ and $\mathbf{r}(0) = x_0\mathbf{i} + y_0\mathbf{j} \Rightarrow [v_0(0)\cos\alpha]\mathbf{i} + \left[v_0(0)\sin\alpha - \frac{1}{2}g(0)^2\right]\mathbf{j} + \mathbf{C}_2$ $= x_0\mathbf{i} + y_0\mathbf{j} \Rightarrow \mathbf{C}_2 = x_0\mathbf{i} + y_0\mathbf{j} \Rightarrow \mathbf{r} = (x_0 + v_0 t \cos\alpha)\mathbf{i} + \left(y_0 + v_0 t \sin\alpha - \frac{1}{2}gt^2\right)\mathbf{j} \Rightarrow x = x_0 + v_0 t \cos\alpha$ and $y = y_0 + v_0 t \sin\alpha - \frac{1}{2}gt^2$

31. (a) At the time t when the projectile hits the line OR we have $\tan\beta = \frac{y}{x}$; $x = [v_0 \cos(\alpha - \beta)]t$ and $y = [v_0 \sin(\alpha - \beta)]t - \frac{1}{2}gt^2 < 0$ since R is below level ground. Therefore let $|y| = \frac{1}{2}gt^2 - [v_0 \sin(\alpha - \beta)]t > 0$ so that $\tan\beta = \frac{\left[\frac{1}{2}gt^2 (v_0 \sin(\alpha-\beta))t\right]}{[v_0 \cos(\alpha-\beta)]t} = \frac{\left[\frac{1}{2}gt - v_0 \sin(\alpha-\beta)\right]}{v_0 \cos(\alpha-\beta)}$

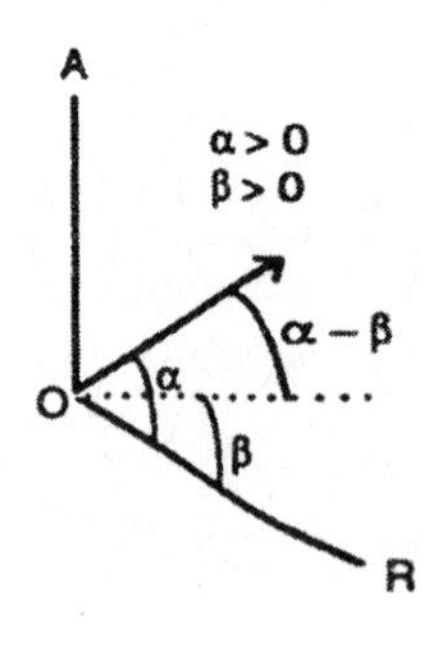

$\Rightarrow v_0 \cos(\alpha - \beta)\tan\beta = \frac{1}{2}gt - v_0 \sin(\alpha - \beta)$

$\Rightarrow t = \frac{2v_0 \sin(\alpha-\beta) + 2v_0 \cos(\alpha-\beta)\tan\beta}{g}$, which is the time when the projectile hits the downhill slope. Therefore,

$x = [v_0 \cos(\alpha - \beta)]\left[\frac{2v_0 \sin(\alpha-\beta) + 2v_0 \cos(\alpha-\beta)\tan\beta}{g}\right] = \frac{2v_0^2}{g}\left[\cos^2(\alpha - \beta)\tan\beta + \sin(\alpha - \beta)\cos(\alpha - \beta)\right]$. If x is maximized, then OR is maximized: $\frac{dx}{d\alpha} = \frac{2v_0^2}{g}[-\sin 2(\alpha - \beta)\tan\beta + \cos 2(\alpha - \beta)] = 0$

$\Rightarrow -\sin 2(\alpha - \beta)\tan\beta + \cos 2(\alpha - \beta) = 0 \Rightarrow \tan\beta = \cot 2(\alpha - \beta) \Rightarrow 2(\alpha - \beta) = 90° - \beta$

$\Rightarrow \alpha - \beta = \frac{1}{2}(90° - \beta) \Rightarrow \alpha = \frac{1}{2}(90° + \beta) = \frac{1}{2}$ of $\angle AOR$.

(b) At the time t when the projectile hits OR we have $\tan\beta = \frac{y}{x}$; $x = [v_0 \cos(\alpha+\beta)]t$ and $y = [v_0 \sin(\alpha+\beta)]t - \frac{1}{2}gt^2$

$\Rightarrow \tan\beta = \frac{[v_0 \sin(\alpha+\beta)]t - \frac{1}{2}gt^2}{[v_0 \cos(\alpha+\beta)]t} = \frac{[v_0 \sin(\alpha+\beta) - \frac{1}{2}gt]}{v_0 \cos(\alpha+\beta)}$

$\Rightarrow v_0 \cos(\alpha+\beta)\tan\beta = v_0 \sin(\alpha+\beta) - \frac{1}{2}gt$

$\Rightarrow t = \frac{2v_0 \sin(\alpha+\beta) - 2v_0 \cos(\alpha+\beta)\tan\beta}{g}$, which is the time

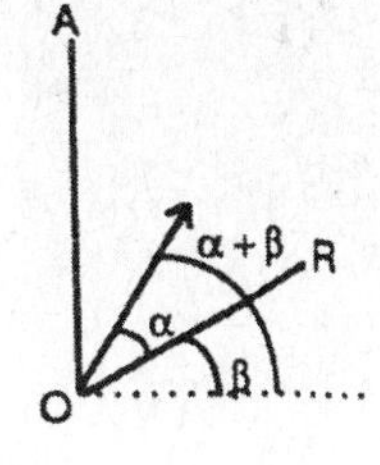

when the projectile hits the uphill slope. Therefore,

$x = [v_0 \cos(\alpha+\beta)]\left[\frac{2v_0 \sin(\alpha+\beta) - 2v_0 \cos(\alpha+\beta)\tan\beta}{g}\right] = \frac{2v_0^2}{g}[\sin(\alpha+\beta)\cos(\alpha+\beta) - \cos^2(\alpha+\beta)\tan\beta]$. If x is maximized, then OR is maximized: $\frac{dx}{d\alpha} = \frac{2v_0^2}{g}[\cos 2(\alpha+\beta) + \sin 2(\alpha+\beta)\tan\beta] = 0$

$\Rightarrow \cos 2(\alpha+\beta) + \sin 2(\alpha+\beta)\tan\beta = 0 \Rightarrow \cot 2(\alpha+\beta) + \tan\beta = 0 \Rightarrow \cot 2(\alpha+\beta) = -\tan\beta$ $= \tan(-\beta) \Rightarrow 2(\alpha+\beta) = 90° - (-\beta) = 90° + \beta \Rightarrow \alpha = \frac{1}{2}(90° - \beta) = \frac{1}{2}$ of $\angle AOR$. Therefore v_0 would bisect $\angle AOR$ for maximum range uphill.

33. (a) (Assuming that "x" is zero at the point of impact:) $\mathbf{r}(t) = (x(t))\mathbf{i} + (y(t))\mathbf{j}$; where $x(t) = (35\cos 27°)t$ and $y(t) = 4 + (35\sin 27°)t - 16t^2$.

(b) $y_{max} = \frac{(v_0 \sin\alpha)^2}{2g} + 4 = \frac{(35\sin 27°)^2}{64} + 4 \approx 7.945$ feet, which is reached at $t = \frac{v_0 \sin\alpha}{g} = \frac{35\sin 27°}{32} \approx 0.497$ seconds.

(c) For the time, solve $y = 4 + (35\sin 27°)t - 16t^2 = 0$ for t, using the quadratic formula $t = \frac{35\sin 27° + \sqrt{(-35\sin 27°)^2 + 256}}{32} \approx 1.201$ sec. Then the range is about $x(1.201) = (35\cos 27°)(1.201) \approx 37.453$ feet.

(d) For the time, solve $y = 4 + (35\sin 27°)t - 16t^2 = 7$ for t, using the quadratic formula $t = \frac{35\sin 27° \pm \sqrt{(-35\sin 27°)^2 - 192}}{32} \approx 0.254$ and 0.740 seconds. At those times the ball is about $x(0.254) = (35\cos 27°)(0.254) \approx 7.921$ feet and $x(0.740) = (35\cos 27°)(0.740) \approx 23.077$ feet the impact point, or about $37.453 - 7.921 \approx 29.532$ feet and $37.453 - 23.077 \approx 14.376$ feet from the landing spot.

(e) Yes. It changes things because the ball won't clear the net ($y_{max} \approx 7.945$).

35. Flight time = 1 sec and the measure of the angle of elevation is about 64° (using a protractor) so that $t = \frac{2v_0 \sin\alpha}{g}$ $\Rightarrow 1 = \frac{2v_0 \sin 64°}{32} \Rightarrow v_0 \approx 17.80$ ft/sec. Then $y_{max} = \frac{(17.80\sin 64°)^2}{2(32)} \approx 4.00$ ft and $R = \frac{v_0^2}{g}\sin 2\alpha \Rightarrow R = \frac{(17.80)^2}{32}\sin 128°$ ≈ 7.80 ft $\Rightarrow$ the engine traveled about 7.80 ft in 1 sec $\Rightarrow$ the engine velocity was about 7.80 ft/sec

37. $\frac{d^2\mathbf{r}}{dt^2} + k\frac{d\mathbf{r}}{dt} = -g\mathbf{j} \Rightarrow P(t) = k$ and $\mathbf{Q}(t) = -g\mathbf{j} \Rightarrow \int P(t)\,dt = kt \Rightarrow v(t) = e^{\int P(t)\,dt} = e^{kt} \Rightarrow \frac{d\mathbf{r}}{dt} = \frac{1}{v(t)}\int v(t)\,\mathbf{Q}(t)\,dt$ $= -ge^{-kt}\int e^{kt}\,\mathbf{j}\,dt = -ge^{-kt}\left[\frac{e^{kt}}{k}\mathbf{j} + \mathbf{C}_1\right] = -\frac{g}{k}\mathbf{j} + \mathbf{C}e^{-kt}$, where $\mathbf{C} = -g\mathbf{C}_1$; apply the initial condition:

$\left.\frac{d\mathbf{r}}{dt}\right|_{t=0} = (v_0\cos\alpha)\mathbf{i} + (v_0\sin\alpha)\mathbf{j} = -\frac{g}{k}\mathbf{j} + \mathbf{C} \Rightarrow \mathbf{C} = (v_0\cos\alpha)\mathbf{i} + \left(\frac{g}{k} + v_0\sin\alpha\right)\mathbf{j}$

$\Rightarrow \frac{d\mathbf{r}}{dt} = (v_0e^{-kt}\cos\alpha)\mathbf{i} + \left(-\frac{g}{k} + e^{-kt}\left(\frac{g}{k} + v_0\sin\alpha\right)\right)\mathbf{j}$, $\mathbf{r} = \int\left[(v_0e^{-kt}\cos\alpha)\mathbf{i} + \left(-\frac{g}{k} + e^{-kt}\left(\frac{g}{k} + v_0\sin\alpha\right)\right)\mathbf{j}\right]dt$

$= \left(-\frac{v_0}{k}e^{-kt}\cos\alpha\right)\mathbf{i} + \left(-\frac{gt}{k} - \frac{e^{-kt}}{k}\left(\frac{g}{k} + v_0\sin\alpha\right)\right)\mathbf{j} + \mathbf{C}_2$; apply the initial condition:

$\mathbf{r}(0) = \mathbf{0} = \left(-\frac{v_0}{k}\cos\alpha\right)\mathbf{i} + \left(-\frac{g}{k^2} - \frac{v_0\sin\alpha}{k}\right)\mathbf{j} + \mathbf{C}_2 \Rightarrow \mathbf{C}_2 = \left(\frac{v_0}{k}\cos\alpha\right)\mathbf{i} + \left(\frac{g}{k^2} + \frac{v_0\sin\alpha}{k}\right)\mathbf{j}$

$\Rightarrow \mathbf{r}(t) = \left(\frac{v_0}{k}(1 - e^{-kt})\cos\alpha\right)\mathbf{i} + \left(\frac{v_0}{k}(1 - e^{-kt})\sin\alpha + \frac{g}{k^2}(1 - kt - e^{-kt})\right)\mathbf{j}$

39. (a) $\int_a^b k\mathbf{r}(t)\,dt = \int_a^b [kf(t)\mathbf{i} + kg(t)\mathbf{j} + kh(t)\mathbf{k}]\,dt = \int_a^b [kf(t)]\,dt\,\mathbf{i} + \int_a^b [kg(t)]\,dt\,\mathbf{j} + \int_a^b [kh(t)]\,dt\,\mathbf{k}$

$= k\left(\int_a^b f(t)\,dt\,\mathbf{i} + \int_a^b g(t)\,dt\,\mathbf{j} + \int_a^b h(t)\,dt\,\mathbf{k}\right) = k\int_a^b \mathbf{r}(t)\,dt$

(b) $\int_a^b [\mathbf{r}_1(t) \pm \mathbf{r}_2(t)]\,dt = \int_a^b ([f_1(t)\mathbf{i} + g_1(t)\mathbf{j} + h_1(t)\mathbf{k}] \pm [f_2(t)\mathbf{i} + g_2(t)\mathbf{j} + h_2(t)\mathbf{k}])\,dt$

$= \int_a^b ([f_1(t) \pm f_2(t)]\,\mathbf{i} + [g_1(t) \pm g_2(t)]\,\mathbf{j} + [h_1(t) \pm h_2(t)]\,\mathbf{k})\,dt$

$= \int_a^b [f_1(t) \pm f_2(t)]\,dt\,\mathbf{i} + \int_a^b [g_1(t) \pm g_2(t)]\,dt\,\mathbf{j} + \int_a^b [h_1(t) \pm h_2(t)]\,dt\,\mathbf{k}$

$= \left[\int_a^b f_1(t)\,dt\,\mathbf{i} \pm \int_a^b f_2(t)\,dt\,\mathbf{i}\right] + \left[\int_a^b g_1(t)\,dt\,\mathbf{j} \pm \int_a^b g_2(t)\,dt\,\mathbf{j}\right] + \left[\int_a^b h_1(t)\,dt\,\mathbf{k} \pm \int_a^b h_2(t)\,dt\,\mathbf{k}\right]$

$= \int_a^b \mathbf{r}_1(t)\,dt \pm \int_a^b \mathbf{r}_2(t)\,dt$

(c) Let $\mathbf{C} = c_1\mathbf{i} + c_2\mathbf{j} + c_3\mathbf{k}$. Then $\int_a^b \mathbf{C}\cdot\mathbf{r}(t)\,dt = \int_a^b [c_1 f(t) + c_2 g(t) + c_3 h(t)]\,dt$

$= c_1 \int_a^b f(t)\,dt + c_2 \int_a^b g(t)\,dt + c_3 \int_a^b h(t)\,dt = \mathbf{C}\cdot\int_a^b \mathbf{r}(t)\,dt;$

$\int_a^b \mathbf{C}\times\mathbf{r}(t)\,dt = \int_a^b [c_2 h(t) - c_3 g(t)]\,\mathbf{i} + [c_3 f(t) - c_1 h(t)]\,\mathbf{j} + [c_1 g(t) - c_2 f(t)]\,\mathbf{k}\,dt$

$= \left[c_2 \int_a^b h(t)\,dt - c_3 \int_a^b g(t)\,dt\right]\mathbf{i} + \left[c_3 \int_a^b f(t)\,dt - c_1 \int_a^b h(t)\,dt\right]\mathbf{j} + \left[c_1 \int_a^b g(t)\,dt - c_2 \int_a^b f(t)\,dt\right]\mathbf{k}$

$= \mathbf{C}\times\int_a^b \mathbf{r}(t)\,dt$

41. (a) If $\mathbf{R}_1(t)$ and $\mathbf{R}_2(t)$ have identical derivatives on I, then $\frac{d\mathbf{R}_1}{dt} = \frac{df_1}{dt}\mathbf{i} + \frac{dg_1}{dt}\mathbf{j} + \frac{dh_1}{dt}\mathbf{k} = \frac{df_2}{dt}\mathbf{i} + \frac{dg_2}{dt}\mathbf{j} + \frac{dh_2}{dt}\mathbf{k}$ $= \frac{d\mathbf{R}_2}{dt} \Rightarrow \frac{df_1}{dt} = \frac{df_2}{dt}, \frac{dg_1}{dt} = \frac{dg_2}{dt}, \frac{dh_1}{dt} = \frac{dh_2}{dt} \Rightarrow f_1(t) = f_2(t) + c_1,\ g_1(t) = g_2(t) + c_2,\ h_1(t) = h_2(t) + c_3$ $\Rightarrow f_1(t)\mathbf{i} + g_1(t)\mathbf{j} + h_1(t)\mathbf{k} = [f_2(t) + c_1]\mathbf{i} + [g_2(t) + c_2]\mathbf{j} + [h_2(t) + c_3]\mathbf{k} \Rightarrow \mathbf{R}_1(t) = \mathbf{R}_2(t) + \mathbf{C}$, where $\mathbf{C} = c_1\mathbf{i} + c_2\mathbf{j} + c_3\mathbf{k}$.

(b) Let $\mathbf{R}(t)$ be an antiderivative of $\mathbf{r}(t)$ on I. Then $\mathbf{R}'(t) = \mathbf{r}(t)$. If $\mathbf{U}(t)$ is an antiderivative of $\mathbf{r}(t)$ on I, then $\mathbf{U}'(t) = \mathbf{r}(t)$. Thus $\mathbf{U}'(t) = \mathbf{R}'(t)$ on I $\Rightarrow \mathbf{U}(t) = \mathbf{R}(t) + \mathbf{C}$.

43. (a) $\mathbf{r}(t) = (x(t))\mathbf{i} + (y(t))\mathbf{j}$; where $x(t) = \left(\frac{1}{0.08}\right)(1 - e^{-0.08t})(152\cos 20° - 17.6)$ and $y(t) = 3 + \left(\frac{152}{0.08}\right)(1 - e^{-0.08t})(\sin 20°) + \left(\frac{32}{0.08^2}\right)(1 - 0.08t - e^{-0.08t})$

(b) Solve graphically using a calculator or CAS: At $t \approx 1.527$ seconds the ball reaches a maximum height of about 41.893 feet.

(c) Use a graphing calculator or CAS to find that $y = 0$ when the ball has traveled for ≈ 3.181 seconds. The range is about $x(3.181) = \left(\frac{1}{0.08}\right)\left(1 - e^{-0.08(3.181)}\right)(152\cos 20° - 17.6) \approx 351.734$ feet.

(d) Use a graphing calculator or CAS to find that $y = 35$ for $t \approx 0.877$ and 2.190 seconds, at which times the ball is about $x(0.877) \approx 106.028$ feet and $x(2.190) \approx 251.530$ feet from home plate.

(e) No; the range is less than 380 feet. To find the wind needed for a home run, first use the method of part (d) to find that $y = 20$ at $t \approx 0.376$ and 2.716 seconds. Then define $x(w) = \left(\frac{1}{0.08}\right)\left(1 - e^{-0.08(2.716)}\right)(152\cos 20° + w)$, and solve $x(w) = 380$ to find $w \approx 12.846$ ft/sec.

13.3 ARC LENGTH IN SPACE

1. $\mathbf{r} = (2\cos t)\mathbf{i} + (2\sin t)\mathbf{j} + \sqrt{5}t\mathbf{k} \Rightarrow \mathbf{v} = (-2\sin t)\mathbf{i} + (2\cos t)\mathbf{j} + \sqrt{5}\mathbf{k}$ $\Rightarrow |\mathbf{v}| = \sqrt{(-2\sin t)^2 + (2\cos t)^2 + \left(\sqrt{5}\right)^2} = \sqrt{4\sin^2 t + 4\cos^2 t + 5} = 3;\ \mathbf{T} = \frac{\mathbf{v}}{|\mathbf{v}|}$ $= \left(-\frac{2}{3}\sin t\right)\mathbf{i} + \left(\frac{2}{3}\cos t\right)\mathbf{j} + \frac{\sqrt{5}}{3}\mathbf{k}$ and Length $= \int_0^\pi |\mathbf{v}|\,dt = \int_0^\pi 3\,dt = [3t]_0^\pi = 3\pi$

3. $\mathbf{r} = t\mathbf{i} + \frac{2}{3}t^{3/2}\mathbf{k} \Rightarrow \mathbf{v} = \mathbf{i} + t^{1/2}\mathbf{k} \Rightarrow |\mathbf{v}| = \sqrt{1^2 + (t^{1/2})^2} = \sqrt{1+t};\ \mathbf{T} = \frac{\mathbf{v}}{|\mathbf{v}|} = \frac{1}{\sqrt{1+t}}\mathbf{i} + \frac{\sqrt{t}}{\sqrt{1+t}}\mathbf{k}$ and Length $= \int_0^8 \sqrt{1+t}\,dt = \left[\frac{2}{3}(1+t)^{3/2}\right]_0^8 = \frac{52}{3}$

5. $\mathbf{r} = (\cos^3 t)\mathbf{j} + (\sin^3 t)\mathbf{k} \Rightarrow \mathbf{v} = (-3\cos^2 t\sin t)\mathbf{j} + (3\sin^2 t\cos t)\mathbf{k} \Rightarrow |\mathbf{v}|$ $= \sqrt{(-3\cos^2 t\sin t)^2 + (3\sin^2 t\cos t)^2} = \sqrt{(9\cos^2 t\sin^2 t)(\cos^2 t + \sin^2 t)} = 3|\cos t\sin t|;$

$\mathbf{T} = \frac{\mathbf{v}}{|\mathbf{v}|} = \frac{-3\cos^2 t \sin t}{3|\cos t \sin t|}\mathbf{j} + \frac{3\sin^2 t \cos t}{3|\cos t \sin t|}\mathbf{k} = (-\cos t)\mathbf{j} + (\sin t)\mathbf{k}$, if $0 \le t \le \frac{\pi}{2}$, and

$\text{Length} = \int_0^{\pi/2} 3|\cos t \sin t|\, dt = \int_0^{\pi/2} 3\cos t \sin t\, dt = \int_0^{\pi/2} \frac{3}{2}\sin 2t\, dt = \left[-\frac{3}{4}\cos 2t\right]_0^{\pi/2} = \frac{3}{2}$

7. $\mathbf{r} = (t\cos t)\mathbf{i} + (t\sin t)\mathbf{j} + \frac{2\sqrt{2}}{3}t^{3/2}\mathbf{k} \Rightarrow \mathbf{v} = (\cos t - t\sin t)\mathbf{i} + (\sin t + t\cos t)\mathbf{j} + \left(\sqrt{2}t^{1/2}\right)\mathbf{k}$

$\Rightarrow |\mathbf{v}| = \sqrt{(\cos t - t\sin t)^2 + (\sin t + t\cos t)^2 + \left(\sqrt{2t}\right)^2} = \sqrt{1 + t^2 + 2t} = \sqrt{(t+1)^2} = |t+1| = t+1$, if $t \ge 0$;

$\mathbf{T} = \frac{\mathbf{v}}{|\mathbf{v}|} = \left(\frac{\cos t - t\sin t}{t+1}\right)\mathbf{i} + \left(\frac{\sin t + t\cos t}{t+1}\right)\mathbf{j} + \left(\frac{\sqrt{2}t^{1/2}}{t+1}\right)\mathbf{k}$ and $\text{Length} = \int_0^{\pi}(t+1)\,dt = \left[\frac{t^2}{2} + t\right]_0^{\pi} = \frac{\pi^2}{2} + \pi$

9. Let $P(t_0)$ denote the point. Then $\mathbf{v} = (5\cos t)\mathbf{i} - (5\sin t)\mathbf{j} + 12\mathbf{k}$ and $26\pi = \int_0^{t_0}\sqrt{25\cos^2 t + 25\sin^2 t + 144}\,dt$

$= \int_0^{t_0} 13\,dt = 13t_0 \Rightarrow t_0 = 2\pi$, and the point is $P(2\pi) = (5\sin 2\pi, 5\cos 2\pi, 24\pi) = (0, 5, 24\pi)$

11. $\mathbf{r} = (4\cos t)\mathbf{i} + (4\sin t)\mathbf{j} + 3t\mathbf{k} \Rightarrow \mathbf{v} = (-4\sin t)\mathbf{i} + (4\cos t)\mathbf{j} + 3\mathbf{k} \Rightarrow |\mathbf{v}| = \sqrt{(-4\sin t)^2 + (4\cos t)^2 + 3^2}$

$= \sqrt{25} = 5 \Rightarrow s(t) = \int_0^t 5\,d\tau = 5t \Rightarrow \text{Length} = s\left(\frac{\pi}{2}\right) = \frac{5\pi}{2}$

13. $\mathbf{r} = (e^t\cos t)\mathbf{i} + (e^t\sin t)\mathbf{j} + e^t\mathbf{k} \Rightarrow \mathbf{v} = (e^t\cos t - e^t\sin t)\mathbf{i} + (e^t\sin t + e^t\cos t)\mathbf{j} + e^t\mathbf{k}$

$\Rightarrow |\mathbf{v}| = \sqrt{(e^t\cos t - e^t\sin t)^2 + (e^t\sin t + e^t\cos t)^2 + (e^t)^2} = = \sqrt{3e^{2t}} = \sqrt{3}e^t \Rightarrow s(t) = \int_0^t \sqrt{3}e^{\tau}\,d\tau$

$= \sqrt{3}e^t - \sqrt{3} \Rightarrow \text{Length} = s(0) - s(-\ln 4) = 0 - \left(\sqrt{3}e^{-\ln 4} - \sqrt{3}\right) = \frac{3\sqrt{3}}{4}$

15. $\mathbf{r} = \left(\sqrt{2}t\right)\mathbf{i} + \left(\sqrt{2}t\right)\mathbf{j} + (1 - t^2)\mathbf{k} \Rightarrow \mathbf{v} = \sqrt{2}\mathbf{i} + \sqrt{2}\mathbf{j} - 2t\mathbf{k} \Rightarrow |\mathbf{v}| = \sqrt{\left(\sqrt{2}\right)^2 + \left(\sqrt{2}\right)^2 + (-2t)^2} = \sqrt{4 + 4t^2}$

$= 2\sqrt{1+t^2} \Rightarrow \text{Length} = \int_0^1 2\sqrt{1+t^2}\,dt = \left[2\left(\frac{t}{2}\sqrt{1+t^2} + \frac{1}{2}\ln\left(t + \sqrt{1+t^2}\right)\right)\right]_0^1 = \sqrt{2} + \ln\left(1 + \sqrt{2}\right)$

17. (a) $\mathbf{r} = (\cos t)\mathbf{i} + (\sin t)\mathbf{j} + (1 - \cos t)\mathbf{k}$, $0 \le t \le 2\pi \Rightarrow x = \cos t$, $y = \sin t$, $z = 1 - \cos t \Rightarrow x^2 + y^2$ $= \cos^2 t + \sin^2 t = 1$, a right circular cylinder with the z-axis as the axis and radius $= 1$. Therefore $P(\cos t, \sin t, 1 - \cos t)$ lies on the cylinder $x^2 + y^2 = 1$; $t = 0 \Rightarrow P(1, 0, 0)$ is on the curve; $t = \frac{\pi}{2} \Rightarrow Q(0, 1, 1)$ is on the curve; $t = \pi \Rightarrow R(-1, 0, 2)$ is on the curve. Then $\overrightarrow{PQ} = -\mathbf{i} + \mathbf{j} + \mathbf{k}$ and $\overrightarrow{PR} = -2\mathbf{i} + 2\mathbf{k}$

$\Rightarrow \overrightarrow{PQ} \times \overrightarrow{PR} = \begin{vmatrix} \mathbf{i} & \mathbf{j} & \mathbf{k} \\ -1 & 1 & 1 \\ -2 & 0 & 2 \end{vmatrix} = 2\mathbf{i} + 2\mathbf{k}$ is a vector normal to the plane of P, Q, and R. Then the

plane containing P, Q, and R has an equation $2x + 2z = 2(1) + 2(0)$ or $x + z = 1$. Any point on the curve will satisfy this equation since $x + z = \cos t + (1 - \cos t) = 1$. Therefore, any point on the curve lies on the intersection of the cylinder $x^2 + y^2 = 1$ and the plane $x + z = 1 \Rightarrow$ the curve is an ellipse.

(b) $\mathbf{v} = (-\sin t)\mathbf{i} + (\cos t)\mathbf{j} + (\sin t)\mathbf{k} \Rightarrow |\mathbf{v}| = \sqrt{\sin^2 t + \cos^2 t + \sin^2 t} = \sqrt{1 + \sin^2 t} \Rightarrow \mathbf{T} = \frac{\mathbf{v}}{|\mathbf{v}|}$

$= \frac{(-\sin t)\mathbf{i} + (\cos t)\mathbf{j} + (\sin t)\mathbf{k}}{\sqrt{1+\sin^2 t}} \Rightarrow \mathbf{T}(0) = \mathbf{j}$, $\mathbf{T}\left(\frac{\pi}{2}\right) = \frac{-\mathbf{i}+\mathbf{k}}{\sqrt{2}}$, $\mathbf{T}(\pi) = -\mathbf{j}$, $\mathbf{T}\left(\frac{3\pi}{2}\right) = \frac{\mathbf{i}-\mathbf{k}}{\sqrt{2}}$

(c) $\mathbf{a} = (-\cos t)\mathbf{i} - (\sin t)\mathbf{j} + (\cos t)\mathbf{k}$; $\mathbf{n} = \mathbf{i} + \mathbf{k}$ is normal to the plane $x + z = 1 \Rightarrow \mathbf{n} \cdot \mathbf{a} = -\cos t + \cos t$ $= 0 \Rightarrow \mathbf{a}$ is orthogonal to $\mathbf{n} \Rightarrow \mathbf{a}$ is parallel to the plane; $\mathbf{a}(0) = -\mathbf{i} + \mathbf{k}$, $\mathbf{a}\left(\frac{\pi}{2}\right) = -\mathbf{j}$, $\mathbf{a}(\pi) = \mathbf{i} - \mathbf{k}$, $\mathbf{a}\left(\frac{3\pi}{2}\right) = \mathbf{j}$

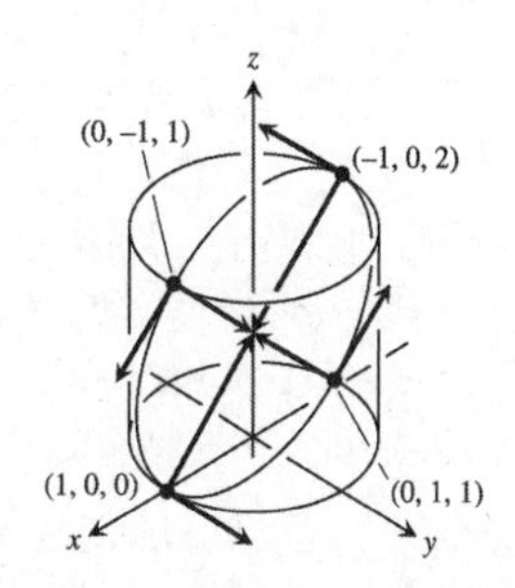

(d) $|\mathbf{v}| = \sqrt{1 + \sin^2 t}$ (See part (b) $\Rightarrow$ $L = \int_0^{2\pi} \sqrt{1 + \sin^2 t}\, dt$

(e) $L \approx 7.64$ (by *Mathematica*)

19. $\angle PQB = \angle QOB = t$ and $PQ = \text{arc}(AQ) = t$ since PQ = length of the unwound string = length of arc (AQ); thus $x = OB + BC = OB + DP = \cos t + t \sin t$, and $y = PC = QB - QD = \sin t - t \cos t$

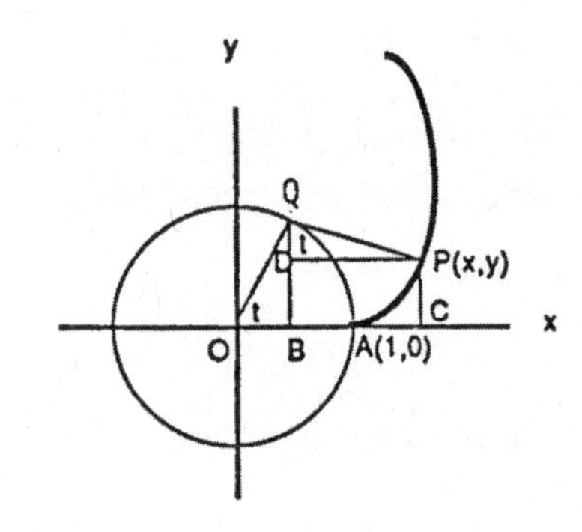

21. $\mathbf{v} = \frac{d}{dt}(x_0 + tu_1)\mathbf{i} + \frac{d}{dt}(y_0 + tu_2)\mathbf{j} + \frac{d}{dt}(z_0 + tu_3)\mathbf{k} = u_1\mathbf{i} + u_2\mathbf{j} + u_3\mathbf{k} = \mathbf{u}$, so $s(t) = \int_0^t |\mathbf{v}|dt = \int_0^t |\mathbf{u}|d\tau = \int_0^t 1\, d\tau = t$

13.4 CURVATURE AND NORMAL VECTORS OF A CURVE

1. $\mathbf{r} = t\mathbf{i} + \ln(\cos t)\mathbf{j} \Rightarrow \mathbf{v} = \mathbf{i} + \left(\frac{-\sin t}{\cos t}\right)\mathbf{j} = \mathbf{i} - (\tan t)\mathbf{j} \Rightarrow |\mathbf{v}| = \sqrt{1^2 + (-\tan t)^2} = \sqrt{\sec^2 t} = |\sec t| = \sec t$, since $-\frac{\pi}{2} < t < \frac{\pi}{2} \Rightarrow \mathbf{T} = \frac{\mathbf{v}}{|\mathbf{v}|} = \left(\frac{1}{\sec t}\right)\mathbf{i} - \left(\frac{\tan t}{\sec t}\right)\mathbf{j} = (\cos t)\mathbf{i} - (\sin t)\mathbf{j}$; $\frac{d\mathbf{T}}{dt} = (-\sin t)\mathbf{i} - (\cos t)\mathbf{j}$ $\Rightarrow \left|\frac{d\mathbf{T}}{dt}\right| = \sqrt{(-\sin t)^2 + (-\cos t)^2} = 1 \Rightarrow \mathbf{N} = \frac{\left(\frac{d\mathbf{T}}{dt}\right)}{\left|\frac{d\mathbf{T}}{dt}\right|} = (-\sin t)\mathbf{i} - (\cos t)\mathbf{j}$; $\kappa = \frac{1}{|\mathbf{v}|} \cdot \left|\frac{d\mathbf{T}}{dt}\right| = \frac{1}{\sec t} \cdot 1 = \cos t$.

3. $\mathbf{r} = (2t + 3)\mathbf{i} + (5 - t^2)\mathbf{j} \Rightarrow \mathbf{v} = 2\mathbf{i} - 2t\mathbf{j} \Rightarrow |\mathbf{v}| = \sqrt{2^2 + (-2t)^2} = 2\sqrt{1 + t^2} \Rightarrow \mathbf{T} = \frac{\mathbf{v}}{|\mathbf{v}|} = \frac{2}{2\sqrt{1+t^2}}\mathbf{i} + \frac{-2t}{2\sqrt{1+t^2}}\mathbf{j}$ $= \frac{1}{\sqrt{1+t^2}}\mathbf{i} - \frac{t}{\sqrt{1+t^2}}\mathbf{j}$; $\frac{d\mathbf{T}}{dt} = \frac{-t}{\left(\sqrt{1+t^2}\right)^3}\mathbf{i} - \frac{1}{\left(\sqrt{1+t^2}\right)^3}\mathbf{j} \Rightarrow \left|\frac{d\mathbf{T}}{dt}\right| = \sqrt{\left(\frac{-t}{\left(\sqrt{1+t^2}\right)^3}\right)^2 + \left(-\frac{1}{\left(\sqrt{1+t^2}\right)^3}\right)^2}$ $= \sqrt{\frac{1}{(1+t^2)^2}} = \frac{1}{1+t^2} \Rightarrow \mathbf{N} = \frac{\left(\frac{d\mathbf{T}}{dt}\right)}{\left|\frac{d\mathbf{T}}{dt}\right|} = \frac{-t}{\sqrt{1+t^2}}\mathbf{i} - \frac{1}{\sqrt{1+t^2}}\mathbf{j}$; $\kappa = \frac{1}{|\mathbf{v}|} \cdot \left|\frac{d\mathbf{T}}{dt}\right| = \frac{1}{2\sqrt{1+t^2}} \cdot \frac{1}{1+t^2} = \frac{1}{2(1+t^2)^{3/2}}$

5. (a) $\kappa(x) = \frac{1}{|\mathbf{v}(x)|} \cdot \left|\frac{d\mathbf{T}(x)}{dt}\right|$. Now, $\mathbf{v} = \mathbf{i} + f'(x)\mathbf{j} \Rightarrow |\mathbf{v}(x)| = \sqrt{1 + [f'(x)]^2} \Rightarrow \mathbf{T} = \frac{\mathbf{v}}{|\mathbf{v}|}$ $= \left(1 + [f'(x)]^2\right)^{-1/2}\mathbf{i} + f'(x)\left(1 + [f'(x)]^2\right)^{-1/2}\mathbf{j}$. Thus $\frac{d\mathbf{T}}{dt}(x) = \frac{-f'(x)f''(x)}{\left(1+[f'(x)]^2\right)^{3/2}}\mathbf{i} + \frac{f''(x)}{\left(1+[f'(x)]^2\right)^{3/2}}\mathbf{j}$

$\Rightarrow \left|\frac{d\mathbf{T}(x)}{dt}\right| = \sqrt{\left[\frac{-f'(x)f''(x)}{\left(1+[f'(x)]^2\right)^{3/2}}\right]^2 + \left(\frac{f''(x)}{\left(1+[f'(x)]^2\right)^{3/2}}\right)^2} = \sqrt{\frac{[f''(x)]^2\left(1+[f'(x)]^2\right)}{\left(1+[f'(x)]^2\right)^3}} = \frac{|f''(x)|}{\left|1+[f'(x)]^2\right|}$

Thus $\kappa(x) = \frac{1}{(1+[f'(x)]^2)^{1/2}} \cdot \frac{|f''(x)|}{|1+[f'(x)]^2|} = \frac{|f''(x)|}{\left(1+[f'(x)]^2\right)^{3/2}}$

(b) $y = \ln(\cos x) \Rightarrow \frac{dy}{dx} = \left(\frac{1}{\cos x}\right)(-\sin x) = -\tan x \Rightarrow \frac{d^2y}{dx^2} = -\sec^2 x \Rightarrow \kappa = \frac{|-\sec^2 x|}{[1+(-\tan x)^2]^{3/2}} = \frac{\sec^2 x}{|\sec^3 x|}$ $= \frac{1}{\sec x} = \cos x$, since $-\frac{\pi}{2} < x < \frac{\pi}{2}$

(c) Note that $f''(x) = 0$ at an inflection point.

7. (a) $\mathbf{r}(t) = f(t)\mathbf{i} + g(t)\mathbf{j} \Rightarrow \mathbf{v} = f'(t)\mathbf{i} + g'(t)\mathbf{j}$ is tangent to the curve at the point $(f(t), g(t))$; $\mathbf{n} \cdot \mathbf{v} = [-g'(t)\mathbf{i} + f'(t)\mathbf{j}] \cdot [f'(t)\mathbf{i} + g'(t)\mathbf{j}] = -g'(t)f'(t) + f'(t)g'(t) = 0$; $-\mathbf{n} \cdot \mathbf{v} = -(\mathbf{n} \cdot \mathbf{v}) = 0$; thus, $\mathbf{n}$ and $-\mathbf{n}$ are both normal to the curve at the point

(b) $\mathbf{r}(t) = t\mathbf{i} + e^{2t}\mathbf{j} \Rightarrow \mathbf{v} = \mathbf{i} + 2e^{2t}\mathbf{j} \Rightarrow \mathbf{n} = -2e^{2t}\mathbf{i} + \mathbf{j}$ points toward the concave side of the curve; $\mathbf{N} = \frac{\mathbf{n}}{|\mathbf{n}|}$ and $|\mathbf{n}| = \sqrt{4e^{4t} + 1} \Rightarrow \mathbf{N} = \frac{-2e^{2t}}{\sqrt{1+4e^{4t}}}\mathbf{i} + \frac{1}{\sqrt{1+4e^{4t}}}\mathbf{j}$

(c) $\mathbf{r}(t) = \sqrt{4-t^2}\,\mathbf{i} + t\mathbf{j} \Rightarrow \mathbf{v} = \frac{-t}{\sqrt{4-t^2}}\,\mathbf{i} + \mathbf{j} \Rightarrow \mathbf{n} = -\mathbf{i} - \frac{t}{\sqrt{4-t^2}}\,\mathbf{j}$ points toward the concave side of the curve;
$\mathbf{N} = \frac{\mathbf{n}}{|\mathbf{n}|}$ and $|\mathbf{n}| = \sqrt{1 + \frac{t^2}{4-t^2}} = \frac{2}{\sqrt{4-t^2}} \Rightarrow \mathbf{N} = -\frac{1}{2}\left(\sqrt{4-t^2}\,\mathbf{i} + t\mathbf{j}\right)$

9. $\mathbf{r} = (3 \sin t)\mathbf{i} + (3 \cos t)\mathbf{j} + 4t\mathbf{k} \Rightarrow \mathbf{v} = (3 \cos t)\mathbf{i} + (-3 \sin t)\mathbf{j} + 4\mathbf{k} \Rightarrow |\mathbf{v}| = \sqrt{(3 \cos t)^2 + (-3 \sin t)^2 + 4^2} = \sqrt{25}$
$= 5 \Rightarrow \mathbf{T} = \frac{\mathbf{v}}{|\mathbf{v}|} = \left(\frac{3}{5} \cos t\right)\mathbf{i} - \left(\frac{3}{5} \sin t\right)\mathbf{j} + \frac{4}{5}\mathbf{k} \Rightarrow \frac{d\mathbf{T}}{dt} = \left(-\frac{3}{5} \sin t\right)\mathbf{i} - \left(\frac{3}{5} \cos t\right)\mathbf{j}$
$\Rightarrow \left|\frac{d\mathbf{T}}{dt}\right| = \sqrt{\left(-\frac{3}{5} \sin t\right)^2 + \left(-\frac{3}{5} \cos t\right)^2} = \frac{3}{5} \Rightarrow \mathbf{N} = \frac{\left(\frac{d\mathbf{T}}{dt}\right)}{\left|\frac{d\mathbf{T}}{dt}\right|} = (-\sin t)\mathbf{i} - (\cos t)\mathbf{j}\,;\ \kappa = \frac{1}{5} \cdot \frac{3}{5} = \frac{3}{25}$

11. $\mathbf{r} = (e^t \cos t)\,\mathbf{i} + (e^t \sin t)\,\mathbf{j} + 2\mathbf{k} \Rightarrow \mathbf{v} = (e^t \cos t - e^t \sin t)\,\mathbf{i} + (e^t \sin t + e^t \cos t)\,\mathbf{j} \Rightarrow$
$|\mathbf{v}| = \sqrt{(e^t \cos t - e^t \sin t)^2 + (e^t \sin t + e^t \cos t)^2} = \sqrt{2e^{2t}} = e^t\sqrt{2}\,;$
$\mathbf{T} = \frac{\mathbf{v}}{|\mathbf{v}|} = \left(\frac{\cos t - \sin t}{\sqrt{2}}\right)\mathbf{i} + \left(\frac{\sin t + \cos t}{\sqrt{2}}\right)\mathbf{j} \Rightarrow \frac{d\mathbf{T}}{dt} = \left(\frac{-\sin t - \cos t}{\sqrt{2}}\right)\mathbf{i} + \left(\frac{\cos t - \sin t}{\sqrt{2}}\right)\mathbf{j}$
$\Rightarrow \left|\frac{d\mathbf{T}}{dt}\right| = \sqrt{\left(\frac{-\sin t - \cos t}{\sqrt{2}}\right)^2 + \left(\frac{\cos t - \sin t}{\sqrt{2}}\right)^2} = 1 \Rightarrow \mathbf{N} = \frac{\left(\frac{d\mathbf{T}}{dt}\right)}{\left|\frac{d\mathbf{T}}{dt}\right|} = \left(\frac{-\cos t - \sin t}{\sqrt{2}}\right)\mathbf{i} + \left(\frac{-\sin t + \cos t}{\sqrt{2}}\right)\mathbf{j}\,;$
$\kappa = \frac{1}{|\mathbf{v}|} \cdot \left|\frac{d\mathbf{T}}{dt}\right| = \frac{1}{e^t\sqrt{2}} \cdot 1 = \frac{1}{e^t\sqrt{2}}$

13. $\mathbf{r} = \left(\frac{t^3}{3}\right)\mathbf{i} + \left(\frac{t^2}{2}\right)\mathbf{j},\ t > 0 \Rightarrow \mathbf{v} = t^2\mathbf{i} + t\mathbf{j} \Rightarrow |\mathbf{v}| = \sqrt{t^4 + t^2} = t\sqrt{t^2+1}$, since $t > 0 \Rightarrow \mathbf{T} = \frac{\mathbf{v}}{|\mathbf{v}|}$
$= \frac{t}{\sqrt{t^2+t}}\,\mathbf{i} + \frac{1}{\sqrt{t^2+1}}\,\mathbf{j} \Rightarrow \frac{d\mathbf{T}}{dt} = \frac{1}{(t^2+1)^{3/2}}\,\mathbf{i} - \frac{t}{(t^2+1)^{3/2}}\,\mathbf{j} \Rightarrow \left|\frac{d\mathbf{T}}{dt}\right| = \sqrt{\left(\frac{1}{(t^2+1)^{3/2}}\right)^2 + \left(\frac{-t}{(t^2+1)^{3/2}}\right)^2}$
$= \sqrt{\frac{1+t^2}{(t^2+1)^3}} = \frac{1}{t^2+1} \Rightarrow \mathbf{N} = \frac{\left(\frac{d\mathbf{T}}{dt}\right)}{\left|\frac{d\mathbf{T}}{dt}\right|} = \frac{1}{\sqrt{t^2+1}}\,\mathbf{i} - \frac{t}{\sqrt{t^2+1}}\,\mathbf{j}\,;\ \kappa = \frac{1}{|\mathbf{v}|} \cdot \left|\frac{d\mathbf{T}}{dt}\right| = \frac{1}{t\sqrt{t^2+1}} \cdot \frac{1}{t^2+1} = \frac{1}{t\,(t^2+1)^{3/2}}.$

15. $\mathbf{r} = t\mathbf{i} + \left(a \cosh \frac{t}{a}\right)\mathbf{j},\ a > 0 \Rightarrow \mathbf{v} = \mathbf{i} + \left(\sinh \frac{t}{a}\right)\mathbf{j} \Rightarrow |\mathbf{v}| = \sqrt{1 + \sinh^2\left(\frac{t}{a}\right)} = \sqrt{\cosh^2\left(\frac{t}{a}\right)} = \cosh \frac{t}{a}$
$\Rightarrow \mathbf{T} = \frac{\mathbf{v}}{|\mathbf{v}|} = \left(\operatorname{sech} \frac{t}{a}\right)\mathbf{i} + \left(\tanh \frac{t}{a}\right)\mathbf{j} \Rightarrow \frac{d\mathbf{T}}{dt} = \left(-\frac{1}{a} \operatorname{sech} \frac{t}{a} \tanh \frac{t}{a}\right)\mathbf{i} + \left(\frac{1}{a} \operatorname{sech}^2 \frac{t}{a}\right)\mathbf{j}$
$\Rightarrow \left|\frac{d\mathbf{T}}{dt}\right| = \sqrt{\frac{1}{a^2} \operatorname{sech}^2\left(\frac{t}{a}\right) \tanh^2\left(\frac{t}{a}\right) + \frac{1}{a^2} \operatorname{sech}^4\left(\frac{t}{a}\right)} = \frac{1}{a} \operatorname{sech}\left(\frac{t}{a}\right) \Rightarrow \mathbf{N} = \frac{\left(\frac{d\mathbf{T}}{dt}\right)}{\left|\frac{d\mathbf{T}}{dt}\right|} = \left(-\tanh \frac{t}{a}\right)\mathbf{i} + \left(\operatorname{sech} \frac{t}{a}\right)\mathbf{j}\,;$
$\kappa = \frac{1}{|\mathbf{v}|} \cdot \left|\frac{d\mathbf{T}}{dt}\right| = \frac{1}{\cosh \frac{t}{a}} \cdot \frac{1}{a} \operatorname{sech}\left(\frac{t}{a}\right) = \frac{1}{a} \operatorname{sech}^2\left(\frac{t}{a}\right).$

17. $y = ax^2 \Rightarrow y' = 2ax \Rightarrow y'' = 2a$; from Exercise 5(a), $\kappa(x) = \frac{|2a|}{(1+4a^2x^2)^{3/2}} = |2a|\,(1 + 4a^2x^2)^{-3/2}$
$\Rightarrow \kappa'(x) = -\frac{3}{2}\,|2a|\,(1 + 4a^2x^2)^{-5/2}\,(8a^2x)$; thus, $\kappa'(x) = 0 \Rightarrow x = 0$. Now, $\kappa'(x) > 0$ for $x < 0$ and $\kappa'(x) < 0$ for $x > 0$ so that $\kappa(x)$ has an absolute maximum at $x = 0$ which is the vertex of the parabola. Since $x = 0$ is the only critical point for $\kappa(x)$, the curvature has no minimum value.

19. $\kappa = \frac{a}{a^2+b^2} \Rightarrow \frac{d\kappa}{da} = \frac{-a^2+b^2}{(a^2+b^2)^2}\,;\ \frac{d\kappa}{da} = 0 \Rightarrow -a^2 + b^2 = 0 \Rightarrow a = \pm b \Rightarrow a = b$ since $a, b \geq 0$. Now, $\frac{d\kappa}{da} > 0$ if $a < b$ and $\frac{d\kappa}{da} < 0$ if $a > b \Rightarrow \kappa$ is at a maximum for $a = b$ and $\kappa(b) = \frac{b}{b^2+b^2} = \frac{1}{2b}$ is the maximum value of κ.

21. $\mathbf{r} = t\mathbf{i} + (\sin t)\mathbf{j} \Rightarrow \mathbf{v} = \mathbf{i} + (\cos t)\mathbf{j} \Rightarrow |\mathbf{v}| = \sqrt{1^2 + (\cos t)^2} = \sqrt{1 + \cos^2 t} \Rightarrow \left|\mathbf{v}\left(\frac{\pi}{2}\right)\right| = \sqrt{1 + \cos^2\left(\frac{\pi}{2}\right)} = 1;\ \mathbf{T} = \frac{\mathbf{v}}{|\mathbf{v}|}$
$= \frac{\mathbf{i} + \cos t\,\mathbf{j}}{\sqrt{1+\cos^2 t}} \Rightarrow \frac{d\mathbf{T}}{dt} = \frac{\sin t \cos t}{(1+\cos^2 t)^{3/2}}\mathbf{i} + \frac{-\sin t}{(1+\cos^2 t)^{3/2}}\mathbf{j} \Rightarrow \left|\frac{d\mathbf{T}}{dt}\right| = \frac{|\sin t|}{1+\cos^2 t};\ \left|\frac{d\mathbf{T}}{dt}\right|_{t=\frac{\pi}{2}} = \frac{\left|\sin \frac{\pi}{2}\right|}{1+\cos^2\left(\frac{\pi}{2}\right)} = \frac{1}{1} = 1$. Thus $\kappa\left(\frac{\pi}{2}\right) = \frac{1}{1} \cdot 1 = 1$
$\Rightarrow \rho = \frac{1}{1} = 1$ and the center is $\left(\frac{\pi}{2}, 0\right) \Rightarrow \left(x - \frac{\pi}{2}\right)^2 + y^2 = 1$

23. $y = x^2 \Rightarrow f'(x) = 2x$ and $f''(x) = 2$

$\Rightarrow \kappa = \frac{|2|}{(1+(2x)^2)^{3/2}} = \frac{2}{(1+4x^2)^{3/2}}$

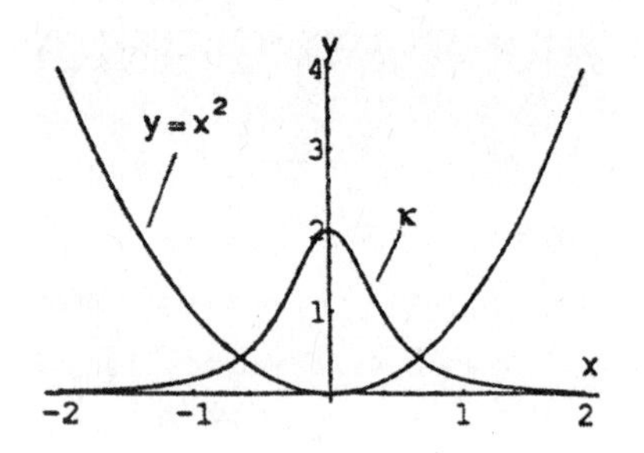

25. $y = \sin x \Rightarrow f'(x) = \cos x$ and $f''(x) = -\sin x$

$\Rightarrow \kappa = \frac{|-\sin x|}{(1+\cos^2 x)^{3/2}} = \frac{|\sin x|}{(1+\cos^2 x)^{3/2}}$

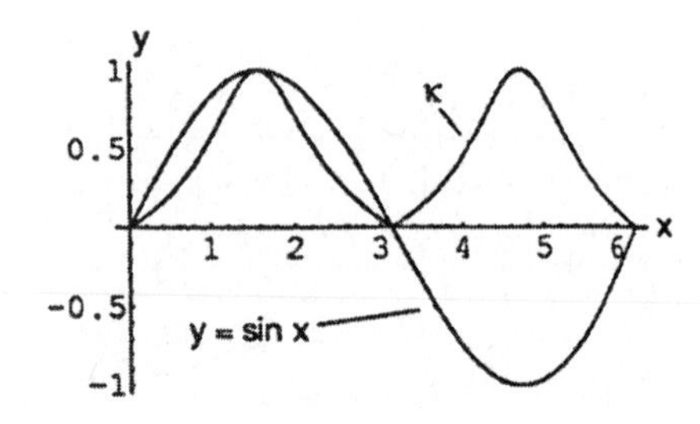

13.5 TANGENTIAL AND NORMAL COMPONENTS OF ACCELERATION

1. $\mathbf{r} = (a\cos t)\mathbf{i} + (a\sin t)\mathbf{j} + bt\mathbf{k} \Rightarrow \mathbf{v} = (-a\sin t)\mathbf{i} + (a\cos t)\mathbf{j} + b\mathbf{k} \Rightarrow |\mathbf{v}| = \sqrt{(-a\sin t)^2 + (a\cos t)^2 + b^2}$
$= \sqrt{a^2+b^2} \Rightarrow a_T = \frac{d}{dt}|\mathbf{v}| = 0;\ \mathbf{a} = (-a\cos t)\mathbf{i} + (-a\sin t)\mathbf{j} \Rightarrow |\mathbf{a}| = \sqrt{(-a\cos t)^2 + (-a\sin t)^2} = \sqrt{a^2} = |a|$
$\Rightarrow a_N = \sqrt{|\mathbf{a}|^2 - a_T^2} = \sqrt{|\mathbf{a}|^2 - 0^2} = |\mathbf{a}| = |a| \Rightarrow \mathbf{a} = (0)\mathbf{T} + |a|\,\mathbf{N} = |a|\,\mathbf{N}$

3. $\mathbf{r} = (t+1)\mathbf{i} + 2t\mathbf{j} + t^2\mathbf{k} \Rightarrow \mathbf{v} = \mathbf{i} + 2\mathbf{j} + 2t\mathbf{k} \Rightarrow |\mathbf{v}| = \sqrt{1^2+2^2+(2t)^2} = \sqrt{5+4t^2} \Rightarrow a_T = \frac{1}{2}(5+4t^2)^{-1/2}(8t)$
$= 4t(5+4t^2)^{-1/2} \Rightarrow a_T(1) = \frac{4}{\sqrt{9}} = \frac{4}{3};\ \mathbf{a} = 2\mathbf{k} \Rightarrow \mathbf{a}(1) = 2\mathbf{k} \Rightarrow |\mathbf{a}(1)| = 2 \Rightarrow a_N = \sqrt{|\mathbf{a}|^2 - a_T^2} = \sqrt{2^2 - \left(\frac{4}{3}\right)^2}$
$= \sqrt{\frac{20}{9}} = \frac{2\sqrt{5}}{3} \Rightarrow \mathbf{a}(1) = \frac{4}{3}\mathbf{T} + \frac{2\sqrt{5}}{3}\mathbf{N}$

5. $\mathbf{r} = t^2\mathbf{i} + \left(t + \frac{1}{3}t^3\right)\mathbf{j} + \left(t - \frac{1}{3}t^3\right)\mathbf{k} \Rightarrow \mathbf{v} = 2t\mathbf{i} + (1+t^2)\mathbf{j} + (1-t^2)\mathbf{k} \Rightarrow |\mathbf{v}| = \sqrt{(2t)^2 + (1+t^2)^2 + (1-t^2)^2}$
$= \sqrt{2(t^4+2t^2+1)} = \sqrt{2}(1+t^2) \Rightarrow a_T = 2t\sqrt{2} \Rightarrow a_T(0) = 0;\ \mathbf{a} = 2\mathbf{i} + 2t\mathbf{j} - 2t\mathbf{k} \Rightarrow \mathbf{a}(0) = 2\mathbf{i} \Rightarrow |\mathbf{a}(0)| = 2$
$\Rightarrow a_N = \sqrt{|\mathbf{a}|^2 - a_T^2} = \sqrt{2^2 - 0^2} = 2 \Rightarrow \mathbf{a}(0) = (0)\mathbf{T} + 2\mathbf{N} = 2\mathbf{N}$

7. $\mathbf{r} = (\cos t)\mathbf{i} + (\sin t)\mathbf{j} - \mathbf{k} \Rightarrow \mathbf{v} = (-\sin t)\mathbf{i} + (\cos t)\mathbf{j} \Rightarrow |\mathbf{v}| = \sqrt{(-\sin t)^2 + (\cos t)^2} = 1 \Rightarrow \mathbf{T} = \frac{\mathbf{v}}{|\mathbf{v}|}$
$= (-\sin t)\mathbf{i} + (\cos t)\mathbf{j} \Rightarrow \mathbf{T}\left(\frac{\pi}{4}\right) = -\frac{\sqrt{2}}{2}\mathbf{i} + \frac{\sqrt{2}}{2}\mathbf{j};\ \frac{d\mathbf{T}}{dt} = (-\cos t)\mathbf{i} - (\sin t)\mathbf{j} \Rightarrow \left|\frac{d\mathbf{T}}{dt}\right| = \sqrt{(-\cos t)^2 + (-\sin t)^2}$
$= 1 \Rightarrow \mathbf{N} = \frac{\left(\frac{d\mathbf{T}}{dt}\right)}{\left|\frac{d\mathbf{T}}{dt}\right|} = (-\cos t)\mathbf{i} - (\sin t)\mathbf{j} \Rightarrow \mathbf{N}\left(\frac{\pi}{4}\right) = -\frac{\sqrt{2}}{2}\mathbf{i} - \frac{\sqrt{2}}{2}\mathbf{j};\ \mathbf{B} = \mathbf{T}\times\mathbf{N} = \begin{vmatrix} \mathbf{i} & \mathbf{j} & \mathbf{k} \\ -\sin t & \cos t & 0 \\ -\cos t & -\sin t & 0 \end{vmatrix} = \mathbf{k}$
$\Rightarrow \mathbf{B}\left(\frac{\pi}{4}\right) = \mathbf{k}$, the normal to the osculating plane; $\mathbf{r}\left(\frac{\pi}{4}\right) = \frac{\sqrt{2}}{2}\mathbf{i} + \frac{\sqrt{2}}{2}\mathbf{j} - \mathbf{k} \Rightarrow P = \left(\frac{\sqrt{2}}{2}, \frac{\sqrt{2}}{2}, -1\right)$ lies on the osculating plane $\Rightarrow 0\left(x - \frac{\sqrt{2}}{2}\right) + 0\left(y - \frac{\sqrt{2}}{2}\right) + (z-(-1)) = 0 \Rightarrow z = -1$ is the osculating plane; $\mathbf{T}$ is normal to the normal plane $\Rightarrow \left(-\frac{\sqrt{2}}{2}\right)\left(x - \frac{\sqrt{2}}{2}\right) + \left(\frac{\sqrt{2}}{2}\right)\left(y - \frac{\sqrt{2}}{2}\right) + 0(z-(-1)) = 0 \Rightarrow -\frac{\sqrt{2}}{2}x + \frac{\sqrt{2}}{2}y = 0$
$\Rightarrow -x + y = 0$ is the normal plane; $\mathbf{N}$ is normal to the rectifying plane
$\Rightarrow \left(-\frac{\sqrt{2}}{2}\right)\left(x - \frac{\sqrt{2}}{2}\right) + \left(-\frac{\sqrt{2}}{2}\right)\left(y - \frac{\sqrt{2}}{2}\right) + 0(z-(-1)) = 0 \Rightarrow -\frac{\sqrt{2}}{2}x - \frac{\sqrt{2}}{2}y = -1 \Rightarrow x + y = \sqrt{2}$ is the rectifying plane

9. By Exercise 9 in Section 13.4, $\mathbf{T} = \left(\frac{3}{5}\cos t\right)\mathbf{i} + \left(-\frac{3}{5}\sin t\right)\mathbf{j} + \frac{4}{5}\mathbf{k}$ and $\mathbf{N} = (-\sin t)\mathbf{i} - (\cos t)\mathbf{j}$ so that $\mathbf{B} = \mathbf{T}\times\mathbf{N}$ $= \begin{vmatrix} \mathbf{i} & \mathbf{j} & \mathbf{k} \\ \frac{3}{5}\cos t & -\frac{3}{5}\sin t & \frac{4}{5} \\ -\sin t & -\cos t & 0 \end{vmatrix} = \left(\frac{4}{5}\cos t\right)\mathbf{i} - \left(\frac{4}{5}\sin t\right)\mathbf{j} - \frac{3}{5}\mathbf{k}$. Also $\mathbf{v} = (3\cos t)\mathbf{i} + (-3\sin t)\mathbf{j} + 4\mathbf{k}$ $\Rightarrow \mathbf{a} = (-3\sin t)\mathbf{i} + (-3\cos t)\mathbf{j} \Rightarrow \frac{d\mathbf{a}}{dt} = (-3\cos t)\mathbf{i} + (3\sin t)\mathbf{j}$ and $\mathbf{v}\times\mathbf{a} = \begin{vmatrix} \mathbf{i} & \mathbf{j} & \mathbf{k} \\ 3\cos t & -3\sin t & 4 \\ -3\sin t & -3\cos t & 0 \end{vmatrix}$ $= (12\cos t)\mathbf{i} - (12\sin t)\mathbf{j} - 9\mathbf{k} \Rightarrow |\mathbf{v}\times\mathbf{a}|^2 = (12\cos t)^2 + (-12\sin t)^2 + (-9)^2 = 225$. Thus

$$\tau = \frac{\begin{vmatrix} 3\cos t & -3\sin t & 4 \\ -3\sin t & -3\sin t & 0 \\ -3\cos t & 3\sin t & 0 \end{vmatrix}}{225} = \frac{4\cdot(-9\sin^2 t - 9\cos^2 t)}{225} = \frac{-36}{225} = -\frac{4}{25}$$

11. By Exercise 11 in Section 13.4, $\mathbf{T} = \left(\frac{\cos t - \sin t}{\sqrt{2}}\right)\mathbf{i} + \left(\frac{\sin t + \cos t}{\sqrt{2}}\right)\mathbf{j}$ and $\mathbf{N} = \left(\frac{-\cos t - \sin t}{\sqrt{2}}\right)\mathbf{i} + \left(\frac{-\sin t + \cos t}{\sqrt{2}}\right)\mathbf{j}$; Thus

$$\mathbf{B} = \mathbf{T}\times\mathbf{N} = \begin{vmatrix} \mathbf{i} & \mathbf{j} & \mathbf{k} \\ \frac{\cos t - \sin t}{\sqrt{2}} & \frac{\sin t + \cos t}{\sqrt{2}} & 0 \\ \frac{-\cos t - \sin t}{\sqrt{2}} & \frac{-\sin t + \cos t}{\sqrt{2}} & 0 \end{vmatrix} = \left[\left(\frac{\cos^2 t - 2\cos t\sin t + \sin^2 t}{2}\right) + \left(\frac{\sin^2 t + 2\sin t\cos t + \cos^2 t}{2}\right)\right]\mathbf{k}$$

$= \left[\left(\frac{1-\sin(2t)}{2}\right) + \left(\frac{1+\sin(2t)}{2}\right)\right]\mathbf{k} = \mathbf{k}$. Also, $\mathbf{v} = (e^t\cos t - e^t\sin t)\,\mathbf{i} + (e^t\sin t + e^t\cos t)\,\mathbf{j}$ $\Rightarrow \mathbf{a} = \left[e^t(-\sin t - \cos t) + e^t(\cos t - \sin t)\right]\mathbf{i} + \left[e^t(\cos t - \sin t) + e^t(\sin t + \cos t)\right]\mathbf{j} = (-2e^t\sin t)\,\mathbf{i} + (2e^t\cos t)\,\mathbf{j}$ $\Rightarrow \frac{d\mathbf{a}}{dt} = -2e^t(\cos t + \sin t)\,\mathbf{i} + 2e^t(-\sin t + \cos t)\mathbf{j}$. Thus $\mathbf{v}\times\mathbf{a} = \begin{vmatrix} \mathbf{i} & \mathbf{j} & \mathbf{k} \\ e^t(\cos t - \sin t) & e^t(\sin t + \cos t) & 0 \\ -2e^t\sin t & 2e^t\cos t & 0 \end{vmatrix} = 2e^{2t}\mathbf{k}$

$$\Rightarrow |\mathbf{v}\times\mathbf{a}|^2 = (2e^{2t})^2 = 4e^{4t}. \text{ Thus } \tau = \frac{\begin{vmatrix} e^t(\cos t - \sin t) & e^t(\sin t + \cos t) & 0 \\ -2e^t\sin t & 2e^t\cos t & 0 \\ -2e^t(\cos t + \sin t) & 2e^t(-\sin t + \cos t) & 0 \end{vmatrix}}{4e^{4t}} = 0$$

13. By Exercise 13 in Section 13.4, $\mathbf{T} = \frac{t}{(t^2+1)^{1/2}}\mathbf{i} + \frac{1}{(t^2+1)^{1/2}}\mathbf{j}$ and $\mathbf{N} = \frac{1}{\sqrt{t^2+1}}\mathbf{i} - \frac{t}{\sqrt{t^2+1}}\mathbf{j}$ so that $\mathbf{B} = \mathbf{T}\times\mathbf{N}$

$$= \begin{vmatrix} \mathbf{i} & \mathbf{j} & \mathbf{k} \\ \frac{t}{\sqrt{t^2+1}} & \frac{1}{\sqrt{t^2+1}} & 0 \\ \frac{1}{\sqrt{t^2+1}} & \frac{-t}{\sqrt{t^2+1}} & 0 \end{vmatrix} = -\mathbf{k}. \text{ Also, } \mathbf{v} = t^2\mathbf{i} + t\mathbf{j} \Rightarrow \mathbf{a} = 2t\mathbf{i} + \mathbf{j} \Rightarrow \frac{d\mathbf{a}}{dt} = 2\mathbf{i} \text{ so that } \begin{vmatrix} t^2 & t & 0 \\ 2t & 1 & 0 \\ 2 & 0 & 0 \end{vmatrix} = 0 \Rightarrow \tau = 0$$

15. By Exercise 15 in Section 13.4, $\mathbf{T} = \frac{\mathbf{v}}{|\mathbf{v}|} = \left(\operatorname{sech}\frac{t}{a}\right)\mathbf{i} + \left(\tanh\frac{t}{a}\right)\mathbf{j}$ and $\mathbf{N} = \left(-\tanh\frac{t}{a}\right)\mathbf{i} + \left(\operatorname{sech}\frac{t}{a}\right)\mathbf{j}$ so that $\mathbf{B} = \mathbf{T}\times\mathbf{N}$

$= \begin{vmatrix} \mathbf{i} & \mathbf{j} & \mathbf{k} \\ \operatorname{sech}\left(\frac{t}{a}\right) & \tanh\left(\frac{t}{a}\right) & 0 \\ -\tanh\left(\frac{t}{a}\right) & \operatorname{sech}\left(\frac{t}{a}\right) & 0 \end{vmatrix} = \mathbf{k}$. Also, $\mathbf{v} = \mathbf{i} + \left(\sinh\frac{t}{a}\right)\mathbf{j} \Rightarrow \mathbf{a} = \left(\frac{1}{a}\cosh\frac{t}{a}\right)\mathbf{j} \Rightarrow \frac{d\mathbf{a}}{dt} = \frac{1}{a^2}\sinh\left(\frac{t}{a}\right)\mathbf{j}$ so that

$$\begin{vmatrix} 1 & \sinh\left(\frac{t}{a}\right) & 0 \\ 0 & \frac{1}{a}\cosh\left(\frac{t}{a}\right) & 0 \\ 0 & \frac{1}{a^2}\sinh\left(\frac{t}{a}\right) & 0 \end{vmatrix} = 0 \Rightarrow \tau = 0$$

17. Yes. If the car is moving along a curved path, then $\kappa \neq 0$ and $a_N = \kappa\,|\mathbf{v}|^2 \neq 0 \Rightarrow \mathbf{a} = a_T\mathbf{T} + a_N\mathbf{N} \neq \mathbf{0}$.

19. $\mathbf{a}\perp\mathbf{v} \Rightarrow \mathbf{a}\perp\mathbf{T} \Rightarrow a_T = 0 \Rightarrow \frac{d}{dt}|\mathbf{v}| = 0 \Rightarrow |\mathbf{v}|$ is constant

21. By $\mathbf{a} = a_T\mathbf{T} + a_N\mathbf{N}$ we have $\mathbf{v}\times\mathbf{a} = \left(\frac{ds}{dt}\mathbf{T}\right)\times\left[\frac{d^2s}{dt^2}\mathbf{T} + \kappa\left(\frac{ds}{dt}\right)^2\mathbf{N}\right] = \left(\frac{ds}{dt}\frac{d^2s}{dt^2}\right)(\mathbf{T}\times\mathbf{T}) + \kappa\left(\frac{ds}{dt}\right)^3(\mathbf{T}\times\mathbf{N})$ $= \kappa\left(\frac{ds}{dt}\right)^3\mathbf{B}$. It follows that $|\mathbf{v}\times\mathbf{a}| = \kappa\left|\frac{ds}{dt}\right|^3|\mathbf{B}| = \kappa\,|\mathbf{v}|^3 \Rightarrow \kappa = \frac{|\mathbf{v}\times\mathbf{a}|}{|\mathbf{v}|^3}$

23. From Example 1, $|\mathbf{v}| = t$ and $a_N = t$ so that $a_N = \kappa\,|\mathbf{v}|^2 \Rightarrow \kappa = \frac{a_N}{|\mathbf{v}|^2} = \frac{t}{t^2} = \frac{1}{t}, t \neq 0 \Rightarrow \rho = \frac{1}{\kappa} = t$

25. If a plane curve is sufficiently differentiable the torsion is zero as the following argument shows:
$\mathbf{r} = f(t)\mathbf{i} + g(t)\mathbf{j} \Rightarrow \mathbf{v} = f'(t)\mathbf{i} + g'(t)\mathbf{j} \Rightarrow \mathbf{a} = f''(t)\mathbf{i} + g''(t)\mathbf{j} \Rightarrow \frac{d\mathbf{a}}{dt} = f'''(t)\mathbf{i} + g'''(t)\mathbf{j}$

$$\Rightarrow \tau = \frac{\begin{vmatrix} f'(t) & g'(t) & 0 \\ f''(t) & g''(t) & 0 \\ f'''(t) & g'''(t) & 0 \end{vmatrix}}{|\mathbf{v}\times\mathbf{a}|^2} = 0$$

27. $\mathbf{r}(t) = f(t)\mathbf{i} + g(t)\mathbf{j} + h(t)\mathbf{k} \Rightarrow \mathbf{v} = f'(t)\mathbf{i} + g'(t)\mathbf{j} + h'(t)\mathbf{k};\ \mathbf{v}\cdot\mathbf{k} = 0 \Rightarrow h'(t) = 0 \Rightarrow h(t) = C$
$\Rightarrow \mathbf{r}(t) = f(t)\mathbf{i} + g(t)\mathbf{j} + C\mathbf{k}$ and $\mathbf{r}(a) = f(a)\mathbf{i} + g(a)\mathbf{j} + C\mathbf{k} = \mathbf{0} \Rightarrow f(a) = 0, g(a) = 0$ and $C = 0 \Rightarrow h(t) = 0.$

13.6 VELOCITY AND ACCELERATION IN POLAR COORDINATES

1. $\frac{d\theta}{dt} = 3 = \dot{\theta} \Rightarrow \ddot{\theta} = 0, r = a(1-\cos\theta) \Rightarrow \dot{r} = a\sin\theta\frac{d\theta}{dt} = 3a\sin\theta \Rightarrow \ddot{r} = 3a\cos\theta\frac{d\theta}{dt} = 9a\cos\theta$
$\mathbf{v} = (3a\sin\theta)\mathbf{u}_r + (a(1-\cos\theta))(3)\mathbf{u}_\theta = (3a\sin\theta)\mathbf{u}_r + 3a(1-\cos\theta)\mathbf{u}_\theta$
$\mathbf{a} = \left(9a\cos\theta - a(1-\cos\theta)(3)^2\right)\mathbf{u}_r + (a(1-\cos\theta)\cdot 0 + 2(3a\sin\theta)(3))\mathbf{u}_\theta$
$= (9a\cos\theta - 9a + 9a\cos\theta)\mathbf{u}_r + (18a\sin\theta)\mathbf{u}_\theta = 9a(2\cos\theta - 1)\mathbf{u}_r + (18a\sin\theta)\mathbf{u}_\theta$

3. $\frac{d\theta}{dt} = 2 = \dot{\theta} \Rightarrow \ddot{\theta} = 0, r = e^{a\theta} \Rightarrow \dot{r} = e^{a\theta}\cdot a\frac{d\theta}{dt} = 2a\,e^{a\theta} \Rightarrow \ddot{r} = 2a\,e^{a\theta}\cdot a\frac{d\theta}{dt} = 4a^2\,e^{a\theta}$
$\mathbf{v} = \left(2a\,e^{a\theta}\right)\mathbf{u}_r + \left(e^{a\theta}\right)(2)\mathbf{u}_\theta = \left(2a\,e^{a\theta}\right)\mathbf{u}_r + \left(2e^{a\theta}\right)\mathbf{u}_\theta$
$\mathbf{a} = \left[\left(4a^2\,e^{a\theta}\right) - \left(e^{a\theta}\right)(2)^2\right]\mathbf{u}_r + \left[\left(e^{a\theta}\right)(0) + 2\left(2a\,e^{a\theta}\right)(2)\right]\mathbf{u}_\theta = \left[4a^2\,e^{a\theta} - 4e^{a\theta}\right]\mathbf{u}_r + \left[0 + 8a\,e^{a\theta}\right]\mathbf{u}_\theta$
$= 4e^{a\theta}(a^2-1)\mathbf{u}_r + \left(8a\,e^{a\theta}\right)\mathbf{u}_\theta$

5. $\theta = 2t \Rightarrow \dot{\theta} = 2 \Rightarrow \ddot{\theta} = 0, r = 2\cos 4t \Rightarrow \dot{r} = -8\sin 4t \Rightarrow \ddot{r} = -32\cos 4t$
$\mathbf{v} = (-8\sin 4t)\mathbf{u}_r + (2\cos 4t)(2)\mathbf{u}_\theta = -8(\sin 4t)\mathbf{u}_r + 4(\cos 4t)\mathbf{u}_\theta$
$\mathbf{a} = \left((-32\cos 4t) - (2\cos 4t)(2)^2\right)\mathbf{u}_r + ((2\cos 4t)\cdot 0 + 2(-8\sin 4t)(2))\mathbf{u}_\theta$
$= (-32\cos 4t - 8\cos 4t)\mathbf{u}_r + (0 - 32\sin 4t)\mathbf{u}_\theta = -40(\cos 4t)\mathbf{u}_r - 32(\sin 4t)\mathbf{u}_\theta$

7. $r = \frac{GM}{v^2} \Rightarrow v^2 = \frac{GM}{r} \Rightarrow v = \sqrt{\frac{GM}{r}}$ which is constant since G, M, and r (the radius of orbit) are constant

9. $T = \left(\frac{2\pi a^2}{r_0 v_0}\right)\sqrt{1-e^2} \Rightarrow T^2 = \left(\frac{4\pi^2 a^4}{r_0^2 v_0^2}\right)(1-e^2) = \left(\frac{4\pi^2 a^4}{r_0^2 v_0^2}\right)\left[1 - \left(\frac{r_0 v_0^2}{GM} - 1\right)^2\right]$ (from Equation 5)
$= \left(\frac{4\pi^2 a^4}{r_0^2 v_0^2}\right)\left[-\frac{r_0^2 v_0^4}{G^2M^2} + 2\left(\frac{r_0 v_0^2}{GM}\right)\right] = \left(\frac{4\pi^2 a^4}{r_0^2 v_0^2}\right)\left[\frac{2GMr_0v_0^2 - r_0^2 v_0^4}{G^2M^2}\right] = \frac{(4\pi^2a^4)\,(2GM - r_0v_0^2)}{r_0G^2M^2}$
$= (4\pi^2 a^4)\left(\frac{2GM - r_0v_0^2}{2r_0GM}\right)\left(\frac{2}{GM}\right) = (4\pi^2a^4)\left(\frac{1}{2a}\right)\left(\frac{2}{GM}\right)$ (from Equation 10) $\Rightarrow T^2 = \frac{4\pi^2a^3}{GM} \Rightarrow \frac{T^2}{a^3} = \frac{4\pi^2}{GM}$

CHAPTER 13 PRACTICE EXERCISES

1. $\mathbf{r}(t) = (4\cos t)\mathbf{i} + \left(\sqrt{2}\sin t\right)\mathbf{j} \Rightarrow x = 4\cos t$
and $y = \sqrt{2}\sin t \Rightarrow \frac{x^2}{16} + \frac{y^2}{2} = 1;$
$\mathbf{v} = (-4\sin t)\mathbf{i} + \left(\sqrt{2}\cos t\right)\mathbf{j}$ and
$\mathbf{a} = (-4\cos t)\mathbf{i} - \left(\sqrt{2}\sin t\right)\mathbf{j};\ \mathbf{r}(0) = 4\,\mathbf{i},\ \mathbf{v}(0) = \sqrt{2}\mathbf{j},$
$\mathbf{a}(0) = -4\mathbf{i};\ \mathbf{r}\left(\frac{\pi}{4}\right) = 2\sqrt{2}\mathbf{i} + \mathbf{j},\ \mathbf{v}\left(\frac{\pi}{4}\right) = -2\sqrt{2}\mathbf{i} + \mathbf{j},\ \mathbf{a}\left(\frac{\pi}{4}\right) = -2\sqrt{2}\mathbf{i} - \mathbf{j};\ |\mathbf{v}| = \sqrt{16\sin^2 t + 2\cos^2 t}$

$\Rightarrow a_T = \frac{d}{dt}|\mathbf{v}| = \frac{14 \sin t \cos t}{\sqrt{16 \sin^2 t + 2 \cos^2 t}}$; at t = 0: $a_T = 0$, $a_N = \sqrt{|\mathbf{a}|^2 - 0} = 4$, $\mathbf{a} = 0\mathbf{T} + 4\mathbf{N} = 4\mathbf{N}$, $\kappa = \frac{a_N}{|\mathbf{v}|^2} = \frac{4}{2} = 2$;

at $t = \frac{\pi}{4}$: $a_T = \frac{7}{\sqrt{8+1}} = \frac{7}{3}$, $a_N = \sqrt{9 - \frac{49}{9}} = \frac{4\sqrt{2}}{3}$, $\mathbf{a} = \frac{7}{3}\mathbf{T} + \frac{4\sqrt{2}}{3}\mathbf{N}$, $\kappa = \frac{a_N}{|\mathbf{v}|^2} = \frac{4\sqrt{2}}{27}$

3. $\mathbf{r} = \frac{1}{\sqrt{1+t^2}}\mathbf{i} + \frac{t}{\sqrt{1+t^2}}\mathbf{j} \Rightarrow \mathbf{v} = -t(1+t^2)^{-3/2}\mathbf{i} + (1+t^2)^{-3/2}\mathbf{j} \Rightarrow |\mathbf{v}| = \sqrt{\left[-t(1+t^2)^{-3/2}\right]^2 + \left[(1+t^2)^{-3/2}\right]^2}$
$= \frac{1}{1+t^2}$. We want to maximize $|\mathbf{v}|$: $\frac{d|\mathbf{v}|}{dt} = \frac{-2t}{(1+t^2)^2}$ and $\frac{d|\mathbf{v}|}{dt} = 0 \Rightarrow \frac{-2t}{(1+t^2)^2} = 0 \Rightarrow t = 0$. For $t < 0$, $\frac{-2t}{(1+t^2)^2} > 0$; for $t > 0$, $\frac{-2t}{(1+t^2)^2} < 0 \Rightarrow |\mathbf{v}|_{max}$ occurs when $t = 0 \Rightarrow |\mathbf{v}|_{max} = 1$

5. $\mathbf{v} = 3\mathbf{i} + 4\mathbf{j}$ and $\mathbf{a} = 5\mathbf{i} + 15\mathbf{j} \Rightarrow \mathbf{v} \times \mathbf{a} = \begin{vmatrix} \mathbf{i} & \mathbf{j} & \mathbf{k} \\ 3 & 4 & 0 \\ 5 & 15 & 0 \end{vmatrix} = 25\mathbf{k} \Rightarrow |\mathbf{v} \times \mathbf{a}| = 25$; $|\mathbf{v}| = \sqrt{3^2 + 4^2} = 5$
$\Rightarrow \kappa = \frac{|\mathbf{v} \times \mathbf{a}|}{|\mathbf{v}|^3} = \frac{25}{5^3} = \frac{1}{5}$

7. $\mathbf{r} = x\mathbf{i} + y\mathbf{j} \Rightarrow \mathbf{v} = \frac{dx}{dt}\mathbf{i} + \frac{dy}{dt}\mathbf{j}$ and $\mathbf{v} \cdot \mathbf{i} = y \Rightarrow \frac{dx}{dt} = y$. Since the particle moves around the unit circle $x^2 + y^2 = 1$, $2x\frac{dx}{dt} + 2y\frac{dy}{dt} = 0 \Rightarrow \frac{dy}{dt} = -\frac{x}{y}\frac{dx}{dt} \Rightarrow \frac{dy}{dt} = -\frac{x}{y}(y) = -x$. Since $\frac{dx}{dt} = y$ and $\frac{dy}{dt} = -x$, we have $\mathbf{v} = y\mathbf{i} - x\mathbf{j} \Rightarrow$ at $(1, 0)$, $\mathbf{v} = -\mathbf{j}$ and the motion is clockwise.

9. $\frac{d\mathbf{r}}{dt}$ orthogonal to $\mathbf{r} \Rightarrow 0 = \frac{d\mathbf{r}}{dt} \cdot \mathbf{r} = \frac{1}{2}\frac{d\mathbf{r}}{dt} \cdot \mathbf{r} + \frac{1}{2}\mathbf{r} \cdot \frac{d\mathbf{r}}{dt} = \frac{1}{2}\frac{d}{dt}(\mathbf{r} \cdot \mathbf{r}) \Rightarrow \mathbf{r} \cdot \mathbf{r} = K$, a constant. If $\mathbf{r} = x\mathbf{i} + y\mathbf{j}$, where x and y are differentiable functions of t, then $\mathbf{r} \cdot \mathbf{r} = x^2 + y^2 \Rightarrow x^2 + y^2 = K$, which is the equation of a circle centered at the origin.

11. $y = y_0 + (v_0 \sin\alpha)t - \frac{1}{2}gt^2 \Rightarrow y = 6.5 + (44 \text{ ft/sec})(\sin 45°)(3 \text{ sec}) - \frac{1}{2}(32 \text{ ft/sec}^2)(3 \text{ sec})^2 = 6.5 + 66\sqrt{2} - 144$
≈ -44.16 ft $\Rightarrow$ the shot put is on the ground. Now, $y = 0 \Rightarrow 6.5 + 22\sqrt{2}t - 16t^2 = 0 \Rightarrow t \approx 2.13$ sec (the positive root) $\Rightarrow x \approx (44 \text{ ft/sec})(\cos 45°)(2.13 \text{ sec}) \approx 66.27$ ft or about 66 ft, 3 in. from the stopboard

13. $x = (v_0 \cos\alpha)t$ and $y = (v_0 \sin\alpha)t - \frac{1}{2}gt^2 \Rightarrow \tan\phi = \frac{y}{x} = \frac{(v_0 \sin\alpha)t - \frac{1}{2}gt^2}{(v_0 \cos\alpha)t} = \frac{(v_0 \sin\alpha) - \frac{1}{2}gt}{v_0 \cos\alpha}$
$\Rightarrow v_0 \cos\alpha \tan\phi = v_0 \sin\alpha - \frac{1}{2}gt \Rightarrow t = \frac{2v_0 \sin\alpha - 2v_0 \cos\alpha \tan\phi}{g}$, which is the time when the golf ball hits the upward slope. At this time $x = (v_0 \cos\alpha)\left(\frac{2v_0 \sin\alpha - 2v_0 \cos\alpha \tan\phi}{g}\right) = \left(\frac{2}{g}\right)(v_0^2 \sin\alpha \cos\alpha - v_0^2 \cos^2\alpha \tan\phi)$.

Now $OR = \frac{x}{\cos\phi} \Rightarrow OR = \left(\frac{2}{g}\right)\left(\frac{v_0^2 \sin\alpha \cos\alpha - v_0^2 \cos^2\alpha \tan\phi}{\cos\phi}\right)$
$= \left(\frac{2v_0^2 \cos\alpha}{g}\right)\left(\frac{\sin\alpha}{\cos\phi} - \frac{\cos\alpha \tan\phi}{\cos\phi}\right)$
$= \left(\frac{2v_0^2 \cos\alpha}{g}\right)\left(\frac{\sin\alpha \cos\phi - \cos\alpha \sin\phi}{\cos^2\phi}\right)$
$= \left(\frac{2v_0^2 \cos\alpha}{g \cos^2\phi}\right)[\sin(\alpha - \phi)]$. The distance OR is maximized when x is maximized:

y R α ϕ O x

$\frac{dx}{d\alpha} = \left(\frac{2v_0^2}{g}\right)(\cos 2\alpha + \sin 2\alpha \tan\phi) = 0$ 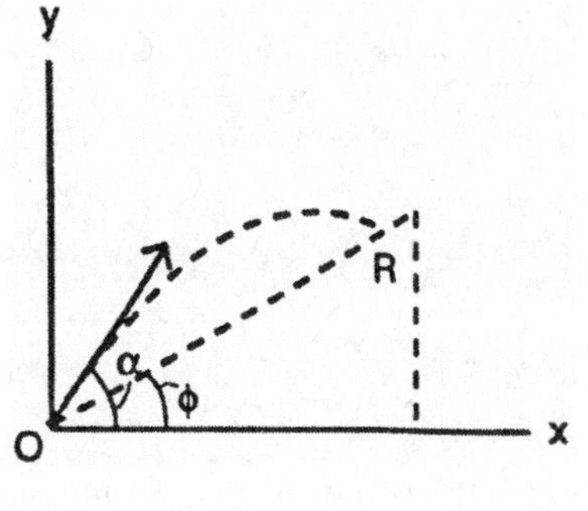
$\Rightarrow (\cos 2\alpha + \sin 2\alpha \tan\phi) = 0 \Rightarrow \cot 2\alpha + \tan\phi = 0 \Rightarrow \cot 2\alpha = \tan(-\phi) \Rightarrow 2\alpha = \frac{\pi}{2} + \phi \Rightarrow \alpha = \frac{\phi}{2} + \frac{\pi}{4}$

15. $\mathbf{r} = (2\cos t)\mathbf{i} + (2\sin t)\mathbf{j} + t^2\mathbf{k} \Rightarrow \mathbf{v} = (-2\sin t)\mathbf{i} + (2\cos t)\mathbf{j} + 2t\mathbf{k} \Rightarrow |\mathbf{v}| = \sqrt{(-2\sin t)^2 + (2\cos t)^2 + (2t)^2}$
$= 2\sqrt{1+t^2} \Rightarrow \text{Length} = \int_0^{\pi/4} 2\sqrt{1+t^2}\,dt = \left[t\sqrt{1+t^2} + \ln\left|t + \sqrt{1+t^2}\right|\right]_0^{\pi/4} = \frac{\pi}{4}\sqrt{1+\frac{\pi^2}{16}} + \ln\left(\frac{\pi}{4} + \sqrt{1+\frac{\pi^2}{16}}\right)$

17. $\mathbf{r} = \frac{4}{9}(1+t)^{3/2}\mathbf{i} + \frac{4}{9}(1-t)^{3/2}\mathbf{j} + \frac{1}{3}t\mathbf{k} \Rightarrow \mathbf{v} = \frac{2}{3}(1+t)^{1/2}\mathbf{i} - \frac{2}{3}(1-t)^{1/2}\mathbf{j} + \frac{1}{3}\mathbf{k}$

$\Rightarrow |\mathbf{v}| = \sqrt{\left[\frac{2}{3}(1+t)^{1/2}\right]^2 + \left[-\frac{2}{3}(1-t)^{1/2}\right]^2 + \left(\frac{1}{3}\right)^2} = 1 \Rightarrow \mathbf{T} = \frac{2}{3}(1+t)^{1/2}\mathbf{i} - \frac{2}{3}(1-t)^{1/2}\mathbf{j} + \frac{1}{3}\mathbf{k}$

$\Rightarrow \mathbf{T}(0) = \frac{2}{3}\mathbf{i} - \frac{2}{3}\mathbf{j} + \frac{1}{3}\mathbf{k};\ \frac{d\mathbf{T}}{dt} = \frac{1}{3}(1+t)^{-1/2}\mathbf{i} + \frac{1}{3}(1-t)^{-1/2}\mathbf{j} \Rightarrow \frac{d\mathbf{T}}{dt}(0) = \frac{1}{3}\mathbf{i} + \frac{1}{3}\mathbf{j} \Rightarrow \left|\frac{d\mathbf{T}}{dt}(0)\right| = \frac{\sqrt{2}}{3}$

$\Rightarrow \mathbf{N}(0) = \frac{1}{\sqrt{2}}\mathbf{i} + \frac{1}{\sqrt{2}}\mathbf{j};\ \mathbf{B}(0) = \mathbf{T}(0) \times \mathbf{N}(0) = \begin{vmatrix} \mathbf{i} & \mathbf{j} & \mathbf{k} \\ \frac{2}{3} & -\frac{2}{3} & \frac{1}{3} \\ \frac{1}{\sqrt{2}} & \frac{1}{\sqrt{2}} & 0 \end{vmatrix} = -\frac{1}{3\sqrt{2}}\mathbf{i} + \frac{1}{3\sqrt{2}}\mathbf{j} + \frac{4}{3\sqrt{2}}\mathbf{k};$

$\mathbf{a} = \frac{1}{3}(1+t)^{-1/2}\mathbf{i} + \frac{1}{3}(1-t)^{-1/2}\mathbf{j} \Rightarrow \mathbf{a}(0) = \frac{1}{3}\mathbf{i} + \frac{1}{3}\mathbf{j}$ and $\mathbf{v}(0) = \frac{2}{3}\mathbf{i} - \frac{2}{3}\mathbf{j} + \frac{1}{3}\mathbf{k} \Rightarrow \mathbf{v}(0) \times \mathbf{a}(0)$

$= \begin{vmatrix} \mathbf{i} & \mathbf{j} & \mathbf{k} \\ \frac{2}{3} & -\frac{2}{3} & \frac{1}{3} \\ \frac{1}{3} & \frac{1}{3} & 0 \end{vmatrix} = -\frac{1}{9}\mathbf{i} + \frac{1}{9}\mathbf{j} + \frac{4}{9}\mathbf{k} \Rightarrow |\mathbf{v} \times \mathbf{a}| = \frac{\sqrt{2}}{3} \Rightarrow \kappa(0) = \frac{|\mathbf{v}\times\mathbf{a}|}{|\mathbf{v}|^3} = \frac{\left(\frac{\sqrt{2}}{3}\right)}{1^3} = \frac{\sqrt{2}}{3};$

$\dot{\mathbf{a}} = -\frac{1}{6}(1+t)^{-3/2}\mathbf{i} + \frac{1}{6}(1-t)^{-3/2}\mathbf{j} \Rightarrow \dot{\mathbf{a}}(0) = -\frac{1}{6}\mathbf{i} + \frac{1}{6}\mathbf{j} \Rightarrow \tau(0) = \frac{\begin{vmatrix} \frac{2}{3} & -\frac{2}{3} & \frac{1}{3} \\ \frac{1}{3} & \frac{1}{3} & 0 \\ -\frac{1}{6} & \frac{1}{6} & 0 \end{vmatrix}}{|\mathbf{v}\times\mathbf{a}|^2} = \frac{\left(\frac{1}{3}\right)\left(\frac{2}{18}\right)}{\left(\frac{\sqrt{2}}{3}\right)^2} = \frac{1}{6}$

19. $\mathbf{r} = t\mathbf{i} + \frac{1}{2}e^{2t}\mathbf{j} \Rightarrow \mathbf{v} = \mathbf{i} + e^{2t}\mathbf{j} \Rightarrow |\mathbf{v}| = \sqrt{1+e^{4t}} \Rightarrow \mathbf{T} = \frac{1}{\sqrt{1+e^{4t}}}\mathbf{i} + \frac{e^{2t}}{\sqrt{1+e^{4t}}}\mathbf{j} \Rightarrow \mathbf{T}(\ln 2) = \frac{1}{\sqrt{17}}\mathbf{i} + \frac{4}{\sqrt{17}}\mathbf{j};$

$\frac{d\mathbf{T}}{dt} = \frac{-2e^{4t}}{(1+e^{4t})^{3/2}}\mathbf{i} + \frac{2e^{2t}}{(1+e^{4t})^{3/2}}\mathbf{j} \Rightarrow \frac{d\mathbf{T}}{dt}(\ln 2) = \frac{-32}{17\sqrt{17}}\mathbf{i} + \frac{8}{17\sqrt{17}}\mathbf{j} \Rightarrow \mathbf{N}(\ln 2) = -\frac{4}{\sqrt{17}}\mathbf{i} + \frac{1}{\sqrt{17}}\mathbf{j};$

$\mathbf{B}(\ln 2) = \mathbf{T}(\ln 2) \times \mathbf{N}(\ln 2) = \begin{vmatrix} \mathbf{i} & \mathbf{j} & \mathbf{k} \\ \frac{1}{\sqrt{17}} & \frac{4}{\sqrt{17}} & 0 \\ -\frac{4}{\sqrt{17}} & \frac{1}{\sqrt{17}} & 0 \end{vmatrix} = \mathbf{k};\ \mathbf{a} = 2e^{2t}\mathbf{j} \Rightarrow \mathbf{a}(\ln 2) = 8\mathbf{j}$ and $\mathbf{v}(\ln 2) = \mathbf{i} + 4\mathbf{j}$

$\Rightarrow \mathbf{v}(\ln 2) \times \mathbf{a}(\ln 2) = \begin{vmatrix} \mathbf{i} & \mathbf{j} & \mathbf{k} \\ 1 & 4 & 0 \\ 0 & 8 & 0 \end{vmatrix} = 8\mathbf{k} \Rightarrow |\mathbf{v} \times \mathbf{a}| = 8$ and $|\mathbf{v}(\ln 2)| = \sqrt{17} \Rightarrow \kappa(\ln 2) = \frac{8}{17\sqrt{17}};\ \dot{\mathbf{a}} = 4e^{2t}\mathbf{j}$

$\Rightarrow \dot{\mathbf{a}}(\ln 2) = 16\mathbf{j} \Rightarrow \tau(\ln 2) = \frac{\begin{vmatrix} 1 & 4 & 0 \\ 0 & 8 & 0 \\ 0 & 16 & 0 \end{vmatrix}}{|\mathbf{v}\times\mathbf{a}|^2} = 0$

21. $\mathbf{r} = (2+3t+3t^2)\mathbf{i} + (4t+4t^2)\mathbf{j} - (6\cos t)\mathbf{k} \Rightarrow \mathbf{v} = (3+6t)\mathbf{i} + (4+8t)\mathbf{j} + (6\sin t)\mathbf{k}$

$\Rightarrow |\mathbf{v}| = \sqrt{(3+6t)^2 + (4+8t)^2 + (6\sin t)^2} = \sqrt{25 + 100t + 100t^2 + 36\sin^2 t}$

$\Rightarrow \frac{d|\mathbf{v}|}{dt} = \frac{1}{2}(25 + 100t + 100t^2 + 36\sin^2 t)^{-1/2}(100 + 200t + 72\sin t\cos t) \Rightarrow a_T(0) = \frac{d|\mathbf{v}|}{dt}(0) = 10;$

$\mathbf{a} = 6\mathbf{i} + 8\mathbf{j} + (6\cos t)\mathbf{k} \Rightarrow |\mathbf{a}| = \sqrt{6^2 + 8^2 + (6\cos t)^2} = \sqrt{100 + 36\cos^2 t} \Rightarrow |\mathbf{a}(0)| = \sqrt{136}$

$\Rightarrow a_N = \sqrt{|\mathbf{a}|^2 - a_T^2} = \sqrt{136 - 10^2} = \sqrt{36} = 6 \Rightarrow \mathbf{a}(0) = 10\mathbf{T} + 6\mathbf{N}$

23. $\mathbf{r} = (\sin t)\mathbf{i} + \left(\sqrt{2}\cos t\right)\mathbf{j} + (\sin t)\mathbf{k} \Rightarrow \mathbf{v} = (\cos t)\mathbf{i} - \left(\sqrt{2}\sin t\right)\mathbf{j} + (\cos t)\mathbf{k}$

$\Rightarrow |\mathbf{v}| = \sqrt{(\cos t)^2 + \left(-\sqrt{2}\sin t\right)^2 + (\cos t)^2} = \sqrt{2} \Rightarrow \mathbf{T} = \frac{\mathbf{v}}{|\mathbf{v}|} = \left(\frac{1}{\sqrt{2}}\cos t\right)\mathbf{i} - (\sin t)\mathbf{j} + \left(\frac{1}{\sqrt{2}}\cos t\right)\mathbf{k};$

$\frac{d\mathbf{T}}{dt} = \left(-\frac{1}{\sqrt{2}}\sin t\right)\mathbf{i} - (\cos t)\mathbf{j} - \left(\frac{1}{\sqrt{2}}\sin t\right)\mathbf{k} \Rightarrow \left|\frac{d\mathbf{T}}{dt}\right| = \sqrt{\left(-\frac{1}{\sqrt{2}}\sin t\right)^2 + (-\cos t)^2 + \left(-\frac{1}{\sqrt{2}}\sin t\right)^2} = 1$

$\Rightarrow \mathbf{N} = \frac{\left(\frac{d\mathbf{T}}{dt}\right)}{\left|\frac{d\mathbf{T}}{dt}\right|} = \left(-\frac{1}{\sqrt{2}}\sin t\right)\mathbf{i} - (\cos t)\mathbf{j} - \left(\frac{1}{\sqrt{2}}\sin t\right)\mathbf{k};\ \mathbf{B} = \mathbf{T} \times \mathbf{N} = \begin{vmatrix} \mathbf{i} & \mathbf{j} & \mathbf{k} \\ \frac{1}{\sqrt{2}}\cos t & -\sin t & \frac{1}{\sqrt{2}}\cos t \\ -\frac{1}{\sqrt{2}}\sin t & -\cos t & -\frac{1}{\sqrt{2}}\sin t \end{vmatrix}$

$= \frac{1}{\sqrt{2}}\mathbf{i} - \frac{1}{\sqrt{2}}\mathbf{k};\ \mathbf{a} = (-\sin t)\mathbf{i} - \left(\sqrt{2}\cos t\right)\mathbf{j} - (\sin t)\mathbf{k} \Rightarrow \mathbf{v} \times \mathbf{a} = \begin{vmatrix} \mathbf{i} & \mathbf{j} & \mathbf{k} \\ \cos t & -\sqrt{2}\sin t & \cos t \\ -\sin t & -\sqrt{2}\cos t & -\sin t \end{vmatrix}$

$= \sqrt{2}\mathbf{i} - \sqrt{2}\mathbf{k} \Rightarrow |\mathbf{v}\times\mathbf{a}| = \sqrt{4} = 2 \Rightarrow \kappa = \frac{|\mathbf{v}\times\mathbf{a}|}{|\mathbf{v}|^3} = \frac{2}{\left(\sqrt{2}\right)^3} = \frac{1}{\sqrt{2}}$; $\dot{\mathbf{a}} = (-\cos t)\mathbf{i} + \left(\sqrt{2}\sin t\right)\mathbf{j} - (\cos t)\mathbf{k}$

$$\Rightarrow \tau = \frac{\begin{vmatrix} \cos t & -\sqrt{2}\sin t & \cos t \\ -\sin t & -\sqrt{2}\cos t & -\sin t \\ -\cos t & \sqrt{2}\sin t & -\cos t \end{vmatrix}}{|\mathbf{v}\times\mathbf{a}|^2} = \frac{(\cos t)\left(\sqrt{2}\right) - \left(\sqrt{2}\sin t\right)(0) + (\cos t)\left(-\sqrt{2}\right)}{4} = 0$$

25. $\mathbf{r} = 2\mathbf{i} + \left(4\sin\frac{t}{2}\right)\mathbf{j} + \left(3 - \frac{t}{\pi}\right)\mathbf{k} \Rightarrow 0 = \mathbf{r}\cdot(\mathbf{i}-\mathbf{j}) = 2(1) + \left(4\sin\frac{t}{2}\right)(-1) \Rightarrow 0 = 2 - 4\sin\frac{t}{2} \Rightarrow \sin\frac{t}{2} = \frac{1}{2} \Rightarrow \frac{t}{2} = \frac{\pi}{6}$
$\Rightarrow t = \frac{\pi}{3}$ (for the first time)

27. $\mathbf{r} = e^t\mathbf{i} + (\sin t)\mathbf{j} + \ln(1-t)\mathbf{k} \Rightarrow \mathbf{v} = e^t\mathbf{i} + (\cos t)\mathbf{j} - \left(\frac{1}{1-t}\right)\mathbf{k} \Rightarrow \mathbf{v}(0) = \mathbf{i} + \mathbf{j} - \mathbf{k}$; $\mathbf{r}(0) = \mathbf{i} \Rightarrow (1,0,0)$ is on the line
$\Rightarrow x = 1 + t$, $y = t$, and $z = -t$ are parametric equations of the line

29. $x^2 = \left(v_0^2\cos^2\alpha\right)t^2$ and $\left(y + \frac{1}{2}gt^2\right)^2 = \left(v_0^2\sin^2\alpha\right)t^2 \Rightarrow x^2 + \left(y + \frac{1}{2}gt^2\right)^2 = v_0^2t^2$

31. $s = a\theta \Rightarrow \theta = \frac{s}{a} \Rightarrow \phi = \frac{s}{a} + \frac{\pi}{2} \Rightarrow \frac{d\phi}{ds} = \frac{1}{a} \Rightarrow \kappa = \left|\frac{1}{a}\right| = \frac{1}{a}$ since $a > 0$

CHAPTER 13 ADDITIONAL AND ADVANCED EXERCISES

1. (a) $\mathbf{r}(\theta) = (a\cos\theta)\mathbf{i} + (a\sin\theta)\mathbf{j} + b\theta\mathbf{k} \Rightarrow \frac{d\mathbf{r}}{dt} = [(-a\sin\theta)\mathbf{i} + (a\cos\theta)\mathbf{j} + b\mathbf{k}]\frac{d\theta}{dt}$; $|\mathbf{v}| = \sqrt{2gz} = \left|\frac{d\mathbf{r}}{dt}\right|$
$= \sqrt{a^2+b^2}\,\frac{d\theta}{dt} \Rightarrow \frac{d\theta}{dt} = \sqrt{\frac{2gz}{a^2+b^2}} = \sqrt{\frac{2gb\theta}{a^2+b^2}} \Rightarrow \left.\frac{d\theta}{dt}\right|_{\theta=2\pi} = \sqrt{\frac{4\pi gb}{a^2+b^2}} = 2\sqrt{\frac{\pi gb}{a^2+b^2}}$

(b) $\frac{d\theta}{dt} = \sqrt{\frac{2gb\theta}{a^2+b^2}} \Rightarrow \frac{d\theta}{\sqrt{\theta}} = \sqrt{\frac{2gb}{a^2+b^2}}\,dt \Rightarrow 2\theta^{1/2} = \sqrt{\frac{2gb}{a^2+b^2}}\,t + C$; $t = 0 \Rightarrow \theta = 0 \Rightarrow C = 0$
$\Rightarrow 2\theta^{1/2} = \sqrt{\frac{2gb}{a^2+b^2}}\,t \Rightarrow \theta = \frac{gbt^2}{2(a^2+b^2)}$; $z = b\theta \Rightarrow z = \frac{gb^2t^2}{2(a^2+b^2)}$

(c) $\mathbf{v}(t) = \frac{d\mathbf{r}}{dt} = [(-a\sin\theta)\mathbf{i} + (a\cos\theta)\mathbf{j} + b\mathbf{k}]\frac{d\theta}{dt} = [(-a\sin\theta)\mathbf{i} + (a\cos\theta)\mathbf{j} + b\mathbf{k}]\left(\frac{gbt}{a^2+b^2}\right)$, from part (b)
$\Rightarrow \mathbf{v}(t) = \left[\frac{(-a\sin\theta)\mathbf{i} + (a\cos\theta)\mathbf{j} + b\mathbf{k}}{\sqrt{a^2+b^2}}\right]\left(\frac{gbt}{\sqrt{a^2+b^2}}\right) = \frac{gbt}{\sqrt{a^2+b^2}}\mathbf{T}$;
$\frac{d^2\mathbf{r}}{dt^2} = [(-a\cos\theta)\mathbf{i} - (a\sin\theta)\mathbf{j}]\left(\frac{d\theta}{dt}\right)^2 + [(-a\sin\theta)\mathbf{i} + (a\cos\theta)\mathbf{j} + b\mathbf{k}]\frac{d^2\theta}{dt^2}$
$= \left(\frac{gbt}{a^2+b^2}\right)^2[(-a\cos\theta)\mathbf{i} - (a\sin\theta)\mathbf{j}] + [(-a\sin\theta)\mathbf{i} + (a\cos\theta)\mathbf{j} + b\mathbf{k}]\left(\frac{gb}{a^2+b^2}\right)$
$= \left[\frac{(-a\sin\theta)\mathbf{i} + (a\cos\theta)\mathbf{j} + b\mathbf{k}}{\sqrt{a^2+b^2}}\right]\left(\frac{gb}{\sqrt{a^2+b^2}}\right) + a\left(\frac{gbt}{a^2+b^2}\right)^2[(-\cos\theta)\mathbf{i} - (\sin\theta)\mathbf{j}]$
$= \frac{gb}{\sqrt{a^2+b^2}}\mathbf{T} + a\left(\frac{gbt}{a^2+b^2}\right)^2\mathbf{N}$ (there is no component in the direction of $\mathbf{B}$).

3. $r = \frac{(1+e)r_0}{1+e\cos\theta} \Rightarrow \frac{dr}{d\theta} = \frac{(1+e)r_0(e\sin\theta)}{(1+e\cos\theta)^2}$; $\frac{dr}{d\theta} = 0 \Rightarrow \frac{(1+e)r_0(e\sin\theta)}{(1+e\cos\theta)^2} = 0 \Rightarrow (1+e)r_0(e\sin\theta) = 0$
$\Rightarrow \sin\theta = 0 \Rightarrow \theta = 0$ or π. Note that $\frac{dr}{d\theta} > 0$ when $\sin\theta > 0$ and $\frac{dr}{d\theta} < 0$ when $\sin\theta < 0$. Since $\sin\theta < 0$ on $-\pi < \theta < 0$ and $\sin\theta > 0$ on $0 < \theta < \pi$; r is a minimum when $\theta = 0$ and $r(0) = \frac{(1+e)r_0}{1+e\cos 0} = r_0$

5. (a) $\mathbf{v} = \dot{x}\mathbf{i} + \dot{y}\mathbf{j}$ and $\mathbf{v} = \dot{r}\mathbf{u}_r + r\dot{\theta}\mathbf{u}_\theta = (\dot{r})[(\cos\theta)\mathbf{i} + (\sin\theta)\mathbf{j}] + \left(r\dot{\theta}\right)[(-\sin\theta)\mathbf{i} + (\cos\theta)\mathbf{j}] \Rightarrow \mathbf{v}\cdot\mathbf{i} = \dot{x}$ and
$\mathbf{v}\cdot\mathbf{i} = \dot{r}\cos\theta - r\dot{\theta}\sin\theta \Rightarrow \dot{x} = \dot{r}\cos\theta - r\dot{\theta}\sin\theta$; $\mathbf{v}\cdot\mathbf{j} = \dot{y}$ and $\mathbf{v}\cdot\mathbf{j} = \dot{r}\sin\theta + r\dot{\theta}\cos\theta$
$\Rightarrow \dot{y} = \dot{r}\sin\theta + r\dot{\theta}\cos\theta$

(b) $\mathbf{u}_r = (\cos\theta)\mathbf{i} + (\sin\theta)\mathbf{j} \Rightarrow \mathbf{v}\cdot\mathbf{u}_r = \dot{x}\cos\theta + \dot{y}\sin\theta$
$= \left(\dot{r}\cos\theta - r\dot{\theta}\sin\theta\right)(\cos\theta) + \left(\dot{r}\sin\theta + r\dot{\theta}\cos\theta\right)(\sin\theta)$ by part (a),
$\Rightarrow \mathbf{v}\cdot\mathbf{u}_r = \dot{r}$; therefore, $\dot{r} = \dot{x}\cos\theta + \dot{y}\sin\theta$;
$\mathbf{u}_\theta = -(\sin\theta)\mathbf{i} + (\cos\theta)\mathbf{j} \Rightarrow \mathbf{v}\cdot\mathbf{u}_\theta = -\dot{x}\sin\theta + \dot{y}\cos\theta$
$= \left(\dot{r}\cos\theta - r\dot{\theta}\sin\theta\right)(-\sin\theta) + \left(\dot{r}\sin\theta + r\dot{\theta}\cos\theta\right)(\cos\theta)$ by part (a) $\Rightarrow \mathbf{v}\cdot\mathbf{u}_\theta = r\dot{\theta}$;
therefore, $r\dot{\theta} = -\dot{x}\sin\theta + \dot{y}\cos\theta$

7. (a) Let $r = 2 - t$ and $\theta = 3t \Rightarrow \frac{dr}{dt} = -1$ and $\frac{d\theta}{dt} = 3 \Rightarrow \frac{d^2r}{dt^2} = \frac{d^2\theta}{dt^2} = 0$. The halfway point is $(1, 3) \Rightarrow t = 1$;
$\mathbf{v} = \frac{dr}{dt}\mathbf{u}_r + r\frac{d\theta}{dt}\mathbf{u}_\theta \Rightarrow \mathbf{v}(1) = -\mathbf{u}_r + 3\mathbf{u}_\theta$; $\mathbf{a} = \left[\frac{d^2r}{dt^2} - r\left(\frac{d\theta}{dt}\right)^2\right]\mathbf{u}_r + \left[r\frac{d^2\theta}{dt^2} + 2\frac{dr}{dt}\frac{d\theta}{dt}\right]\mathbf{u}_\theta \Rightarrow \mathbf{a}(1) = -9\mathbf{u}_r - 6\mathbf{u}_\theta$

(b) It takes the beetle 2 min to crawl to the origin $\Rightarrow$ the rod has revolved 6 radians
$\Rightarrow L = \int_0^6 \sqrt{[f(\theta)]^2 + [f'(\theta)]^2}\, d\theta = \int_0^6 \sqrt{\left(2 - \frac{\theta}{3}\right)^2 + \left(-\frac{1}{3}\right)^2}\, d\theta = \int_0^6 \sqrt{4 - \frac{4\theta}{3} + \frac{\theta^2}{9} + \frac{1}{9}}\, d\theta$
$= \int_0^6 \sqrt{\frac{37 - 12\theta + \theta^2}{9}}\, d\theta = \frac{1}{3}\int_0^6 \sqrt{(\theta - 6)^2 + 1}\, d\theta = \frac{1}{3}\left[\frac{(\theta-6)}{2}\sqrt{(\theta - 6)^2 + 1} + \frac{1}{2}\ln\left|\theta - 6 + \sqrt{(\theta - 6)^2 + 1}\right|\right]_0^6$
$= \sqrt{37} - \frac{1}{6}\ln\left(\sqrt{37} - 6\right) \approx 6.5$ in.

9. (a) $\mathbf{u}_r \times \mathbf{u}_\theta = \begin{vmatrix} \mathbf{i} & \mathbf{j} & \mathbf{k} \\ \cos\theta & \sin\theta & 0 \\ -\sin\theta & \cos\theta & 0 \end{vmatrix} = \mathbf{k} \Rightarrow$ a right-handed frame of unit vectors

(b) $\frac{d\mathbf{u}_r}{d\theta} = (-\sin\theta)\mathbf{i} + (\cos\theta)\mathbf{j} = \mathbf{u}_\theta$ and $\frac{d\mathbf{u}_\theta}{d\theta} = (-\cos\theta)\mathbf{i} - (\sin\theta)\mathbf{j} = -\mathbf{u}_r$

(c) From Eq. (7), $\mathbf{v} = \dot{r}\mathbf{u}_r + r\dot{\theta}\mathbf{u}_\theta + \dot{z}\mathbf{k} \Rightarrow \mathbf{a} = \dot{\mathbf{v}} = (\ddot{r}\mathbf{u}_r + \dot{r}\dot{\mathbf{u}}_r) + \left(\dot{r}\dot{\theta}\mathbf{u}_\theta + r\ddot{\theta}\mathbf{u}_\theta + r\dot{\theta}\dot{\mathbf{u}}_\theta\right) + \ddot{z}\mathbf{k}$
$= \left(\ddot{r} - r\dot{\theta}^2\right)\mathbf{u}_r + \left(r\ddot{\theta} + 2\dot{r}\dot{\theta}\right)\mathbf{u}_\theta + \ddot{z}\mathbf{k}$

CHAPTER 14 PARTIAL DERIVATIVES

14.1 FUNCTIONS OF SEVERAL VARIABLES

1. (a) $f(0, 0) = 0$ (b) $f(-1, 1) = 0$ (c) $f(2, 3) = 58$
   (d) $f(-3, -2) = 33$

3. (a) $f(3, -1, 2) = \frac{4}{5}$ (b) $f\left(1, \frac{1}{2}, -\frac{1}{4}\right) = \frac{8}{5}$ (c) $f\left(0, -\frac{1}{3}, 0\right) = 3$
   (d) $f(2, 2, 100) = 0$

5. Domain: all points (x, y) on or above the line $y = x + 2$

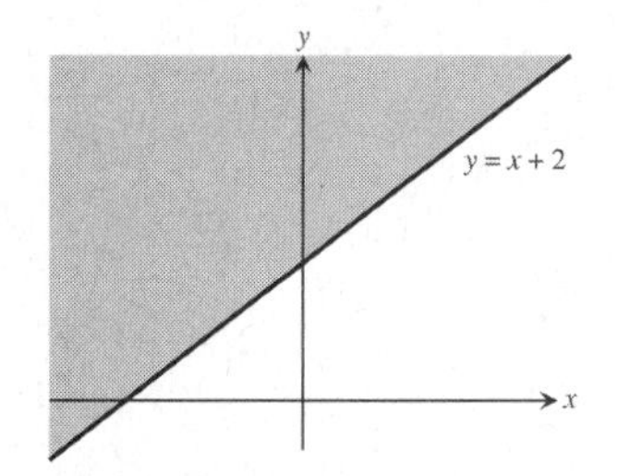

7. Domain: all points (x, y) not liying on the graph of $y = x$ or $y = x^3$

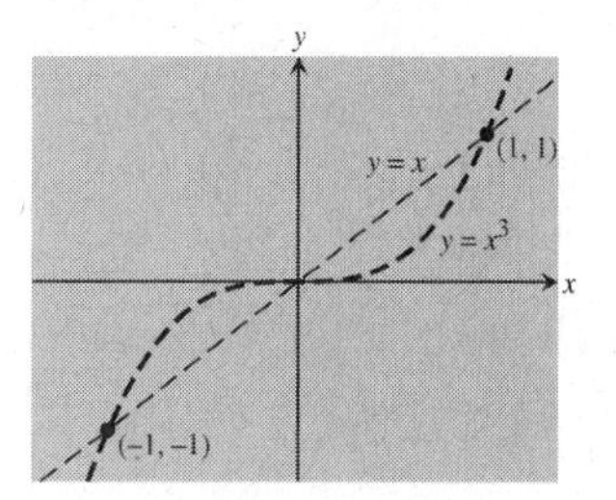

9. Domain: all points (x, y) satisfying $x^2 - 1 \le y \le x^2 + 1$

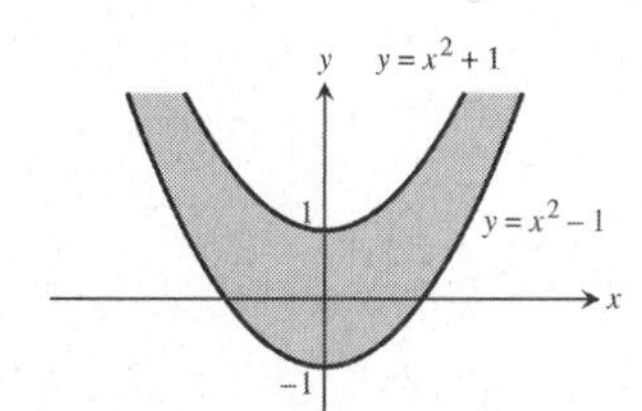

11. Domain: all points (x, y) satisfying $(x - 2)(x + 2)(y - 3)(y + 3) \ge 0$

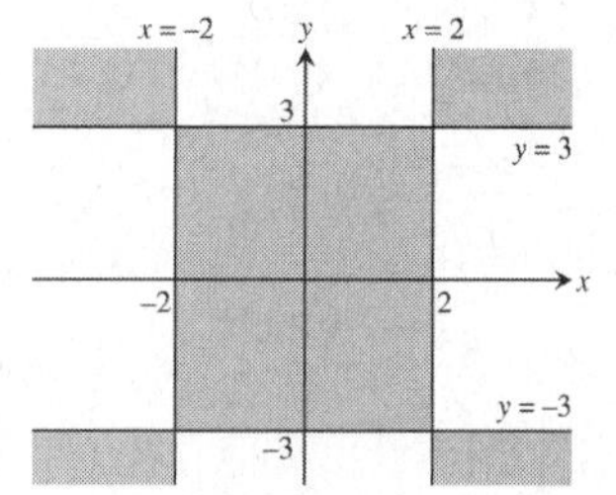

13.

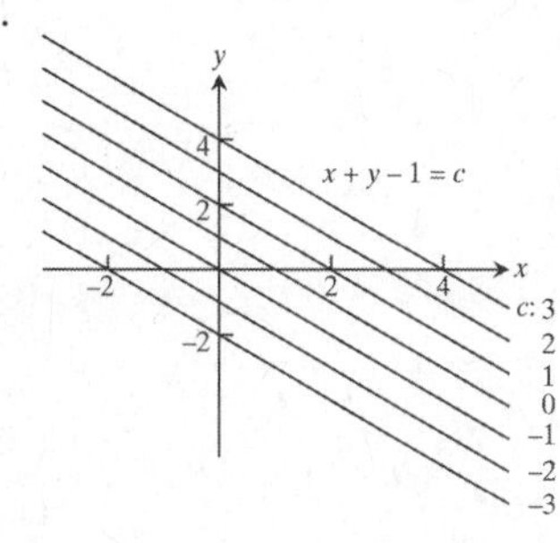

15.

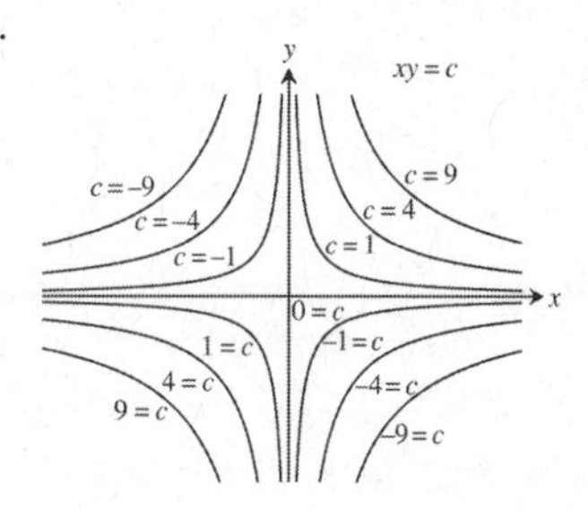

17. (a) Domain: all points in the xy-plane
    (b) Range: all real numbers
    (c) level curves are straight lines $y - x = c$ parallel to the line $y = x$
    (d) no boundary points
    (e) both open and closed
    (f) unbounded

19. (a) Domain: all points in the xy-plane
    (b) Range: $z \geq 0$
    (c) level curves: for $f(x, y) = 0$, the origin; for $f(x, y) = c > 0$, ellipses with center $(0, 0)$ and major and minor axes along the x- and y-axes, respectively
    (d) no boundary points
    (e) both open and closed
    (f) unbounded

21. (a) Domain: all points in the xy-plane
    (b) Range: all real numbers
    (c) level curves are hyperbolas with the x- and y-axes as asymptotes when $f(x, y) \neq 0$, and the x- and y-axes when $f(x, y) = 0$
    (d) no boundary points
    (e) both open and closed
    (f) unbounded

23. (a) Domain: all (x, y) satisfying $x^2 + y^2 < 16$
    (b) Range: $z \geq \frac{1}{4}$
    (c) level curves are circles centered at the origin with radii $r < 4$
    (d) boundary is the circle $x^2 + y^2 = 16$
    (e) open
    (f) bounded

25. (a) Domain: $(x, y) \neq (0, 0)$
    (b) Range: all real numbers
    (c) level curves are circles with center $(0, 0)$ and radii $r > 0$
    (d) boundary is the single point $(0, 0)$
    (e) open
    (f) unbounded

27. (a) Domain: all (x, y) satisfying $-1 \leq y - x \leq 1$
    (b) Range: $-\frac{\pi}{2} \leq z \leq \frac{\pi}{2}$
    (c) level curves are straight lines of the form $y - x = c$ where $-1 \leq c \leq 1$
    (d) boundary is the two straight lines $y = 1 + x$ and $y = -1 + x$
    (e) closed
    (f) unbounded

29. (a) Domain: all points (x, y) outside the circle $x^2 + y^2 = 1$
    (b) Range: all reals
    (c) Circles centered ar the origin with radii $r > 1$
    (d) Boundary: the cricle $x^2 + y^2 = 1$
    (e) open
    (f) unbounded

31. f

33. a

35. d

37. (a)

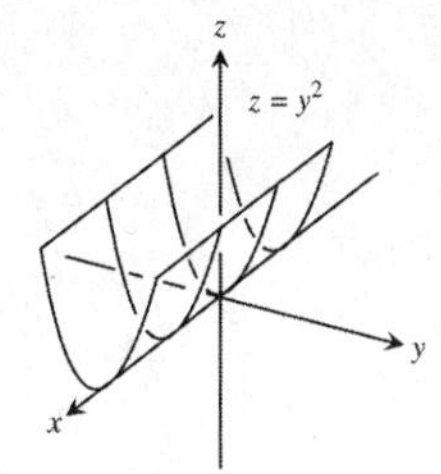

(b)

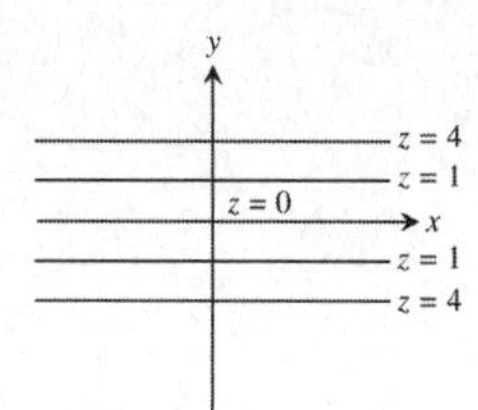

39. (a)

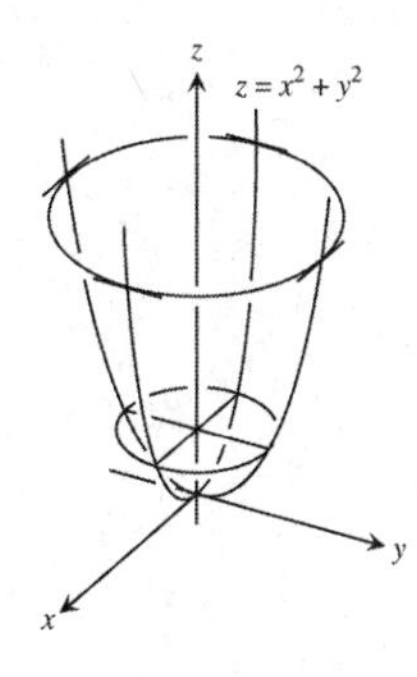

(b)

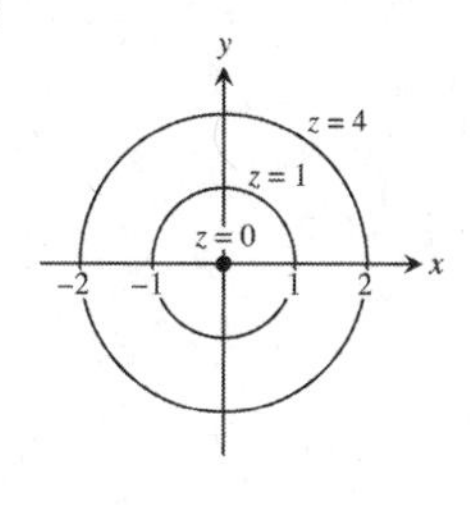

41. (a)

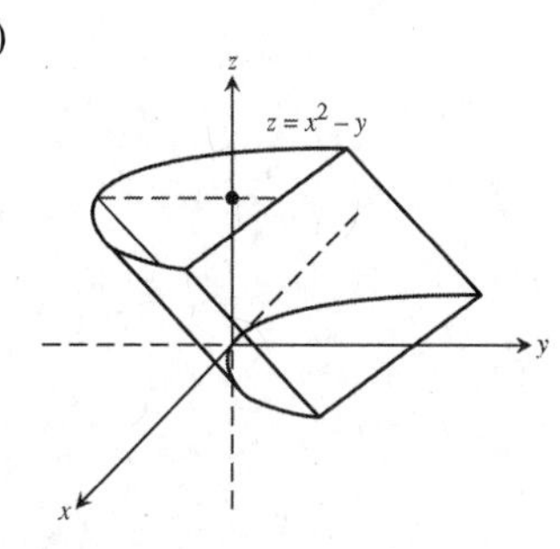

(b)

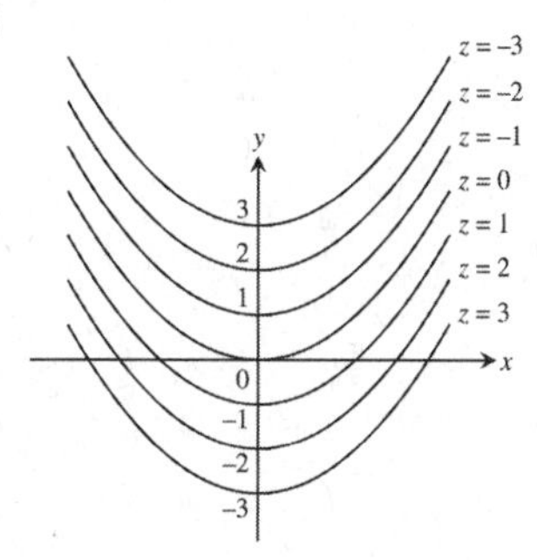

43. (a)

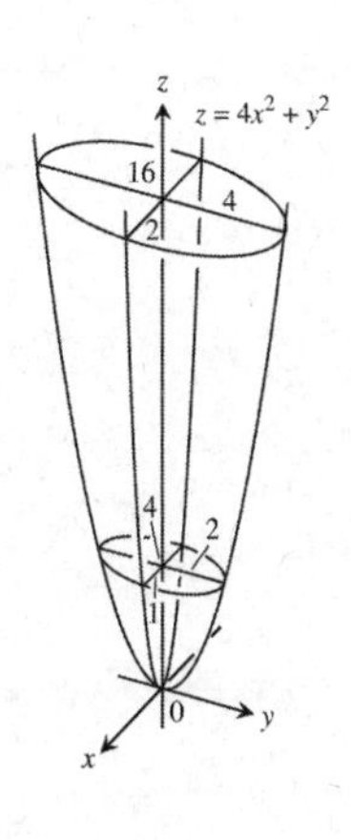

(b)

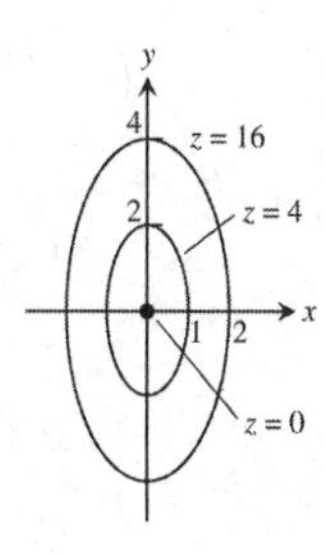

45. (a)

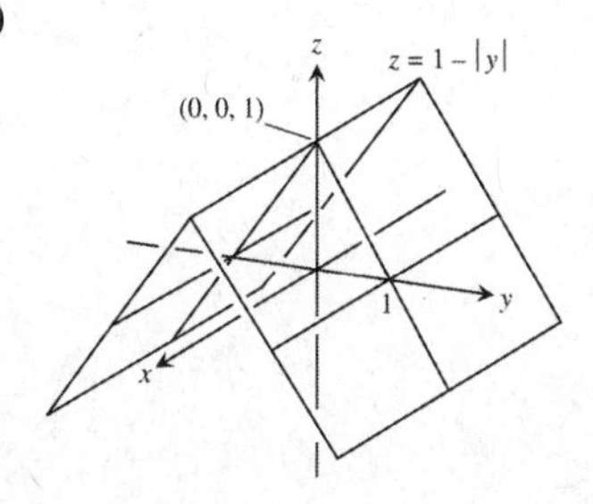

(b)

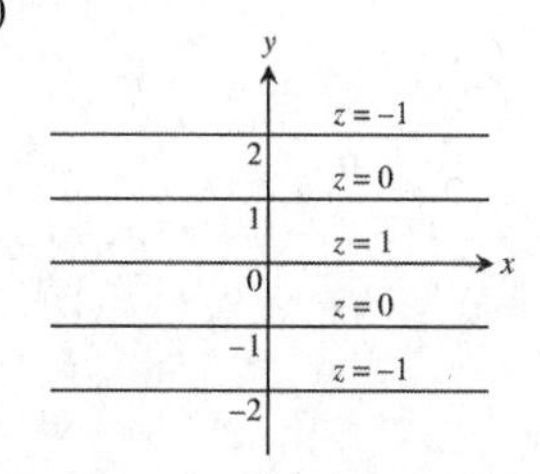

47. (a) 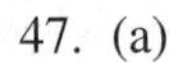

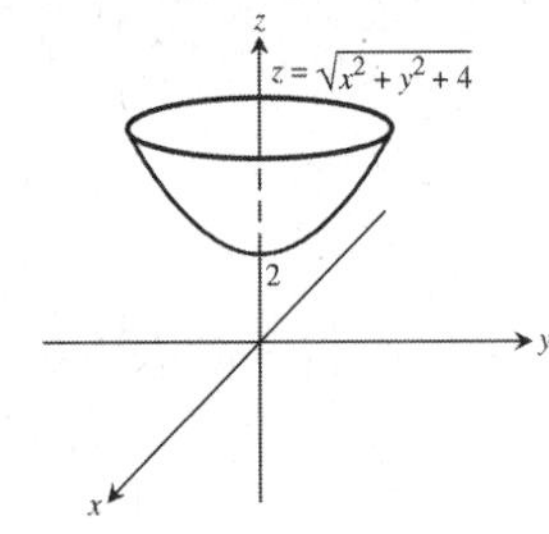

(b)

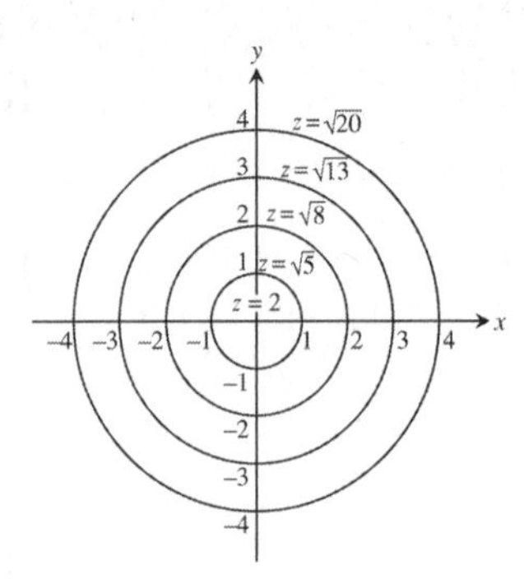

49. $f(x, y) = 16 - x^2 - y^2$ and $\left(2\sqrt{2}, \sqrt{2}\right) \Rightarrow z = 16 - \left(2\sqrt{2}\right)^2 - \left(\sqrt{2}\right)^2 = 6 \Rightarrow 6 = 16 - x^2 - y^2 \Rightarrow x^2 + y^2 = 10$

51. $f(x, y) = \sqrt{x + y^2 - 3}$ and $(3, -1) \Rightarrow z = \sqrt{3 + (-1)^2 - 3} = 1 \Rightarrow x + y^2 - 3 = 1 \Rightarrow x + y^2 = 4$

53.

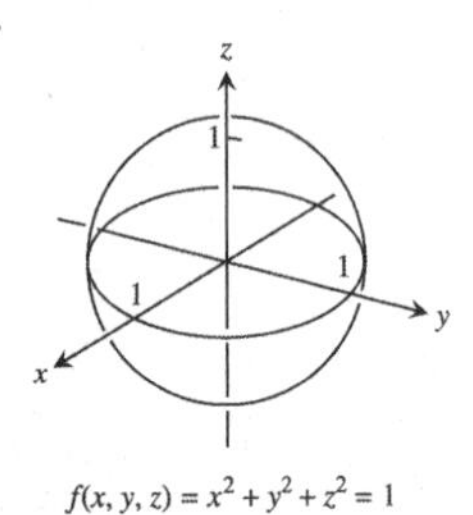

55.

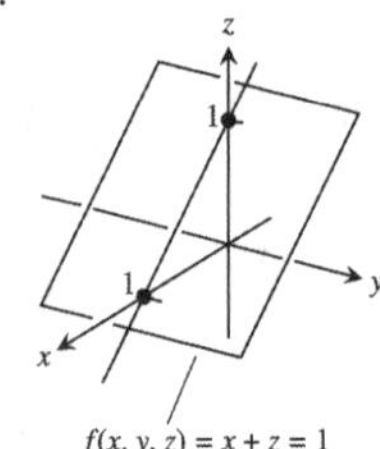

57.

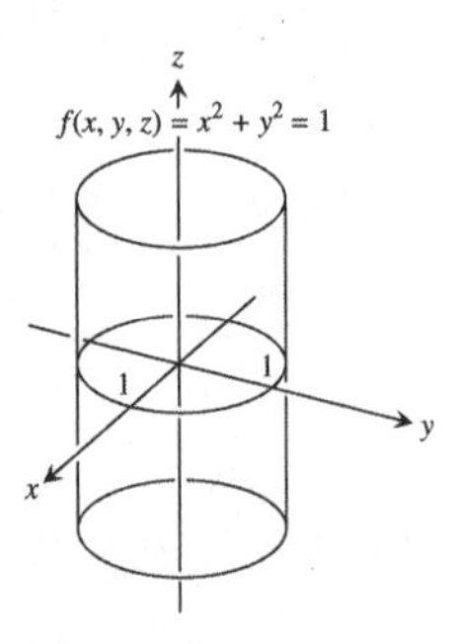

59.

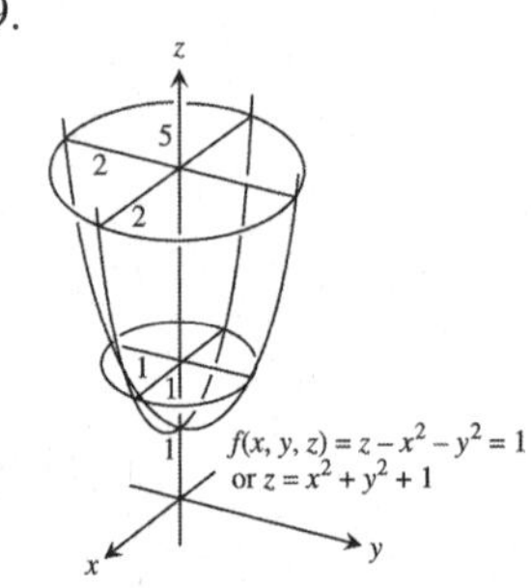

61. $f(x, y, z) = \sqrt{x - y} - \ln z$ at $(3, -1, 1) \Rightarrow w = \sqrt{x - y} - \ln z$; at $(3, -1, 1) \Rightarrow w = \sqrt{3 - (-1)} - \ln 1 = 2$
$\Rightarrow \sqrt{x - y} - \ln z = 2$

63. $g(x, y, z) = \sqrt{x^2 + y^2 + z^2}$ at $\left(1, -1, \sqrt{2}\right) \Rightarrow w = \sqrt{x^2 + y^2 + z^2}$; at $\left(1, -1, \sqrt{2}\right) \Rightarrow w = \sqrt{1^2 + (-1)^2 + \left(\sqrt{2}\right)^2}$
$= 2 \Rightarrow 2 = \sqrt{x^2 + y^2 + z^2} \Rightarrow x^2 + y^2 + z^2 = 4$

65. $f(x, y) = \sum_{n=0}^{\infty} \left(\frac{x}{y}\right)^n = \frac{1}{1 - \left(\frac{x}{y}\right)} = \frac{y}{y - x}$ for
$\left|\frac{x}{y}\right| < 1 \Rightarrow$ Domain: all points (x, y) satisfying $|x| < |y|$;
at $(1, 2) \Rightarrow$ since $\left|\frac{1}{2}\right| < 1 \Rightarrow z = \frac{2}{2-1} = 2$
$\Rightarrow \frac{y}{y - x} = 2 \Rightarrow y = 2x$

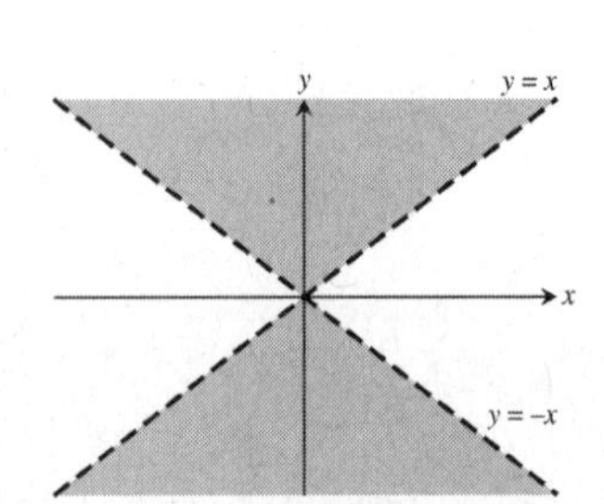

67. $f(x, y) = \int_x^y \frac{d\theta}{\sqrt{1-\theta^2}} = \sin^{-1}y - \sin^{-1}x \Rightarrow$ Domain: all points (x, y) satisfying $-1 \le x \le 1$ and $-1 \le y \le 1$; at $(0, 1) \Rightarrow \sin^{-1}1 - \sin^{-1}0 = \frac{\pi}{2} \Rightarrow \sin^{-1}y - \sin^{-1}x = \frac{\pi}{2}$. Since $-\frac{\pi}{2} \le \sin^{-1}y \le \frac{\pi}{2}$ and $-\frac{\pi}{2} \le \sin^{-1}x \le \frac{\pi}{2}$, in order for $\sin^{-1}y - \sin^{-1}x$ to equal $\frac{\pi}{2}$, $0 \le \sin^{-1}y \le \frac{\pi}{2}$ and $-\frac{\pi}{2} \le \sin^{-1}x \le 0$; that is $0 \le y \le 1$ and $-1 \le x \le 0$. Thus $y = \sin\left(\frac{\pi}{2} + \sin^{-1}x\right) = \sqrt{1 - x^2}, x \le 0$

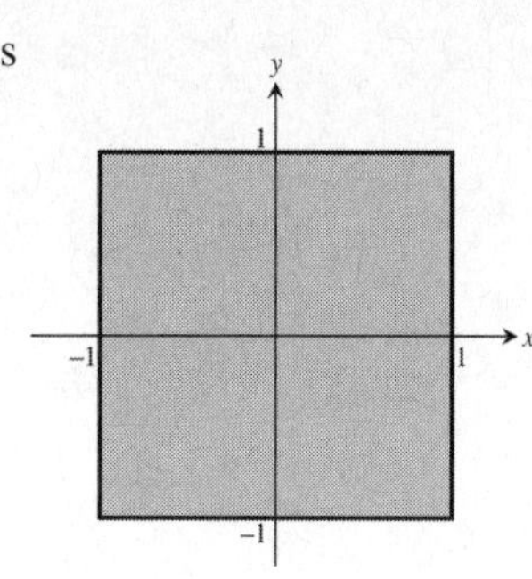

14.2 LIMITS AND CONTINUITY IN HIGHER DIMENSIONS

1. $\lim\limits_{(x,y)\to(0,0)} \frac{3x^2 - y^2 + 5}{x^2 + y^2 + 2} = \frac{3(0)^2 - 0^2 + 5}{0^2 + 0^2 + 2} = \frac{5}{2}$

3. $\lim\limits_{(x,y)\to(3,4)} \sqrt{x^2 + y^2 - 1} = \sqrt{3^2 + 4^2 - 1} = \sqrt{24} = 2\sqrt{6}$

5. $\lim\limits_{(x,y)\to\left(0,\frac{\pi}{4}\right)} \sec x \tan y = (\sec 0)\left(\tan \frac{\pi}{4}\right) = (1)(1) = 1$

7. $\lim\limits_{(x,y)\to(0,\ln 2)} e^{x-y} = e^{0 - \ln 2} = e^{\ln\left(\frac{1}{2}\right)} = \frac{1}{2}$

9. $\lim\limits_{(x,y)\to(0,0)} \frac{e^y \sin x}{x} = \lim\limits_{(x,y)\to(0,0)} (e^y)\left(\frac{\sin x}{x}\right) = e^0 \cdot \lim\limits_{x\to 0} \left(\frac{\sin x}{x}\right) = 1 \cdot 1 = 1$

11. $\lim\limits_{(x,y)\to(1,\pi/6)} \frac{x \sin y}{x^2 + 1} = \frac{1 \cdot \sin\left(\frac{\pi}{6}\right)}{1^2 + 1} = \frac{1/2}{2} = \frac{1}{4}$

13. $\lim\limits_{\substack{(x,y)\to(1,1)\\ x \ne y}} \frac{x^2 - 2xy + y^2}{x - y} = \lim\limits_{(x,y)\to(1,1)} \frac{(x-y)^2}{x - y} = \lim\limits_{(x,y)\to(1,1)} (x - y) = (1 - 1) = 0$

15. $\lim\limits_{\substack{(x,y)\to(1,1)\\ x \ne 1}} \frac{xy - y - 2x + 2}{x - 1} = \lim\limits_{\substack{(x,y)\to(1,1)\\ x \ne 1}} \frac{(x-1)(y-2)}{x - 1} = \lim\limits_{(x,y)\to(1,1)} (y - 2) = (1 - 2) = -1$

17. $\lim\limits_{\substack{(x,y)\to(0,0)\\ x \ne y}} \frac{x - y + 2\sqrt{x} - 2\sqrt{y}}{\sqrt{x} - \sqrt{y}} = \lim\limits_{\substack{(x,y)\to(0,0)\\ x \ne y}} \frac{\left(\sqrt{x} - \sqrt{y}\right)\left(\sqrt{x} + \sqrt{y} + 2\right)}{\sqrt{x} - \sqrt{y}} = \lim\limits_{(x,y)\to(0,0)} \left(\sqrt{x} + \sqrt{y} + 2\right)$
$= \left(\sqrt{0} + \sqrt{0} + 2\right) = 2$

Note: (x, y) must approach $(0, 0)$ through the first quadrant only with $x \ne y$.

19. $\lim\limits_{\substack{(x,y)\to(2,0)\\ 2x - y \ne 4}} \frac{\sqrt{2x - y} - 2}{2x - y - 4} = \lim\limits_{\substack{(x,y)\to(2,0)\\ 2x - y \ne 4}} \frac{\sqrt{2x - y} - 2}{\left(\sqrt{2x - y} + 2\right)\left(\sqrt{2x - y} - 2\right)} = \lim\limits_{(x,y)\to(2,0)} \frac{1}{\sqrt{2x - y} + 2}$
$= \frac{1}{\sqrt{(2)(2) - 0} + 2} = \frac{1}{2 + 2} = \frac{1}{4}$

21. $\lim\limits_{(x,y)\to(0,0)} \frac{\sin(x^2 + y^2)}{x^2 + y^2} = \lim\limits_{r\to 0} \frac{\sin(r^2)}{r^2} = \lim\limits_{r\to 0} \frac{2r \cdot \cos(r^2)}{2r} = \lim\limits_{r\to 0} \cos(r^2) = 1$

23. $\lim\limits_{(x,y)\to(1,-1)} \frac{x^3 + y^3}{x + y} = \lim\limits_{(x,y)\to(1,-1)} \frac{(x + y)(x^2 - xy + y^2)}{x + y} = \lim\limits_{(x,y)\to(1,-1)} (x^2 - xy + y^2) = \left(1^2 - (1)(-1) + (-1)^2\right) = 3$

25. $\lim\limits_{P \to (1,3,4)} \left(\frac{1}{x} + \frac{1}{y} + \frac{1}{z}\right) = \frac{1}{1} + \frac{1}{3} + \frac{1}{4} = \frac{12+4+3}{12} = \frac{19}{12}$

27. $\lim\limits_{P \to (3,3,0)} (\sin^2 x + \cos^2 y + \sec^2 z) = (\sin^2 3 + \cos^2 3) + \sec^2 0 = 1 + 1^2 = 2$

29. $\lim\limits_{P \to (\pi,0,3)} ze^{-2y} \cos 2x = 3e^{-2(0)} \cos 2\pi = (3)(1)(1) = 3$

31. (a) All (x, y) (b) All (x, y) except $(0, 0)$

33. (a) All (x, y) except where $x = 0$ or $y = 0$ (b) All (x, y)

35. (a) All (x, y, z) (b) All (x, y, z) except the interior of the cylinder $x^2 + y^2 = 1$

37. (a) All (x, y, z) with $z \neq 0$ (b) All (x, y, z) with $x^2 + z^2 \neq 1$

39. (a) All (x, y, z) such that $z > x^2 + y^2 + 1$ (b) All (x, y, z) such that $z \neq \sqrt{x^2 + y^2}$

41. $\lim\limits_{\substack{(x,y) \to (0,0) \\ \text{along } y = x \\ x > 0}} -\frac{x}{\sqrt{x^2+y^2}} = \lim\limits_{x \to 0^+} -\frac{x}{\sqrt{x^2+x^2}} = \lim\limits_{x \to 0^+} -\frac{x}{\sqrt{2}\,|x|} = \lim\limits_{x \to 0^+} -\frac{x}{\sqrt{2}\,x} = \lim\limits_{x \to 0^+} -\frac{1}{\sqrt{2}} = -\frac{1}{\sqrt{2}}$;

$\lim\limits_{\substack{(x,y) \to (0,0) \\ \text{along } y = x \\ x < 0}} -\frac{x}{\sqrt{x^2+y^2}} = \lim\limits_{x \to 0^-} -\frac{x}{\sqrt{2}\,|x|} = \lim\limits_{x \to 0^-} -\frac{x}{\sqrt{2}(-x)} = \lim\limits_{x \to 0^-} \frac{1}{\sqrt{2}} = \frac{1}{\sqrt{2}}$

43. $\lim\limits_{\substack{(x,y) \to (0,0) \\ \text{along } y = kx^2}} \frac{x^4 - y^2}{x^4 + y^2} = \lim\limits_{x \to 0} \frac{x^4 - (kx^2)^2}{x^4 + (kx^2)^2} = \lim\limits_{x \to 0} \frac{x^4 - k^2x^4}{x^4 + k^2x^4} = \frac{1-k^2}{1+k^2} \Rightarrow$ different limits for different values of k

45. $\lim\limits_{\substack{(x,y) \to (0,0) \\ \text{along } y = kx \\ k \neq -1}} \frac{x-y}{x+y} = \lim\limits_{x \to 0} \frac{x - kx}{x + kx} = \frac{1-k}{1+k} \Rightarrow$ different limits for different values of k, $k \neq -1$

47. $\lim\limits_{\substack{(x,y) \to (0,0) \\ \text{along } y = kx^2 \\ k \neq 0}} \frac{x^2+y}{y} = \lim\limits_{x \to 0} \frac{x^2 + kx^2}{kx^2} = \frac{1+k}{k} \Rightarrow$ different limits for different values of k, $k \neq 0$

49. $\lim\limits_{\substack{(x,y) \to (1,1) \\ \text{along } x = 1}} \frac{xy^2 - 1}{y - 1} = \lim\limits_{y \to 1} \frac{y^2 - 1}{y - 1} = \lim\limits_{y \to 1} (y + 1) = 2$; $\lim\limits_{\substack{(x,y) \to (1,1) \\ \text{along } y = x}} \frac{xy^2 - 1}{y - 1} = \lim\limits_{y \to 1} \frac{y^3 - 1}{y - 1} = \lim\limits_{y \to 1} (y^2 + y + 1) = 3$

51. $f(x, y) = \begin{cases} 1 & \text{if } y \geq x^4 \\ 1 & \text{if } y \leq 0 \\ 0 & \text{otherwise} \end{cases}$

(a) $\lim\limits_{(x,y) \to (0,1)} f(x, y) = 1$ since any path through $(0, 1)$ that is close to $(0, 1)$ satisfies $y \geq x^4$

(b) $\lim\limits_{(x,y) \to (2,3)} f(x, y) = 0$ since any path through $(2, 3)$ that is close to $(2, 3)$ does not satisfiy either $y \geq x^4$ or $y \leq 0$

(c) $\lim\limits_{\substack{(x,y) \to (0,0) \\ \text{along } x = 0}} f(x, y) = 1$ and $\lim\limits_{\substack{(x,y) \to (0,0) \\ \text{along } y = x^2}} f(x, y) = 0 \Rightarrow \lim\limits_{(x,y) \to (0,0)} f(x, y)$ does not exist

53. First consider the vertical line $x = 0 \Rightarrow \lim\limits_{\substack{(x,y)\to(0,0) \\ \text{along } x=0}} \frac{2x^2y}{x^4+y^2} = \lim\limits_{y\to 0} \frac{2(0)^2y}{(0)^4+y^2} = \lim\limits_{y\to 0} 0 = 0$. Now consider any nonvertical through (0, 0). The equation of any line through (0, 0) is of the form $y = mx \Rightarrow \lim\limits_{\substack{(x,y)\to(0,0) \\ \text{along } y=mx}} f(x,y) = \lim\limits_{\substack{(x,y)\to(0,0) \\ \text{along } y=mx}} \frac{2x^2y}{x^4+y^2}$ $= \lim\limits_{x\to 0} \frac{2x^2(mx)}{x^4+(mx)^2} = \lim\limits_{x\to 0} \frac{2mx^3}{x^4+m^2x^2} = \lim\limits_{x\to 0} \frac{2mx^3}{x^2(x^2+m^2)} = \lim\limits_{x\to 0} \frac{2mx}{(x^2+m^2)} = 0$. Thus $\lim\limits_{\substack{(x,y)\to(0,0) \\ \text{any line though } (0,0)}} \frac{2x^2y}{x^4+y^2} = 0$.

55. $\lim\limits_{(x,y)\to(0,0)} \left(1 - \frac{x^2y^2}{3}\right) = 1$ and $\lim\limits_{(x,y)\to(0,0)} 1 = 1 \Rightarrow \lim\limits_{(x,y)\to(0,0)} \frac{\tan^{-1} xy}{xy} = 1$, by the Sandwich Theorem

57. The limit is 0 since $\left|\sin\left(\frac{1}{x}\right)\right| \le 1 \Rightarrow -1 \le \sin\left(\frac{1}{x}\right) \le 1 \Rightarrow -y \le y\sin\left(\frac{1}{x}\right) \le y$ for $y \ge 0$, and $-y \ge y\sin\left(\frac{1}{x}\right) \ge y$ for $y \le 0$. Thus as $(x,y) \to (0,0)$, both $-y$ and y approach $0 \Rightarrow y\sin\left(\frac{1}{x}\right) \to 0$, by the Sandwich Theorem.

59. (a) $f(x,y)|_{y=mx} = \frac{2m}{1+m^2} = \frac{2\tan\theta}{1+\tan^2\theta} = \sin 2\theta$. The value of $f(x,y) = \sin 2\theta$ varies with θ, which is the line's angle of inclination.

(b) Since $f(x,y)|_{y=mx} = \sin 2\theta$ and since $-1 \le \sin 2\theta \le 1$ for every θ, $\lim\limits_{(x,y)\to(0,0)} f(x,y)$ varies from -1 to 1 along $y = mx$.

61. $\lim\limits_{(x,y)\to(0,0)} \frac{x^3-xy^2}{x^2+y^2} = \lim\limits_{r\to 0} \frac{r^3\cos^3\theta-(r\cos\theta)(r^2\sin^2\theta)}{r^2\cos^2\theta+r^2\sin^2\theta} = \lim\limits_{r\to 0} \frac{r(\cos^3\theta-\cos\theta\sin^2\theta)}{1} = 0$

63. $\lim\limits_{(x,y)\to(0,0)} \frac{y^2}{x^2+y^2} = \lim\limits_{r\to 0} \frac{r^2\sin^2\theta}{r^2} = \lim\limits_{r\to 0} (\sin^2\theta) = \sin^2\theta$; the limit does not exist since $\sin^2\theta$ is between 0 and 1 depending on θ

65. $\lim\limits_{(x,y)\to(0,0)} \tan^{-1}\left[\frac{|x|+|y|}{x^2+y^2}\right] = \lim\limits_{r\to 0} \tan^{-1}\left[\frac{|r\cos\theta|+|r\sin\theta|}{r^2}\right] = \lim\limits_{r\to 0} \tan^{-1}\left[\frac{|r|(|\cos\theta|+|\sin\theta|)}{r^2}\right]$; if $r \to 0^+$, then $\lim\limits_{r\to 0^+} \tan^{-1}\left[\frac{|r|(|\cos\theta|+|\sin\theta|)}{r^2}\right] = \lim\limits_{r\to 0^+} \tan^{-1}\left[\frac{|\cos\theta|+|\sin\theta|}{r}\right] = \frac{\pi}{2}$; if $r \to 0^-$, then $\lim\limits_{r\to 0^-} \tan^{-1}\left[\frac{|r|(|\cos\theta|+|\sin\theta|)}{r^2}\right] = \lim\limits_{r\to 0^-} \tan^{-1}\left(\frac{|\cos\theta|+|\sin\theta|}{-r}\right) = \frac{\pi}{2} \Rightarrow$ the limit is $\frac{\pi}{2}$

67. $\lim\limits_{(x,y)\to(0,0)} \ln\left(\frac{3x^2-x^2y^2+3y^2}{x^2+y^2}\right) = \lim\limits_{r\to 0} \ln\left(\frac{3r^2\cos^2\theta-r^4\cos^2\theta\sin^2\theta+3r^2\sin^2\theta}{r^2}\right)$ $= \lim\limits_{r\to 0} \ln(3-r^2\cos^2\theta\sin^2\theta) = \ln 3 \Rightarrow$ define $f(0,0) = \ln 3$

69. Let $\delta = 0.1$. Then $\sqrt{x^2+y^2} < \delta \Rightarrow \sqrt{x^2+y^2} < 0.1 \Rightarrow x^2+y^2 < 0.01 \Rightarrow |x^2+y^2-0| < 0.01$ $\Rightarrow |f(x,y)-f(0,0)| < 0.01 = \epsilon$.

71. Let $\delta = 0.005$. Then $|x| < \delta$ and $|y| < \delta \Rightarrow |f(x,y)-f(0,0)| = \left|\frac{x+y}{x^2+1} - 0\right| = \left|\frac{x+y}{x^2+1}\right| \le |x+y| < |x|+|y|$ $< 0.005 + 0.005 = 0.01 = \epsilon$.

73. Let $\delta = 0.04$. Since $y^2 \le x^2+y^2 \Rightarrow \frac{y^2}{x^2+y^2} \le 1 \Rightarrow \frac{|x|y^2}{x^2+y^2} \le |x| = \sqrt{x^2} \le \sqrt{x^2+y^2} < \delta \Rightarrow |f(x,y)-f(0,0)|$ $= \left|\frac{xy^2}{x^2+y^2} - 0\right| < 0.04 = \epsilon$.

75. Let $\delta = \sqrt{0.015}$. Then $\sqrt{x^2+y^2+z^2} < \delta \Rightarrow |f(x,y,z)-f(0,0,0)| = |x^2+y^2+z^2-0| = |x^2+y^2+z^2|$ $= \left(\sqrt{x^2+t^2+x^2}\right)^2 < \left(\sqrt{0.015}\right)^2 = 0.015 = \epsilon$.

77. Let $\delta = 0.005$. Then $|x| < \delta$, $|y| < \delta$, and $|z| < \delta \Rightarrow |f(x,y,z) - f(0,0,0)| = \left|\frac{x+y+z}{x^2+y^2+z^2+1} - 0\right|$ $= \left|\frac{x+y+z}{x^2+y^2+z^2+1}\right| \le |x+y+z| \le |x| + |y| + |z| < 0.005 + 0.005 + 0.005 = 0.015 = \epsilon.$

79. $\lim_{(x,y,z) \to (x_0,y_0,z_0)} f(x,y,z) = \lim_{(x,y,z) \to (x_0,y_0,z_0)} (x+y-z) = x_0 + y_0 - z_0 = f(x_0,y_0,z_0) \Rightarrow$ f is continuous at every (x_0, y_0, z_0)

14.3 PARTIAL DERIVATIVES

1. $\frac{\partial f}{\partial x} = 4x,\ \frac{\partial f}{\partial y} = -3$

3. $\frac{\partial f}{\partial x} = 2x(y+2),\ \frac{\partial f}{\partial y} = x^2 - 1$

5. $\frac{\partial f}{\partial x} = 2y(xy-1),\ \frac{\partial f}{\partial y} = 2x(xy-1)$

7. $\frac{\partial f}{\partial x} = \frac{x}{\sqrt{x^2+y^2}},\ \frac{\partial f}{\partial y} = \frac{y}{\sqrt{x^2+y^2}}$

9. $\frac{\partial f}{\partial x} = -\frac{1}{(x+y)^2} \cdot \frac{\partial}{\partial x}(x+y) = -\frac{1}{(x+y)^2},\ \frac{\partial f}{\partial y} = -\frac{1}{(x+y)^2} \cdot \frac{\partial}{\partial y}(x+y) = -\frac{1}{(x+y)^2}$

11. $\frac{\partial f}{\partial x} = = \frac{(xy-1)(1)-(x+y)(y)}{(xy-1)^2} = \frac{-y^2-1}{(xy-1)^2},\ \frac{\partial f}{\partial y} = \frac{(xy-1)(1)-(x+y)(x)}{(xy-1)^2} = \frac{-x^2-1}{(xy-1)^2}$

13. $\frac{\partial f}{\partial x} = e^{(x+y+1)} \cdot \frac{\partial}{\partial x}(x+y+1) = e^{(x+y+1)},\ \frac{\partial f}{\partial y} = e^{(x+y+1)} \cdot \frac{\partial}{\partial y}(x+y+1) = e^{(x+y+1)}$

15. $\frac{\partial f}{\partial x} = \frac{1}{x+y} \cdot \frac{\partial}{\partial x}(x+y) = \frac{1}{x+y},\ \frac{\partial f}{\partial y} = \frac{1}{x+y} \cdot \frac{\partial}{\partial y}(x+y) = \frac{1}{x+y}$

17. $\frac{\partial f}{\partial x} = 2\sin(x-3y) \cdot \frac{\partial}{\partial x}\sin(x-3y) = 2\sin(x-3y)\cos(x-3y) \cdot \frac{\partial}{\partial x}(x-3y) = 2\sin(x-3y)\cos(x-3y),$
$\frac{\partial f}{\partial y} = 2\sin(x-3y) \cdot \frac{\partial}{\partial y}\sin(x-3y) = 2\sin(x-3y)\cos(x-3y) \cdot \frac{\partial}{\partial y}(x-3y) = -6\sin(x-3y)\cos(x-3y)$

19. $\frac{\partial f}{\partial x} = yx^{y-1},\ \frac{\partial f}{\partial y} = x^y \ln x$

21. $\frac{\partial f}{\partial x} = -g(x),\ \frac{\partial f}{\partial y} = g(y)$

23. $f_x = y^2,\ f_y = 2xy,\ f_z = -4z$

25. $f_x = 1,\ f_y = -\frac{y}{\sqrt{y^2+z^2}},\ f_z = -\frac{z}{\sqrt{y^2+z^2}}$

27. $f_x = \frac{yz}{\sqrt{1-x^2y^2z^2}},\ f_y = \frac{xz}{\sqrt{1-x^2y^2z^2}},\ f_z = \frac{xy}{\sqrt{1-x^2y^2z^2}}$

29. $f_x = \frac{1}{x+2y+3z},\ f_y = \frac{2}{x+2y+3z},\ f_z = \frac{3}{x+2y+3z}$

31. $f_x = -2xe^{-(x^2+y^2+z^2)},\ f_y = -2ye^{-(x^2+y^2+z^2)},\ f_z = -2ze^{-(x^2+y^2+z^2)}$

33. $f_x = \operatorname{sech}^2(x+2y+3z),\ f_y = 2\operatorname{sech}^2(x+2y+3z),\ f_z = 3\operatorname{sech}^2(x+2y+3z)$

35. $\frac{\partial f}{\partial t} = -2\pi\sin(2\pi t - \alpha),\ \frac{\partial f}{\partial \alpha} = \sin(2\pi t - \alpha)$

37. $\frac{\partial h}{\partial \rho} = \sin\phi\cos\theta,\ \frac{\partial h}{\partial \phi} = \rho\cos\phi\cos\theta,\ \frac{\partial h}{\partial \theta} = -\rho\sin\phi\sin\theta$

39. $W_p = V,\ W_v = P + \frac{\delta v^2}{2g},\ W_\delta = \frac{Vv^2}{2g},\ W_v = \frac{2V\delta v}{2g} = \frac{V\delta v}{g},\ W_g = -\frac{V\delta v^2}{2g^2}$

41. $\frac{\partial f}{\partial x} = 1+y,\ \frac{\partial f}{\partial y} = 1+x,\ \frac{\partial^2 f}{\partial x^2} = 0,\ \frac{\partial^2 f}{\partial y^2} = 0,\ \frac{\partial^2 f}{\partial y \partial x} = \frac{\partial^2 f}{\partial x \partial y} = 1$

43. $\frac{\partial g}{\partial x} = 2xy + y\cos x$, $\frac{\partial g}{\partial y} = x^2 - \sin y + \sin x$, $\frac{\partial^2 g}{\partial x^2} = 2y - y\sin x$, $\frac{\partial^2 g}{\partial y^2} = -\cos y$, $\frac{\partial^2 g}{\partial y \partial x} = \frac{\partial^2 g}{\partial x \partial y} = 2x + \cos x$

45. $\frac{\partial r}{\partial x} = \frac{1}{x+y}$, $\frac{\partial r}{\partial y} = \frac{1}{x+y}$, $\frac{\partial^2 r}{\partial x^2} = \frac{-1}{(x+y)^2}$, $\frac{\partial^2 r}{\partial y^2} = \frac{-1}{(x+y)^2}$, $\frac{\partial^2 r}{\partial y \partial x} = \frac{\partial^2 r}{\partial x \partial y} = \frac{-1}{(x+y)^2}$

47. $\frac{\partial w}{\partial x} = 2x\tan(xy) + x^2\sec^2(xy) \cdot y = 2x\tan(xy) + x^2 y\sec^2(xy)$, $\frac{\partial w}{\partial y} = x^2\sec^2(xy) \cdot x = x^3\sec^2(xy)$,
$\frac{\partial^2 w}{\partial x^2} = 2\tan(xy) + 2x\sec^2(xy) \cdot y + 2xy\sec^2(xy) + x^2 y\,(2\sec(xy)\sec(xy)\tan(xy) \cdot y)$
$= 2\tan(xy) + 4xy\sec^2(xy) + 2x^2y^2\sec^2(xy)\tan(xy)$, $\frac{\partial^2 w}{\partial y^2} = x^3(2\sec(xy)\sec(xy)\tan(xy) \cdot x) = 2x^4\sec^2(xy)\tan(xy)$
$\frac{\partial^2 w}{\partial y \partial x} = \frac{\partial^2 w}{\partial x \partial y} = 3x^2\sec^2(xy) + x^3(2\sec(xy)\sec(xy)\tan(xy) \cdot y) = 3x^2\sec^2(xy) + x^3 y\sec^2(xy)\tan(xy)$

49. $\frac{\partial w}{\partial x} = \sin(x^2y) + x\cos(x^2y) \cdot 2xy = \sin(x^2y) + 2x^2y\cos(x^2y)$, $\frac{\partial w}{\partial y} = x\cos(x^2y) \cdot x^2 = x^3\cos(x^2y)$,
$\frac{\partial^2 w}{\partial x^2} = \cos(x^2y) \cdot 2xy + 4xy\cos(x^2y) - 2x^2y\sin(x^2y) \cdot 2xy = 6xy\cos(x^2y) - 4x^3y^2\sin(x^2y)$,
$\frac{\partial^2 w}{\partial y^2} = -x^3\sin(x^2y) \cdot x^2 = -x^5\sin(x^2y)$, $\frac{\partial^2 w}{\partial y \partial x} = \frac{\partial^2 w}{\partial x \partial y} = 3x^2\cos(x^2y) - x^3\sin(x^2y) \cdot 2xy = 3x^2\cos(x^2y) - 2x^4y\sin(x^2y)$

51. $\frac{\partial w}{\partial x} = \frac{2}{2x+3y}$, $\frac{\partial w}{\partial y} = \frac{3}{2x+3y}$, $\frac{\partial^2 w}{\partial y \partial x} = \frac{-6}{(2x+3y)^2}$, and $\frac{\partial^2 w}{\partial x \partial y} = \frac{-6}{(2x+3y)^2}$

53. $\frac{\partial w}{\partial x} = y^2 + 2xy^3 + 3x^2y^4$, $\frac{\partial w}{\partial y} = 2xy + 3x^2y^2 + 4x^3y^3$, $\frac{\partial^2 w}{\partial y \partial x} = 2y + 6xy^2 + 12x^2y^3$, and $\frac{\partial^2 w}{\partial x \partial y} = 2y + 6xy^2 + 12x^2y^3$

55. (a) x first (b) y first (c) x first (d) x first (e) y first (f) y first

57. $f_x(1,2) = \lim_{h \to 0} \frac{f(1+h,2) - f(1,2)}{h} = \lim_{h \to 0} \frac{[1-(1+h)+2-6(1+h)^2] - (2-6)}{h} = \lim_{h \to 0} \frac{-h-6(1+2h+h^2)+6}{h}$
$= \lim_{h \to 0} \frac{-13h-6h^2}{h} = \lim_{h \to 0} (-13-6h) = -13$,
$f_y(1,2) = \lim_{h \to 0} \frac{f(1,2+h) - f(1,2)}{h} = \lim_{h \to 0} \frac{[1-1+(2+h)-3(2+h)] - (2-6)}{h} = \lim_{h \to 0} \frac{(2-6-2h)-(2-6)}{h}$
$= \lim_{h \to 0} (-2) = -2$

59. $f_x(-2,3) = \lim_{h \to 0} \frac{f(-2+h,3) - f(-2,3)}{h} = \lim_{h \to 0} \frac{\sqrt{2(-2+h)+9-1} - \sqrt{-4+9-1}}{h}$
$= \lim_{h \to 0} \frac{\sqrt{2h+4}-2}{h} = \lim_{h \to 0} \left(\frac{\sqrt{2h+4}-2}{h}\,\frac{\sqrt{2h+4}+2}{\sqrt{2h+4}+2}\right) = \lim_{h \to 0} \frac{2}{\sqrt{2h+4}+2} = \frac{1}{2}$,
$f_y(-2,3) = \lim_{h \to 0} \frac{f(-2,3+h) - f(-2,3)}{h} = \lim_{h \to 0} \frac{\sqrt{-4+3(3+h)-1} - \sqrt{-4+9-1}}{h}$
$= \lim_{h \to 0} \frac{\sqrt{3h+4}-2}{h} = \lim_{h \to 0} \left(\frac{\sqrt{3h+4}-2}{h}\,\frac{\sqrt{3h+4}+2}{\sqrt{3h+4}+2}\right) = \lim_{h \to 0} \frac{3}{\sqrt{2h+4}+2} = \frac{3}{4}$

61. (a) In the plane $x = 2 \Rightarrow f_y(x,y) = 3 \Rightarrow f_y(2,-1) = 3 \Rightarrow m = 3$
(b) In the plane $y = -1 \Rightarrow f_x(x,y) = 2 \Rightarrow f_y(2,-1) = 2 \Rightarrow m = 2$

63. $f_z(x_0,y_0,z_0) = \lim_{h \to 0} \frac{f(x_0,y_0,z_0+h) - f(x_0,y_0,z_0)}{h}$;
$f_z(1,2,3) = \lim_{h \to 0} \frac{f(1,2,3+h) - f(1,2,3)}{h} = \lim_{h \to 0} \frac{2(3+h)^2 - 2(9)}{h} = \lim_{h \to 0} \frac{12h+2h^2}{h} = \lim_{h \to 0} (12+2h) = 12$

65. $y + \left(3z^2\,\frac{\partial z}{\partial x}\right)x + z^3 - 2y\,\frac{\partial z}{\partial x} = 0 \Rightarrow (3xz^2 - 2y)\,\frac{\partial z}{\partial x} = -y - z^3 \Rightarrow$ at $(1,1,1)$ we have $(3-2)\,\frac{\partial z}{\partial x} = -1-1$ or $\frac{\partial z}{\partial x} = -2$

67. $a^2 = b^2 + c^2 - 2bc\cos A \Rightarrow 2a = (2bc\sin A)\,\frac{\partial A}{\partial a} \Rightarrow \frac{\partial A}{\partial a} = \frac{a}{bc\sin A}$; also $0 = 2b - 2c\cos A + (2bc\sin A)\,\frac{\partial A}{\partial b}$
$\Rightarrow 2c\cos A - 2b = (2bc\sin A)\,\frac{\partial A}{\partial b} \Rightarrow \frac{\partial A}{\partial b} = \frac{c\cos A - b}{bc\sin A}$

69. Differentiating each equation implicitly gives $1 = v_x \ln u + \left(\frac{v}{u}\right) u_x$ and $0 = u_x \ln v + \left(\frac{u}{v}\right) v_x$ or

$$\left.\begin{aligned}(\ln u)\, v_x \quad + \left(\tfrac{v}{u}\right) u_x &= 1 \\ \left(\tfrac{u}{v}\right) v_x + (\ln v)\, u_x &= 0\end{aligned}\right\} \Rightarrow v_x = \frac{\begin{vmatrix} 1 & \frac{v}{u} \\ 0 & \ln v \end{vmatrix}}{\begin{vmatrix} \ln u & \frac{v}{u} \\ \frac{u}{v} & \ln v \end{vmatrix}} = \frac{\ln v}{(\ln u)(\ln v) - 1}$$

71. $f_x(x, y) = \begin{cases} 0 & \text{if } y \ge 0 \\ 0 & \text{if } y < 0 \end{cases} \Rightarrow f_x(x, y) = 0$ for all points (x, y); at $y = 0$, $f_y(x, 0) = \lim\limits_{h\to 0} \frac{f(x, 0+h) - f(x, 0)}{h} = \lim\limits_{h\to 0} \frac{f(x, h) - 0}{h}$

$= \lim\limits_{h\to 0} \frac{f(x, h)}{h} = 0$ because $\lim\limits_{h\to 0^-} \frac{f(x, h)}{h} = \lim\limits_{h\to 0^-} \frac{h^3}{h} = 0$ and $\lim\limits_{h\to 0^+} \frac{f(x, h)}{h} = \lim\limits_{h\to 0^+} \frac{h^2}{h} = 0 \Rightarrow f_y(x, y) = \begin{cases} 3y^2 & \text{if } y \ge 0 \\ -2y & \text{if } y < 0 \end{cases}$;

$f_{yx}(x, y) = f_{xy}(x, y) = 0$ for all points (x, y)

73. $\frac{\partial f}{\partial x} = 2x,\ \frac{\partial f}{\partial y} = 2y,\ \frac{\partial f}{\partial z} = -4z \Rightarrow \frac{\partial^2 f}{\partial x^2} = 2,\ \frac{\partial^2 f}{\partial y^2} = 2,\ \frac{\partial^2 f}{\partial z^2} = -4 \Rightarrow \frac{\partial^2 f}{\partial x^2} + \frac{\partial^2 f}{\partial y^2} + \frac{\partial^2 f}{\partial z^2} = 2 + 2 + (-4) = 0$

75. $\frac{\partial f}{\partial x} = -2e^{-2y} \sin 2x,\ \frac{\partial f}{\partial y} = -2e^{-2y} \cos 2x,\ \frac{\partial^2 f}{\partial x^2} = -4e^{-2y} \cos 2x,\ \frac{\partial^2 f}{\partial y^2} = 4e^{-2y} \cos 2x \Rightarrow \frac{\partial^2 f}{\partial x^2} + \frac{\partial^2 f}{\partial y^2}$

$= -4e^{-2y} \cos 2x + 4e^{-2y} \cos 2x = 0$

77. $\frac{\partial f}{\partial x} = 3,\ \frac{\partial f}{\partial y} = 2,\ \frac{\partial^2 f}{\partial x^2} = 0,\ \frac{\partial^2 f}{\partial y^2} = 0 \Rightarrow \frac{\partial^2 f}{\partial x^2} + \frac{\partial^2 f}{\partial y^2} = 0 + 0 = 0$

79. $\frac{\partial f}{\partial x} = -\frac{1}{2}\left(x^2 + y^2 + z^2\right)^{-3/2}(2x) = -x\left(x^2 + y^2 + z^2\right)^{-3/2},\ \frac{\partial f}{\partial y} = -\frac{1}{2}\left(x^2 + y^2 + z^2\right)^{-3/2}(2y)$

$= -y\left(x^2 + y^2 + z^2\right)^{-3/2},\ \frac{\partial f}{\partial z} = -\frac{1}{2}\left(x^2 + y^2 + z^2\right)^{-3/2}(2z) = -z\left(x^2 + y^2 + z^2\right)^{-3/2}$;

$\frac{\partial^2 f}{\partial x^2} = -\left(x^2 + y^2 + z^2\right)^{-3/2} + 3x^2\left(x^2 + y^2 + z^2\right)^{-5/2},\ \frac{\partial^2 f}{\partial y^2} = -\left(x^2 + y^2 + z^2\right)^{-3/2} + 3y^2\left(x^2 + y^2 + z^2\right)^{-5/2}$,

$\frac{\partial^2 f}{\partial z^2} = -\left(x^2 + y^2 + z^2\right)^{-3/2} + 3z^2\left(x^2 + y^2 + z^2\right)^{-5/2} \Rightarrow \frac{\partial^2 f}{\partial x^2} + \frac{\partial^2 f}{\partial y^2} + \frac{\partial^2 f}{\partial z^2}$

$= \left[-\left(x^2 + y^2 + z^2\right)^{-3/2} + 3x^2\left(x^2 + y^2 + z^2\right)^{-5/2}\right] + \left[-\left(x^2 + y^2 + z^2\right)^{-3/2} + 3y^2\left(x^2 + y^2 + z^2\right)^{-5/2}\right]$

$+ \left[-\left(x^2 + y^2 + z^2\right)^{-3/2} + 3z^2\left(x^2 + y^2 + z^2\right)^{-5/2}\right] = -3\left(x^2 + y^2 + z^2\right)^{-3/2} + \left(3x^2 + 3y^2 + 3z^2\right)\left(x^2 + y^2 + z^2\right)^{-5/2} = 0$

81. $\frac{\partial w}{\partial x} = \cos(x + ct),\ \frac{\partial w}{\partial t} = c\cos(x + ct);\ \frac{\partial^2 w}{\partial x^2} = -\sin(x + ct),\ \frac{\partial^2 w}{\partial t^2} = -c^2 \sin(x + ct) \Rightarrow \frac{\partial^2 w}{\partial t^2} = c^2[-\sin(x + ct)] = c^2 \frac{\partial^2 w}{\partial x^2}$

83. $\frac{\partial w}{\partial x} = \cos(x + ct) - 2\sin(2x + 2ct),\ \frac{\partial w}{\partial t} = c\cos(x + ct) - 2c\sin(2x + 2ct)$;

$\frac{\partial^2 w}{\partial x^2} = -\sin(x + ct) - 4\cos(2x + 2ct),\ \frac{\partial^2 w}{\partial t^2} = -c^2\sin(x + ct) - 4c^2\cos(2x + 2ct)$

$\Rightarrow \frac{\partial^2 w}{\partial t^2} = c^2[-\sin(x + ct) - 4\cos(2x + 2ct)] = c^2 \frac{\partial^2 w}{\partial x^2}$

85. $\frac{\partial w}{\partial x} = 2\sec^2(2x - 2ct),\ \frac{\partial w}{\partial t} = -2c\sec^2(2x - 2ct);\ \frac{\partial^2 w}{\partial x^2} = 8\sec^2(2x - 2ct)\tan(2x - 2ct)$,

$\frac{\partial^2 w}{\partial t^2} = 8c^2\sec^2(2x - 2ct)\tan(2x - 2ct) \Rightarrow ux\frac{\partial^2 w}{\partial t^2} = c^2[8\sec^2(2x - 2ct)\tan(2x - 2ct)] = c^2 \frac{\partial^2 w}{\partial x^2}$

87. $\frac{\partial w}{\partial t} = \frac{\partial f}{\partial u}\frac{\partial u}{\partial t} = \frac{\partial f}{\partial u}(ac) \Rightarrow \frac{\partial^2 w}{\partial t^2} = (ac)\left(\frac{\partial^2 f}{\partial u^2}\right)(ac) = a^2c^2\frac{\partial^2 f}{\partial u^2};\ \frac{\partial w}{\partial x} = \frac{\partial f}{\partial u}\frac{\partial u}{\partial x} = \frac{\partial f}{\partial u} \cdot a \Rightarrow \frac{\partial^2 w}{\partial x^2} = \left(a\frac{\partial^2 f}{\partial u^2}\right) \cdot a$

$= a^2\frac{\partial^2 f}{\partial u^2} \Rightarrow \frac{\partial^2 w}{\partial t^2} = a^2c^2\frac{\partial^2 f}{\partial u^2} = c^2\left(a^2\frac{\partial^2 f}{\partial u^2}\right) = c^2\frac{\partial^2 w}{\partial x^2}$

89. Yes, since f_{xx}, f_{yy}, f_{xy}, and f_{yx} are all continuous on R, use the same reasoning as in Exercise 76 with

$f_x(x, y) = f_x(x_0, y_0) + f_{xx}(x_0, y_0)\,\Delta x + f_{xy}(x_0, y_0)\,\Delta y + \epsilon_1\Delta x + \epsilon_2\Delta y$ and

$f_y(x, y) = f_y(x_0, y_0) + f_{yx}(x_0, y_0)\,\Delta x + f_{yy}(x_0, y_0)\,\Delta y + \hat{\epsilon}_1\Delta x + \hat{\epsilon}_2\Delta y$. Then $\lim\limits_{(x, y)\to(x_0, y_0)} f_x(x, y) = f_x(x_0, y_0)$

and $\lim\limits_{(x, y)\to(x_0, y_0)} f_y(x, y) = f_y(x_0, y_0)$.

91. $f_x(0,0) = \lim\limits_{h \to 0} \frac{f(0+h,0)-f(0,0)}{h} = \lim\limits_{h \to 0} \frac{\frac{h \cdot 0^2}{h^2+0^4} - 0}{h} = \lim\limits_{h \to 0} \frac{0}{h} = 0$; $f_y(0,0) = \lim\limits_{h \to 0} \frac{f(0,0+h)-f(0,0)}{h} = \lim\limits_{h \to 0} \frac{\frac{0 \cdot h^2}{0^2+h^4} - 0}{h} = \lim\limits_{h \to 0} \frac{0}{h} = 0$;

$\lim\limits_{\substack{(x,y) \to (0,0) \\ \text{along } x = ky^2}} f(x,y) = \lim\limits_{y \to 0} \frac{(ky^2)y^2}{(ky^2)^2 + y^4} = \lim\limits_{y \to 0} \frac{ky^4}{k^2y^4 + y^4} = \lim\limits_{y \to 0} \frac{k}{k^2+1} = \frac{k}{k^2+1} \Rightarrow$ different limits for different values of $k \Rightarrow \lim\limits_{(x,y) \to (0,0)} f(x,y)$ does not exist $\Rightarrow f(x,y)$ is not continuous at $(0,0) \Rightarrow$ by Theorem 4, $f(x,y)$ is not differentiable at $(0,0)$.

14.4 THE CHAIN RULE

1. (a) $\frac{\partial w}{\partial x} = 2x$, $\frac{\partial w}{\partial y} = 2y$, $\frac{dx}{dt} = -\sin t$, $\frac{dy}{dt} = \cos t \Rightarrow \frac{dw}{dt} = -2x \sin t + 2y \cos t = -2 \cos t \sin t + 2 \sin t \cos t$ $= 0$; $w = x^2 + y^2 = \cos^2 t + \sin^2 t = 1 \Rightarrow \frac{dw}{dt} = 0$

(b) $\frac{dw}{dt}(\pi) = 0$

3. (a) $\frac{\partial w}{\partial x} = \frac{1}{z}$, $\frac{\partial w}{\partial y} = \frac{1}{z}$, $\frac{\partial w}{\partial z} = \frac{-(x+y)}{z^2}$, $\frac{dx}{dt} = -2 \cos t \sin t$, $\frac{dy}{dt} = 2 \sin t \cos t$, $\frac{dz}{dt} = -\frac{1}{t^2}$

$\Rightarrow \frac{dw}{dt} = -\frac{2}{z} \cos t \sin t + \frac{2}{z} \sin t \cos t + \frac{x+y}{z^2 t^2} = \frac{\cos^2 t + \sin^2 t}{\left(\frac{1}{t^2}\right)(t^2)} = 1$; $w = \frac{x}{z} + \frac{y}{z} = \frac{\cos^2 t}{\left(\frac{1}{t}\right)} + \frac{\sin^2 t}{\left(\frac{1}{t}\right)} = t \Rightarrow \frac{dw}{dt} = 1$

(b) $\frac{dw}{dt}(3) = 1$

5. (a) $\frac{\partial w}{\partial x} = 2ye^x$, $\frac{\partial w}{\partial y} = 2e^x$, $\frac{\partial w}{\partial z} = -\frac{1}{z}$, $\frac{dx}{dt} = \frac{2t}{t^2+1}$, $\frac{dy}{dt} = \frac{1}{t^2+1}$, $\frac{dz}{dt} = e^t \Rightarrow \frac{dw}{dt} = \frac{4yte^x}{t^2+1} + \frac{2e^x}{t^2+1} - \frac{e^t}{z}$

$= \frac{(4t)(\tan^{-1} t)(t^2+1)}{t^2+1} + \frac{2(t^2+1)}{t^2+1} - \frac{e^t}{e^t} = 4t \tan^{-1} t + 1$; $w = 2ye^x - \ln z = (2 \tan^{-1} t)(t^2+1) - t$

$\Rightarrow \frac{dw}{dt} = \left(\frac{2}{t^2+1}\right)(t^2+1) + (2 \tan^{-1} t)(2t) - 1 = 4t \tan^{-1} t + 1$

(b) $\frac{dw}{dt}(1) = (4)(1)\left(\frac{\pi}{4}\right) + 1 = \pi + 1$

7. (a) $\frac{\partial z}{\partial u} = \frac{\partial z}{\partial x}\frac{\partial x}{\partial u} + \frac{\partial z}{\partial y}\frac{\partial y}{\partial u} = (4e^x \ln y)\left(\frac{\cos v}{u \cos v}\right) + \left(\frac{4e^x}{y}\right)(\sin v) = \frac{4e^x \ln y}{u} + \frac{4e^x \sin v}{y}$

$= \frac{4(u \cos v) \ln (u \sin v)}{u} + \frac{4(u \cos v)(\sin v)}{u \sin v} = (4 \cos v) \ln (u \sin v) + 4 \cos v$;

$\frac{\partial z}{\partial v} = \frac{\partial z}{\partial x}\frac{\partial x}{\partial v} + \frac{\partial z}{\partial y}\frac{\partial y}{\partial v} = (4e^x \ln y)\left(\frac{-u \sin v}{u \cos v}\right) + \left(\frac{4e^x}{y}\right)(u \cos v) = -(4e^x \ln y)(\tan v) + \frac{4e^x u \cos v}{y}$

$= [-4(u \cos v) \ln (u \sin v)](\tan v) + \frac{4(u \cos v)(u \cos v)}{u \sin v} = (-4u \sin v) \ln (u \sin v) + \frac{4u \cos^2 v}{\sin v}$;

$z = 4e^x \ln y = 4(u \cos v) \ln (u \sin v) \Rightarrow \frac{\partial z}{\partial u} = (4 \cos v) \ln (u \sin v) + 4(u \cos v)\left(\frac{\sin v}{u \sin v}\right)$

$= (4 \cos v) \ln (u \sin v) + 4 \cos v$; also $\frac{\partial z}{\partial v} = (-4u \sin v) \ln (u \sin v) + 4(u \cos v)\left(\frac{u \cos v}{u \sin v}\right)$

$= (-4u \sin v) \ln (u \sin v) + \frac{4u \cos^2 v}{\sin v}$

(b) At $\left(2, \frac{\pi}{4}\right)$: $\frac{\partial z}{\partial u} = 4 \cos \frac{\pi}{4} \ln \left(2 \sin \frac{\pi}{4}\right) + 4 \cos \frac{\pi}{4} = 2\sqrt{2} \ln \sqrt{2} + 2\sqrt{2} = \sqrt{2}(\ln 2 + 2)$;

$\frac{\partial z}{\partial v} = (-4)(2) \sin \frac{\pi}{4} \ln \left(2 \sin \frac{\pi}{4}\right) + \frac{(4)(2)\left(\cos^2 \frac{\pi}{4}\right)}{\left(\sin \frac{\pi}{4}\right)} = -4\sqrt{2} \ln \sqrt{2} + 4\sqrt{2} = -2\sqrt{2} \ln 2 + 4\sqrt{2}$

9. (a) $\frac{\partial w}{\partial u} = \frac{\partial w}{\partial x}\frac{\partial x}{\partial u} + \frac{\partial w}{\partial y}\frac{\partial y}{\partial u} + \frac{\partial w}{\partial z}\frac{\partial z}{\partial u} = (y+z)(1) + (x+z)(1) + (y+x)(v) = x + y + 2z + v(y+x)$

$= (u+v) + (u-v) + 2uv + v(2u) = 2u + 4uv$; $\frac{\partial w}{\partial v} = \frac{\partial w}{\partial x}\frac{\partial x}{\partial v} + \frac{\partial w}{\partial y}\frac{\partial y}{\partial v} + \frac{\partial w}{\partial z}\frac{\partial z}{\partial v}$

$= (y+z)(1) + (x+z)(-1) + (y+x)(u) = y - x + (y+x)u = -2v + (2u)u = -2v + 2u^2$;

$w = xy + yz + xz = (u^2 - v^2) + (u^2 v - uv^2) + (u^2 v + uv^2) = u^2 - v^2 + 2u^2 v \Rightarrow \frac{\partial w}{\partial u} = 2u + 4uv$ and $\frac{\partial w}{\partial v} = -2v + 2u^2$

(b) At $\left(\frac{1}{2}, 1\right)$: $\frac{\partial w}{\partial u} = 2\left(\frac{1}{2}\right) + 4\left(\frac{1}{2}\right)(1) = 3$ and $\frac{\partial w}{\partial v} = -2(1) + 2\left(\frac{1}{2}\right)^2 = -\frac{3}{2}$

11. (a) $\frac{\partial u}{\partial x} = \frac{\partial u}{\partial p}\frac{\partial p}{\partial x} + \frac{\partial u}{\partial q}\frac{\partial q}{\partial x} + \frac{\partial u}{\partial r}\frac{\partial r}{\partial x} = \frac{1}{q-r} + \frac{r-p}{(q-r)^2} + \frac{p-q}{(q-r)^2} = \frac{q-r+r-p+p-q}{(q-r)^2} = 0;$

$\frac{\partial u}{\partial y} = \frac{\partial u}{\partial p}\frac{\partial p}{\partial y} + \frac{\partial u}{\partial q}\frac{\partial q}{\partial y} + \frac{\partial u}{\partial r}\frac{\partial r}{\partial y} = \frac{1}{q-r} - \frac{r-p}{(q-r)^2} + \frac{p-q}{(q-r)^2} = \frac{q-r-r+p+p-q}{(q-r)^2} = \frac{2p-2r}{(q-r)^2}$

$= \frac{(2x+2y+2z)-(2x+2y-2z)}{(2z-2y)^2} = \frac{z}{(z-y)^2}$; $\frac{\partial u}{\partial z} = \frac{\partial u}{\partial p}\frac{\partial p}{\partial z} + \frac{\partial u}{\partial q}\frac{\partial q}{\partial z} + \frac{\partial u}{\partial r}\frac{\partial r}{\partial z}$

$= \frac{1}{q-r} + \frac{r-p}{(q-r)^2} - \frac{p-q}{(q-r)^2} = \frac{q-r+r-p-p+q}{(q-r)^2} = \frac{2q-2p}{(q-r)^2} = \frac{-4y}{(2z-2y)^2} = -\frac{y}{(z-y)^2}$;

$u = \frac{p-q}{q-r} = \frac{2y}{2z-2y} = \frac{y}{z-y} \Rightarrow \frac{\partial u}{\partial x} = 0$, $\frac{\partial u}{\partial y} = \frac{(z-y)-y(-1)}{(z-y)^2} = \frac{z}{(z-y)^2}$, and $\frac{\partial u}{\partial z} = \frac{(z-y)(0)-y(1)}{(z-y)^2}$

$= -\frac{y}{(z-y)^2}$

(b) At $\left(\sqrt{3}, 2, 1\right)$: $\frac{\partial u}{\partial x} = 0$, $\frac{\partial u}{\partial y} = \frac{1}{(1-2)^2} = 1$, and $\frac{\partial u}{\partial z} = \frac{-2}{(1-2)^2} = -2$

13. $\frac{dz}{dt} = \frac{\partial z}{\partial x}\frac{dx}{dt} + \frac{\partial z}{\partial y}\frac{dy}{dt}$

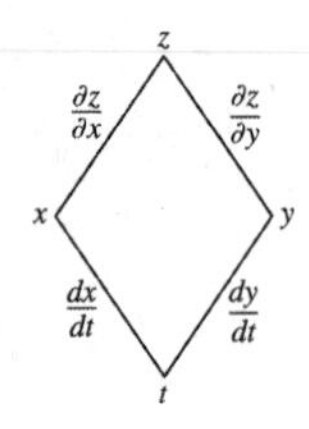

15. $\frac{\partial w}{\partial u} = \frac{\partial w}{\partial x}\frac{\partial x}{\partial u} + \frac{\partial w}{\partial y}\frac{\partial y}{\partial u} + \frac{\partial w}{\partial z}\frac{\partial z}{\partial u}$ $\qquad$ $\frac{\partial w}{\partial v} = \frac{\partial w}{\partial x}\frac{\partial x}{\partial v} + \frac{\partial w}{\partial y}\frac{\partial y}{\partial v} + \frac{\partial w}{\partial z}\frac{\partial z}{\partial v}$

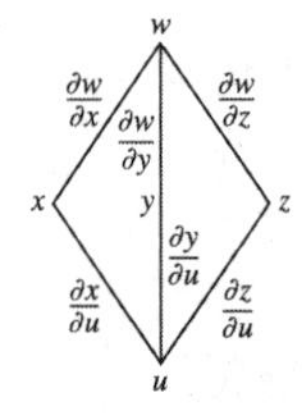

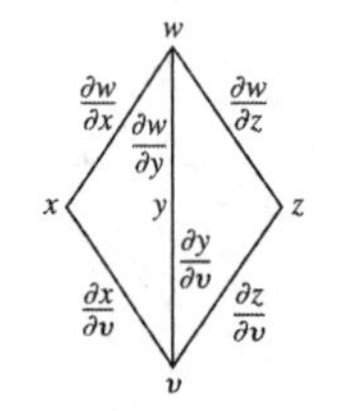

17. $\frac{\partial w}{\partial u} = \frac{\partial w}{\partial x}\frac{\partial x}{\partial u} + \frac{\partial w}{\partial y}\frac{\partial y}{\partial u}$ $\qquad$ $\frac{\partial w}{\partial v} = \frac{\partial w}{\partial x}\frac{\partial x}{\partial v} + \frac{\partial w}{\partial y}\frac{\partial y}{\partial v}$

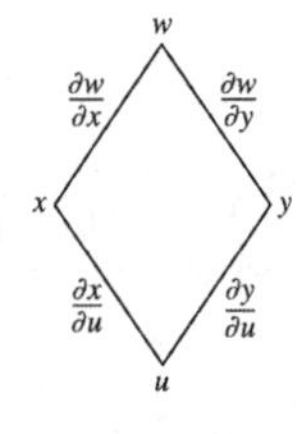

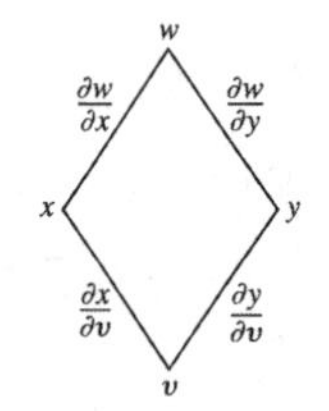

19. $\frac{\partial z}{\partial t} = \frac{\partial z}{\partial x}\frac{\partial x}{\partial t} + \frac{\partial z}{\partial y}\frac{\partial y}{\partial t}$ $\qquad$ $\frac{\partial z}{\partial s} = \frac{\partial z}{\partial x}\frac{\partial x}{\partial s} + \frac{\partial z}{\partial y}\frac{\partial y}{\partial s}$

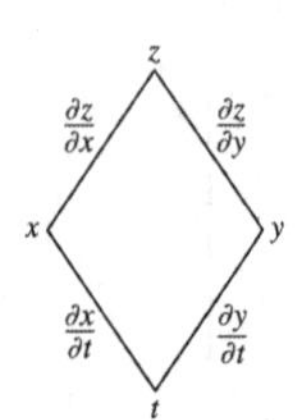

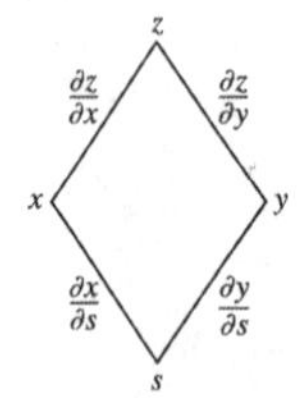

21. $\frac{\partial w}{\partial s} = \frac{dw}{du}\frac{\partial u}{\partial s}$ $\quad \frac{\partial w}{\partial t} = \frac{dw}{du}\frac{\partial u}{\partial t}$

23. $\frac{\partial w}{\partial r} = \frac{\partial w}{\partial x}\frac{dx}{dr} + \frac{\partial w}{\partial y}\frac{dy}{dr} = \frac{\partial w}{\partial x}\frac{dx}{dr}$ since $\frac{dy}{dr} = 0$ $\quad \frac{\partial w}{\partial s} = \frac{\partial w}{\partial x}\frac{dx}{ds} + \frac{\partial w}{\partial y}\frac{dy}{ds} = \frac{\partial w}{\partial y}\frac{dy}{ds}$ since $\frac{dx}{ds} = 0$

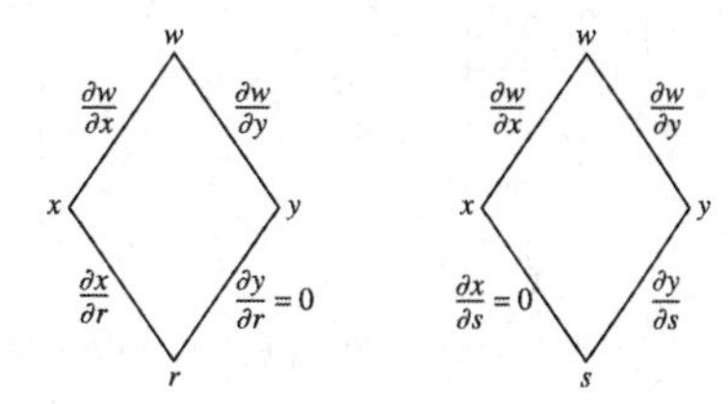

25. Let $F(x, y) = x^3 - 2y^2 + xy = 0 \Rightarrow F_x(x, y) = 3x^2 + y$ and $F_y(x, y) = -4y + x \Rightarrow \frac{dy}{dx} = -\frac{F_x}{F_y} = -\frac{3x^2+y}{(-4y+x)}$
$\Rightarrow \frac{dy}{dx}(1, 1) = \frac{4}{3}$

27. Let $F(x, y) = x^2 + xy + y^2 - 7 = 0 \Rightarrow F_x(x, y) = 2x + y$ and $F_y(x, y) = x + 2y \Rightarrow \frac{dy}{dx} = -\frac{F_x}{F_y} = -\frac{2x+y}{x+2y}$
$\Rightarrow \frac{dy}{dx}(1, 2) = -\frac{4}{5}$

29. Let $F(x, y, z) = z^3 - xy + yz + y^3 - 2 = 0 \Rightarrow F_x(x, y, z) = -y$, $F_y(x, y, z) = -x + z + 3y^2$, $F_z(x, y, z) = 3z^2 + y$
$\Rightarrow \frac{\partial z}{\partial x} = -\frac{F_x}{F_z} = -\frac{-y}{3z^2+y} = \frac{y}{3z^2+y} \Rightarrow \frac{\partial z}{\partial x}(1, 1, 1) = \frac{1}{4}$; $\frac{\partial z}{\partial y} = -\frac{F_y}{F_z} = -\frac{-x+z+3y^2}{3z^2+y} = \frac{x-z-3y^2}{3z^2+y}$
$\Rightarrow \frac{\partial z}{\partial y}(1, 1, 1) = -\frac{3}{4}$

31. Let $F(x, y, z) = \sin(x + y) + \sin(y + z) + \sin(x + z) = 0 \Rightarrow F_x(x, y, z) = \cos(x + y) + \cos(x + z)$,
$F_y(x, y, z) = \cos(x + y) + \cos(y + z)$, $F_z(x, y, z) = \cos(y + z) + \cos(x + z) \Rightarrow \frac{\partial z}{\partial x} = -\frac{F_x}{F_z}$
$= -\frac{\cos(x+y)+\cos(x+z)}{\cos(y+z)+\cos(x+z)} \Rightarrow \frac{\partial z}{\partial x}(\pi, \pi, \pi) = -1$; $\frac{\partial z}{\partial y} = -\frac{F_y}{F_z} = -\frac{\cos(x+y)+\cos(y+z)}{\cos(y+z)+\cos(x+z)} \Rightarrow \frac{\partial z}{\partial y}(\pi, \pi, \pi) = -1$

33. $\frac{\partial w}{\partial r} = \frac{\partial w}{\partial x}\frac{\partial x}{\partial r} + \frac{\partial w}{\partial y}\frac{\partial y}{\partial r} + \frac{\partial w}{\partial z}\frac{\partial z}{\partial r} = 2(x + y + z)(1) + 2(x + y + z)[-\sin(r + s)] + 2(x + y + z)[\cos(r + s)]$
$= 2(x + y + z)[1 - \sin(r + s) + \cos(r + s)] = 2[r - s + \cos(r + s) + \sin(r + s)][1 - \sin(r + s) + \cos(r + s)]$
$\Rightarrow \left.\frac{\partial w}{\partial r}\right|_{r=1,s=-1} = 2(3)(2) = 12$

35. $\frac{\partial w}{\partial v} = \frac{\partial w}{\partial x}\frac{\partial x}{\partial v} + \frac{\partial w}{\partial y}\frac{\partial y}{\partial v} = \left(2x - \frac{y}{x^2}\right)(-2) + \left(\frac{1}{x}\right)(1) = \left[2(u - 2v + 1) - \frac{2u+v-2}{(u-2v+1)^2}\right](-2) + \frac{1}{u-2v+1}$
$\Rightarrow \left.\frac{\partial w}{\partial v}\right|_{u=0,v=0} = -7$

37. $\frac{\partial z}{\partial u} = \frac{dz}{dx}\frac{\partial x}{\partial u} = \left(\frac{5}{1+x^2}\right)e^u = \left[\frac{5}{1+(e^u+\ln v)^2}\right]e^u \Rightarrow \left.\frac{\partial z}{\partial u}\right|_{u=\ln 2,v=1} = \left[\frac{5}{1+(2)^2}\right](2) = 2$;
$\frac{\partial z}{\partial v} = \frac{dz}{dx}\frac{\partial x}{\partial v} = \left(\frac{5}{1+x^2}\right)\left(\frac{1}{v}\right) = \left[\frac{5}{1+(e^u+\ln v)^2}\right]\left(\frac{1}{v}\right) \Rightarrow \left.\frac{\partial z}{\partial v}\right|_{u=\ln 2,v=1} = \left[\frac{5}{1+(2)^2}\right](1) = 1$

39. Let $x = s^3 + t^2 \Rightarrow w = f(s^3 + t^2) = f(x) \Rightarrow \frac{\partial w}{\partial s} = \frac{dw}{dx}\frac{\partial x}{\partial s} = f'(x) \cdot 3s^2 = 3s^2 e^{s^3+t^2}$, $\frac{\partial w}{\partial t} = \frac{dw}{dx}\frac{\partial x}{\partial t} = f'(x) \cdot 2t = 2t\,e^{s^3+t^2}$

41. $V = IR \Rightarrow \frac{\partial V}{\partial I} = R$ and $\frac{\partial V}{\partial R} = I$; $\frac{dV}{dt} = \frac{\partial V}{\partial I}\frac{dI}{dt} + \frac{\partial V}{\partial R}\frac{dR}{dt} = R\frac{dI}{dt} + I\frac{dR}{dt} \Rightarrow -0.01$ volts/sec
$= (600 \text{ ohms})\frac{dI}{dt} + (0.04 \text{ amps})(0.5 \text{ ohms/sec}) \Rightarrow \frac{dI}{dt} = -0.00005$ amps/sec

43. $\frac{\partial f}{\partial x} = \frac{\partial f}{\partial u}\frac{\partial u}{\partial x} + \frac{\partial f}{\partial v}\frac{\partial v}{\partial x} + \frac{\partial f}{\partial w}\frac{\partial w}{\partial x} = \frac{\partial f}{\partial u}(1) + \frac{\partial f}{\partial v}(0) + \frac{\partial f}{\partial w}(-1) = \frac{\partial f}{\partial u} - \frac{\partial f}{\partial w}$,
$\frac{\partial f}{\partial y} = \frac{\partial f}{\partial u}\frac{\partial u}{\partial y} + \frac{\partial f}{\partial v}\frac{\partial v}{\partial y} + \frac{\partial f}{\partial w}\frac{\partial w}{\partial y} = \frac{\partial f}{\partial u}(-1) + \frac{\partial f}{\partial v}(1) + \frac{\partial f}{\partial w}(0) = -\frac{\partial f}{\partial u} + \frac{\partial f}{\partial v}$, and
$\frac{\partial f}{\partial z} = \frac{\partial f}{\partial u}\frac{\partial u}{\partial z} + \frac{\partial f}{\partial v}\frac{\partial v}{\partial z} + \frac{\partial f}{\partial w}\frac{\partial w}{\partial z} = \frac{\partial f}{\partial u}(0) + \frac{\partial f}{\partial v}(-1) + \frac{\partial f}{\partial w}(1) = -\frac{\partial f}{\partial v} + \frac{\partial f}{\partial w} \Rightarrow \frac{\partial f}{\partial x} + \frac{\partial f}{\partial y} + \frac{\partial f}{\partial z} = 0$

45. $w_x = \frac{\partial w}{\partial x} = \frac{\partial w}{\partial u}\frac{\partial u}{\partial x} + \frac{\partial w}{\partial v}\frac{\partial v}{\partial x} = x\frac{\partial w}{\partial u} + y\frac{\partial w}{\partial v} \Rightarrow w_{xx} = \frac{\partial w}{\partial u} + x\frac{\partial}{\partial x}\left(\frac{\partial w}{\partial u}\right) + y\frac{\partial}{\partial x}\left(\frac{\partial w}{\partial v}\right)$
$= \frac{\partial w}{\partial u} + x\left(\frac{\partial^2 w}{\partial u^2}\frac{\partial u}{\partial x} + \frac{\partial^2 w}{\partial v \partial u}\frac{\partial v}{\partial x}\right) + y\left(\frac{\partial^2 w}{\partial u \partial v}\frac{\partial u}{\partial x} + \frac{\partial^2 w}{\partial v^2}\frac{\partial v}{\partial x}\right) = \frac{\partial w}{\partial u} + x\left(x\frac{\partial^2 w}{\partial u^2} + y\frac{\partial^2 w}{\partial v \partial u}\right) + y\left(x\frac{\partial^2 w}{\partial u \partial v} + y\frac{\partial^2 w}{\partial v^2}\right)$
$= \frac{\partial w}{\partial u} + x^2\frac{\partial^2 w}{\partial u^2} + 2xy\frac{\partial^2 w}{\partial v \partial u} + y^2\frac{\partial^2 w}{\partial v^2}$; $w_y = \frac{\partial w}{\partial y} = \frac{\partial w}{\partial u}\frac{\partial u}{\partial y} + \frac{\partial w}{\partial v}\frac{\partial v}{\partial y} = -y\frac{\partial w}{\partial u} + x\frac{\partial w}{\partial v}$
$\Rightarrow w_{yy} = -\frac{\partial w}{\partial u} - y\left(\frac{\partial^2 w}{\partial u^2}\frac{\partial u}{\partial y} + \frac{\partial^2 w}{\partial v \partial u}\frac{\partial v}{\partial y}\right) + x\left(\frac{\partial^2 w}{\partial u \partial v}\frac{\partial u}{\partial y} + \frac{\partial^2 w}{\partial v^2}\frac{\partial v}{\partial y}\right)$
$= -\frac{\partial w}{\partial u} - y\left(-y\frac{\partial^2 w}{\partial u^2} + x\frac{\partial^2 w}{\partial v \partial u}\right) + x\left(-y\frac{\partial^2 w}{\partial u \partial v} + x\frac{\partial^2 w}{\partial v^2}\right) = -\frac{\partial w}{\partial u} + y^2\frac{\partial^2 w}{\partial u^2} - 2xy\frac{\partial^2 w}{\partial v \partial u} + x^2\frac{\partial^2 w}{\partial v^2}$; thus
$w_{xx} + w_{yy} = (x^2 + y^2)\frac{\partial^2 w}{\partial u^2} + (x^2 + y^2)\frac{\partial^2 w}{\partial v^2} = (x^2 + y^2)(w_{uu} + w_{vv}) = 0$, since $w_{uu} + w_{vv} = 0$

47. $f_x(x, y, z) = \cos t$, $f_y(x, y, z) = \sin t$, and $f_z(x, y, z) = t^2 + t - 2 \Rightarrow \frac{df}{dt} = \frac{\partial f}{\partial x}\frac{dx}{dt} + \frac{\partial f}{\partial y}\frac{dy}{dt} + \frac{\partial f}{\partial z}\frac{dz}{dt}$
$= (\cos t)(-\sin t) + (\sin t)(\cos t) + (t^2 + t - 2)(1) = t^2 + t - 2$; $\frac{df}{dt} = 0 \Rightarrow t^2 + t - 2 = 0 \Rightarrow t = -2$
or $t = 1$; $t = -2 \Rightarrow x = \cos(-2)$, $y = \sin(-2)$, $z = -2$ for the point $(\cos(-2), \sin(-2), -2)$; $t = 1 \Rightarrow x = \cos 1$,
$y = \sin 1$, $z = 1$ for the point $(\cos 1, \sin 1, 1)$

49. (a) $\frac{\partial T}{\partial x} = 8x - 4y$ and $\frac{\partial T}{\partial y} = 8y - 4x \Rightarrow \frac{dT}{dt} = \frac{\partial T}{\partial x}\frac{dx}{dt} + \frac{\partial T}{\partial y}\frac{dy}{dt} = (8x - 4y)(-\sin t) + (8y - 4x)(\cos t)$
$= (8\cos t - 4\sin t)(-\sin t) + (8\sin t - 4\cos t)(\cos t) = 4\sin^2 t - 4\cos^2 t \Rightarrow \frac{d^2T}{dt^2} = 16\sin t\cos t$;
$\frac{dT}{dt} = 0 \Rightarrow 4\sin^2 t - 4\cos^2 t = 0 \Rightarrow \sin^2 t = \cos^2 t \Rightarrow \sin t = \cos t$ or $\sin t = -\cos t \Rightarrow t = \frac{\pi}{4}, \frac{5\pi}{4}, \frac{3\pi}{4}, \frac{7\pi}{4}$ on the interval $0 \le t \le 2\pi$;
$\left.\frac{d^2T}{dt^2}\right|_{t=\frac{\pi}{4}} = 16\sin\frac{\pi}{4}\cos\frac{\pi}{4} > 0 \Rightarrow T$ has a minimum at $(x, y) = \left(\frac{\sqrt{2}}{2}, \frac{\sqrt{2}}{2}\right)$;
$\left.\frac{d^2T}{dt^2}\right|_{t=\frac{3\pi}{4}} = 16\sin\frac{3\pi}{4}\cos\frac{3\pi}{4} < 0 \Rightarrow T$ has a maximum at $(x, y) = \left(-\frac{\sqrt{2}}{2}, \frac{\sqrt{2}}{2}\right)$;
$\left.\frac{d^2T}{dt^2}\right|_{t=\frac{5\pi}{4}} = 16\sin\frac{5\pi}{4}\cos\frac{5\pi}{4} > 0 \Rightarrow T$ has a minimum at $(x, y) = \left(-\frac{\sqrt{2}}{2}, -\frac{\sqrt{2}}{2}\right)$;
$\left.\frac{d^2T}{dt^2}\right|_{t=\frac{7\pi}{4}} = 16\sin\frac{7\pi}{4}\cos\frac{7\pi}{4} < 0 \Rightarrow T$ has a maximum at $(x, y) = \left(\frac{\sqrt{2}}{2}, -\frac{\sqrt{2}}{2}\right)$

(b) $T = 4x^2 - 4xy + 4y^2 \Rightarrow \frac{\partial T}{\partial x} = 8x - 4y$, and $\frac{\partial T}{\partial y} = 8y - 4x$ so the extreme values occur at the four points found in part (a): $T\left(-\frac{\sqrt{2}}{2}, \frac{\sqrt{2}}{2}\right) = T\left(\frac{\sqrt{2}}{2}, -\frac{\sqrt{2}}{2}\right) = 4\left(\frac{1}{2}\right) - 4\left(-\frac{1}{2}\right) + 4\left(\frac{1}{2}\right) = 6$, the maximum and
$T\left(\frac{\sqrt{2}}{2}, \frac{\sqrt{2}}{2}\right) = T\left(-\frac{\sqrt{2}}{2}, -\frac{\sqrt{2}}{2}\right) = 4\left(\frac{1}{2}\right) - 4\left(\frac{1}{2}\right) + 4\left(\frac{1}{2}\right) = 2$, the minimum

51. $G(u, x) = \int_a^u g(t, x)\,dt$ where $u = f(x) \Rightarrow \frac{dG}{dx} = \frac{\partial G}{\partial u}\frac{du}{dx} + \frac{\partial G}{\partial x}\frac{dx}{dx} = g(u, x)f'(x) + \int_a^u g_x(t, x)\,dt$; thus
$F(x) = \int_0^{x^2}\sqrt{t^4 + x^3}\,dt \Rightarrow F'(x) = \sqrt{(x^2)^4 + x^3}\,(2x) + \int_0^{x^2}\frac{\partial}{\partial x}\sqrt{t^4 + x^3}\,dt = 2x\sqrt{x^8 + x^3} + \int_0^{x^2}\frac{3x^2}{2\sqrt{t^4 + x^3}}\,dt$

14.5 DIRECTIONAL DERIVATIVES AND GRADIENT VECTORS

1. $\frac{\partial f}{\partial x} = -1$, $\frac{\partial f}{\partial y} = 1 \Rightarrow \nabla f = -\mathbf{i} + \mathbf{j}$; $f(2,1) = -1$
   $\Rightarrow -1 = y - x$ is the level curve

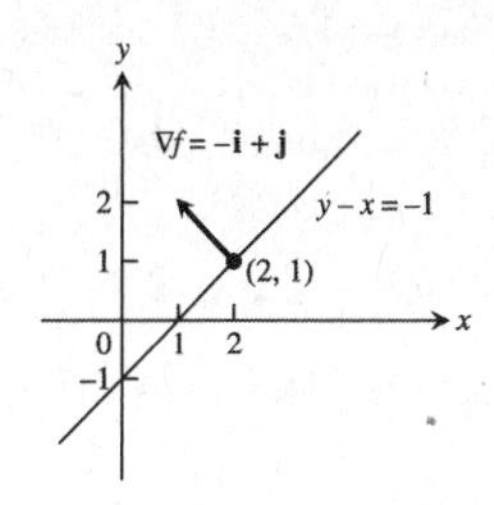

3. $\frac{\partial g}{\partial x} = y^2 \Rightarrow \frac{\partial g}{\partial x}(2,-1) = 1$; $\frac{\partial g}{\partial y} = 2xy \Rightarrow \frac{\partial g}{\partial x}(2,-1) = -4$;
   $\Rightarrow \nabla g = \mathbf{i} - 4\mathbf{j}$; $g(2,-1) = 2 \Rightarrow x = \frac{2}{y^2}$ is the level curve

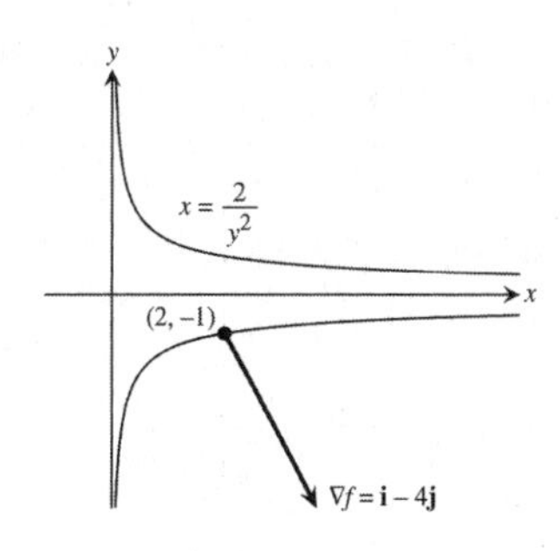

5. $\frac{\partial f}{\partial x} = \frac{1}{\sqrt{2x+3y}} \Rightarrow \frac{\partial f}{\partial x}(-1,2) = \frac{1}{2}$; $\frac{\partial f}{\partial y} = \frac{3}{2\sqrt{2x+3y}}$
   $\Rightarrow \frac{\partial f}{\partial x}(-1,2) = \frac{3}{4}$; $\Rightarrow \nabla f = \frac{1}{2}\mathbf{i} + \frac{3}{4}\mathbf{j}$; $f(-1,2) = 2$
   $\Rightarrow 4 = 2x + 3y$ is the level curve

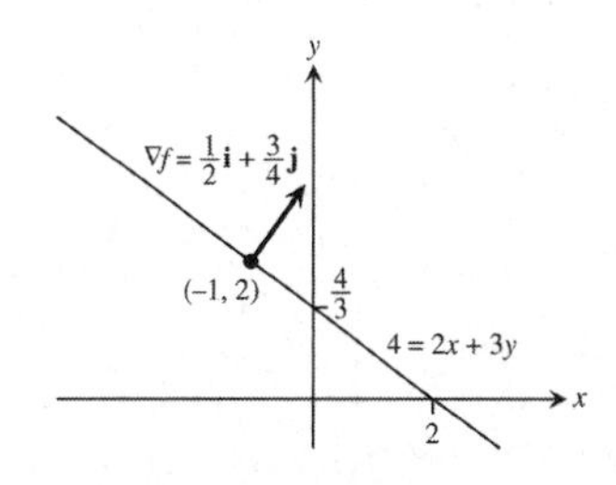

7. $\frac{\partial f}{\partial x} = 2x + \frac{z}{x} \Rightarrow \frac{\partial f}{\partial x}(1,1,1) = 3$; $\frac{\partial f}{\partial y} = 2y \Rightarrow \frac{\partial f}{\partial y}(1,1,1) = 2$; $\frac{\partial f}{\partial z} = -4z + \ln x \Rightarrow \frac{\partial f}{\partial z}(1,1,1) = -4$;
   thus $\nabla f = 3\mathbf{i} + 2\mathbf{j} - 4\mathbf{k}$

9. $\frac{\partial f}{\partial x} = -\frac{x}{(x^2+y^2+z^2)^{3/2}} + \frac{1}{x} \Rightarrow \frac{\partial f}{\partial x}(-1,2,-2) = -\frac{26}{27}$; $\frac{\partial f}{\partial y} = -\frac{y}{(x^2+y^2+z^2)^{3/2}} + \frac{1}{y} \Rightarrow \frac{\partial f}{\partial y}(-1,2,-2) = \frac{23}{54}$;
   $\frac{\partial f}{\partial z} = -\frac{z}{(x^2+y^2+z^2)^{3/2}} + \frac{1}{z} \Rightarrow \frac{\partial f}{\partial z}(-1,2,-2) = -\frac{23}{54}$; thus $\nabla f = -\frac{26}{27}\mathbf{i} + \frac{23}{54}\mathbf{j} - \frac{23}{54}\mathbf{k}$

11. $\mathbf{u} = \frac{\mathbf{A}}{|\mathbf{A}|} = \frac{4\mathbf{i}+3\mathbf{j}}{\sqrt{4^2+3^2}} = \frac{4}{5}\mathbf{i} + \frac{3}{5}\mathbf{j}$; $f_x(x,y) = 2y \Rightarrow f_x(5,5) = 10$; $f_y(x,y) = 2x - 6y \Rightarrow f_y(5,5) = -20$
   $\Rightarrow \nabla f = 10\mathbf{i} - 20\mathbf{j} \Rightarrow (D_{\mathbf{u}}f)_{P_0} = \nabla f \cdot \mathbf{u} = 10\left(\frac{4}{5}\right) - 20\left(\frac{3}{5}\right) = -4$

13. $\mathbf{u} = \frac{\mathbf{A}}{|\mathbf{A}|} = \frac{12\mathbf{i}+5\mathbf{j}}{\sqrt{12^2+5^2}} = \frac{12}{13}\mathbf{i} + \frac{5}{13}\mathbf{j}$; $g_x(x,y) = \frac{y^2+2}{(xy+2)^2} \Rightarrow g_x(1,-1) = 3$; $g_y(x,y) = -\frac{x^2+2}{(xy+2)^2} \Rightarrow g_y(1,-1) = -3$
   $\Rightarrow \nabla g = 3\mathbf{i} - 3\mathbf{j} \Rightarrow (D_{\mathbf{u}}g)_{P_0} = \nabla g \cdot \mathbf{u} = \frac{36}{13} - \frac{15}{13} = \frac{21}{13}$

15. $\mathbf{u} = \frac{\mathbf{A}}{|\mathbf{A}|} = \frac{3\mathbf{i}+6\mathbf{j}-2\mathbf{k}}{\sqrt{3^2+6^2+(-2)^2}} = \frac{3}{7}\mathbf{i} + \frac{6}{7}\mathbf{j} - \frac{2}{7}\mathbf{k}$; $f_x(x,y,z) = y + z \Rightarrow f_x(1,-1,2) = 1$; $f_y(x,y,z) = x + z$
   $\Rightarrow f_y(1,-1,2) = 3$; $f_z(x,y,z) = y + x \Rightarrow f_z(1,-1,2) = 0 \Rightarrow \nabla f = \mathbf{i} + 3\mathbf{j} \Rightarrow (D_{\mathbf{u}}f)_{P_0} = \nabla f \cdot \mathbf{u} = \frac{3}{7} + \frac{18}{7} = 3$

17. $\mathbf{u} = \frac{\mathbf{A}}{|\mathbf{A}|} = \frac{2\mathbf{i}+\mathbf{j}-2\mathbf{k}}{\sqrt{2^2+1^2+(-2)^2}} = \frac{2}{3}\mathbf{i} + \frac{1}{3}\mathbf{j} - \frac{2}{3}\mathbf{k}$; $g_x(x,y,z) = 3e^x \cos yz \Rightarrow g_x(0,0,0) = 3$; $g_y(x,y,z) = -3ze^x \sin yz$
   $\Rightarrow g_y(0,0,0) = 0$; $g_z(x,y,z) = -3ye^x \sin yz \Rightarrow g_z(0,0,0) = 0 \Rightarrow \nabla g = 3\mathbf{i} \Rightarrow (D_{\mathbf{u}}g)_{P_0} = \nabla g \cdot \mathbf{u} = 2$

19. $\nabla f = (2x+y)\mathbf{i} + (x+2y)\mathbf{j} \Rightarrow \nabla f(-1,1) = -\mathbf{i}+\mathbf{j} \Rightarrow \mathbf{u} = \frac{\nabla f}{|\nabla f|} = \frac{-\mathbf{i}+\mathbf{j}}{\sqrt{(-1)^2+1^2}} = -\frac{1}{\sqrt{2}}\mathbf{i} + \frac{1}{\sqrt{2}}\mathbf{j}$; f increases most rapidly in the direction $\mathbf{u} = -\frac{1}{\sqrt{2}}\mathbf{i} + \frac{1}{\sqrt{2}}\mathbf{j}$ and decreases most rapidly in the direction $-\mathbf{u} = \frac{1}{\sqrt{2}}\mathbf{i} - \frac{1}{\sqrt{2}}\mathbf{j}$; $(D_{\mathbf{u}}f)_{P_0} = \nabla f \cdot \mathbf{u} = |\nabla f| = \sqrt{2}$ and $(D_{-\mathbf{u}}f)_{P_0} = -\sqrt{2}$

21. $\nabla f = \frac{1}{y}\mathbf{i} - \left(\frac{x}{y^2}+z\right)\mathbf{j} - y\mathbf{k} \Rightarrow \nabla f(4,1,1) = \mathbf{i} - 5\mathbf{j} - \mathbf{k} \Rightarrow \mathbf{u} = \frac{\nabla f}{|\nabla f|} = \frac{\mathbf{i}-5\mathbf{j}-\mathbf{k}}{\sqrt{1^2+(-5)^2+(-1)^2}}$ $= \frac{1}{3\sqrt{3}}\mathbf{i} - \frac{5}{3\sqrt{3}}\mathbf{j} - \frac{1}{3\sqrt{3}}\mathbf{k}$; f increases most rapidly in the direction of $\mathbf{u} = \frac{1}{3\sqrt{3}}\mathbf{i} - \frac{5}{3\sqrt{3}}\mathbf{j} - \frac{1}{3\sqrt{3}}\mathbf{k}$ and decreases most rapidly in the direction $-\mathbf{u} = -\frac{1}{3\sqrt{3}}\mathbf{i} + \frac{5}{3\sqrt{3}}\mathbf{j} + \frac{1}{3\sqrt{3}}\mathbf{k}$; $(D_{\mathbf{u}}f)_{P_0} = \nabla f \cdot \mathbf{u} = |\nabla f| = 3\sqrt{3}$ and $(D_{-\mathbf{u}}f)_{P_0} = -3\sqrt{3}$

23. $\nabla f = \left(\frac{1}{x}+\frac{1}{x}\right)\mathbf{i} + \left(\frac{1}{y}+\frac{1}{y}\right)\mathbf{j} + \left(\frac{1}{z}+\frac{1}{z}\right)\mathbf{k} \Rightarrow \nabla f(1,1,1) = 2\mathbf{i}+2\mathbf{j}+2\mathbf{k} \Rightarrow \mathbf{u} = \frac{\nabla f}{|\nabla f|} = \frac{1}{\sqrt{3}}\mathbf{i} + \frac{1}{\sqrt{3}}\mathbf{j} + \frac{1}{\sqrt{3}}\mathbf{k}$; f increases most rapidly in the direction $\mathbf{u} = \frac{1}{\sqrt{3}}\mathbf{i} + \frac{1}{\sqrt{3}}\mathbf{j} + \frac{1}{\sqrt{3}}\mathbf{k}$ and decreases most rapidly in the direction $-\mathbf{u} = -\frac{1}{\sqrt{3}}\mathbf{i} - \frac{1}{\sqrt{3}}\mathbf{j} - \frac{1}{\sqrt{3}}\mathbf{k}$; $(D_{\mathbf{u}}f)_{P_0} = \nabla f \cdot \mathbf{u} = |\nabla f| = 2\sqrt{3}$ and $(D_{-\mathbf{u}}f)_{P_0} = -2\sqrt{3}$

25. $\nabla f = 2x\mathbf{i} + 2y\mathbf{j} \Rightarrow \nabla f\left(\sqrt{2},\sqrt{2}\right) = 2\sqrt{2}\mathbf{i} + 2\sqrt{2}\mathbf{j}$
$\Rightarrow$ Tangent line: $2\sqrt{2}\left(x-\sqrt{2}\right) + 2\sqrt{2}\left(y-\sqrt{2}\right) = 0$
$\Rightarrow \sqrt{2}x + \sqrt{2}y = 4$

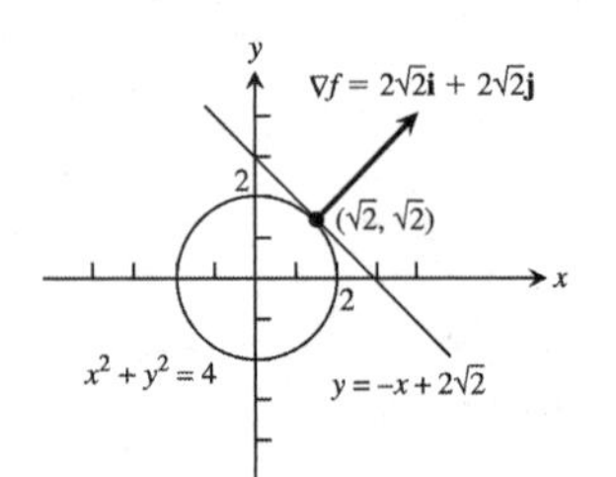

27. $\nabla f = y\mathbf{i} + x\mathbf{j} \Rightarrow \nabla f(2,-2) = -2\mathbf{i} + 2\mathbf{j}$
$\Rightarrow$ Tangent line: $-2(x-2) + 2(y+2) = 0$
$\Rightarrow y = x - 4$

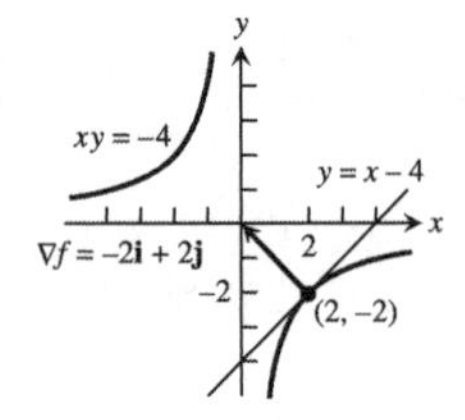

29. $\nabla f = (2x-y)\mathbf{i} + (-x+2y-1)\mathbf{j}$
(a) $\nabla f(1,-1) = 3\mathbf{i} - 4\mathbf{j} \Rightarrow |\nabla f(1,-1)| = 5 \Rightarrow D_{\mathbf{u}}f(1,-1) = 5$ in the direction of $\mathbf{u} = \frac{3}{5}\mathbf{i} - \frac{4}{5}\mathbf{j}$
(b) $-\nabla f(1,-1) = -3\mathbf{i} + 4\mathbf{j} \Rightarrow |\nabla f(1,-1)| = 5 \Rightarrow D_{\mathbf{u}}f(1,-1) = -5$ in the direction of $\mathbf{u} = -\frac{3}{5}\mathbf{i} + \frac{4}{5}\mathbf{j}$
(c) $D_{\mathbf{u}}f(1,-1) = 0$ in the direction of $\mathbf{u} = \frac{4}{5}\mathbf{i} + \frac{3}{5}\mathbf{j}$ or $\mathbf{u} = -\frac{4}{5}\mathbf{i} - \frac{3}{5}\mathbf{j}$
(d) Let $\mathbf{u} = u_1\mathbf{i} + u_2\mathbf{j} \Rightarrow |\mathbf{u}| = \sqrt{u_1^2+u_2^2} = 1 \Rightarrow u_1^2 + u_2^2 = 1$; $D_{\mathbf{u}}f(1,-1) = \nabla f(1,-1)\cdot\mathbf{u} = (3\mathbf{i}-4\mathbf{j})\cdot(u_1\mathbf{i}+u_2\mathbf{j})$ $= 3u_1 - 4u_2 = 4 \Rightarrow u_2 = \frac{3}{4}u_1 - 1 \Rightarrow u_1^2 + \left(\frac{3}{4}u_1 - 1\right)^2 = 1 \Rightarrow \frac{25}{16}u_1^2 - \frac{3}{2}u_1 = 0 \Rightarrow u_1 = 0$ or $u_1 = \frac{24}{25}$; $u_1 = 0 \Rightarrow u_2 = -1 \Rightarrow \mathbf{u} = -\mathbf{j}$, or $u_1 = \frac{24}{25} \Rightarrow u_2 = -\frac{7}{25} \Rightarrow \mathbf{u} = \frac{24}{25}\mathbf{i} - \frac{7}{25}\mathbf{j}$
(e) Let $\mathbf{u} = u_1\mathbf{i} + u_2\mathbf{j} \Rightarrow |\mathbf{u}| = \sqrt{u_1^2+u_2^2} = 1 \Rightarrow u_1^2 + u_2^2 = 1$; $D_{\mathbf{u}}f(1,-1) = \nabla f(1,-1)\cdot\mathbf{u} = (3\mathbf{i}-4\mathbf{j})\cdot(u_1\mathbf{i}+u_2\mathbf{j})$ $= 3u_1 - 4u_2 = -3 \Rightarrow u_1 = \frac{4}{3}u_2 - 1 \Rightarrow \left(\frac{4}{3}u_2 - 1\right)^2 + u_2^2 = 1 \Rightarrow \frac{25}{9}u_2^2 - \frac{8}{3}u_2 = 0 \Rightarrow u_2 = 0$ or $u_2 = \frac{24}{25}$; $u_2 = 0 \Rightarrow u_1 = -1 \Rightarrow \mathbf{u} = -\mathbf{i}$, or $u_2 = \frac{24}{25} \Rightarrow u_2 = \frac{7}{25} \Rightarrow \mathbf{u} = \frac{7}{25}\mathbf{i} + \frac{24}{25}\mathbf{j}$

31. $\nabla f = y\mathbf{i} + (x+2y)\mathbf{j} \Rightarrow \nabla f(3,2) = 2\mathbf{i} + 7\mathbf{j}$; a vector orthogonal to ∇f is $\mathbf{v} = 7\mathbf{i} - 2\mathbf{j} \Rightarrow \mathbf{u} = \frac{\mathbf{v}}{|\mathbf{v}|} = \frac{7\mathbf{i}-2\mathbf{j}}{\sqrt{7^2+(-2)^2}}$ $= \frac{7}{\sqrt{53}}\mathbf{i} - \frac{2}{\sqrt{53}}\mathbf{j}$ and $-\mathbf{u} = -\frac{7}{\sqrt{53}}\mathbf{i} + \frac{2}{\sqrt{53}}\mathbf{j}$ are the directions where the derivative is zero

33. $\nabla f = (2x-3y)\mathbf{i} + (-3x+8y)\mathbf{j} \Rightarrow \nabla f(1,2) = -4\mathbf{i} + 13\mathbf{j} \Rightarrow |\nabla f(1,2)| = \sqrt{(-4)^2+(13)^2} = \sqrt{185}$; no, the maximum rate of change is $\sqrt{185} < 14$

35. $\nabla f = f_x(1,2)\mathbf{i} + f_y(1,2)\mathbf{j}$ and $\mathbf{u}_1 = \frac{\mathbf{i}+\mathbf{j}}{\sqrt{1^2+1^2}} = \frac{1}{\sqrt{2}}\mathbf{i} + \frac{1}{\sqrt{2}}\mathbf{j} \Rightarrow (D_{\mathbf{u}_1}f)(1,2) = f_x(1,2)\left(\frac{1}{\sqrt{2}}\right) + f_y(1,2)\left(\frac{1}{\sqrt{2}}\right)$ $= 2\sqrt{2} \Rightarrow f_x(1,2) + f_y(1,2) = 4$; $\mathbf{u}_2 = -\mathbf{j} \Rightarrow (D_{\mathbf{u}_2}f)(1,2) = f_x(1,2)(0) + f_y(1,2)(-1) = -3 \Rightarrow -f_y(1,2) = -3$ $\Rightarrow f_y(1,2) = 3$; then $f_x(1,2) + 3 = 4 \Rightarrow f_x(1,2) = 1$; thus $\nabla f(1,2) = \mathbf{i} + 3\mathbf{j}$ and $\mathbf{u} = \frac{\mathbf{v}}{|\mathbf{v}|} = \frac{-\mathbf{i}-2\mathbf{j}}{\sqrt{(-1)^2+(-2)^2}}$ $= -\frac{1}{\sqrt{5}}\mathbf{i} - \frac{2}{\sqrt{5}}\mathbf{j} \Rightarrow (D_{\mathbf{u}}f)_{P_0} = \nabla f \cdot \mathbf{u} = -\frac{1}{\sqrt{5}} - \frac{6}{\sqrt{5}} = -\frac{7}{\sqrt{5}}$

37. The directional derivative is the scalar component. With ∇f evaluated at P_0, the scalar component of ∇f in the direction of $\mathbf{u}$ is $\nabla f \cdot \mathbf{u} = (D_{\mathbf{u}}f)_{P_0}$.

39. If (x, y) is a point on the line, then $\mathbf{T}(x, y) = (x - x_0)\mathbf{i} + (y - y_0)\mathbf{j}$ is a vector parallel to the line $\Rightarrow \mathbf{T} \cdot \mathbf{N} = 0$ $\Rightarrow A(x - x_0) + B(y - y_0) = 0$, as claimed.

14.6 TANGENT PLANES AND DIFFERENTIALS

1. (a) $\nabla f = 2x\mathbf{i} + 2y\mathbf{j} + 2z\mathbf{k} \Rightarrow \nabla f(1,1,1) = 2\mathbf{i} + 2\mathbf{j} + 2\mathbf{k} \Rightarrow$ Tangent plane: $2(x-1) + 2(y-1) + 2(z-1) = 0$ $\Rightarrow x + y + z = 3$;
   (b) Normal line: $x = 1 + 2t$, $y = 1 + 2t$, $z = 1 + 2t$

3. (a) $\nabla f = -2x\mathbf{i} + 2\mathbf{k} \Rightarrow \nabla f(2,0,2) = -4\mathbf{i} + 2\mathbf{k} \Rightarrow$ Tangent plane: $-4(x-2) + 2(z-2) = 0$ $\Rightarrow -4x + 2z + 4 = 0 \Rightarrow -2x + z + 2 = 0$;
   (b) Normal line: $x = 2 - 4t$, $y = 0$, $z = 2 + 2t$

5. (a) $\nabla f = (-\pi \sin \pi x - 2xy + ze^{xz})\,\mathbf{i} + (-x^2 + z)\,\mathbf{j} + (xe^{xz} + y)\,\mathbf{k} \Rightarrow \nabla f(0,1,2) = 2\mathbf{i} + 2\mathbf{j} + \mathbf{k} \Rightarrow$ Tangent plane: $2(x-0) + 2(y-1) + 1(z-2) = 0 \Rightarrow 2x + 2y + z - 4 = 0$;
   (b) Normal line: $x = 2t$, $y = 1 + 2t$, $z = 2 + t$

7. (a) $\nabla f = \mathbf{i} + \mathbf{j} + \mathbf{k}$ for all points $\Rightarrow \nabla f(0,1,0) = \mathbf{i} + \mathbf{j} + \mathbf{k} \Rightarrow$ Tangent plane: $1(x-0) + 1(y-1) + 1(z-0) = 0$ $\Rightarrow x + y + z - 1 = 0$;
   (b) Normal line: $x = t$, $y = 1 + t$, $z = t$

9. $z = f(x,y) = \ln(x^2 + y^2) \Rightarrow f_x(x,y) = \frac{2x}{x^2+y^2}$ and $f_y(x,y) = \frac{2y}{x^2+y^2} \Rightarrow f_x(1,0) = 2$ and $f_y(1,0) = 0 \Rightarrow$ from Eq. (4) the tangent plane at $(1,0,0)$ is $2(x-1) - z = 0$ or $2x - z - 2 = 0$

11. $z = f(x,y) = \sqrt{y - x} \Rightarrow f_x(x,y) = -\frac{1}{2}(y-x)^{-1/2}$ and $f_y(x,y) = \frac{1}{2}(y-x)^{-1/2} \Rightarrow f_x(1,2) = -\frac{1}{2}$ and $f_y(1,2) = \frac{1}{2}$ $\Rightarrow$ from Eq. (4) the tangent plane at $(1,2,1)$ is $-\frac{1}{2}(x-1) + \frac{1}{2}(y-2) - (z-1) = 0 \Rightarrow x - y + 2z - 1 = 0$

13. $\nabla f = \mathbf{i} + 2y\mathbf{j} + 2\mathbf{k} \Rightarrow \nabla f(1,1,1) = \mathbf{i} + 2\mathbf{j} + 2\mathbf{k}$ and $\nabla g = \mathbf{i}$ for all points; $\mathbf{v} = \nabla f \times \nabla g$

$$\Rightarrow \mathbf{v} = \begin{vmatrix} \mathbf{i} & \mathbf{j} & \mathbf{k} \\ 1 & 2 & 2 \\ 1 & 0 & 0 \end{vmatrix} = 2\mathbf{j} - 2\mathbf{k} \Rightarrow \text{Tangent line: } x = 1,\ y = 1 + 2t,\ z = 1 - 2t$$

15. $\nabla f = 2x\mathbf{i} + 2\mathbf{j} + 2\mathbf{k} \Rightarrow \nabla f\left(1, 1, \frac{1}{2}\right) = 2\mathbf{i} + 2\mathbf{j} + 2\mathbf{k}$ and $\nabla g = \mathbf{j}$ for all points; $\mathbf{v} = \nabla f \times \nabla g$

$$\Rightarrow \mathbf{v} = \begin{vmatrix} \mathbf{i} & \mathbf{j} & \mathbf{k} \\ 2 & 2 & 2 \\ 0 & 1 & 0 \end{vmatrix} = -2\mathbf{i} + 2\mathbf{k} \Rightarrow \text{Tangent line: } x = 1 - 2t,\ y = 1,\ z = \tfrac{1}{2} + 2t$$

17. $\nabla f = (3x^2 + 6xy^2 + 4y)\mathbf{i} + (6x^2y + 3y^2 + 4x)\mathbf{j} - 2z\mathbf{k} \Rightarrow \nabla f(1,1,3) = 13\mathbf{i} + 13\mathbf{j} - 6\mathbf{k}$; $\nabla g = 2x\mathbf{i} + 2y\mathbf{j} + 2z\mathbf{k}$

$\Rightarrow \nabla g(1,1,3) = 2\mathbf{i} + 2\mathbf{j} + 6\mathbf{k}$; $\mathbf{v} = \nabla f \times \nabla g \Rightarrow \mathbf{v} = \begin{vmatrix} \mathbf{i} & \mathbf{j} & \mathbf{k} \\ 13 & 13 & -6 \\ 2 & 2 & 6 \end{vmatrix} = 90\mathbf{i} - 90\mathbf{j} \Rightarrow$ Tangent line:

$x = 1 + 90t,\ y = 1 - 90t,\ z = 3$

19. $\nabla f = \left(\frac{x}{x^2+y^2+z^2}\right)\mathbf{i} + \left(\frac{y}{x^2+y^2+z^2}\right)\mathbf{j} + \left(\frac{z}{x^2+y^2+z^2}\right)\mathbf{k} \Rightarrow \nabla f(3,4,12) = \frac{3}{169}\mathbf{i} + \frac{4}{169}\mathbf{j} + \frac{12}{169}\mathbf{k}$;

$\mathbf{u} = \frac{\mathbf{v}}{|\mathbf{v}|} = \frac{3\mathbf{i}+6\mathbf{j}-2\mathbf{k}}{\sqrt{3^2+6^2+(-2)^2}} = \frac{3}{7}\mathbf{i} + \frac{6}{7}\mathbf{j} - \frac{2}{7}\mathbf{k} \Rightarrow \nabla f \cdot \mathbf{u} = \frac{9}{1183}$ and $df = (\nabla f \cdot \mathbf{u})\,ds = \left(\frac{9}{1183}\right)(0.1) \approx 0.0008$

21. $\nabla g = (1 + \cos z)\mathbf{i} + (1 - \sin z)\mathbf{j} + (-x \sin z - y \cos z)\mathbf{k} \Rightarrow \nabla g(2,-1,0) = 2\mathbf{i} + \mathbf{j} + \mathbf{k}$; $\mathbf{A} = \overrightarrow{P_0P_1} = -2\mathbf{i} + 2\mathbf{j} + 2\mathbf{k}$

$\Rightarrow \mathbf{u} = \frac{\mathbf{v}}{|\mathbf{v}|} = \frac{-2\mathbf{i}+2\mathbf{j}+2\mathbf{k}}{\sqrt{(-2)^2+2^2+2^2}} = -\frac{1}{\sqrt{3}}\mathbf{i} + \frac{1}{\sqrt{3}}\mathbf{j} + \frac{1}{\sqrt{3}}\mathbf{k} \Rightarrow \nabla g \cdot \mathbf{u} = 0$ and $dg = (\nabla g \cdot \mathbf{u})\,ds = (0)(0.2) = 0$

23. (a) The unit tangent vector at $\left(\frac{1}{2}, \frac{\sqrt{3}}{2}\right)$ in the direction of motion is $\mathbf{u} = \frac{\sqrt{3}}{2}\mathbf{i} - \frac{1}{2}\mathbf{j}$;

$\nabla T = (\sin 2y)\mathbf{i} + (2x \cos 2y)\mathbf{j} \Rightarrow \nabla T\left(\frac{1}{2}, \frac{\sqrt{3}}{2}\right) = \left(\sin \sqrt{3}\right)\mathbf{i} + \left(\cos \sqrt{3}\right)\mathbf{j} \Rightarrow D_{\mathbf{u}}T\left(\frac{1}{2}, \frac{\sqrt{3}}{2}\right) = \nabla T \cdot \mathbf{u}$

$= \frac{\sqrt{3}}{2} \sin \sqrt{3} - \frac{1}{2} \cos \sqrt{3} \approx 0.935°$ C/ft

(b) $\mathbf{r}(t) = (\sin 2t)\mathbf{i} + (\cos 2t)\mathbf{j} \Rightarrow \mathbf{v}(t) = (2 \cos 2t)\mathbf{i} - (2 \sin 2t)\mathbf{j}$ and $|\mathbf{v}| = 2$; $\frac{dT}{dt} = \frac{\partial T}{\partial x}\frac{dx}{dt} + \frac{\partial T}{\partial y}\frac{dy}{dt}$

$= \nabla T \cdot \mathbf{v} = \left(\nabla T \cdot \frac{\mathbf{v}}{|\mathbf{v}|}\right)|\mathbf{v}| = (D_{\mathbf{u}}T)\,|\mathbf{v}|$, where $\mathbf{u} = \frac{\mathbf{v}}{|\mathbf{v}|}$; at $\left(\frac{1}{2}, \frac{\sqrt{3}}{2}\right)$ we have $\mathbf{u} = \frac{\sqrt{3}}{2}\mathbf{i} - \frac{1}{2}\mathbf{j}$ from part (a)

$\Rightarrow \frac{dT}{dt} = \left(\frac{\sqrt{3}}{2} \sin \sqrt{3} - \frac{1}{2} \cos \sqrt{3}\right) \cdot 2 = \sqrt{3} \sin \sqrt{3} - \cos \sqrt{3} \approx 1.87°$ C/sec

25. (a) $f(0,0) = 1,\ f_x(x,y) = 2x \Rightarrow f_x(0,0) = 0,\ f_y(x,y) = 2y \Rightarrow f_y(0,0) = 0 \Rightarrow L(x,y) = 1 + 0(x-0) + 0(y-0) = 1$

(b) $f(1,1) = 3,\ f_x(1,1) = 2,\ f_y(1,1) = 2 \Rightarrow L(x,y) = 3 + 2(x-1) + 2(y-1) = 2x + 2y - 1$

27. (a) $f(0,0) = 5,\ f_x(x,y) = 3$ for all (x,y), $f_y(x,y) = -4$ for all $(x,y) \Rightarrow L(x,y) = 5 + 3(x-0) - 4(y-0) = 3x - 4y + 5$

(b) $f(1,1) = 4,\ f_x(1,1) = 3,\ f_y(1,1) = -4 \Rightarrow L(x,y) = 4 + 3(x-1) - 4(y-1) = 3x - 4y + 5$

29. (a) $f(0,0) = 1,\ f_x(x,y) = e^x \cos y \Rightarrow f_x(0,0) = 1,\ f_y(x,y) = -e^x \sin y \Rightarrow f_y(0,0) = 0$

$\Rightarrow L(x,y) = 1 + 1(x-0) + 0(y-0) = x + 1$

(b) $f\left(0, \frac{\pi}{2}\right) = 0,\ f_x\left(0, \frac{\pi}{2}\right) = 0,\ f_y\left(0, \frac{\pi}{2}\right) = -1 \Rightarrow L(x,y) = 0 + 0(x-0) - 1\left(y - \frac{\pi}{2}\right) = -y + \frac{\pi}{2}$

31. (a) $W(20,25) = 11°F$; $W(30,-10) = -39°F$; $W(15,15) = 0°F$

(b) $W(10,-40) = -65.5°F$; $W(50,-40) = -88°F$; $W(60,30) = 10.2°F$;

(c) $W(25,5) = -17.4088°F$; $\frac{\partial W}{\partial V} = -\frac{5.72}{V^{0.84}} + \frac{0.0684t}{V^{0.84}} \Rightarrow \frac{\partial W}{\partial V}(25,5) = -0.36$; $\frac{\partial W}{\partial T} = 0.6215 + 0.4275V^{0.16}$

$\Rightarrow \frac{\partial W}{\partial T}(25,5) = 1.3370 \Rightarrow L(V,T) = -17.4088 - 0.36(V-25) + 1.337(T-5) = 1.337T - 0.36V - 15.0938$

(d) i) $W(24,6) \approx L(24,6) = -15.7118 \approx -15.7°F$

ii) $W(27,2) \approx L(27,2) = -22.1398 \approx -22.1°F$

ii) $W(5,-10) \approx L(5,-10) = -30.2638 \approx -30.2°F$ This value is very different because the point $(5,-10)$ is not close to the point $(25,5)$.

33. $f(2,1) = 3,\ f_x(x,y) = 2x - 3y \Rightarrow f_x(2,1) = 1,\ f_y(x,y) = -3x \Rightarrow f_y(2,1) = -6 \Rightarrow L(x,y) = 3 + 1(x-2) - 6(y-1)$

$= 7 + x - 6y$; $f_{xx}(x,y) = 2,\ f_{yy}(x,y) = 0,\ f_{xy}(x,y) = -3 \Rightarrow M = 3$; thus $|E(x,y)| \le \left(\frac{1}{2}\right)(3)\left(|x-2| + |y-1|\right)^2$

$\le \left(\frac{3}{2}\right)(0.1 + 0.1)^2 = 0.06$

35. $f(0,0) = 1$, $f_x(x,y) = \cos y \Rightarrow f_x(0,0) = 1$, $f_y(x,y) = 1 - x \sin y \Rightarrow f_y(0,0) = 1$
$\Rightarrow L(x,y) = 1 + 1(x-0) + 1(y-0) = x + y + 1$; $f_{xx}(x,y) = 0$, $f_{yy}(x,y) = -x\cos y$, $f_{xy}(x,y) = -\sin y \Rightarrow M = 1$;
thus $|E(x,y)| \le \left(\frac{1}{2}\right)(1)\left(|x| + |y|\right)^2 \le \left(\frac{1}{2}\right)(0.2 + 0.2)^2 = 0.08$

37. $f(0,0) = 1$, $f_x(x,y) = e^x \cos y \Rightarrow f_x(0,0) = 1$, $f_y(x,y) = -e^x \sin y \Rightarrow f_y(0,0) = 0$
$\Rightarrow L(x,y) = 1 + 1(x-0) + 0(y-0) = 1 + x$; $f_{xx}(x,y) = e^x \cos y$, $f_{yy}(x,y) = -e^x \cos y$, $f_{xy}(x,y) = -e^x \sin y$;
$|x| \le 0.1 \Rightarrow -0.1 \le x \le 0.1$ and $|y| \le 0.1 \Rightarrow -0.1 \le y \le 0.1$; thus the max of $|f_{xx}(x,y)|$ on R is $e^{0.1} \cos(0.1)$
≤ 1.11, the max of $|f_{yy}(x,y)|$ on R is $e^{0.1}\cos(0.1) \le 1.11$, and the max of $|f_{xy}(x,y)|$ on R is $e^{0.1}\sin(0.1)$
$\le 0.12 \Rightarrow M = 1.11$; thus $|E(x,y)| \le \left(\frac{1}{2}\right)(1.11)\left(|x| + |y|\right)^2 \le (0.555)(0.1 + 0.1)^2 = 0.0222$

39. (a) $f(1,1,1) = 3$, $f_x(1,1,1) = y + z|_{(1,1,1)} = 2$, $f_y(1,1,1) = x + z|_{(1,1,1)} = 2$, $f_z(1,1,1) = y + x|_{(1,1,1)} = 2$
$\Rightarrow L(x,y,z) = 3 + 2(x-1) + 2(y-1) + 2(z-1) = 2x + 2y + 2z - 3$
(b) $f(1,0,0) = 0$, $f_x(1,0,0) = 0$, $f_y(1,0,0) = 1$, $f_z(1,0,0) = 1 \Rightarrow L(x,y,z) = 0 + 0(x-1) + (y-0) + (z-0) = y + z$
(c) $f(0,0,0) = 0$, $f_x(0,0,0) = 0$, $f_y(0,0,0) = 0$, $f_z(0,0,0) = 0 \Rightarrow L(x,y,z) = 0$

41. (a) $f(1,0,0) = 1$, $f_x(1,0,0) = \left.\frac{x}{\sqrt{x^2+y^2+z^2}}\right|_{(1,0,0)} = 1$, $f_y(1,0,0) = \left.\frac{y}{\sqrt{x^2+y^2+z^2}}\right|_{(1,0,0)} = 0$,
$f_z(1,0,0) = \left.\frac{z}{\sqrt{x^2+y^2+z^2}}\right|_{(1,0,0)} = 0 \Rightarrow L(x,y,z) = 1 + 1(x-1) + 0(y-0) + 0(z-0) = x$
(b) $f(1,1,0) = \sqrt{2}$, $f_x(1,1,0) = \frac{1}{\sqrt{2}}$, $f_y(1,1,0) = \frac{1}{\sqrt{2}}$, $f_z(1,1,0) = 0$
$\Rightarrow L(x,y,z) = \sqrt{2} + \frac{1}{\sqrt{2}}(x-1) + \frac{1}{\sqrt{2}}(y-1) + 0(z-0) = \frac{1}{\sqrt{2}}x + \frac{1}{\sqrt{2}}y$
(c) $f(1,2,2) = 3$, $f_x(1,2,2) = \frac{1}{3}$, $f_y(1,2,2) = \frac{2}{3}$, $f_z(1,2,2) = \frac{2}{3} \Rightarrow L(x,y,z) = 3 + \frac{1}{3}(x-1) + \frac{2}{3}(y-2) + \frac{2}{3}(z-2)$
$= \frac{1}{3}x + \frac{2}{3}y + \frac{2}{3}z$

43. (a) $f(0,0,0) = 2$, $f_x(0,0,0) = e^x|_{(0,0,0)} = 1$, $f_y(0,0,0) = -\sin(y+z)|_{(0,0,0)} = 0$,
$f_z(0,0,0) = -\sin(y+z)|_{(0,0,0)} = 0 \Rightarrow L(x,y,z) = 2 + 1(x-0) + 0(y-0) + 0(z-0) = 2 + x$
(b) $f\left(0,\frac{\pi}{2},0\right) = 1$, $f_x\left(0,\frac{\pi}{2},0\right) = 1$, $f_y\left(0,\frac{\pi}{2},0\right) = -1$, $f_z\left(0,\frac{\pi}{2},0\right) = -1 \Rightarrow L(x,y,z)$
$= 1 + 1(x-0) - 1\left(y - \frac{\pi}{2}\right) - 1(z-0) = x - y - z + \frac{\pi}{2} + 1$
(c) $f\left(0,\frac{\pi}{4},\frac{\pi}{4}\right) = 1$, $f_x\left(0,\frac{\pi}{4},\frac{\pi}{4}\right) = 1$, $f_y\left(0,\frac{\pi}{4},\frac{\pi}{4}\right) = -1$, $f_z\left(0,\frac{\pi}{4},\frac{\pi}{4}\right) = -1 \Rightarrow L(x,y,z)$
$= 1 + 1(x-0) - 1\left(y - \frac{\pi}{4}\right) - 1\left(z - \frac{\pi}{4}\right) = x - y - z + \frac{\pi}{2} + 1$

45. $f(x,y,z) = xz - 3yz + 2$ at $P_0(1,1,2) \Rightarrow f(1,1,2) = -2$; $f_x = z$, $f_y = -3z$, $f_z = x - 3y \Rightarrow L(x,y,z)$
$= -2 + 2(x-1) - 6(y-1) - 2(z-2) = 2x - 6y - 2z + 6$; $f_{xx} = 0$, $f_{yy} = 0$, $f_{zz} = 0$, $f_{xy} = 0$, $f_{yz} = -3$
$\Rightarrow M = 3$; thus, $|E(x,y,z)| \le \left(\frac{1}{2}\right)(3)(0.01 + 0.01 + 0.02)^2 = 0.0024$

47. $f(x,y,z) = xy + 2yz - 3xz$ at $P_0(1,1,0) \Rightarrow f(1,1,0) = 1$; $f_x = y - 3z$, $f_y = x + 2z$, $f_z = 2y - 3x$
$\Rightarrow L(x,y,z) = 1 + (x-1) + (y-1) - (z-0) = x + y - z - 1$; $f_{xx} = 0$, $f_{yy} = 0$, $f_{zz} = 0$, $f_{xy} = 1$, $f_{xz} = -3$,
$f_{yz} = 2 \Rightarrow M = 3$; thus $|E(x,y,z)| \le \left(\frac{1}{2}\right)(3)(0.01 + 0.01 + 0.01)^2 = 0.00135$

49. $T_x(x,y) = e^y + e^{-y}$ and $T_y(x,y) = x\left(e^y - e^{-y}\right) \Rightarrow dT = T_x(x,y)\,dx + T_y(x,y)\,dy$
$= \left(e^y + e^{-y}\right)dx + x\left(e^y - e^{-y}\right)dy \Rightarrow dT|_{(2,\ln 2)} = 2.5\,dx + 3.0\,dy$. If $|dx| \le 0.1$ and $|dy| \le 0.02$, then the maximum possible error in the computed value of T is $(2.5)(0.1) + (3.0)(0.02) = 0.31$ in magnitude.

51. $\frac{dx}{x} \le 0.02$, $\frac{dy}{y} \le 0.03$

(a) $S = 2x^2 + 4xy \Rightarrow dS = (4x + 4y)dx + 4x\,dy = (4x^2 + 4xy)\frac{dx}{x} + 4xy\,\frac{dy}{y} \le (4x^2 + 4xy)(0.02) + (4xy)(0.03)$
$= 0.04(2x^2) + 0.05(4xy) \le 0.05(2x^2) + 0.05(4xy) = (0.05)(2x^2 + 4xy) = 0.05S$

(b) $V = x^2y \Rightarrow dV = 2xy\,dx + x^2dy = 2x^2y\,\frac{dx}{x} + x^2y\frac{dy}{y} \le (2x^2y)(0.02) + (x^2y)(0.03) = 0.07(x^2y) = 0.07V$

53. $V_r = 2\pi rh$ and $V_h = \pi r^2 \Rightarrow dV = V_r\,dr + V_h\,dh \Rightarrow dV = 2\pi rh\,dr + \pi r^2\,dh \Rightarrow dV|_{(5,12)} = 120\pi\,dr + 25\pi\,dh$;
$|dr| \le 0.1$ cm and $|dh| \le 0.1$ cm $\Rightarrow dV \le (120\pi)(0.1) + (25\pi)(0.1) = 14.5\pi\text{ cm}^3$; $V(5, 12) = 300\pi\text{ cm}^3$
$\Rightarrow$ maximum percentage error is $\pm\,\frac{14.5\pi}{300\pi} \times 100 = \pm\,4.83\%$

55. $A = xy \Rightarrow dA = x\,dy + y\,dx$; if $x > y$ then a 1-unit change in y gives a greater change in dA than a 1-unit change in x. Thus, pay more attention to y which is the smaller of the two dimensions.

57. (a) $r^2 = x^2 + y^2 \Rightarrow 2r\,dr = 2x\,dx + 2y\,dy \Rightarrow dr = \frac{x}{r}\,dx + \frac{y}{r}\,dy \Rightarrow dr|_{(3,4)} = \left(\frac{3}{5}\right)(\pm\,0.01) + \left(\frac{4}{5}\right)(\pm\,0.01)$
$= \pm\,\frac{0.07}{5} = \pm\,0.014 \Rightarrow \left|\frac{dr}{r} \times 100\right| = \left|\pm\,\frac{0.014}{5} \times 100\right| = 0.28\%$; $d\theta = \frac{\left(-\frac{y}{x^2}\right)}{\left(\frac{y}{x}\right)^2+1}\,dx + \frac{\left(\frac{1}{x}\right)}{\left(\frac{y}{x}\right)^2+1}\,dy$
$= \frac{-y}{y^2+x^2}\,dx + \frac{x}{y^2+x^2}\,dy \Rightarrow d\theta|_{(3,4)} = \left(\frac{-4}{25}\right)(\pm\,0.01) + \left(\frac{3}{25}\right)(\pm\,0.01) = \frac{\mp 0.04}{25} + \frac{\pm 0.03}{25}$
$\Rightarrow$ maximum change in $d\theta$ occurs when dx and dy have opposite signs ($dx = 0.01$ and $dy = -0.01$ or vice versa) $\Rightarrow d\theta = \frac{\pm 0.07}{25} \approx \pm\,0.0028$; $\theta = \tan^{-1}\left(\frac{4}{3}\right) \approx 0.927255218 \Rightarrow \left|\frac{d\theta}{\theta} \times 100\right| = \left|\frac{\pm 0.0028}{0.927255218} \times 100\right|$
$\approx 0.30\%$

(b) the radius r is more sensitive to changes in y, and the angle θ is more sensitive to changes in x

59. $f(a, b, c, d) = \begin{vmatrix} a & b \\ c & d \end{vmatrix} = ad - bc \Rightarrow f_a = d,\ f_b = -c,\ f_c = -b,\ f_d = a \Rightarrow df = d\,da - c\,db - b\,dc + a\,dd$; since $|a|$ is much greater than $|b|$, $|c|$, and $|d|$, the function f is most sensitive to a change in d.

61. $Q_K = \frac{1}{2}\left(\frac{2KM}{h}\right)^{-1/2}\left(\frac{2M}{h}\right)$, $Q_M = \frac{1}{2}\left(\frac{2KM}{h}\right)^{-1/2}\left(\frac{2K}{h}\right)$, and $Q_h = \frac{1}{2}\left(\frac{2KM}{h}\right)^{-1/2}\left(\frac{-2KM}{h^2}\right)$
$\Rightarrow dQ = \frac{1}{2}\left(\frac{2KM}{h}\right)^{-1/2}\left(\frac{2M}{h}\right)dK + \frac{1}{2}\left(\frac{2KM}{h}\right)^{-1/2}\left(\frac{2K}{h}\right)dM + \frac{1}{2}\left(\frac{2KM}{h}\right)^{-1/2}\left(\frac{-2KM}{h^2}\right)dh$
$= \frac{1}{2}\left(\frac{2KM}{h}\right)^{-1/2}\left[\frac{2M}{h}\,dK + \frac{2K}{h}\,dM - \frac{2KM}{h^2}\,dh\right] \Rightarrow dQ|_{(2,20,0.0.05)}$
$= \frac{1}{2}\left[\frac{(2)(2)(20)}{0.05}\right]^{-1/2}\left[\frac{(2)(20)}{0.05}\,dK + \frac{(2)(2)}{0.05}\,dM - \frac{(2)(2)(20)}{(0.05)^2}\,dh\right] = (0.0125)(800\,dK + 80\,dM - 32{,}000\,dh)$
$\Rightarrow$ Q is most sensitive to changes in h

63. $z = f(x, y) \Rightarrow g(x, y, z) = f(x, y) - z = 0 \Rightarrow g_x(x, y, z) = f_x(x, y)$, $g_y(x, y, z) = f_y(x, y)$ and $g_z(x, y, z) = -1$
$\Rightarrow g_x(x_0, y_0, f(x_0, y_0)) = f_x(x_0, y_0)$, $g_y(x_0, y_0, f(x_0, y_0)) = f_y(x_0, y_0)$ and $g_z(x_0, y_0, f(x_0, y_0)) = -1 \Rightarrow$ the tangent plane at the point P_0 is $f_x(x_0, y_0)(x - x_0) + f_y(x_0, y_0)(y - y_0) - [z - f(x_0, y_0)] = 0$ or
$z = f_x(x_0, y_0)(x - x_0) + f_y(x_0, y_0)(y - y_0) + f(x_0, y_0)$

65. $\nabla f = 2x\mathbf{i} + 2y\mathbf{j} + 2z\mathbf{k} = (2\cos t)\mathbf{i} + (2\sin t)\mathbf{j} + 2t\mathbf{k}$ and $\mathbf{v} = (-\sin t)\mathbf{i} + (\cos t)\mathbf{j} + \mathbf{k} \Rightarrow \mathbf{u} = \frac{\mathbf{v}}{|\mathbf{v}|}$
$= \frac{(-\sin t)\mathbf{i} + (\cos t)\mathbf{j} + \mathbf{k}}{\sqrt{(\sin t)^2 + (\cos t)^2 + 1^2}} = \left(\frac{-\sin t}{\sqrt{2}}\right)\mathbf{i} + \left(\frac{\cos t}{\sqrt{2}}\right)\mathbf{j} + \frac{1}{\sqrt{2}}\mathbf{k} \Rightarrow (D_{\mathbf{u}}f)_{P_0} = \nabla f \cdot \mathbf{u}$
$= (2\cos t)\left(\frac{-\sin t}{\sqrt{2}}\right) + (2\sin t)\left(\frac{\cos t}{\sqrt{2}}\right) + (2t)\left(\frac{1}{\sqrt{2}}\right) = \frac{2t}{\sqrt{2}} \Rightarrow (D_{\mathbf{u}}f)\left(\frac{-\pi}{4}\right) = \frac{-\pi}{2\sqrt{2}}$, $(D_{\mathbf{u}}f)(0) = 0$ and
$(D_{\mathbf{u}}f)\left(\frac{\pi}{4}\right) = \frac{\pi}{2\sqrt{2}}$

67. $\mathbf{r} = \sqrt{t}\mathbf{i} + \sqrt{t}\mathbf{j} + (2t-1)\mathbf{k} \Rightarrow \mathbf{v} = \frac{1}{2}t^{-1/2}\mathbf{i} + \frac{1}{2}t^{-1/2}\mathbf{j} + 2\mathbf{k}$; $t = 1 \Rightarrow x = 1, y = 1, z = 1 \Rightarrow P_0 = (1,1,1)$ and $\mathbf{v}(1) = \frac{1}{2}\mathbf{i} + \frac{1}{2}\mathbf{j} + 2\mathbf{k}$; $f(x,y,z) = x^2 + y^2 - z - 1 = 0 \Rightarrow \nabla f = 2x\mathbf{i} + 2y\mathbf{j} - \mathbf{k} \Rightarrow \nabla f(1,1,1) = 2\mathbf{i} + 2\mathbf{j} - \mathbf{k}$; now $\mathbf{v}(1) \cdot \nabla f(1,1,1) = 0$, thus the curve is tangent to the surface when $t = 1$

14.7 EXTREME VALUES AND SADDLE POINTS

1. $f_x(x,y) = 2x + y + 3 = 0$ and $f_y(x,y) = x + 2y - 3 = 0 \Rightarrow x = -3$ and $y = 3 \Rightarrow$ critical point is $(-3,3)$; $f_{xx}(-3,3) = 2, f_{yy}(-3,3) = 2, f_{xy}(-3,3) = 1 \Rightarrow f_{xx}f_{yy} - f_{xy}^2 = 3 > 0$ and $f_{xx} > 0 \Rightarrow$ local minimum of $f(-3,3) = -5$

3. $f_x(x,y) = 2x + y + 3 = 0$ and $f_y(x,y) = x + 2 = 0 \Rightarrow x = -2$ and $y = 1 \Rightarrow$ critical point is $(-2,1)$; $f_{xx}(-2,1) = 2, f_{yy}(-2,1) = 0, f_{xy}(-2,1) = 1 \Rightarrow f_{xx}f_{yy} - f_{xy}^2 = -1 < 0 \Rightarrow$ saddle point

5. $f_x(x,y) = 2y - 2x + 3 = 0$ and $f_y(x,y) = 2x - 4y = 0 \Rightarrow x = 3$ and $y = \frac{3}{2} \Rightarrow$ critical point is $\left(3, \frac{3}{2}\right)$; $f_{xx}\left(3, \frac{3}{2}\right) = -2, f_{yy}\left(3, \frac{3}{2}\right) = -4, f_{xy}\left(3, \frac{3}{2}\right) = 2 \Rightarrow f_{xx}f_{yy} - f_{xy}^2 = 4 > 0$ and $f_{xx} < 0 \Rightarrow$ local maximum of $f\left(3, \frac{3}{2}\right) = \frac{17}{2}$

7. $f_x(x,y) = 4x + 3y - 5 = 0$ and $f_y(x,y) = 3x + 8y + 2 = 0 \Rightarrow x = 2$ and $y = -1 \Rightarrow$ critical point is $(2,-1)$; $f_{xx}(2,-1) = 4, f_{yy}(2,-1) = 8, f_{xy}(2,-1) = 3 \Rightarrow f_{xx}f_{yy} - f_{xy}^2 = 23 > 0$ and $f_{xx} > 0 \Rightarrow$ local minimum of $f(2,-1) = -6$

9. $f_x(x,y) = 2x - 2 = 0$ and $f_y(x,y) = -2y + 4 = 0 \Rightarrow x = 1$ and $y = 2 \Rightarrow$ critical point is $(1,2)$; $f_{xx}(1,2) = 2$, $f_{yy}(1,2) = -2, f_{xy}(1,2) = 0 \Rightarrow f_{xx}f_{yy} - f_{xy}^2 = -4 < 0 \Rightarrow$ saddle point

11. $f_x(x,y) = \frac{112x - 8x}{\sqrt{56x^2 - 8y^2 - 16x - 31}} - 8 = 0$ and $f_y(x,y) = \frac{-8y}{\sqrt{56x^2 - 8y^2 - 16x - 31}} = 0 \Rightarrow$ critical point is $\left(\frac{16}{7}, 0\right)$; $f_{xx}\left(\frac{16}{7}, 0\right) = -\frac{8}{15}, f_{yy}\left(\frac{16}{7}, 0\right) = -\frac{8}{15}, f_{xy}\left(\frac{16}{7}, 0\right) = 0 \Rightarrow f_{xx}f_{yy} - f_{xy}^2 = \frac{64}{225} > 0$ and $f_{xx} < 0 \Rightarrow$ local maximum of $f\left(\frac{16}{7}, 0\right) = -\frac{16}{7}$

13. $f_x(x,y) = 3x^2 - 2y = 0$ and $f_y(x,y) = -3y^2 - 2x = 0 \Rightarrow x = 0$ and $y = 0$, or $x = -\frac{2}{3}$ and $y = \frac{2}{3} \Rightarrow$ critical points are $(0,0)$ and $\left(-\frac{2}{3}, \frac{2}{3}\right)$; for $(0,0)$: $f_{xx}(0,0) = 6x|_{(0,0)} = 0, f_{yy}(0,0) = -6y|_{(0,0)} = 0, f_{xy}(0,0) = -2$ $\Rightarrow f_{xx}f_{yy} - f_{xy}^2 = -4 < 0 \Rightarrow$ saddle point; for $\left(-\frac{2}{3}, \frac{2}{3}\right)$: $f_{xx}\left(-\frac{2}{3}, \frac{2}{3}\right) = -4, f_{yy}\left(-\frac{2}{3}, \frac{2}{3}\right) = -4, f_{xy}\left(-\frac{2}{3}, \frac{2}{3}\right) = -2$ $\Rightarrow f_{xx}f_{yy} - f_{xy}^2 = 12 > 0$ and $f_{xx} < 0 \Rightarrow$ local maximum of $f\left(-\frac{2}{3}, \frac{2}{3}\right) = \frac{170}{27}$

15. $f_x(x,y) = 12x - 6x^2 + 6y = 0$ and $f_y(x,y) = 6y + 6x = 0 \Rightarrow x = 0$ and $y = 0$, or $x = 1$ and $y = -1 \Rightarrow$ critical points are $(0,0)$ and $(1,-1)$; for $(0,0)$: $f_{xx}(0,0) = 12 - 12x|_{(0,0)} = 12, f_{yy}(0,0) = 6, f_{xy}(0,0) = 6 \Rightarrow f_{xx}f_{yy} - f_{xy}^2$ $= 36 > 0$ and $f_{xx} > 0 \Rightarrow$ local minimum of $f(0,0) = 0$; for $(1,-1)$: $f_{xx}(1,-1) = 0, f_{yy}(1,-1) = 6$, $f_{xy}(1,-1) = 6 \Rightarrow f_{xx}f_{yy} - f_{xy}^2 = -36 < 0 \Rightarrow$ saddle point

17. $f_x(x,y) = 3x^2 + 3y^2 - 15 = 0$ and $f_y(x,y) = 6xy + 3y^2 - 15 = 0 \Rightarrow$ critical points are $(2,1), (-2,-1), \left(0, \sqrt{5}\right)$, and $\left(0, -\sqrt{5}\right)$; for $(2,1)$: $f_{xx}(2,1) = 6x|_{(2,1)} = 12,\ f_{yy}(2,1) = (6x + 6y)|_{(2,1)} = 18,\ f_{xy}(2,1) = 6y|_{(2,1)} = 6$ $\Rightarrow f_{xx}f_{yy} - f_{xy}^2 = 180 > 0$ and $f_{xx} > 0 \Rightarrow$ local minimum of $f(2,1) = -30$; for $(-2,-1)$: $f_{xx}(-2,-1) = 6x|_{(-2,-1)}$ $= -12,\ f_{yy}(-2,-1) = (6x + 6y)|_{(-2,-1)} = -18,\ f_{xy}(-2,-1) = 6y|_{(-2,-1)} = -6 \Rightarrow f_{xx}f_{yy} - f_{xy}^2 = 180 > 0$ and $f_{xx} < 0 \Rightarrow$ local maximum of $f(-2,-1) = 30$; for $\left(0, \sqrt{5}\right)$: $f_{xx}\left(0, \sqrt{5}\right) = 6x\Big|_{\left(0,\sqrt{5}\right)} = 0, f_{yy}\left(0, \sqrt{5}\right)$

$= (6x+6y)|_{(0,\sqrt{5})} = 6\sqrt{5}$, $f_{xy}\left(0,\sqrt{5}\right) = 6y\big|_{(0,\sqrt{5})} = 6\sqrt{5} \Rightarrow f_{xx}f_{yy} - f_{xy}^2 = -180 < 0 \Rightarrow$ saddle point;

for $\left(0,-\sqrt{5}\right)$: $f_{xx}\left(0,-\sqrt{5}\right) = 6x\big|_{(0,-\sqrt{5})} = 0$, $f_{yy}\left(0,-\sqrt{5}\right) = (6x+6y)|_{(0,-\sqrt{5})} = -6\sqrt{5}$,

$f_{xy}\left(0,-\sqrt{5}\right) = 6y\big|_{(0,-\sqrt{5})} = -6\sqrt{5} \Rightarrow f_{xx}f_{yy} - f_{xy}^2 = -180 < 0 \Rightarrow$ saddle point.

19. $f_x(x,y) = 4y - 4x^3 = 0$ and $f_y(x,y) = 4x - 4y^3 = 0 \Rightarrow x = y \Rightarrow x(1-x^2) = 0 \Rightarrow x = 0, 1, -1 \Rightarrow$ the critical points are $(0,0)$, $(1,1)$, and $(-1,-1)$; for $(0,0)$: $f_{xx}(0,0) = -12x^2|_{(0,0)} = 0$, $f_{yy}(0,0) = -12y^2|_{(0,0)} = 0$, $f_{xy}(0,0) = 4 \Rightarrow f_{xx}f_{yy} - f_{xy}^2 = -16 < 0 \Rightarrow$ saddle point; for $(1,1)$: $f_{xx}(1,1) = -12$, $f_{yy}(1,1) = -12$, $f_{xy}(1,1) = 4$ $\Rightarrow f_{xx}f_{yy} - f_{xy}^2 = 128 > 0$ and $f_{xx} < 0 \Rightarrow$ local maximum of $f(1,1) = 2$; for $(-1,-1)$: $f_{xx}(-1,-1) = -12$, $f_{yy}(-1,-1) = -12$, $f_{xy}(-1,-1) = 4 \Rightarrow f_{xx}f_{yy} - f_{xy}^2 = 128 > 0$ and $f_{xx} < 0 \Rightarrow$ local maximum of $f(-1,-1) = 2$

21. $f_x(x,y) = \frac{-2x}{(x^2+y^2-1)^2} = 0$ and $f_y(x,y) = \frac{-2y}{(x^2+y^2-1)^2} = 0 \Rightarrow x = 0$ and $y = 0 \Rightarrow$ the critical point is $(0,0)$; $f_{xx} = \frac{4x^2-2y^2+2}{(x^2+y^2-1)^3}$, $f_{yy} = \frac{-2x^2+4y^2+2}{(x^2+y^2-1)^3}$, $f_{xy} = \frac{8xy}{(x^2+y^2-1)^3}$; $f_{xx}(0,0) = -2$, $f_{yy}(0,0) = -2$, $f_{xy}(0,0) = 0$ $\Rightarrow f_{xx}f_{yy} - f_{xy}^2 = 4 > 0$ and $f_{xx} < 0 \Rightarrow$ local maximum of $f(0,0) = -1$

23. $f_x(x,y) = y\cos x = 0$ and $f_y(x,y) = \sin x = 0 \Rightarrow x = n\pi$, n an integer, and $y = 0 \Rightarrow$ the critical points are $(n\pi, 0)$, n an integer (Note: $\cos x$ and $\sin x$ cannot both be 0 for the same x, so $\sin x$ must be 0 and $y = 0$); $f_{xx} = -y\sin x$, $f_{yy} = 0$, $f_{xy} = \cos x$; $f_{xx}(n\pi,0) = 0$, $f_{yy}(n\pi,0) = 0$, $f_{xy}(n\pi,0) = 1$ if n is even and $f_{xy}(n\pi,0) = -1$ if n is odd $\Rightarrow f_{xx}f_{yy} - f_{xy}^2 = -1 < 0 \Rightarrow$ saddle point.

25. $f_x(x,y) = (2x-4)e^{x^2+y^2-4x} = 0$ and $f_y(x,y) = 2ye^{x^2+y^2-4x} = 0 \Rightarrow$ critical point is $(2,0)$; $f_{xx}(2,0) = \frac{2}{e^4}$, $f_{xy}(2,0) = 0$, $f_{yy}(2,0) = \frac{2}{e^4} \Rightarrow f_{xx}f_{yy} - f_{xy}^2 = \frac{4}{e^8} > 0$ and $f_{xx} > 0 \Rightarrow$ local mimimum of $f(2,0) = \frac{1}{e^4}$

27. $f_x(x,y) = 2xe^{-y} = 0$ and $f_y(x,y) = 2ye^{-y} - e^{-y}(x^2+y^2) = 0 \Rightarrow$ critical points are $(0,0)$ and $(0,2)$; for $(0,0)$: $f_{xx}(0,0) = 2e^{-y}|_{(0,0)} = 2$, $f_{yy}(0,0) = (2e^{-y} - 4ye^{-y} + e^{-y}(x^2+y^2))|_{(0,0)} = 2$, $f_{xy}(0,0) = -2xe^{-y}|_{(0,0)} = 0$ $\Rightarrow f_{xx}f_{yy} - f_{xy}^2 = 4 > 0$ and $f_{xx} > 0 \Rightarrow$ local mimimum of $f(0,0) = 0$; for $(0,2)$: $f_{xx}(0,2) = 2e^{-y}|_{(0,2)} = \frac{2}{e^2}$, $f_{yy}(0,2) = (2e^{-y} - 4ye^{-y} + e^{-y}(x^2+y^2))|_{(0,2)} = -\frac{2}{e^2}$, $f_{xy}(0,2) = -2xe^{-y}|_{(0,2)} = 0 \Rightarrow f_{xx}f_{yy} - f_{xy}^2 = -\frac{4}{e^4} < 0$ $\Rightarrow$ saddle point

29. $f_x(x,y) = -4 + \frac{2}{x} = 0$ and $f_y(x,y) = -1 + \frac{1}{y} = 0 \Rightarrow$ critical point is $\left(\frac{1}{2},1\right)$; $f_{xx}\left(\frac{1}{2},1\right) = -8$, $f_{yy}\left(\frac{1}{2},1\right) = -1$, $f_{xy}\left(\frac{1}{2},1\right) = 0 \Rightarrow f_{xx}f_{yy} - f_{xy}^2 = 8 > 0$ and $f_{xx} < 0 \Rightarrow$ local maximum of $f\left(\frac{1}{2},1\right) = -3 - 2\ln 2$

31. (i) On OA, $f(x,y) = f(0,y) = y^2 - 4y + 1$ on $0 \le y \le 2$;
$f'(0,y) = 2y - 4 = 0 \Rightarrow y = 2$;
$f(0,0) = 1$ and $f(0,2) = -3$

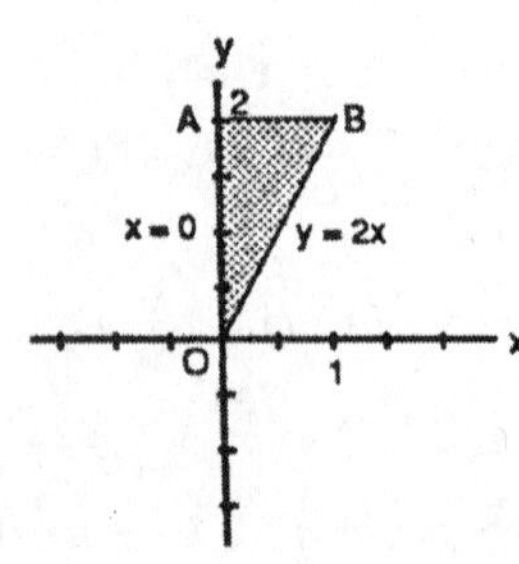

(ii) On AB, $f(x,y) = f(x,2) = 2x^2 - 4x - 3$ on $0 \le x \le 1$;
$f'(x,2) = 4x - 4 = 0 \Rightarrow x = 1$;
$f(0,2) = -3$ and $f(1,2) = -5$

(iii) On OB, $f(x,y) = f(x,2x) = 6x^2 - 12x + 1$ on $0 \le x \le 1$; endpoint values have been found above;
$f'(x,2x) = 12x - 12 = 0 \Rightarrow x = 1$ and $y = 2$, but $(1,2)$ is not an interior point of OB

(iv) For interior points of the triangular region, $f_x(x,y) = 4x - 4 = 0$ and $f_y(x,y) = 2y - 4 = 0$ $\Rightarrow x = 1$ and $y = 2$, but $(1,2)$ is not an interior point of the region. Therefore, the absolute maximum is 1 at $(0,0)$ and the absolute minimum is -5 at $(1,2)$.

33. (i) On OA, $f(x, y) = f(0, y) = y^2$ on $0 \le y \le 2$; $f'(0, y) = 2y = 0 \Rightarrow y = 0$ and $x = 0$; $f(0, 0) = 0$ and $f(0, 2) = 4$

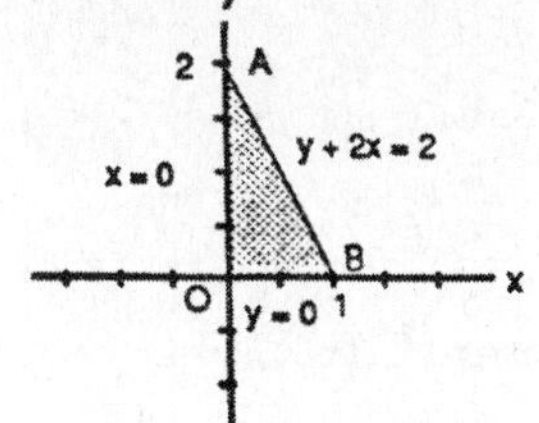

(ii) On OB, $f(x, y) = f(x, 0) = x^2$ on $0 \le x \le 1$; $f'(x, 0) = 2x = 0 \Rightarrow x = 0$ and $y = 0$; $f(0, 0) = 0$ and $f(1, 0) = 1$

(iii) On AB, $f(x, y) = f(x, -2x + 2) = 5x^2 - 8x + 4$ on $0 \le x \le 1$; $f'(x, -2x + 2) = 10x - 8 = 0 \Rightarrow x = \frac{4}{5}$ and $y = \frac{2}{5}$; $f\left(\frac{4}{5}, \frac{2}{5}\right) = \frac{4}{5}$; endpoint values have been found above.

(iv) For interior points of the triangular region, $f_x(x, y) = 2x = 0$ and $f_y(x, y) = 2y = 0 \Rightarrow x = 0$ and $y = 0$, but $(0, 0)$ is not an interior point of the region. Therefore the absolute maximum is 4 at $(0, 2)$ and the absolute minimum is 0 at $(0, 0)$.

35. (i) On OC, $T(x, y) = T(x, 0) = x^2 - 6x + 2$ on $0 \le x \le 5$; $T'(x, 0) = 2x - 6 = 0 \Rightarrow x = 3$ and $y = 0$; $T(3, 0) = -7$, $T(0, 0) = 2$, and $T(5, 0) = -3$

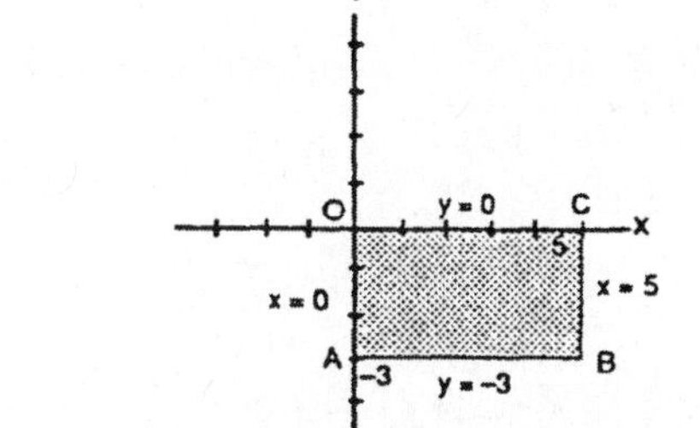

(ii) On CB, $T(x, y) = T(5, y) = y^2 + 5y - 3$ on $-3 \le y \le 0$; $T'(5, y) = 2y + 5 = 0 \Rightarrow y = -\frac{5}{2}$ and $x = 5$; $T\left(5, -\frac{5}{2}\right) = -\frac{37}{4}$ and $T(5, -3) = -9$

(iii) On AB, $T(x, y) = T(x, -3) = x^2 - 9x + 11$ on $0 \le x \le 5$; $T'(x, -3) = 2x - 9 = 0 \Rightarrow x = \frac{9}{2}$ and $y = -3$; $T\left(\frac{9}{2}, -3\right) = -\frac{37}{4}$ and $T(0, -3) = 11$

(iv) On AO, $T(x, y) = T(0, y) = y^2 + 2$ on $-3 \le y \le 0$; $T'(0, y) = 2y = 0 \Rightarrow y = 0$ and $x = 0$, but $(0, 0)$ is not an interior point of AO

(v) For interior points of the rectangular region, $T_x(x, y) = 2x + y - 6 = 0$ and $T_y(x, y) = x + 2y = 0 \Rightarrow x = 4$ and $y = -2$, an interior critical point with $T(4, -2) = -10$. Therefore the absolute maximum is 11 at $(0, -3)$ and the absolute minimum is -10 at $(4, -2)$.

37. (i) On AB, $f(x, y) = f(1, y) = 3 \cos y$ on $-\frac{\pi}{4} \le y \le \frac{\pi}{4}$; $f'(1, y) = -3 \sin y = 0 \Rightarrow y = 0$ and $x = 1$; $f(1, 0) = 3$, $f\left(1, -\frac{\pi}{4}\right) = \frac{3\sqrt{2}}{2}$, and $f\left(1, \frac{\pi}{4}\right) = \frac{3\sqrt{2}}{2}$

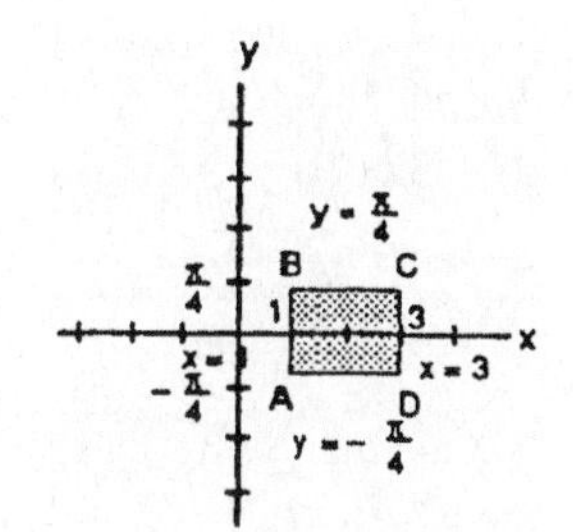

(ii) On CD, $f(x, y) = f(3, y) = 3 \cos y$ on $-\frac{\pi}{4} \le y \le \frac{\pi}{4}$; $f'(3, y) = -3 \sin y = 0 \Rightarrow y = 0$ and $x = 3$; $f(3, 0) = 3$, $f\left(3, -\frac{\pi}{4}\right) = \frac{3\sqrt{2}}{2}$ and $f\left(3, \frac{\pi}{4}\right) = \frac{3\sqrt{2}}{2}$

(iii) On BC, $f(x, y) = f\left(x, \frac{\pi}{4}\right) = \frac{\sqrt{2}}{2}(4x - x^2)$ on $1 \le x \le 3$; $f'\left(x, \frac{\pi}{4}\right) = \sqrt{2}(2 - x) = 0 \Rightarrow x = 2$ and $y = \frac{\pi}{4}$; $f\left(2, \frac{\pi}{4}\right) = 2\sqrt{2}$, $f\left(1, \frac{\pi}{4}\right) = \frac{3\sqrt{2}}{2}$, and $f\left(3, \frac{\pi}{4}\right) = \frac{3\sqrt{2}}{2}$

(iv) On AD, $f(x, y) = f\left(x, -\frac{\pi}{4}\right) = \frac{\sqrt{2}}{2}(4x - x^2)$ on $1 \le x \le 3$; $f'\left(x, -\frac{\pi}{4}\right) = \sqrt{2}(2 - x) = 0 \Rightarrow x = 2$ and $y = -\frac{\pi}{4}$; $f\left(2, -\frac{\pi}{4}\right) = 2\sqrt{2}$, $f\left(1, -\frac{\pi}{4}\right) = \frac{3\sqrt{2}}{2}$, and $f\left(3, -\frac{\pi}{4}\right) = \frac{3\sqrt{2}}{2}$

(v) For interior points of the region, $f_x(x, y) = (4 - 2x)\cos y = 0$ and $f_y(x, y) = -(4x - x^2)\sin y = 0 \Rightarrow x = 2$ and $y = 0$, which is an interior critical point with $f(2, 0) = 4$. Therefore the absolute maximum is 4 at $(2, 0)$ and the absolute minimum is $\frac{3\sqrt{2}}{2}$ at $\left(3, -\frac{\pi}{4}\right)$, $\left(3, \frac{\pi}{4}\right)$, $\left(1, -\frac{\pi}{4}\right)$, and $\left(1, \frac{\pi}{4}\right)$.

39. Let $F(a, b) = \int_a^b (6 - x - x^2)\, dx$ where $a \le b$. The boundary of the domain of F is the line $a = b$ in the ab-plane, and $F(a, a) = 0$, so F is identically 0 on the boundary of its domain. For interior critical points we have:

$\frac{\partial F}{\partial a} = -(6 - a - a^2) = 0 \Rightarrow a = -3, 2$ and $\frac{\partial F}{\partial b} = (6 - b - b^2) = 0 \Rightarrow b = -3, 2$. Since $a \le b$, there is only one interior critical point $(-3, 2)$ and $F(-3, 2) = \int_{-3}^{2} (6 - x - x^2)\,dx$ gives the area under the parabola $y = 6 - x - x^2$ that is above the x-axis. Therefore, $a = -3$ and $b = 2$.

41. $T_x(x, y) = 2x - 1 = 0$ and $T_y(x, y) = 4y = 0 \Rightarrow x = \frac{1}{2}$ and $y = 0$ with $T\left(\frac{1}{2}, 0\right) = -\frac{1}{4}$; on the boundary $x^2 + y^2 = 1$: $T(x, y) = -x^2 - x + 2$ for $-1 \le x \le 1 \Rightarrow T'(x, y) = -2x - 1 = 0 \Rightarrow x = -\frac{1}{2}$ and $y = \pm\frac{\sqrt{3}}{2}$; $T\left(-\frac{1}{2}, \frac{\sqrt{3}}{2}\right) = \frac{9}{4}$, $T\left(-\frac{1}{2}, -\frac{\sqrt{3}}{2}\right) = \frac{9}{4}$, $T(-1, 0) = 2$, and $T(1, 0) = 0 \Rightarrow$ the hottest is $2\frac{1}{4}°$ at $\left(-\frac{1}{2}, \frac{\sqrt{3}}{2}\right)$ and $\left(-\frac{1}{2}, -\frac{\sqrt{3}}{2}\right)$; the coldest is $-\frac{1}{4}°$ at $\left(\frac{1}{2}, 0\right)$.

43. (a) $f_x(x, y) = 2x - 4y = 0$ and $f_y(x, y) = 2y - 4x = 0 \Rightarrow x = 0$ and $y = 0$; $f_{xx}(0, 0) = 2$, $f_{yy}(0, 0) = 2$, $f_{xy}(0, 0) = -4 \Rightarrow f_{xx}f_{yy} - f_{xy}^2 = -12 < 0 \Rightarrow$ saddle point at $(0, 0)$

(b) $f_x(x, y) = 2x - 2 = 0$ and $f_y(x, y) = 2y - 4 = 0 \Rightarrow x = 1$ and $y = 2$; $f_{xx}(1, 2) = 2$, $f_{yy}(1, 2) = 2$, $f_{xy}(1, 2) = 0 \Rightarrow f_{xx}f_{yy} - f_{xy}^2 = 4 > 0$ and $f_{xx} > 0 \Rightarrow$ local minimum at $(1, 2)$

(c) $f_x(x, y) = 9x^2 - 9 = 0$ and $f_y(x, y) = 2y + 4 = 0 \Rightarrow x = \pm 1$ and $y = -2$; $f_{xx}(1, -2) = 18x|_{(1,-2)} = 18$, $f_{yy}(1, -2) = 2$, $f_{xy}(1, -2) = 0 \Rightarrow f_{xx}f_{yy} - f_{xy}^2 = 36 > 0$ and $f_{xx} > 0 \Rightarrow$ local minimum at $(1, -2)$; $f_{xx}(-1, -2) = -18$, $f_{yy}(-1, -2) = 2$, $f_{xy}(-1, -2) = 0 \Rightarrow f_{xx}f_{yy} - f_{xy}^2 = -36 < 0 \Rightarrow$ saddle point at $(-1, -2)$

45. If $k = 0$, then $f(x, y) = x^2 + y^2 \Rightarrow f_x(x, y) = 2x = 0$ and $f_y(x, y) = 2y = 0 \Rightarrow x = 0$ and $y = 0 \Rightarrow (0, 0)$ is the only critical point. If $k \ne 0$, $f_x(x, y) = 2x + ky = 0 \Rightarrow y = -\frac{2}{k}x$; $f_y(x, y) = kx + 2y = 0 \Rightarrow kx + 2\left(-\frac{2}{k}x\right) = 0 \Rightarrow kx - \frac{4x}{k} = 0 \Rightarrow \left(k - \frac{4}{k}\right)x = 0 \Rightarrow x = 0$ or $k = \pm 2 \Rightarrow y = \left(-\frac{2}{k}\right)(0) = 0$ or $y = \pm x$; in any case $(0, 0)$ is a critical point.

47. No; for example $f(x, y) = xy$ has a saddle point at $(a, b) = (0, 0)$ where $f_x = f_y = 0$.

49. We want the point on $z = 10 - x^2 - y^2$ where the tangent plane is parallel to the plane $x + 2y + 3z = 0$. To find a normal vector to $z = 10 - x^2 - y^2$ let $w = z + x^2 + y^2 - 10$. Then $\nabla w = 2x\mathbf{i} + 2y\mathbf{j} + \mathbf{k}$ is normal to $z = 10 - x^2 - y^2$ at (x, y). The vector ∇w is parallel to $\mathbf{i} + 2\mathbf{j} + 3\mathbf{k}$ which is normal to the plane $x + 2y + 3z = 0$ if $6x\mathbf{i} + 6y\mathbf{j} + 3\mathbf{k} = \mathbf{i} + 2\mathbf{j} + 3\mathbf{k}$ or $x = \frac{1}{6}$ and $y = \frac{1}{3}$. Thus the point is $\left(\frac{1}{6}, \frac{1}{3}, 10 - \frac{1}{36} - \frac{1}{9}\right)$ or $\left(\frac{1}{6}, \frac{1}{3}, \frac{355}{36}\right)$.

51. $d(x, y, z) = \sqrt{(x - 0)^2 + (y - 0)^2 + (z - 0)^2} \Rightarrow$ we can minimize $d(x, y, z)$ by minimizing $D(x, y, z) = x^2 + y^2 + z^2$; $3x + 2y + z = 6 \Rightarrow z = 6 - 3x - 2y \Rightarrow D(x, y) = x^2 + y^2 + (6 - 3x - 2y)^2 \Rightarrow D_x(x, y) = 2x - 6(6 - 3x - 2y) = 0$ and $D_y(x, y) = 2y - 4(6 - 3x - 2y) = 0 \Rightarrow$ critical point is $\left(\frac{9}{7}, \frac{6}{7}\right) \Rightarrow z = \frac{3}{7}$; $D_{xx}\left(\frac{9}{7}, \frac{6}{7}\right) = 20$, $D_{yy}\left(\frac{1}{2}, 1\right) = 10$, $D_{xy}\left(\frac{1}{2}, 1\right) = 12 \Rightarrow D_{xx}D_{yy} - D_{xy}^2 = 56 > 0$ and $D_{xx} > 0 \Rightarrow$ local minimum of $d\left(\frac{9}{7}, \frac{6}{7}, \frac{3}{7}\right) = \frac{3\sqrt{14}}{7}$

53. $s(x, y, z) = x^2 + y^2 + z^2$; $x + y + z = 9 \Rightarrow z = 9 - x - y \Rightarrow s(x, y) = x^2 + y^2 + (9 - x - y)^2 \Rightarrow s_x(x, y) = 2x - 2(9 - x - y) = 0$ and $s_y(x, y) = 2y - 2(9 - x - y) = 0 \Rightarrow$ critical point is $(3, 3) \Rightarrow z = 3$; $s_{xx}(3, 3) = 4$, $s_{yy}(3, 3) = 4$, $s_{xy}(3, 3) = 2 \Rightarrow s_{xx}s_{yy} - s_{xy}^2 = 12 > 0$ and $s_{xx} > 0 \Rightarrow$ local minimum of $s(3, 3, 3) = 27$

55. $s(x, y, z) = xy + yz + xz$; $x + y + z = 6 \Rightarrow z = 6 - x - y \Rightarrow s(x, y) = xy + y(6 - x - y) + x(6 - x - y) = 6x + 6y - xy - x^2 - y^2 \Rightarrow s_x(x, y) = 6 - 2x - y = 0$ and $s_y(x, y) = 6 - x - 2y = 0 \Rightarrow$ critical point is $(2, 2) \Rightarrow z = 2$; $s_{xx}(2, 2) = -2$, $s_{yy}(2, 2) = -2$, $s_{xy}(2, 2) = -1 \Rightarrow s_{xx}s_{yy} - s_{xy}^2 = 3 > 0$ and $s_{xx} < 0 \Rightarrow$ local maximum of $s(2, 2, 2) = 12$

57. $V(x, y, z) = (2x)(2y)(2z) = 8xyz;\ x^2 + y^2 + z^2 = 4 \Rightarrow z = \sqrt{4 - x^2 - y^2} \Rightarrow V(x, y) = 8xy\sqrt{4 - x^2 - y^2}$, $x \geq 0$ and $y \geq 0 \Rightarrow V_x(x, y) = \frac{32y - 16x^2y - 8y^3}{\sqrt{4 - x^2 - y^2}} = 0$ and $V_y(x, y) = \frac{32x - 16xy^2 - 8x^3}{\sqrt{4 - x^2 - y^2}} = 0 \Rightarrow$ critical points are $(0, 0), \left(\frac{2}{\sqrt{3}}, \frac{2}{\sqrt{3}}\right), \left(\frac{2}{\sqrt{3}}, -\frac{2}{\sqrt{3}}\right), \left(-\frac{2}{\sqrt{3}}, \frac{2}{\sqrt{3}}\right)$, and $\left(-\frac{2}{\sqrt{3}}, -\frac{2}{\sqrt{3}}\right)$. Only $(0, 0)$ and $\left(\frac{2}{\sqrt{3}}, \frac{2}{\sqrt{3}}\right)$ satisfy $x \geq 0$ and $y \geq 0$ $V(0, 0) = 0$ and $V\left(\frac{2}{\sqrt{3}}, \frac{2}{\sqrt{3}}\right) = \frac{64}{3\sqrt{3}}$; On $x = 0,\ 0 \leq y \leq 2 \Rightarrow V(0, y) = 8(0)y\sqrt{4 - 0^2 - y^2} = 0$, no critical points, $V(0, 0) = 0,\ V(0, 2) = 0$; On $y = 0,\ 0 \leq x \leq 2 \Rightarrow V(x, 0) = 8x(0)\sqrt{4 - x^2 - 0^2} = 0$, no critical points, $V(0, 0) = 0$, $V(0, 2) = 0$; On $y = \sqrt{4 - x^2},\ 0 \leq x \leq 2 \Rightarrow V\left(x, \sqrt{4 - x^2}\right) = 8x\sqrt{4 - x^2}\sqrt{4 - x^2 - \left(\sqrt{4 - x^2}\right)^2} = 0$ no critical points, $V(0, 2) = 0,\ V(2, 0) = 0$. Thus, there is a maximum volume of $\frac{64}{3\sqrt{3}}$ if the box is $\frac{2}{\sqrt{3}} \times \frac{2}{\sqrt{3}} \times \frac{2}{\sqrt{3}}$.

59. Let x = height of the box, y = width, and z = length, cut out squares of length x from corner of the material See diagram at right. Fold along the dashed lines to form the box. From the diagram we see that the length of the material is $2x + y$ and the width is $2x + z$. Thus $(2x + y)(2x + z) = 12$ $\Rightarrow z = \frac{2(6 - 2x^2 + xy)}{2x + y}$. Since $V(x, y, z) = xyz$ $\Rightarrow V(x, y) = \frac{2xy(6 - 2x^2 + xy)}{2x + y}$, where $x > 0,\ y > 0$. $V_x(x, y) = \frac{4(3y^2 - 4x^3y - 4x^2y^2 - xy^3)}{(2x + y)^2} = 0$ and $V_y(x, y) = \frac{2(12x^2 - 4x^4 - 4x^3y - x^2y^2)}{(2x + y)^2} = 0 \Rightarrow$ critical points are $\left(\sqrt{3}, 0\right), \left(-\sqrt{3}, 0\right), \left(\frac{1}{\sqrt{3}}, \frac{4}{\sqrt{3}}\right)$, and $\left(-\frac{1}{\sqrt{3}}, -\frac{4}{\sqrt{3}}\right)$. Only $\left(\sqrt{3}, 0\right)$ and $\left(\frac{1}{\sqrt{3}}, \frac{4}{\sqrt{3}}\right)$ satisfy $x > 0$ and $y > 0$. For $\left(\sqrt{3}, 0\right)$: $z = 0$; $V_{xx}\left(\sqrt{3}, 0\right) = 0$, $V_{yy}\left(\sqrt{3}, 0\right) = -2\sqrt{3},\ V_{xy}\left(\sqrt{3}, 0\right) = -4\sqrt{3} \Rightarrow V_{xx}V_{yy} - V_{xy}^2 = -48 < 0 \Rightarrow$ saddle point. For $\left(\frac{1}{\sqrt{3}}, \frac{4}{\sqrt{3}}\right)$: $z = \frac{4}{\sqrt{3}}$; $V_{xx}\left(\frac{1}{\sqrt{3}}, \frac{4}{\sqrt{3}}\right) = -\frac{80}{3\sqrt{3}},\ V_{yy}\left(\frac{1}{\sqrt{3}}, \frac{4}{\sqrt{3}}\right) = -\frac{2}{3\sqrt{3}},\ V_{xy}\left(\frac{1}{\sqrt{3}}, \frac{4}{\sqrt{3}}\right) = -\frac{4}{3\sqrt{3}} \Rightarrow V_{xx}V_{yy} - V_{xy}^2 = \frac{16}{3} > 0$ and $V_{xx} < 0 \Rightarrow$ local maximum of $V\left(\frac{1}{\sqrt{3}}, \frac{4}{\sqrt{3}}, \frac{4}{\sqrt{3}}\right) = \frac{16}{3\sqrt{3}}$

61. (a) $\frac{df}{dt} = \frac{\partial f}{\partial x}\frac{dx}{dt} + \frac{\partial f}{\partial y}\frac{dy}{dt} = \frac{dx}{dt} + \frac{dy}{dt} = -2\sin t + 2\cos t = 0 \Rightarrow \cos t = \sin t \Rightarrow x = y$

(i) On the semicircle $x^2 + y^2 = 4,\ y \geq 0$, we have $t = \frac{\pi}{4}$ and $x = y = \sqrt{2} \Rightarrow f\left(\sqrt{2}, \sqrt{2}\right) = 2\sqrt{2}$. At the endpoints, $f(-2, 0) = -2$ and $f(2, 0) = 2$. Therefore the absolute minimum is $f(-2, 0) = -2$ when $t = \pi$; the absolute maximum is $f\left(\sqrt{2}, \sqrt{2}\right) = 2\sqrt{2}$ when $t = \frac{\pi}{4}$.

(ii) On the quartercircle $x^2 + y^2 = 4,\ x \geq 0$ and $y \geq 0$, the endpoints give $f(0, 2) = 2$ and $f(2, 0) = 2$. Therefore the absolute minimum is $f(2, 0) = 2$ and $f(0, 2) = 2$ when $t = 0, \frac{\pi}{2}$ respectively; the absolute maximum is $f\left(\sqrt{2}, \sqrt{2}\right) = 2\sqrt{2}$ when $t = \frac{\pi}{4}$.

(b) $\frac{dg}{dt} = \frac{\partial g}{\partial x}\frac{dx}{dt} + \frac{\partial g}{\partial y}\frac{dy}{dt} = y\frac{dx}{dt} + x\frac{dy}{dt} = -4\sin^2 t + 4\cos^2 t = 0 \Rightarrow \cos t = \pm\sin t \Rightarrow x = \pm y$.

(i) On the semicircle $x^2 + y^2 = 4,\ y \geq 0$, we obtain $x = y = \sqrt{2}$ at $t = \frac{\pi}{4}$ and $x = -\sqrt{2},\ y = \sqrt{2}$ at $t = \frac{3\pi}{4}$. Then $g\left(\sqrt{2}, \sqrt{2}\right) = 2$ and $g\left(-\sqrt{2}, \sqrt{2}\right) = -2$. At the endpoints, $g(-2, 0) = g(2, 0) = 0$. Therefore the absolute minimum is $g\left(-\sqrt{2}, \sqrt{2}\right) = -2$ when $t = \frac{3\pi}{4}$; the absolute maximum is $g\left(\sqrt{2}, \sqrt{2}\right) = 2$ when $t = \frac{\pi}{4}$.

(ii) On the quartercircle $x^2 + y^2 = 4,\ x \geq 0$ and $y \geq 0$, the endpoints give $g(0, 2) = 0$ and $g(2, 0) = 0$. Therefore the absolute minimum is $g(2, 0) = 0$ and $g(0, 2) = 0$ when $t = 0, \frac{\pi}{2}$ respectively; the absolute maximum is $g\left(\sqrt{2}, \sqrt{2}\right) = 2$ when $t = \frac{\pi}{4}$.

(c) $\frac{dh}{dt} = \frac{\partial h}{\partial x}\frac{dx}{dt} + \frac{\partial h}{\partial y}\frac{dy}{dt} = 4x\frac{dx}{dt} + 2y\frac{dy}{dt} = (8\cos t)(-2\sin t) + (4\sin t)(2\cos t) = -8\cos t\sin t = 0$
$\Rightarrow t = 0, \frac{\pi}{2}, \pi$ yielding the points $(2,0)$, $(0,2)$ for $0 \le t \le \pi$.

(i) On the semicircle $x^2 + y^2 = 4$, $y \ge 0$ we have $h(2,0) = 8$, $h(0,2) = 4$, and $h(-2,0) = 8$. Therefore, the absolute minimum is $h(0,2) = 4$ when $t = \frac{\pi}{2}$; the absolute maximum is $h(2,0) = 8$ and $h(-2,0) = 8$ when $t = 0, \pi$ respectively.

(ii) On the quartercircle $x^2 + y^2 = 4$, $x \ge 0$ and $y \ge 0$ the absolute minimum is $h(0,2) = 4$ when $t = \frac{\pi}{2}$; the absolute maximum is $h(2,0) = 8$ when $t = 0$.

63. $\frac{df}{dt} = \frac{\partial f}{\partial x}\frac{dx}{dt} + \frac{\partial f}{\partial y}\frac{dy}{dt} = y\frac{dx}{dt} + x\frac{dy}{dt}$

(i) $x = 2t$ and $y = t + 1 \Rightarrow \frac{df}{dt} = (t+1)(2) + (2t)(1) = 4t + 2 = 0 \Rightarrow t = -\frac{1}{2} \Rightarrow x = -1$ and $y = \frac{1}{2}$ with $f\left(-1, \frac{1}{2}\right) = -\frac{1}{2}$. The absolute minimum is $f\left(-1, \frac{1}{2}\right) = -\frac{1}{2}$ when $t = -\frac{1}{2}$; there is no absolute maximum.

(ii) For the endpoints: $t = -1 \Rightarrow x = -2$ and $y = 0$ with $f(-2,0) = 0$; $t = 0 \Rightarrow x = 0$ and $y = 1$ with $f(0,1) = 0$. The absolute minimum is $f\left(-1, \frac{1}{2}\right) = -\frac{1}{2}$ when $t = -\frac{1}{2}$; the absolute maximum is $f(0,1) = 0$ and $f(-2,0) = 0$ when $t = -1, 0$ respectively.

(iii) There are no interior critical points. For the endpoints: $t = 0 \Rightarrow x = 0$ and $y = 1$ with $f(0,1) = 0$; $t = 1 \Rightarrow x = 2$ and $y = 2$ with $f(2,2) = 4$. The absolute minimum is $f(0,1) = 0$ when $t = 0$; the absolute maximum is $f(2,2) = 4$ when $t = 1$.

65. $w = (mx_1 + b - y_1)^2 + (mx_2 + b - y_2)^2 + \cdots + (mx_n + b - y_n)^2$
$\Rightarrow \frac{\partial w}{\partial m} = 2(mx_1 + b - y_1)(x_1) + 2(mx_2 + b - y_2)(x_2) + \cdots + 2(mx_n + b - y_n)(x_n)$
$\Rightarrow \frac{\partial w}{\partial b} = 2(mx_1 + b - y_1)(1) + 2(mx_2 + b - y_2)(1) + \cdots + 2(mx_n + b - y_n)(1)$
$\frac{\partial w}{\partial m} = 0 \Rightarrow 2\left[(mx_1 + b - y_1)(x_1) + (mx_2 + b - y_2)(x_2) + \cdots + (mx_n + b - y_n)(x_n)\right] = 0$
$\Rightarrow mx_1^2 + bx_1 - x_1y_1 + mx_2^2 + bx_2 - x_2y_2 + \cdots + mx_n^2 + bx_n - x_ny_n = 0$
$\Rightarrow m(x_1^2 + x_2^2 + \cdots + x_n^2) + b(x_1 + x_2 + \cdots + x_n) - (x_1y_1 + x_2y_2 + \cdots + x_ny_n) = 0$
$\Rightarrow m\sum_{k=1}^{n}(x_k^2) + b\sum_{k=1}^{n}x_k - \sum_{k=1}^{n}(x_ky_k) = 0$
$\frac{\partial w}{\partial b} = 0 \Rightarrow 2\left[(mx_1 + b - y_1) + (mx_2 + b - y_2) + \cdots + (mx_n + b - y_n)\right] = 0$
$\Rightarrow mx_1 + b - y_1 + mx_2 + b - y_2 + \cdots + mx_n + b - y_n = 0$
$\Rightarrow m(x_1 + x_2 + \cdots + x_n) + (b + b + \cdots + b) - (y_1 + y_2 + \cdots + y_n) = 0$
$\Rightarrow m\sum_{k=1}^{n}x_k + b\sum_{k=1}^{n}1 - \sum_{k=1}^{n}y_k = 0 \Rightarrow m\sum_{k=1}^{n}x_k + bn - \sum_{k=1}^{n}y_k = 0 \Rightarrow b = \frac{1}{n}\left(\sum_{k=1}^{n}y_k - m\sum_{k=1}^{n}x_k\right)$.

Substituting for b in the equation obtained for $\frac{\partial w}{\partial m}$ we get $m\sum_{k=1}^{n}(x_k^2) + \frac{1}{n}\left(\sum_{k=1}^{n}y_k - m\sum_{k=1}^{n}x_k\right)\sum_{k=1}^{n}x_k - \sum_{k=1}^{n}(x_ky_k) = 0$.

Multiply both sides by n to obtain $mn\sum_{k=1}^{n}(x_k^2) + \left(\sum_{k=1}^{n}y_k - m\sum_{k=1}^{n}x_k\right)\sum_{k=1}^{n}x_k - n\sum_{k=1}^{n}(x_ky_k) = 0$

$\Rightarrow mn\sum_{k=1}^{n}(x_k^2) + \left(\sum_{k=1}^{n}x_k\right)\left(\sum_{k=1}^{n}y_k\right) - m\left(\sum_{k=1}^{n}x_k\right)^2 - n\sum_{k=1}^{n}(x_ky_k) = 0$

$\Rightarrow mn\sum_{k=1}^{n}(x_k^2) - m\left(\sum_{k=1}^{n}x_k\right)^2 = n\sum_{k=1}^{n}(x_ky_k) - \left(\sum_{k=1}^{n}x_k\right)\left(\sum_{k=1}^{n}y_k\right)$

$\Rightarrow m\left[n\sum_{k=1}^{n}(x_k^2) - \left(\sum_{k=1}^{n}x_k\right)^2\right] = n\sum_{k=1}^{n}(x_ky_k) - \left(\sum_{k=1}^{n}x_k\right)\left(\sum_{k=1}^{n}y_k\right)$

$$\Rightarrow m = \frac{n\sum_{k=1}^{n}(x_ky_k) - \left(\sum_{k=1}^{n}x_k\right)\left(\sum_{k=1}^{n}y_k\right)}{n\sum_{k=1}^{n}(x_k^2) - \left(\sum_{k=1}^{n}x_k\right)^2} = \frac{\left(\sum_{k=1}^{n}x_k\right)\left(\sum_{k=1}^{n}y_k\right) - n\sum_{k=1}^{n}(x_ky_k)}{\left(\sum_{k=1}^{n}x_k\right)^2 - n\sum_{k=1}^{n}(x_k^2)}$$

To show that these values for m and b minimize the sum of the squares of the distances, use second derivative test.

$\frac{\partial^2 w}{\partial m^2} = 2x_1^2 + 2x_2^2 + \cdots + 2x_n^2 = 2\sum_{k=1}^{n}(x_k^2)$; $\frac{\partial^2 w}{\partial m\,\partial b} = 2x_1 + 2x_2 + \cdots + 2x_n = 2\sum_{k=1}^{n} x_k$; $\frac{\partial^2 w}{\partial b^2} = 2 + 2 + \cdots + 2 = 2n$

The discriminant is: $\left(\frac{\partial^2 w}{\partial m^2}\right)\left(\frac{\partial^2 w}{\partial b^2}\right) - \left(\frac{\partial^2 w}{\partial m\,\partial b}\right)^2 = \left[2\sum_{k=1}^{n}(x_k^2)\right](2n) - \left[2\sum_{k=1}^{n} x_k\right]^2 = 4\left[n\sum_{k=1}^{n}(x_k^2) - \left(\sum_{k=1}^{n} x_k\right)^2\right]$.

Now, $n\sum_{k=1}^{n}(x_k^2) - \left(\sum_{k=1}^{n} x_k\right)^2 = n(x_1^2 + x_2^2 + \cdots + x_n^2) - (x_1 + x_2 + \cdots + x_n)(x_1 + x_2 + \cdots + x_n)$

$= nx_1^2 + nx_2^2 + \cdots + nx_n^2 - x_1^2 - x_1x_2 - \cdots - x_1x_n - x_2x_1 - x_2^2 - \cdots - x_2x_n - x_nx_1 - x_nx_2 - \cdots - x_n^2$

$= (n-1)x_1^2 + (n-1)x_2^2 + \cdots + (n-1)x_n^2 - 2x_1x_2 - 2x_1x_3 - \cdots - 2x_1x_n - 2x_2x_3 - \cdots - 2x_2x_n - \cdots - 2x_{n-1}x_n$

$= (x_1^2 - 2x_1x_2 + x_2^2) + (x_1^2 - 2x_1x_3 + x_3^2) + \cdots + (x_1^2 - 2x_1x_n + x_n^2) + (x_2^2 - 2x_2x_3 + x_3^2) + \cdots + (x_2^2 - 2x_2x_n + x_n^2)$
$+ \cdots + (x_{n-1}^2 - 2x_{n-1}x_n + x_n^2)$

$= (x_1 - x_2)^2 + (x_1 - x_3)^2 + \cdots + (x_1 - x_n)^2 + (x_2 - x_3)^2 + \cdots + (x_2 - x_n)^2 + \cdots + (x_{n-1} - x_n)^2 \ge 0.$

Thus we have : $\left(\frac{\partial^2 w}{\partial m^2}\right)\left(\frac{\partial^2 w}{\partial b^2}\right) - \left(\frac{\partial^2 w}{\partial m\,\partial b}\right)^2 = 4\left[n\sum_{k=1}^{n}(x_k^2) - \left(\sum_{k=1}^{n} x_k\right)^2\right] \ge 4(0) = 0$. If $x_1 = x_2 = \cdots = x_n$ then

$\left(\frac{\partial^2 w}{\partial m^2}\right)\left(\frac{\partial^2 w}{\partial b^2}\right) - \left(\frac{\partial^2 w}{\partial m\,\partial b}\right)^2 = 0$. Also, $\frac{\partial^2 w}{\partial m^2} = 2\sum_{k=1}^{n}(x_k^2) \ge 0$. If $x_1 = x_2 = \cdots = x_n = 0$, then $\frac{\partial^2 w}{\partial m^2} = 0$.

Provided that at least one x_i is nonzero and different from the rest of x_j, $j \ne i$, then $\left(\frac{\partial^2 w}{\partial m^2}\right)\left(\frac{\partial^2 w}{\partial b^2}\right) - \left(\frac{\partial^2 w}{\partial m\,\partial b}\right)^2 > 0$ and $\frac{\partial^2 w}{\partial m^2} > 0 \Rightarrow$ the values given above for m and b minimize w.

67. $m = \frac{(2)(-1) - 3(-14)}{(2)^2 - 3(10)} = -\frac{20}{13}$ and
$b = \frac{1}{3}\left[-1 - \left(-\frac{20}{13}\right)(2)\right] = \frac{9}{13}$
$\Rightarrow y = -\frac{20}{13}x + \frac{9}{13}$; $y\big|_{x=4} = -\frac{71}{13}$

k	x_k	y_k	x_k^2	$x_k y_k$
1	−1	2	1	−2
2	0	1	0	0
3	3	−4	9	−12
Σ	2	−1	10	−14

14.8 LAGRANGE MULTIPLIERS

1. $\nabla f = y\mathbf{i} + x\mathbf{j}$ and $\nabla g = 2x\mathbf{i} + 4y\mathbf{j}$ so that $\nabla f = \lambda \nabla g \Rightarrow y\mathbf{i} + x\mathbf{j} = \lambda(2x\mathbf{i} + 4y\mathbf{j}) \Rightarrow y = 2x\lambda$ and $x = 4y\lambda$ $\Rightarrow x = 8x\lambda^2 \Rightarrow \lambda = \pm\frac{\sqrt{2}}{4}$ or $x = 0$.
CASE 1: If $x = 0$, then $y = 0$. But $(0, 0)$ is not on the ellipse so $x \ne 0$.
CASE 2: $x \ne 0 \Rightarrow \lambda = \pm\frac{\sqrt{2}}{4} \Rightarrow x = \pm\sqrt{2}y \Rightarrow \left(\pm\sqrt{2}y\right)^2 + 2y^2 = 1 \Rightarrow y = \pm\frac{1}{2}$.
Therefore f takes on its extreme values at $\left(\pm\frac{\sqrt{2}}{2}, \frac{1}{2}\right)$ and $\left(\pm\frac{\sqrt{2}}{2}, -\frac{1}{2}\right)$. The extreme values of f on the ellipse are $\pm\frac{\sqrt{2}}{2}$.

3. $\nabla f = -2x\mathbf{i} - 2y\mathbf{j}$ and $\nabla g = \mathbf{i} + 3\mathbf{j}$ so that $\nabla f = \lambda \nabla g \Rightarrow -2x\mathbf{i} - 2y\mathbf{j} = \lambda(\mathbf{i} + 3\mathbf{j}) \Rightarrow x = -\frac{\lambda}{2}$ and $y = -\frac{3\lambda}{2}$ $\Rightarrow \left(-\frac{\lambda}{2}\right) + 3\left(-\frac{3\lambda}{2}\right) = 10 \Rightarrow \lambda = -2 \Rightarrow x = 1$ and $y = 3 \Rightarrow$ f takes on its extreme value at $(1, 3)$ on the line. The extreme value is $f(1, 3) = 49 - 1 - 9 = 39$.

5. We optimize $f(x, y) = x^2 + y^2$, the square of the distance to the origin, subject to the constraint $g(x, y) = xy^2 - 54 = 0$. Thus $\nabla f = 2x\mathbf{i} + 2y\mathbf{j}$ and $\nabla g = y^2\mathbf{i} + 2xy\mathbf{j}$ so that $\nabla f = \lambda \nabla g \Rightarrow 2x\mathbf{i} + 2y\mathbf{j}$ $= \lambda\left(y^2\mathbf{i} + 2xy\mathbf{j}\right) \Rightarrow 2x = \lambda y^2$ and $2y = 2\lambda xy$.
CASE 1: If $y = 0$, then $x = 0$. But $(0, 0)$ does not satisfy the constraint $xy^2 = 54$ so $y \ne 0$.
CASE 2: If $y \ne 0$, then $2 = 2\lambda x \Rightarrow x = \frac{1}{\lambda} \Rightarrow 2\left(\frac{1}{\lambda}\right) = \lambda y^2 \Rightarrow y^2 = \frac{2}{\lambda^2}$. Then $xy^2 = 54 \Rightarrow \left(\frac{1}{\lambda}\right)\left(\frac{2}{\lambda^2}\right) = 54$ $\Rightarrow \lambda^3 = \frac{1}{27} \Rightarrow \lambda = \frac{1}{3} \Rightarrow x = 3$ and $y^2 = 18 \Rightarrow x = 3$ and $y = \pm 3\sqrt{2}$.

Therefore $\left(3, \pm 3\sqrt{2}\right)$ are the points on the curve $xy^2 = 54$ nearest the origin (since $xy^2 = 54$ has points increasingly far away as y gets close to 0, no points are farthest away).

7. (a) $\nabla f = \mathbf{i} + \mathbf{j}$ and $\nabla g = y\mathbf{i} + x\mathbf{j}$ so that $\nabla f = \lambda \nabla g \Rightarrow \mathbf{i} + \mathbf{j} = \lambda(y\mathbf{i} + x\mathbf{j}) \Rightarrow 1 = \lambda y$ and $1 = \lambda x \Rightarrow y = \frac{1}{\lambda}$ and $x = \frac{1}{\lambda} \Rightarrow \frac{1}{\lambda^2} = 16 \Rightarrow \lambda = \pm\frac{1}{4}$. Use $\lambda = \frac{1}{4}$ since $x > 0$ and $y > 0$. Then $x = 4$ and $y = 4 \Rightarrow$ the minimum value is 8 at the point $(4, 4)$. Now, $xy = 16$, $x > 0$, $y > 0$ is a branch of a hyperbola in the first quadrant with the x-and y-axes as asymptotes. The equations $x + y = c$ give a family of parallel lines with $m = -1$. As these lines move away from the origin, the number c increases. Thus the minimum value of c occurs where $x + y = c$ is tangent to the hyperbola's branch.

(b) $\nabla f = y\mathbf{i} + x\mathbf{j}$ and $\nabla g = \mathbf{i} + \mathbf{j}$ so that $\nabla f = \lambda \nabla g \Rightarrow y\mathbf{i} + x\mathbf{j} = \lambda(\mathbf{i} + \mathbf{j}) \Rightarrow y = \lambda = x\ y + y = 16 \Rightarrow y = 8$ $\Rightarrow x = 8 \Rightarrow f(8, 8) = 64$ is the maximum value. The equations $xy = c$ ($x > 0$ and $y > 0$ or $x < 0$ and $y < 0$ to get a maximum value) give a family of hyperbolas in the first and third quadrants with the x- and y-axes as asymptotes. The maximum value of c occurs where the hyperbola $xy = c$ is tangent to the line $x + y = 16$.

9. $V = \pi r^2 h \Rightarrow 16\pi = \pi r^2 h \Rightarrow 16 = r^2 h \Rightarrow g(r, h) = r^2 h - 16$; $S = 2\pi rh + 2\pi r^2 \Rightarrow \nabla S = (2\pi h + 4\pi r)\mathbf{i} + 2\pi r\mathbf{j}$ and $\nabla g = 2rh\mathbf{i} + r^2\mathbf{j}$ so that $\nabla S = \lambda \nabla g \Rightarrow (2\pi rh + 4\pi r)\mathbf{i} + 2\pi r\mathbf{j} = \lambda\left(2rh\mathbf{i} + r^2\mathbf{j}\right) \Rightarrow 2\pi rh + 4\pi r = 2rh\lambda$ and $2\pi r = \lambda r^2$ $\Rightarrow r = 0$ or $\lambda = \frac{2\pi}{r}$. But $r = 0$ gives no physical can, so $r \neq 0 \Rightarrow \lambda = \frac{2\pi}{r} \Rightarrow 2\pi h + 4\pi r = 2rh\left(\frac{2\pi}{r}\right) \Rightarrow 2r = h$ $\Rightarrow 16 = r^2(2r) \Rightarrow r = 2 \Rightarrow h = 4$; thus $r = 2$ cm and $h = 4$ cm give the only extreme surface area of 24π cm^2. Since $r = 4$ cm and $h = 1$ cm $\Rightarrow V = 16\pi$ cm^3 and $S = 40\pi$ cm^2, which is a larger surface area, then 24π cm^2 must be the minimum surface area.

11. $A = (2x)(2y) = 4xy$ subject to $g(x, y) = \frac{x^2}{16} + \frac{y^2}{9} - 1 = 0$; $\nabla A = 4y\mathbf{i} + 4x\mathbf{j}$ and $\nabla g = \frac{x}{8}\mathbf{i} + \frac{2y}{9}\mathbf{j}$ so that ∇A $= \lambda \nabla g \Rightarrow 4y\mathbf{i} + 4x\mathbf{j} = \lambda\left(\frac{x}{8}\mathbf{i} + \frac{2y}{9}\mathbf{j}\right) \Rightarrow 4y = \left(\frac{x}{8}\right)\lambda$ and $4x = \left(\frac{2y}{9}\right)\lambda \Rightarrow \lambda = \frac{32y}{x}$ and $4x = \left(\frac{2y}{9}\right)\left(\frac{32y}{x}\right)$ $\Rightarrow y = \pm\frac{3}{4}x \Rightarrow \frac{x^2}{16} + \frac{\left(\frac{\pm 3}{4}x\right)^2}{9} = 1 \Rightarrow x^2 = 8 \Rightarrow x = \pm 2\sqrt{2}$. We use $x = 2\sqrt{2}$ since x represents distance. Then $y = \frac{3}{4}\left(2\sqrt{2}\right) = \frac{3\sqrt{2}}{2}$, so the length is $2x = 4\sqrt{2}$ and the width is $2y = 3\sqrt{2}$.

13. $\nabla f = 2x\mathbf{i} + 2y\mathbf{j}$ and $\nabla g = (2x - 2)\mathbf{i} + (2y - 4)\mathbf{j}$ so that $\nabla f = \lambda \nabla g = 2x\mathbf{i} + 2y\mathbf{j} = \lambda[(2x - 2)\mathbf{i} + (2y - 4)\mathbf{j}]$ $\Rightarrow 2x = \lambda(2x - 2)$ and $2y = \lambda(2y - 4) \Rightarrow x = \frac{\lambda}{\lambda - 1}$ and $y = \frac{2\lambda}{\lambda - 1}$, $\lambda \neq 1 \Rightarrow y = 2x \Rightarrow x^2 - 2x + (2x)^2 - 4(2x) = 0$ $\Rightarrow x = 0$ and $y = 0$, or $x = 2$ and $y = 4$. Therefore $f(0, 0) = 0$ is the minimum value and $f(2, 4) = 20$ is the maximum value. (Note that $\lambda = 1$ gives $2x = 2x - 2$ or $0 = -2$, which is impossible.)

15. $\nabla T = (8x - 4y)\mathbf{i} + (-4x + 2y)\mathbf{j}$ and $g(x, y) = x^2 + y^2 - 25 = 0 \Rightarrow \nabla g = 2x\mathbf{i} + 2y\mathbf{j}$ so that $\nabla T = \lambda \nabla g$ $\Rightarrow (8x - 4y)\mathbf{i} + (-4x + 2y)\mathbf{j} = \lambda(2x\mathbf{i} + 2y\mathbf{j}) \Rightarrow 8x - 4y = 2\lambda x$ and $-4x + 2y = 2\lambda y \Rightarrow y = \frac{-2x}{\lambda - 1}$, $\lambda \neq 1$ $\Rightarrow 8x - 4\left(\frac{-2x}{\lambda - 1}\right) = 2\lambda x \Rightarrow x = 0$, or $\lambda = 0$, or $\lambda = 5$.

CASE 1: $x = 0 \Rightarrow y = 0$; but $(0, 0)$ is not on $x^2 + y^2 = 25$ so $x \neq 0$.

CASE 2: $\lambda = 0 \Rightarrow y = 2x \Rightarrow x^2 + (2x)^2 = 25 \Rightarrow x = \pm\sqrt{5}$ and $y = 2x$.

CASE 3: $\lambda = 5 \Rightarrow y = \frac{-2x}{4} = -\frac{x}{2} \Rightarrow x^2 + \left(-\frac{x}{2}\right)^2 = 25 \Rightarrow x = \pm 2\sqrt{5} \Rightarrow x = 2\sqrt{5}$ and $y = -\sqrt{5}$, or $x = -2\sqrt{5}$ and $y = \sqrt{5}$.

Therefore $T\left(\sqrt{5}, 2\sqrt{5}\right) = 0° = T\left(-\sqrt{5}, -2\sqrt{5}\right)$ is the minimum value and $T\left(2\sqrt{5}, -\sqrt{5}\right) = 125°$ $= T\left(-2\sqrt{5}, \sqrt{5}\right)$ is the maximum value. (Note: $\lambda = 1 \Rightarrow x = 0$ from the equation $-4x + 2y = 2\lambda y$; but we found $x \neq 0$ in CASE 1.)

17. Let $f(x, y, z) = (x - 1)^2 + (y - 1)^2 + (z - 1)^2$ be the square of the distance from $(1, 1, 1)$. Then $\nabla f = 2(x - 1)\mathbf{i} + 2(y - 1)\mathbf{j} + 2(z - 1)\mathbf{k}$ and $\nabla g = \mathbf{i} + 2\mathbf{j} + 3\mathbf{k}$ so that $\nabla f = \lambda \nabla g$ $\Rightarrow 2(x - 1)\mathbf{i} + 2(y - 1)\mathbf{j} + 2(z - 1)\mathbf{k} = \lambda(\mathbf{i} + 2\mathbf{j} + 3\mathbf{k}) \Rightarrow 2(x - 1) = \lambda$, $2(y - 1) = 2\lambda$, $2(z - 1) = 3\lambda$

$\Rightarrow 2(y-1) = 2[2(x-1)]$ and $2(z-1) = 3[2(x-1)] \Rightarrow x = \frac{y+1}{2} \Rightarrow z+2 = 3\left(\frac{y+1}{2}\right)$ or $z = \frac{3y-1}{2}$; thus $\frac{y+1}{2} + 2y + 3\left(\frac{3y-1}{2}\right) - 13 = 0 \Rightarrow y = 2 \Rightarrow x = \frac{3}{2}$ and $z = \frac{5}{2}$. Therefore the point $\left(\frac{3}{2}, 2, \frac{5}{2}\right)$ is closest (since no point on the plane is farthest from the point $(1, 1, 1)$).

19. Let $f(x, y, z) = x^2 + y^2 + z^2$ be the square of the distance from the origin. Then $\nabla f = 2x\mathbf{i} + 2y\mathbf{j} + 2z\mathbf{k}$ and $\nabla g = 2x\mathbf{i} - 2y\mathbf{j} - 2z\mathbf{k}$ so that $\nabla f = \lambda \nabla g \Rightarrow 2x\mathbf{i} + 2y\mathbf{j} + 2z\mathbf{k} = \lambda(2x\mathbf{i} - 2y\mathbf{j} - 2z\mathbf{k}) \Rightarrow 2x = 2x\lambda, 2y = 2y\lambda$, and $2z = -2z\lambda \Rightarrow x = 0$ or $\lambda = 1$.
CASE 1: $\lambda = 1 \Rightarrow 2y = -2y \Rightarrow y = 0; 2z = -2z \Rightarrow z = 0 \Rightarrow x^2 - 1 = 0 \Rightarrow x^2 - 1 = 0 \Rightarrow x = \pm 1$ and $y = z = 0$.
CASE 2: $x = 0 \Rightarrow y^2 - z^2 = 1$, which has no solution.
Therefore the points on the unit circle $x^2 + y^2 = 1$, are the points on the surface $x^2 + y^2 - z^2 = 1$ closest to the origin. The minimum distance is 1.

21. Let $f(x, y, z) = x^2 + y^2 + z^2$ be the square of the distance to the origin. Then $\nabla f = 2x\mathbf{i} + 2y\mathbf{j} + 2z\mathbf{k}$ and $\nabla g = -y\mathbf{i} - x\mathbf{j} + 2z\mathbf{k}$ so that $\nabla f = \lambda \nabla g \Rightarrow 2x\mathbf{i} + 2y\mathbf{j} + 2z\mathbf{k} = \lambda(-y\mathbf{i} - x\mathbf{j} + 2z\mathbf{k}) \Rightarrow 2x = -y\lambda, 2y = -x\lambda$, and $2z = 2z\lambda \Rightarrow \lambda = 1$ or $z = 0$.
CASE 1: $\lambda = 1 \Rightarrow 2x = -y$ and $2y = -x \Rightarrow y = 0$ and $x = 0 \Rightarrow z^2 - 4 = 0 \Rightarrow z = \pm 2$ and $x = y = 0$.
CASE 2: $z = 0 \Rightarrow -xy - 4 = 0 \Rightarrow y = -\frac{4}{x}$. Then $2x = \frac{4}{x}\lambda \Rightarrow \lambda = \frac{x^2}{2}$, and $-\frac{8}{x} = -x\lambda \Rightarrow -\frac{8}{x} = -x\left(\frac{x^2}{2}\right)$
$\Rightarrow x^4 = 16 \Rightarrow x = \pm 2$. Thus, $x = 2$ and $y = -2$, or $x = -2$ and $y = 2$.
Therefore we get four points: $(2, -2, 0)$, $(-2, 2, 0)$, $(0, 0, 2)$ and $(0, 0, -2)$. But the points $(0, 0, 2)$ and $(0, 0, -2)$ are closest to the origin since they are 2 units away and the others are $2\sqrt{2}$ units away.

23. $\nabla f = \mathbf{i} - 2\mathbf{j} + 5\mathbf{k}$ and $\nabla g = 2x\mathbf{i} + 2y\mathbf{j} + 2z\mathbf{k}$ so that $\nabla f = \lambda \nabla g \Rightarrow \mathbf{i} - 2\mathbf{j} + 5\mathbf{k} = \lambda(2x\mathbf{i} + 2y\mathbf{j} + 2z\mathbf{k}) \Rightarrow 1 = 2x\lambda$, $-2 = 2y\lambda$, and $5 = 2z\lambda \Rightarrow x = \frac{1}{2\lambda}, y = -\frac{1}{\lambda} = -2x$, and $z = \frac{5}{2\lambda} = 5x \Rightarrow x^2 + (-2x)^2 + (5x)^2 = 30 \Rightarrow x = \pm 1$. Thus, $x = 1, y = -2, z = 5$ or $x = -1, y = 2, z = -5$. Therefore $f(1, -2, 5) = 30$ is the maximum value and $f(-1, 2, -5) = -30$ is the minimum value.

25. $f(x, y, z) = x^2 + y^2 + z^2$ and $g(x, y, z) = x + y + z - 9 = 0 \Rightarrow \nabla f = 2x\mathbf{i} + 2y\mathbf{j} + 2z\mathbf{k}$ and $\nabla g = \mathbf{i} + \mathbf{j} + \mathbf{k}$ so that $\nabla f = \lambda \nabla g \Rightarrow 2x\mathbf{i} + 2y\mathbf{j} + 2z\mathbf{k} = \lambda(\mathbf{i} + \mathbf{j} + \mathbf{k}) \Rightarrow 2x = \lambda, 2y = \lambda$, and $2z = \lambda \Rightarrow x = y = z \Rightarrow x + x + x - 9 = 0$ $\Rightarrow x = 3, y = 3$, and $z = 3$.

27. $V = xyz$ and $g(x, y, z) = x^2 + y^2 + z^2 - 1 = 0 \Rightarrow \nabla V = yz\mathbf{i} + xz\mathbf{j} + xy\mathbf{k}$ and $\nabla g = 2x\mathbf{i} + 2y\mathbf{j} + 2z\mathbf{k}$ so that $\nabla V = \lambda \nabla g \Rightarrow yz = \lambda x, xz = \lambda y$, and $xy = \lambda z \Rightarrow xyz = \lambda x^2$ and $xyz = \lambda y^2 \Rightarrow y = \pm x \Rightarrow z = \pm x$ $\Rightarrow x^2 + x^2 + x^2 = 1 \Rightarrow x = \frac{1}{\sqrt{3}}$ since $x > 0 \Rightarrow$ the dimensions of the box are $\frac{1}{\sqrt{3}}$ by $\frac{1}{\sqrt{3}}$ by $\frac{1}{\sqrt{3}}$ for maximum volume. (Note that there is no minimum volume since the box could be made arbitrarily thin.)

29. $\nabla T = 16x\mathbf{i} + 4z\mathbf{j} + (4y - 16)\mathbf{k}$ and $\nabla g = 8x\mathbf{i} + 2y\mathbf{j} + 8z\mathbf{k}$ so that $\nabla T = \lambda \nabla g \Rightarrow 16x\mathbf{i} + 4z\mathbf{j} + (4y - 16)\mathbf{k}$ $= \lambda(8x\mathbf{i} + 2y\mathbf{j} + 8z\mathbf{k}) \Rightarrow 16x = 8x\lambda, 4z = 2y\lambda$, and $4y - 16 = 8z\lambda \Rightarrow \lambda = 2$ or $x = 0$.
CASE 1: $\lambda = 2 \Rightarrow 4z = 2y(2) \Rightarrow z = y$. Then $4z - 16 = 16z \Rightarrow z = -\frac{4}{3} \Rightarrow y = -\frac{4}{3}$. Then
$$4x^2 + \left(-\tfrac{4}{3}\right)^2 + 4\left(-\tfrac{4}{3}\right)^2 = 16 \Rightarrow x = \pm\tfrac{4}{3}.$$
CASE 2: $x = 0 \Rightarrow \lambda = \frac{2z}{y} \Rightarrow 4y - 16 = 8z\left(\frac{2z}{y}\right) \Rightarrow y^2 - 4y = 4z^2 \Rightarrow 4(0)^2 + y^2 + (y^2 - 4y) - 16 = 0$
$\Rightarrow y^2 - 2y - 8 = 0 \Rightarrow (y - 4)(y + 2) = 0 \Rightarrow y = 4$ or $y = -2$. Now $y = 4 \Rightarrow 4z^2 = 4^2 - 4(4)$
$\Rightarrow z = 0$ and $y = -2 \Rightarrow 4z^2 = (-2)^2 - 4(-2) \Rightarrow z = \pm\sqrt{3}$.
The temperatures are $T\left(\pm\frac{4}{3}, -\frac{4}{3}, -\frac{4}{3}\right) = 642\frac{2}{3}^\circ$, $T(0, 4, 0) = 600^\circ$, $T\left(0, -2, \sqrt{3}\right) = \left(600 - 24\sqrt{3}\right)^\circ$, and $T\left(0, -2, -\sqrt{3}\right) = \left(600 + 24\sqrt{3}\right)^\circ \approx 641.6^\circ$. Therefore $\left(\pm\frac{4}{3}, -\frac{4}{3}, -\frac{4}{3}\right)$ are the hottest points on the space probe.

31. $\nabla U = (y+2)\mathbf{i} + x\mathbf{j}$ and $\nabla g = 2\mathbf{i} + \mathbf{j}$ so that $\nabla U = \lambda \nabla g \Rightarrow (y+2)\mathbf{i} + x\mathbf{j} = \lambda(2\mathbf{i}+\mathbf{j}) \Rightarrow y+2 = 2\lambda$ and $x = \lambda \Rightarrow y+2 = 2x \Rightarrow y = 2x-2 \Rightarrow 2x + (2x-2) = 30 \Rightarrow x = 8$ and $y = 14$. Therefore $U(8,14) = \$128$ is the maximum value of U under the constraint.

33. Let $g_1(x,y,z) = 2x - y = 0$ and $g_2(x,y,z) = y + z = 0 \Rightarrow \nabla g_1 = 2\mathbf{i} - \mathbf{j}$, $\nabla g_2 = \mathbf{j} + \mathbf{k}$, and $\nabla f = 2x\mathbf{i} + 2\mathbf{j} - 2z\mathbf{k}$ so that $\nabla f = \lambda \nabla g_1 + \mu \nabla g_2 \Rightarrow 2x\mathbf{i} + 2\mathbf{j} - 2z\mathbf{k} = \lambda(2\mathbf{i} - \mathbf{j}) + \mu(\mathbf{j}+\mathbf{k}) \Rightarrow 2x\mathbf{i} + 2\mathbf{j} - 2z\mathbf{k} = 2\lambda\mathbf{i} + (\mu - \lambda)\mathbf{j} + \mu\mathbf{k}$ $\Rightarrow 2x = 2\lambda$, $2 = \mu - \lambda$, and $-2z = \mu \Rightarrow x = \lambda$. Then $2 = -2z - x \Rightarrow x = -2z - 2$ so that $2x - y = 0$ $\Rightarrow 2(-2z-2) - y = 0 \Rightarrow -4z - 4 - y = 0$. This equation coupled with $y + z = 0$ implies $z = -\frac{4}{3}$ and $y = \frac{4}{3}$. Then $x = \frac{2}{3}$ so that $\left(\frac{2}{3}, \frac{4}{3}, -\frac{4}{3}\right)$ is the point that gives the maximum value $f\left(\frac{2}{3}, \frac{4}{3}, -\frac{4}{3}\right) = \left(\frac{2}{3}\right)^2 + 2\left(\frac{4}{3}\right) - \left(-\frac{4}{3}\right)^2 = \frac{4}{3}$.

35. Let $f(x,y,z) = x^2 + y^2 + z^2$ be the square of the distance from the origin. We want to minimize $f(x,y,z)$ subject to the constraints $g_1(x,y,z) = y + 2z - 12 = 0$ and $g_2(x,y,z) = x + y - 6 = 0$. Thus $\nabla f = 2x\mathbf{i} + 2y\mathbf{j} + 2z\mathbf{k}$, $\nabla g_1 = \mathbf{j} + 2\mathbf{k}$, and $\nabla g_2 = \mathbf{i} + \mathbf{j}$ so that $\nabla f = \lambda \nabla g_1 + \mu \nabla g_2 \Rightarrow 2x = \mu$, $2y = \lambda + \mu$, and $2z = 2\lambda$. Then $0 = y + 2z - 12$ $= \left(\frac{\lambda}{2} + \frac{\mu}{2}\right) + 2\lambda - 12 \Rightarrow \frac{5}{2}\lambda + \frac{1}{2}\mu = 12 \Rightarrow 5\lambda + \mu = 24$; $0 = x + y - 6 = \frac{\mu}{2} + \left(\frac{\lambda}{2} + \frac{\mu}{2}\right) - 6 \Rightarrow \frac{1}{2}\lambda + \mu = 6$ $\Rightarrow \lambda + 2\mu = 12$. Solving these two equations for λ and μ gives $\lambda = 4$ and $\mu = 4 \Rightarrow x = \frac{\mu}{2} = 2$, $y = \frac{\lambda+\mu}{2} = 4$, and $z = \lambda = 4$. The point $(2,4,4)$ on the line of intersection is closest to the origin. (There is no maximum distance from the origin since points on the line can be arbitrarily far away.)

37. Let $g_1(x,y,z) = z - 1 = 0$ and $g_2(x,y,z) = x^2 + y^2 + z^2 - 10 = 0 \Rightarrow \nabla g_1 = \mathbf{k}$, $\nabla g_2 = 2x\mathbf{i} + 2y\mathbf{j} + 2z\mathbf{k}$, and $\nabla f = 2xyz\mathbf{i} + x^2z\mathbf{j} + x^2y\mathbf{k}$ so that $\nabla f = \lambda \nabla g_1 + \mu \nabla g_2 \Rightarrow 2xyz\mathbf{i} + x^2z\mathbf{j} + x^2y\mathbf{k} = \lambda(\mathbf{k}) + \mu(2x\mathbf{i} + 2y\mathbf{j} + 2z\mathbf{k})$ $\Rightarrow 2xyz = 2x\mu$, $x^2z = 2y\mu$, and $x^2y = 2z\mu + \lambda \Rightarrow xyz = x\mu \Rightarrow x = 0$ or $yz = \mu \Rightarrow \mu = y$ since $z = 1$.

CASE 1: $x = 0$ and $z = 1 \Rightarrow y^2 - 9 = 0$ (from g_2) $\Rightarrow y = \pm 3$ yielding the points $(0, \pm 3, 1)$.

CASE 2: $\mu = y \Rightarrow x^2z = 2y^2 \Rightarrow x^2 = 2y^2$ (since $z = 1$) $\Rightarrow 2y^2 + y^2 + 1 - 10 = 0$ (from g_2) $\Rightarrow 3y^2 - 9 = 0$

$\Rightarrow y = \pm\sqrt{3} \Rightarrow x^2 = 2\left(\pm\sqrt{3}\right)^2 \Rightarrow x = \pm\sqrt{6}$ yielding the points $\left(\pm\sqrt{6}, \pm\sqrt{3}, 1\right)$.

Now $f(0, \pm 3, 1) = 1$ and $f\left(\pm\sqrt{6}, \pm\sqrt{3}, 1\right) = 6\left(\pm\sqrt{3}\right) + 1 = 1 \pm 6\sqrt{3}$. Therefore the maximum of f is $1 + 6\sqrt{3}$ at $\left(\pm\sqrt{6}, \sqrt{3}, 1\right)$, and the minimum of f is $1 - 6\sqrt{3}$ at $\left(\pm\sqrt{6}, -\sqrt{3}, 1\right)$.

39. Let $g_1(x,y,z) = y - x = 0$ and $g_2(x,y,z) = x^2 + y^2 + z^2 - 4 = 0$. Then $\nabla f = y\mathbf{i} + x\mathbf{j} + 2z\mathbf{k}$, $\nabla g_1 = -\mathbf{i} + \mathbf{j}$, and $\nabla g_2 = 2x\mathbf{i} + 2y\mathbf{j} + 2z\mathbf{k}$ so that $\nabla f = \lambda \nabla g_1 + \mu \nabla g_2 \Rightarrow y\mathbf{i} + x\mathbf{j} + 2z\mathbf{k} = \lambda(-\mathbf{i} + \mathbf{j}) + \mu(2x\mathbf{i} + 2y\mathbf{j} + 2z\mathbf{k})$ $\Rightarrow y = -\lambda + 2x\mu$, $x = \lambda + 2y\mu$, and $2z = 2z\mu \Rightarrow z = 0$ or $\mu = 1$.

CASE 1: $z = 0 \Rightarrow x^2 + y^2 - 4 = 0 \Rightarrow 2x^2 - 4 = 0$ (since $x = y$) $\Rightarrow x = \pm\sqrt{2}$ and $y = \pm\sqrt{2}$ yielding the points $\left(\pm\sqrt{2}, \pm\sqrt{2}, 0\right)$.

CASE 2: $\mu = 1 \Rightarrow y = -\lambda + 2x$ and $x = \lambda + 2y \Rightarrow x + y = 2(x+y) \Rightarrow 2x = 2(2x)$ since $x = y \Rightarrow x = 0 \Rightarrow y = 0$ $\Rightarrow z^2 - 4 = 0 \Rightarrow z = \pm 2$ yielding the points $(0, 0, \pm 2)$.

Now, $f(0,0,\pm 2) = 4$ and $f\left(\pm\sqrt{2}, \pm\sqrt{2}, 0\right) = 2$. Therefore the maximum value of f is 4 at $(0,0,\pm 2)$ and the minimum value of f is 2 at $\left(\pm\sqrt{2}, \pm\sqrt{2}, 0\right)$.

41. $\nabla f = \mathbf{i} + \mathbf{j}$ and $\nabla g = y\mathbf{i} + x\mathbf{j}$ so that $\nabla f = \lambda \nabla g \Rightarrow \mathbf{i} + \mathbf{j} = \lambda(y\mathbf{i} + x\mathbf{j}) \Rightarrow 1 = y\lambda$ and $1 = x\lambda \Rightarrow y = x$ $\Rightarrow y^2 = 16 \Rightarrow y = \pm 4 \Rightarrow (4,4)$ and $(-4,-4)$ are candidates for the location of extreme values. But as $x \to \infty$, $y \to \infty$ and $f(x,y) \to \infty$; as $x \to -\infty$, $y \to 0$ and $f(x,y) \to -\infty$. Therefore no maximum or minimum value exists subject to the constraint.

43. (a) Maximize $f(a, b, c) = a^2b^2c^2$ subject to $a^2 + b^2 + c^2 = r^2$. Thus $\nabla f = 2ab^2c^2\mathbf{i} + 2a^2bc^2\mathbf{j} + 2a^2b^2c\mathbf{k}$ and $\nabla g = 2a\mathbf{i} + 2b\mathbf{j} + 2c\mathbf{k}$ so that $\nabla f = \lambda \nabla g \Rightarrow 2ab^2c^2 = 2a\lambda,\ 2a^2bc^2 = 2b\lambda$, and $2a^2b^2c = 2c\lambda$ $\Rightarrow 2a^2b^2c^2 = 2a^2\lambda = 2b^2\lambda = 2c^2\lambda \Rightarrow \lambda = 0$ or $a^2 = b^2 = c^2$.

CASE 1: $\lambda = 0 \Rightarrow a^2b^2c^2 = 0$.

CASE 2: $a^2 = b^2 = c^2 \Rightarrow f(a, b, c) = a^2a^2a^2$ and $3a^2 = r^2 \Rightarrow f(a, b, c) = \left(\frac{r^2}{3}\right)^3$ is the maximum value.

(b) The point $\left(\sqrt{a}, \sqrt{b}, \sqrt{c}\right)$ is on the sphere if $a + b + c = r^2$. Moreover, by part (a), $abc = f\left(\sqrt{a}, \sqrt{b}, \sqrt{c}\right)$ $\le \left(\frac{r^2}{3}\right)^3 \Rightarrow (abc)^{1/3} \le \frac{r^2}{3} = \frac{a+b+c}{3}$, as claimed.

14.9 TAYLOR'S FORMULA FOR TWO VARIABLES

1. $f(x, y) = xe^y \Rightarrow f_x = e^y,\ f_y = xe^y,\ f_{xx} = 0,\ f_{xy} = e^y,\ f_{yy} = xe^y$
$\Rightarrow f(x, y) \approx f(0, 0) + xf_x(0, 0) + yf_y(0, 0) + \frac{1}{2}\left[x^2f_{xx}(0, 0) + 2xyf_{xy}(0, 0) + y^2f_{yy}(0, 0)\right]$
$= 0 + x \cdot 1 + y \cdot 0 + \frac{1}{2}\left(x^2 \cdot 0 + 2xy \cdot 1 + y^2 \cdot 0\right) = x + xy$ quadratic approximation;
$f_{xxx} = 0,\ f_{xxy} = 0,\ f_{xyy} = e^y,\ f_{yyy} = xe^y$
$\Rightarrow f(x, y) \approx \text{quadratic} + \frac{1}{6}\left[x^3f_{xxx}(0, 0) + 3x^2yf_{xxy}(0, 0) + 3xy^2f_{xyy}(0, 0) + y^3f_{yyy}(0, 0)\right]$
$= x + xy + \frac{1}{6}\left(x^3 \cdot 0 + 3x^2y \cdot 0 + 3xy^2 \cdot 1 + y^3 \cdot 0\right) = x + xy + \frac{1}{2}xy^2$, cubic approximation

3. $f(x, y) = y \sin x \Rightarrow f_x = y \cos x,\ f_y = \sin x,\ f_{xx} = -y \sin x,\ f_{xy} = \cos x,\ f_{yy} = 0$
$\Rightarrow f(x, y) \approx f(0, 0) + xf_x(0, 0) + yf_y(0, 0) + \frac{1}{2}\left[x^2f_{xx}(0, 0) + 2xyf_{xy}(0, 0) + y^2f_{yy}(0, 0)\right]$
$= 0 + x \cdot 0 + y \cdot 0 + \frac{1}{2}\left(x^2 \cdot 0 + 2xy \cdot 1 + y^2 \cdot 0\right) = xy$, quadratic approximation;
$f_{xxx} = -y \cos x,\ f_{xxy} = -\sin x,\ f_{xyy} = 0,\ f_{yyy} = 0$
$\Rightarrow f(x, y) \approx \text{quadratic} + \frac{1}{6}\left[x^3f_{xxx}(0, 0) + 3x^2yf_{xxy}(0, 0) + 3xy^2f_{xyy}(0, 0) + y^3f_{yyy}(0, 0)\right]$
$= xy + \frac{1}{6}\left(x^3 \cdot 0 + 3x^2y \cdot 0 + 3xy^2 \cdot 0 + y^3 \cdot 0\right) = xy$, cubic approximation

5. $f(x, y) = e^x \ln(1 + y) \Rightarrow f_x = e^x \ln(1 + y),\ f_y = \frac{e^x}{1+y},\ f_{xx} = e^x \ln(1 + y),\ f_{xy} = \frac{e^x}{1+y},\ f_{yy} = -\frac{e^x}{(1+y)^2}$
$\Rightarrow f(x, y) \approx f(0, 0) + xf_x(0, 0) + yf_y(0, 0) + \frac{1}{2}\left[x^2f_{xx}(0, 0) + 2xyf_{xy}(0, 0) + y^2f_{yy}(0, 0)\right]$
$= 0 + x \cdot 0 + y \cdot 1 + \frac{1}{2}\left[x^2 \cdot 0 + 2xy \cdot 1 + y^2 \cdot (-1)\right] = y + \frac{1}{2}\left(2xy - y^2\right)$, quadratic approximation;
$f_{xxx} = e^x \ln(1 + y),\ f_{xxy} = \frac{e^x}{1+y},\ f_{xyy} = -\frac{e^x}{(1+y)^2},\ f_{yyy} = \frac{2e^x}{(1+y)^3}$
$\Rightarrow f(x, y) \approx \text{quadratic} + \frac{1}{6}\left[x^3f_{xxx}(0, 0) + 3x^2yf_{xxy}(0, 0) + 3xy^2f_{xyy}(0, 0) + y^3f_{yyy}(0, 0)\right]$
$= y + \frac{1}{2}\left(2xy - y^2\right) + \frac{1}{6}\left[x^3 \cdot 0 + 3x^2y \cdot 1 + 3xy^2 \cdot (-1) + y^3 \cdot 2\right]$
$= y + \frac{1}{2}\left(2xy - y^2\right) + \frac{1}{6}\left(3x^2y - 3xy^2 + 2y^3\right)$, cubic approximation

7. $f(x, y) = \sin\left(x^2 + y^2\right) \Rightarrow f_x = 2x \cos\left(x^2 + y^2\right),\ f_y = 2y \cos\left(x^2 + y^2\right),\ f_{xx} = 2 \cos\left(x^2 + y^2\right) - 4x^2 \sin\left(x^2 + y^2\right)$,
$f_{xy} = -4xy \sin\left(x^2 + y^2\right),\ f_{yy} = 2 \cos\left(x^2 + y^2\right) - 4y^2 \sin\left(x^2 + y^2\right)$
$\Rightarrow f(x, y) \approx f(0, 0) + xf_x(0, 0) + yf_y(0, 0) + \frac{1}{2}\left[x^2f_{xx}(0, 0) + 2xyf_{xy}(0, 0) + y^2f_{yy}(0, 0)\right]$
$= 0 + x \cdot 0 + y \cdot 0 + \frac{1}{2}\left(x^2 \cdot 2 + 2xy \cdot 0 + y^2 \cdot 2\right) = x^2 + y^2$, quadratic approximation;
$f_{xxx} = -12x \sin\left(x^2 + y^2\right) - 8x^3 \cos\left(x^2 + y^2\right),\ f_{xxy} = -4y \sin\left(x^2 + y^2\right) - 8x^2y \cos\left(x^2 + y^2\right)$,
$f_{xyy} = -4x \sin\left(x^2 + y^2\right) - 8xy^2 \cos\left(x^2 + y^2\right),\ f_{yyy} = -12y \sin\left(x^2 + y^2\right) - 8y^3 \cos\left(x^2 + y^2\right)$
$\Rightarrow f(x, y) \approx \text{quadratic} + \frac{1}{6}\left[x^3f_{xxx}(0, 0) + 3x^2yf_{xxy}(0, 0) + 3xy^2f_{xyy}(0, 0) + y^3f_{yyy}(0, 0)\right]$
$= x^2 + y^2 + \frac{1}{6}\left(x^3 \cdot 0 + 3x^2y \cdot 0 + 3xy^2 \cdot 0 + y^3 \cdot 0\right) = x^2 + y^2$, cubic approximation

9. $f(x, y) = \frac{1}{1-x-y} \Rightarrow f_x = \frac{1}{(1-x-y)^2} = f_y,\ f_{xx} = \frac{2}{(1-x-y)^3} = f_{xy} = f_{yy}$
$\Rightarrow f(x, y) \approx f(0, 0) + xf_x(0, 0) + yf_y(0, 0) + \frac{1}{2}\left[x^2f_{xx}(0, 0) + 2xyf_{xy}(0, 0) + y^2f_{yy}(0, 0)\right]$
$= 1 + x \cdot 1 + y \cdot 1 + \frac{1}{2}\left(x^2 \cdot 2 + 2xy \cdot 2 + y^2 \cdot 2\right) = 1 + (x + y) + \left(x^2 + 2xy + y^2\right)$

$= 1 + (x+y) + (x+y)^2$, quadratic approximation; $f_{xxx} = \frac{6}{(1-x-y)^4} = f_{xxy} = f_{xyy} = f_{yyy}$

$\Rightarrow f(x,y) \approx \text{quadratic} + \frac{1}{6}\left[x^3 f_{xxx}(0,0) + 3x^2 y f_{xxy}(0,0) + 3xy^2 f_{xyy}(0,0) + y^3 f_{yyy}(0,0)\right]$

$= 1 + (x+y) + (x+y)^2 + \frac{1}{6}\left(x^3 \cdot 6 + 3x^2 y \cdot 6 + 3xy^2 \cdot 6 + y^3 \cdot 6\right)$

$= 1 + (x+y) + (x+y)^2 + (x^3 + 3x^2 y + 3xy^2 + y^3) = 1 + (x+y) + (x+y)^2 + (x+y)^3$, cubic approximation

11. $f(x,y) = \cos x \cos y \Rightarrow f_x = -\sin x \cos y,\ f_y = -\cos x \sin y,\ f_{xx} = -\cos x \cos y,\ f_{xy} = \sin x \sin y,$
$f_{yy} = -\cos x \cos y \Rightarrow f(x,y) \approx f(0,0) + x f_x(0,0) + y f_y(0,0) + \frac{1}{2}\left[x^2 f_{xx}(0,0) + 2xy f_{xy}(0,0) + y^2 f_{yy}(0,0)\right]$
$= 1 + x \cdot 0 + y \cdot 0 + \frac{1}{2}\left[x^2 \cdot (-1) + 2xy \cdot 0 + y^2 \cdot (-1)\right] = 1 - \frac{x^2}{2} - \frac{y^2}{2}$, quadratic approximation. Since all partial derivatives of f are products of sines and cosines, the absolute value of these derivatives is less than or equal to 1 $\Rightarrow E(x,y) \le \frac{1}{6}\left[(0.1)^3 + 3(0.1)^3 + 3(0.1)^3 + 0.1)^3\right] \le 0.00134.$

14.10 PARTIAL DERIVATIVES WITH CONSTRAINED VARIABLES

1. $w = x^2 + y^2 + z^2$ and $z = x^2 + y^2$:

(a) $\begin{pmatrix} y \\ z \end{pmatrix} \to \begin{pmatrix} x = x(y,z) \\ y = y \\ z = z \end{pmatrix} \to w \Rightarrow \left(\frac{\partial w}{\partial y}\right)_z = \frac{\partial w}{\partial x}\frac{\partial x}{\partial y} + \frac{\partial w}{\partial y}\frac{\partial y}{\partial y} + \frac{\partial w}{\partial z}\frac{\partial z}{\partial y};\ \frac{\partial z}{\partial y} = 0$ and $\frac{\partial z}{\partial y} = 2x\frac{\partial x}{\partial y} + 2y\frac{\partial y}{\partial y}$

$= 2x\frac{\partial x}{\partial y} + 2y \Rightarrow 0 = 2x\frac{\partial x}{\partial y} + 2y \Rightarrow \frac{\partial x}{\partial y} = -\frac{y}{x} \Rightarrow \left(\frac{\partial w}{\partial y}\right)_z = (2x)\left(-\frac{y}{x}\right) + (2y)(1) + (2z)(0) = -2y + 2y = 0$

(b) $\begin{pmatrix} x \\ z \end{pmatrix} \to \begin{pmatrix} x = x \\ y = y(x,z) \\ z = z \end{pmatrix} \to w \Rightarrow \left(\frac{\partial w}{\partial z}\right)_x = \frac{\partial w}{\partial x}\frac{\partial x}{\partial z} + \frac{\partial w}{\partial y}\frac{\partial y}{\partial z} + \frac{\partial w}{\partial z}\frac{\partial z}{\partial z};\ \frac{\partial x}{\partial z} = 0$ and $\frac{\partial z}{\partial z} = 2x\frac{\partial x}{\partial z} + 2y\frac{\partial y}{\partial z}$

$\Rightarrow 1 = 2y\frac{\partial y}{\partial z} \Rightarrow \frac{\partial y}{\partial z} = \frac{1}{2y} \Rightarrow \left(\frac{\partial w}{\partial z}\right)_x = (2x)(0) + (2y)\left(\frac{1}{2y}\right) + (2z)(1) = 1 + 2z$

(c) $\begin{pmatrix} y \\ z \end{pmatrix} \to \begin{pmatrix} x = x(y,z) \\ y = y \\ z = z \end{pmatrix} \to w \Rightarrow \left(\frac{\partial w}{\partial z}\right)_y = \frac{\partial w}{\partial x}\frac{\partial x}{\partial z} + \frac{\partial w}{\partial y}\frac{\partial y}{\partial z} + \frac{\partial w}{\partial z}\frac{\partial z}{\partial z};\ \frac{\partial y}{\partial z} = 0$ and $\frac{\partial z}{\partial z} = 2x\frac{\partial x}{\partial z} + 2y\frac{\partial y}{\partial z}$

$\Rightarrow 1 = 2x\frac{\partial x}{\partial z} \Rightarrow \frac{\partial x}{\partial z} = \frac{1}{2x} \Rightarrow \left(\frac{\partial w}{\partial z}\right)_y = (2x)\left(\frac{1}{2x}\right) + (2y)(0) + (2z)(1) = 1 + 2z$

3. $U = f(P,V,T)$ and $PV = nRT$

(a) $\begin{pmatrix} P \\ V \end{pmatrix} \to \begin{pmatrix} P = P \\ V = V \\ T = \frac{PV}{nR} \end{pmatrix} \to U \Rightarrow \left(\frac{\partial U}{\partial P}\right)_V = \frac{\partial U}{\partial P}\frac{\partial P}{\partial P} + \frac{\partial U}{\partial V}\frac{\partial V}{\partial P} + \frac{\partial U}{\partial T}\frac{\partial T}{\partial P} = \frac{\partial U}{\partial P} + \left(\frac{\partial U}{\partial V}\right)(0) + \left(\frac{\partial U}{\partial T}\right)\left(\frac{V}{nR}\right)$

$= \frac{\partial U}{\partial P} + \left(\frac{\partial U}{\partial T}\right)\left(\frac{V}{nR}\right)$

(b) $\begin{pmatrix} V \\ T \end{pmatrix} \to \begin{pmatrix} P = \frac{nRT}{V} \\ V = V \\ T = T \end{pmatrix} \to U \Rightarrow \left(\frac{\partial U}{\partial T}\right)_V = \frac{\partial U}{\partial P}\frac{\partial P}{\partial T} + \frac{\partial U}{\partial V}\frac{\partial V}{\partial T} + \frac{\partial U}{\partial T}\frac{\partial T}{\partial T} = \left(\frac{\partial U}{\partial P}\right)\left(\frac{nR}{V}\right) + \left(\frac{\partial U}{\partial V}\right)(0) + \frac{\partial U}{\partial T}$

$= \left(\frac{\partial U}{\partial P}\right)\left(\frac{nR}{V}\right) + \frac{\partial U}{\partial T}$

5. $w = x^2y^2 + yz - z^3$ and $x^2 + y^2 + z^2 = 6$

(a) $\begin{pmatrix} x \\ y \end{pmatrix} \to \begin{pmatrix} x = x \\ y = y \\ z = z(x,y) \end{pmatrix} \to w \Rightarrow \left(\frac{\partial w}{\partial y}\right)_x = \frac{\partial w}{\partial x}\frac{\partial x}{\partial y} + \frac{\partial w}{\partial y}\frac{\partial y}{\partial y} + \frac{\partial w}{\partial z}\frac{\partial z}{\partial y}$

$= (2xy^2)(0) + (2x^2y + z)(1) + (y - 3z^2)\frac{\partial z}{\partial y} = 2x^2y + z + (y - 3z^2)\frac{\partial z}{\partial y}$. Now $(2x)\frac{\partial x}{\partial y} + 2y + (2z)\frac{\partial z}{\partial y} = 0$ and $\frac{\partial x}{\partial y} = 0 \Rightarrow 2y + (2z)\frac{\partial z}{\partial y} = 0 \Rightarrow \frac{\partial z}{\partial y} = -\frac{y}{z}$. At $(w,x,y,z) = (4,2,1,-1)$, $\frac{\partial z}{\partial y} = -\frac{1}{-1} = 1 \Rightarrow \left(\frac{\partial w}{\partial y}\right)_x\Big|_{(4,2,1,-1)}$
$= \left[(2)(2)^2(1) + (-1)\right] + \left[1 - 3(-1)^2\right](1) = 5$

(b) $\begin{pmatrix} y \\ z \end{pmatrix} \to \begin{pmatrix} x = x(y,z) \\ y = y \\ z = z \end{pmatrix} \to w \Rightarrow \left(\frac{\partial w}{\partial y}\right)_z = \frac{\partial w}{\partial x}\frac{\partial x}{\partial y} + \frac{\partial w}{\partial y}\frac{\partial y}{\partial y} + \frac{\partial w}{\partial z}\frac{\partial z}{\partial y}$

$= (2xy^2)\frac{\partial x}{\partial y} + (2x^2y + z)(1) + (y - 3z^2)(0) = (2x^2y)\frac{\partial x}{\partial y} + 2x^2y + z$. Now $(2x)\frac{\partial x}{\partial y} + 2y + (2z)\frac{\partial z}{\partial y} = 0$ and $\frac{\partial z}{\partial y} = 0 \Rightarrow (2x)\frac{\partial x}{\partial y} + 2y = 0 \Rightarrow \frac{\partial x}{\partial y} = -\frac{y}{x}$. At $(w,x,y,z) = (4,2,1,-1)$, $\frac{\partial x}{\partial y} = -\frac{1}{2} \Rightarrow \left(\frac{\partial w}{\partial y}\right)_z\Big|_{(4,2,1,-1)}$
$= (2)(2)(1)^2\left(-\frac{1}{2}\right) + (2)(2)^2(1) + (-1) = 5$

7. $\begin{pmatrix} r \\ \theta \end{pmatrix} \to \begin{pmatrix} x = r\cos\theta \\ y = r\sin\theta \end{pmatrix} \Rightarrow \left(\frac{\partial x}{\partial r}\right)_\theta = \cos\theta$; $x^2 + y^2 = r^2 \Rightarrow 2x + 2y\frac{\partial y}{\partial x} = 2r\frac{\partial r}{\partial x}$ and $\frac{\partial y}{\partial x} = 0 \Rightarrow 2x = 2r\frac{\partial r}{\partial x}$
$\Rightarrow \frac{\partial r}{\partial x} = \frac{x}{r} \Rightarrow \left(\frac{\partial r}{\partial x}\right)_y = \frac{x}{\sqrt{x^2+y^2}}$

9. If x is a differentiable function of y and z, then $f(x,y,z) = 0 \Rightarrow \frac{\partial f}{\partial x}\frac{\partial x}{\partial x} + \frac{\partial f}{\partial y}\frac{\partial y}{\partial x} + \frac{\partial f}{\partial z}\frac{\partial z}{\partial x} = 0 \Rightarrow \frac{\partial f}{\partial x} + \frac{\partial f}{\partial y}\frac{\partial y}{\partial x} = 0$
$\Rightarrow \left(\frac{\partial x}{\partial y}\right)_z = -\frac{\partial f/\partial y}{\partial f/\partial z}$. Similarly, if y is a differentiable function of x and z, $\left(\frac{\partial y}{\partial z}\right)_x = -\frac{\partial f/\partial z}{\partial f/\partial x}$ and if z is a differentiable function of x and y, $\left(\frac{\partial z}{\partial x}\right)_y = -\frac{\partial f/\partial x}{\partial f/\partial y}$. Then $\left(\frac{\partial x}{\partial y}\right)_z\left(\frac{\partial y}{\partial z}\right)_x\left(\frac{\partial z}{\partial x}\right)_y$
$= \left(-\frac{\partial f/\partial y}{\partial f/\partial z}\right)\left(-\frac{\partial f/\partial z}{\partial f/\partial x}\right)\left(-\frac{\partial f/\partial x}{\partial f/\partial y}\right) = -1.$

11. If x and y are independent, then $g(x,y,z) = 0 \Rightarrow \frac{\partial g}{\partial x}\frac{\partial x}{\partial y} + \frac{\partial g}{\partial y}\frac{\partial y}{\partial y} + \frac{\partial g}{\partial z}\frac{\partial z}{\partial y} = 0$ and $\frac{\partial x}{\partial y} = 0 \Rightarrow \frac{\partial g}{\partial y} + \frac{\partial g}{\partial z}\frac{\partial z}{\partial y} = 0$
$\Rightarrow \left(\frac{\partial z}{\partial y}\right)_x = -\frac{\partial g/\partial y}{\partial g/\partial z}$, as claimed.

CHAPTER 14 PRACTICE EXERCISES

1. Domain: All points in the xy-plane
Range: $z \geq 0$

Level curves are ellipses with major axis along the y-axis and minor axis along the x-axis.

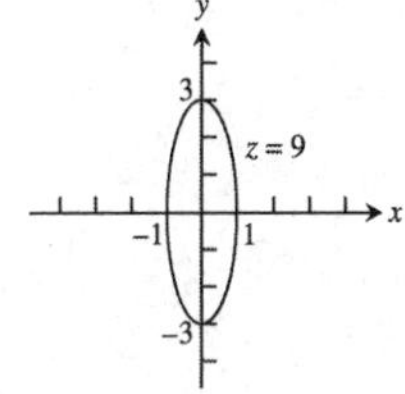

3. Domain: All (x,y) such that $x \neq 0$ and $y \neq 0$
Range: $z \neq 0$

Level curves are hyperbolas with the x- and y-axes as asymptotes.

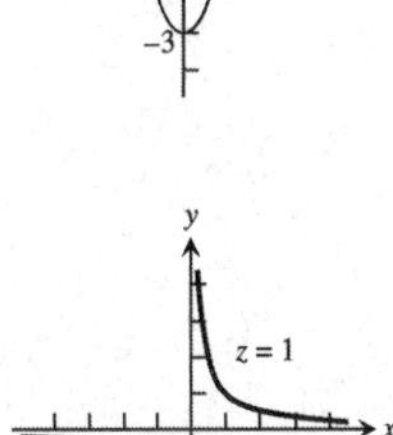

5. Domain: All points (x,y,z) in space
Range: All real numbers

Level surfaces are paraboloids of revolution with the z-axis as axis.

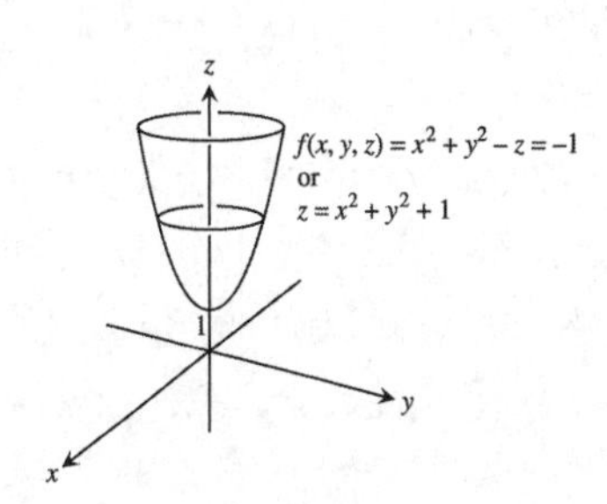

7. Domain: All (x, y, z) such that $(x, y, z) \neq (0, 0, 0)$
Range: Positive real numbers

Level surfaces are spheres with center $(0, 0, 0)$ and radius $r > 0$.

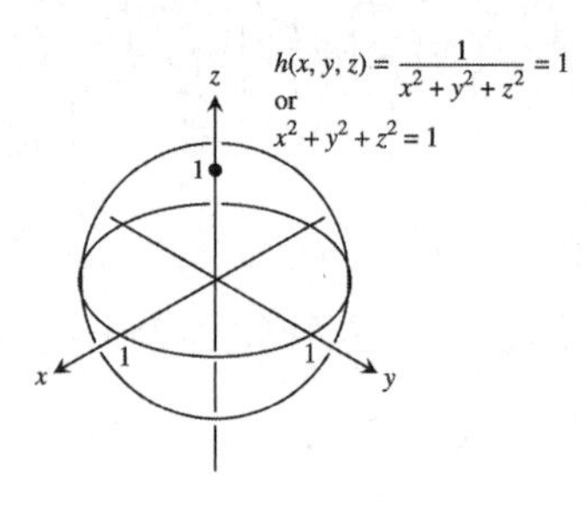

9. $\lim_{(x, y) \to (\pi, \ln 2)} e^y \cos x = e^{\ln 2} \cos \pi = (2)(-1) = -2$

11. $\lim_{\substack{(x, y) \to (1, 1) \\ x \neq \pm y}} \frac{x - y}{x^2 - y^2} = \lim_{\substack{(x, y) \to (1, 1) \\ x \neq \pm y}} \frac{x - y}{(x - y)(x + y)} = \lim_{(x, y) \to (1, 1)} \frac{1}{x + y} = \frac{1}{1 + 1} = \frac{1}{2}$

13. $\lim_{P \to (1, -1, e)} \ln |x + y + z| = \ln |1 + (-1) + e| = \ln e = 1$

15. Let $y = kx^2$, $k \neq 1$. Then $\lim_{\substack{(x, y) \to (0, 0) \\ y \neq x^2}} \frac{y}{x^2 - y} = \lim_{(x, kx^2) \to (0, 0)} \frac{kx^2}{x^2 - kx^2} = \frac{k}{1 - k^2}$ which gives different limits for different values of k $\Rightarrow$ the limit does not exist.

17. Let $y = kx$. Then $\lim_{(x, y) \to (0, 0)} \frac{x^2 - y^2}{x^2 + y^2} = \frac{x^2 - k^2x^2}{x^2 + k^2x^2} = \frac{1 - k^2}{1 + k^2}$ which gives different limits for different values of k $\Rightarrow$ the limit does not exist so $f(0, 0)$ cannot be defined in a way that makes f continuous at the origin.

19. $\frac{\partial g}{\partial r} = \cos \theta + \sin \theta$, $\frac{\partial g}{\partial \theta} = -r \sin \theta + r \cos \theta$

21. $\frac{\partial f}{\partial R_1} = -\frac{1}{R_1^2}$, $\frac{\partial f}{\partial R_2} = -\frac{1}{R_2^2}$, $\frac{\partial f}{\partial R_3} = -\frac{1}{R_3^2}$

23. $\frac{\partial P}{\partial n} = \frac{RT}{V}$, $\frac{\partial P}{\partial R} = \frac{nT}{V}$, $\frac{\partial P}{\partial T} = \frac{nR}{V}$, $\frac{\partial P}{\partial V} = -\frac{nRT}{V^2}$

25. $\frac{\partial g}{\partial x} = \frac{1}{y}$, $\frac{\partial g}{\partial y} = 1 - \frac{x}{y^2} \Rightarrow \frac{\partial^2 g}{\partial x^2} = 0$, $\frac{\partial^2 g}{\partial y^2} = \frac{2x}{y^3}$, $\frac{\partial^2 g}{\partial y \partial x} = \frac{\partial^2 g}{\partial x \partial y} = -\frac{1}{y^2}$

27. $\frac{\partial f}{\partial x} = 1 + y - 15x^2 + \frac{2x}{x^2 + 1}$, $\frac{\partial f}{\partial y} = x \Rightarrow \frac{\partial^2 f}{\partial x^2} = -30x + \frac{2 - 2x^2}{(x^2 + 1)^2}$, $\frac{\partial^2 f}{\partial y^2} = 0$, $\frac{\partial^2 f}{\partial y \partial x} = \frac{\partial^2 f}{\partial x \partial y} = 1$

29. $\frac{\partial w}{\partial x} = y \cos (xy + \pi)$, $\frac{\partial w}{\partial y} = x \cos (xy + \pi)$, $\frac{dx}{dt} = e^t$, $\frac{dy}{dt} = \frac{1}{t + 1}$
$\Rightarrow \frac{dw}{dt} = [y \cos (xy + \pi)]e^t + [x \cos (xy + \pi)] \left(\frac{1}{t+1}\right)$; $t = 0 \Rightarrow x = 1$ and $y = 0$
$\Rightarrow \left.\frac{dw}{dt}\right|_{t=0} = 0 \cdot 1 + [1 \cdot (-1)] \left(\frac{1}{0+1}\right) = -1$

31. $\frac{\partial w}{\partial x} = 2 \cos (2x - y)$, $\frac{\partial w}{\partial y} = -\cos (2x - y)$, $\frac{\partial x}{\partial r} = 1$, $\frac{\partial x}{\partial s} = \cos s$, $\frac{\partial y}{\partial r} = s$, $\frac{\partial y}{\partial s} = r$
$\Rightarrow \frac{\partial w}{\partial r} = [2 \cos (2x - y)](1) + [-\cos (2x - y)](s)$; $r = \pi$ and $s = 0 \Rightarrow x = \pi$ and $y = 0$
$\Rightarrow \left.\frac{\partial w}{\partial r}\right|_{(\pi,0)} = (2 \cos 2\pi) - (\cos 2\pi)(0) = 2$; $\frac{\partial w}{\partial s} = [2 \cos (2x - y)](\cos s) + [-\cos (2x - y)](r)$
$\Rightarrow \left.\frac{\partial w}{\partial s}\right|_{(\pi,0)} = (2 \cos 2\pi)(\cos 0) - (\cos 2\pi)(\pi) = 2 - \pi$

33. $\frac{\partial f}{\partial x} = y + z$, $\frac{\partial f}{\partial y} = x + z$, $\frac{\partial f}{\partial z} = y + x$, $\frac{dx}{dt} = -\sin t$, $\frac{dy}{dt} = \cos t$, $\frac{dz}{dt} = -2 \sin 2t$
$\Rightarrow \frac{df}{dt} = -(y + z)(\sin t) + (x + z)(\cos t) - 2(y + x)(\sin 2t)$; $t = 1 \Rightarrow x = \cos 1$, $y = \sin 1$, and $z = \cos 2$
$\Rightarrow \left.\frac{df}{dt}\right|_{t=1} = -(\sin 1 + \cos 2)(\sin 1) + (\cos 1 + \cos 2)(\cos 1) - 2(\sin 1 + \cos 1)(\sin 2)$

35. $F(x, y) = 1 - x - y^2 - \sin xy \Rightarrow F_x = -1 - y\cos xy$ and $F_y = -2y - x\cos xy \Rightarrow \frac{dy}{dx} = -\frac{F_x}{F_y} = -\frac{-1 - y\cos xy}{-2y - x\cos xy}$ $= \frac{1 + y\cos xy}{-2y - x\cos xy} \Rightarrow$ at $(x, y) = (0, 1)$ we have $\left.\frac{dy}{dx}\right|_{(0,1)} = \frac{1+1}{-2} = -1$

37. $\nabla f = (-\sin x \cos y)\mathbf{i} - (\cos x \sin y)\mathbf{j} \Rightarrow \nabla f|_{(\frac{\pi}{4},\frac{\pi}{4})} = -\frac{1}{2}\mathbf{i} - \frac{1}{2}\mathbf{j} \Rightarrow |\nabla f| = \sqrt{\left(-\frac{1}{2}\right)^2 + \left(-\frac{1}{2}\right)^2} = \frac{1}{\sqrt{2}} = \frac{\sqrt{2}}{2}$; $\mathbf{u} = \frac{\nabla f}{|\nabla f|} = -\frac{\sqrt{2}}{2}\mathbf{i} - \frac{\sqrt{2}}{2}\mathbf{j} \Rightarrow$ f increases most rapidly in the direction $\mathbf{u} = -\frac{\sqrt{2}}{2}\mathbf{i} - \frac{\sqrt{2}}{2}\mathbf{j}$ and decreases most rapidly in the direction $-\mathbf{u} = \frac{\sqrt{2}}{2}\mathbf{i} + \frac{\sqrt{2}}{2}\mathbf{j}$; $(D_{\mathbf{u}}f)_{P_0} = |\nabla f| = \frac{\sqrt{2}}{2}$ and $(D_{-\mathbf{u}}f)_{P_0} = -\frac{\sqrt{2}}{2}$; $\mathbf{u}_1 = \frac{\mathbf{v}}{|\mathbf{v}|} = \frac{3\mathbf{i} + 4\mathbf{j}}{\sqrt{3^2 + 4^2}} = \frac{3}{5}\mathbf{i} + \frac{4}{5}\mathbf{j} \Rightarrow (D_{\mathbf{u}_1}f)_{P_0} = \nabla f \cdot \mathbf{u}_1 = \left(-\frac{1}{2}\right)\left(\frac{3}{5}\right) + \left(-\frac{1}{2}\right)\left(\frac{4}{5}\right) = -\frac{7}{10}$

39. $\nabla f = \left(\frac{2}{2x + 3y + 6z}\right)\mathbf{i} + \left(\frac{3}{2x + 3y + 6z}\right)\mathbf{j} + \left(\frac{6}{2x + 3y + 6z}\right)\mathbf{k} \Rightarrow \nabla f|_{(-1,-1,1)} = 2\mathbf{i} + 3\mathbf{j} + 6\mathbf{k}$; $\mathbf{u} = \frac{\nabla f}{|\nabla f|} = \frac{2\mathbf{i} + 3\mathbf{j} + 6\mathbf{k}}{\sqrt{2^2 + 3^2 + 6^2}} = \frac{2}{7}\mathbf{i} + \frac{3}{7}\mathbf{j} + \frac{6}{7}\mathbf{k} \Rightarrow$ f increases most rapidly in the direction $\mathbf{u} = \frac{2}{7}\mathbf{i} + \frac{3}{7}\mathbf{j} + \frac{6}{7}\mathbf{k}$ and decreases most rapidly in the direction $-\mathbf{u} = -\frac{2}{7}\mathbf{i} - \frac{3}{7}\mathbf{j} - \frac{6}{7}\mathbf{k}$; $(D_{\mathbf{u}}f)_{P_0} = |\nabla f| = 7$, $(D_{-\mathbf{u}}f)_{P_0} = -7$; $\mathbf{u}_1 = \frac{\mathbf{v}}{|\mathbf{v}|} = \frac{2}{7}\mathbf{i} + \frac{3}{7}\mathbf{j} + \frac{6}{7}\mathbf{k} \Rightarrow (D_{\mathbf{u}_1}f)_{P_0} = (D_{\mathbf{u}}f)_{P_0} = 7$

41. $\mathbf{r} = (\cos 3t)\mathbf{i} + (\sin 3t)\mathbf{j} + 3t\mathbf{k} \Rightarrow \mathbf{v}(t) = (-3\sin 3t)\mathbf{i} + (3\cos 3t)\mathbf{j} + 3\mathbf{k} \Rightarrow \mathbf{v}\left(\frac{\pi}{3}\right) = -3\mathbf{j} + 3\mathbf{k}$ $\Rightarrow \mathbf{u} = -\frac{1}{\sqrt{2}}\mathbf{j} + \frac{1}{\sqrt{2}}\mathbf{k}$; $f(x, y, z) = xyz \Rightarrow \nabla f = yz\mathbf{i} + xz\mathbf{j} + xy\mathbf{k}$; $t = \frac{\pi}{3}$ yields the point on the helix $(-1, 0, \pi)$ $\Rightarrow \nabla f|_{(-1,0,\pi)} = -\pi\mathbf{j} \Rightarrow \nabla f \cdot \mathbf{u} = (-\pi\mathbf{j}) \cdot \left(-\frac{1}{\sqrt{2}}\mathbf{j} + \frac{1}{\sqrt{2}}\mathbf{k}\right) = \frac{\pi}{\sqrt{2}}$

43. (a) Let $\nabla f = a\mathbf{i} + b\mathbf{j}$ at $(1, 2)$. The direction toward $(2, 2)$ is determined by $\mathbf{v}_1 = (2 - 1)\mathbf{i} + (2 - 2)\mathbf{j} = \mathbf{i} = \mathbf{u}$ so that $\nabla f \cdot \mathbf{u} = 2 \Rightarrow a = 2$. The direction toward $(1, 1)$ is determined by $\mathbf{v}_2 = (1 - 1)\mathbf{i} + (1 - 2)\mathbf{j} = -\mathbf{j} = \mathbf{u}$ so that $\nabla f \cdot \mathbf{u} = -2 \Rightarrow -b = -2 \Rightarrow b = 2$. Therefore $\nabla f = 2\mathbf{i} + 2\mathbf{j}$; $f_x(1, 2) = f_y(1, 2) = 2$.

(b) The direction toward $(4, 6)$ is determined by $\mathbf{v}_3 = (4 - 1)\mathbf{i} + (6 - 2)\mathbf{j} = 3\mathbf{i} + 4\mathbf{j} \Rightarrow \mathbf{u} = \frac{3}{5}\mathbf{i} + \frac{4}{5}\mathbf{j}$ $\Rightarrow \nabla f \cdot \mathbf{u} = \frac{14}{5}$.

45. $\nabla f = 2x\mathbf{i} + \mathbf{j} + 2z\mathbf{k} \Rightarrow$
$\nabla f|_{(0,-1,-1)} = \mathbf{j} - 2\mathbf{k}$,
$\nabla f|_{(0,0,0)} = \mathbf{j}$,
$\nabla f|_{(0,-1,1)} = \mathbf{j} + 2\mathbf{k}$

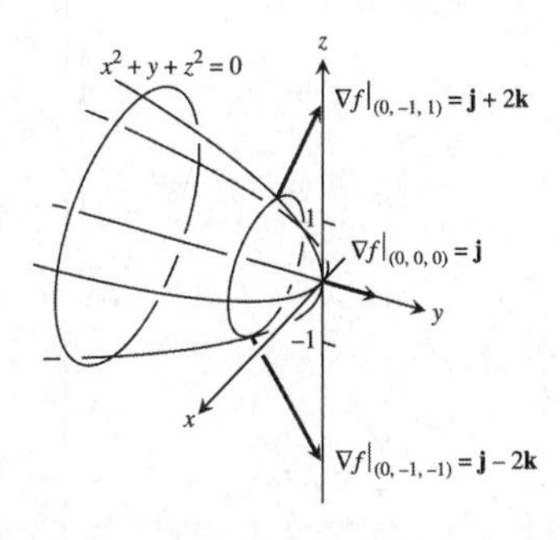

47. $\nabla f = 2x\mathbf{i} - \mathbf{j} - 5\mathbf{k} \Rightarrow \nabla f|_{(2,-1,1)} = 4\mathbf{i} - \mathbf{j} - 5\mathbf{k} \Rightarrow$ Tangent Plane: $4(x - 2) - (y + 1) - 5(z - 1) = 0$ $\Rightarrow 4x - y - 5z = 4$; Normal Line: $x = 2 + 4t$, $y = -1 - t$, $z = 1 - 5t$

49. $\frac{\partial z}{\partial x} = \frac{2x}{x^2 + y^2} \Rightarrow \left.\frac{\partial z}{\partial x}\right|_{(0,1,0)} = 0$ and $\frac{\partial z}{\partial y} = \frac{2y}{x^2 + y^2} \Rightarrow \left.\frac{\partial z}{\partial y}\right|_{(0,1,0)} = 2$; thus the tangent plane is $2(y - 1) - (z - 0) = 0$ or $2y - z - 2 = 0$

51. $\nabla f = (-\cos x)\mathbf{i} + \mathbf{j} \Rightarrow \nabla f|_{(\pi,1)} = \mathbf{i} + \mathbf{j} \Rightarrow$ the tangent line is $(x - \pi) + (y - 1) = 0 \Rightarrow x + y = \pi + 1$; the normal line is $y - 1 = 1(x - \pi) \Rightarrow y = x - \pi + 1$

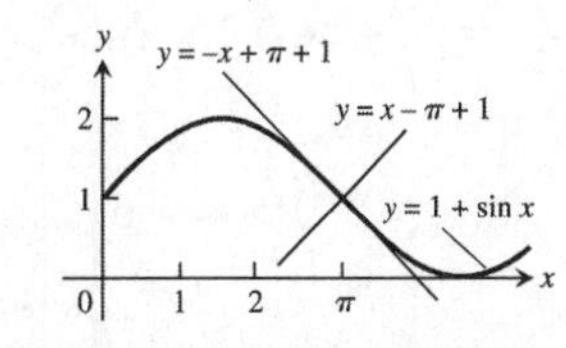

53. Let $f(x, y, z) = x^2 + 2y + 2z - 4$ and $g(x, y, z) = y - 1$. Then $\nabla f = 2x\mathbf{i} + 2\mathbf{j} + 2\mathbf{k}|_{(1,1,\frac{1}{2})} = 2\mathbf{i} + 2\mathbf{j} + 2\mathbf{k}$ and $\nabla g = \mathbf{j} \Rightarrow \nabla f \times \nabla g = \begin{vmatrix} \mathbf{i} & \mathbf{j} & \mathbf{k} \\ 2 & 2 & 2 \\ 0 & 1 & 0 \end{vmatrix} = -2\mathbf{i} + 2\mathbf{k} \Rightarrow$ the line is $x = 1 - 2t$, $y = 1$, $z = \frac{1}{2} + 2t$

55. $f\left(\frac{\pi}{4}, \frac{\pi}{4}\right) = \frac{1}{2}$, $f_x\left(\frac{\pi}{4}, \frac{\pi}{4}\right) = \cos x \cos y|_{(\pi/4,\pi/4)} = \frac{1}{2}$, $f_y\left(\frac{\pi}{4}, \frac{\pi}{4}\right) = -\sin x \sin y|_{(\pi/4,\pi/4)} = -\frac{1}{2}$ $\Rightarrow L(x, y) = \frac{1}{2} + \frac{1}{2}\left(x - \frac{\pi}{4}\right) - \frac{1}{2}\left(y - \frac{\pi}{4}\right) = \frac{1}{2} + \frac{1}{2}x - \frac{1}{2}y$; $f_{xx}(x, y) = -\sin x \cos y$, $f_{yy}(x, y) = -\sin x \cos y$, and $f_{xy}(x, y) = -\cos x \sin y$. Thus an upper bound for E depends on the bound M used for $|f_{xx}|$, $|f_{xy}|$, and $|f_{yy}|$. With $M = \frac{\sqrt{2}}{2}$ we have $|E(x, y)| \le \frac{1}{2}\left(\frac{\sqrt{2}}{2}\right)\left(\left|x - \frac{\pi}{4}\right| + \left|y - \frac{\pi}{4}\right|\right)^2 \le \frac{\sqrt{2}}{4}(0.2)^2 \le 0.0142$; with $M = 1$, $|E(x, y)| \le \frac{1}{2}(1)\left(\left|x - \frac{\pi}{4}\right| + \left|y - \frac{\pi}{4}\right|\right)^2 = \frac{1}{2}(0.2)^2 = 0.02$.

57. $f(1, 0, 0) = 0$, $f_x(1, 0, 0) = y - 3z|_{(1,0,0)} = 0$, $f_y(1, 0, 0) = x + 2z|_{(1,0,0)} = 1$, $f_z(1, 0, 0) = 2y - 3x|_{(1,0,0)} = -3$ $\Rightarrow L(x, y, z) = 0(x - 1) + (y - 0) - 3(z - 0) = y - 3z$; $f(1, 1, 0) = 1$, $f_x(1, 1, 0) = 1$, $f_y(1, 1, 0) = 1$, $f_z(1, 1, 0) = -1$ $\Rightarrow L(x, y, z) = 1 + (x - 1) + (y - 1) - 1(z - 0) = x + y - z - 1$

59. $V = \pi r^2 h \Rightarrow dV = 2\pi rh\,dr + \pi r^2\,dh \Rightarrow dV|_{(1.5,5280)} = 2\pi(1.5)(5280)\,dr + \pi(1.5)^2\,dh = 15{,}840\pi\,dr + 2.25\pi\,dh$. You should be more careful with the diameter since it has a greater effect on dV.

61. $dI = \frac{1}{R}\,dV - \frac{V}{R^2}\,dR \Rightarrow dI|_{(24,100)} = \frac{1}{100}\,dV - \frac{24}{100^2}\,dR \Rightarrow dI|_{dV=-1,dR=-20} = -0.01 + (480)(.0001) = 0.038$, or increases by 0.038 amps; % change in $V = (100)\left(-\frac{1}{24}\right) \approx -4.17\%$; % change in $R = \left(-\frac{20}{100}\right)(100) = -20\%$; $I = \frac{24}{100} = 0.24 \Rightarrow$ estimated % change in $I = \frac{dI}{I} \times 100 = \frac{0.038}{0.24} \times 100 \approx 15.83\% \Rightarrow$ more sensitive to voltage change.

63. (a) $y = uv \Rightarrow dy = v\,du + u\,dv$; percentage change in $u \le 2\% \Rightarrow |du| \le 0.02$, and percentage change in $v \le 3\%$ $\Rightarrow |dv| \le 0.03$; $\frac{dy}{y} = \frac{v\,du + u\,dv}{uv} = \frac{du}{u} + \frac{dv}{v} \Rightarrow \left|\frac{dy}{y} \times 100\right| = \left|\frac{du}{u} \times 100 + \frac{dv}{v} \times 100\right| \le \left|\frac{du}{u} \times 100\right| + \left|\frac{dv}{v} \times 100\right|$ $\le 2\% + 3\% = 5\%$

(b) $z = u + v \Rightarrow \frac{dz}{z} = \frac{du + dv}{u + v} = \frac{du}{u + v} + \frac{dv}{u + v} \le \frac{du}{u} + \frac{dv}{v}$ (since $u > 0$, $v > 0$) $\Rightarrow \left|\frac{dz}{z} \times 100\right| \le \left|\frac{du}{u} \times 100 + \frac{dv}{v} \times 100\right| = \left|\frac{dy}{y} \times 100\right|$

65. $f_x(x, y) = 2x - y + 2 = 0$ and $f_y(x, y) = -x + 2y + 2 = 0 \Rightarrow x = -2$ and $y = -2 \Rightarrow (-2, -2)$ is the critical point; $f_{xx}(-2, -2) = 2$, $f_{yy}(-2, -2) = 2$, $f_{xy}(-2, -2) = -1 \Rightarrow f_{xx}f_{yy} - f_{xy}^2 = 3 > 0$ and $f_{xx} > 0 \Rightarrow$ local minimum value of $f(-2, -2) = -8$

67. $f_x(x, y) = 6x^2 + 3y = 0$ and $f_y(x, y) = 3x + 6y^2 = 0 \Rightarrow y = -2x^2$ and $3x + 6(4x^4) = 0 \Rightarrow x(1 + 8x^3) = 0$ $\Rightarrow x = 0$ and $y = 0$, or $x = -\frac{1}{2}$ and $y = -\frac{1}{2} \Rightarrow$ the critical points are $(0, 0)$ and $\left(-\frac{1}{2}, -\frac{1}{2}\right)$. For $(0, 0)$: $f_{xx}(0, 0) = 12x|_{(0,0)} = 0$, $f_{yy}(0, 0) = 12y|_{(0,0)} = 0$, $f_{xy}(0, 0) = 3 \Rightarrow f_{xx}f_{yy} - f_{xy}^2 = -9 < 0 \Rightarrow$ saddle point with $f(0, 0) = 0$. For $\left(-\frac{1}{2}, -\frac{1}{2}\right)$: $f_{xx} = -6$, $f_{yy} = -6$, $f_{xy} = 3 \Rightarrow f_{xx}f_{yy} - f_{xy}^2 = 27 > 0$ and $f_{xx} < 0 \Rightarrow$ local maximum value of $f\left(-\frac{1}{2}, -\frac{1}{2}\right) = \frac{1}{4}$

69. $f_x(x, y) = 3x^2 + 6x = 0$ and $f_y(x, y) = 3y^2 - 6y = 0 \Rightarrow x(x + 2) = 0$ and $y(y - 2) = 0 \Rightarrow x = 0$ or $x = -2$ and $y = 0$ or $y = 2 \Rightarrow$ the critical points are $(0, 0)$, $(0, 2)$, $(-2, 0)$, and $(-2, 2)$. For $(0, 0)$: $f_{xx}(0, 0) = 6x + 6|_{(0,0)}$ $= 6$, $f_{yy}(0, 0) = 6y - 6|_{(0,0)} = -6$, $f_{xy}(0, 0) = 0 \Rightarrow f_{xx}f_{yy} - f_{xy}^2 = -36 < 0 \Rightarrow$ saddle point with $f(0, 0) = 0$. For $(0, 2)$: $f_{xx}(0, 2) = 6$, $f_{yy}(0, 2) = 6$, $f_{xy}(0, 2) = 0 \Rightarrow f_{xx}f_{yy} - f_{xy}^2 = 36 > 0$ and $f_{xx} > 0 \Rightarrow$ local minimum value of $f(0, 2) = -4$. For $(-2, 0)$: $f_{xx}(-2, 0) = -6$, $f_{yy}(-2, 0) = -6$, $f_{xy}(-2, 0) = 0 \Rightarrow f_{xx}f_{yy} - f_{xy}^2 = 36 > 0$ and $f_{xx} < 0$

$\Rightarrow$ local maximum value of $f(-2, 0) = 4$. For $(-2, 2)$: $f_{xx}(-2, 2) = -6$, $f_{yy}(-2, 2) = 6$, $f_{xy}(-2, 2) = 0$
$\Rightarrow f_{xx}f_{yy} - f_{xy}^2 = -36 < 0 \Rightarrow$ saddle point with $f(-2, 2) = 0$.

71. (i) On OA, $f(x, y) = f(0, y) = y^2 + 3y$ for $0 \le y \le 4$
$\Rightarrow f'(0, y) = 2y + 3 = 0 \Rightarrow y = -\frac{3}{2}$. But $\left(0, -\frac{3}{2}\right)$
is not in the region.

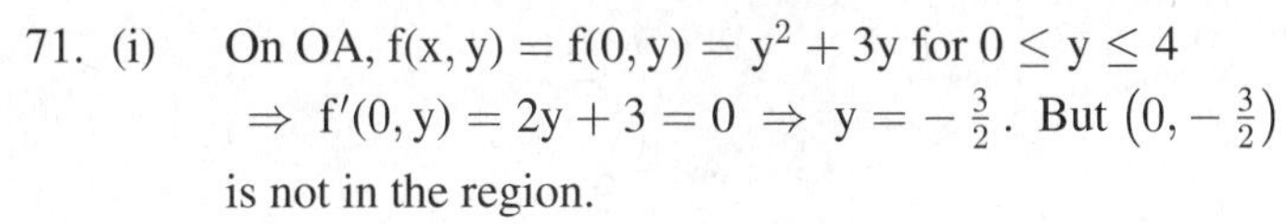

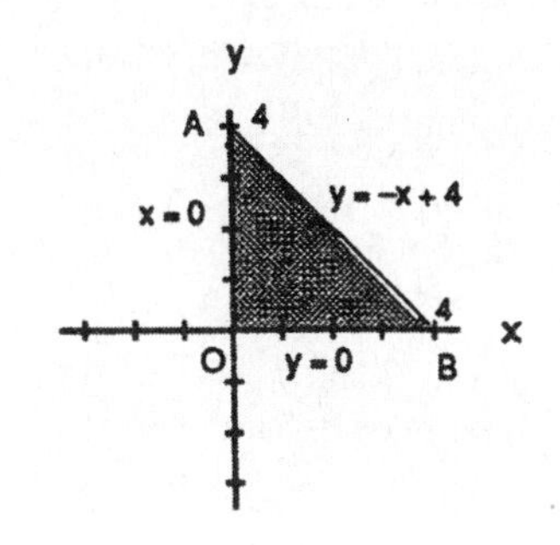

Endpoints: $f(0, 0) = 0$ and $f(0, 4) = 28$.

(ii) On AB, $f(x, y) = f(x, -x + 4) = x^2 - 10x + 28$
for $0 \le x \le 4 \Rightarrow f'(x, -x + 4) = 2x - 10 = 0$
$\Rightarrow x = 5, y = -1$. But $(5, -1)$ is not in the region.

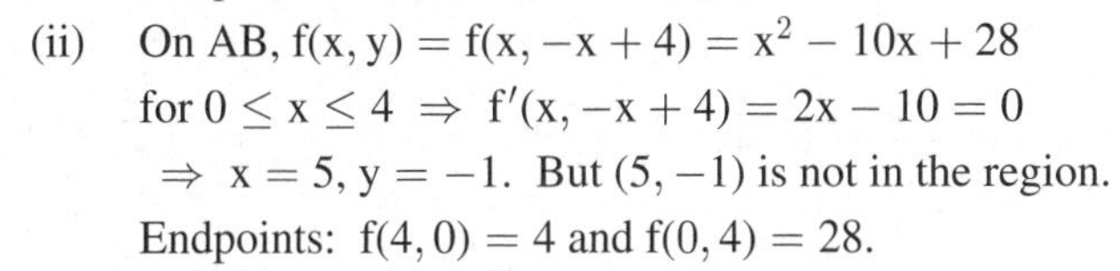

Endpoints: $f(4, 0) = 4$ and $f(0, 4) = 28$.

(iii) On OB, $f(x, y) = f(x, 0) = x^2 - 3x$ for $0 \le x \le 4 \Rightarrow f'(x, 0) = 2x - 3 \Rightarrow x = \frac{3}{2}$ and $y = 0 \Rightarrow \left(\frac{3}{2}, 0\right)$ is a critical point with $f\left(\frac{3}{2}, 0\right) = -\frac{9}{4}$.
Endpoints: $f(0, 0) = 0$ and $f(4, 0) = 4$.

(iv) For the interior of the triangular region, $f_x(x, y) = 2x + y - 3 = 0$ and $f_y(x, y) = x + 2y + 3 = 0 \Rightarrow x = 3$ and $y = -3$. But $(3, -3)$ is not in the region. Therefore the absolute maximum is 28 at $(0, 4)$ and the absolute minimum is $-\frac{9}{4}$ at $\left(\frac{3}{2}, 0\right)$.

73. (i) On AB, $f(x, y) = f(-2, y) = y^2 - y - 4$ for
$-2 \le y \le 2 \Rightarrow f'(-2, y) = 2y - 1 \Rightarrow y = \frac{1}{2}$ and
$x = -2 \Rightarrow \left(-2, \frac{1}{2}\right)$ is an interior critical point in AB
with $f\left(-2, \frac{1}{2}\right) = -\frac{17}{4}$. Endpoints: $f(-2, -2) = 2$ and
$f(2, 2) = -2$.

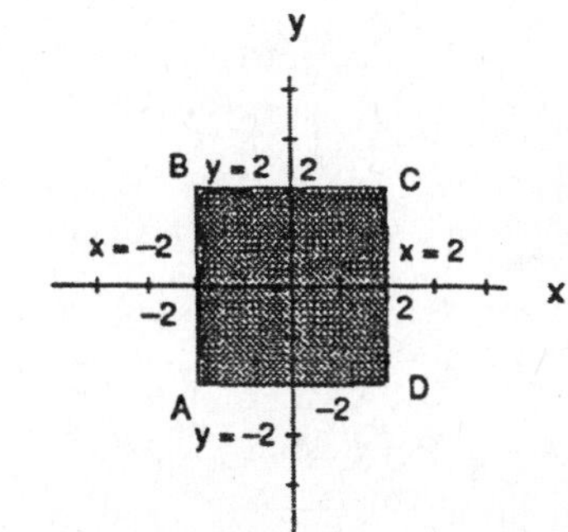

(ii) On BC, $f(x, y) = f(x, 2) = -2$ for $-2 \le x \le 2$
$\Rightarrow f'(x, 2) = 0 \Rightarrow$ no critical points in the interior of
BC. Endpoints: $f(-2, 2) = -2$ and $f(2, 2) = -2$.

(iii) On CD, $f(x, y) = f(2, y) = y^2 - 5y + 4$ for
$-2 \le y \le 2 \Rightarrow f'(2, y) = 2y - 5 = 0 \Rightarrow y = \frac{5}{2}$ and $x = 2$. But $\left(2, \frac{5}{2}\right)$ is not in the region.
Endpoints: $f(2, -2) = 18$ and $f(2, 2) = -2$.

(iv) On AD, $f(x, y) = f(x, -2) = 4x + 10$ for $-2 \le x \le 2 \Rightarrow f'(x, -2) = 4 \Rightarrow$ no critical points in the interior of AD. Endpoints: $f(-2, -2) = 2$ and $f(2, -2) = 18$.

(v) For the interior of the square, $f_x(x, y) = -y + 2 = 0$ and $f_y(x, y) = 2y - x - 3 = 0 \Rightarrow y = 2$ and $x = 1$
$\Rightarrow (1, 2)$ is an interior critical point of the square with $f(1, 2) = -2$. Therefore the absolute maximum is 18 at $(2, -2)$ and the absolute minimum is $-\frac{17}{4}$ at $\left(-2, \frac{1}{2}\right)$.

75. (i) On AB, $f(x, y) = f(x, x + 2) = -2x + 4$ for
$-2 \le x \le 2 \Rightarrow f'(x, x + 2) = -2 = 0 \Rightarrow$ no critical
points in the interior of AB. Endpoints: $f(-2, 0) = 8$
and $f(2, 4) = 0$.

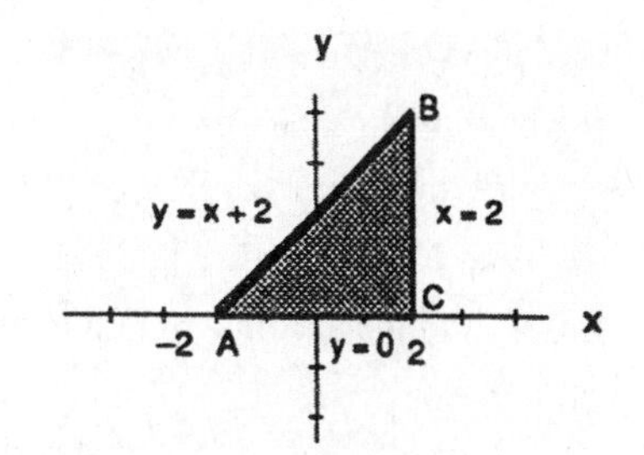

(ii) On BC, $f(x, y) = f(2, y) = -y^2 + 4y$ for $0 \le y \le 4$
$\Rightarrow f'(2, y) = -2y + 4 = 0 \Rightarrow y = 2$ and $x = 2$
$\Rightarrow (2, 2)$ is an interior critical point of BC with
$f(2, 2) = 4$. Endpoints: $f(2, 0) = 0$ and $f(2, 4) = 0$.

(iii) On AC, $f(x, y) = f(x, 0) = x^2 - 2x$ for $-2 \le x \le 2 \Rightarrow f'(x, 0) = 2x - 2 \Rightarrow x = 1$ and $y = 0 \Rightarrow (1, 0)$ is an interior critical point of AC with $f(1, 0) = -1$. Endpoints: $f(-2, 0) = 8$ and $f(2, 0) = 0$.

(iv) For the interior of the triangular region, $f_x(x, y) = 2x - 2 = 0$ and $f_y(x, y) = -2y + 4 = 0 \Rightarrow x = 1$ and $y = 2 \Rightarrow (1, 2)$ is an interior critical point of the region with $f(1, 2) = 3$. Therefore the absolute maximum is 8 at $(-2, 0)$ and the absolute minimum is -1 at $(1, 0)$.

77. (i) On AB, $f(x, y) = f(-1, y) = y^3 - 3y^2 + 2$ for $-1 \le y \le 1 \Rightarrow f'(-1, y) = 3y^2 - 6y = 0 \Rightarrow y = 0$ and $x = -1$, or $y = 2$ and $x = -1 \Rightarrow (-1, 0)$ is an interior critical point of AB with $f(-1, 0) = 2$; $(-1, 2)$ is outside the boundary. Endpoints: $f(-1, -1) = -2$ and $f(-1, 1) = 0$.

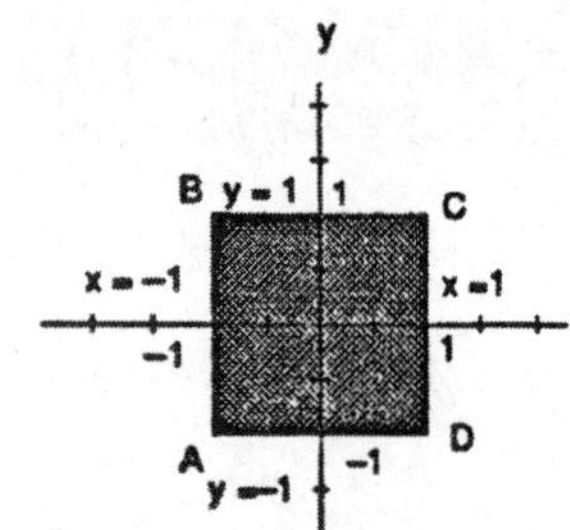

(ii) On BC, $f(x, y) = f(x, 1) = x^3 + 3x^2 - 2$ for $-1 \le x \le 1 \Rightarrow f'(x, 1) = 3x^2 + 6x = 0 \Rightarrow x = 0$ and $y = 1$, or $x = -2$ and $y = 1 \Rightarrow (0, 1)$ is an interior critical point of BC with $f(0, 1) = -2$; $(-2, 1)$ is outside the boundary. Endpoints: $f(-1, 1) = 0$ and $f(1, 1) = 2$.

(iii) On CD, $f(x, y) = f(1, y) = y^3 - 3y^2 + 4$ for $-1 \le y \le 1 \Rightarrow f'(1, y) = 3y^2 - 6y = 0 \Rightarrow y = 0$ and $x = 1$, or $y = 2$ and $x = 1 \Rightarrow (1, 0)$ is an interior critical point of CD with $f(1, 0) = 4$; $(1, 2)$ is outside the boundary. Endpoints: $f(1, 1) = 2$ and $f(1, -1) = 0$.

(iv) On AD, $f(x, y) = f(x, -1) = x^3 + 3x^2 - 4$ for $-1 \le x \le 1 \Rightarrow f'(x, -1) = 3x^2 + 6x = 0 \Rightarrow x = 0$ and $y = -1$, or $x = -2$ and $y = -1 \Rightarrow (0, -1)$ is an interior point of AD with $f(0, -1) = -4$; $(-2, -1)$ is outside the boundary. Endpoints: $f(-1, -1) = -2$ and $f(1, -1) = 0$.

(v) For the interior of the square, $f_x(x, y) = 3x^2 + 6x = 0$ and $f_y(x, y) = 3y^2 - 6y = 0 \Rightarrow x = 0$ or $x = -2$, and $y = 0$ or $y = 2 \Rightarrow (0, 0)$ is an interior critical point of the square region with $f(0, 0) = 0$; the points $(0, 2)$, $(-2, 0)$, and $(-2, 2)$ are outside the region. Therefore the absolute maximum is 4 at $(1, 0)$ and the absolute minimum is -4 at $(0, -1)$.

79. $\nabla f = 3x^2\mathbf{i} + 2y\mathbf{j}$ and $\nabla g = 2x\mathbf{i} + 2y\mathbf{j}$ so that $\nabla f = \lambda \nabla g \Rightarrow 3x^2\mathbf{i} + 2y\mathbf{j} = \lambda(2x\mathbf{i} + 2y\mathbf{j}) \Rightarrow 3x^2 = 2x\lambda$ and $2y = 2y\lambda \Rightarrow \lambda = 1$ or $y = 0$.

CASE 1: $\lambda = 1 \Rightarrow 3x^2 = 2x \Rightarrow x = 0$ or $x = \frac{2}{3}$; $x = 0 \Rightarrow y = \pm 1$ yielding the points $(0, 1)$ and $(0, -1)$; $x = \frac{2}{3}$ $\Rightarrow y = \pm \frac{\sqrt{5}}{3}$ yielding the points $\left(\frac{2}{3}, \frac{\sqrt{5}}{3}\right)$ and $\left(\frac{2}{3}, -\frac{\sqrt{5}}{3}\right)$.

CASE 2: $y = 0 \Rightarrow x^2 - 1 = 0 \Rightarrow x = \pm 1$ yielding the points $(1, 0)$ and $(-1, 0)$.

Evaluations give $f(0, \pm 1) = 1$, $f\left(\frac{2}{3}, \pm \frac{\sqrt{5}}{3}\right) = \frac{23}{27}$, $f(1, 0) = 1$, and $f(-1, 0) = -1$. Therefore the absolute maximum is 1 at $(0, \pm 1)$ and $(1, 0)$, and the absolute minimum is -1 at $(-1, 0)$.

81. (i) $f(x, y) = x^2 + 3y^2 + 2y$ on $x^2 + y^2 = 1 \Rightarrow \nabla f = 2x\mathbf{i} + (6y + 2)\mathbf{j}$ and $\nabla g = 2x\mathbf{i} + 2y\mathbf{j}$ so that $\nabla f = \lambda \nabla g$ $\Rightarrow 2x\mathbf{i} + (6y + 2)\mathbf{j} = \lambda(2x\mathbf{i} + 2y\mathbf{j}) \Rightarrow 2x = 2x\lambda$ and $6y + 2 = 2y\lambda \Rightarrow \lambda = 1$ or $x = 0$.

CASE 1: $\lambda = 1 \Rightarrow 6y + 2 = 2y \Rightarrow y = -\frac{1}{2}$ and $x = \pm \frac{\sqrt{3}}{2}$ yielding the points $\left(\pm \frac{\sqrt{3}}{2}, -\frac{1}{2}\right)$.

CASE 2: $x = 0 \Rightarrow y^2 = 1 \Rightarrow y = \pm 1$ yielding the points $(0, \pm 1)$.

Evaluations give $f\left(\pm \frac{\sqrt{3}}{2}, -\frac{1}{2}\right) = \frac{1}{2}$, $f(0, 1) = 5$, and $f(0, -1) = 1$. Therefore $\frac{1}{2}$ and 5 are the extreme values on the boundary of the disk.

(ii) For the interior of the disk, $f_x(x, y) = 2x = 0$ and $f_y(x, y) = 6y + 2 = 0 \Rightarrow x = 0$ and $y = -\frac{1}{3}$ $\Rightarrow \left(0, -\frac{1}{3}\right)$ is an interior critical point with $f\left(0, -\frac{1}{3}\right) = -\frac{1}{3}$. Therefore the absolute maximum of f on the disk is 5 at $(0, 1)$ and the absolute minimum of f on the disk is $-\frac{1}{3}$ at $\left(0, -\frac{1}{3}\right)$.

83. $\nabla f = \mathbf{i} - \mathbf{j} + \mathbf{k}$ and $\nabla g = 2x\mathbf{i} + 2y\mathbf{j} + 2z\mathbf{k}$ so that $\nabla f = \lambda \nabla g \Rightarrow \mathbf{i} - \mathbf{j} + \mathbf{k} = \lambda(2x\mathbf{i} + 2y\mathbf{j} + 2z\mathbf{k}) \Rightarrow 1 = 2x\lambda$, $-1 = 2y\lambda$, $1 = 2z\lambda \Rightarrow x = -y = z = \frac{1}{\lambda}$. Thus $x^2 + y^2 + z^2 = 1 \Rightarrow 3x^2 = 1 \Rightarrow x = \pm \frac{1}{\sqrt{3}}$ yielding the points $\left(\frac{1}{\sqrt{3}}, -\frac{1}{\sqrt{3}}, \frac{1}{\sqrt{3}}\right)$ and $\left(-\frac{1}{\sqrt{3}}, \frac{1}{\sqrt{3}}, -\frac{1}{\sqrt{3}}\right)$. Evaluations give the absolute maximum value of $f\left(\frac{1}{\sqrt{3}}, -\frac{1}{\sqrt{3}}, \frac{1}{\sqrt{3}}\right) = \frac{3}{\sqrt{3}} = \sqrt{3}$ and the absolute minimum value of $f\left(-\frac{1}{\sqrt{3}}, \frac{1}{\sqrt{3}}, -\frac{1}{\sqrt{3}}\right) = -\sqrt{3}$.

85. The cost is $f(x,y,z) = 2axy + 2bxz + 2cyz$ subject to the constraint $xyz = V$. Then $\nabla f = \lambda \nabla g$ $\Rightarrow 2ay + 2bz = \lambda yz,\ 2ax + 2cz = \lambda xz$, and $2bx + 2cy = \lambda xy \Rightarrow 2axy + 2bxz = \lambda xyz,\ 2axy + 2cyz = \lambda xyz$, and $2bxz + 2cyz = \lambda xyz \Rightarrow 2axy + 2bxz = 2axy + 2cyz \Rightarrow y = \left(\frac{b}{c}\right)x$. Also $2axy + 2bxz = 2bxz + 2cyz \Rightarrow z = \left(\frac{a}{c}\right)x$. Then $x\left(\frac{b}{c}x\right)\left(\frac{a}{c}x\right) = V \Rightarrow x^3 = \frac{c^2V}{ab} \Rightarrow$ width $= x = \left(\frac{c^2V}{ab}\right)^{1/3}$, Depth $= y = \left(\frac{b}{c}\right)\left(\frac{c^2V}{ab}\right)^{1/3} = \left(\frac{b^2V}{ac}\right)^{1/3}$, and Height $= z = \left(\frac{a}{c}\right)\left(\frac{c^2V}{ab}\right)^{1/3} = \left(\frac{a^2V}{bc}\right)^{1/3}$.

87. $\nabla f = (y+z)\mathbf{i} + x\mathbf{j} + x\mathbf{k}$, $\nabla g = 2x\mathbf{i} + 2y\mathbf{j}$, and $\nabla h = z\mathbf{i} + x\mathbf{k}$ so that $\nabla f = \lambda \nabla g + \mu \nabla h$ $\Rightarrow (y+z)\mathbf{i} + x\mathbf{j} + x\mathbf{k} = \lambda(2x\mathbf{i} + 2y\mathbf{j}) + \mu(z\mathbf{i} + x\mathbf{k}) \Rightarrow y + z = 2\lambda x + \mu z,\ x = 2\lambda y,\ x = \mu x \Rightarrow x = 0$ or $\mu = 1$.
CASE 1: $x = 0$ which is impossible since $xz = 1$.
CASE 2: $\mu = 1 \Rightarrow y + z = 2\lambda x + z \Rightarrow y = 2\lambda x$ and $x = 2\lambda y \Rightarrow y = (2\lambda)(2\lambda y) \Rightarrow y = 0$ or $4\lambda^2 = 1$. If $y = 0$, then $x^2 = 1 \Rightarrow x = \pm 1$ so with $xz = 1$ we obtain the points $(1,0,1)$ and $(-1,0,-1)$. If $4\lambda^2 = 1$, then $\lambda = \pm\frac{1}{2}$. For $\lambda = -\frac{1}{2}$, $y = -x$ so $x^2 + y^2 = 1 \Rightarrow x^2 = \frac{1}{2}$ $\Rightarrow x = \pm\frac{1}{\sqrt{2}}$ with $xz = 1 \Rightarrow z = \pm\sqrt{2}$, and we obtain the points $\left(\frac{1}{\sqrt{2}}, -\frac{1}{\sqrt{2}}, \sqrt{2}\right)$ and $\left(-\frac{1}{\sqrt{2}}, \frac{1}{\sqrt{2}}, -\sqrt{2}\right)$. For $\lambda = \frac{1}{2}$, $y = x \Rightarrow x^2 = \frac{1}{2} \Rightarrow x = \pm\frac{1}{\sqrt{2}}$ with $xz = 1 \Rightarrow z = \pm\sqrt{2}$, and we obtain the points $\left(\frac{1}{\sqrt{2}}, \frac{1}{\sqrt{2}}, \sqrt{2}\right)$ and $\left(-\frac{1}{\sqrt{2}}, -\frac{1}{\sqrt{2}}, -\sqrt{2}\right)$.
Evaluations give $f(1,0,1) = 1$, $f(-1,0,-1) = 1$, $f\left(\frac{1}{\sqrt{2}}, -\frac{1}{\sqrt{2}}, \sqrt{2}\right) = \frac{1}{2}$, $f\left(-\frac{1}{\sqrt{2}}, \frac{1}{\sqrt{2}}, -\sqrt{2}\right) = \frac{1}{2}$, $f\left(\frac{1}{\sqrt{2}}, \frac{1}{\sqrt{2}}, \sqrt{2}\right) = \frac{3}{2}$, and $f\left(-\frac{1}{\sqrt{2}}, -\frac{1}{\sqrt{2}}, -\sqrt{2}\right) = \frac{3}{2}$. Therefore the absolute maximum is $\frac{3}{2}$ at $\left(\frac{1}{\sqrt{2}}, \frac{1}{\sqrt{2}}, \sqrt{2}\right)$ and $\left(-\frac{1}{\sqrt{2}}, -\frac{1}{\sqrt{2}}, -\sqrt{2}\right)$, and the absolute minimum is $\frac{1}{2}$ at $\left(-\frac{1}{\sqrt{2}}, \frac{1}{\sqrt{2}}, -\sqrt{2}\right)$ and $\left(\frac{1}{\sqrt{2}}, -\frac{1}{\sqrt{2}}, \sqrt{2}\right)$.

89. (a) y, z are independent with $w = x^2e^{yz}$ and $z = x^2 - y^2 \Rightarrow \frac{\partial w}{\partial y} = \frac{\partial w}{\partial x}\frac{\partial x}{\partial y} + \frac{\partial w}{\partial y}\frac{\partial y}{\partial y} + \frac{\partial w}{\partial z}\frac{\partial z}{\partial y}$ $= (2xe^{yz})\frac{\partial x}{\partial y} + (zx^2e^{yz})(1) + (yx^2e^{yz})(0)$; $z = x^2 - y^2 \Rightarrow 0 = 2x\frac{\partial x}{\partial y} - 2y \Rightarrow \frac{\partial x}{\partial y} = \frac{y}{x}$; therefore, $\left(\frac{\partial w}{\partial y}\right)_z = (2xe^{yz})\left(\frac{y}{x}\right) + zx^2e^{yz} = (2y + zx^2)e^{yz}$

(b) z, x are independent with $w = x^2e^{yz}$ and $z = x^2 - y^2 \Rightarrow \frac{\partial w}{\partial z} = \frac{\partial w}{\partial x}\frac{\partial x}{\partial z} + \frac{\partial w}{\partial y}\frac{\partial y}{\partial z} + \frac{\partial w}{\partial z}\frac{\partial z}{\partial z}$ $= (2xe^{yz})(0) + (zx^2e^{yz})\frac{\partial y}{\partial z} + (yx^2e^{yz})(1)$; $z = x^2 - y^2 \Rightarrow 1 = 0 - 2y\frac{\partial y}{\partial z} \Rightarrow \frac{\partial y}{\partial z} = -\frac{1}{2y}$; therefore, $\left(\frac{\partial w}{\partial z}\right)_x = (zx^2e^{yz})\left(-\frac{1}{2y}\right) + yx^2e^{yz} = x^2e^{yz}\left(y - \frac{z}{2y}\right)$

(c) z, y are independent with $w = x^2e^{yz}$ and $z = x^2 - y^2 \Rightarrow \frac{\partial w}{\partial z} = \frac{\partial w}{\partial x}\frac{\partial x}{\partial z} + \frac{\partial w}{\partial y}\frac{\partial y}{\partial z} + \frac{\partial w}{\partial z}\frac{\partial z}{\partial z}$ $= (2xe^{yz})\frac{\partial x}{\partial z} + (zx^2e^{yz})(0) + (yx^2e^{yz})(1)$; $z = x^2 - y^2 \Rightarrow 1 = 2x\frac{\partial x}{\partial z} - 0 \Rightarrow \frac{\partial x}{\partial z} = \frac{1}{2x}$; therefore, $\left(\frac{\partial w}{\partial z}\right)_y = (2xe^{yz})\left(\frac{1}{2x}\right) + yx^2e^{yz} = (1 + x^2y)e^{yz}$

91. Note that $x = r\cos\theta$ and $y = r\sin\theta \Rightarrow r = \sqrt{x^2 + y^2}$ and $\theta = \tan^{-1}\left(\frac{y}{x}\right)$. Thus,

$$\frac{\partial w}{\partial x} = \frac{\partial w}{\partial r}\frac{\partial r}{\partial x} + \frac{\partial w}{\partial \theta}\frac{\partial \theta}{\partial x} = \left(\frac{\partial w}{\partial r}\right)\left(\frac{x}{\sqrt{x^2+y^2}}\right) + \left(\frac{\partial w}{\partial \theta}\right)\left(\frac{-y}{x^2+y^2}\right) = (\cos\theta)\frac{\partial w}{\partial r} - \left(\frac{\sin\theta}{r}\right)\frac{\partial w}{\partial \theta};$$

$$\frac{\partial w}{\partial y} = \frac{\partial w}{\partial r}\frac{\partial r}{\partial y} + \frac{\partial w}{\partial \theta}\frac{\partial \theta}{\partial y} = \left(\frac{\partial w}{\partial r}\right)\left(\frac{y}{\sqrt{x^2+y^2}}\right) + \left(\frac{\partial w}{\partial \theta}\right)\left(\frac{x}{x^2+y^2}\right) = (\sin\theta)\frac{\partial w}{\partial r} + \left(\frac{\cos\theta}{r}\right)\frac{\partial w}{\partial \theta}$$

93. $\frac{\partial u}{\partial y} = b$ and $\frac{\partial u}{\partial x} = a \Rightarrow \frac{\partial w}{\partial x} = \frac{dw}{du}\frac{\partial u}{\partial x} = a\frac{dw}{du}$ and $\frac{\partial w}{\partial y} = \frac{dw}{du}\frac{\partial u}{\partial y} = b\frac{dw}{du} \Rightarrow \frac{1}{a}\frac{\partial w}{\partial x} = \frac{dw}{du}$ and $\frac{1}{b}\frac{\partial w}{\partial y} = \frac{dw}{du}$ $\Rightarrow \frac{1}{a}\frac{\partial w}{\partial x} = \frac{1}{b}\frac{\partial w}{\partial y} \Rightarrow b\frac{\partial w}{\partial x} = a\frac{\partial w}{\partial y}$

95. $e^u \cos v - x = 0 \Rightarrow (e^u \cos v)\frac{\partial u}{\partial x} - (e^u \sin v)\frac{\partial v}{\partial x} = 1$; $e^u \sin v - y = 0 \Rightarrow (e^u \sin v)\frac{\partial u}{\partial x} + (e^u \cos v)\frac{\partial v}{\partial x} = 0$. Solving this system yields $\frac{\partial u}{\partial x} = e^{-u}\cos v$ and $\frac{\partial v}{\partial x} = -e^{-u}\sin v$. Similarly, $e^u \cos v - x = 0$ $\Rightarrow (e^u \cos v)\frac{\partial u}{\partial y} - (e^u \sin v)\frac{\partial v}{\partial y} = 0$ and $e^u \sin v - y = 0 \Rightarrow (e^u \sin v)\frac{\partial u}{\partial y} + (e^u \cos v)\frac{\partial v}{\partial y} = 1$. Solving this second system yields $\frac{\partial u}{\partial y} = e^{-u}\sin v$ and $\frac{\partial v}{\partial y} = e^{-u}\cos v$. Therefore $\left(\frac{\partial u}{\partial x}\mathbf{i} + \frac{\partial u}{\partial y}\mathbf{j}\right)\cdot\left(\frac{\partial v}{\partial x}\mathbf{i} + \frac{\partial v}{\partial y}\mathbf{j}\right)$ $= [(e^{-u}\cos v)\mathbf{i} + (e^{-u}\sin v)\mathbf{j}]\cdot[(-e^{-u}\sin v)\mathbf{i} + (e^{-u}\cos v)\mathbf{j}] = 0 \Rightarrow$ the vectors are orthogonal $\Rightarrow$ the angle between the vectors is the constant $\frac{\pi}{2}$.

97. $(y+z)^2 + (z-x)^2 = 16 \Rightarrow \nabla f = -2(z-x)\mathbf{i} + 2(y+z)\mathbf{j} + 2(y+2z-x)\mathbf{k}$; if the normal line is parallel to the yz-plane, then x is constant $\Rightarrow \frac{\partial f}{\partial x} = 0 \Rightarrow -2(z-x) = 0 \Rightarrow z = x \Rightarrow (y+z)^2 + (z-z)^2 = 16 \Rightarrow y+z = \pm 4$. Let $x = t \Rightarrow z = t \Rightarrow y = -t \pm 4$. Therefore the points are $(t, -t \pm 4, t)$, t a real number.

99. $\nabla f = \lambda(x\mathbf{i} + y\mathbf{j} + z\mathbf{k}) \Rightarrow \frac{\partial f}{\partial x} = \lambda x \Rightarrow f(x,y,z) = \frac{1}{2}\lambda x^2 + g(y,z)$ for some function $g \Rightarrow \lambda y = \frac{\partial f}{\partial y} = \frac{\partial g}{\partial y}$ $\Rightarrow g(y,z) = \frac{1}{2}\lambda y^2 + h(z)$ for some function $h \Rightarrow \lambda z = \frac{\partial f}{\partial z} = \frac{\partial g}{\partial z} = h'(z) \Rightarrow h(z) = \frac{1}{2}\lambda z^2 + C$ for some arbitrary constant $C \Rightarrow g(y,z) = \frac{1}{2}\lambda y^2 + \left(\frac{1}{2}\lambda z^2 + C\right) \Rightarrow f(x,y,z) = \frac{1}{2}\lambda x^2 + \frac{1}{2}\lambda y^2 + \frac{1}{2}\lambda z^2 + C \Rightarrow f(0,0,a) = \frac{1}{2}\lambda a^2 + C$ and $f(0,0,-a) = \frac{1}{2}\lambda(-a)^2 + C \Rightarrow f(0,0,a) = f(0,0,-a)$ for any constant a, as claimed.

101. Let $f(x,y,z) = xy + z - 2 \Rightarrow \nabla f = y\mathbf{i} + x\mathbf{j} + \mathbf{k}$. At $(1,1,1)$, we have $\nabla f = \mathbf{i} + \mathbf{j} + \mathbf{k} \Rightarrow$ the normal line is $x = 1+t, y = 1+t, z = 1+t$, so at $t = -1 \Rightarrow x = 0, y = 0, z = 0$ and the normal line passes through the origin.

CHAPTER 14 ADDITIONAL AND ADVANCED EXERCISES

1. By definition, $f_{xy}(0,0) = \lim_{h\to 0}\frac{f_x(0,h) - f_x(0,0)}{h}$ so we need to calculate the first partial derivatives in the numerator. For $(x,y) \neq (0,0)$ we calculate $f_x(x,y)$ by applying the differentiation rules to the formula for $f(x,y)$: $f_x(x,y) = \frac{x^2y - y^3}{x^2+y^2} + (xy)\frac{(x^2+y^2)(2x) - (x^2-y^2)(2x)}{(x^2+y^2)^2} = \frac{x^2y-y^3}{x^2+y^2} + \frac{4x^2y^3}{(x^2+y^2)^2} \Rightarrow f_x(0,h) = -\frac{h^3}{h^2} = -h$. For $(x,y) = (0,0)$ we apply the definition: $f_x(0,0) = \lim_{h\to 0}\frac{f(h,0) - f(0,0)}{h} = \lim_{h\to 0}\frac{0-0}{h} = 0$. Then by definition $f_{xy}(0,0) = \lim_{h\to 0}\frac{-h-0}{h} = -1$. Similarly, $f_{yx}(0,0) = \lim_{h\to 0}\frac{f_y(h,0) - f_y(0,0)}{h}$, so for $(x,y) \neq (0,0)$ we have $f_y(x,y) = \frac{x^3 - xy^2}{x^2+y^2} - \frac{4x^3y^2}{(x^2+y^2)^2} \Rightarrow f_y(h,0) = \frac{h^3}{h^2} = h$; for $(x,y) = (0,0)$ we obtain $f_y(0,0) = \lim_{h\to 0}\frac{f(0,h) - f(0,0)}{h}$ $= \lim_{h\to 0}\frac{0-0}{h} = 0$. Then by definition $f_{yx}(0,0) = \lim_{h\to 0}\frac{h-0}{h} = 1$. Note that $f_{xy}(0,0) \neq f_{yx}(0,0)$ in this case.

3. Substitution of $u + u(x)$ and $v = v(x)$ in $g(u,v)$ gives $g(u(x), v(x))$ which is a function of the independent variable x. Then, $g(u,v) = \int_u^v f(t)\,dt \Rightarrow \frac{dg}{dx} = \frac{\partial g}{\partial u}\frac{du}{dx} + \frac{\partial g}{\partial v}\frac{dv}{dx} = \left(\frac{\partial}{\partial u}\int_u^v f(t)\,dt\right)\frac{du}{dx} + \left(\frac{\partial}{\partial v}\int_u^v f(t)\,dt\right)\frac{dv}{dx}$ $= \left(-\frac{\partial}{\partial u}\int_v^u f(t)\,dt\right)\frac{du}{dx} + \left(\frac{\partial}{\partial v}\int_u^v f(t)\,dt\right)\frac{dv}{dx} = -f(u(x))\frac{du}{dx} + f(v(x))\frac{dv}{dx} = f(v(x))\frac{dv}{dx} - f(u(x))\frac{du}{dx}$

5. (a) Let $u = tx$, $v = ty$, and $w = f(u,v) = f(u(t,x), v(t,y)) = f(tx,ty) = t^n f(x,y)$, where t, x, and y are independent variables. Then $nt^{n-1}f(x,y) = \frac{\partial w}{\partial t} = \frac{\partial w}{\partial u}\frac{\partial u}{\partial t} + \frac{\partial w}{\partial v}\frac{\partial v}{\partial t} = x\frac{\partial w}{\partial u} + y\frac{\partial w}{\partial v}$. Now, $\frac{\partial w}{\partial x} = \frac{\partial w}{\partial u}\frac{\partial u}{\partial x} + \frac{\partial w}{\partial v}\frac{\partial v}{\partial x} = \left(\frac{\partial w}{\partial u}\right)(t) + \left(\frac{\partial w}{\partial v}\right)(0) = t\frac{\partial w}{\partial u} \Rightarrow \frac{\partial w}{\partial u} = \left(\frac{1}{t}\right)\left(\frac{\partial w}{\partial x}\right)$. Likewise, $\frac{\partial w}{\partial y} = \frac{\partial w}{\partial u}\frac{\partial u}{\partial y} + \frac{\partial w}{\partial v}\frac{\partial v}{\partial y} = \left(\frac{\partial w}{\partial u}\right)(0) + \left(\frac{\partial w}{\partial v}\right)(t) \Rightarrow \frac{\partial w}{\partial v} = \left(\frac{1}{t}\right)\left(\frac{\partial w}{\partial y}\right)$. Therefore, $nt^{n-1}f(x,y) = x\frac{\partial w}{\partial u} + y\frac{\partial w}{\partial v} = \left(\frac{x}{t}\right)\left(\frac{\partial w}{\partial x}\right) + \left(\frac{y}{t}\right)\left(\frac{\partial w}{\partial y}\right)$. When $t = 1$, $u = x$, $v = y$, and $w = f(x,y)$ $\Rightarrow \frac{\partial w}{\partial x} = \frac{\partial f}{\partial x}$ and $\frac{\partial w}{\partial y} = \frac{\partial f}{\partial x} \Rightarrow nf(x,y) = x\frac{\partial f}{\partial x} + y\frac{\partial f}{\partial y}$, as claimed.

(b) From part (a), $nt^{n-1}f(x,y) = x\frac{\partial w}{\partial u} + y\frac{\partial w}{\partial v}$. Differentiating with respect to t again we obtain $n(n-1)t^{n-2}f(x,y) = x\frac{\partial^2 w}{\partial u^2}\frac{\partial u}{\partial t} + x\frac{\partial^2 w}{\partial v\partial u}\frac{\partial v}{\partial t} + y\frac{\partial^2 w}{\partial u\partial v}\frac{\partial u}{\partial t} + y\frac{\partial^2 w}{\partial v^2}\frac{\partial v}{\partial t} = x^2\frac{\partial^2 w}{\partial u^2} + 2xy\frac{\partial^2 w}{\partial u\partial v} + y^2\frac{\partial^2 w}{\partial v^2}$. Also from part (a), $\frac{\partial^2 w}{\partial x^2} = \frac{\partial}{\partial x}\left(\frac{\partial w}{\partial x}\right) = \frac{\partial}{\partial x}\left(t\frac{\partial w}{\partial u}\right) = t\frac{\partial^2 w}{\partial u^2}\frac{\partial u}{\partial x} + t\frac{\partial^2 w}{\partial v\partial u}\frac{\partial v}{\partial x} = t^2\frac{\partial^2 w}{\partial u^2}$, $\frac{\partial^2 w}{\partial y^2} = \frac{\partial}{\partial y}\left(\frac{\partial w}{\partial y}\right)$ $= \frac{\partial}{\partial y}\left(t\frac{\partial w}{\partial v}\right) = t\frac{\partial^2 w}{\partial u\partial v}\frac{\partial u}{\partial y} + t\frac{\partial^2 w}{\partial v^2}\frac{\partial v}{\partial y} = t^2\frac{\partial^2 w}{\partial v^2}$, and $\frac{\partial^2 w}{\partial y\partial x} = \frac{\partial}{\partial y}\left(\frac{\partial w}{\partial x}\right) = \frac{\partial}{\partial y}\left(t\frac{\partial w}{\partial u}\right) = t\frac{\partial^2 w}{\partial u^2}\frac{\partial u}{\partial y} + t\frac{\partial^2 w}{\partial v\partial u}\frac{\partial v}{\partial y}$ $= t^2\frac{\partial^2 w}{\partial v\partial u} \Rightarrow \left(\frac{1}{t^2}\right)\frac{\partial^2 w}{\partial x^2} = \frac{\partial^2 w}{\partial u^2}$, $\left(\frac{1}{t^2}\right)\frac{\partial^2 w}{\partial y^2} = \frac{\partial^2 w}{\partial v^2}$, and $\left(\frac{1}{t^2}\right)\frac{\partial^2 w}{\partial y\partial x} = \frac{\partial^2 w}{\partial v\partial u}$ $\Rightarrow n(n-1)t^{n-2}f(x,y) = \left(\frac{x^2}{t^2}\right)\left(\frac{\partial^2 w}{\partial x^2}\right) + \left(\frac{2xy}{t^2}\right)\left(\frac{\partial^2 w}{\partial y\partial x}\right) + \left(\frac{y^2}{t^2}\right)\left(\frac{\partial^2 w}{\partial y^2}\right)$ for $t \neq 0$. When $t = 1$, $w = f(x,y)$ and we have $n(n-1)f(x,y) = x^2\left(\frac{\partial^2 f}{\partial x^2}\right) + 2xy\left(\frac{\partial^2 f}{\partial x\partial y}\right) + y^2\left(\frac{\partial^2 f}{\partial y^2}\right)$ as claimed.

7. (a) $\mathbf{r} = x\mathbf{i} + y\mathbf{j} + z\mathbf{k} \Rightarrow r = |\mathbf{r}| = \sqrt{x^2+y^2+z^2}$ and $\nabla r = \frac{x}{\sqrt{x^2+y^2+z^2}}\mathbf{i} + \frac{y}{\sqrt{x^2+y^2+z^2}}\mathbf{j} + \frac{z}{\sqrt{x^2+y^2+z^2}}\mathbf{k} = \frac{\mathbf{r}}{r}$

(b) $r^n = \left(\sqrt{x^2+y^2+z^2}\right)^n$ $\Rightarrow \nabla(r^n) = nx(x^2+y^2+z^2)^{(n/2)-1}\mathbf{i} + ny(x^2+y^2+z^2)^{(n/2)-1}\mathbf{j} + nz(x^2+y^2+z^2)^{(n/2)-1}\mathbf{k} = nr^{n-2}\mathbf{r}$

(c) Let $n = 2$ in part (b). Then $\frac{1}{2}\nabla(r^2) = \mathbf{r} \Rightarrow \nabla\left(\frac{1}{2}r^2\right) = \mathbf{r} \Rightarrow \frac{r^2}{2} = \frac{1}{2}(x^2+y^2+z^2)$ is the function.

(d) $d\mathbf{r} = dx\mathbf{i} + dy\mathbf{j} + dz\mathbf{k} \Rightarrow \mathbf{r}\cdot d\mathbf{r} = x\,dx + y\,dy + z\,dz$, and $dr = r_x\,dx + r_y\,dy + r_z\,dz = \frac{x}{r}\,dx + \frac{y}{r}\,dy + \frac{z}{r}\,dz$ $\Rightarrow r\,dr = x\,dx + y\,dy + z\,dz = \mathbf{r}\cdot d\mathbf{r}$

(e) $\mathbf{A} = a\mathbf{i} + b\mathbf{j} + c\mathbf{k} \Rightarrow \mathbf{A}\cdot\mathbf{r} = ax + by + cz \Rightarrow \nabla(\mathbf{A}\cdot\mathbf{r}) = a\mathbf{i} + b\mathbf{j} + c\mathbf{k} = \mathbf{A}$

9. $f(x,y,z) = xz^2 - yz + \cos xy - 1 \Rightarrow \nabla f = (z^2 - y\sin xy)\mathbf{i} + (-z - x\sin xy)\mathbf{j} + (2xz - y)\mathbf{k} \Rightarrow \nabla f(0,0,1) = \mathbf{i} - \mathbf{j}$ $\Rightarrow$ the tangent plane is $x - y = 0$; $\mathbf{r} = (\ln t)\mathbf{i} + (t\ln t)\mathbf{j} + t\mathbf{k} \Rightarrow \mathbf{r}' = \left(\frac{1}{t}\right)\mathbf{i} + (\ln t + 1)\mathbf{j} + \mathbf{k}$; $x = y = 0$, $z = 1$ $\Rightarrow t = 1 \Rightarrow \mathbf{r}'(1) = \mathbf{i} + \mathbf{j} + \mathbf{k}$. Since $(\mathbf{i}+\mathbf{j}+\mathbf{k})\cdot(\mathbf{i}-\mathbf{j}) = \mathbf{r}'(1)\cdot\nabla f = 0$, $\mathbf{r}$ is parallel to the plane, and $\mathbf{r}(1) = 0\mathbf{i} + 0\mathbf{j} + \mathbf{k} \Rightarrow \mathbf{r}$ is contained in the plane.

11. $\frac{\partial z}{\partial x} = 3x^2 - 9y = 0$ and $\frac{\partial z}{\partial y} = 3y^2 - 9x = 0 \Rightarrow y = \frac{1}{3}x^2$ and $3\left(\frac{1}{3}x^2\right)^2 - 9x = 0 \Rightarrow \frac{1}{3}x^4 - 9x = 0$ $\Rightarrow x(x^3 - 27) = 0 \Rightarrow x = 0$ or $x = 3$. Now $x = 0 \Rightarrow y = 0$ or $(0,0)$ and $x = 3 \Rightarrow y = 3$ or $(3,3)$. Next $\frac{\partial^2 z}{\partial x^2} = 6x$, $\frac{\partial^2 z}{\partial y^2} = 6y$, and $\frac{\partial^2 z}{\partial x\partial y} = -9$. For $(0,0)$, $\frac{\partial^2 z}{\partial x^2}\frac{\partial^2 z}{\partial y^2} - \left(\frac{\partial^2 z}{\partial x\partial y}\right)^2 = -81 \Rightarrow$ no extremum (a saddle point), and for $(3,3)$, $\frac{\partial^2 z}{\partial x^2}\frac{\partial^2 z}{\partial y^2} - \left(\frac{\partial^2 z}{\partial x\partial y}\right)^2 = 243 > 0$ and $\frac{\partial^2 z}{\partial x^2} = 18 > 0 \Rightarrow$ a local minimum.

13. Let $f(x,y,z) = \frac{x^2}{a^2} + \frac{y^2}{b^2} + \frac{z^2}{c^2} - 1 \Rightarrow \nabla f = \frac{2x}{a^2}\mathbf{i} + \frac{2y}{b^2}\mathbf{j} + \frac{2z}{c^2}\mathbf{k} \Rightarrow$ an equation of the plane tangent at the point $P_0(x_0, y_0, y_0)$ is $\left(\frac{2x_0}{a^2}\right)x + \left(\frac{2y_0}{b^2}\right)y + \left(\frac{2z_0}{c^2}\right)z = \frac{2x_0^2}{a^2} + \frac{2y_0^2}{b^2} + \frac{2z_0^2}{c^2} = 2$ or $\left(\frac{x_0}{a^2}\right)x + \left(\frac{y_0}{b^2}\right)y + \left(\frac{z_0}{c^2}\right)z = 1$. The intercepts of the plane are $\left(\frac{a^2}{x_0}, 0, 0\right)$, $\left(0, \frac{b^2}{y_0}, 0\right)$ and $\left(0, 0, \frac{c^2}{z_0}\right)$. The volume of the tetrahedron formed by the plane and the coordinate planes is $V = \left(\frac{1}{3}\right)\left(\frac{1}{2}\right)\left(\frac{a^2}{x_0}\right)\left(\frac{b^2}{y_0}\right)\left(\frac{c^2}{z_0}\right) \Rightarrow$ we need to maximize $V(x,y,z) = \frac{(abc)^2}{6}(xyz)^{-1}$ subject to the constraint $f(x,y,z) = \frac{x^2}{a^2} + \frac{y^2}{b^2} + \frac{z^2}{c^2} = 1$. Thus, $\left[-\frac{(abc)^2}{6}\right]\left(\frac{1}{x^2yz}\right) = \frac{2x}{a^2}\lambda$, $\left[-\frac{(abc)^2}{6}\right]\left(\frac{1}{xy^2z}\right) = \frac{2y}{b^2}\lambda$, and $\left[-\frac{(abc)^2}{6}\right]\left(\frac{1}{xyz^2}\right) = \frac{2z}{c^2}\lambda$. Multiply the first equation by a^2yz, the second by b^2xz, and the third by c^2xy. Then equate the first and second $\Rightarrow a^2y^2 = b^2x^2 \Rightarrow y = \frac{b}{a}x$, $x > 0$; equate the first and third $\Rightarrow a^2z^2 = c^2x^2 \Rightarrow z = \frac{c}{a}x$, $x > 0$; substitute into $f(x,y,z) = 0 \Rightarrow x = \frac{a}{\sqrt{3}} \Rightarrow y = \frac{b}{\sqrt{3}} \Rightarrow z = \frac{c}{\sqrt{3}} \Rightarrow V = \frac{\sqrt{3}}{2}abc$.

15. Let (x_0, y_0) be any point in R. We must show $\lim_{(x,y)\to(x_0,y_0)} f(x,y) = f(x_0,y_0)$ or, equivalently that $\lim_{(h,k)\to(0,0)} |f(x_0+h, y_0+k) - f(x_0,y_0)| = 0$. Consider $f(x_0+h, y_0+k) - f(x_0,y_0)$ $= [f(x_0+h, y_0+k) - f(x_0, y_0+k)] + [f(x_0, y_0+k) - f(x_0,y_0)]$. Let $F(x) = f(x, y_0+k)$ and apply the Mean Value Theorem: there exists ξ with $x_0 < \xi < x_0 + h$ such that $F'(\xi)h = F(x_0+h) - F(x_0) \Rightarrow hf_x(\xi, y_0+k)$ $= f(x_0+h, y_0+k) - f(x_0, y_0+k)$. Similarly, $k\,f_y(x_0,\eta) = f(x_0, y_0+k) - f(x_0,y_0)$ for some η with

$y_0 < \eta < y_0 + k$. Then $|f(x_0 + h, y_0 + k) - f(x_0, y_0)| \le |hf_x(\xi, y_0 + k)| + |kf_y(x_0, \eta)|$. If M, N are positive real numbers such that $|f_x| \le M$ and $|f_y| \le N$ for all (x, y) in the xy-plane, then $|f(x_0 + h, y_0 + k) - f(x_0, y_0)|$ $\le M|h| + N|k|$. As $(h, k) \to 0$, $|f(x_0 + h, y_0 + k) - f(x_0, y_0)| \to 0 \Rightarrow \lim_{(h,k) \to (0,0)} |f(x_0 + h, y_0 + k) - f(x_0, y_0)|$ $= 0 \Rightarrow$ f is continuous at (x_0, y_0).

17. $\frac{\partial f}{\partial x} = 0 \Rightarrow f(x, y) = h(y)$ is a function of y only. Also, $\frac{\partial g}{\partial y} = \frac{\partial f}{\partial x} = 0 \Rightarrow g(x, y) = k(x)$ is a function of x only. Moreover, $\frac{\partial f}{\partial y} = \frac{\partial g}{\partial x} \Rightarrow h'(y) = k'(x)$ for all x and y. This can happen only if $h'(y) = k'(x) = c$ is a constant. Integration gives $h(y) = cy + c_1$ and $k(x) = cx + c_2$, where c_1 and c_2 are constants. Therefore $f(x, y) = cy + c_1$ and $g(x, y) = cx + c_2$. Then $f(1, 2) = g(1, 2) = 5 \Rightarrow 5 = 2c + c_1 = c + c_2$, and $f(0, 0) = 4 \Rightarrow c_1 = 4 \Rightarrow c = \frac{1}{2}$ $\Rightarrow c_2 = \frac{9}{2}$. Thus, $f(x, y) = \frac{1}{2}y + 4$ and $g(x, y) = \frac{1}{2}x + \frac{9}{2}$.

19. Since the particle is heat-seeking, at each point (x, y) it moves in the direction of maximal temperature increase, that is in the direction of $\nabla T(x, y) = (e^{-2y}\sin x)\mathbf{i} + (2e^{-2y}\cos x)\mathbf{j}$. Since $\nabla T(x, y)$ is parallel to the particle's velocity vector, it is tangent to the path $y = f(x)$ of the particle $\Rightarrow f'(x) = \frac{2e^{-2y}\cos x}{e^{-2y}\sin x} = 2\cot x$. Integration gives $f(x) = 2\ln|\sin x| + C$ and $f\left(\frac{\pi}{4}\right) = 0 \Rightarrow 0 = 2\ln\left|\sin\frac{\pi}{4}\right| + C \Rightarrow C = -2\ln\frac{\sqrt{2}}{2} = \ln\left(\frac{2}{\sqrt{2}}\right)^2$ $= \ln 2$. Therefore, the path of the particle is the graph of $y = 2\ln|\sin x| + \ln 2$.

21. (a) $\mathbf{k}$ is a vector normal to $z = 10 - x^2 - y^2$ at the point (0, 0, 10). So directions tangential to S at (0, 0, 10) will be unit vectors $\mathbf{u} = a\mathbf{i} + b\mathbf{j}$. Also, $\nabla T(x, y, z) = (2xy + 4)\mathbf{i} + (x^2 + 2yz + 14)\mathbf{j} + (y^2 + 1)\mathbf{k}$ $\Rightarrow \nabla T(0, 0, 10) = 4\mathbf{i} + 14\mathbf{j} + \mathbf{k}$. We seek the unit vector $\mathbf{u} = a\mathbf{i} + b\mathbf{j}$ such that $D_{\mathbf{u}}T(0, 0, 10)$ $= (4\mathbf{i} + 14\mathbf{j} + \mathbf{k}) \cdot (a\mathbf{i} + b\mathbf{j}) = (4\mathbf{i} + 14\mathbf{j}) \cdot (a\mathbf{i} + b\mathbf{j})$ is a maximum. The maximum will occur when $a\mathbf{i} + b\mathbf{j}$ has the same direction as $4\mathbf{i} + 14\mathbf{j}$, or $\mathbf{u} = \frac{1}{\sqrt{53}}(2\mathbf{i} + 7\mathbf{j})$.

(b) A vector normal to S at (1, 1, 8) is $\mathbf{n} = 2\mathbf{i} + 2\mathbf{j} + \mathbf{k}$. Now, $\nabla T(1, 1, 8) = 6\mathbf{i} + 31\mathbf{j} + 2\mathbf{k}$ and we seek the unit vector $\mathbf{u}$ such that $D_{\mathbf{u}}T(1, 1, 8) = \nabla T \cdot \mathbf{u}$ has its largest value. Now write $\nabla T = \mathbf{v} + \mathbf{w}$, where $\mathbf{v}$ is parallel to ∇T and $\mathbf{w}$ is orthogonal to ∇T. Then $D_{\mathbf{u}}T = \nabla T \cdot \mathbf{u} = (\mathbf{v} + \mathbf{w}) \cdot \mathbf{u} = \mathbf{v} \cdot \mathbf{u} + \mathbf{w} \cdot \mathbf{u} = \mathbf{w} \cdot \mathbf{u}$. Thus $D_{\mathbf{u}}T(1, 1, 8)$ is a maximum when $\mathbf{u}$ has the same direction as $\mathbf{w}$. Now, $\mathbf{w} = \nabla T - \left(\frac{\nabla T \cdot \mathbf{n}}{|\mathbf{n}|^2}\right)\mathbf{n}$ $= (6\mathbf{i} + 31\mathbf{j} + 2\mathbf{k}) - \left(\frac{12 + 62 + 2}{4 + 4 + 1}\right)(2\mathbf{i} + 2\mathbf{j} + \mathbf{k}) = \left(6 - \frac{152}{9}\right)\mathbf{i} + \left(31 - \frac{152}{9}\right)\mathbf{j} + \left(2 - \frac{76}{9}\right)\mathbf{k}$ $= -\frac{98}{9}\mathbf{i} + \frac{127}{9}\mathbf{j} - \frac{58}{9}\mathbf{k} \Rightarrow \mathbf{u} = \frac{\mathbf{w}}{|\mathbf{w}|} = -\frac{1}{\sqrt{29{,}097}}(98\mathbf{i} - 127\mathbf{j} + 58\mathbf{k})$.

23. $w = e^{rt}\sin \pi x \Rightarrow w_t = re^{rt}\sin \pi x$ and $w_x = \pi e^{rt}\cos \pi x \Rightarrow w_{xx} = -\pi^2 e^{rt}\sin \pi x$; $w_{xx} = \frac{1}{c^2}w_t$, where c^2 is the positive constant determined by the material of the rod $\Rightarrow -\pi^2 e^{rt}\sin \pi x = \frac{1}{c^2}(re^{rt}\sin \pi x)$ $\Rightarrow (r + c^2\pi^2)e^{rt}\sin \pi x = 0 \Rightarrow r = -c^2\pi^2 \Rightarrow w = e^{-c^2\pi^2 t}\sin \pi x$

CHAPTER 15 MULTIPLE INTEGRALS

15.1 DOUBLE AND ITERATED INTEGRALS OVER RECTANGLES

1. $\int_1^2 \int_0^4 2xy\,dy\,dx = \int_1^2 \left[x\,y^2\right]_0^4 dx = \int_1^2 16x\,dx = \left[8\,x^2\right]_1^2 = 24$

3. $\int_{-1}^0 \int_{-1}^1 (x+y+1)\,dx\,dy = \int_{-1}^0 \left[\frac{x^2}{2} + yx + x\right]_{-1}^1 dy = \int_{-1}^0 (2y+2)\,dy = \left[y^2+2y\right]_{-1}^0 = 1$

5. $\int_0^3 \int_0^2 (4-y^2)\,dy\,dx = \int_0^3 \left[4y - \frac{y^3}{3}\right]_0^2 dx = \int_0^3 \frac{16}{3}\,dx = \left[\frac{16}{3}\,x\right]_0^3 = 16$

7. $\int_0^1 \int_0^1 \frac{y}{1+xy}\,dx\,dy = \int_0^1 \left[\ln|1+x\,y|\right]_0^1 dy = \int_0^1 \ln|1+y|dy = \left[y\ln|1+y| - y + \ln|1+y|\right]_0^1 = 2\ln 2 - 1$

9. $\int_0^{\ln 2} \int_1^{\ln 5} e^{2x+y}\,dy\,dx = \int_0^{\ln 2} \left[e^{2x+y}\right]_1^{\ln 5} dx = \int_0^{\ln 2} (5e^{2x} - e^{2x+1})\,dx = \left[\frac{5}{2}e^{2x} - \frac{1}{2}e^{2x+1}\right]_0^{\ln 2} = \frac{3}{2}(5-e)$

11. $\int_{-1}^2 \int_0^{\pi/2} y\sin x\,dx\,dy = \int_{-1}^2 \left[-y\cos x\right]_0^{\pi/2} dy = \int_{-1}^2 y\,dy = \left[\frac{1}{2}y^2\right]_{-1}^2 = \frac{3}{2}$

13. $\iint_R (6\,y^2 - 2\,x)dA = \int_0^1 \int_0^2 (6\,y^2 - 2\,x)\,dy\,dx = \int_0^1 \left[2\,y^3 - 2\,x\,y\right]_0^2 dx = \int_0^1 (16 - 4\,x)\,dx = \left[16\,x - 2\,x^2\right]_0^1 = 14$

15. $\iint_R x\,y\cos y\,dA = \int_{-1}^1 \int_0^\pi x\,y\cos y\,dy\,dx = \int_{-1}^1 \left[x\,y\sin y + x\cos y\right]_0^\pi dx = \int_{-1}^1 (-2x)\,dx = \left[-x^2\right]_{-1}^1 = 0$

17. $\iint_R e^{x-y}dA = \int_0^{\ln 2} \int_0^{\ln 2} e^{x-y}\,dy\,dx = \int_0^{\ln 2} \left[-e^{x-y}\right]_0^{\ln 2} dx = \int_0^{\ln 2} (-e^{x-\ln 2} + e^x)\,dx = \left[-e^{x-\ln 2} + e^x\right]_0^{\ln 2} = \frac{1}{2}$

19. $\iint_R \frac{x\,y^3}{x^2+1}dA = \int_0^1 \int_0^2 \frac{x\,y^3}{x^2+1}\,dy\,dx = \int_0^1 \left[\frac{x\,y^4}{4(x^2+1)}\right]_0^2 dx = \int_0^1 \frac{4\,x}{x^2+1}\,dx = \left[2\ln|x^2+1|\right]_0^1 = 2\ln 2$

21. $\int_1^2 \int_1^2 \frac{1}{xy}\,dy\,dx = \int_1^2 \frac{1}{x}(\ln 2 - \ln 1)\,dx = (\ln 2)\int_1^2 \frac{1}{x}\,dx = (\ln 2)^2$

23. $V = \iint_R f(x,y)\,dA = \int_{-1}^1 \int_{-1}^1 (x^2+y^2)\,dy\,dx = \int_{-1}^1 \left[x^2\,y + \frac{1}{3}y^3\right]_{-1}^1 dx = \int_{-1}^1 \left(2\,x^2 + \frac{2}{3}\right) dx = \left[\frac{2}{3}x^3 + \frac{2}{3}x\right]_{-1}^1 = \frac{8}{3}$

25. $V = \iint_R f(x,y)\,dA = \int_0^1 \int_0^1 (2-x-y)\,dy\,dx = \int_0^1 \left[2\,y - x\,y - \frac{1}{2}y^2\right]_0^1 dx = \int_0^1 \left(\frac{3}{2} - x\right) dx = \left[\frac{3}{2}x - \frac{1}{2}x^2\right]_0^1 = 1$

27. $V = \iint_R f(x,y)\,dA = \int_0^{\pi/2} \int_0^{\pi/4} 2\sin x\cos y\,dy\,dx = \int_0^{\pi/2} \left[2\sin x\sin y\right]_0^{\pi/4} dx = \int_0^{\pi/2} \left(\sqrt{2}\sin x\right) dx = \left[-\sqrt{2}\cos x\right]_0^{\pi/2}$
$= \sqrt{2}$

15.2 DOUBLE INTEGRALS OVER GENERAL REGIONS

1.

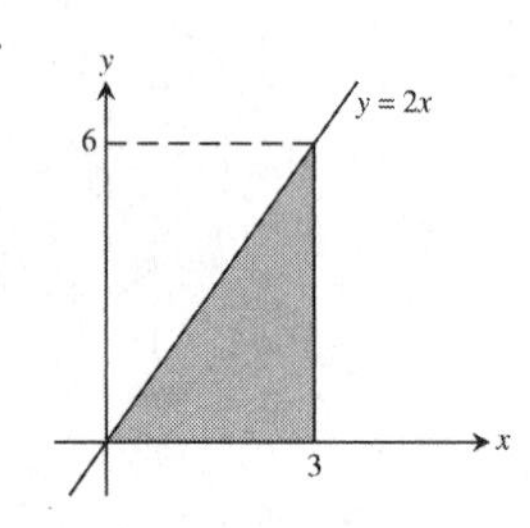

3.

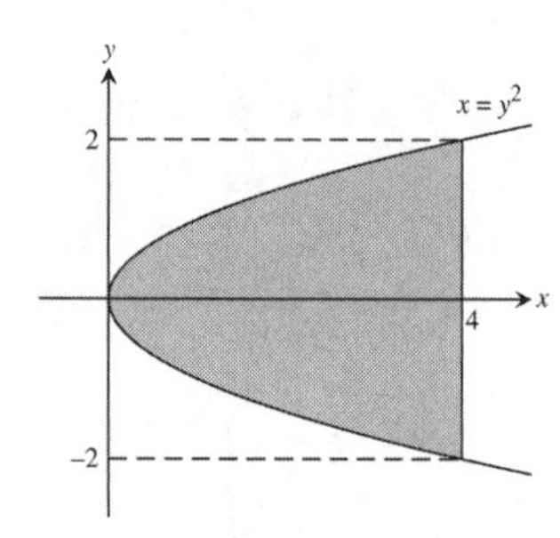

5.

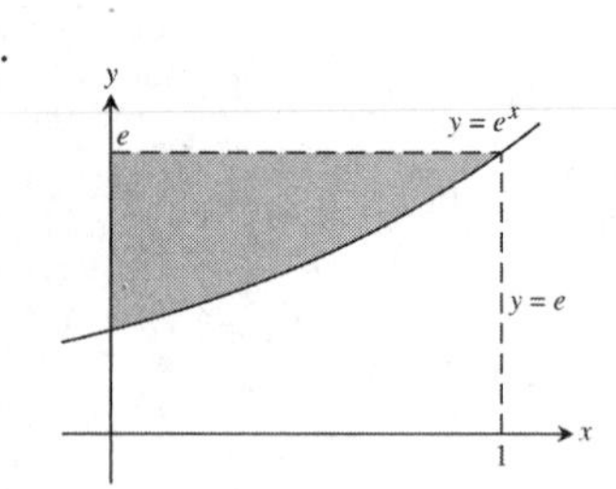

7.

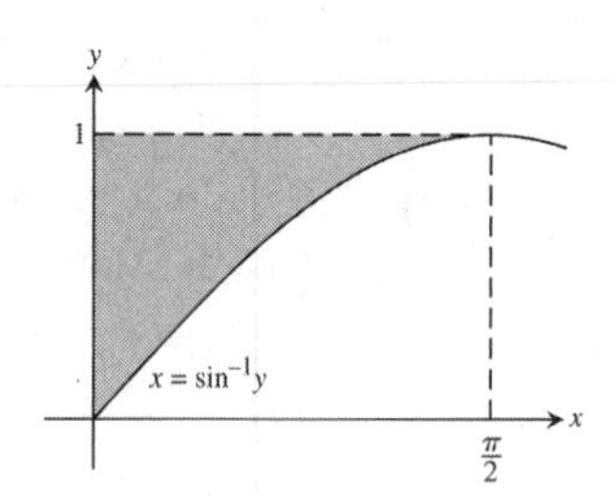

9. (a) $\int_0^2 \int_{x^3}^8 dy\,dx$ (b) $\int_0^8 \int_0^{y^{1/3}} dx\,dy$

11. (a) $\int_0^3 \int_{x^2}^{3x} dy\,dx$ (b) $\int_0^9 \int_{y/3}^{\sqrt{y}} dx\,dy$

13. (a) $\int_0^9 \int_0^{\sqrt{x}} dy\,dx$
(b) $\int_0^3 \int_{y^2}^9 dx\,dy$

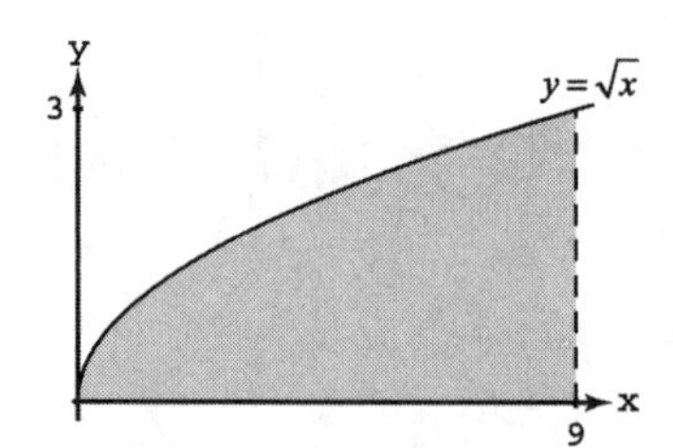

15. (a) $\int_0^{\ln 3} \int_{e^{-x}}^1 dy\,dx$
(b) $\int_{1/3}^1 \int_{-\ln y}^{\ln 3} dx\,dy$

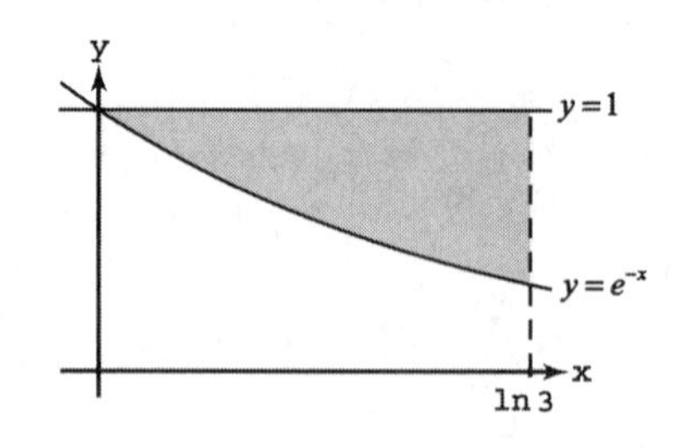

17. (a) $\int_0^1 \int_x^{3-2x} dy\,dx$
(b) $\int_0^1 \int_0^y dx\,dy + \int_1^3 \int_0^{(3-y)/2} dx\,dy$

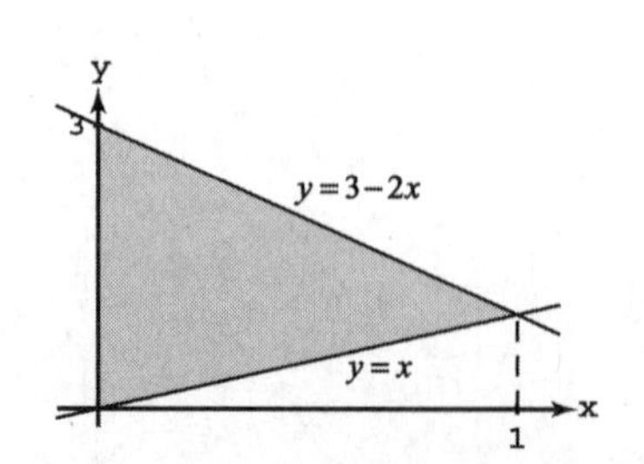

19. $\int_0^{\pi}\int_0^{x}(x\sin y)\,dy\,dx = \int_0^{\pi}[-x\cos y]_0^x\,dx$
$= \int_0^{\pi}(x - x\cos x)\,dx = \left[\frac{x^2}{2} - (\cos x + x\sin x)\right]_0^{\pi}$
$= \frac{\pi^2}{2} + 2$

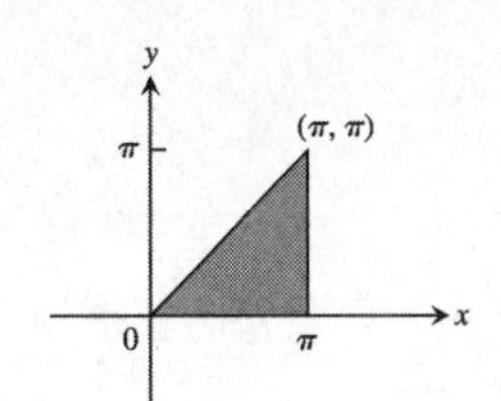

21. $\int_1^{\ln 8}\int_0^{\ln y} e^{x+y}\,dx\,dy = \int_1^{\ln 8}\left[e^{x+y}\right]_0^{\ln y}\,dy = \int_1^{\ln 8}(ye^y - e^y)\,dy$
$= \left[(y-1)e^y - e^y\right]_1^{\ln 8} = 8(\ln 8 - 1) - 8 + e$
$= 8\ln 8 - 16 + e$

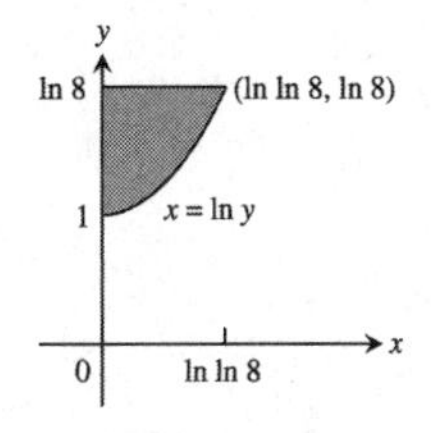

23. $\int_0^1\int_0^{y^2} 3y^3e^{xy}\,dx\,dy = \int_0^1\left[3y^2e^{xy}\right]_0^{y^2}\,dy$
$= \int_0^1\left(3y^2e^{y^3} - 3y^2\right)dy = \left[e^{y^3} - y^3\right]_0^1 = e - 2$

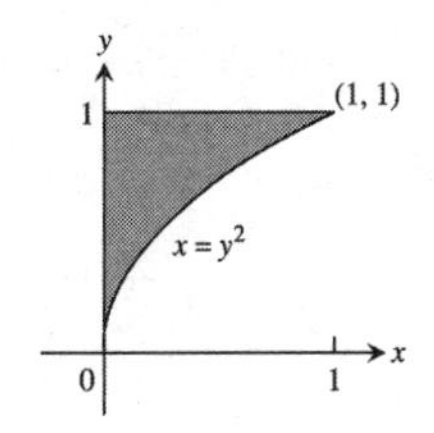

25. $\int_1^2\int_x^{2x}\frac{x}{y}\,dy\,dx = \int_1^2[x\ln y]_x^{2x}\,dx = (\ln 2)\int_1^2 x\,dx = \frac{3}{2}\ln 2$

27. $\int_0^1\int_0^{1-u}\left(v - \sqrt{u}\right)dv\,du = \int_0^1\left[\frac{v^2}{2} - v\sqrt{u}\right]_0^{1-u}du = \int_0^1\left[\frac{1-2u+u^2}{2} - \sqrt{u}(1-u)\right]du$
$= \int_0^1\left(\frac{1}{2} - u + \frac{u^2}{2} - u^{1/2} + u^{3/2}\right)du = \left[\frac{u}{2} - \frac{u^2}{2} + \frac{u^3}{6} - \frac{2}{3}u^{3/2} + \frac{2}{5}u^{5/2}\right]_0^1 = \frac{1}{2} - \frac{1}{2} + \frac{1}{6} - \frac{2}{3} + \frac{2}{5} = -\frac{1}{2} + \frac{2}{5} = -\frac{1}{10}$

29. $\int_{-2}^0\int_v^{-v} 2\,dp\,dv = 2\int_{-2}^0[p]_v^{-v}\,dv = 2\int_{-2}^0 -2v\,dv$
$= -2\left[v^2\right]_{-2}^0 = 8$

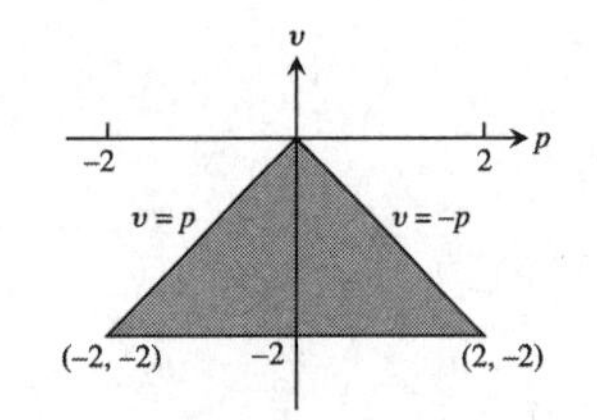

31. $\int_{-\pi/3}^{\pi/3}\int_0^{\sec t} 3\cos t\,du\,dt = \int_{-\pi/3}^{\pi/3}[(3\cos t)u]_0^{\sec t}$
$= \int_{-\pi/3}^{\pi/3} 3\,dt = 2\pi$

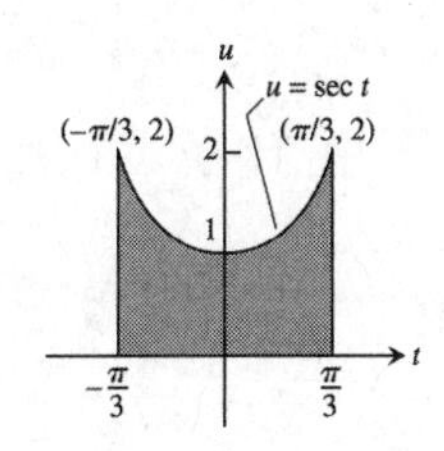

33. $\int_2^4\int_0^{(4-y)/2} dx\,dy$

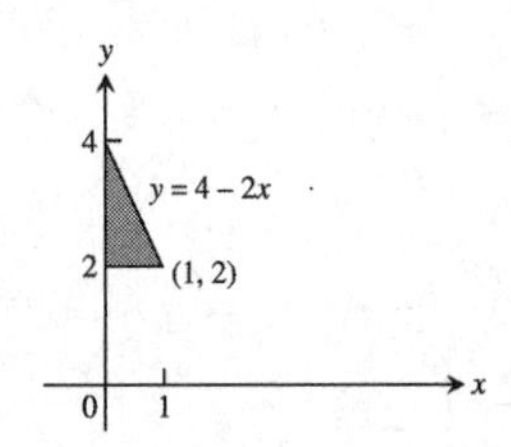

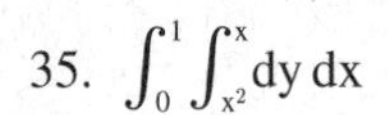

35. $\int_0^1 \int_{x^2}^{x} dy\, dx$

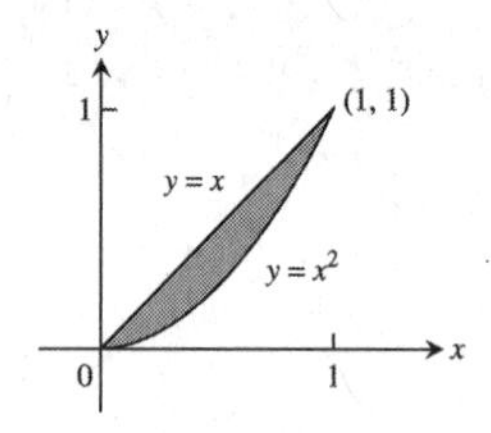

37. $\int_1^e \int_{\ln y}^{1} dx\, dy$

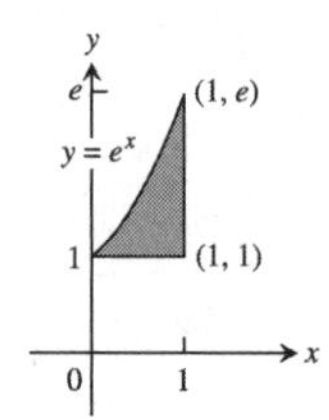

39. $\int_0^9 \int_0^{\frac{1}{2}\sqrt{9-y}} 16x\, dx\, dy$

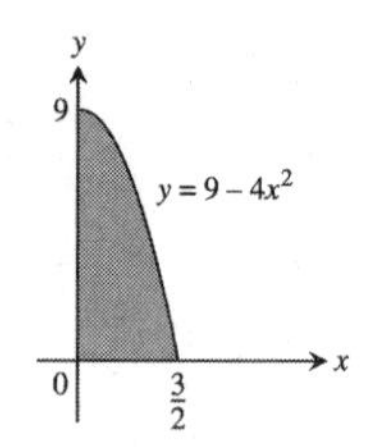

41. $\int_{-1}^1 \int_0^{\sqrt{1-x^2}} 3y\, dy\, dx$

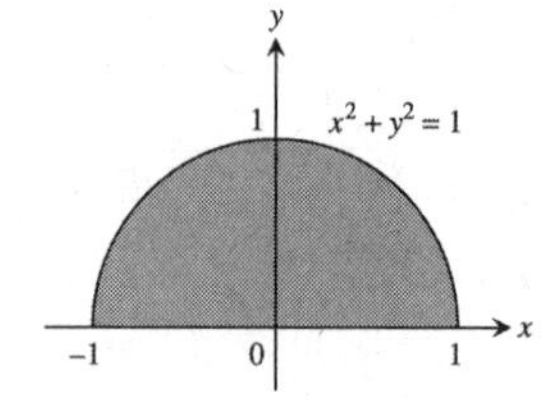

43. $\int_0^1 \int_{e^y}^{e} x\, y\, dx\, dy$

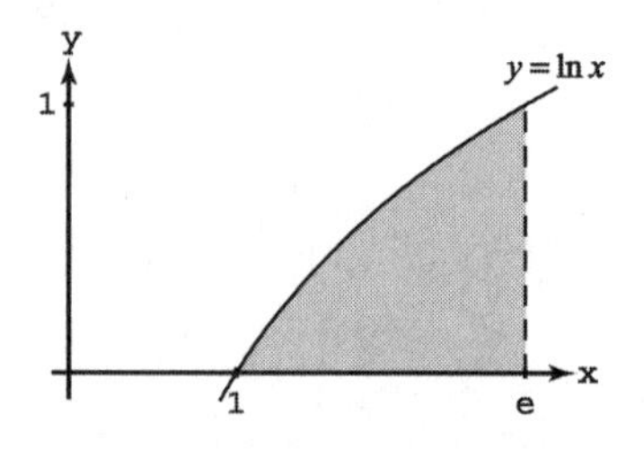

45. $\int_1^{e^3} \int_{\ln x}^{3} (x + y) dy\, dx$

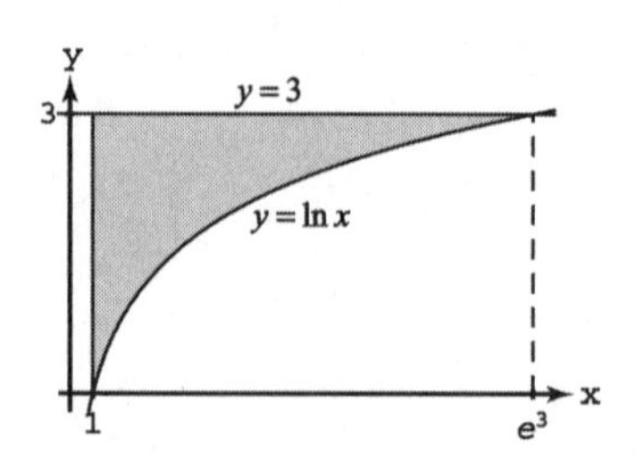

47. $\int_0^\pi \int_x^\pi \frac{\sin y}{y}\, dy\, dx = \int_0^\pi \int_0^y \frac{\sin y}{y}\, dx\, dy = \int_0^\pi \sin y\, dy = 2$

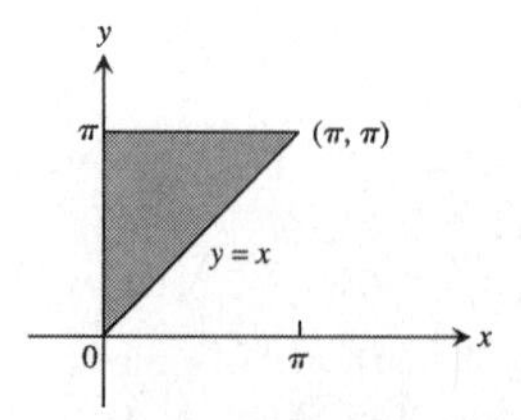

49. $\int_0^1 \int_y^1 x^2 e^{xy}\,dx\,dy = \int_0^1 \int_0^x x^2 e^{xy}\,dy\,dx = \int_0^1 \left[x e^{xy}\right]_0^x dx$
$= \int_0^1 \left(x e^{x^2} - x\right) dx = \left[\frac{1}{2} e^{x^2} - \frac{x^2}{2}\right]_0^1 = \frac{e-2}{2}$

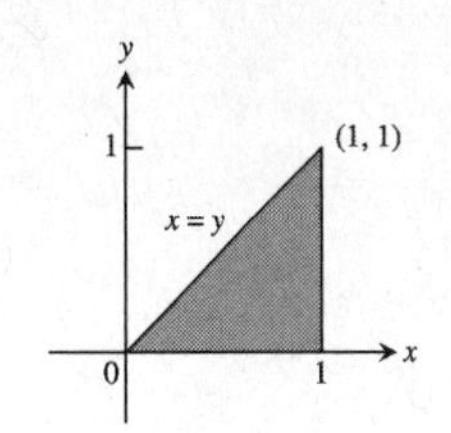

51. $\int_0^{2\sqrt{\ln 3}} \int_{y/2}^{\sqrt{\ln 3}} e^{x^2}\,dx\,dy = \int_0^{\sqrt{\ln 3}} \int_0^{2x} e^{x^2}\,dy\,dx$
$= \int_0^{\sqrt{\ln 3}} 2x e^{x^2}\,dx = \left[e^{x^2}\right]_0^{\sqrt{\ln 3}} = e^{\ln 3} - 1 = 2$

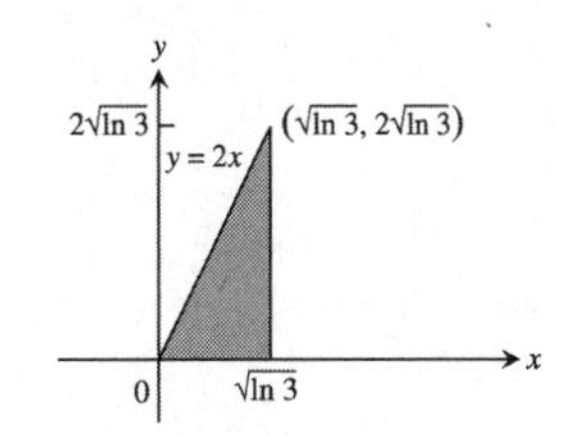

53. $\int_0^{1/16} \int_{y^{1/4}}^{1/2} \cos(16\pi x^5)\,dx\,dy = \int_0^{1/2} \int_0^{x^4} \cos(16\pi x^5)\,dy\,dx$
$= \int_0^{1/2} x^4 \cos(16\pi x^5)\,dx = \left[\frac{\sin(16\pi x^5)}{80\pi}\right]_0^{1/2} = \frac{1}{80\pi}$

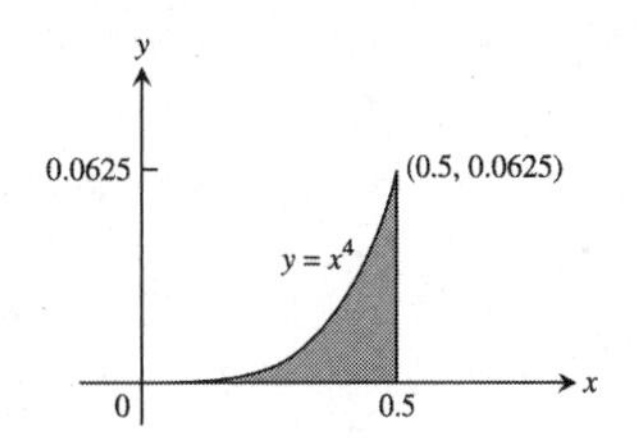

55. $\iint_R (y - 2x^2)\,dA$
$= \int_{-1}^0 \int_{-x-1}^{x+1} (y - 2x^2)\,dy\,dx + \int_0^1 \int_{x-1}^{1-x} (y - 2x^2)\,dy\,dx$
$= \int_{-1}^0 \left[\frac{1}{2} y^2 - 2x^2 y\right]_{-x-1}^{x+1} dx + \int_0^1 \left[\frac{1}{2} y^2 - 2x^2 y\right]_{x-1}^{1-x} dx$
$= \int_{-1}^0 \left[\frac{1}{2}(x+1)^2 - 2x^2(x+1) - \frac{1}{2}(-x-1)^2 + 2x^2(-x-1)\right] dx$
$+ \int_0^1 \left[\frac{1}{2}(1-x)^2 - 2x^2(1-x) - \frac{1}{2}(x-1)^2 + 2x^2(x-1)\right] dx$
$= -4 \int_{-1}^0 (x^3 + x^2)\,dx + 4 \int_0^1 (x^3 - x^2)\,dx$
$= -4 \left[\frac{x^4}{4} + \frac{x^3}{3}\right]_{-1}^0 + 4 \left[\frac{x^4}{4} - \frac{x^3}{3}\right]_0^1 = 4 \left[\frac{(-1)^4}{4} + \frac{(-1)^3}{3}\right] + 4\left(\frac{1}{4} - \frac{1}{3}\right) = 8\left(\frac{3}{12} - \frac{4}{12}\right) = -\frac{8}{12} = -\frac{2}{3}$

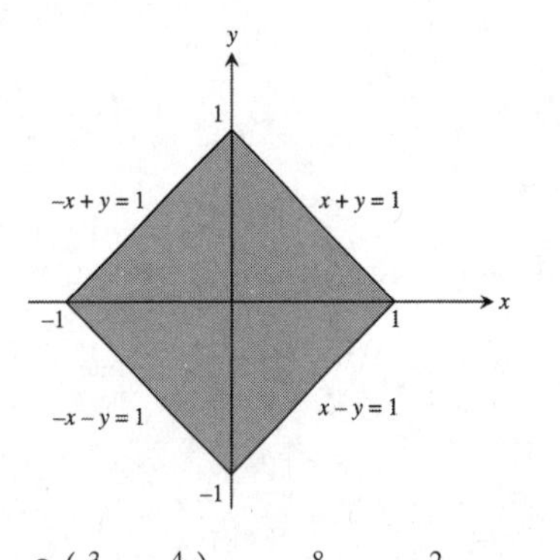

57. $V = \int_0^1 \int_x^{2-x} (x^2 + y^2)\,dy\,dx = \int_0^1 \left[x^2 y + \frac{y^3}{3}\right]_x^{2-x} dx = \int_0^1 \left[2x^2 - \frac{7x^3}{3} + \frac{(2-x)^3}{3}\right] dx = \left[\frac{2x^3}{3} - \frac{7x^4}{12} - \frac{(2-x)^4}{12}\right]_0^1$
$= \left(\frac{2}{3} - \frac{7}{12} - \frac{1}{12}\right) - \left(0 - 0 - \frac{16}{12}\right) = \frac{4}{3}$

59. $V = \int_{-4}^1 \int_{3x}^{4-x^2} (x + 4)\,dy\,dx = \int_{-4}^1 \left[xy + 4y\right]_{3x}^{4-x^2} dx = \int_{-4}^1 \left[x(4 - x^2) + 4(4 - x^2) - 3x^2 - 12x\right] dx$
$= \int_{-4}^1 (-x^3 - 7x^2 - 8x + 16)\,dx = \left[-\frac{1}{4}x^4 - \frac{7}{3}x^3 - 4x^2 + 16x\right]_{-4}^1 = \left(-\frac{1}{4} - \frac{7}{3} + 12\right) - \left(\frac{64}{3} - 64\right) = \frac{157}{3} - \frac{1}{4} = \frac{625}{12}$

61. $V = \int_0^2 \int_0^3 (4 - y^2)\,dx\,dy = \int_0^2 \left[4x - y^2 x\right]_0^3 dy = \int_0^2 (12 - 3y^2)\,dy = \left[12y - y^3\right]_0^2 = 24 - 8 = 16$

63. $V = \int_0^2 \int_0^{2-x} (12 - 3y^2)\,dy\,dx = \int_0^2 \left[12y - y^3\right]_0^{2-x} dx = \int_0^2 \left[24 - 12x - (2 - x)^3\right] dx = \left[24x - 6x^2 + \frac{(2-x)^4}{4}\right]_0^2 = 20$

65. $V = \int_1^2 \int_{-1/x}^{1/x} (x + 1)\,dy\,dx = \int_1^2 \left[xy + y\right]_{-1/x}^{1/x} dx = \int_1^2 \left[1 + \frac{1}{x} - \left(-1 - \frac{1}{x}\right)\right] dx = 2 \int_1^2 \left(1 + \frac{1}{x}\right) dx = 2\left[x + \ln x\right]_1^2$
$= 2(1 + \ln 2)$

67.

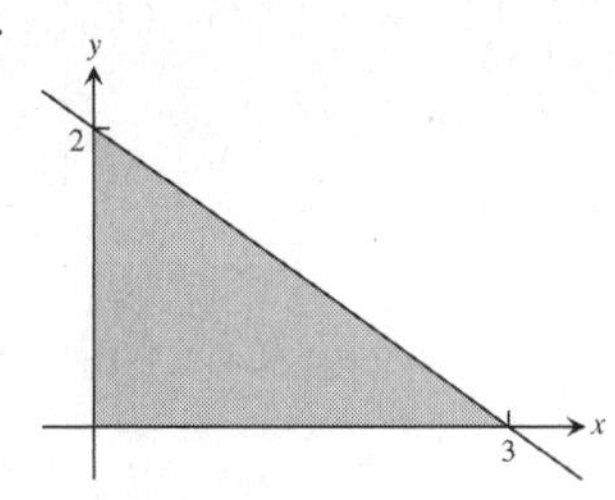

69. $\int_1^{\infty}\int_{e^{-x}}^{1} \frac{1}{x^3 y}\, dy\, dx = \int_1^{\infty} \left[\frac{\ln y}{x^3}\right]_{e^{-x}}^{1} dx = \int_1^{\infty} -\left(\frac{-x}{x^3}\right) dx = -\lim_{b \to \infty} \left[\frac{1}{x}\right]_1^b = -\lim_{b \to \infty} \left(\frac{1}{b} - 1\right) = 1$

71. $\int_{-\infty}^{\infty}\int_{-\infty}^{\infty} \frac{1}{(x^2+1)(y^2+1)} dx\, dy = 2\int_0^{\infty} \left(\frac{2}{y^2+1}\right)\left(\lim_{b \to \infty} \tan^{-1} b - \tan^{-1} 0\right) dy = 2\pi \lim_{b \to \infty} \int_0^b \frac{1}{y^2+1}\, dy$
$= 2\pi \left(\lim_{b \to \infty} \tan^{-1} b - \tan^{-1} 0\right) = (2\pi)\left(\frac{\pi}{2}\right) = \pi^2$

73. $\iint_R f(x,y)\, dA \approx \frac{1}{4} f\left(-\frac{1}{2}, 0\right) + \frac{1}{8} f(0,0) + \frac{1}{8} f\left(\frac{1}{4}, 0\right) = \frac{1}{4}\left(-\frac{1}{2}\right) + \frac{1}{8}\left(0 + \frac{1}{4}\right) = -\frac{3}{32}$

75. The ray $\theta = \frac{\pi}{6}$ meets the circle $x^2 + y^2 = 4$ at the point $\left(\sqrt{3}, 1\right)$ $\Rightarrow$ the ray is represented by the line $y = \frac{x}{\sqrt{3}}$. Thus,
$\iint_R f(x,y)\, dA = \int_0^{\sqrt{3}}\int_{x/\sqrt{3}}^{\sqrt{4-x^2}} \sqrt{4-x^2}\, dy\, dx = \int_0^{\sqrt{3}} \left[(4 - x^2) - \frac{x}{\sqrt{3}}\sqrt{4 - x^2}\right] dx = \left[4x - \frac{x^3}{3} + \frac{(4-x^2)^{3/2}}{3\sqrt{3}}\right]_0^{\sqrt{3}} = \frac{20\sqrt{3}}{9}$

77. $V = \int_0^1\int_x^{2-x} (x^2 + y^2)\, dy\, dx = \int_0^1 \left[x^2 y + \frac{y^3}{3}\right]_x^{2-x} dx$
$= \int_0^1 \left[2x^2 - \frac{7x^3}{3} + \frac{(2-x)^3}{3}\right] dx = \left[\frac{2x^3}{3} - \frac{7x^4}{12} - \frac{(2-x)^4}{12}\right]_0^1$
$= \left(\frac{2}{3} - \frac{7}{12} - \frac{1}{12}\right) - \left(0 - 0 - \frac{16}{12}\right) = \frac{4}{3}$

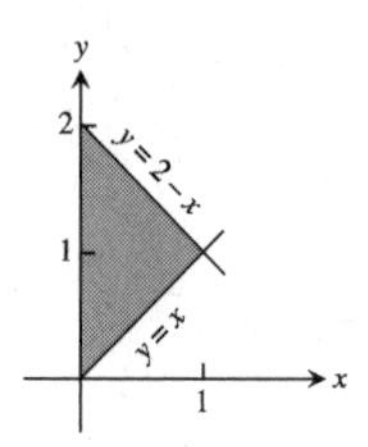

79. To maximize the integral, we want the domain to include all points where the integrand is positive and to exclude all points where the integrand is negative. These criteria are met by the points (x, y) such that $4 - x^2 - 2y^2 \ge 0$ or $x^2 + 2y^2 \le 4$, which is the ellipse $x^2 + 2y^2 = 4$ together with its interior.

81. No, it is not possible. By Fubini's theorem, the two orders of integration must give the same result.

83. $\int_{-b}^{b}\int_{-b}^{b} e^{-x^2-y^2}\, dx\, dy = \int_{-b}^{b}\int_{-b}^{b} e^{-y^2} e^{-x^2}\, dx\, dy = \int_{-b}^{b} e^{-y^2}\left(\int_{-b}^{b} e^{-x^2}\, dx\right) dy = \left(\int_{-b}^{b} e^{-x^2}\, dx\right)\left(\int_{-b}^{b} e^{-y^2}\, dy\right)$
$= \left(\int_{-b}^{b} e^{-x^2}\, dx\right)^2 = \left(2\int_0^b e^{-x^2}\, dx\right)^2 = 4\left(\int_0^b e^{-x^2}\, dx\right)^2$; taking limits as $b \to \infty$ gives the stated result.

85. $\int_1^3\int_1^x \frac{1}{xy}\, dy\, dx \approx 0.603$

87. $\int_0^1\int_0^1 \tan^{-1} xy\, dy\, dx \approx 0.233$

89. Evaluate the integrals:

$\int_0^1 \int_{2y}^4 e^{x^2}\, dx\, dy$
$= \int_0^2 \int_0^{x/2} e^{x^2}\, dy\, dx + \int_2^4 \int_0^1 e^{x^2}\, dy\, dx$
$= -\frac{1}{4} + \frac{1}{4}\left(e^4 - 2\sqrt{\pi}\, \text{erfi}(2) + 2\sqrt{\pi}\, \text{erfi}(4)\right)$
$\approx 1.1494 \times 10^6$

The following graph was generated using Mathematica.

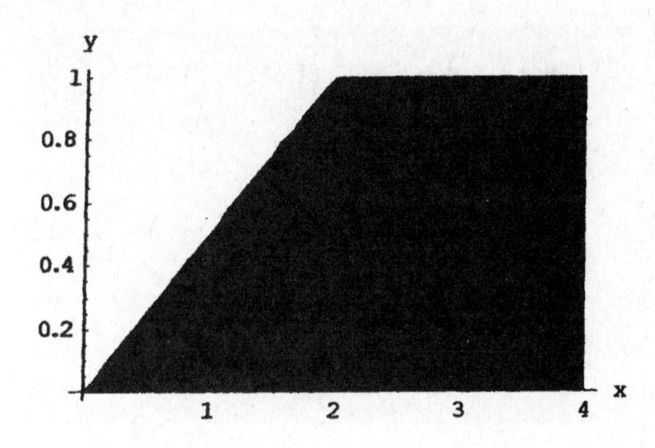

91. Evaluate the integrals:

$\int_0^2 \int_{y^3}^{4\sqrt{2y}} (x^2y - xy^2)dx\, dy = \int_0^8 \int_{x^2/32}^{\sqrt[3]{x}} (x^2y - xy^2)dy\, dx$
$= \frac{67,520}{693} \approx 97.4315$

The following graph was generated using Mathematica.

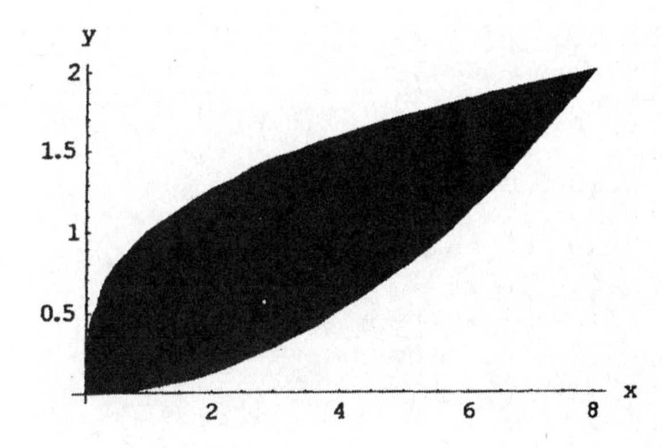

93. Evaluate the integrals:

$\int_1^2 \int_0^{x^2} \frac{1}{x+y}\, dy\, dx$
$= \int_0^1 \int_1^2 \frac{1}{x+y}\, dx\, dy + \int_1^4 \int_{\sqrt{y}}^2 \frac{1}{x+y}\, dx\, dy$
$-1 + \ln\left(\frac{27}{4}\right) \approx 0.909543$

The following graph was generated using Mathematica.

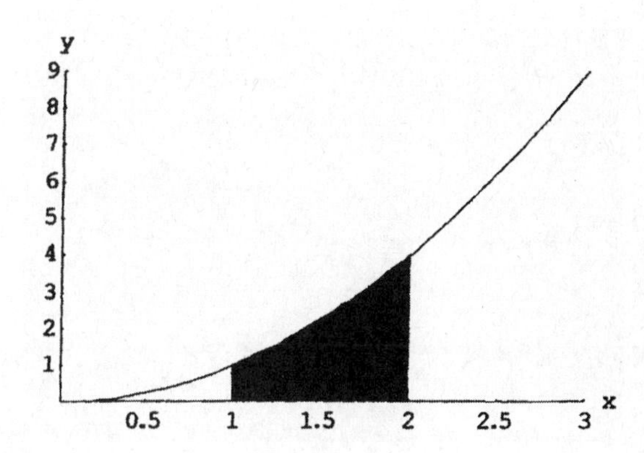

15.3 AREA BY DOUBLE INTEGRATION

1. $\int_0^2 \int_0^{2-x} dy\, dx = \int_0^2 (2-x)\, dx = \left[2x - \frac{x^2}{2}\right]_0^2 = 2,$
or $\int_0^2 \int_0^{2-y} dx\, dy = \int_0^2 (2-y)\, dy = 2$

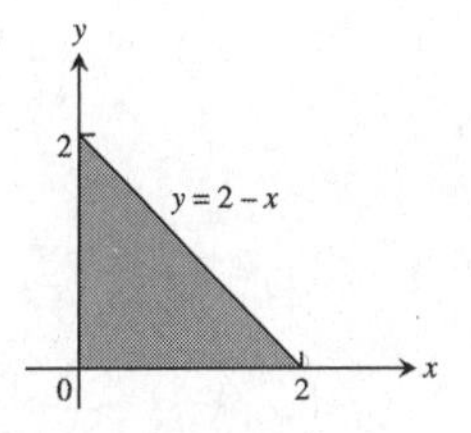

3. $\int_{-2}^{1}\int_{y-2}^{-y^2} dx\,dy = \int_{-2}^{1}(-y^2 - y + 2)\,dy$
$= \left[-\frac{y^3}{3} - \frac{y^2}{2} + 2y\right]_{-2}^{1}$
$= \left(-\frac{1}{3} - \frac{1}{2} + 2\right) - \left(\frac{8}{3} - 2 - 4\right) = \frac{9}{2}$

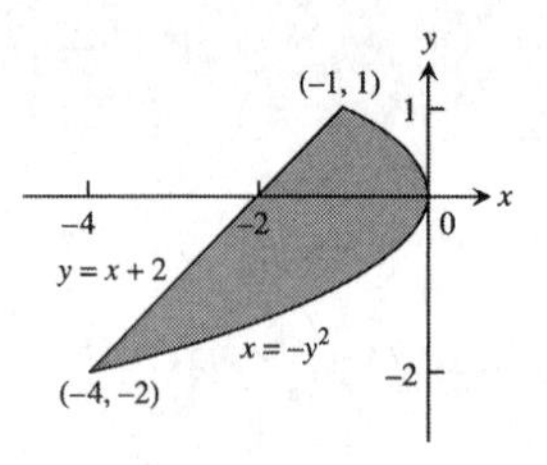

5. $\int_{0}^{\ln 2}\int_{0}^{e^x} dy\,dx = \int_{0}^{\ln 2} e^x\,dx = \left[e^x\right]_0^{\ln 2} = 2 - 1 = 1$

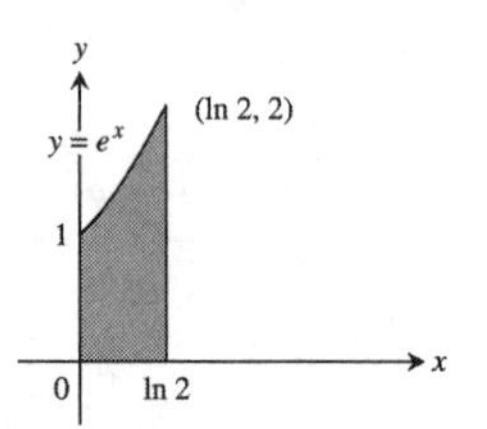

7. $\int_{0}^{1}\int_{y^2}^{2y-y^2} dx\,dy = \int_{0}^{1}(2y - 2y^2)\,dy = \left[y^2 - \frac{2}{3}y^3\right]_0^1$
$= \frac{1}{3}$

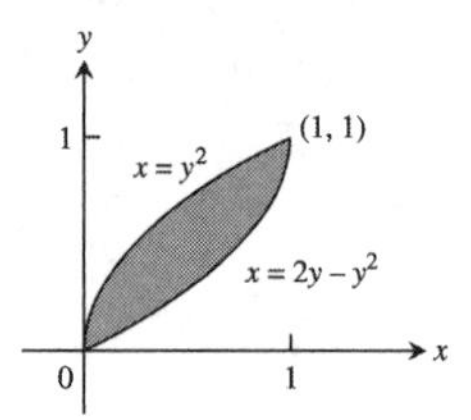

9. $\int_{0}^{2}\int_{y}^{3y} 1\,dx\,dy = \int_{0}^{2}[x]_y^{3y}\,dy$
$= \int_{0}^{2}(2y)\,dy = [y^2]_0^2 = 4$

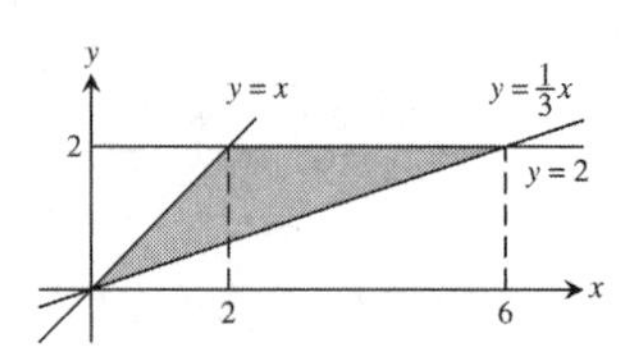

11. $\int_{0}^{1}\int_{x/2}^{2x} 1\,dy\,dx + \int_{1}^{2}\int_{x/2}^{3-x} 1\,dy\,dx$
$= \int_{0}^{1}[y]_{x/2}^{2x}dx + \int_{1}^{2}[y]_{x/2}^{3-x}dx$
$= \int_{0}^{1}\left(\frac{3}{2}x\right)dx + \int_{1}^{2}\left(3 - \frac{3}{2}x\right)dx$
$= \left[\frac{3}{4}x^2\right]_0^1 + \left[3x - \frac{3}{4}x^2\right]_1^2 = \frac{3}{2}$

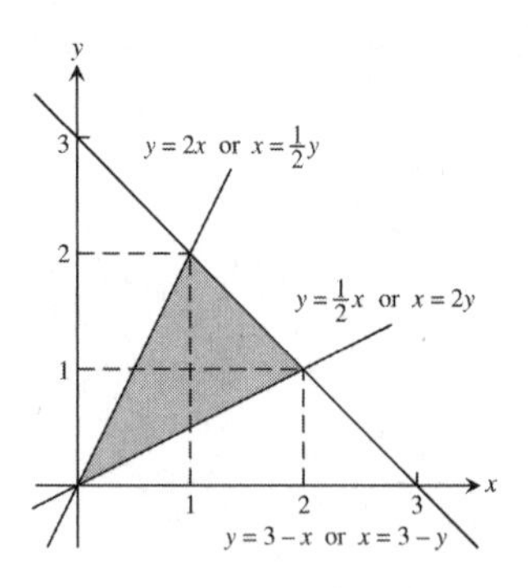

13. $\int_{0}^{6}\int_{y^2/3}^{2y} dx\,dy = \int_{0}^{6}\left(2y - \frac{y^2}{3}\right)dy = \left[y^2 - \frac{y^3}{9}\right]_0^6$
$= 36 - \frac{216}{9} = 12$

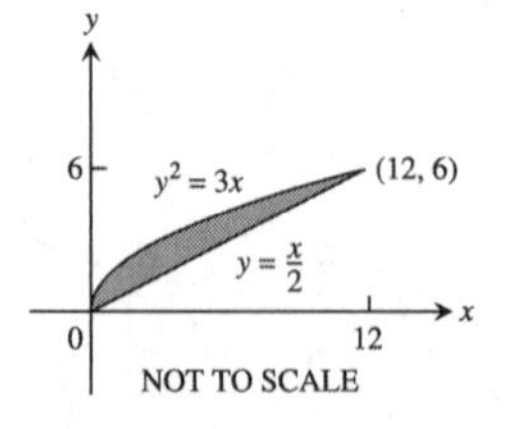

15. $\int_{0}^{\pi/4}\int_{\sin x}^{\cos x} dy\,dx$
$= \int_{0}^{\pi/4}(\cos x - \sin x)\,dx = [\sin x + \cos x]_0^{\pi/4}$
$= \left(\frac{\sqrt{2}}{2} + \frac{\sqrt{2}}{2}\right) - (0 + 1) = \sqrt{2} - 1$

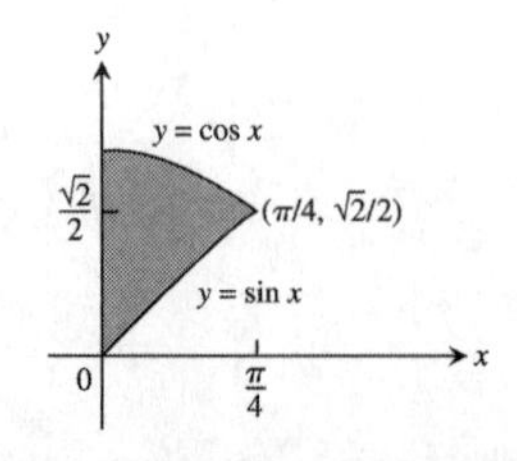

17. $\int_{-1}^{0}\int_{-2x}^{1-x} dy\,dx + \int_{0}^{2}\int_{-x/2}^{1-x} dy\,dx$
$= \int_{-1}^{0}(1+x)\,dx + \int_{0}^{2}\left(1-\frac{x}{2}\right)dx$
$= \left[x+\frac{x^2}{2}\right]_{-1}^{0} + \left[x-\frac{x^2}{4}\right]_{0}^{2} = -\left(-1+\frac{1}{2}\right)+(2-1) = \frac{3}{2}$

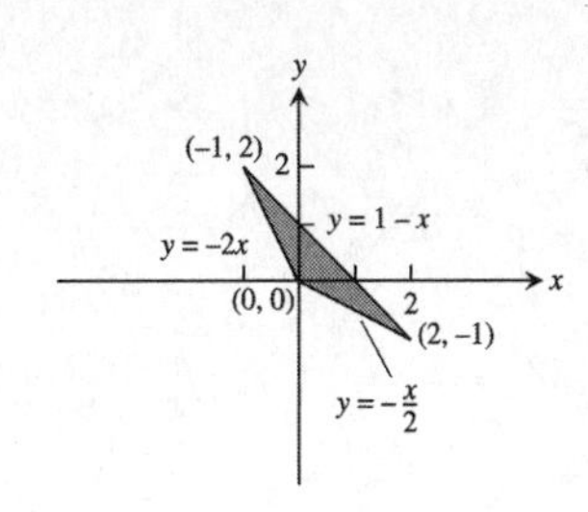

19. (a) average $= \frac{1}{\pi^2}\int_{0}^{\pi}\int_{0}^{\pi}\sin(x+y)\,dy\,dx = \frac{1}{\pi^2}\int_{0}^{\pi}[-\cos(x+y)]_{0}^{\pi}\,dx = \frac{1}{\pi^2}\int_{0}^{\pi}[-\cos(x+\pi)+\cos x]\,dx$
$= \frac{1}{\pi^2}[-\sin(x+\pi)+\sin x]_{0}^{\pi} = \frac{1}{\pi^2}[(-\sin 2\pi+\sin\pi)-(-\sin\pi+\sin 0)] = 0$

(b) average $= \frac{1}{\left(\frac{\pi^2}{2}\right)}\int_{0}^{\pi}\int_{0}^{\pi/2}\sin(x+y)\,dy\,dx = \frac{2}{\pi^2}\int_{0}^{\pi}[-\cos(x+y)]_{0}^{\pi/2}\,dx = \frac{2}{\pi^2}\int_{0}^{\pi}\left[-\cos\left(x+\frac{\pi}{2}\right)+\cos x\right]dx$
$= \frac{2}{\pi^2}\left[-\sin\left(x+\frac{\pi}{2}\right)+\sin x\right]_{0}^{\pi} = \frac{2}{\pi^2}\left[\left(-\sin\frac{3\pi}{2}+\sin\pi\right)-\left(-\sin\frac{\pi}{2}+\sin 0\right)\right] = \frac{4}{\pi^2}$

21. average height $= \frac{1}{4}\int_{0}^{2}\int_{0}^{2}(x^2+y^2)\,dy\,dx = \frac{1}{4}\int_{0}^{2}\left[x^2y+\frac{y^3}{3}\right]_{0}^{2}dx = \frac{1}{4}\int_{0}^{2}\left(2x^2+\frac{8}{3}\right)dx = \frac{1}{2}\left[\frac{x^3}{3}+\frac{4x}{3}\right]_{0}^{2} = \frac{8}{3}$

23. $\int_{-5}^{5}\int_{-2}^{0}\frac{10{,}000e^y}{1+\frac{|x|}{2}}\,dy\,dx = 10{,}000\left(1-e^{-2}\right)\int_{-5}^{5}\frac{dx}{1+\frac{|x|}{2}} = 10{,}000\left(1-e^{-2}\right)\left[\int_{-5}^{0}\frac{dx}{1-\frac{x}{2}}+\int_{0}^{5}\frac{dx}{1+\frac{x}{2}}\right]$
$= 10{,}000\left(1-e^{-2}\right)\left[-2\ln\left(1-\frac{x}{2}\right)\right]_{-5}^{0} + 10{,}000\left(1-e^{-2}\right)\left[2\ln\left(1+\frac{x}{2}\right)\right]_{0}^{5}$
$= 10{,}000\left(1-e^{-2}\right)\left[2\ln\left(1+\frac{5}{2}\right)\right] + 10{,}000\left(1-e^{-2}\right)\left[2\ln\left(1+\frac{5}{2}\right)\right] = 40{,}000\left(1-e^{-2}\right)\ln\left(\frac{7}{2}\right) \approx 43{,}329$

25. Let (x_i, y_i) be the location of the weather station in county i for $i = 1, \ldots, 254$. The average temperature in Texas at time t_0 is approximately $\frac{\sum_{i=1}^{254} T(x_i,y_i)\,\Delta_i A}{A}$, where $T(x_i, y_i)$ is the temperature at time t_0 at the weather station in county i, $\Delta_i A$ is the area of county i, and A is the area of Texas.

15.4 DOUBLE INTEGRALS IN POLAR FORM

1. $x^2+y^2 = 9^2 \Rightarrow r = 9 \Rightarrow \frac{\pi}{2} \le \theta \le 2\pi,\ 0 \le r \le 9$

3. $y = x \Rightarrow \theta = \frac{\pi}{4},\ y = -x \Rightarrow \theta = \frac{3\pi}{4},\ y = 1 \Rightarrow r = \csc\theta \Rightarrow \frac{\pi}{4} \le \theta \le \frac{3\pi}{4},\ 0 \le r \le \csc\theta$

5. $x^2+y^2 = 1^2 \Rightarrow r = 1,\ x = 2\sqrt{3} \Rightarrow r = 2\sqrt{3}\sec\theta,\ y = 2 \Rightarrow r = 2\csc\theta;\ 2\sqrt{3}\sec\theta = 2\csc\theta \Rightarrow \theta = \frac{\pi}{6}$
$\Rightarrow 0 \le \theta \le \frac{\pi}{6},\ 1 \le r \le 2\sqrt{3}\sec\theta;\ \frac{\pi}{6} \le \theta \le \frac{\pi}{2},\ 1 \le r \le 2\sqrt{3}\csc\theta$

7. $x^2+y^2 = 2x \Rightarrow r = 2\cos\theta \Rightarrow -\frac{\pi}{2} \le \theta \le \frac{\pi}{2},\ 0 \le r \le 2\cos\theta$

9. $\int_{-1}^{1}\int_{0}^{\sqrt{1-x^2}} dy\,dx = \int_{0}^{\pi}\int_{0}^{1} r\,dr\,d\theta = \frac{1}{2}\int_{0}^{\pi} d\theta = \frac{\pi}{2}$

11. $\int_{0}^{2}\int_{0}^{\sqrt{4-y^2}}(x^2+y^2)\,dx\,dy = \int_{0}^{\pi/2}\int_{0}^{2} r^3\,dr\,d\theta = 4\int_{0}^{\pi/2} d\theta = 2\pi$

13. $\int_{0}^{6}\int_{0}^{y} x\,dx\,dy = \int_{\pi/4}^{\pi/2}\int_{0}^{6\csc\theta} r^2\cos\theta\,dr\,d\theta = 72\int_{\pi/4}^{\pi/2}\cot\theta\csc^2\theta\,d\theta = -36\left[\cot^2\theta\right]_{\pi/4}^{\pi/2} = 36$

15. $\int_{1}^{\sqrt{3}}\int_{1}^{x} dy\,dx = \int_{\pi/6}^{\pi/4}\int_{\csc\theta}^{\sqrt{3}\sec\theta} r\,dr\,d\theta = \int_{\pi/6}^{\pi/4}\left(\frac{3}{2}\sec^2\theta-\frac{1}{2}\csc^2\theta\right)d\theta = \left[\frac{3}{2}\tan\theta+\frac{1}{2}\cot\theta\right]_{\pi/6}^{\pi/4} = 2-\sqrt{3}$

17. $\int_{-1}^{0}\int_{-\sqrt{1-x^2}}^{0}\frac{2}{1+\sqrt{x^2+y^2}}\,dy\,dx = \int_{\pi}^{3\pi/2}\int_{0}^{1}\frac{2r}{1+r}\,dr\,d\theta = 2\int_{\pi}^{3\pi/2}\int_{0}^{1}\left(1-\frac{1}{1+r}\right)dr\,d\theta = 2\int_{\pi}^{3\pi/2}(1-\ln 2)\,d\theta$
$= (1-\ln 2)\pi$

19. $\int_{0}^{\ln 2}\int_{0}^{\sqrt{(\ln 2)^2-y^2}} e^{\sqrt{x^2+y^2}}\,dx\,dy = \int_{0}^{\pi/2}\int_{0}^{\ln 2} re^{r}\,dr\,d\theta = \int_{0}^{\pi/2}(2\ln 2-1)\,d\theta = \frac{\pi}{2}(2\ln 2-1)$

21. $\int_{0}^{1}\int_{x}^{\sqrt{2-x^2}}(x+2y)\,dy\,dx = \int_{\pi/4}^{\pi/2}\int_{0}^{\sqrt{2}}(r\cos\theta+2r\sin\theta)\,r\,dr\,d\theta = \int_{\pi/4}^{\pi/2}\left[\frac{r^3}{3}\cos\theta+\frac{2r^3}{3}\sin\theta\right]_{0}^{\sqrt{2}}d\theta$
$= \int_{\pi/4}^{\pi/2}\left(\frac{2\sqrt{2}}{3}\cos\theta+\frac{4\sqrt{2}}{3}\sin\theta\right)d\theta = \left[\frac{2\sqrt{2}}{3}\sin\theta-\frac{4\sqrt{2}}{3}\cos\theta\right]_{\pi/4}^{\pi/2} = \frac{2(1+\sqrt{2})}{3}$

23. $\int_{0}^{1}\int_{0}^{\sqrt{1-x^2}} x\,y\,dy\,dx$ or
$\int_{0}^{1}\int_{0}^{\sqrt{1-y^2}} x\,y\,dx\,dy$

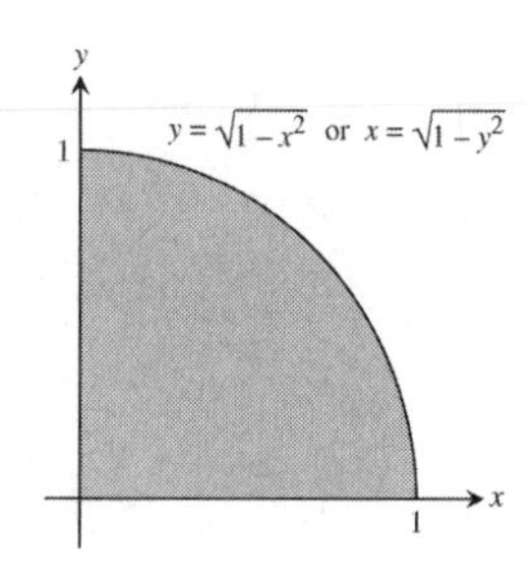

25. $\int_{0}^{2}\int_{0}^{x} y^2(x^2+y^2)\,dy\,dx$ or
$\int_{0}^{2}\int_{y}^{2} y^2(x^2+y^2)\,dx\,dy$

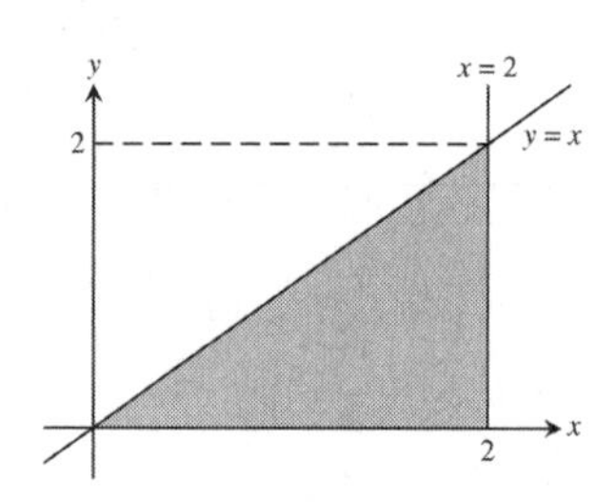

27. $\int_{0}^{\pi/2}\int_{0}^{2\sqrt{2-\sin 2\theta}} r\,dr\,d\theta = 2\int_{0}^{\pi/2}(2-\sin 2\theta)\,d\theta = 2(\pi-1)$

29. $A = 2\int_{0}^{\pi/6}\int_{0}^{12\cos 3\theta} r\,dr\,d\theta = 144\int_{0}^{\pi/6}\cos^2 3\theta\,d\theta = 12\pi$

31. $A = \int_{0}^{\pi/2}\int_{0}^{1+\sin\theta} r\,dr\,d\theta = \frac{1}{2}\int_{0}^{\pi/2}\left(\frac{3}{2}+2\sin\theta-\frac{\cos 2\theta}{2}\right)d\theta = \frac{3\pi}{8}+1$

33. $\text{average} = \frac{4}{\pi a^2}\int_{0}^{\pi/2}\int_{0}^{a} r\sqrt{a^2-r^2}\,dr\,d\theta = \frac{4}{3\pi a^2}\int_{0}^{\pi/2} a^3\,d\theta = \frac{2a}{3}$

35. $\text{average} = \frac{1}{\pi a^2}\int_{-a}^{a}\int_{-\sqrt{a^2-x^2}}^{\sqrt{a^2-x^2}}\sqrt{x^2+y^2}\,dy\,dx = \frac{1}{\pi a^2}\int_{0}^{2\pi}\int_{0}^{a} r^2\,dr\,d\theta = \frac{a}{3\pi}\int_{0}^{2\pi}d\theta = \frac{2a}{3}$

37. $\int_{0}^{2\pi}\int_{1}^{\sqrt{e}}\left(\frac{\ln r^2}{r}\right)r\,dr\,d\theta = \int_{0}^{2\pi}\int_{1}^{\sqrt{e}} 2\ln r\,dr\,d\theta = 2\int_{0}^{2\pi}[r\ln r-r]_{1}^{e^{1/2}}\,d\theta = 2\int_{0}^{2\pi}\sqrt{e}\left[\left(\frac{1}{2}-1\right)+1\right]d\theta = 2\pi\left(2-\sqrt{e}\right)$

39. $V = 2\int_{0}^{\pi/2}\int_{1}^{1+\cos\theta} r^2\cos\theta\,dr\,d\theta = \frac{2}{3}\int_{0}^{\pi/2}(3\cos^2\theta+3\cos^3\theta+\cos^4\theta)\,d\theta$
$= \frac{2}{3}\left[\frac{15\theta}{8}+\sin 2\theta+3\sin\theta-\sin^3\theta+\frac{\sin 4\theta}{32}\right]_{0}^{\pi/2} = \frac{4}{3}+\frac{5\pi}{8}$

41. (a) $I^2 = \int_0^\infty \int_0^\infty e^{-(x^2+y^2)}\,dx\,dy = \int_0^{\pi/2} \int_0^\infty \left(e^{-r^2}\right) r\,dr\,d\theta = \int_0^{\pi/2} \left[\lim_{b\to\infty} \int_0^b re^{-r^2}\,dr\right] d\theta$

$= -\frac{1}{2}\int_0^{\pi/2} \lim_{b\to\infty} \left(e^{-b^2} - 1\right) d\theta = \frac{1}{2}\int_0^{\pi/2} d\theta = \frac{\pi}{4} \Rightarrow I = \frac{\sqrt{\pi}}{2}$

(b) $\lim_{x\to\infty} \int_0^x \frac{2e^{-t^2}}{\sqrt{\pi}}\,dt = \frac{2}{\sqrt{\pi}} \int_0^\infty e^{-t^2}\,dt = \left(\frac{2}{\sqrt{\pi}}\right)\left(\frac{\sqrt{\pi}}{2}\right) = 1$, from part (a)

43. Over the disk $x^2 + y^2 \le \frac{3}{4}$: $\iint\limits_R \frac{1}{1-x^2-y^2}\,dA = \int_0^{2\pi} \int_0^{\sqrt{3}/2} \frac{r}{1-r^2}\,dr\,d\theta = \int_0^{2\pi} \left[-\frac{1}{2}\ln(1-r^2)\right]_0^{\sqrt{3}/2} d\theta$

$= \int_0^{2\pi} \left(-\frac{1}{2}\ln\frac{1}{4}\right) d\theta = (\ln 2)\int_0^{2\pi} d\theta = \pi \ln 4$

Over the disk $x^2 + y^2 \le 1$: $\iint\limits_R \frac{1}{1-x^2-y^2}\,dA = \int_0^{2\pi} \int_0^1 \frac{r}{1-r^2}\,dr\,d\theta = \int_0^{2\pi} \left[\lim_{a\to 1^-} \int_0^a \frac{r}{1-r^2}\,dr\right] d\theta$

$= \int_0^{2\pi} \lim_{a\to 1^-} \left[-\frac{1}{2}\ln(1-a^2)\right] d\theta = 2\pi \cdot \lim_{a\to 1^-} \left[-\frac{1}{2}\ln(1-a^2)\right] = 2\pi\cdot\infty$, so the integral does not exist over $x^2 + y^2 \le 1$

45. average $= \frac{1}{\pi a^2} \int_0^{2\pi} \int_0^a \left[(r\cos\theta - h)^2 + r^2\sin^2\theta\right] r\,dr\,d\theta = \frac{1}{\pi a^2} \int_0^{2\pi} \int_0^a (r^3 - 2r^2 h\cos\theta + rh^2)\,dr\,d\theta$

$= \frac{1}{\pi a^2} \int_0^{2\pi} \left(\frac{a^4}{4} - \frac{2a^3 h\cos\theta}{3} + \frac{a^2h^2}{2}\right) d\theta = \frac{1}{\pi} \int_0^{2\pi} \left(\frac{a^2}{4} - \frac{2ah\cos\theta}{3} + \frac{h^2}{2}\right) d\theta = \frac{1}{\pi}\left[\frac{a^2\theta}{4} - \frac{2ah\sin\theta}{3} + \frac{h^2\theta}{2}\right]_0^{2\pi}$

$= \frac{1}{2}(a^2 + 2h^2)$

15.5 TRIPLE INTEGRALS IN RECTANGULAR COORDINATES

1. $\int_0^1 \int_0^{1-x} \int_{x+z}^1 F(x,y,z)\,dy\,dz\,dx = \int_0^1 \int_0^{1-x} \int_{x+z}^1 dy\,dz\,dx = \int_0^1 \int_0^{1-x} (1-x-z)\,dz\,dx$

$= \int_0^1 \left[(1-x) - x(1-x) - \frac{(1-x)^2}{2}\right] dx = \int_0^1 \frac{(1-x)^2}{2}\,dx = \left[-\frac{(1-x)^3}{6}\right]_0^1 = \frac{1}{6}$

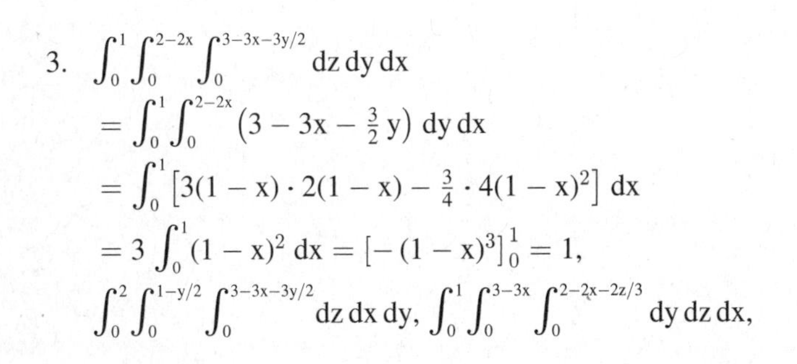

3. $\int_0^1 \int_0^{2-2x} \int_0^{3-3x-3y/2} dz\,dy\,dx$

$= \int_0^1 \int_0^{2-2x} \left(3 - 3x - \frac{3}{2}y\right) dy\,dx$

$= \int_0^1 \left[3(1-x)\cdot 2(1-x) - \frac{3}{4}\cdot 4(1-x)^2\right] dx$

$= 3\int_0^1 (1-x)^2\,dx = \left[-(1-x)^3\right]_0^1 = 1,$

$\int_0^2 \int_0^{1-y/2} \int_0^{3-3x-3y/2} dz\,dx\,dy, \int_0^1 \int_0^{3-3x} \int_0^{2-2x-2z/3} dy\,dz\,dx,$

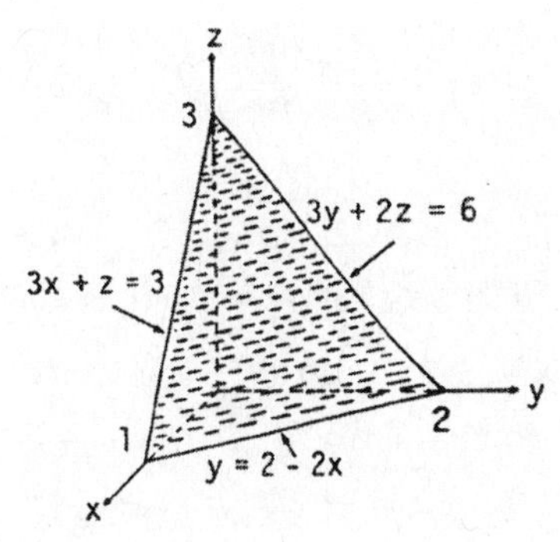

$\int_0^3 \int_0^{1-z/3} \int_0^{2-2x-2z/3} dy\,dx\,dz, \int_0^2 \int_0^{3-3y/2} \int_0^{1-y/2-z/3} dx\,dz\,dy, \int_0^3 \int_0^{2-2z/3} \int_0^{1-y/2-z/3} dx\,dy\,dz$

5. $\int_{-2}^{2}\int_{-\sqrt{4-x^2}}^{\sqrt{4-x^2}}\int_{x^2+y^2}^{8-x^2-y^2} dz\,dy\,dx = 4\int_{0}^{2}\int_{0}^{\sqrt{4-x^2}}\int_{x^2+y^2}^{8-x^2-y^2} dz\,dy\,dx$
$= 4\int_{0}^{2}\int_{0}^{\sqrt{4-x^2}} [8-2(x^2+y^2)]\,dy\,dx$
$= 8\int_{0}^{2}\int_{0}^{\sqrt{4-x^2}} (4-x^2-y^2)\,dy\,dx$
$= 8\int_{0}^{\pi/2}\int_{0}^{2}(4-r^2)\,r\,dr\,d\theta = 8\int_{0}^{\pi/2}\left[2r^2-\frac{r^4}{4}\right]_0^2 d\theta$
$= 32\int_{0}^{\pi/2} d\theta = 32\left(\frac{\pi}{2}\right) = 16\pi,$
$\int_{-2}^{2}\int_{-\sqrt{4-y^2}}^{\sqrt{4-y^2}}\int_{x^2+y^2}^{8-x^2-y^2} dz\,dx\,dy,$
$\int_{-2}^{2}\int_{y^2}^{4}\int_{-\sqrt{z-y^2}}^{\sqrt{z-y^2}} dx\,dz\,dy + \int_{-2}^{2}\int_{4}^{8-y^2}\int_{-\sqrt{8-z-y^2}}^{\sqrt{8-z-y^2}} dx\,dz\,dy,$
$\int_{0}^{4}\int_{-\sqrt{z}}^{\sqrt{z}}\int_{-\sqrt{z-y^2}}^{\sqrt{z-y^2}} dx\,dy\,dz + \int_{4}^{8}\int_{-\sqrt{8-z}}^{\sqrt{8-z}}\int_{-\sqrt{8-z-y^2}}^{\sqrt{8-z-y^2}} dx\,dy\,dz,$ $\int_{-2}^{2}\int_{x^2}^{4}\int_{-\sqrt{z-x^2}}^{\sqrt{z-x^2}} dy\,dz\,dx + \int_{-2}^{2}\int_{4}^{8-x^2}\int_{-\sqrt{8-z-x^2}}^{\sqrt{8-z-x^2}} dy\,dz\,dx,$
$\int_{0}^{4}\int_{-\sqrt{z}}^{\sqrt{z}}\int_{-\sqrt{z-x^2}}^{\sqrt{z-x^2}} dy\,dx\,dz + \int_{4}^{8}\int_{-\sqrt{8-z}}^{\sqrt{8-z}}\int_{-\sqrt{8-z-x^2}}^{\sqrt{8-z-x^2}} dy\,dx\,dz$

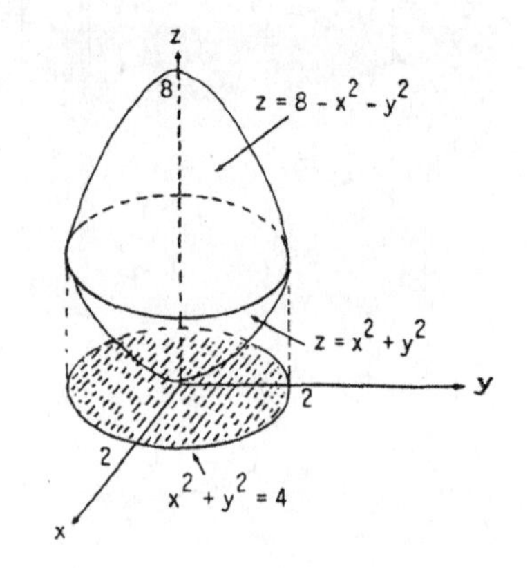

7. $\int_{0}^{1}\int_{0}^{1}\int_{0}^{1} (x^2+y^2+z^2)\,dz\,dy\,dx = \int_{0}^{1}\int_{0}^{1}\left(x^2+y^2+\frac{1}{3}\right)dy\,dx = \int_{0}^{1}\left(x^2+\frac{2}{3}\right)dx = 1$

9. $\int_{1}^{e}\int_{1}^{e^2}\int_{1}^{e^3} \frac{1}{xyz}\,dx\,dy\,dz = \int_{1}^{e}\int_{1}^{e^2}\left[\frac{\ln x}{yz}\right]_1^{e^3} dy\,dz = \int_{1}^{e}\int_{1}^{e^2}\frac{3}{yz}\,dy\,dz = \int_{1}^{e}\left[\frac{\ln y}{z}\right]_1^{e^2} dz = \int_{1}^{e}\frac{6}{z}\,dz = 6$

11. $\int_{0}^{\pi/6}\int_{0}^{1}\int_{-2}^{3} y\sin z\,dx\,dy\,dz = \int_{0}^{\pi/6}\int_{0}^{1} 5y\sin z\,dy\,dz = \frac{5}{2}\int_{0}^{\pi/6}\sin z\,dz = \frac{5\left(2-\sqrt{3}\right)}{4}$

13. $\int_{0}^{3}\int_{0}^{\sqrt{9-x^2}}\int_{0}^{\sqrt{9-x^2}} dz\,dy\,dx = \int_{0}^{3}\int_{0}^{\sqrt{9-x^2}}\sqrt{9-x^2}\,dy\,dx = \int_{0}^{3}(9-x^2)\,dx = \left[9x-\frac{x^3}{3}\right]_0^3 = 18$

15. $\int_{0}^{1}\int_{0}^{2-x}\int_{0}^{2-x-y} dz\,dy\,dx = \int_{0}^{1}\int_{0}^{2-x}(2-x-y)\,dy\,dx = \int_{0}^{1}\left[(2-x)^2-\frac{1}{2}(2-x)^2\right]dx = \frac{1}{2}\int_{0}^{1}(2-x)^2\,dx$
$= \left[-\frac{1}{6}(2-x)^3\right]_0^1 = -\frac{1}{6}+\frac{8}{6} = \frac{7}{6}$

17. $\int_{0}^{\pi}\int_{0}^{\pi}\int_{0}^{\pi}\cos(u+v+w)\,du\,dv\,dw = \int_{0}^{\pi}\int_{0}^{\pi}[\sin(w+v+\pi)-\sin(w+v)]\,dv\,dw$
$= \int_{0}^{\pi}[(-\cos(w+2\pi)+\cos(w+\pi))+(\cos(w+\pi)-\cos w)]\,dw$
$= [-\sin(w+2\pi)+\sin(w+\pi)-\sin w+\sin(w+\pi)]_0^{\pi} = 0$

19. $\int_{0}^{\pi/4}\int_{0}^{\ln\sec v}\int_{-\infty}^{2t} e^x\,dx\,dt\,dv = \int_{0}^{\pi/4}\int_{0}^{\ln\sec v}\lim_{b\to-\infty}(e^{2t}-e^{b})\,dt\,dv = \int_{0}^{\pi/4}\int_{0}^{\ln\sec v} e^{2t}\,dt\,dv = \int_{0}^{\pi/4}\left(\frac{1}{2}e^{2\ln\sec v}-\frac{1}{2}\right)dv$
$= \int_{0}^{\pi/4}\left(\frac{\sec^2 v}{2}-\frac{1}{2}\right)dv = \left[\frac{\tan v}{2}-\frac{v}{2}\right]_0^{\pi/4} = \frac{1}{2}-\frac{\pi}{8}$

21. (a) $\int_{-1}^{1}\int_{0}^{1-x^2}\int_{x^2}^{1-z} dy\,dz\,dx$ (b) $\int_{0}^{1}\int_{-\sqrt{1-z}}^{\sqrt{1-z}}\int_{x^2}^{1-z} dy\,dx\,dz$ (c) $\int_{0}^{1}\int_{0}^{1-z}\int_{-\sqrt{y}}^{\sqrt{y}} dx\,dy\,dz$
(d) $\int_{0}^{1}\int_{0}^{1-y}\int_{-\sqrt{y}}^{\sqrt{y}} dx\,dz\,dy$ (e) $\int_{0}^{1}\int_{-\sqrt{y}}^{\sqrt{y}}\int_{0}^{1-y} dz\,dx\,dy$

23. $V = \int_{0}^{1}\int_{-1}^{1}\int_{0}^{y^2} dz\,dy\,dx = \int_{0}^{1}\int_{-1}^{1} y^2\,dy\,dx = \frac{2}{3}\int_{0}^{1} dx = \frac{2}{3}$

25. $V = \int_0^4 \int_0^{\sqrt{4-x}} \int_0^{2-y} dz\,dy\,dx = \int_0^4 \int_0^{\sqrt{4-x}} (2-y)\,dy\,dx = \int_0^4 \left[2\sqrt{4-x} - \left(\frac{4-x}{2}\right)\right] dx$
$= \left[-\frac{4}{3}(4-x)^{3/2} + \frac{1}{4}(4-x)^2\right]_0^4 = \frac{4}{3}(4)^{3/2} - \frac{1}{4}(16) = \frac{32}{3} - 4 = \frac{20}{3}$

27. $V = \int_0^1 \int_0^{2-2x} \int_0^{3-3x-3y/2} dz\,dy\,dx = \int_0^1 \int_0^{2-2x} \left(3 - 3x - \frac{3}{2}y\right) dy\,dx = \int_0^1 \left[6(1-x)^2 - \frac{3}{4} \cdot 4(1-x)^2\right] dx$
$= \int_0^1 3(1-x)^2\,dx = \left[-(1-x)^3\right]_0^1 = 1$

29. $V = 8 \int_0^1 \int_0^{\sqrt{1-x^2}} \int_0^{\sqrt{1-x^2}} dz\,dy\,dx = 8 \int_0^1 \int_0^{\sqrt{1-x^2}} \sqrt{1-x^2}\,dy\,dx = 8 \int_0^1 (1-x^2)\,dx = \frac{16}{3}$

31. $V = \int_0^4 \int_0^{(\sqrt{16-y^2})/2} \int_0^{4-y} dx\,dz\,dy = \int_0^4 \int_0^{(\sqrt{16-y^2})/2} (4-y)\,dz\,dy = \int_0^4 \frac{\sqrt{16-y^2}}{2}(4-y)\,dy$
$= \int_0^4 2\sqrt{16-y^2}\,dy - \frac{1}{2}\int_0^4 y\sqrt{16-y^2}\,dy = \left[y\sqrt{16-y^2} + 16\sin^{-1}\frac{y}{4}\right]_0^4 + \left[\frac{1}{6}(16-y^2)^{3/2}\right]_0^4$
$= 16\left(\frac{\pi}{2}\right) - \frac{1}{6}(16)^{3/2} = 8\pi - \frac{32}{3}$

33. $\int_0^2 \int_0^{2-x} \int_{(2-x-y)/2}^{4-2x-2y} dz\,dy\,dx = \int_0^2 \int_0^{2-x} \left(3 - \frac{3x}{2} - \frac{3y}{2}\right) dy\,dx$
$= \int_0^2 \left[3\left(1 - \frac{x}{2}\right)(2-x) - \frac{3}{4}(2-x)^2\right] dx$
$= \int_0^2 \left[6 - 6x + \frac{3x^2}{2} - \frac{3(2-x)^2}{4}\right] dx$
$= \left[6x - 3x^2 + \frac{x^3}{2} + \frac{(2-x)^3}{4}\right]_0^2 = (12 - 12 + 4 + 0) - \frac{2^3}{4} = 2$

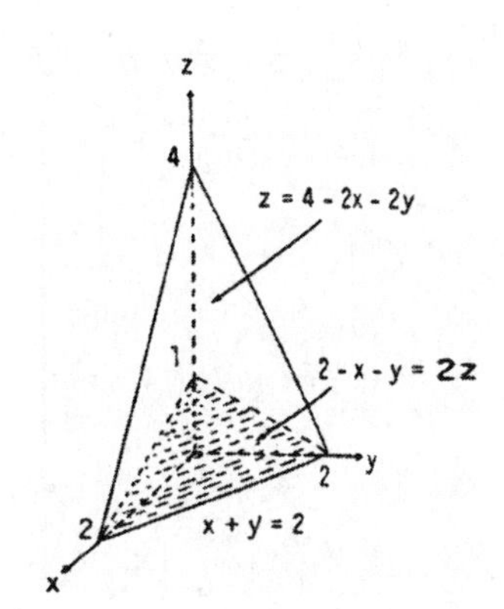

35. $V = 2\int_{-2}^2 \int_0^{\sqrt{4-x^2}/2} \int_0^{x+2} dz\,dy\,dx = 2\int_{-2}^2 \int_0^{\sqrt{4-x^2}/2} (x+2)\,dy\,dx = \int_{-2}^2 (x+2)\sqrt{4-x^2}\,dx$
$= \int_{-2}^2 2\sqrt{4-x^2}\,dx + \int_{-2}^2 x\sqrt{4-x^2}\,dx = \left[x\sqrt{4-x^2} + 4\sin^{-1}\frac{x}{2}\right]_{-2}^2 + \left[-\frac{1}{3}(4-x^2)^{3/2}\right]_{-2}^2$
$= 4\left(\frac{\pi}{2}\right) - 4\left(-\frac{\pi}{2}\right) = 4\pi$

37. $\text{average} = \frac{1}{8}\int_0^2 \int_0^2 \int_0^2 (x^2+9)\,dz\,dy\,dx = \frac{1}{8}\int_0^2 \int_0^2 (2x^2+18)\,dy\,dx = \frac{1}{8}\int_0^2 (4x^2+36)\,dx = \frac{31}{3}$

39. $\text{average} = \int_0^1 \int_0^1 \int_0^1 (x^2+y^2+z^2)\,dz\,dy\,dx = \int_0^1 \int_0^1 \left(x^2+y^2+\frac{1}{3}\right) dy\,dx = \int_0^1 \left(x^2+\frac{2}{3}\right) dx = 1$

41. $\int_0^4 \int_0^1 \int_{2y}^2 \frac{4\cos(x^2)}{2\sqrt{z}}\,dx\,dy\,dz = \int_0^4 \int_0^2 \int_0^{x/2} \frac{4\cos(x^2)}{2\sqrt{z}}\,dy\,dx\,dz = \int_0^4 \int_0^2 \frac{x\cos(x^2)}{\sqrt{z}}\,dx\,dz = \int_0^4 \left(\frac{\sin 4}{2}\right) z^{-1/2}\,dz$
$= \left[(\sin 4)z^{1/2}\right]_0^4 = 2\sin 4$

43. $\int_0^1 \int_{\sqrt[3]{z}}^1 \int_0^{\ln 3} \frac{\pi e^{2x}\sin(\pi y^2)}{y^2}\,dx\,dy\,dz = \int_0^1 \int_{\sqrt[3]{z}}^1 \frac{4\pi\sin(\pi y^2)}{y^2}\,dy\,dz = \int_0^1 \int_0^{y^3} \frac{4\pi\sin(\pi y^2)}{y^2}\,dz\,dy$
$= \int_0^1 4\pi y\sin(\pi y^2)\,dy = \left[-2\cos(\pi y^2)\right]_0^1 = -2(-1) + 2(1) = 4$

45. $\int_0^1 \int_0^{4-a-x^2} \int_a^{4-x^2-y} dz\,dy\,dx = \frac{4}{15} \Rightarrow \int_0^1 \int_0^{4-a-x^2} (4-x^2-y-a)\,dy\,dx = \frac{4}{15}$
$\Rightarrow \int_0^1 \left[(4-a-x^2)^2 - \frac{1}{2}(4-a-x^2)^2\right] dx = \frac{4}{15} \Rightarrow \frac{1}{2}\int_0^1 (4-a-x^2)^2\,dx = \frac{4}{15} \Rightarrow \int_0^1 \left[(4-a)^2 - 2x^2(4-a) + x^4\right] dx$

$= \frac{8}{15} \Rightarrow \left[(4-a)^2 x - \frac{2}{3}x^3(4-a) + \frac{x^5}{5}\right]_0^1 = \frac{8}{15} \Rightarrow (4-a)^2 - \frac{2}{3}(4-a) + \frac{1}{5} = \frac{8}{15} \Rightarrow 15(4-a)^2 - 10(4-a) - 5 = 0$
$\Rightarrow 3(4-a)^2 - 2(4-a) - 1 = 0 \Rightarrow [3(4-a)+1][(4-a)-1] = 0 \Rightarrow 4-a = -\frac{1}{3}$ or $4-a = 1 \Rightarrow a = \frac{13}{3}$ or $a = 3$

47. To minimize the integral, we want the domain to include all points where the integrand is negative and to exclude all points where it is positive. These criteria are met by the points (x, y, z) such that $4x^2 + 4y^2 + z^2 - 4 \le 0$ or $4x^2 + 4y^2 + z^2 \le 4$, which is a solid ellipsoid centered at the origin.

15.6 MOMENTS AND CENTERS OF MASS

1. $M = \int_0^1 \int_x^{2-x^2} 3\,dy\,dx = 3\int_0^1 (2 - x^2 - x)\,dx = \frac{7}{2}$; $M_y = \int_0^1 \int_x^{2-x^2} 3x\,dy\,dx = 3\int_0^1 [xy]_x^{2-x^2}\,dx$
$= 3\int_0^1 (2x - x^3 - x^2)\,dx = \frac{5}{4}$; $M_x = \int_0^1 \int_x^{2-x^2} 3y\,dy\,dx = \frac{3}{2}\int_0^1 [y^2]_x^{2-x^2}\,dx = \frac{3}{2}\int_0^1 (4 - 5x^2 + x^4)\,dx = \frac{19}{5}$
$\Rightarrow \bar{x} = \frac{5}{14}$ and $\bar{y} = \frac{38}{35}$

3. $M = \int_0^2 \int_{y^2/2}^{4-y} dx\,dy = \int_0^2 \left(4 - y - \frac{y^2}{2}\right) dy = \frac{14}{3}$; $M_y = \int_0^2 \int_{y^2/2}^{4-y} x\,dx\,dy = \frac{1}{2}\int_0^2 [x^2]_{y^2/2}^{4-y}\,dy$
$= \frac{1}{2}\int_0^2 \left(16 - 8y + y^2 - \frac{y^4}{4}\right) dy = \frac{128}{15}$; $M_x = \int_0^2 \int_{y^2/2}^{4-y} y\,dx\,dy = \int_0^2 \left(4y - y^2 - \frac{y^3}{2}\right) dy = \frac{10}{3}$
$\Rightarrow \bar{x} = \frac{64}{35}$ and $\bar{y} = \frac{5}{7}$

5. $M = \int_0^a \int_0^{\sqrt{a^2-x^2}} dy\,dx = \frac{\pi a^2}{4}$; $M_y = \int_0^a \int_0^{\sqrt{a^2-x^2}} x\,dy\,dx = \int_0^a [xy]_0^{\sqrt{a^2-x^2}}\,dx = \int_0^a x\sqrt{a^2-x^2}\,dx = \frac{a^3}{3}$
$\Rightarrow \bar{x} = \bar{y} = \frac{4a}{3\pi}$, by symmetry

7. $I_x = \int_{-2}^2 \int_{-\sqrt{4-x^2}}^{\sqrt{4-x^2}} y^2\,dy\,dx = \int_{-2}^2 \left[\frac{y^3}{3}\right]_{-\sqrt{4-x^2}}^{\sqrt{4-x^2}} dx = \frac{2}{3}\int_{-2}^2 (4-x^2)^{3/2}\,dx = 4\pi$; $I_y = 4\pi$, by symmetry;
$I_o = I_x + I_y = 8\pi$

9. $M = \int_{-\infty}^0 \int_0^{e^x} dy\,dx = \int_{-\infty}^0 e^x\,dx = \lim_{b \to -\infty} \int_b^0 e^x\,dx = 1 - \lim_{b \to -\infty} e^b = 1$; $M_y = \int_{-\infty}^0 \int_0^{e^x} x\,dy\,dx = \int_{-\infty}^0 xe^x\,dx$
$= \lim_{b \to -\infty} \int_b^0 xe^x\,dx = \lim_{b \to -\infty} [xe^x - e^x]_b^0 = -1 - \lim_{b \to -\infty} (be^b - e^b) = -1$; $M_x = \int_{-\infty}^0 \int_0^{e^x} y\,dy\,dx$
$= \frac{1}{2}\int_{-\infty}^0 e^{2x}\,dx = \frac{1}{2}\lim_{b \to -\infty} \int_b^0 e^{2x}\,dx = \frac{1}{4} \Rightarrow \bar{x} = -1$ and $\bar{y} = \frac{1}{4}$

11. $M = \int_0^2 \int_{-y}^{y-y^2} (x+y)\,dx\,dy = \int_0^2 \left[\frac{x^2}{2} + xy\right]_{-y}^{y-y^2} dy = \int_0^2 \left(\frac{y^4}{2} - 2y^3 + 2y^2\right) dy = \left[\frac{y^5}{10} - \frac{y^4}{2} + \frac{2y^3}{3}\right]_0^2 = \frac{8}{15}$;
$I_x = \int_0^2 \int_{-y}^{y-y^2} y^2(x+y)\,dx\,dy = \int_0^2 \left[\frac{x^2y^2}{2} + xy^3\right]_{-y}^{y-y^2} dy = \int_0^2 \left(\frac{y^6}{2} - 2y^5 + 2y^4\right) dy = \frac{64}{105}$;

13. $M = \int_0^1 \int_x^{2-x} (6x + 3y + 3)\,dy\,dx = \int_0^1 \left[6xy + \frac{3}{2}y^2 + 3y\right]_x^{2-x} dx = \int_0^1 (12 - 12x^2)\,dx = 8$;
$M_y = \int_0^1 \int_x^{2-x} x(6x + 3y + 3)\,dy\,dx = \int_0^1 (12x - 12x^3)\,dx = 3$; $M_x = \int_0^1 \int_x^{2-x} y(6x + 3y + 3)\,dy\,dx$
$= \int_0^1 (14 - 6x - 6x^2 - 2x^3)\,dx = \frac{17}{2} \Rightarrow \bar{x} = \frac{3}{8}$ and $\bar{y} = \frac{17}{16}$

15. $M = \int_0^1 \int_0^6 (x + y + 1)\,dx\,dy = \int_0^1 (6y + 24)\,dy = 27$; $M_x = \int_0^1 \int_0^6 y(x + y + 1)\,dx\,dy = \int_0^1 y(6y + 24)\,dy = 14$;
$M_y = \int_0^1 \int_0^6 x(x + y + 1)\,dx\,dy = \int_0^1 (18y + 90)\,dy = 99 \Rightarrow \bar{x} = \frac{11}{3}$ and $\bar{y} = \frac{14}{27}$; $I_y = \int_0^1 \int_0^6 x^2(x + y + 1)\,dx\,dy$
$= 216\int_0^1 \left(\frac{y}{3} + \frac{11}{6}\right) dy = 432$

17. $M = \int_{-1}^{1}\int_{0}^{x^2} (7y+1)\,dy\,dx = \int_{-1}^{1}\left(\frac{7x^4}{2}+x^2\right)dx = \frac{31}{15}$; $M_x = \int_{-1}^{1}\int_{0}^{x^2} y(7y+1)\,dy\,dx = \int_{-1}^{1}\left(\frac{7x^6}{3}+\frac{x^4}{2}\right)dx = \frac{13}{15}$;
$M_y = \int_{-1}^{1}\int_{0}^{x^2} x(7y+1)\,dy\,dx = \int_{-1}^{1}\left(\frac{7x^5}{2}+x^3\right)dx = 0 \Rightarrow \bar{x} = 0$ and $\bar{y} = \frac{13}{31}$; $I_y = \int_{-1}^{1}\int_{0}^{x^2} x^2(7y+1)\,dy\,dx$
$= \int_{-1}^{1}\left(\frac{7x^6}{2}+x^4\right)dx = \frac{7}{5}$

19. $M = \int_{0}^{1}\int_{-y}^{y} (y+1)\,dx\,dy = \int_{0}^{1}(2y^2+2y)\,dy = \frac{5}{3}$; $M_x = \int_{0}^{1}\int_{-y}^{y} y(y+1)\,dx\,dy = 2\int_{0}^{1}(y^3+y^2)\,dy = \frac{7}{6}$;
$M_y = \int_{0}^{1}\int_{-y}^{y} x(y+1)\,dx\,dy = \int_{0}^{1} 0\,dy = 0 \Rightarrow \bar{x} = 0$ and $\bar{y} = \frac{7}{10}$; $I_x = \int_{0}^{1}\int_{-y}^{y} y^2(y+1)\,dx\,dy = \int_{0}^{1}(2y^4+2y^3)\,dy$
$= \frac{9}{10}$; $I_y = \int_{0}^{1}\int_{-y}^{y} x^2(y+1)\,dx\,dy = \frac{1}{3}\int_{0}^{1}(2y^4+2y^3)\,dy = \frac{3}{10} \Rightarrow I_o = I_x + I_y = \frac{6}{5}$

21. $I_x = \int_{0}^{a}\int_{0}^{b}\int_{0}^{c}(y^2+z^2)\,dz\,dy\,dx = \int_{0}^{a}\int_{0}^{b}\left(cy^2+\frac{c^3}{3}\right)dy\,dx = \int_{0}^{a}\left(\frac{cb^3}{3}+\frac{c^3b}{3}\right)dx = \frac{abc\,(b^2+c^2)}{3}$
$= \frac{M}{3}(b^2+c^2)$ where $M = abc$; $I_y = \frac{M}{3}(a^2+c^2)$ and $I_z = \frac{M}{3}(a^2+b^2)$, by symmetry

23. $M = 4\int_{0}^{1}\int_{0}^{1}\int_{4y^2}^{4} dz\,dy\,dx = 4\int_{0}^{1}\int_{0}^{1}(4-4y^2)\,dy\,dx = 16\int_{0}^{1}\frac{2}{3}\,dx = \frac{32}{3}$; $M_{xy} = 4\int_{0}^{1}\int_{0}^{1}\int_{4y^2}^{4} z\,dz\,dy\,dx$
$= 2\int_{0}^{1}\int_{0}^{1}(16-16y^4)\,dy\,dx = \frac{128}{5}\int_{0}^{1}dx = \frac{128}{5} \Rightarrow \bar{z} = \frac{12}{5}$, and $\bar{x} = \bar{y} = 0$, by symmetry;
$I_x = 4\int_{0}^{1}\int_{0}^{1}\int_{4y^2}^{4}(y^2+z^2)\,dz\,dy\,dx = 4\int_{0}^{1}\int_{0}^{1}\left[\left(4y^2+\frac{64}{3}\right)-\left(4y^4+\frac{64y^6}{3}\right)\right]dy\,dx = 4\int_{0}^{1}\frac{1976}{105}\,dx = \frac{7904}{105}$;
$I_y = 4\int_{0}^{1}\int_{0}^{1}\int_{4y^2}^{4}(x^2+z^2)\,dz\,dy\,dx = 4\int_{0}^{1}\int_{0}^{1}\left[\left(4x^2+\frac{64}{3}\right)-\left(4x^2y^2+\frac{64y^6}{3}\right)\right]dy\,dx = 4\int_{0}^{1}\left(\frac{8}{3}x^2+\frac{128}{7}\right)dx$
$= \frac{4832}{63}$; $I_z = 4\int_{0}^{1}\int_{0}^{1}\int_{4y^2}^{4}(x^2+y^2)\,dz\,dy\,dx = 16\int_{0}^{1}\int_{0}^{1}(x^2-x^2y^2+y^2-y^4)\,dy\,dx$
$= 16\int_{0}^{1}\left(\frac{2x^2}{3}+\frac{2}{15}\right)dx = \frac{256}{45}$

25. (a) $M = 4\int_{0}^{2}\int_{0}^{\sqrt{4-x^2}}\int_{x^2+y^2}^{4} dz\,dy\,dx = 4\int_{0}^{\pi/2}\int_{0}^{2}\int_{r^2}^{4} r\,dz\,dr\,d\theta = 4\int_{0}^{\pi/2}\int_{0}^{2}(4r-r^3)\,dr\,d\theta = 4\int_{0}^{\pi/2}4\,d\theta = 8\pi$;
$M_{xy} = \int_{0}^{2\pi}\int_{0}^{2}\int_{r^2}^{4} zr\,dz\,dr\,d\theta = \int_{0}^{2\pi}\int_{0}^{2}\frac{r}{2}(16-r^4)\,dr\,d\theta = \frac{32}{3}\int_{0}^{2\pi}d\theta = \frac{64\pi}{3} \Rightarrow \bar{z} = \frac{8}{3}$, and $\bar{x} = \bar{y} = 0$, by symmetry
(b) $M = 8\pi \Rightarrow 4\pi = \int_{0}^{2\pi}\int_{0}^{\sqrt{c}}\int_{r^2}^{c} r\,dz\,dr\,d\theta = \int_{0}^{2\pi}\int_{0}^{\sqrt{c}}(cr-r^3)\,dr\,d\theta = \int_{0}^{2\pi}\frac{c^2}{4}\,d\theta = \frac{c^2\pi}{2} \Rightarrow c^2 = 8 \Rightarrow c = 2\sqrt{2}$,
since $c > 0$

27. The plane $y + 2z = 2$ is the top of the wedge $\Rightarrow I_L = \int_{-2}^{2}\int_{-2}^{4}\int_{-1}^{(2-y)/2}[(y-6)^2+z^2]\,dz\,dy\,dx$
$= \int_{-2}^{2}\int_{-2}^{4}\left[\frac{(y-6)^2(4-y)}{2}+\frac{(2-y)^3}{24}+\frac{1}{3}\right]dy\,dx$; let $t = 2-y \Rightarrow I_L = 4\int_{-2}^{4}\left(\frac{13t^3}{24}+5t^2+16t+\frac{49}{3}\right)dt = 1386$;
$M = \frac{1}{2}(3)(6)(4) = 36$

29. (a) $M = \int_{0}^{2}\int_{0}^{2-x}\int_{0}^{2-x-y} 2x\,dz\,dy\,dx = \int_{0}^{2}\int_{0}^{2-x}(4x-2x^2-2xy)\,dy\,dx = \int_{0}^{2}(x^3-4x^2+4x)\,dx = \frac{4}{3}$
(b) $M_{xy} = \int_{0}^{2}\int_{0}^{2-x}\int_{0}^{2-x-y} 2xz\,dz\,dy\,dx = \int_{0}^{2}\int_{0}^{2-x} x(2-x-y)^2\,dy\,dx = \int_{0}^{2}\frac{x(2-x)^3}{3}\,dx = \frac{8}{15}$; $M_{xz} = \frac{8}{15}$ by symmetry; $M_{yz} = \int_{0}^{2}\int_{0}^{2-x}\int_{0}^{2-x-y} 2x^2\,dz\,dy\,dx = \int_{0}^{2}\int_{0}^{2-x} 2x^2(2-x-y)\,dy\,dx = \int_{0}^{2}(2x-x^2)^2\,dx = \frac{16}{15}$
$\Rightarrow \bar{x} = \frac{4}{5}$, and $\bar{y} = \bar{z} = \frac{2}{5}$

31. (a) $M = \int_{0}^{1}\int_{0}^{1}\int_{0}^{1}(x+y+z+1)\,dz\,dy\,dx = \int_{0}^{1}\int_{0}^{1}\left(x+y+\frac{3}{2}\right)dy\,dx = \int_{0}^{1}(x+2)\,dx = \frac{5}{2}$
(b) $M_{xy} = \int_{0}^{1}\int_{0}^{1}\int_{0}^{1} z(x+y+z+1)\,dz\,dy\,dx = \frac{1}{2}\int_{0}^{1}\int_{0}^{1}\left(x+y+\frac{5}{3}\right)dy\,dx = \frac{1}{2}\int_{0}^{1}\left(x+\frac{13}{6}\right)dx = \frac{4}{3}$
$\Rightarrow M_{xy} = M_{yz} = M_{xz} = \frac{4}{3}$, by symmetry $\Rightarrow \bar{x} = \bar{y} = \bar{z} = \frac{8}{15}$

(c) $I_z = \int_0^1\int_0^1\int_0^1 (x^2+y^2)(x+y+z+1)\,dz\,dy\,dx = \int_0^1\int_0^1 (x^2+y^2)\left(x+y+\frac{3}{2}\right)dy\,dx$
$= \int_0^1 \left(x^3+2x^2+\frac{1}{3}x+\frac{3}{4}\right)dx = \frac{11}{6} \Rightarrow I_x = I_y = I_z = \frac{11}{6}$, by symmetry

33. $M = \int_0^1\int_{z-1}^{1-z}\int_0^{\sqrt{z}} (2y+5)\,dy\,dx\,dz = \int_0^1\int_{z-1}^{1-z}\left(z+5\sqrt{z}\right)dx\,dz = \int_0^1 2\left(z+5\sqrt{z}\right)(1-z)\,dz$
$= 2\int_0^1 \left(5z^{1/2}+z-5z^{3/2}-z^2\right)dz = 2\left[\frac{10}{3}z^{3/2}+\frac{1}{2}z^2-2z^{5/2}-\frac{1}{3}z^3\right]_0^1 = 2\left(\frac{9}{3}-\frac{3}{2}\right) = 3$

35. (a) $\overline{x} = \frac{M_{yz}}{M} = 0 \Rightarrow \iiint\limits_R x\delta(x,y,z)\,dx\,dy\,dz = 0 \Rightarrow M_{yz} = 0$

(b) $I_L = \iiint\limits_D |\mathbf{v}-h\mathbf{i}|^2\,dm = \iiint\limits_D |(x-h)\mathbf{i}+y\mathbf{j}|^2\,dm = \iiint\limits_D (x^2-2xh+h^2+y^2)\,dm$
$= \iiint\limits_D (x^2+y^2)\,dm - 2h\iiint\limits_D x\,dm + h^2\iiint\limits_D dm = I_x - 0 + h^2m = I_{c.m.} + h^2m$

37. (a) $(\overline{x},\overline{y},\overline{z}) = \left(\frac{a}{2},\frac{b}{2},\frac{c}{2}\right) \Rightarrow I_z = I_{c.m.} + abc\left(\sqrt{\frac{a^2}{4}+\frac{b^2}{4}}\right)^2 \Rightarrow I_{c.m.} = I_z - \frac{abc\,(a^2+b^2)}{4}$
$= \frac{abc\,(a^2+b^2)}{3} - \frac{abc\,(a^2+b^2)}{4} = \frac{abc\,(a^2+b^2)}{12}$; $R_{c.m.} = \sqrt{\frac{I_{c.m.}}{M}} = \sqrt{\frac{a^2+b^2}{12}}$

(b) $I_L = I_{c.m.} + abc\left(\sqrt{\frac{a^2}{4}+\left(\frac{b}{2}-2b\right)^2}\right)^2 = \frac{abc\,(a^2+b^2)}{12} + \frac{abc\,(a^2+9b^2)}{4} = \frac{abc\,(4a^2+28b^2)}{12}$
$= \frac{abc\,(a^2+7b^2)}{3}$; $R_L = \sqrt{\frac{I_L}{M}} = \sqrt{\frac{a^2+7b^2}{3}}$

15.7 TRIPLE INTEGRALS IN CYLINDRICAL AND SPHERICAL COORDINATES

1. $\int_0^{2\pi}\int_0^1\int_r^{\sqrt{2-r^2}} dz\,r\,dr\,d\theta = \int_0^{2\pi}\int_0^1 \left[r\left(2-r^2\right)^{1/2}-r^2\right]dr\,d\theta = \int_0^{2\pi}\left[-\frac{1}{3}\left(2-r^2\right)^{3/2}-\frac{r^3}{3}\right]_0^1 d\theta$
$= \int_0^{2\pi}\left(\frac{2^{3/2}}{3}-\frac{2}{3}\right)d\theta = \frac{4\pi\left(\sqrt{2}-1\right)}{3}$

3. $\int_0^{2\pi}\int_0^{\theta/2\pi}\int_0^{3+24r^2} dz\,r\,dr\,d\theta = \int_0^{2\pi}\int_0^{\theta/2\pi}(3r+24r^3)\,dr\,d\theta = \int_0^{2\pi}\left[\frac{3}{2}r^2+6r^4\right]_0^{\theta/2\pi}d\theta = \frac{3}{2}\int_0^{2\pi}\left(\frac{\theta^2}{4\pi^2}+\frac{4\theta^4}{16\pi^4}\right)d\theta$
$= \frac{3}{2}\left[\frac{\theta^3}{12\pi^2}+\frac{\theta^5}{20\pi^4}\right]_0^{2\pi} = \frac{17\pi}{5}$

5. $\int_0^{2\pi}\int_0^1\int_r^{(2-r^2)^{-1/2}} 3\,dz\,r\,dr\,d\theta = 3\int_0^{2\pi}\int_0^1\left[r\left(2-r^2\right)^{-1/2}-r^2\right]dr\,d\theta = 3\int_0^{2\pi}\left[-\left(2-r^2\right)^{1/2}-\frac{r^3}{3}\right]_0^1 d\theta$
$= 3\int_0^{2\pi}\left(\sqrt{2}-\frac{4}{3}\right)d\theta = \pi\left(6\sqrt{2}-8\right)$

7. $\int_0^{2\pi}\int_0^3\int_0^{z/3} r^3\,dr\,dz\,d\theta = \int_0^{2\pi}\int_0^3 \frac{z^4}{324}\,dz\,d\theta = \int_0^{2\pi}\frac{3}{20}\,d\theta = \frac{3\pi}{10}$

9. $\int_0^1\int_0^{\sqrt{z}}\int_0^{2\pi} (r^2\cos^2\theta+z^2)\,r\,d\theta\,dr\,dz = \int_0^1\int_0^{\sqrt{z}}\left[\frac{r^2\theta}{2}+\frac{r^2\sin 2\theta}{4}+z^2\theta\right]_0^{2\pi} r\,dr\,dz = \int_0^1\int_0^{\sqrt{z}}(\pi r^3+2\pi rz^2)\,dr\,dz$
$= \int_0^1\left[\frac{\pi r^4}{4}+\pi r^2z^2\right]_0^{\sqrt{z}}dz = \int_0^1\left(\frac{\pi z^2}{4}+\pi z^3\right)dz = \left[\frac{\pi z^3}{12}+\frac{\pi z^4}{4}\right]_0^1 = \frac{\pi}{3}$

11. (a) $\int_0^{2\pi}\int_0^1\int_0^{\sqrt{4-r^2}} dz\, r\, dr\, d\theta$

(b) $\int_0^{2\pi}\int_0^{\sqrt{3}}\int_0^1 r\, dr\, dz\, d\theta + \int_0^{2\pi}\int_{\sqrt{3}}^2\int_0^{\sqrt{4-z^2}} r\, dr\, dz\, d\theta$

(c) $\int_0^1\int_0^{\sqrt{4-r^2}}\int_0^{2\pi} r\, d\theta\, dz\, dr$

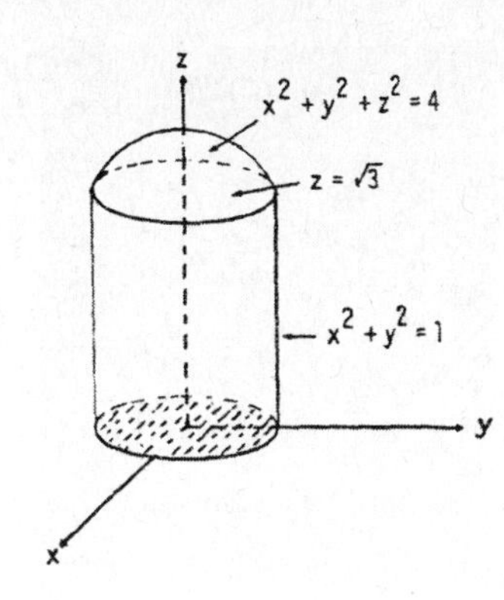

13. $\int_{-\pi/2}^{\pi/2}\int_0^{\cos\theta}\int_0^{3r^2} f(r,\theta,z)\, dz\, r\, dr\, d\theta$

15. $\int_0^{\pi}\int_0^{2\sin\theta}\int_0^{4-r\sin\theta} f(r,\theta,z)\, dz\, r\, dr\, d\theta$

17. $\int_{-\pi/2}^{\pi/2}\int_1^{1+\cos\theta}\int_0^4 f(r,\theta,z)\, dz\, r\, dr\, d\theta$

19. $\int_0^{\pi/4}\int_0^{\sec\theta}\int_0^{2-r\sin\theta} f(r,\theta,z)\, dz\, r\, dr\, d\theta$

21. $\int_0^{\pi}\int_0^{\pi}\int_0^{2\sin\phi} \rho^2\sin\phi\, d\rho\, d\phi\, d\theta = \frac{8}{3}\int_0^{\pi}\int_0^{\pi}\sin^4\phi\, d\phi\, d\theta = \frac{8}{3}\int_0^{\pi}\left(\left[-\frac{\sin^3\phi\cos\phi}{4}\right]_0^{\pi} + \frac{3}{4}\int_0^{\pi}\sin^2\phi\, d\phi\right) d\theta$
$= 2\int_0^{\pi}\int_0^{\pi}\sin^2\phi\, d\phi\, d\theta = \int_0^{\pi}\left[\theta - \frac{\sin 2\theta}{2}\right]_0^{\pi} d\theta = \int_0^{\pi}\pi\, d\theta = \pi^2$

23. $\int_0^{2\pi}\int_0^{\pi}\int_0^{(1-\cos\phi)/2} \rho^2\sin\phi\, d\rho\, d\phi\, d\theta = \frac{1}{24}\int_0^{2\pi}\int_0^{\pi}(1-\cos\phi)^3\sin\phi\, d\phi\, d\theta = \frac{1}{96}\int_0^{2\pi}\left[(1-\cos\phi)^4\right]_0^{\pi} d\theta$
$= \frac{1}{96}\int_0^{2\pi}(2^4-0)\, d\theta = \frac{16}{96}\int_0^{2\pi} d\theta = \frac{1}{6}(2\pi) = \frac{\pi}{3}$

25. $\int_0^{2\pi}\int_0^{\pi/3}\int_{\sec\phi}^2 3\rho^2\sin\phi\, d\rho\, d\phi\, d\theta = \int_0^{2\pi}\int_0^{\pi/3}(8-\sec^3\phi)\sin\phi\, d\phi\, d\theta = \int_0^{2\pi}\left[-8\cos\phi - \frac{1}{2}\sec^2\phi\right]_0^{\pi/3} d\theta$
$= \int_0^{2\pi}\left[(-4-2) - \left(-8-\frac{1}{2}\right)\right] d\theta = \frac{5}{2}\int_0^{2\pi} d\theta = 5\pi$

27. $\int_0^2\int_{-\pi}^0\int_{\pi/4}^{\pi/2}\rho^3\sin 2\phi\, d\phi\, d\theta\, d\rho = \int_0^2\int_{-\pi}^0\rho^3\left[-\frac{\cos 2\phi}{2}\right]_{\pi/4}^{\pi/2} d\theta\, d\rho = \int_0^2\int_{-\pi}^0\frac{\rho^3}{2}\, d\theta\, d\rho = \int_0^2\frac{\rho^3\pi}{2}\, d\rho = \left[\frac{\pi\rho^4}{8}\right]_0^2 = 2\pi$

29. $\int_0^1\int_0^{\pi}\int_0^{\pi/4} 12\rho\sin^3\phi\, d\phi\, d\theta\, d\rho = \int_0^1\int_0^{\pi}\left(12\rho\left[\frac{-\sin^2\phi\cos\phi}{3}\right]_0^{\pi/4} + 8\rho\int_0^{\pi/4}\sin\phi\, d\phi\right) d\theta\, d\rho$
$= \int_0^1\int_0^{\pi}\left(-\frac{2\rho}{\sqrt{2}} - 8\rho\left[\cos\phi\right]_0^{\pi/4}\right) d\theta\, d\rho = \int_0^1\int_0^{\pi}\left(8\rho - \frac{10\rho}{\sqrt{2}}\right) d\theta\, d\rho = \pi\int_0^1\left(8\rho - \frac{10\rho}{\sqrt{2}}\right) d\rho = \pi\left[4\rho^2 - \frac{5\rho^2}{\sqrt{2}}\right]_0^1$
$= \frac{\left(4\sqrt{2}-5\right)\pi}{\sqrt{2}}$

31. (a) $x^2+y^2=1 \Rightarrow \rho^2\sin^2\phi = 1$, and $\rho\sin\phi = 1 \Rightarrow \rho = \csc\phi$; thus
$\int_0^{2\pi}\int_0^{\pi/6}\int_0^2\rho^2\sin\phi\, d\rho\, d\phi\, d\theta + \int_0^{2\pi}\int_{\pi/6}^{\pi/2}\int_0^{\csc\phi}\rho^2\sin\phi\, d\rho\, d\phi\, d\theta$

(b) $\int_0^{2\pi}\int_1^2\int_{\pi/6}^{\sin^{-1}(1/\rho)}\rho^2\sin\phi\, d\phi\, d\rho\, d\theta + \int_0^{2\pi}\int_0^2\int_0^{\pi/6}\rho^2\sin\phi\, d\phi\, d\rho\, d\theta$

33. $V = \int_0^{2\pi}\int_0^{\pi/2}\int_{\cos\phi}^2\rho^2\sin\phi\, d\rho\, d\phi\, d\theta = \frac{1}{3}\int_0^{2\pi}\int_0^{\pi/2}(8-\cos^3\phi)\sin\phi\, d\phi\, d\theta$
$= \frac{1}{3}\int_0^{2\pi}\left[-8\cos\phi + \frac{\cos^4\phi}{4}\right]_0^{\pi/2} d\theta = \frac{1}{3}\int_0^{2\pi}\left(8-\frac{1}{4}\right) d\theta = \left(\frac{31}{12}\right)(2\pi) = \frac{31\pi}{6}$

35. $V = \int_0^{2\pi}\int_0^{\pi}\int_0^{1-\cos\phi} \rho^2 \sin\phi \, d\rho \, d\phi \, d\theta = \frac{1}{3}\int_0^{2\pi}\int_0^{\pi} (1-\cos\phi)^3 \sin\phi \, d\phi \, d\theta = \frac{1}{3}\int_0^{2\pi}\left[\frac{(1-\cos\phi)^4}{4}\right]_0^{\pi} d\theta$
$= \frac{1}{12}(2)^4 \int_0^{2\pi} d\theta = \frac{4}{3}(2\pi) = \frac{8\pi}{3}$

37. $V = \int_0^{2\pi}\int_{\pi/4}^{\pi/2}\int_0^{2\cos\phi} \rho^2 \sin\phi \, d\rho \, d\phi \, d\theta = \frac{8}{3}\int_0^{2\pi}\int_{\pi/4}^{\pi/2} \cos^3\phi \sin\phi \, d\phi \, d\theta = \frac{8}{3}\int_0^{2\pi}\left[-\frac{\cos^4\phi}{4}\right]_{\pi/4}^{\pi/2} d\theta$
$= \left(\frac{8}{3}\right)\left(\frac{1}{16}\right)\int_0^{2\pi} d\theta = \frac{1}{6}(2\pi) = \frac{\pi}{3}$

39. (a) $8\int_0^{\pi/2}\int_0^{\pi/2}\int_0^2 \rho^2 \sin\phi \, d\rho \, d\phi \, d\theta$ (b) $8\int_0^{\pi/2}\int_0^2\int_0^{\sqrt{4-r^2}} dz \, r \, dr \, d\theta$
(c) $8\int_0^2\int_0^{\sqrt{4-x^2}}\int_0^{\sqrt{4-x^2-y^2}} dz \, dy \, dx$

41. (a) $V = \int_0^{2\pi}\int_0^{\pi/3}\int_{\sec\phi}^2 \rho^2 \sin\phi \, d\rho \, d\phi \, d\theta$ (b) $V = \int_0^{2\pi}\int_0^{\sqrt{3}}\int_1^{\sqrt{4-r^2}} dz \, r \, dr \, d\theta$
(c) $V = \int_{-\sqrt{3}}^{\sqrt{3}}\int_{-\sqrt{3-x^2}}^{\sqrt{3-x^2}}\int_1^{\sqrt{4-x^2-y^2}} dz \, dy \, dx$
(d) $V = \int_0^{2\pi}\int_0^{\sqrt{3}}\left[r(4-r^2)^{1/2} - r\right] dr \, d\theta = \int_0^{2\pi}\left[-\frac{(4-r^2)^{3/2}}{3} - \frac{r^2}{2}\right]_0^{\sqrt{3}} d\theta = \int_0^{2\pi}\left(-\frac{1}{3} - \frac{3}{2} + \frac{4^{3/2}}{3}\right) d\theta$
$= \frac{5}{6}\int_0^{2\pi} d\theta = \frac{5\pi}{3}$

43. $V = 4\int_0^{\pi/2}\int_0^1\int_{r^4-1}^{4-4r^2} dz \, r \, dr \, d\theta = 4\int_0^{\pi/2}\int_0^1 (5r - 4r^3 - r^5) \, dr \, d\theta = 4\int_0^{\pi/2}\left(\frac{5}{2} - 1 - \frac{1}{6}\right) d\theta = 4\int_0^{\pi/2} d\theta = \frac{8\pi}{3}$

45. $V = \int_{3\pi/2}^{2\pi}\int_0^{3\cos\theta}\int_0^{-r\sin\theta} dz \, r \, dr \, d\theta = \int_{3\pi/2}^{2\pi}\int_0^{3\cos\theta} -r^2 \sin\theta \, dr \, d\theta = \int_{3\pi/2}^{2\pi} (-9\cos^3\theta)(\sin\theta) \, d\theta = \left[\frac{9}{4}\cos^4\theta\right]_{3\pi/2}^{2\pi}$
$= \frac{9}{4} - 0 = \frac{9}{4}$

47. $V = \int_0^{\pi/2}\int_0^{\sin\theta}\int_0^{\sqrt{1-r^2}} dz \, r \, dr \, d\theta = \int_0^{\pi/2}\int_0^{\sin\theta} r\sqrt{1-r^2} \, dr \, d\theta = \int_0^{\pi/2}\left[-\frac{1}{3}(1-r^2)^{3/2}\right]_0^{\sin\theta} d\theta$
$= -\frac{1}{3}\int_0^{\pi/2}\left[(1-\sin^2\theta)^{3/2} - 1\right] d\theta = -\frac{1}{3}\int_0^{\pi/2}(\cos^3\theta - 1) \, d\theta = -\frac{1}{3}\left(\left[\frac{\cos^2\theta \sin\theta}{3}\right]_0^{\pi/2} + \frac{2}{3}\int_0^{\pi/2}\cos\theta \, d\theta\right) + \left[\frac{\theta}{3}\right]_0^{\pi/2}$
$= -\frac{2}{9}\left[\sin\theta\right]_0^{\pi/2} + \frac{\pi}{6} = \frac{-4+3\pi}{18}$

49. $V = \int_0^{2\pi}\int_{\pi/3}^{2\pi/3}\int_0^a \rho^2 \sin\phi \, d\rho \, d\phi \, d\theta = \int_0^{2\pi}\int_{\pi/3}^{2\pi/3} \frac{a^3}{3}\sin\phi \, d\phi \, d\theta = \frac{a^3}{3}\int_0^{2\pi}\left[-\cos\phi\right]_{\pi/3}^{2\pi/3} d\theta = \frac{a^3}{3}\int_0^{2\pi}\left(\frac{1}{2} + \frac{1}{2}\right) d\theta = \frac{2\pi a^3}{3}$

51. $V = \int_0^{2\pi}\int_0^{\pi/3}\int_{\sec\phi}^2 \rho^2 \sin\phi \, d\rho \, d\phi \, d\theta$
$= \frac{1}{3}\int_0^{2\pi}\int_0^{\pi/3}(8\sin\phi - \tan\phi \sec^2\phi) \, d\phi \, d\theta$
$= \frac{1}{3}\int_0^{2\pi}\left[-8\cos\phi - \frac{1}{2}\tan^2\phi\right]_0^{\pi/3} d\theta$
$= \frac{1}{3}\int_0^{2\pi}\left[-4 - \frac{1}{2}(3) + 8\right] d\theta = \frac{1}{3}\int_0^{2\pi}\frac{5}{2} \, d\theta = \frac{5}{6}(2\pi) = \frac{5\pi}{3}$

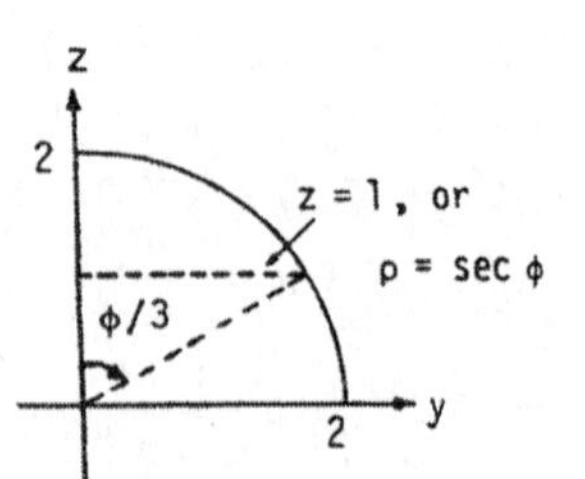

53. $V = 4\int_0^{\pi/2}\int_0^1\int_0^{r^2} dz \, r \, dr \, d\theta = 4\int_0^{\pi/2}\int_0^1 r^3 \, dr \, d\theta = \int_0^{\pi/2} d\theta = \frac{\pi}{2}$

55. $V = 8\int_0^{\pi/2}\int_1^{\sqrt{2}}\int_0^r dz \, r \, dr \, d\theta = 8\int_0^{\pi/2}\int_1^{\sqrt{2}} r^2 \, dr \, d\theta = 8\left(\frac{2\sqrt{2}-1}{3}\right)\int_0^{\pi/2} d\theta = \frac{4\pi\left(2\sqrt{2}-1\right)}{3}$

57. $V = \int_0^{2\pi}\int_0^2\int_0^{4-r\sin\theta} dz\, r\, dr\, d\theta = \int_0^{2\pi}\int_0^2 (4r - r^2\sin\theta)\, dr\, d\theta = 8\int_0^{2\pi}\left(1 - \frac{\sin\theta}{3}\right) d\theta = 16\pi$

59. The paraboloids intersect when $4x^2 + 4y^2 = 5 - x^2 - y^2 \Rightarrow x^2 + y^2 = 1$ and $z = 4$
$\Rightarrow V = 4\int_0^{\pi/2}\int_0^1\int_{4r^2}^{5-r^2} dz\, r\, dr\, d\theta = 4\int_0^{\pi/2}\int_0^1 (5r - 5r^3)\, dr\, d\theta = 20\int_0^{\pi/2}\left[\frac{r^2}{2} - \frac{r^4}{4}\right]_0^1 d\theta = 5\int_0^{\pi/2} d\theta = \frac{5\pi}{2}$

61. $V = 8\int_0^{2\pi}\int_0^1\int_0^{\sqrt{4-r^2}} dz\, r\, dr\, d\theta = 8\int_0^{2\pi}\int_0^1 r\,(4 - r^2)^{1/2}\, dr\, d\theta = 8\int_0^{2\pi}\left[-\frac{1}{3}(4 - r^2)^{3/2}\right]_0^1 d\theta$
$= -\frac{8}{3}\int_0^{2\pi}\left(3^{3/2} - 8\right) d\theta = \frac{4\pi\left(8 - 3\sqrt{3}\right)}{3}$

63. $\text{average} = \frac{1}{2\pi}\int_0^{2\pi}\int_0^1\int_{-1}^1 r^2\, dz\, dr\, d\theta = \frac{1}{2\pi}\int_0^{2\pi}\int_0^1 2r^2\, dr\, d\theta = \frac{1}{3\pi}\int_0^{2\pi} d\theta = \frac{2}{3}$

65. $\text{average} = \frac{1}{\left(\frac{4\pi}{3}\right)}\int_0^{2\pi}\int_0^{\pi}\int_0^1 \rho^3\sin\phi\, d\rho\, d\phi\, d\theta = \frac{3}{16\pi}\int_0^{2\pi}\int_0^{\pi}\sin\phi\, d\phi\, d\theta = \frac{3}{8\pi}\int_0^{2\pi} d\theta = \frac{3}{4}$

67. $M = 4\int_0^{\pi/2}\int_0^1\int_0^r dz\, r\, dr\, d\theta = 4\int_0^{\pi/2}\int_0^1 r^2\, dr\, d\theta = \frac{4}{3}\int_0^{\pi/2} d\theta = \frac{2\pi}{3}$; $M_{xy} = \int_0^{2\pi}\int_0^1\int_0^r z\, dz\, r\, dr\, d\theta$
$= \frac{1}{2}\int_0^{2\pi}\int_0^1 r^3\, dr\, d\theta = \frac{1}{8}\int_0^{2\pi} d\theta = \frac{\pi}{4} \Rightarrow \overline{z} = \frac{M_{xy}}{M} = \left(\frac{\pi}{4}\right)\left(\frac{3}{2\pi}\right) = \frac{3}{8}$, and $\overline{x} = \overline{y} = 0$, by symmetry

69. $M = \frac{8\pi}{3}$; $M_{xy} = \int_0^{2\pi}\int_{\pi/3}^{\pi/2}\int_0^2 z\rho^2\sin\phi\, d\rho\, d\phi\, d\theta = \int_0^{2\pi}\int_{\pi/3}^{\pi/2}\int_0^2 \rho^3\cos\phi\sin\phi\, d\rho\, d\phi\, d\theta = 4\int_0^{2\pi}\int_{\pi/3}^{\pi/2}\cos\phi\sin\phi\, d\phi\, d\theta$
$= 4\int_0^{2\pi}\left[\frac{\sin^2\phi}{2}\right]_{\pi/3}^{\pi/2} d\theta = 4\int_0^{2\pi}\left(\frac{1}{2} - \frac{3}{8}\right) d\theta = \frac{1}{2}\int_0^{2\pi} d\theta = \pi \Rightarrow \overline{z} = \frac{M_{xy}}{M} = (\pi)\left(\frac{3}{8\pi}\right) = \frac{3}{8}$, and $\overline{x} = \overline{y} = 0$, by symmetry

71. $M = \int_0^{2\pi}\int_0^4\int_0^{\sqrt{r}} dz\, r\, dr\, d\theta = \int_0^{2\pi}\int_0^4 r^{3/2}\, dr\, d\theta = \frac{64}{5}\int_0^{2\pi} d\theta = \frac{128\pi}{5}$; $M_{xy} = \int_0^{2\pi}\int_0^4\int_0^{\sqrt{r}} z\, dz\, r\, dr\, d\theta$
$= \frac{1}{2}\int_0^{2\pi}\int_0^4 r^2\, dr\, d\theta = \frac{32}{3}\int_0^{2\pi} d\theta = \frac{64\pi}{3} \Rightarrow \overline{z} = \frac{M_{xy}}{M} = \frac{5}{6}$, and $\overline{x} = \overline{y} = 0$, by symmetry

73. We orient the cone with its vertex at the origin and axis along the z-axis $\Rightarrow \phi = \frac{\pi}{4}$. We use the the x-axis which is through the vertex and parallel to the base of the cone $\Rightarrow I_x = \int_0^{2\pi}\int_0^1\int_r^1 \left(r^2\sin^2\theta + z^2\right) dz\, r\, dr\, d\theta$
$= \int_0^{2\pi}\int_0^1 \left(r^3\sin^2\theta - r^4\sin^2\theta + \frac{r}{3} - \frac{r^4}{3}\right) dr\, d\theta = \int_0^{2\pi}\left(\frac{\sin^2\theta}{20} + \frac{1}{10}\right) d\theta = \left[\frac{\theta}{40} - \frac{\sin 2\theta}{80} + \frac{\theta}{10}\right]_0^{2\pi} = \frac{\pi}{20} + \frac{\pi}{5} = \frac{\pi}{4}$

75. $I_z = \int_0^{2\pi}\int_0^a\int_{\left(\frac{h}{a}\right)r}^h \left(x^2 + y^2\right) dz\, r\, dr\, d\theta = \int_0^{2\pi}\int_0^a \int_{\frac{hr}{a}}^{h} r^3\, dz\, dr\, d\theta = \int_0^{2\pi}\int_0^a \left(hr^3 - \frac{hr^4}{a}\right) dr\, d\theta = \int_0^{2\pi} h\left[\frac{r^4}{4} - \frac{r^5}{5a}\right]_0^a d\theta$
$= \int_0^{2\pi} h\left(\frac{a^4}{4} - \frac{a^5}{5a}\right) d\theta = \frac{ha^4}{20}\int_0^{2\pi} d\theta = \frac{\pi ha^4}{10}$

77. (a) $M = \int_0^{2\pi}\int_0^1\int_r^1 z\, dz\, r\, dr\, d\theta = \frac{1}{2}\int_0^{2\pi}\int_0^1 (r - r^3)\, dr\, d\theta = \frac{1}{8}\int_0^{2\pi} d\theta = \frac{\pi}{4}$; $M_{xy} = \int_0^{2\pi}\int_0^1\int_r^1 z^2\, dz\, r\, dr\, d\theta$
$= \frac{1}{3}\int_0^{2\pi}\int_0^1 (r - r^4)\, dr\, d\theta = \frac{1}{10}\int_0^{2\pi} d\theta = \frac{\pi}{5} \Rightarrow \overline{z} = \frac{4}{5}$, and $\overline{x} = \overline{y} = 0$, by symmetry; $I_z = \int_0^{2\pi}\int_0^1\int_r^1 zr^3\, dz\, dr\, d\theta$
$= \frac{1}{2}\int_0^{2\pi}\int_0^1 (r^3 - r^5)\, dr\, d\theta = \frac{1}{24}\int_0^{2\pi} d\theta = \frac{\pi}{12}$

(b) $M = \int_0^{2\pi}\int_0^1\int_r^1 z^2\, dz\, r\, dr\, d\theta = \frac{\pi}{5}$ from part (a); $M_{xy} = \int_0^{2\pi}\int_0^1\int_r^1 z^3\, dz\, r\, dr\, d\theta = \frac{1}{4}\int_0^{2\pi}\int_0^1 (r - r^5)\, dr\, d\theta$
$= \frac{1}{12}\int_0^{2\pi} d\theta = \frac{\pi}{6} \Rightarrow \overline{z} = \frac{5}{6}$, and $\overline{x} = \overline{y} = 0$, by symmetry; $I_z = \int_0^{2\pi}\int_0^1\int_r^1 z^2r^3\, dz\, dr\, d\theta = \frac{1}{3}\int_0^{2\pi}\int_0^1 (r^3 - r^6)\, dr\, d\theta$
$= \frac{1}{28}\int_0^{2\pi} d\theta = \frac{\pi}{14}$

79. $M = \int_0^{2\pi}\int_0^a\int_0^{\frac{h}{a}\sqrt{a^2-r^2}} dz\, r\, dr\, d\theta = \int_0^{2\pi}\int_0^a \frac{h}{a} r\sqrt{a^2-r^2}\, dr\, d\theta = \frac{h}{a}\int_0^{2\pi}\left[-\frac{1}{3}(a^2-r^2)^{3/2}\right]_0^a d\theta$
$= \frac{h}{a}\int_0^{2\pi}\frac{a^3}{3}\, d\theta = \frac{2ha^2\pi}{3}$; $M_{xy} = \int_0^{2\pi}\int_0^a\int_0^{\frac{h}{a}\sqrt{a^2-r^2}} z\, dz\, r\, dr\, d\theta = \frac{h^2}{2a^2}\int_0^{2\pi}\int_0^a (a^2r - r^3)\, dr\, d\theta$
$= \frac{h^2}{2a^2}\int_0^{2\pi}\left(\frac{a^4}{2} - \frac{a^4}{4}\right) d\theta = \frac{a^2h^2\pi}{4} \Rightarrow \bar{z} = \left(\frac{\pi a^2h^2}{4}\right)\left(\frac{3}{2ha^2\pi}\right) = \frac{3}{8}h$, and $\bar{x} = \bar{y} = 0$, by symmetry

81. The density distribution function is linear so it has the form $\delta(\rho) = k\rho + C$, where ρ is the distance from the center of the planet. Now, $\delta(R) = 0 \Rightarrow kR + C = 0$, and $\delta(\rho) = k\rho - kR$. It remains to determine the constant k: $M = \int_0^{2\pi}\int_0^{\pi}\int_0^R (k\rho - kR)\,\rho^2 \sin\phi\, d\rho\, d\phi\, d\theta = \int_0^{2\pi}\int_0^{\pi}\left[k\frac{\rho^4}{4} - kR\frac{\rho^3}{3}\right]_0^R \sin\phi\, d\phi\, d\theta$
$= \int_0^{2\pi}\int_0^{\pi} k\left(\frac{R^4}{4} - \frac{R^4}{3}\right)\sin\phi\, d\phi\, d\theta = \int_0^{2\pi} -\frac{k}{12}R^4\left[-\cos\phi\right]_0^{\pi} d\theta = \int_0^{2\pi} -\frac{k}{6}R^4\, d\theta = -\frac{k\pi R^4}{3} \Rightarrow k = -\frac{3M}{\pi R^4}$
$\Rightarrow \delta(\rho) = -\frac{3M}{\pi R^4}\rho + \frac{3M}{\pi R^4}R$. At the center of the planet $\rho = 0 \Rightarrow \delta(0) = \left(\frac{3M}{\pi R^4}\right)R = \frac{3M}{\pi R^3}$.

83. (a) A plane perpendicular to the x-axis has the form $x = a$ in rectangular coordinates $\Rightarrow r\cos\theta = a \Rightarrow r = \frac{a}{\cos\theta}$ $\Rightarrow r = a\sec\theta$, in cylindrical coordinates.
(b) A plane perpendicular to the y-axis has the form $y = b$ in rectangular coordinates $\Rightarrow r\sin\theta = b \Rightarrow r = \frac{b}{\sin\theta}$ $\Rightarrow r = b\csc\theta$, in cylindrical coordinates.

85. The equation $r = f(z)$ implies that the point (r, θ, z) $= (f(z), \theta, z)$ will lie on the surface for all θ. In particular $(f(z), \theta + \pi, z)$ lies on the surface whenever $(f(z), \theta, z)$ does $\Rightarrow$ the surface is symmetric with respect to the z-axis.

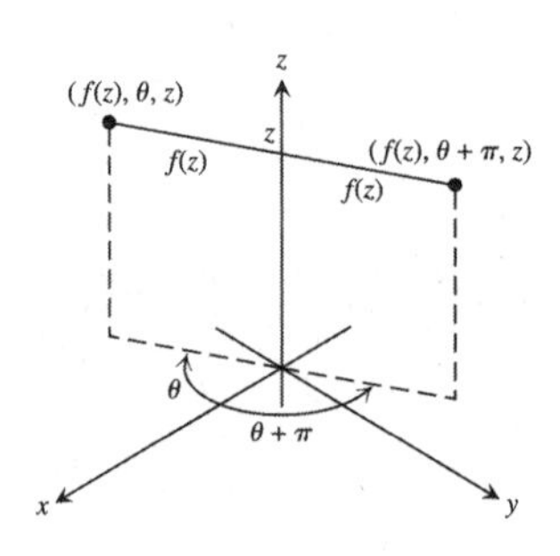

15.8 SUBSTITUTIONS IN MULTIPLE INTEGRALS

1. (a) $x - y = u$ and $2x + y = v \Rightarrow 3x = u + v$ and $y = x - u \Rightarrow x = \frac{1}{3}(u + v)$ and $y = \frac{1}{3}(-2u + v)$;
$\frac{\partial(x,y)}{\partial(u,v)} = \begin{vmatrix} \frac{1}{3} & \frac{1}{3} \\ -\frac{2}{3} & \frac{1}{3} \end{vmatrix} = \frac{1}{9} + \frac{2}{9} = \frac{1}{3}$
(b) The line segment $y = x$ from $(0, 0)$ to $(1, 1)$ is $x - y = 0$ $\Rightarrow u = 0$; the line segment $y = -2x$ from $(0, 0)$ to $(1, -2)$ is $2x + y = 0 \Rightarrow v = 0$; the line segment $x = 1$ from $(1, 1)$ to $(1, -2)$ is $(x - y) + (2x + y) = 3$ $\Rightarrow u + v = 3$. The transformed region is sketched at the right.

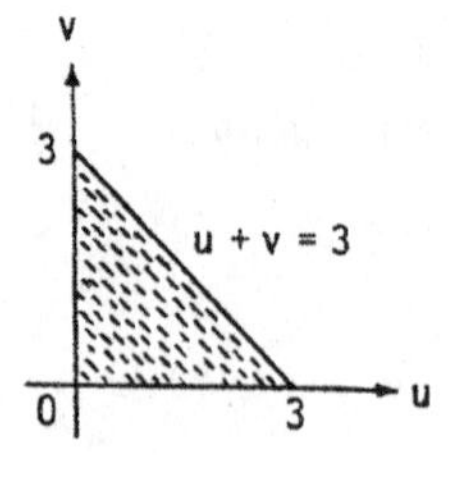

3. (a) $3x + 2y = u$ and $x + 4y = v \Rightarrow -5x = -2u + v$ and $y = \frac{1}{2}(u - 3x) \Rightarrow x = \frac{1}{5}(2u - v)$ and $y = \frac{1}{10}(3v - u)$;
$\frac{\partial(x,y)}{\partial(u,v)} = \begin{vmatrix} \frac{2}{5} & -\frac{1}{5} \\ -\frac{1}{10} & \frac{3}{10} \end{vmatrix} = \frac{6}{50} - \frac{1}{50} = \frac{1}{10}$

(b) The x-axis $y = 0 \Rightarrow u = 3v$; the y-axis $x = 0$ $\Rightarrow v = 2u$; the line $x + y = 1$ $\Rightarrow \frac{1}{5}(2u - v) + \frac{1}{10}(3v - u) = 1$ $\Rightarrow 2(2u - v) + (3v - u) = 10 \Rightarrow 3u + v = 10$. The transformed region is sketched at the right.

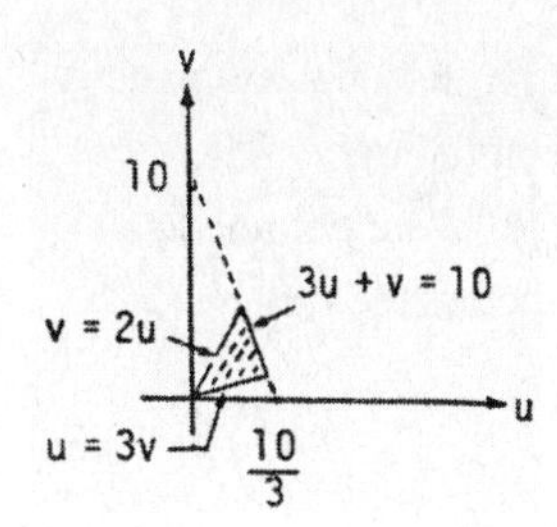

5. $\int_0^4 \int_{y/2}^{(y/2)+1} \left(x - \frac{y}{2}\right) dx\, dy = \int_0^4 \left[\frac{x^2}{2} - \frac{xy}{2}\right]_{\frac{y}{2}}^{\frac{y}{2}+1} dy = \frac{1}{2}\int_0^4 \left[\left(\frac{y}{2}+1\right)^2 - \left(\frac{y}{2}\right)^2 - \left(\frac{y}{2}+1\right)y + \left(\frac{y}{2}\right)y\right] dy$

$= \frac{1}{2}\int_0^4 (y + 1 - y)\, dy = \frac{1}{2}\int_0^4 dy = \frac{1}{2}(4) = 2$

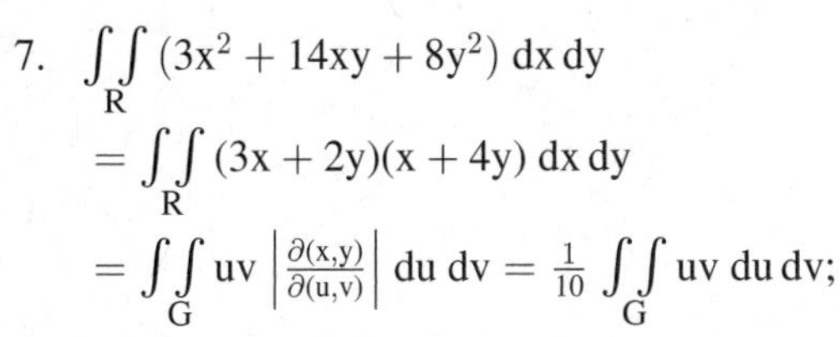

7. $\iint_R (3x^2 + 14xy + 8y^2)\, dx\, dy$

$= \iint_R (3x + 2y)(x + 4y)\, dx\, dy$

$= \iint_G uv \left|\frac{\partial(x,y)}{\partial(u,v)}\right| du\, dv = \frac{1}{10}\iint_G uv\, du\, dv;$

We find the boundaries of G from the boundaries of R, shown in the accompanying figure:

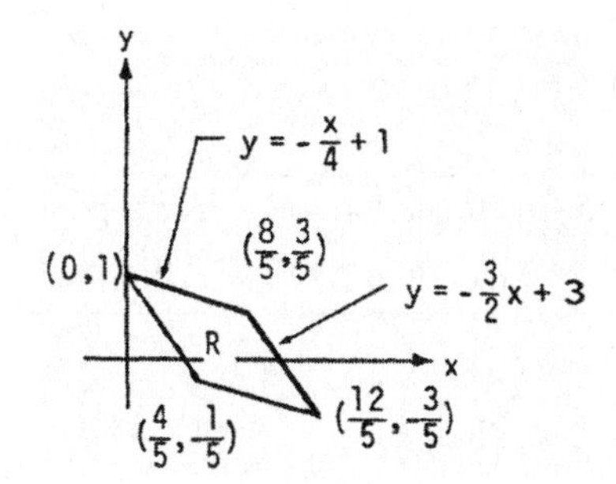

xy-equations for the boundary of R	Corresponding uv-equations for the boundary of G	Simplified uv-equations
$y = -\frac{3}{2}x + 1$	$\frac{1}{10}(3v - u) = -\frac{3}{10}(2u - v) + 1$	$u = 2$
$y = -\frac{3}{2}x + 3$	$\frac{1}{10}(3v - u) = -\frac{3}{10}(2u - v) + 3$	$u = 6$
$y = -\frac{1}{4}x$	$\frac{1}{10}(3v - u) = -\frac{1}{20}(2u - v)$	$v = 0$
$y = -\frac{1}{4}x + 1$	$\frac{1}{10}(3v - u) = -\frac{1}{20}(2u - v) + 1$	$v = 4$

$\Rightarrow \frac{1}{10}\iint_G uv\, du\, dv = \frac{1}{10}\int_2^6 \int_0^4 uv\, dv\, du = \frac{1}{10}\int_2^6 u\left[\frac{v^2}{2}\right]_0^4 du = \frac{4}{5}\int_2^6 u\, du = \left(\frac{4}{5}\right)\left[\frac{u^2}{2}\right]_2^6 = \left(\frac{4}{5}\right)(18 - 2) = \frac{64}{5}$

9. $x = \frac{u}{v}$ and $y = uv \Rightarrow \frac{y}{x} = v^2$ and $xy = u^2$; $\frac{\partial(x,y)}{\partial(u,v)} = J(u,v) = \begin{vmatrix} v^{-1} & -uv^{-2} \\ v & u \end{vmatrix} = v^{-1}u + v^{-1}u = \frac{2u}{v}$;

$y = x \Rightarrow uv = \frac{u}{v} \Rightarrow v = 1$, and $y = 4x \Rightarrow v = 2$; $xy = 1 \Rightarrow u = 1$, and $xy = 9 \Rightarrow u = 3$; thus

$\iint_R \left(\sqrt{\frac{y}{x}} + \sqrt{xy}\right) dx\, dy = \int_1^3 \int_1^2 (v + u)\left(\frac{2u}{v}\right) dv\, du = \int_1^3 \int_1^2 \left(2u + \frac{2u^2}{v}\right) dv\, du = \int_1^3 \left[2uv + 2u^2 \ln v\right]_1^2 du$

$= \int_1^3 (2u + 2u^2 \ln 2)\, du = \left[u^2 + \frac{2}{3}u^2 \ln 2\right]_1^3 = 8 + \frac{2}{3}(26)(\ln 2) = 8 + \frac{52}{3}(\ln 2)$

11. $x = ar\cos\theta$ and $y = ar\sin\theta \Rightarrow \frac{\partial(x,y)}{\partial(r,\theta)} = J(r,\theta) = \begin{vmatrix} a\cos\theta & -ar\sin\theta \\ b\sin\theta & br\cos\theta \end{vmatrix} = abr\cos^2\theta + abr\sin^2\theta = abr;$

$I_0 = \iint_R (x^2 + y^2)\, dA = \int_0^{2\pi}\int_0^1 r^2(a^2\cos^2\theta + b^2\sin^2\theta)\,|J(r,\theta)|\, dr\, d\theta = \int_0^{2\pi}\int_0^1 abr^3(a^2\cos^2\theta + b^2\sin^2\theta)\, dr\, d\theta$

$= \frac{ab}{4}\int_0^{2\pi}(a^2\cos^2\theta + b^2\sin^2\theta)\, d\theta = \frac{ab}{4}\left[\frac{a^2\theta}{2} + \frac{a^2\sin 2\theta}{4} + \frac{b^2\theta}{2} - \frac{b^2\sin 2\theta}{4}\right]_0^{2\pi} = \frac{ab\pi(a^2 + b^2)}{4}$

13. The region of integration R in the xy-plane is sketched in the figure at the right. The boundaries of the image G are obtained as follows, with G sketched at the right:

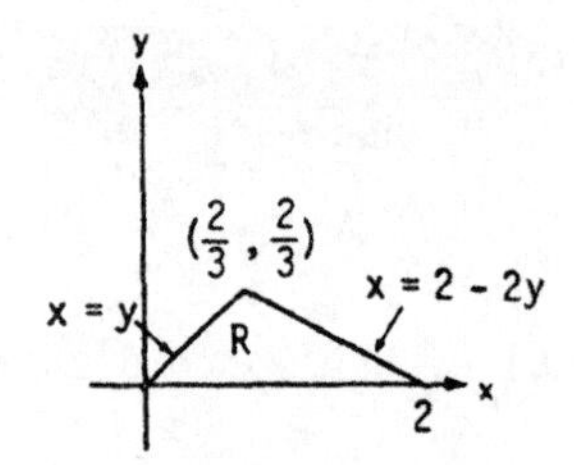

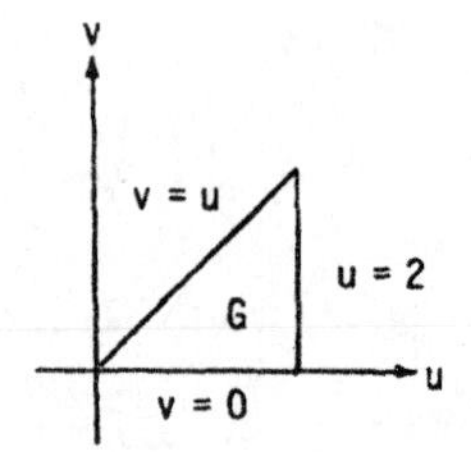

xy-equations for the boundary of R	Corresponding uv-equations for the boundary of G	Simplified uv-equations
$x = y$	$\frac{1}{3}(u + 2v) = \frac{1}{3}(u - v)$	$v = 0$
$x = 2 - 2y$	$\frac{1}{3}(u + 2v) = 2 - \frac{2}{3}(u - v)$	$u = 2$
$y = 0$	$0 = \frac{1}{3}(u - v)$	$v = u$

Also, from Exercise 2, $\frac{\partial(x,y)}{\partial(u,v)} = J(u,v) = -\frac{1}{3} \Rightarrow \int_0^{2/3}\int_y^{2-2y} (x + 2y)\, e^{(y-x)}\, dx\, dy = \int_0^2 \int_0^u ue^{-v} \left|-\frac{1}{3}\right| dv\, du$
$= \frac{1}{3}\int_0^2 u\left[-e^{-v}\right]_0^u du = \frac{1}{3}\int_0^2 u\left(1 - e^{-u}\right) du = \frac{1}{3}\left[u\left(u + e^{-u}\right) - \frac{u^2}{2} + e^{-u}\right]_0^2 = \frac{1}{3}\left[2\left(2 + e^{-2}\right) - 2 + e^{-2} - 1\right]$
$= \frac{1}{3}\left(3e^{-2} + 1\right) \approx 0.4687$

15. $x = \frac{u}{v}$ and $y = uv \Rightarrow \frac{y}{x} = v^2$ and $xy = u^2$; $\frac{\partial(x,y)}{\partial(u,v)} = J(u,v) = \begin{vmatrix} v^{-1} & -uv^{-2} \\ v & u \end{vmatrix} = v^{-1}u + v^{-1}u = \frac{2u}{v}$;
$y = x \Rightarrow uv = \frac{u}{v} \Rightarrow v = 1$, and $y = 4x \Rightarrow v = 2$; $xy = 1 \Rightarrow u = 1$, and $xy = 4 \Rightarrow u = 2$; thus
$\int_1^2\int_{1/y}^{y} \left(x^2 + y^2\right) dx\, dy + \int_2^4 \int_{y/4}^{4/y} \left(x^2 + y^2\right) dx\, dy = \int_1^2\int_1^2 \left(\frac{u^2}{v^2} + u^2v^2\right)\left(\frac{2u}{v}\right) du\, dv = \int_1^2\int_1^2 \left(\frac{2u^3}{v^3} + 2u^3v\right) du\, dv$
$= \int_1^2 \left[\frac{u^4}{2v^3} + \frac{1}{2}u^4v\right]_1^2 dv = \int_1^2 \left(\frac{15}{2v^3} + \frac{15v}{2}\right) dv = \left[-\frac{15}{4v^2} + \frac{15v^2}{4}\right]_1^2 = \frac{225}{16}$

17. (a) $x = u\cos v$ and $y = u \sin v \Rightarrow \frac{\partial(x,y)}{\partial(u,v)} = \begin{vmatrix} \cos v & -u\sin v \\ \sin v & u\cos v \end{vmatrix} = u\cos^2 v + u\sin^2 v = u$

(b) $x = u\sin v$ and $y = u\cos v \Rightarrow \frac{\partial(x,y)}{\partial(u,v)} = \begin{vmatrix} \sin v & u\cos v \\ \cos v & -u\sin v \end{vmatrix} = -u\sin^2 v - u\cos^2 v = -u$

19. $\begin{vmatrix} \sin\phi\cos\theta & \rho\cos\phi\cos\theta & -\rho\sin\phi\sin\theta \\ \sin\phi\sin\theta & \rho\cos\phi\sin\theta & \rho\sin\phi\cos\theta \\ \cos\phi & -\rho\sin\phi & 0 \end{vmatrix}$
$= (\cos\phi)\begin{vmatrix} \rho\cos\phi\cos\theta & -\rho\sin\phi\sin\theta \\ \rho\cos\phi\sin\theta & \rho\sin\phi\cos\theta \end{vmatrix} + (\rho\sin\phi)\begin{vmatrix} \sin\phi\cos\theta & -\rho\sin\phi\sin\theta \\ \sin\phi\sin\theta & \rho\sin\phi\cos\theta \end{vmatrix}$
$= \left(\rho^2\cos\phi\right)\left(\sin\phi\cos\phi\cos^2\theta + \sin\phi\cos\phi\sin^2\theta\right) + \left(\rho^2\sin\phi\right)\left(\sin^2\phi\cos^2\theta + \sin^2\phi\sin^2\theta\right)$
$= \rho^2\sin\phi\cos^2\phi + \rho^2\sin^3\phi = \left(\rho^2\sin\phi\right)\left(\cos^2\phi + \sin^2\phi\right) = \rho^2\sin\phi$

21. $\int_0^3 \int_0^4 \int_{y/2}^{1+(y/2)} \left(\frac{2x-y}{2} + \frac{z}{3}\right) dx\,dy\,dz = \int_0^3 \int_0^4 \left[\frac{x^2}{2} - \frac{xy}{2} + \frac{xz}{3}\right]_{y/2}^{1+(y/2)} dy\,dz = \int_0^3 \int_0^4 \left[\frac{1}{2}(y+1) - \frac{y}{2} + \frac{z}{3}\right] dy\,dz$

$= \int_0^3 \left[\frac{(y+1)^2}{4} - \frac{y^2}{4} + \frac{yz}{3}\right]_0^4 dz = \int_0^3 \left(\frac{9}{4} + \frac{4z}{3} - \frac{1}{4}\right) dz = \int_0^3 \left(2 + \frac{4z}{3}\right) dz = \left[2z + \frac{2z^2}{3}\right]_0^3 = 12$

23. $J(u, v, w) = \begin{vmatrix} a & 0 & 0 \\ 0 & b & 0 \\ 0 & 0 & c \end{vmatrix} = abc$; for R and G as in Exercise 22, $\iiint_R |xyz|\,dx\,dy\,dz$

$= \iiint_G a^2b^2c^2 uvw\,dw\,dv\,du = 8a^2b^2c^2 \int_0^{\pi/2} \int_0^{\pi/2} \int_0^1 (\rho \sin\phi \cos\theta)(\rho \sin\phi \sin\theta)(\rho \cos\phi)\,(\rho^2 \sin\phi)\,d\rho\,d\phi\,d\theta$

$= \frac{4a^2b^2c^2}{3} \int_0^{\pi/2} \int_0^{\pi/2} \sin\theta \cos\theta \sin^3\phi \cos\phi\,d\phi\,d\theta = \frac{a^2b^2c^2}{3} \int_0^{\pi/2} \sin\theta \cos\theta\,d\theta = \frac{a^2b^2c^2}{6}$

25. The first moment about the xy-coordinate plane for the semi-ellipsoid, $\frac{x^2}{a^2} + \frac{y^2}{b^2} + \frac{z^2}{c^2} = 1$ using the transformation in Exercise 23 is, $M_{xy} = \iiint_D z\,dz\,dy\,dx = \iiint_G cw\,|J(u, v, w)|\,du\,dv\,dw$

$= abc^2 \iiint_G w\,du\,dv\,dw = (abc^2) \cdot (M_{xy}$ of the hemisphere $x^2 + y^2 + z^2 = 1, z \geq 0) = \frac{abc^2\pi}{4}$;

the mass of the semi-ellipsoid is $\frac{2abc\pi}{3} \Rightarrow \bar{z} = \left(\frac{abc^2\pi}{4}\right)\left(\frac{3}{2abc\pi}\right) = \frac{3}{8}c$

CHAPTER 15 PRACTICE EXERCISES

1. $\int_1^{10} \int_0^{1/y} ye^{xy}\,dx\,dy = \int_1^{10} \left[e^{xy}\right]_0^{1/y} dy$

$= \int_1^{10} (e - 1)\,dy = 9e - 9$

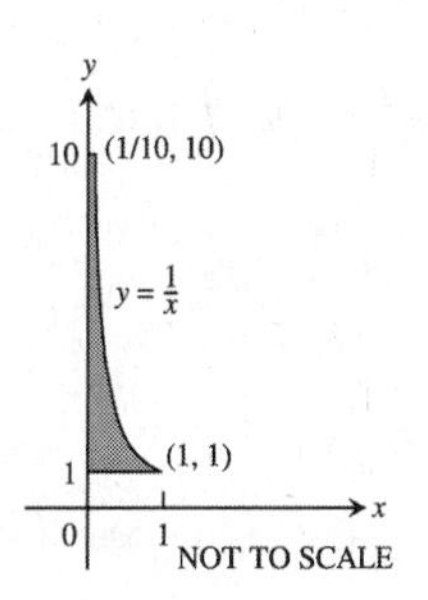

3. $\int_0^{3/2} \int_{-\sqrt{9-4t^2}}^{\sqrt{9-4t^2}} t\,ds\,dt = \int_0^{3/2} \left[ts\right]_{-\sqrt{9-4t^2}}^{\sqrt{9-4t^2}} dt$

$= \int_0^{3/2} 2t\sqrt{9-4t^2}\,dt = \left[-\frac{1}{6}(9-4t^2)^{3/2}\right]_0^{3/2}$

$= -\frac{1}{6}\left(0^{3/2} - 9^{3/2}\right) = \frac{27}{6} = \frac{9}{2}$

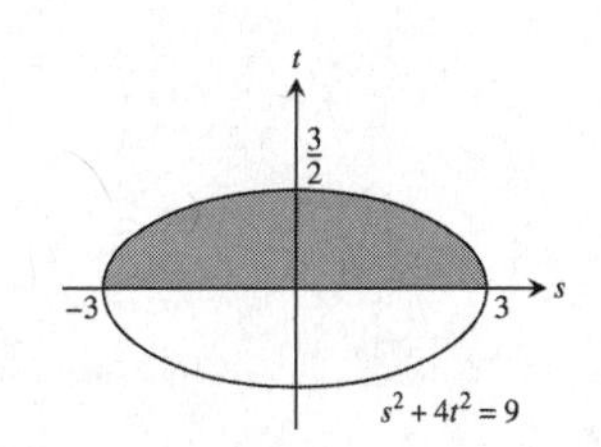

5. $\int_{-2}^{0} \int_{2x+4}^{4-x^2} dy\,dx = \int_{-2}^{0} (-x^2 - 2x)\,dx$

$= \left[-\frac{x^3}{3} - x^2\right]_{-2}^{0} = -\left(\frac{8}{3} - 4\right) = \frac{4}{3}$

$\int_0^4 \int_{-\sqrt{4-y}}^{(y-4)/2} dx\,dy = \int_0^4 \left(\frac{y-4}{2} + \sqrt{4-y}\right) dy$

$= \left[\frac{y^2}{4} - 2y - \frac{2}{3}(4-y)^{3/2}\right]_0^4 = 4 - 8 + \frac{2}{3} \cdot 4^{3/2}$

$= -4 + \frac{16}{3} = \frac{4}{3}$

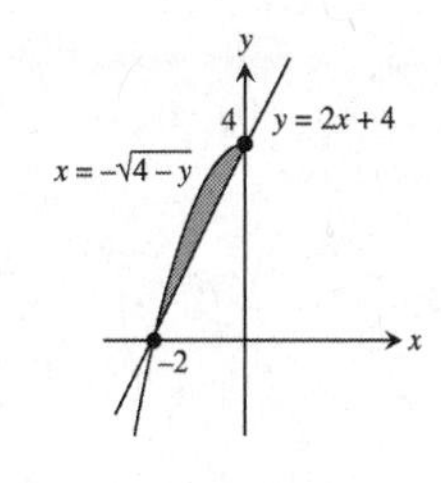

7. $\int_{-3}^{3}\int_{0}^{(1/2)\sqrt{9-x^2}} y\,dy\,dx = \int_{-3}^{3}\left[\frac{y^2}{2}\right]_0^{(1/2)\sqrt{9-x^2}} dx$
$= \int_{-3}^{3}\frac{1}{8}(9-x^2)\,dx = \left[\frac{9x}{8} - \frac{x^3}{24}\right]_{-3}^{3}$
$= \left(\frac{27}{8} - \frac{27}{24}\right) - \left(-\frac{27}{8} + \frac{27}{24}\right) = \frac{27}{6} = \frac{9}{2}$
$\int_0^{3/2}\int_{-\sqrt{9-4y^2}}^{\sqrt{9-4y^2}} y\,dx\,dy = \int_0^{3/2} 2y\sqrt{9-4y^2}\,dy$
$= -\frac{1}{4}\cdot\frac{2}{3}(9-4y^2)^{3/2}\Big|_0^{3/2} = \frac{1}{6}\cdot 9^{3/2} = \frac{27}{6} = \frac{9}{2}$

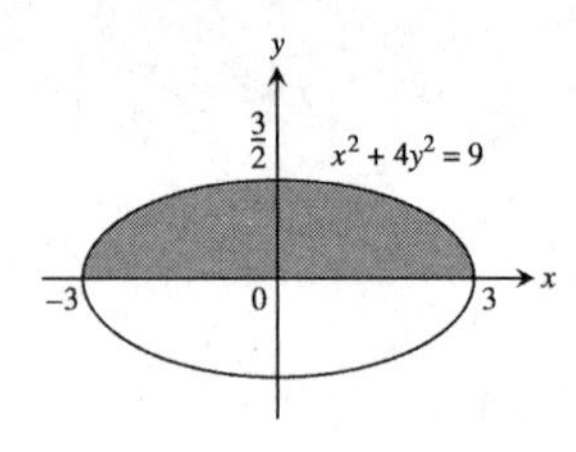

9. $\int_0^1\int_{2y}^2 4\cos(x^2)\,dx\,dy = \int_0^2\int_0^{x/2} 4\cos(x^2)\,dy\,dx = \int_0^2 2x\cos(x^2)\,dx = [\sin(x^2)]_0^2 = \sin 4$

11. $\int_0^8\int_{\sqrt[3]{x}}^2 \frac{1}{y^4+1}\,dy\,dx = \int_0^2\int_0^{y^3}\frac{1}{y^4+1}\,dx\,dy = \frac{1}{4}\int_0^2\frac{4y^3}{y^4+1}\,dy = \frac{\ln 17}{4}$

13. $A = \int_{-2}^{0}\int_{2x+4}^{4-x^2} dy\,dx = \int_{-2}^{0}(-x^2-2x)\,dx = \frac{4}{3}$

15. $V = \int_0^1\int_x^{2-x}(x^2+y^2)\,dy\,dx = \int_0^1\left[x^2y + \frac{y^3}{3}\right]_x^{2-x} dx = \int_0^1\left[2x^2 + \frac{(2-x)^3}{3} - \frac{7x^3}{3}\right] dx = \left[\frac{2x^3}{3} - \frac{(2-x)^4}{12} - \frac{7x^4}{12}\right]_0^1$
$= \left(\frac{2}{3} - \frac{1}{12} - \frac{7}{12}\right) + \frac{2^4}{12} = \frac{4}{3}$

17. average value $= \int_0^1\int_0^1 xy\,dy\,dx = \int_0^1\left[\frac{xy^2}{2}\right]_0^1 dx = \int_0^1\frac{x}{2}\,dx = \frac{1}{4}$

19. $\int_{-1}^{1}\int_{-\sqrt{1-x^2}}^{\sqrt{1-x^2}}\frac{2}{(1+x^2+y^2)^2}\,dy\,dx = \int_0^{2\pi}\int_0^1\frac{2r}{(1+r^2)^2}\,dr\,d\theta = \int_0^{2\pi}\left[-\frac{1}{1+r^2}\right]_0^1 d\theta = \frac{1}{2}\int_0^{2\pi}d\theta = \pi$

21. $(x^2+y^2)^2 - (x^2-y^2) = 0 \Rightarrow r^4 - r^2\cos 2\theta = 0 \Rightarrow r^2 = \cos 2\theta$ so the integral is $\int_{-\pi/4}^{\pi/4}\int_0^{\sqrt{\cos 2\theta}}\frac{r}{(1+r^2)^2}\,dr\,d\theta$
$= \int_{-\pi/4}^{\pi/4}\left[-\frac{1}{2(1+r^2)}\right]_0^{\sqrt{\cos 2\theta}} d\theta = \frac{1}{2}\int_{-\pi/4}^{\pi/4}\left(1 - \frac{1}{1+\cos 2\theta}\right)d\theta = \frac{1}{2}\int_{-\pi/4}^{\pi/4}\left(1 - \frac{1}{2\cos^2\theta}\right)d\theta$
$= \frac{1}{2}\int_{-\pi/4}^{\pi/4}\left(1 - \frac{\sec^2\theta}{2}\right)d\theta = \frac{1}{2}\left[\theta - \frac{\tan\theta}{2}\right]_{-\pi/4}^{\pi/4} = \frac{\pi-2}{4}$

23. $\int_0^{\pi}\int_0^{\pi}\int_0^{\pi}\cos(x+y+z)\,dx\,dy\,dz = \int_0^{\pi}\int_0^{\pi}[\sin(z+y+\pi) - \sin(z+y)]\,dy\,dz$
$= \int_0^{\pi}[-\cos(z+2\pi) + \cos(z+\pi) - \cos z + \cos(z+\pi)]\,dz = 0$

25. $\int_0^1\int_0^{x^2}\int_0^{x+y}(2x-y-z)\,dz\,dy\,dx = \int_0^1\int_0^{x^2}\left(\frac{3x^2}{2} - \frac{3y^2}{2}\right)dy\,dx = \int_0^1\left(\frac{3x^4}{2} - \frac{x^6}{2}\right)dx = \frac{8}{35}$

27. $V = 2\int_0^{\pi/2}\int_{-\cos y}^{0}\int_0^{-2x}dz\,dx\,dy = 2\int_0^{\pi/2}\int_{-\cos y}^{0} -2x\,dx\,dy = 2\int_0^{\pi/2}\cos^2 y\,dy = 2\left[\frac{y}{2} + \frac{\sin 2y}{4}\right]_0^{\pi/2} = \frac{\pi}{2}$

29. average $= \frac{1}{3}\int_0^1\int_0^3\int_0^1 30xz\sqrt{x^2+y}\,dz\,dy\,dx = \frac{1}{3}\int_0^1\int_0^3 15x\sqrt{x^2+y}\,dy\,dx = \frac{1}{3}\int_0^3\int_0^1 15x\sqrt{x^2+y}\,dx\,dy$
$= \frac{1}{3}\int_0^3\left[5(x^2+y)^{3/2}\right]_0^1 dy = \frac{1}{3}\int_0^3\left[5(1+y)^{3/2} - 5y^{3/2}\right]dy = \frac{1}{3}\left[2(1+y)^{5/2} - 2y^{5/2}\right]_0^3 = \frac{1}{3}\left[2(4)^{5/2} - 2(3)^{5/2} - 2\right]$
$= \frac{1}{3}\left[2\left(31 - 3^{5/2}\right)\right]$

31. (a) $\int_{-\sqrt{2}}^{\sqrt{2}}\int_{-\sqrt{2-y^2}}^{\sqrt{2-y^2}}\int_{\sqrt{x^2+y^2}}^{\sqrt{4-x^2-y^2}} 3\,dz\,dx\,dy$

(b) $\int_0^{2\pi}\int_0^{\pi/4}\int_0^{2} 3\rho^2 \sin\phi\,d\rho\,d\phi\,d\theta$

(c) $\int_0^{2\pi}\int_0^{\sqrt{2}}\int_r^{\sqrt{4-r^2}} 3\,dz\,r\,dr\,d\theta = 3\int_0^{2\pi}\int_0^{\sqrt{2}}\left[r(4-r^2)^{1/2}-r^2\right]dr\,d\theta = 3\int_0^{2\pi}\left[-\frac{1}{3}(4-r^2)^{3/2}-\frac{r^3}{3}\right]_0^{\sqrt{2}} d\theta$
$= \int_0^{2\pi}\left(-2^{3/2}-2^{3/2}+4^{3/2}\right)d\theta = \left(8-4\sqrt{2}\right)\int_0^{2\pi} d\theta = 2\pi\left(8-4\sqrt{2}\right)$

33. (a) $\int_0^{2\pi}\int_0^{\pi/4}\int_0^{\sec\phi} \rho^2 \sin\phi\,d\rho\,d\phi\,d\theta$

(b) $\int_0^{2\pi}\int_0^{\pi/4}\int_0^{\sec\phi} \rho^2 \sin\phi\,d\rho\,d\phi\,d\theta = \frac{1}{3}\int_0^{2\pi}\int_0^{\pi/4}(\sec\phi)(\sec\phi\tan\phi)\,d\phi\,d\theta = \frac{1}{3}\int_0^{2\pi}\left[\frac{1}{2}\tan^2\phi\right]_0^{\pi/4} d\theta = \frac{1}{6}\int_0^{2\pi} d\theta = \frac{\pi}{3}$

35. $\int_0^1\int_{\sqrt{1-x^2}}^{\sqrt{3-x^2}}\int_1^{\sqrt{4-x^2-y^2}} z^2yx\,dz\,dy\,dx + \int_1^{\sqrt{3}}\int_0^{\sqrt{3-x^2}}\int_1^{\sqrt{4-x^2-y^2}} z^2yx\,dz\,dy\,dx$

37. (a) $V = \int_0^{2\pi}\int_0^{2}\int_2^{\sqrt{8-r^2}} dz\,r\,dr\,d\theta = \int_0^{2\pi}\int_0^{2}\left(r\sqrt{8-r^2}-2r\right)dr\,d\theta = \int_0^{2\pi}\left[-\frac{1}{3}(8-r^2)^{3/2}-r^2\right]_0^2 d\theta$
$= \int_0^{2\pi}\left[-\frac{1}{3}(4)^{3/2}-4+\frac{1}{3}(8)^{3/2}\right]d\theta = \int_0^{2\pi}\frac{4}{3}\left(-2-3+2\sqrt{8}\right)d\theta = \frac{4}{3}\left(4\sqrt{2}-5\right)\int_0^{2\pi} d\theta = \frac{8\pi\left(4\sqrt{2}-5\right)}{3}$

(b) $V = \int_0^{2\pi}\int_0^{\pi/4}\int_{2\sec\phi}^{\sqrt{8}} \rho^2\sin\phi\,d\rho\,d\phi\,d\theta = \frac{8}{3}\int_0^{2\pi}\int_0^{\pi/4}\left(2\sqrt{2}\sin\phi-\sec^3\phi\sin\phi\right)d\phi\,d\theta$
$= \frac{8}{3}\int_0^{2\pi}\int_0^{\pi/4}\left(2\sqrt{2}\sin\phi-\tan\phi\sec^2\phi\right)d\phi\,d\theta = \frac{8}{3}\int_0^{2\pi}\left[-2\sqrt{2}\cos\phi-\frac{1}{2}\tan^2\phi\right]_0^{\pi/4} d\theta$
$= \frac{8}{3}\int_0^{2\pi}\left(-2-\frac{1}{2}+2\sqrt{2}\right)d\theta = \frac{8}{3}\int_0^{2\pi}\left(\frac{-5+4\sqrt{2}}{2}\right)d\theta = \frac{8\pi\left(4\sqrt{2}-5\right)}{3}$

39. With the centers of the spheres at the origin, $I_z = \int_0^{2\pi}\int_0^{\pi}\int_a^{b} \delta(\rho\sin\phi)^2(\rho^2\sin\phi)\,d\rho\,d\phi\,d\theta$
$= \frac{\delta(b^5-a^5)}{5}\int_0^{2\pi}\int_0^{\pi}\sin^3\phi\,d\phi\,d\theta = \frac{\delta(b^5-a^5)}{5}\int_0^{2\pi}\int_0^{\pi}(\sin\phi-\cos^2\phi\sin\phi)\,d\phi\,d\theta$
$= \frac{\delta(b^5-a^5)}{5}\int_0^{2\pi}\left[-\cos\phi+\frac{\cos^3\phi}{3}\right]_0^{\pi} d\theta = \frac{4\delta(b^5-a^5)}{15}\int_0^{2\pi} d\theta = \frac{8\pi\delta(b^5-a^5)}{15}$

41. $M = \int_1^2\int_{2/x}^2 dy\,dx = \int_1^2\left(2-\frac{2}{x}\right)dx = 2-\ln 4$; $M_y = \int_1^2\int_{2/x}^2 x\,dy\,dx = \int_1^2 x\left(2-\frac{2}{x}\right)dx = 1$;
$M_x = \int_1^2\int_{2/x}^2 y\,dy\,dx = \int_1^2\left(2-\frac{2}{x^2}\right)dx = 1 \Rightarrow \bar{x} = \bar{y} = \frac{1}{2-\ln 4}$

43. $I_o = \int_0^2\int_{2x}^4 (x^2+y^2)(3)\,dy\,dx = 3\int_0^2\left(4x^2+\frac{64}{3}-\frac{14x^3}{3}\right)dx = 104$

45. $M = \delta\int_0^3\int_0^{2x/3} dy\,dx = \delta\int_0^3 \frac{2x}{3}\,dx = 3\delta$; $I_x = \delta\int_0^3\int_0^{2x/3} y^2\,dy\,dx = \frac{8\delta}{81}\int_0^3 x^3\,dx = \left(\frac{8\delta}{81}\right)\left(\frac{3^4}{4}\right) = 2\delta$

47. $M = \int_{-1}^1\int_{-1}^1\left(x^2+y^2+\frac{1}{3}\right)dy\,dx = \int_{-1}^1\left(2x^2+\frac{4}{3}\right)dx = 4$; $M_x = \int_{-1}^1\int_{-1}^1 y\left(x^2+y^2+\frac{1}{3}\right)dy\,dx = \int_{-1}^1 0\,dx = 0$;
$M_y = \int_{-1}^1\int_{-1}^1 x\left(x^2+y^2+\frac{1}{3}\right)dy\,dx = \int_{-1}^1\left(2x^3+\frac{4}{3}x\right)dx = 0$

49. $M = \int_{-\pi/3}^{\pi/3}\int_0^3 r\,dr\,d\theta = \frac{9}{2}\int_{-\pi/3}^{\pi/3} d\theta = 3\pi$; $M_y = \int_{-\pi/3}^{\pi/3}\int_0^3 r^2\cos\theta\,dr\,d\theta = 9\int_{-\pi/3}^{\pi/3}\cos\theta\,d\theta = 9\sqrt{3} \Rightarrow \bar{x} = \frac{3\sqrt{3}}{\pi}$,
and $\bar{y} = 0$ by symmetry

51. (a) $M = 2\int_0^{\pi/2}\int_1^{1+\cos\theta} r\,dr\,d\theta$
$= \int_0^{\pi/2}\left(2\cos\theta + \frac{1+\cos 2\theta}{2}\right)d\theta = \frac{8+\pi}{4}$;
$M_y = \int_{-\pi/2}^{\pi/2}\int_1^{1+\cos\theta} (r\cos\theta)\,r\,dr\,d\theta$
$= \int_{-\pi/2}^{\pi/2}\left(\cos^2\theta + \cos^3\theta + \frac{\cos^4\theta}{3}\right)d\theta$
$= \frac{32+15\pi}{24} \Rightarrow \overline{x} = \frac{15\pi+32}{6\pi+48}$, and
$\overline{y} = 0$ by symmetry

(b)

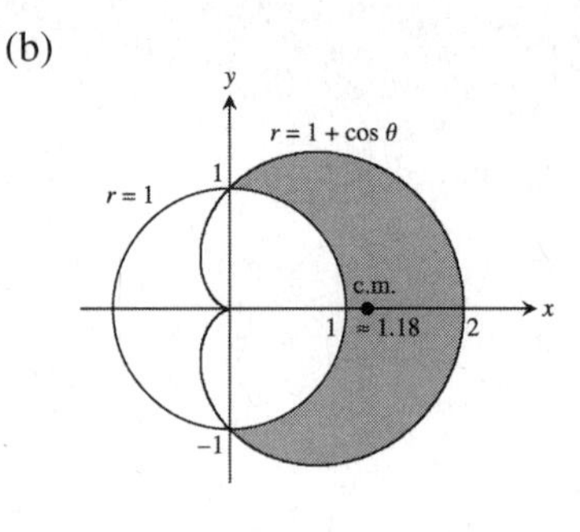

53. $x = u + y$ and $y = v \Rightarrow x = u + v$ and $y = v$
$\Rightarrow J(u, v) = \begin{vmatrix} 1 & 1 \\ 0 & 1 \end{vmatrix} = 1$; the boundary of the image G is obtained from the boundary of R as follows:

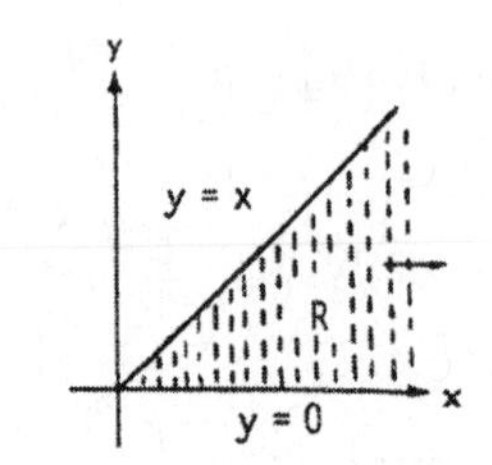

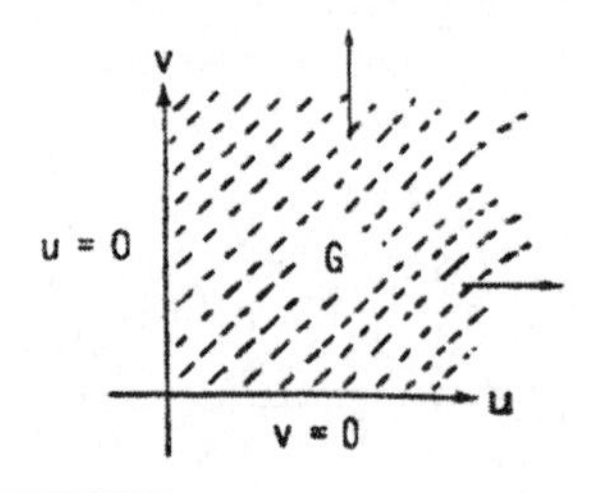

xy-equations for the boundary of R	Corresponding uv-equations for the boundary of G	Simplified uv-equations
$y = x$	$v = u + v$	$u = 0$
$y = 0$	$v = 0$	$v = 0$

$\Rightarrow \int_0^{\infty}\int_0^{x} e^{-sx} f(x-y, y)\,dy\,dx = \int_0^{\infty}\int_0^{\infty} e^{-s(u+v)} f(u, v)\,du\,dv$

CHAPTER 15 ADDITIONAL AND ADVANCED EXERCISES

1. (a) $V = \int_{-3}^{2}\int_x^{6-x^2} x^2\,dy\,dx$ (b) $V = \int_{-3}^{2}\int_x^{6-x^2}\int_0^{x^2} dz\,dy\,dx$
(c) $V = \int_{-3}^{2}\int_x^{6-x^2} x^2\,dy\,dx = \int_{-3}^{2}\int_x^{6-x^2} (6x^2 - x^4 - x^3)\,dx = \left[2x^3 - \frac{x^5}{5} - \frac{x^4}{4}\right]_{-3}^{2} = \frac{125}{4}$

3. Using cylindrical coordinates, $V = \int_0^{2\pi}\int_0^1\int_0^{2-r(\cos\theta+\sin\theta)} dz\,r\,dr\,d\theta = \int_0^{2\pi}\int_0^1 (2r - r^2\cos\theta - r^2\sin\theta)\,dr\,d\theta$
$= \int_0^{2\pi}\left(1 - \frac{1}{3}\cos\theta - \frac{1}{3}\sin\theta\right)d\theta = \left[\theta - \frac{1}{3}\sin\theta + \frac{1}{3}\cos\theta\right]_0^{2\pi} = 2\pi$

5. The surfaces intersect when $3 - x^2 - y^2 = 2x^2 + 2y^2 \Rightarrow x^2 + y^2 = 1$. Thus the volume is
$V = 4\int_0^1\int_0^{\sqrt{1-x^2}}\int_{2x^2+2y^2}^{3-x^2-y^2} dz\,dy\,dx = 4\int_0^{\pi/2}\int_0^1\int_{2r^2}^{3-r^2} dz\,r\,dr\,d\theta = 4\int_0^{\pi/2}\int_0^1 (3r - 3r^3)\,dr\,d\theta = 3\int_0^{\pi/2} d\theta = \frac{3\pi}{2}$

7. (a) The radius of the hole is 1, and the radius of the sphere is 2.

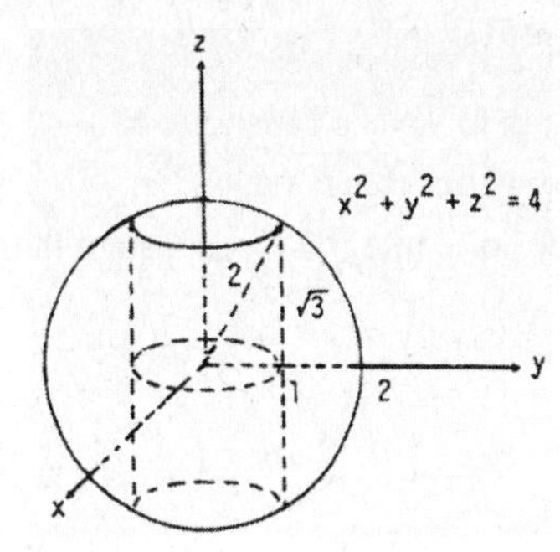

(b) $V = 2\int_0^{2\pi}\int_0^{\sqrt{3}}\int_1^{\sqrt{4-z^2}} r\,dr\,dz\,d\theta = \int_0^{2\pi}\int_0^{\sqrt{3}} (3 - z^2)\,dz\,d\theta = 2\sqrt{3}\int_0^{2\pi} d\theta = 4\sqrt{3}\pi$

9. The surfaces intersect when $x^2 + y^2 = \frac{x^2+y^2+1}{2} \Rightarrow x^2 + y^2 = 1$. Thus the volume in cylindrical coordinates is $V = 4\int_0^{\pi/2}\int_0^1\int_{r^2}^{(r^2+1)/2} dz\,r\,dr\,d\theta = 4\int_0^{\pi/2}\int_0^1 \left(\frac{r}{2} - \frac{r^3}{2}\right) dr\,d\theta = 4\int_0^{\pi/2}\left[\frac{r^2}{4} - \frac{r^4}{8}\right]_0^1 d\theta = \frac{1}{2}\int_0^{\pi/2} d\theta = \frac{\pi}{4}$

11. $\int_0^\infty \frac{e^{-ax}-e^{-bx}}{x}\,dx = \int_0^\infty\int_a^b e^{-xy}\,dy\,dx = \int_a^b\int_0^\infty e^{-xy}\,dx\,dy = \int_a^b \left(\lim_{t\to\infty}\int_0^t e^{-xy}\,dx\right) dy$
$= \int_a^b \lim_{t\to\infty}\left[-\frac{e^{-xy}}{y}\right]_0^t dy = \int_a^b \lim_{t\to\infty}\left(\frac{1}{y} - \frac{e^{-yt}}{y}\right) dy = \int_a^b \frac{1}{y}\,dy = [\ln y]_a^b = \ln\left(\frac{b}{a}\right)$

13. $\int_0^x\int_0^u e^{m(x-t)} f(t)\,dt\,du = \int_0^x\int_t^x e^{m(x-t)} f(t)\,du\,dt = \int_0^x (x-t)e^{m(x-t)} f(t)\,dt$; also
$\int_0^x\int_0^v\int_0^u e^{m(x-t)} f(t)\,dt\,du\,dv = \int_0^x\int_t^x\int_t^v e^{m(x-t)} f(t)\,du\,dv\,dt = \int_0^x\int_t^x (v-t)e^{m(x-t)} f(t)\,dv\,dt$
$= \int_0^x \left[\frac{1}{2}(v-t)^2 e^{m(x-t)} f(t)\right]_t^x dt = \int_0^x \frac{(x-t)^2}{2} e^{m(x-t)} f(t)\,dt$

15. $I_o(a) = \int_0^a\int_0^{x/a^2} (x^2 + y^2)\,dy\,dx = \int_0^a \left[x^2y + \frac{y^3}{3}\right]_0^{x/a^2} dx = \int_0^a \left(\frac{x^3}{a^2} + \frac{x^3}{3a^6}\right) dx = \left[\frac{x^4}{4a^2} + \frac{x^4}{12a^6}\right]_0^a$
$= \frac{a^2}{4} + \frac{1}{12}a^{-2};\ I_o'(a) = \frac{1}{2}a - \frac{1}{6}a^{-3} = 0 \Rightarrow a^4 = \frac{1}{3} \Rightarrow a = \sqrt[4]{\frac{1}{3}} = \frac{1}{\sqrt[4]{3}}$. Since $I_o''(a) = \frac{1}{2} + \frac{1}{2}a^{-4} > 0$, the value of a does provide a minimum for the polar moment of inertia $I_o(a)$.

17. $M = \int_{-\theta}^{\theta}\int_{b\sec\theta}^{a} r\,dr\,d\theta = \int_{-\theta}^{\theta}\left(\frac{a^2}{2} - \frac{b^2}{2}\sec^2\theta\right) d\theta$
$= a^2\theta - b^2\tan\theta = a^2\cos^{-1}\left(\frac{b}{a}\right) - b^2\left(\frac{\sqrt{a^2-b^2}}{b}\right)$
$= a^2\cos^{-1}\left(\frac{b}{a}\right) - b\sqrt{a^2-b^2};\ I_o = \int_{-\theta}^{\theta}\int_{b\sec\theta}^{a} r^3\,dr\,d\theta$
$= \frac{1}{4}\int_{-\theta}^{\theta}(a^4 + b^4\sec^4\theta)\,d\theta$
$= \frac{1}{4}\int_{-\theta}^{\theta}[a^4 + b^4(1+\tan^2\theta)(\sec^2\theta)]\,d\theta$
$= \frac{1}{4}\left[a^4\theta - b^4\tan\theta - \frac{b^4\tan^3\theta}{3}\right]_{-\theta}^{\theta}$
$= \frac{a^4\theta}{2} - \frac{b^4\tan\theta}{2} - \frac{b^4\tan^3\theta}{6}$
$= \frac{1}{2}a^4\cos^{-1}\left(\frac{b}{a}\right) - \frac{1}{2}b^3\sqrt{a^2-b^2} - \frac{1}{6}b^3(a^2-b^2)^{3/2}$

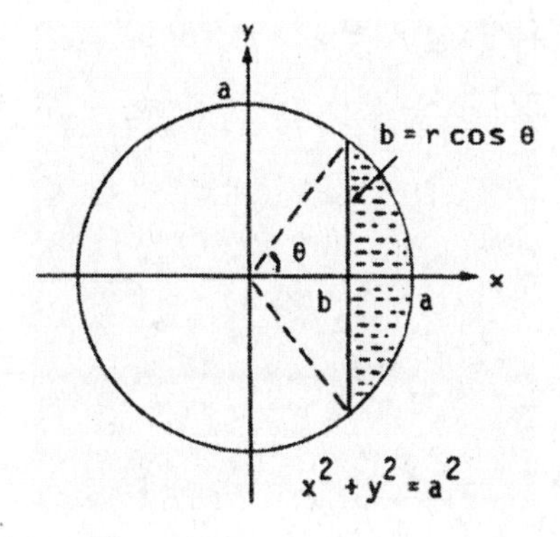

19. $\int_0^a\int_0^b e^{\max(b^2x^2,a^2y^2)}\,dy\,dx = \int_0^a\int_0^{bx/a} e^{b^2x^2}\,dy\,dx + \int_0^b\int_0^{ay/b} e^{a^2y^2}\,dx\,dy$
$= \int_0^a \left(\frac{b}{a}x\right) e^{b^2x^2}\,dx + \int_0^b \left(\frac{a}{b}y\right) e^{a^2y^2}\,dy = \left[\frac{1}{2ab}e^{b^2x^2}\right]_0^a + \left[\frac{1}{2ba}e^{a^2y^2}\right]_0^b = \frac{1}{2ab}\left(e^{b^2a^2} - 1\right) + \frac{1}{2ab}\left(e^{a^2b^2} - 1\right)$
$= \frac{1}{ab}\left(e^{a^2b^2} - 1\right)$

21. (a) (i) Fubini's Theorem
(ii) Treating G(y) as a constant
(iii) Algebraic rearrangement
(iv) The definite integral is a constant number

(b) $\int_0^{\ln 2}\int_0^{\pi/2} e^x \cos y\, dy\, dx = \left(\int_0^{\ln 2} e^x\, dx\right)\left(\int_0^{\pi/2} \cos y\, dy\right) = \left(e^{\ln 2} - e^0\right)\left(\sin\frac{\pi}{2} - \sin 0\right) = (1)(1) = 1$

(c) $\int_1^2\int_{-1}^1 \frac{x}{y^2}\, dx\, dy = \left(\int_1^2 \frac{1}{y^2}\, dy\right)\left(\int_{-1}^1 x\, dx\right) = \left[-\frac{1}{y}\right]_1^2\left[\frac{x^2}{2}\right]_{-1}^1 = \left(-\frac{1}{2}+1\right)\left(\frac{1}{2}-\frac{1}{2}\right) = 0$

23. (a) $I^2 = \int_0^\infty\int_0^\infty e^{-(x^2+y^2)}\, dx\, dy = \int_0^{\pi/2}\int_0^\infty \left(e^{-r^2}\right) r\, dr\, d\theta = \int_0^{\pi/2}\left[\lim_{b\to\infty}\int_0^b re^{-r^2}\, dr\right] d\theta$

$= -\frac{1}{2}\int_0^{\pi/2}\lim_{b\to\infty}\left(e^{-b^2}-1\right) d\theta = \frac{1}{2}\int_0^{\pi/2} d\theta = \frac{\pi}{4} \Rightarrow I = \frac{\sqrt{\pi}}{2}$

(b) $\Gamma\left(\frac{1}{2}\right) = \int_0^\infty t^{-1/2}e^{-t}\, dt = \int_0^\infty \left(y^2\right)^{-1/2} e^{-y^2}(2y)\, dy = 2\int_0^\infty e^{-y^2}\, dy = 2\left(\frac{\sqrt{\pi}}{2}\right) = \sqrt{\pi}$, where $y = \sqrt{t}$

25. For a height h in the bowl the volume of water is $V = \int_{-\sqrt{h}}^{\sqrt{h}}\int_{-\sqrt{h-x^2}}^{\sqrt{h-x^2}}\int_{x^2+y^2}^{h} dz\, dy\, dx$

$= \int_{-\sqrt{h}}^{\sqrt{h}}\int_{-\sqrt{h-x^2}}^{\sqrt{h-x^2}} \left(h - x^2 - y^2\right) dy\, dx = \int_0^{2\pi}\int_0^{\sqrt{h}} \left(h - r^2\right) r\, dr\, d\theta = \int_0^{2\pi}\left[\frac{hr^2}{2} - \frac{r^4}{4}\right]_0^{\sqrt{h}} d\theta = \int_0^{2\pi} \frac{h^2}{4}\, d\theta = \frac{h^2\pi}{2}.$

Since the top of the bowl has area 10π, then we calibrate the bowl by comparing it to a right circular cylinder whose cross sectional area is 10π from $z = 0$ to $z = 10$. If such a cylinder contains $\frac{h^2\pi}{2}$ cubic inches of water to a depth w then we have $10\pi w = \frac{h^2\pi}{2} \Rightarrow w = \frac{h^2}{20}$. So for 1 inch of rain, $w = 1$ and $h = \sqrt{20}$; for 3 inches of rain, $w = 3$ and $h = \sqrt{60}$.

27. The cylinder is given by $x^2 + y^2 = 1$ from $z = 1$ to $\infty \Rightarrow \iiint_D z\left(r^2+z^2\right)^{-5/2} dV$

$= \int_0^{2\pi}\int_0^1\int_1^\infty \frac{z}{\left(r^2+z^2\right)^{5/2}}\, dz\, r\, dr\, d\theta = \lim_{a\to\infty}\int_0^{2\pi}\int_0^1\int_1^a \frac{rz}{\left(r^2+z^2\right)^{5/2}}\, dz\, dr\, d\theta$

$= \lim_{a\to\infty}\int_0^{2\pi}\int_0^1\left[\left(-\frac{1}{3}\right)\frac{r}{\left(r^2+z^2\right)^{3/2}}\right]_1^a dr\, d\theta = \lim_{a\to\infty}\int_0^{2\pi}\int_0^1\left[\left(-\frac{1}{3}\right)\frac{r}{\left(r^2+a^2\right)^{3/2}} + \left(\frac{1}{3}\right)\frac{r}{\left(r^2+1\right)^{3/2}}\right] dr\, d\theta$

$= \lim_{a\to\infty}\int_0^{2\pi}\left[\frac{1}{3}\left(r^2+a^2\right)^{-1/2} - \frac{1}{3}\left(r^2+1\right)^{-1/2}\right]_0^1 d\theta = \lim_{a\to\infty}\int_0^{2\pi}\left[\frac{1}{3}\left(1+a^2\right)^{-1/2} - \frac{1}{3}\left(2^{-1/2}\right) - \frac{1}{3}\left(a^2\right)^{-1/2} + \frac{1}{3}\right] d\theta$

$= \lim_{a\to\infty} 2\pi\left[\frac{1}{3}\left(1+a^2\right)^{-1/2} - \frac{1}{3}\left(\frac{\sqrt{2}}{2}\right) - \frac{1}{3}\left(\frac{1}{a}\right) + \frac{1}{3}\right] = 2\pi\left[\frac{1}{3} - \left(\frac{1}{3}\right)\frac{\sqrt{2}}{2}\right].$

CHAPTER 16 INTEGRATION IN VECTOR FIELDS

16.1 LINE INTEGRALS

1. $\mathbf{r} = t\mathbf{i} + (1-t)\mathbf{j} \Rightarrow x = t$ and $y = 1 - t \Rightarrow y = 1 - x \Rightarrow$ (c)

3. $\mathbf{r} = (2\cos t)\mathbf{i} + (2\sin t)\mathbf{j} \Rightarrow x = 2\cos t$ and $y = 2\sin t \Rightarrow x^2 + y^2 = 4 \Rightarrow$ (g)

5. $\mathbf{r} = t\mathbf{i} + t\mathbf{j} + t\mathbf{k} \Rightarrow x = t, y = t,$ and $z = t \Rightarrow$ (d)

7. $\mathbf{r} = (t^2 - 1)\mathbf{j} + 2t\mathbf{k} \Rightarrow y = t^2 - 1$ and $z = 2t \Rightarrow y = \frac{z^2}{4} - 1 \Rightarrow$ (f)

9. $\mathbf{r}(t) = t\mathbf{i} + (1-t)\mathbf{j},\ 0 \le t \le 1 \Rightarrow \frac{d\mathbf{r}}{dt} = \mathbf{i} - \mathbf{j} \Rightarrow \left|\frac{d\mathbf{r}}{dt}\right| = \sqrt{2}\,\mathbf{j}$; $x = t$ and $y = 1 - t \Rightarrow x + y = t + (1 - t) = 1$
$\Rightarrow \int_C f(x,y,z)\,ds = \int_0^1 f(t, 1-t, 0)\left|\frac{d\mathbf{r}}{dt}\right| dt = \int_0^1 (1)\left(\sqrt{2}\right) dt = \left[\sqrt{2}\,t\right]_0^1 = \sqrt{2}$

11. $\mathbf{r}(t) = 2t\mathbf{i} + t\mathbf{j} + (2 - 2t)\mathbf{k},\ 0 \le t \le 1 \Rightarrow \frac{d\mathbf{r}}{dt} = 2\mathbf{i} + \mathbf{j} - 2\mathbf{k} \Rightarrow \left|\frac{d\mathbf{r}}{dt}\right| = \sqrt{4 + 1 + 4} = 3$; $xy + y + z$
$= (2t)t + t + (2 - 2t) \Rightarrow \int_C f(x,y,z)\,ds = \int_0^1 (2t^2 - t + 2)\,3\,dt = 3\left[\frac{2}{3}t^3 - \frac{1}{2}t^2 + 2t\right]_0^1 = 3\left(\frac{2}{3} - \frac{1}{2} + 2\right) = \frac{13}{2}$

13. $\mathbf{r}(t) = (\mathbf{i} + 2\mathbf{j} + 3\mathbf{k}) + t(-\mathbf{i} - 3\mathbf{j} - 2\mathbf{k}) = (1 - t)\mathbf{i} + (2 - 3t)\mathbf{j} + (3 - 2t)\mathbf{k},\ 0 \le t \le 1 \Rightarrow \frac{d\mathbf{r}}{dt} = -\mathbf{i} - 3\mathbf{j} - 2\mathbf{k}$
$\Rightarrow \left|\frac{d\mathbf{r}}{dt}\right| = \sqrt{1 + 9 + 4} = \sqrt{14}$; $x + y + z = (1 - t) + (2 - 3t) + (3 - 2t) = 6 - 6t \Rightarrow \int_C f(x,y,z)\,ds$
$= \int_0^1 (6 - 6t)\sqrt{14}\,dt = 6\sqrt{14}\left[t - \frac{t^2}{2}\right]_0^1 = \left(6\sqrt{14}\right)\left(\frac{1}{2}\right) = 3\sqrt{14}$

15. C_1: $\mathbf{r}(t) = t\mathbf{i} + t^2\mathbf{j},\ 0 \le t \le 1 \Rightarrow \frac{d\mathbf{r}}{dt} = \mathbf{i} + 2t\mathbf{j} \Rightarrow \left|\frac{d\mathbf{r}}{dt}\right| = \sqrt{1 + 4t^2}$; $x + \sqrt{y} - z^2 = t + \sqrt{t^2} - 0 = t + |t| = 2t$
since $t \ge 0 \Rightarrow \int_{C_1} f(x,y,z)\,ds = \int_0^1 2t\sqrt{1 + 4t^2}\,dt = \left[\frac{1}{6}\left(1 + 4t^2\right)^{3/2}\right]_0^1 = \frac{1}{6}(5)^{3/2} - \frac{1}{6} = \frac{1}{6}\left(5\sqrt{5} - 1\right)$;
C_2: $\mathbf{r}(t) = \mathbf{i} + \mathbf{j} + t\mathbf{k},\ 0 \le t \le 1 \Rightarrow \frac{d\mathbf{r}}{dt} = \mathbf{k} \Rightarrow \left|\frac{d\mathbf{r}}{dt}\right| = 1$; $x + \sqrt{y} - z^2 = 1 + \sqrt{1} - t^2 = 2 - t^2$
$\Rightarrow \int_{C_2} f(x,y,z)\,ds = \int_0^1 (2 - t^2)(1)\,dt = \left[2t - \frac{1}{3}t^3\right]_0^1 = 2 - \frac{1}{3} = \frac{5}{3}$; therefore $\int_C f(x,y,z)\,ds$
$= \int_{C_1} f(x,y,z)\,ds + \int_{C_2} f(x,y,z)\,ds = \frac{5}{6}\sqrt{5} + \frac{3}{2}$

17. $\mathbf{r}(t) = t\mathbf{i} + t\mathbf{j} + t\mathbf{k},\ 0 < a \le t \le b \Rightarrow \frac{d\mathbf{r}}{dt} = \mathbf{i} + \mathbf{j} + \mathbf{k} \Rightarrow \left|\frac{d\mathbf{r}}{dt}\right| = \sqrt{3}$; $\frac{x + y + z}{x^2 + y^2 + z^2} = \frac{t + t + t}{t^2 + t^2 + t^2} = \frac{1}{t}$
$\Rightarrow \int_C f(x,y,z)\,ds = \int_a^b \left(\frac{1}{t}\right)\sqrt{3}\,dt = \left[\sqrt{3}\ln|t|\right]_a^b = \sqrt{3}\ln\left(\frac{b}{a}\right)$, since $0 < a \le b$

19. (a) $\mathbf{r}(t) = t\mathbf{i} + \frac{1}{2}t\mathbf{j},\ 0 \le t \le 4 \Rightarrow \frac{d\mathbf{r}}{dt} = \mathbf{i} + \frac{1}{2}\mathbf{j} \Rightarrow \left|\frac{d\mathbf{r}}{dt}\right| = \frac{\sqrt{5}}{2} \Rightarrow \int_C x\,ds = \int_0^4 t\,\frac{\sqrt{5}}{2}dt = \frac{\sqrt{5}}{2}\int_0^4 t\,dt = \left[\frac{\sqrt{5}}{4}t^2\right]_0^4 = 4\sqrt{5}$
(b) $\mathbf{r}(t) = t\mathbf{i} + t^2\mathbf{j},\ 0 \le t \le 2 \Rightarrow \frac{d\mathbf{r}}{dt} = \mathbf{i} + 2t\mathbf{j} \Rightarrow \left|\frac{d\mathbf{r}}{dt}\right| = \sqrt{1 + 4t^2} \Rightarrow \int_C x\,ds = \int_0^2 t\sqrt{1 + 4t^2}dt$
$= \left[\frac{1}{12}\left(1 + 4t^2\right)^{3/2}\right]_0^2 = \frac{17\sqrt{17} - 1}{12}$

21. $\mathbf{r}(t) = 4t\mathbf{i} - 3t\mathbf{j},\ -1 \le t \le 2 \Rightarrow \frac{d\mathbf{r}}{dt} = 4\mathbf{i} - 3\mathbf{j} \Rightarrow \left|\frac{d\mathbf{r}}{dt}\right| = 5 \Rightarrow \int_C y e^{x^2}\,ds = \int_{-1}^2 (-3t)\,e^{(4t)^2} \cdot 5dt$
$= -15\int_{-1}^2 t e^{16t^2}dt = \left[-\frac{15}{32}e^{16t^2}\right]_{-1}^2 = -\frac{15}{32}e^{64} + \frac{15}{32}e^{16} = \frac{15}{32}\left(e^{16} - e^{64}\right)$

23. $\mathbf{r}(t) = t^2\mathbf{i} + t^3\mathbf{j},\ 1 \le t \le 2 \Rightarrow \frac{d\mathbf{r}}{dt} = 2t\mathbf{i} + 3t^2\mathbf{j} \Rightarrow \left|\frac{d\mathbf{r}}{dt}\right| = \sqrt{(2t)^2 + (3t^2)^2} = t\sqrt{4+9t^2} \Rightarrow \int_C \frac{x^2}{y^{4/3}}\,ds$
$= \int_1^2 \frac{(t^2)^2}{(t^3)^{4/3}} \cdot t\sqrt{4+9t^2}\,dt = \int_1^2 t\sqrt{4+9t^2}\,dt = \left[\frac{1}{27}(4+9t^2)^{3/2}\right]_1^2 = \frac{80\sqrt{10}-13\sqrt{13}}{27}$

25. C_1: $\mathbf{r}(t) = t\mathbf{i} + t^2\mathbf{j},\ 0 \le t \le 1 \Rightarrow \frac{d\mathbf{r}}{dt} = \mathbf{i} + 2t\mathbf{j} \Rightarrow \left|\frac{d\mathbf{r}}{dt}\right| = \sqrt{1+4t^2}$; C_2: $\mathbf{r}(t) = (1-t)\mathbf{i} + (1-t)\mathbf{j},\ 0 \le t \le 1$
$\Rightarrow \frac{d\mathbf{r}}{dt} = -\mathbf{i} - \mathbf{j} \Rightarrow \left|\frac{d\mathbf{r}}{dt}\right| = \sqrt{2} \Rightarrow \int_C \left(x + \sqrt{y}\right)ds = \int_{C_1} \left(x + \sqrt{y}\right)ds + \int_{C_2} \left(x + \sqrt{y}\right)ds$
$= \int_0^1 \left(t + \sqrt{t^2}\right)\sqrt{1+4t^2}\,dt + \int_0^1 \left((1-t) + \sqrt{1-t}\right)\sqrt{2}dt = \int_0^1 2t\sqrt{1+4t^2}\,dt + \int_0^1 \left(1 - t + \sqrt{1-t}\right)\sqrt{2}dt$
$= \left[\frac{1}{6}(1+4t^2)^{3/2}\right]_0^1 + \sqrt{2}\left[t - \frac{1}{2}t^2 - \frac{2}{3}(1-t)^{3/2}\right]_0^1 = \frac{5\sqrt{5}-1}{6} + \frac{7\sqrt{2}}{6} = \frac{5\sqrt{5}+7\sqrt{2}-1}{6}$

27. $\mathbf{r}(x) = x\mathbf{i} + y\mathbf{j} = x\mathbf{i} + \frac{x^2}{2}\mathbf{j},\ 0 \le x \le 2 \Rightarrow \frac{d\mathbf{r}}{dx} = \mathbf{i} + x\mathbf{j} \Rightarrow \left|\frac{d\mathbf{r}}{dx}\right| = \sqrt{1+x^2}$; $f(x,y) = f\left(x, \frac{x^2}{2}\right) = \frac{x^3}{\left(\frac{x^2}{2}\right)} = 2x \Rightarrow \int_C f\,ds$
$= \int_0^2 (2x)\sqrt{1+x^2}\,dx = \left[\frac{2}{3}\left(1+x^2\right)^{3/2}\right]_0^2 = \frac{2}{3}\left(5^{3/2} - 1\right) = \frac{10\sqrt{5}-2}{3}$

29. $\mathbf{r}(t) = (2\cos t)\mathbf{i} + (2\sin t)\mathbf{j},\ 0 \le t \le \frac{\pi}{2} \Rightarrow \frac{d\mathbf{r}}{dt} = (-2\sin t)\mathbf{i} + (2\cos t)\mathbf{j} \Rightarrow \left|\frac{d\mathbf{r}}{dt}\right| = 2$; $f(x,y) = f(2\cos t, 2\sin t)$
$= 2\cos t + 2\sin t \Rightarrow \int_C f\,ds = \int_0^{\pi/2} (2\cos t + 2\sin t)(2)\,dt = [4\sin t - 4\cos t]_0^{\pi/2} = 4 - (-4) = 8$

31. $y = x^2,\ 0 \le x \le 2 \Rightarrow \mathbf{r}(t) = t\mathbf{i} + t^2\mathbf{j},\ 0 \le t \le 2 \Rightarrow \frac{d\mathbf{r}}{dt} = \mathbf{i} + 2t\mathbf{j} \Rightarrow \left|\frac{d\mathbf{r}}{dt}\right| = \sqrt{1+4t^2} \Rightarrow A = \int_C f(x,y)\,ds$
$= \int_C \left(x + \sqrt{y}\right)ds = \int_0^2 \left(t + \sqrt{t^2}\right)\sqrt{1+4t^2}\,dt = \int_0^2 2t\sqrt{1+4t^2}\,dt = \left[\frac{1}{6}(1+4t^2)^{3/2}\right]_0^2 = \frac{17\sqrt{17}-1}{6}$

33. $\mathbf{r}(t) = (t^2-1)\mathbf{j} + 2t\mathbf{k},\ 0 \le t \le 1 \Rightarrow \frac{d\mathbf{r}}{dt} = 2t\mathbf{j} + 2\mathbf{k} \Rightarrow \left|\frac{d\mathbf{r}}{dt}\right| = 2\sqrt{t^2+1}$; $M = \int_C \delta(x,y,z)\,ds = \int_0^1 \delta(t)\left(2\sqrt{t^2+1}\right)dt$
$= \int_0^1 \left(\frac{3}{2}t\right)\left(2\sqrt{t^2+1}\right)dt = \left[\left(t^2+1\right)^{3/2}\right]_0^1 = 2^{3/2} - 1 = 2\sqrt{2} - 1$

35. $\mathbf{r}(t) = \sqrt{2}t\mathbf{i} + \sqrt{2}t\mathbf{j} + (4-t^2)\mathbf{k},\ 0 \le t \le 1 \Rightarrow \frac{d\mathbf{r}}{dt} = \sqrt{2}\mathbf{i} + \sqrt{2}\mathbf{j} - 2t\mathbf{k} \Rightarrow \left|\frac{d\mathbf{r}}{dt}\right| = \sqrt{2+2+4t^2} = 2\sqrt{1+t^2}$;

(a) $M = \int_C \delta\,ds = \int_0^1 (3t)\left(2\sqrt{1+t^2}\right)dt = \left[2\left(1+t^2\right)^{3/2}\right]_0^1 = 2\left(2^{3/2} - 1\right) = 4\sqrt{2} - 2$

(b) $M = \int_C \delta\,ds = \int_0^1 (1)\left(2\sqrt{1+t^2}\right)dt = \left[t\sqrt{1+t^2} + \ln\left(t + \sqrt{1+t^2}\right)\right]_0^1 = \left[\sqrt{2} + \ln\left(1+\sqrt{2}\right)\right] - (0 + \ln 1)$
$= \sqrt{2} + \ln\left(1+\sqrt{2}\right)$

37. Let $x = a\cos t$ and $y = a\sin t$, $0 \le t \le 2\pi$. Then $\frac{dx}{dt} = -a\sin t$, $\frac{dy}{dt} = a\cos t$, $\frac{dz}{dt} = 0$
$\Rightarrow \sqrt{\left(\frac{dx}{dt}\right)^2 + \left(\frac{dy}{dt}\right)^2 + \left(\frac{dz}{dt}\right)^2}\,dt = a\,dt$; $I_z = \int_C (x^2+y^2)\,\delta\,ds = \int_0^{2\pi} (a^2\sin^2 t + a^2\cos^2 t)\,a\delta\,dt$
$= \int_0^{2\pi} a^3\delta\,dt = 2\pi\delta a^3.$

39. $\mathbf{r}(t) = (\cos t)\mathbf{i} + (\sin t)\mathbf{j} + t\mathbf{k},\ 0 \le t \le 2\pi \Rightarrow \frac{d\mathbf{r}}{dt} = (-\sin t)\mathbf{i} + (\cos t)\mathbf{j} + \mathbf{k} \Rightarrow \left|\frac{d\mathbf{r}}{dt}\right| = \sqrt{\sin^2 t + \cos^2 t + 1} = \sqrt{2}$;

(a) $I_z = \int_C (x^2+y^2)\,\delta\,ds = \int_0^{2\pi} (\cos^2 t + \sin^2 t)\,\delta\sqrt{2}\,dt = 2\pi\delta\sqrt{2}$

(b) $I_z = \int_C (x^2+y^2)\,\delta\,ds = \int_0^{4\pi} \delta\sqrt{2}\,dt = 4\pi\delta\sqrt{2}$

41. $\delta(x,y,z) = 2 - z$ and $\mathbf{r}(t) = (\cos t)\mathbf{j} + (\sin t)\mathbf{k},\ 0 \le t \le \pi \Rightarrow M = 2\pi - 2$ as found in Example 3 of the text;
also $\left|\frac{d\mathbf{r}}{dt}\right| = 1$; $I_x = \int_C (y^2+z^2)\,\delta\,ds = \int_0^\pi (\cos^2 t + \sin^2 t)(2 - \sin t)\,dt = \int_0^\pi (2 - \sin t)\,dt = 2\pi - 2$

16.2 VECTOR FIELDS, WORK, CIRCULATION, AND FLUX

1. $f(x, y, z) = (x^2+y^2+z^2)^{-1/2} \Rightarrow \frac{\partial f}{\partial x} = -\frac{1}{2}(x^2+y^2+z^2)^{-3/2}(2x) = -x(x^2+y^2+z^2)^{-3/2}$; similarly, $\frac{\partial f}{\partial y} = -y(x^2+y^2+z^2)^{-3/2}$ and $\frac{\partial f}{\partial z} = -z(x^2+y^2+z^2)^{-3/2} \Rightarrow \nabla f = \frac{-x\mathbf{i}-y\mathbf{j}-z\mathbf{k}}{(x^2+y^2+z^2)^{3/2}}$

3. $g(x, y, z) = e^z - \ln(x^2+y^2) \Rightarrow \frac{\partial g}{\partial x} = -\frac{2x}{x^2+y^2}, \frac{\partial g}{\partial y} = -\frac{2y}{x^2+y^2}$ and $\frac{\partial g}{\partial z} = e^z$
$\Rightarrow \nabla g = \left(\frac{-2x}{x^2+y^2}\right)\mathbf{i} - \left(\frac{2y}{x^2+y^2}\right)\mathbf{j} + e^z\mathbf{k}$

5. $|\mathbf{F}|$ inversely proportional to the square of the distance from (x, y) to the origin $\Rightarrow \sqrt{(M(x,y))^2+(N(x,y))^2}$ $= \frac{k}{x^2+y^2}$, $k > 0$; $\mathbf{F}$ points toward the origin $\Rightarrow$ $\mathbf{F}$ is in the direction of $\mathbf{n} = \frac{-x}{\sqrt{x^2+y^2}}\mathbf{i} - \frac{y}{\sqrt{x^2+y^2}}\mathbf{j}$
$\Rightarrow \mathbf{F} = a\mathbf{n}$, for some constant $a > 0$. Then $M(x,y) = \frac{-ax}{\sqrt{x^2+y^2}}$ and $N(x,y) = \frac{-ay}{\sqrt{x^2+y^2}}$
$\Rightarrow \sqrt{(M(x,y))^2+(N(x,y))^2} = a \Rightarrow a = \frac{k}{x^2+y^2} \Rightarrow \mathbf{F} = \frac{-kx}{(x^2+y^2)^{3/2}}\mathbf{i} - \frac{ky}{(x^2+y^2)^{3/2}}\mathbf{j}$, for any constant $k > 0$

7. Substitute the parametric representations for $\mathbf{r}(t) = x(t)\mathbf{i} + y(t)\mathbf{j} + z(t)\mathbf{k}$ representing each path into the vector field $\mathbf{F}$, and calculate $\int_C \mathbf{F}\cdot\frac{d\mathbf{r}}{dt}$.
(a) $\mathbf{F} = 3t\mathbf{i} + 2t\mathbf{j} + 4t\mathbf{k}$ and $\frac{d\mathbf{r}}{dt} = \mathbf{i}+\mathbf{j}+\mathbf{k} \Rightarrow \mathbf{F}\cdot\frac{d\mathbf{r}}{dt} = 9t \Rightarrow \int_0^1 9t\,dt = \frac{9}{2}$
(b) $\mathbf{F} = 3t^2\mathbf{i} + 2t\mathbf{j} + 4t^4\mathbf{k}$ and $\frac{d\mathbf{r}}{dt} = \mathbf{i}+2t\mathbf{j}+4t^3\mathbf{k} \Rightarrow \mathbf{F}\cdot\frac{d\mathbf{r}}{dt} = 7t^2+16t^7 \Rightarrow \int_0^1 (7t^2+16t^7)\,dt = \left[\frac{7}{3}t^3+2t^8\right]_0^1$ $= \frac{7}{3}+2 = \frac{13}{3}$
(c) $\mathbf{r}_1 = t\mathbf{i}+t\mathbf{j}$ and $\mathbf{r}_2 = \mathbf{i}+\mathbf{j}+t\mathbf{k}$; $\mathbf{F}_1 = 3t\mathbf{i}+2t\mathbf{j}$ and $\frac{d\mathbf{r}_1}{dt} = \mathbf{i}+\mathbf{j} \Rightarrow \mathbf{F}_1\cdot\frac{d\mathbf{r}_1}{dt} = 5t \Rightarrow \int_0^1 5t\,dt = \frac{5}{2}$;
$\mathbf{F}_2 = 3\mathbf{i}+2\mathbf{j}+4t\mathbf{k}$ and $\frac{d\mathbf{r}_2}{dt} = \mathbf{k} \Rightarrow \mathbf{F}_2\cdot\frac{d\mathbf{r}_2}{dt} = 4t \Rightarrow \int_0^1 4t\,dt = 2 \Rightarrow \frac{5}{2}+2 = \frac{9}{2}$

9. Substitute the parametric representation for $\mathbf{r}(t) = x(t)\mathbf{i} + y(t)\mathbf{j} + z(t)\mathbf{k}$ representing each path into the vector field $\mathbf{F}$, and calculate $\int_C \mathbf{F}\cdot\frac{d\mathbf{r}}{dt}$.
(a) $\mathbf{F} = \sqrt{t}\mathbf{i} - 2t\mathbf{j} + \sqrt{t}\mathbf{k}$ and $\frac{d\mathbf{r}}{dt} = \mathbf{i}+\mathbf{j}+\mathbf{k} \Rightarrow \mathbf{F}\cdot\frac{d\mathbf{r}}{dt} = 2\sqrt{t}-2t \Rightarrow \int_0^1 (2\sqrt{t}-2t)\,dt = \left[\frac{4}{3}t^{3/2}-t^2\right]_0^1 = \frac{1}{3}$
(b) $\mathbf{F} = t^2\mathbf{i} - 2t\mathbf{j} + t\mathbf{k}$ and $\frac{d\mathbf{r}}{dt} = \mathbf{i}+2t\mathbf{j}+4t^3\mathbf{k} \Rightarrow \mathbf{F}\cdot\frac{d\mathbf{r}}{dt} = 4t^4-3t^2 \Rightarrow \int_0^1 (4t^4-3t^2)\,dt = \left[\frac{4}{5}t^5-t^3\right]_0^1 = -\frac{1}{5}$
(c) $\mathbf{r}_1 = t\mathbf{i}+t\mathbf{j}$ and $\mathbf{r}_2 = \mathbf{i}+\mathbf{j}+t\mathbf{k}$; $\mathbf{F}_1 = -2t\mathbf{j}+\sqrt{t}\mathbf{k}$ and $\frac{d\mathbf{r}_1}{dt} = \mathbf{i}+\mathbf{j} \Rightarrow \mathbf{F}_1\cdot\frac{d\mathbf{r}_1}{dt} = -2t \Rightarrow \int_0^1 -2t\,dt = -1$;
$\mathbf{F}_2 = \sqrt{t}\mathbf{i}-2\mathbf{j}+\mathbf{k}$ and $\frac{d\mathbf{r}_2}{dt} = \mathbf{k} \Rightarrow \mathbf{F}_2\cdot\frac{d\mathbf{r}_2}{dt} = 1 \Rightarrow \int_0^1 dt = 1 \Rightarrow -1+1 = 0$

11. Substitute the parametric representation for $\mathbf{r}(t) = x(t)\mathbf{i} + y(t)\mathbf{j} + z(t)\mathbf{k}$ representing each path into the vector field $\mathbf{F}$, and calculate $\int_C \mathbf{F}\cdot\frac{d\mathbf{r}}{dt}$.
(a) $\mathbf{F} = (3t^2-3t)\mathbf{i} + 3t\mathbf{j} + \mathbf{k}$ and $\frac{d\mathbf{r}}{dt} = \mathbf{i}+\mathbf{j}+\mathbf{k} \Rightarrow \mathbf{F}\cdot\frac{d\mathbf{r}}{dt} = 3t^2+1 \Rightarrow \int_0^1 (3t^2+1)\,dt = [t^3+t]_0^1 = 2$
(b) $\mathbf{F} = (3t^2-3t)\mathbf{i} + 3t^4\mathbf{j} + \mathbf{k}$ and $\frac{d\mathbf{r}}{dt} = \mathbf{i}+2t\mathbf{j}+4t^3\mathbf{k} \Rightarrow \mathbf{F}\cdot\frac{d\mathbf{r}}{dt} = 6t^5+4t^3+3t^2-3t$
$\Rightarrow \int_0^1 (6t^5+4t^3+3t^2-3t)\,dt = \left[t^6+t^4+t^3-\frac{3}{2}t^2\right]_0^1 = \frac{3}{2}$
(c) $\mathbf{r}_1 = t\mathbf{i}+t\mathbf{j}$ and $\mathbf{r}_2 = \mathbf{i}+\mathbf{j}+t\mathbf{k}$; $\mathbf{F}_1 = (3t^2-3t)\mathbf{i}+\mathbf{k}$ and $\frac{d\mathbf{r}_1}{dt} = \mathbf{i}+\mathbf{j} \Rightarrow \mathbf{F}_1\cdot\frac{d\mathbf{r}_1}{dt} = 3t^2-3t$
$\Rightarrow \int_0^1 (3t^2-3t)\,dt = \left[t^3-\frac{3}{2}t^2\right]_0^1 = -\frac{1}{2}$; $\mathbf{F}_2 = 3t\mathbf{j}+\mathbf{k}$ and $\frac{d\mathbf{r}_2}{dt} = \mathbf{k} \Rightarrow \mathbf{F}_2\cdot\frac{d\mathbf{r}_2}{dt} = 1 \Rightarrow \int_0^1 dt = 1$
$\Rightarrow -\frac{1}{2}+1 = \frac{1}{2}$

13. $x = t$, $y = 2t+1$, $0 \le t \le 3 \Rightarrow dx = dt \Rightarrow \int_C (x-y)\,dx = \int_0^3 (t-(2t+1))\,dt = \int_0^3 (-t-1)\,dt = \left[-\frac{1}{2}t^2-t\right]_0^3 = -\frac{15}{2}$

15. $C_1\colon x = t,\ y = 0,\ 0 \le t \le 3 \Rightarrow dy = 0;\ C_2\colon x = 3,\ y = t,\ 0 \le t \le 3 \Rightarrow dy = dt \Rightarrow \int_C (x^2 + y^2)\,dy$
$= \int_{C_1} (x^2 + y^2)\,dx + \int_{C_2} (x^2 + y^2)\,dx = \int_0^3 (t^2 + 0^2)\cdot 0 + \int_0^3 (3^2 + t^2)\,dt = \int_0^3 (9 + t^2)dt = \left[9t + \frac{1}{3}t^3\right]_0^3 = 36$

17. $\mathbf{r}(t) = t\mathbf{i} - \mathbf{j} + t^2\mathbf{k},\ 0 \le t \le 1 \Rightarrow dx = dt,\ dy = 0,\ dz = 2t\,dt$
(a) $\int_C (x + y - z)\,dx = \int_0^1 (t - 1 - t^2)\,dt = \left[\frac{1}{2}t^2 - t - \frac{1}{3}t^3\right]_0^1 = -\frac{5}{6}$
(b) $\int_C (x + y - z)\,dy = \int_0^1 (t - 1 - t^2)\cdot 0 = 0$
(c) $\int_C (x + y - z)\,dz = \int_0^1 (t - 1 - t^2)\,2t\,dt = \int_0^1 (2t^2 - 2t - 2t^3)\,dt = = \left[\frac{2}{3}t^3 - t^2 - \frac{1}{2}t^4\right]_0^1 = -\frac{5}{6}$

19. $\mathbf{r} = t\mathbf{i} + t^2\mathbf{j} + t\mathbf{k},\ 0 \le t \le 1$, and $\mathbf{F} = xy\mathbf{i} + y\mathbf{j} - yz\mathbf{k} \Rightarrow \mathbf{F} = t^3\mathbf{i} + t^2\mathbf{j} - t^3\mathbf{k}$ and $\frac{d\mathbf{r}}{dt} = \mathbf{i} + 2t\mathbf{j} + \mathbf{k}$
$\Rightarrow \mathbf{F}\cdot\frac{d\mathbf{r}}{dt} = 2t^3 \Rightarrow \text{work} = \int_0^1 2t^3\,dt = \frac{1}{2}$

21. $\mathbf{r} = (\sin t)\mathbf{i} + (\cos t)\mathbf{j} + t\mathbf{k},\ 0 \le t \le 2\pi$, and $\mathbf{F} = z\mathbf{i} + x\mathbf{j} + y\mathbf{k} \Rightarrow \mathbf{F} = t\mathbf{i} + (\sin t)\mathbf{j} + (\cos t)\mathbf{k}$ and
$\frac{d\mathbf{r}}{dt} = (\cos t)\mathbf{i} - (\sin t)\mathbf{j} + \mathbf{k} \Rightarrow \mathbf{F}\cdot\frac{d\mathbf{r}}{dt} = t\cos t - \sin^2 t + \cos t \Rightarrow \text{work} = \int_0^{2\pi} (t\cos t - \sin^2 t + \cos t)\,dt$
$= \left[\cos t + t\sin t - \frac{t}{2} + \frac{\sin 2t}{4} + \sin t\right]_0^{2\pi} = -\pi$

23. $x = t$ and $y = x^2 = t^2 \Rightarrow \mathbf{r} = t\mathbf{i} + t^2\mathbf{j},\ -1 \le t \le 2$, and $\mathbf{F} = xy\mathbf{i} + (x + y)\mathbf{j} \Rightarrow \mathbf{F} = t^3\mathbf{i} + (t + t^2)\,\mathbf{j}$ and
$\frac{d\mathbf{r}}{dt} = \mathbf{i} + 2t\mathbf{j} \Rightarrow \mathbf{F}\cdot\frac{d\mathbf{r}}{dt} = t^3 + (2t^2 + 2t^3) = 3t^3 + 2t^2 \Rightarrow \int_C xy\,dx + (x + y)\,dy = \int_C \mathbf{F}\cdot\frac{d\mathbf{r}}{dt}\,dt = \int_{-1}^2 (3t^3 + 2t^2)\,dt$
$= \left[\frac{3}{4}t^4 + \frac{2}{3}t^3\right]_{-1}^2 = \left(12 + \frac{16}{3}\right) - \left(\frac{3}{4} - \frac{2}{3}\right) = \frac{45}{4} + \frac{18}{3} = \frac{69}{4}$

25. $\mathbf{r} = x\mathbf{i} + y\mathbf{j} = y^2\mathbf{i} + y\mathbf{j},\ 2 \ge y \ge -1$, and $\mathbf{F} = x^2\mathbf{i} - y\mathbf{j} = y^4\mathbf{i} - y\mathbf{j} \Rightarrow \frac{d\mathbf{r}}{dy} = 2y\mathbf{i} + \mathbf{j}$ and $\mathbf{F}\cdot\frac{d\mathbf{r}}{dy} = 2y^5 - y$
$\Rightarrow \int_C \mathbf{F}\cdot\mathbf{T}\,ds = \int_2^{-1} \mathbf{F}\cdot\frac{d\mathbf{r}}{dy}\,dy = \int_2^{-1} (2y^5 - y)\,dy = \left[\frac{1}{3}y^6 - \frac{1}{2}y^2\right]_2^{-1} = \left(\frac{1}{3} - \frac{1}{2}\right) - \left(\frac{64}{3} - \frac{4}{2}\right) = \frac{3}{2} - \frac{63}{3} = -\frac{39}{2}$

27. $\mathbf{r} = (\mathbf{i} + \mathbf{j}) + t(\mathbf{i} + 2\mathbf{j}) = (1 + t)\mathbf{i} + (1 + 2t)\mathbf{j},\ 0 \le t \le 1$, and $\mathbf{F} = xy\mathbf{i} + (y - x)\mathbf{j} \Rightarrow \mathbf{F} = (1 + 3t + 2t^2)\,\mathbf{i} + t\mathbf{j}$ and
$\frac{d\mathbf{r}}{dt} = \mathbf{i} + 2\mathbf{j} \Rightarrow \mathbf{F}\cdot\frac{d\mathbf{r}}{dt} = 1 + 5t + 2t^2 \Rightarrow \text{work} = \int_C \mathbf{F}\cdot\frac{d\mathbf{r}}{dt}\,dt = \int_0^1 (1 + 5t + 2t^2)\,dt = \left[t + \frac{5}{2}t^2 + \frac{2}{3}t^3\right]_0^1 = \frac{25}{6}$

29. (a) $\mathbf{r} = (\cos t)\mathbf{i} + (\sin t)\mathbf{j},\ 0 \le t \le 2\pi,\ \mathbf{F}_1 = x\mathbf{i} + y\mathbf{j}$, and $\mathbf{F}_2 = -y\mathbf{i} + x\mathbf{j} \Rightarrow \frac{d\mathbf{r}}{dt} = (-\sin t)\mathbf{i} + (\cos t)\mathbf{j}$,
$\mathbf{F}_1 = (\cos t)\mathbf{i} + (\sin t)\mathbf{j}$, and $\mathbf{F}_2 = (-\sin t)\mathbf{i} + (\cos t)\mathbf{j} \Rightarrow \mathbf{F}_1\cdot\frac{d\mathbf{r}}{dt} = 0$ and $\mathbf{F}_2\cdot\frac{d\mathbf{r}}{dt} = \sin^2 t + \cos^2 t = 1$
$\Rightarrow \text{Circ}_1 = \int_0^{2\pi} 0\,dt = 0$ and $\text{Circ}_2 = \int_0^{2\pi} dt = 2\pi;\ \mathbf{n} = (\cos t)\mathbf{i} + (\sin t)\mathbf{j} \Rightarrow \mathbf{F}_1\cdot\mathbf{n} = \cos^2 t + \sin^2 t = 1$ and
$\mathbf{F}_2\cdot\mathbf{n} = 0 \Rightarrow \text{Flux}_1 = \int_0^{2\pi} dt = 2\pi$ and $\text{Flux}_2 = \int_0^{2\pi} 0\,dt = 0$
(b) $\mathbf{r} = (\cos t)\mathbf{i} + (4\sin t)\mathbf{j},\ 0 \le t \le 2\pi \Rightarrow \frac{d\mathbf{r}}{dt} = (-\sin t)\mathbf{i} + (4\cos t)\mathbf{j},\ \mathbf{F}_1 = (\cos t)\mathbf{i} + (4\sin t)\mathbf{j}$, and
$\mathbf{F}_2 = (-4\sin t)\mathbf{i} + (\cos t)\mathbf{j} \Rightarrow \mathbf{F}_1\cdot\frac{d\mathbf{r}}{dt} = 15\sin t\cos t$ and $\mathbf{F}_2\cdot\frac{d\mathbf{r}}{dt} = 4 \Rightarrow \text{Circ}_1 = \int_0^{2\pi} 15\sin t\cos t\,dt$
$= \left[\frac{15}{2}\sin^2 t\right]_0^{2\pi} = 0$ and $\text{Circ}_2 = \int_0^{2\pi} 4\,dt = 8\pi;\ \mathbf{n} = \left(\frac{4}{\sqrt{17}}\cos t\right)\mathbf{i} + \left(\frac{1}{\sqrt{17}}\sin t\right)\mathbf{j} \Rightarrow \mathbf{F}_1\cdot\mathbf{n}$
$= \frac{4}{\sqrt{17}}\cos^2 t + \frac{4}{\sqrt{17}}\sin^2 t$ and $\mathbf{F}_2\cdot\mathbf{n} = -\frac{15}{\sqrt{17}}\sin t\cos t \Rightarrow \text{Flux}_1 = \int_0^{2\pi} (\mathbf{F}_1\cdot\mathbf{n})\,|\mathbf{v}|\,dt = \int_0^{2\pi} \left(\frac{4}{\sqrt{17}}\right)\sqrt{17}\,dt$
$= 8\pi$ and $\text{Flux}_2 = \int_0^{2\pi} (\mathbf{F}_2\cdot\mathbf{n})\,|\mathbf{v}|\,dt = \int_0^{2\pi} \left(-\frac{15}{\sqrt{17}}\sin t\cos t\right)\sqrt{17}\,dt = \left[-\frac{15}{2}\sin^2 t\right]_0^{2\pi} = 0$

31. $\mathbf{F}_1 = (a\cos t)\mathbf{i} + (a\sin t)\mathbf{j},\ \frac{d\mathbf{r}_1}{dt} = (-a\sin t)\mathbf{i} + (a\cos t)\mathbf{j} \Rightarrow \mathbf{F}_1\cdot\frac{d\mathbf{r}_1}{dt} = 0 \Rightarrow \text{Circ}_1 = 0;\ M_1 = a\cos t$,
$N_1 = a\sin t,\ dx = -a\sin t\,dt,\ dy = a\cos t\,dt \Rightarrow \text{Flux}_1 = \int_C M_1\,dy - N_1\,dx = \int_0^\pi (a^2\cos^2 t + a^2\sin^2 t)\,dt$
$= \int_0^\pi a^2\,dt = a^2\pi;$

$\mathbf{F}_2 = t\mathbf{i}$, $\frac{d\mathbf{r}_2}{dt} = \mathbf{i} \Rightarrow \mathbf{F}_2 \cdot \frac{d\mathbf{r}_2}{dt} = t \Rightarrow \text{Circ}_2 = \int_{-a}^{a} t\,dt = 0$; $M_2 = t, N_2 = 0, dx = dt, dy = 0 \Rightarrow \text{Flux}_2$
$= \int_C M_2\,dy - N_2\,dx = \int_{-a}^{a} 0\,dt = 0$; therefore, $\text{Circ} = \text{Circ}_1 + \text{Circ}_2 = 0$ and $\text{Flux} = \text{Flux}_1 + \text{Flux}_2 = a^2\pi$

33. $\mathbf{F}_1 = (-a\sin t)\mathbf{i} + (a\cos t)\mathbf{j}$, $\frac{d\mathbf{r}_1}{dt} = (-a\sin t)\mathbf{i} + (a\cos t)\mathbf{j} \Rightarrow \mathbf{F}_1 \cdot \frac{d\mathbf{r}_1}{dt} = a^2\sin^2 t + a^2\cos^2 t = a^2$
$\Rightarrow \text{Circ}_1 = \int_0^{\pi} a^2\,dt = a^2\pi$; $M_1 = -a\sin t$, $N_1 = a\cos t$, $dx = -a\sin t\,dt$, $dy = a\cos t\,dt$
$\Rightarrow \text{Flux}_1 = \int_C M_1\,dy - N_1\,dx = \int_0^{\pi} (-a^2\sin t\cos t + a^2\sin t\cos t)\,dt = 0$; $\mathbf{F}_2 = t\mathbf{j}$, $\frac{d\mathbf{r}_2}{dt} = \mathbf{i} \Rightarrow \mathbf{F}_2 \cdot \frac{d\mathbf{r}_2}{dt} = 0$
$\Rightarrow \text{Circ}_2 = 0$; $M_2 = 0, N_2 = t, dx = dt, dy = 0 \Rightarrow \text{Flux}_2 = \int_C M_2\,dy - N_2\,dx = \int_{-a}^{a} -t\,dt = 0$; therefore,
$\text{Circ} = \text{Circ}_1 + \text{Circ}_2 = a^2\pi$ and $\text{Flux} = \text{Flux}_1 + \text{Flux}_2 = 0$

35. (a) $\mathbf{r} = (\cos t)\mathbf{i} + (\sin t)\mathbf{j}$, $0 \le t \le \pi$, and $\mathbf{F} = (x+y)\mathbf{i} - (x^2+y^2)\mathbf{j} \Rightarrow \frac{d\mathbf{r}}{dt} = (-\sin t)\mathbf{i} + (\cos t)\mathbf{j}$ and
$\mathbf{F} = (\cos t + \sin t)\mathbf{i} - (\cos^2 t + \sin^2 t)\mathbf{j} \Rightarrow \mathbf{F} \cdot \frac{d\mathbf{r}}{dt} = -\sin t\cos t - \sin^2 t - \cos t \Rightarrow \int_C \mathbf{F} \cdot \mathbf{T}\,ds$
$= \int_0^{\pi} (-\sin t\cos t - \sin^2 t - \cos t)\,dt = \left[-\frac{1}{2}\sin^2 t - \frac{t}{2} + \frac{\sin 2t}{4} - \sin t\right]_0^{\pi} = -\frac{\pi}{2}$

(b) $\mathbf{r} = (1-2t)\mathbf{i}$, $0 \le t \le 1$, and $\mathbf{F} = (x+y)\mathbf{i} - (x^2+y^2)\mathbf{j} \Rightarrow \frac{d\mathbf{r}}{dt} = -2\mathbf{i}$ and $\mathbf{F} = (1-2t)\mathbf{i} - (1-2t)^2\mathbf{j} \Rightarrow$
$\mathbf{F} \cdot \frac{d\mathbf{r}}{dt} = 4t - 2 \Rightarrow \int_C \mathbf{F} \cdot \mathbf{T}\,ds = \int_0^1 (4t-2)\,dt = \left[2t^2 - 2t\right]_0^1 = 0$

(c) $\mathbf{r}_1 = (1-t)\mathbf{i} - t\mathbf{j}$, $0 \le t \le 1$, and $\mathbf{F} = (x+y)\mathbf{i} - (x^2+y^2)\mathbf{j} \Rightarrow \frac{d\mathbf{r}_1}{dt} = -\mathbf{i} - \mathbf{j}$ and $\mathbf{F} = (1-2t)\mathbf{i} - (1-2t+2t^2)\mathbf{j}$
$\Rightarrow \mathbf{F} \cdot \frac{d\mathbf{r}_1}{dt} = (2t-1) + (1-2t+2t^2) = 2t^2 \Rightarrow \text{Flow}_1 = \int_{C_1} \mathbf{F} \cdot \frac{d\mathbf{r}_1}{dt} = \int_0^1 2t^2\,dt = \frac{2}{3}$; $\mathbf{r}_2 = -t\mathbf{i} + (t-1)\mathbf{j}$,
$0 \le t \le 1$, and $\mathbf{F} = (x+y)\mathbf{i} - (x^2+y^2)\mathbf{j} \Rightarrow \frac{d\mathbf{r}_2}{dt} = -\mathbf{i} + \mathbf{j}$ and $\mathbf{F} = -\mathbf{i} - (t^2 + t^2 - 2t + 1)\mathbf{j}$
$= -\mathbf{i} - (2t^2 - 2t + 1)\mathbf{j} \Rightarrow \mathbf{F} \cdot \frac{d\mathbf{r}_2}{dt} = 1 - (2t^2 - 2t + 1) = 2t - 2t^2 \Rightarrow \text{Flow}_2 = \int_{C_2} \mathbf{F} \cdot \frac{d\mathbf{r}_2}{dt} = \int_0^1 (2t - 2t^2)\,dt$
$= \left[t^2 - \frac{2}{3}t^3\right]_0^1 = \frac{1}{3} \Rightarrow \text{Flow} = \text{Flow}_1 + \text{Flow}_2 = \frac{2}{3} + \frac{1}{3} = 1$

37. (a) $y = 2x, 0 \le x \le 2 \Rightarrow \mathbf{r}(t) = t\mathbf{i} + 2t\mathbf{j}, 0 \le t \le 2 \Rightarrow \frac{d\mathbf{r}}{dt} = \mathbf{i} + 2\mathbf{j} \Rightarrow \mathbf{F} \cdot \frac{d\mathbf{r}}{dt} = \left((2t)^2\mathbf{i} + 2(t)(2t)\mathbf{j}\right) \cdot (\mathbf{i} + 2\mathbf{j})$
$= 4t^2 + 8t^2 = 12t^2 \Rightarrow \text{Flow} = \int_C \mathbf{F} \cdot \frac{d\mathbf{r}}{dt}\,dt = \int_0^2 12t^2\,dt = \left[4t^3\right]_0^2 = 32$

(b) $y = x^2, 0 \le x \le 2 \Rightarrow \mathbf{r}(t) = t\mathbf{i} + t^2\mathbf{j}, 0 \le t \le 2 \Rightarrow \frac{d\mathbf{r}}{dt} = \mathbf{i} + 2t\mathbf{j} \Rightarrow \mathbf{F} \cdot \frac{d\mathbf{r}}{dt} = \left((t^2)^2\mathbf{i} + 2(t)(t^2)\mathbf{j}\right) \cdot (\mathbf{i} + 2t\mathbf{j})$
$= t^4 + 4t^4 = 5t^4 \Rightarrow \text{Flow} = \int_C \mathbf{F} \cdot \frac{d\mathbf{r}}{dt}\,dt = \int_0^2 5t^4\,dt = \left[t^5\right]_0^2 = 32$

(c) answers will vary, one possible path is $y = \frac{1}{2}x^3, 0 \le x \le 2 \Rightarrow \mathbf{r}(t) = t\mathbf{i} + \frac{1}{2}t^3\mathbf{j}, 0 \le t \le 2 \Rightarrow \frac{d\mathbf{r}}{dt} = \mathbf{i} + 3t^2\mathbf{j}$
$\Rightarrow \mathbf{F} \cdot \frac{d\mathbf{r}}{dt} = \left(\left(\frac{1}{2}t^3\right)^2\mathbf{i} + 2(t)\left(\frac{1}{2}t^3\right)\mathbf{j}\right) \cdot (\mathbf{i} + 3t^2\mathbf{j}) = \frac{1}{4}t^6 + \frac{3}{2}t^6 = \frac{7}{4}t^6 \Rightarrow \text{Flow} = \int_C \mathbf{F} \cdot \frac{d\mathbf{r}}{dt}\,dt = \int_0^2 \frac{7}{4}t^6\,dt = \left[\frac{1}{4}t^7\right]_0^2$
$= 32$

39. $\mathbf{F} = -\frac{y}{\sqrt{x^2+y^2}}\mathbf{i} + \frac{x}{\sqrt{x^2+y^2}}\mathbf{j}$ on $x^2 + y^2 = 4$;
at $(2,0)$, $\mathbf{F} = \mathbf{j}$; at $(0,2)$, $\mathbf{F} = -\mathbf{i}$; at $(-2,0)$,
$\mathbf{F} = -\mathbf{j}$; at $(0,-2)$, $\mathbf{F} = \mathbf{i}$; at $\left(\sqrt{2}, \sqrt{2}\right)$, $\mathbf{F} = -\frac{\sqrt{3}}{2}\mathbf{i} + \frac{1}{2}\mathbf{j}$;
at $\left(\sqrt{2}, -\sqrt{2}\right)$, $\mathbf{F} = \frac{\sqrt{3}}{2}\mathbf{i} + \frac{1}{2}\mathbf{j}$; at $\left(-\sqrt{2}, \sqrt{2}\right)$,
$\mathbf{F} = -\frac{\sqrt{3}}{2}\mathbf{i} - \frac{1}{2}\mathbf{j}$; at $\left(-\sqrt{2}, -\sqrt{2}\right)$, $\mathbf{F} = \frac{\sqrt{3}}{2}\mathbf{i} - \frac{1}{2}\mathbf{j}$

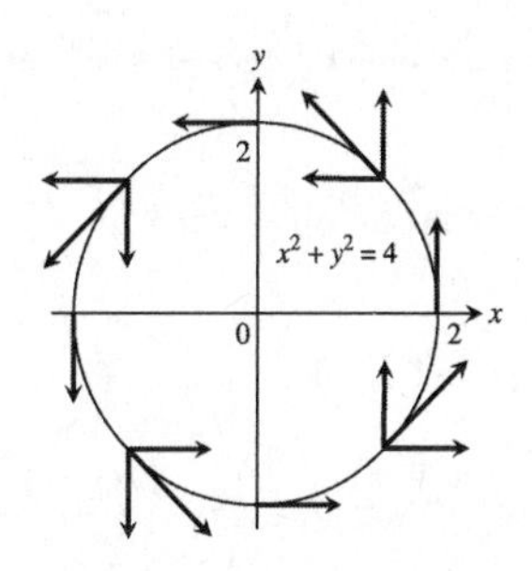

41. (a) $\mathbf{G} = P(x,y)\mathbf{i} + Q(x,y)\mathbf{j}$ is to have a magnitude $\sqrt{a^2+b^2}$ and to be tangent to $x^2 + y^2 = a^2 + b^2$ in a counterclockwise direction. Thus $x^2 + y^2 = a^2 + b^2 \Rightarrow 2x + 2yy' = 0 \Rightarrow y' = -\frac{x}{y}$ is the slope of the tangent line at any point on the circle $\Rightarrow y' = -\frac{a}{b}$ at (a,b). Let $\mathbf{v} = -b\mathbf{i} + a\mathbf{j} \Rightarrow |\mathbf{v}| = \sqrt{a^2+b^2}$, with $\mathbf{v}$ in a

counterclockwise direction and tangent to the circle. Then let $P(x, y) = -y$ and $Q(x, y) = x$
$\Rightarrow \mathbf{G} = -y\mathbf{i} + x\mathbf{j} \Rightarrow$ for (a, b) on $x^2 + y^2 = a^2 + b^2$ we have $\mathbf{G} = -b\mathbf{i} + a\mathbf{j}$ and $|\mathbf{G}| = \sqrt{a^2 + b^2}$.

(b) $\mathbf{G} = \left(\sqrt{x^2 + y^2}\right)\mathbf{F} = \left(\sqrt{a^2 + b^2}\right)\mathbf{F}$.

43. The slope of the line through (x, y) and the origin is $\frac{y}{x} \Rightarrow \mathbf{v} = x\mathbf{i} + y\mathbf{j}$ is a vector parallel to that line and pointing away from the origin $\Rightarrow \mathbf{F} = -\frac{x\mathbf{i} + y\mathbf{j}}{\sqrt{x^2 + y^2}}$ is the unit vector pointing toward the origin.

45. Yes. The work and area have the same numerical value because work $= \int_C \mathbf{F} \cdot d\mathbf{r} = \int_C y\mathbf{i} \cdot d\mathbf{r}$
$= \int_b^a [f(t)\mathbf{i}] \cdot \left[\mathbf{i} + \frac{df}{dt}\mathbf{j}\right] dt$ [On the path, y equals f(t)]
$= \int_a^b f(t)\, dt =$ Area under the curve [because f(t) > 0]

47. $\mathbf{F} = -4t^3\mathbf{i} + 8t^2\mathbf{j} + 2\mathbf{k}$ and $\frac{d\mathbf{r}}{dt} = \mathbf{i} + 2t\mathbf{j} \Rightarrow \mathbf{F} \cdot \frac{d\mathbf{r}}{dt} = 12t^3 \Rightarrow \text{Flow} = \int_0^2 12t^3\, dt = \left[3t^4\right]_0^2 = 48$

49. $\mathbf{F} = (\cos t - \sin t)\mathbf{i} + (\cos t)\mathbf{k}$ and $\frac{d\mathbf{r}}{dt} = (-\sin t)\mathbf{i} + (\cos t)\mathbf{k} \Rightarrow \mathbf{F} \cdot \frac{d\mathbf{r}}{dt} = -\sin t \cos t + 1$
$\Rightarrow \text{Flow} = \int_0^\pi (-\sin t \cos t + 1)\, dt = \left[\frac{1}{2}\cos^2 t + t\right]_0^\pi = \left(\frac{1}{2} + \pi\right) - \left(\frac{1}{2} + 0\right) = \pi$

51. C_1: $\mathbf{r} = (\cos t)\mathbf{i} + (\sin t)\mathbf{j} + t\mathbf{k},\ 0 \le t \le \frac{\pi}{2} \Rightarrow \mathbf{F} = (2\cos t)\mathbf{i} + 2t\mathbf{j} + (2\sin t)\mathbf{k}$ and $\frac{d\mathbf{r}}{dt} = (-\sin t)\mathbf{i} + (\cos t)\mathbf{j} + \mathbf{k}$
$\Rightarrow \mathbf{F} \cdot \frac{d\mathbf{r}}{dt} = -2\cos t \sin t + 2t\cos t + 2\sin t = -\sin 2t + 2t\cos t + 2\sin t$
$\Rightarrow \text{Flow}_1 = \int_0^{\pi/2} (-\sin 2t + 2t\cos t + 2\sin t)\, dt = \left[\frac{1}{2}\cos 2t + 2t\sin t + 2\cos t - 2\cos t\right]_0^{\pi/2} = -1 + \pi;$
C_2: $\mathbf{r} = \mathbf{j} + \frac{\pi}{2}(1 - t)\mathbf{k},\ 0 \le t \le 1 \Rightarrow \mathbf{F} = \pi(1 - t)\mathbf{j} + 2\mathbf{k}$ and $\frac{d\mathbf{r}}{dt} = -\frac{\pi}{2}\mathbf{k} \Rightarrow \mathbf{F} \cdot \frac{d\mathbf{r}}{dt} = -\pi$
$\Rightarrow \text{Flow}_2 = \int_0^1 -\pi\, dt = \left[-\pi t\right]_0^1 = -\pi;$
C_3: $\mathbf{r} = t\mathbf{i} + (1 - t)\mathbf{j},\ 0 \le t \le 1 \Rightarrow \mathbf{F} = 2t\mathbf{i} + 2(1 - t)\mathbf{k}$ and $\frac{d\mathbf{r}}{dt} = \mathbf{i} - \mathbf{j} \Rightarrow \mathbf{F} \cdot \frac{d\mathbf{r}}{dt} = 2t$
$\Rightarrow \text{Flow}_3 = \int_0^1 2t\, dt = \left[t^2\right]_0^1 = 1 \Rightarrow \text{Circulation} = (-1 + \pi) - \pi + 1 = 0$

53. Let $x = t$ be the parameter $\Rightarrow y = x^2 = t^2$ and $z = x = t \Rightarrow \mathbf{r} = t\mathbf{i} + t^2\mathbf{j} + t\mathbf{k},\ 0 \le t \le 1$ from (0, 0, 0) to (1, 1, 1)
$\Rightarrow \frac{d\mathbf{r}}{dt} = \mathbf{i} + 2t\mathbf{j} + \mathbf{k}$ and $\mathbf{F} = xy\mathbf{i} + y\mathbf{j} - yz\mathbf{k} = t^3\mathbf{i} + t^2\mathbf{j} - t^3\mathbf{k} \Rightarrow \mathbf{F} \cdot \frac{d\mathbf{r}}{dt} = t^3 + 2t^3 - t^3 = 2t^3 \Rightarrow \text{Flow} = \int_0^1 2t^3\, dt$
$= \frac{1}{2}$

16.3 PATH INDEPENDENCE, POTENTIAL FUNCTIONS, AND CONSERVATIVE FIELDS

1. $\frac{\partial P}{\partial y} = x = \frac{\partial N}{\partial z},\ \frac{\partial M}{\partial z} = y = \frac{\partial P}{\partial x},\ \frac{\partial N}{\partial x} = z = \frac{\partial M}{\partial y} \Rightarrow$ Conservative

3. $\frac{\partial P}{\partial y} = -1 \ne 1 = \frac{\partial N}{\partial z} \Rightarrow$ Not Conservative

5. $\frac{\partial N}{\partial x} = 0 \ne 1 = \frac{\partial M}{\partial y} \Rightarrow$ Not Conservative

7. $\frac{\partial f}{\partial x} = 2x \Rightarrow f(x, y, z) = x^2 + g(y, z) \Rightarrow \frac{\partial f}{\partial y} = \frac{\partial g}{\partial y} = 3y \Rightarrow g(y, z) = \frac{3y^2}{2} + h(z) \Rightarrow f(x, y, z) = x^2 + \frac{3y^2}{2} + h(z)$
$\Rightarrow \frac{\partial f}{\partial z} = h'(z) = 4z \Rightarrow h(z) = 2z^2 + C \Rightarrow f(x, y, z) = x^2 + \frac{3y^2}{2} + 2z^2 + C$

9. $\frac{\partial f}{\partial x} = e^{y+2z} \Rightarrow f(x, y, z) = xe^{y+2z} + g(y, z) \Rightarrow \frac{\partial f}{\partial y} = xe^{y+2z} + \frac{\partial g}{\partial y} = xe^{y+2z} \Rightarrow \frac{\partial g}{\partial y} = 0 \Rightarrow f(x, y, z)$
$= xe^{y+2z} + h(z) \Rightarrow \frac{\partial f}{\partial z} = 2xe^{y+2z} + h'(z) = 2xe^{y+2z} \Rightarrow h'(z) = 0 \Rightarrow h(z) = C \Rightarrow f(x, y, z) = xe^{y+2z} + C$

11. $\frac{\partial f}{\partial z} = \frac{z}{y^2+z^2} \Rightarrow f(x, y, z) = \frac{1}{2}\ln(y^2+z^2) + g(x, y) \Rightarrow \frac{\partial f}{\partial x} = \frac{\partial g}{\partial x} = \ln x + \sec^2(x+y) \Rightarrow g(x, y)$
$= (x\ln x - x) + \tan(x+y) + h(y) \Rightarrow f(x, y, z) = \frac{1}{2}\ln(y^2+z^2) + (x\ln x - x) + \tan(x+y) + h(y)$
$\Rightarrow \frac{\partial f}{\partial y} = \frac{y}{y^2+z^2} + \sec^2(x+y) + h'(y) = \sec^2(x+y) + \frac{y}{y^2+z^2} \Rightarrow h'(y) = 0 \Rightarrow h(y) = C \Rightarrow f(x, y, z)$
$= \frac{1}{2}\ln(y^2+z^2) + (x\ln x - x) + \tan(x+y) + C$

13. Let $\mathbf{F}(x, y, z) = 2x\mathbf{i} + 2y\mathbf{j} + 2z\mathbf{k} \Rightarrow \frac{\partial P}{\partial y} = 0 = \frac{\partial N}{\partial z}, \frac{\partial M}{\partial z} = 0 = \frac{\partial P}{\partial x}, \frac{\partial N}{\partial x} = 0 = \frac{\partial M}{\partial y} \Rightarrow M\,dx + N\,dy + P\,dz$ is exact; $\frac{\partial f}{\partial x} = 2x \Rightarrow f(x, y, z) = x^2 + g(y, z) \Rightarrow \frac{\partial f}{\partial y} = \frac{\partial g}{\partial y} = 2y \Rightarrow g(y, z) = y^2 + h(z) \Rightarrow f(x, y, z) = x^2 + y^2 = h(z)$
$\Rightarrow \frac{\partial f}{\partial z} = h'(z) = 2z \Rightarrow h(z) = z^2 + C \Rightarrow f(x, y, z) = x^2 + y^2 + z^2 + C \Rightarrow \int_{(0,0,0)}^{(2,3,-6)} 2x\,dx + 2y\,dy + 2z\,dz$
$= f(2, 3, -6) - f(0, 0, 0) = 2^2 + 3^2 + (-6)^2 = 49$

15. Let $\mathbf{F}(x, y, z) = 2xy\mathbf{i} + (x^2 - z^2)\mathbf{j} - 2yz\mathbf{k} \Rightarrow \frac{\partial P}{\partial y} = -2z = \frac{\partial N}{\partial z}, \frac{\partial M}{\partial z} = 0 = \frac{\partial P}{\partial x}, \frac{\partial N}{\partial x} = 2x = \frac{\partial M}{\partial y}$
$\Rightarrow M\,dx + N\,dy + P\,dz$ is exact; $\frac{\partial f}{\partial x} = 2xy \Rightarrow f(x, y, z) = x^2y + g(y, z) \Rightarrow \frac{\partial f}{\partial y} = x^2 + \frac{\partial g}{\partial y} = x^2 - z^2 \Rightarrow \frac{\partial g}{\partial y} = -z^2$
$\Rightarrow g(y, z) = -yz^2 + h(z) \Rightarrow f(x, y, z) = x^2y - yz^2 + h(z) \Rightarrow \frac{\partial f}{\partial z} = -2yz + h'(z) = -2yz \Rightarrow h'(z) = 0 \Rightarrow h(z) = C$
$\Rightarrow f(x, y, z) = x^2y - yz^2 + C \Rightarrow \int_{(0,0,0)}^{(1,2,3)} 2xy\,dx + (x^2 - z^2)\,dy - 2yz\,dz = f(1, 2, 3) - f(0, 0, 0) = 2 - 2(3)^2 = -16$

17. Let $\mathbf{F}(x, y, z) = (\sin y \cos x)\mathbf{i} + (\cos y \sin x)\mathbf{j} + \mathbf{k} \Rightarrow \frac{\partial P}{\partial y} = 0 = \frac{\partial N}{\partial z}, \frac{\partial M}{\partial z} = 0 = \frac{\partial P}{\partial x}, \frac{\partial N}{\partial x} = \cos y \cos x = \frac{\partial M}{\partial y}$
$\Rightarrow M\,dx + N\,dy + P\,dz$ is exact; $\frac{\partial f}{\partial x} = \sin y \cos x \Rightarrow f(x, y, z) = \sin y \sin x + g(y, z) \Rightarrow \frac{\partial f}{\partial y} = \cos y \sin x + \frac{\partial g}{\partial y}$
$= \cos y \sin x \Rightarrow \frac{\partial g}{\partial y} = 0 \Rightarrow g(y, z) = h(z) \Rightarrow f(x, y, z) = \sin y \sin x + h(z) \Rightarrow \frac{\partial f}{\partial z} = h'(z) = 1 \Rightarrow h(z) = z + C$
$\Rightarrow f(x, y, z) = \sin y \sin x + z + C \Rightarrow \int_{(1,0,0)}^{(0,1,1)} \sin y \cos x\,dx + \cos y \sin x\,dy + dz = f(0, 1, 1) - f(1, 0, 0)$
$= (0 + 1) - (0 + 0) = 1$

19. Let $\mathbf{F}(x, y, z) = 3x^2\mathbf{i} + \left(\frac{z^2}{y}\right)\mathbf{j} + (2z \ln y)\mathbf{k} \Rightarrow \frac{\partial P}{\partial y} = \frac{2z}{y} = \frac{\partial N}{\partial z}, \frac{\partial M}{\partial z} = 0 = \frac{\partial P}{\partial x}, \frac{\partial N}{\partial x} = 0 = \frac{\partial M}{\partial y}$
$\Rightarrow M\,dx + N\,dy + P\,dz$ is exact; $\frac{\partial f}{\partial x} = 3x^2 \Rightarrow f(x, y, z) = x^3 + g(y, z) \Rightarrow \frac{\partial f}{\partial y} = \frac{\partial g}{\partial y} = \frac{z^2}{y} \Rightarrow g(y, z) = z^2 \ln y + h(z)$
$\Rightarrow f(x, y, z) = x^3 + z^2 \ln y + h(z) \Rightarrow \frac{\partial f}{\partial z} = 2z \ln y + h'(z) = 2z \ln y \Rightarrow h'(z) = 0 \Rightarrow h(z) = C \Rightarrow f(x, y, z)$
$= x^3 + z^2 \ln y + C \Rightarrow \int_{(1,1,1)}^{(1,2,3)} 3x^2\,dx + \frac{z^2}{y}\,dy + 2z \ln y\,dz = f(1, 2, 3) - f(1, 1, 1)$
$= (1 + 9\ln 2 + C) - (1 + 0 + C) = 9 \ln 2$

21. Let $\mathbf{F}(x, y, z) = \left(\frac{1}{y}\right)\mathbf{i} + \left(\frac{1}{z} - \frac{x}{y^2}\right)\mathbf{j} - \left(\frac{y}{z^2}\right)\mathbf{k} \Rightarrow \frac{\partial P}{\partial y} = -\frac{1}{z^2} = \frac{\partial N}{\partial z}, \frac{\partial M}{\partial z} = 0 = \frac{\partial P}{\partial x}, \frac{\partial N}{\partial x} = -\frac{1}{y^2} = \frac{\partial M}{\partial y}$
$\Rightarrow M\,dx + N\,dy + P\,dz$ is exact; $\frac{\partial f}{\partial x} = \frac{1}{y} \Rightarrow f(x, y, z) = \frac{x}{y} + g(y, z) \Rightarrow \frac{\partial f}{\partial y} = -\frac{x}{y^2} + \frac{\partial g}{\partial y} = \frac{1}{z} - \frac{x}{y^2}$
$\Rightarrow \frac{\partial g}{\partial y} = \frac{1}{z} \Rightarrow g(y, z) = \frac{y}{z} + h(z) \Rightarrow f(x, y, z) = \frac{x}{y} + \frac{y}{z} + h(z) \Rightarrow \frac{\partial f}{\partial z} = -\frac{y}{z^2} + h'(z) = -\frac{y}{z^2} \Rightarrow h'(z) = 0 \Rightarrow h(z) = C$
$\Rightarrow f(x, y, z) = \frac{x}{y} + \frac{y}{z} + C \Rightarrow \int_{(1,1,1)}^{(2,2,2)} \frac{1}{y}\,dx + \left(\frac{1}{z} - \frac{x}{y^2}\right)dy - \frac{y}{z^2}\,dz = f(2, 2, 2) - f(1, 1, 1) = \left(\frac{2}{2} + \frac{2}{2} + C\right) - \left(\frac{1}{1} + \frac{1}{1} + C\right)$
$= 0$

23. $\mathbf{r} = (\mathbf{i} + \mathbf{j} + \mathbf{k}) + t(\mathbf{i} + 2\mathbf{j} - 2\mathbf{k}) = (1 + t)\mathbf{i} + (1 + 2t)\mathbf{j} + (1 - 2t)\mathbf{k}, 0 \le t \le 1 \Rightarrow dx = dt, dy = 2\,dt, dz = -2\,dt$
$\Rightarrow \int_{(1,1,1)}^{(2,3,-1)} y\,dx + x\,dy + 4\,dz = \int_0^1 (2t + 1)\,dt + (t + 1)(2\,dt) + 4(-2)\,dt = \int_0^1 (4t - 5)\,dt = \left[2t^2 - 5t\right]_0^1 = -3$

25. $\frac{\partial P}{\partial y} = 0 = \frac{\partial N}{\partial z}, \frac{\partial M}{\partial z} = 2z = \frac{\partial P}{\partial x}, \frac{\partial N}{\partial x} = 0 = \frac{\partial M}{\partial y} \Rightarrow M\,dx + N\,dy + P\,dz$ is exact $\Rightarrow \mathbf{F}$ is conservative
$\Rightarrow$ path independence

27. $\frac{\partial P}{\partial y} = 0 = \frac{\partial N}{\partial z}$, $\frac{\partial M}{\partial z} = 0 = \frac{\partial P}{\partial x}$, $\frac{\partial N}{\partial x} = -\frac{2x}{y^2} = \frac{\partial M}{\partial y}$ $\Rightarrow$ **F** is conservative $\Rightarrow$ there exists an f so that $\mathbf{F} = \nabla f$;
$\frac{\partial f}{\partial x} = \frac{2x}{y} \Rightarrow f(x,y) = \frac{x^2}{y} + g(y) \Rightarrow \frac{\partial f}{\partial y} = -\frac{x^2}{y^2} + g'(y) = \frac{1-x^2}{y^2} \Rightarrow g'(y) = \frac{1}{y^2} \Rightarrow g(y) = -\frac{1}{y} + C$
$\Rightarrow f(x,y) = \frac{x^2}{y} - \frac{1}{y} + C \Rightarrow \mathbf{F} = \nabla\left(\frac{x^2-1}{y}\right)$

29. $\frac{\partial P}{\partial y} = 0 = \frac{\partial N}{\partial z}$, $\frac{\partial M}{\partial z} = 0 = \frac{\partial P}{\partial x}$, $\frac{\partial N}{\partial x} = 1 = \frac{\partial M}{\partial y}$ $\Rightarrow$ **F** is conservative $\Rightarrow$ there exists an f so that $\mathbf{F} = \nabla f$;
$\frac{\partial f}{\partial x} = x^2 + y \Rightarrow f(x,y,z) = \frac{1}{3}x^3 + xy + g(y,z) \Rightarrow \frac{\partial f}{\partial y} = x + \frac{\partial g}{\partial y} = y^2 + x \Rightarrow \frac{\partial g}{\partial y} = y^2 \Rightarrow g(y,z) = \frac{1}{3}y^3 + h(z)$
$\Rightarrow f(x,y,z) = \frac{1}{3}x^3 + xy + \frac{1}{3}y^3 + h(z) \Rightarrow \frac{\partial f}{\partial z} = h'(z) = ze^z \Rightarrow h(z) = ze^z - e^z + C \Rightarrow f(x,y,z)$
$= \frac{1}{3}x^3 + xy + \frac{1}{3}y^3 + ze^z - e^z + C \Rightarrow \mathbf{F} = \nabla\left(\frac{1}{3}x^3 + xy + \frac{1}{3}y^3 + ze^z - e^z\right)$

(a) work $= \int_A^B \mathbf{F} \cdot \frac{d\mathbf{r}}{dt}\,dt = \int_A^B \mathbf{F} \cdot d\mathbf{r} = \left[\frac{1}{3}x^3 + xy + \frac{1}{3}y^3 + ze^z - e^z\right]_{(1,0,0)}^{(1,0,1)} = \left(\frac{1}{3} + 0 + 0 + e - e\right) - \left(\frac{1}{3} + 0 + 0 - 1\right)$
$= 1$

(b) work $= \int_A^B \mathbf{F} \cdot d\mathbf{r} = \left[\frac{1}{3}x^3 + xy + \frac{1}{3}y^3 + ze^z - e^z\right]_{(1,0,0)}^{(1,0,1)} = 1$

(c) work $= \int_A^B \mathbf{F} \cdot d\mathbf{r} = \left[\frac{1}{3}x^3 + xy + \frac{1}{3}y^3 + ze^z - e^z\right]_{(1,0,0)}^{(1,0,1)} = 1$

Note: Since **F** is conservative, $\int_A^B \mathbf{F} \cdot d\mathbf{r}$ is independent of the path from $(1,0,0)$ to $(1,0,1)$.

31. (a) $\mathbf{F} = \nabla(x^3y^2) \Rightarrow \mathbf{F} = 3x^2y^2\mathbf{i} + 2x^3y\mathbf{j}$; let C_1 be the path from $(-1,1)$ to $(0,0) \Rightarrow x = t-1$ and $y = -t+1,\ 0 \le t \le 1 \Rightarrow \mathbf{F} = 3(t-1)^2(-t+1)^2\mathbf{i} + 2(t-1)^3(-t+1)\mathbf{j} = 3(t-1)^4\mathbf{i} - 2(t-1)^4\mathbf{j}$
and $\mathbf{r}_1 = (t-1)\mathbf{i} + (-t+1)\mathbf{j} \Rightarrow d\mathbf{r}_1 = dt\,\mathbf{i} - dt\,\mathbf{j} \Rightarrow \int_{C_1} \mathbf{F} \cdot d\mathbf{r}_1 = \int_0^1 \left[3(t-1)^4 + 2(t-1)^4\right] dt$
$= \int_0^1 5(t-1)^4\,dt = \left[(t-1)^5\right]_0^1 = 1$; let C_2 be the path from $(0,0)$ to $(1,1) \Rightarrow x = t$ and $y = t$,
$0 \le t \le 1 \Rightarrow \mathbf{F} = 3t^4\mathbf{i} + 2t^4\mathbf{j}$ and $\mathbf{r}_2 = t\mathbf{i} + t\mathbf{j} \Rightarrow d\mathbf{r}_2 = dt\,\mathbf{i} + dt\,\mathbf{j} \Rightarrow \int_{C_2} \mathbf{F} \cdot d\mathbf{r}_2 = \int_0^1 (3t^4 + 2t^4)\,dt$
$= \int_0^1 5t^4\,dt = 1 \Rightarrow \int_C \mathbf{F} \cdot d\mathbf{r} = \int_{C_1} \mathbf{F} \cdot d\mathbf{r}_1 + \int_{C_2} \mathbf{F} \cdot d\mathbf{r}_2 = 2$

(b) Since $f(x,y) = x^3y^2$ is a potential function for **F**, $\int_{(-1,1)}^{(1,1)} \mathbf{F} \cdot d\mathbf{r} = f(1,1) - f(-1,1) = 2$

33. (a) If the differential form is exact, then $\frac{\partial P}{\partial y} = \frac{\partial N}{\partial z} \Rightarrow 2ay = cy$ for all $y \Rightarrow 2a = c$, $\frac{\partial M}{\partial z} = \frac{\partial P}{\partial x} \Rightarrow 2cx = 2cx$ for all x, and $\frac{\partial N}{\partial x} = \frac{\partial M}{\partial y} \Rightarrow by = 2ay$ for all $y \Rightarrow b = 2a$ and $c = 2a$

(b) $\mathbf{F} = \nabla f \Rightarrow$ the differential form with $a = 1$ in part (a) is exact $\Rightarrow b = 2$ and $c = 2$

35. The path will not matter; the work along any path will be the same because the field is conservative.

37. Let the coordinates of points A and B be (x_A, y_A, z_A) and (x_B, y_B, z_B), respectively. The force $\mathbf{F} = a\mathbf{i} + b\mathbf{j} + c\mathbf{k}$ is conservative because all the partial derivatives of M, N, and P are zero. Therefore, the potential function is $f(x,y,z) = ax + by + cz + C$, and the work done by the force in moving a particle along any path from A to B is $f(B) - f(A) = f(x_B, y_B, z_B) - f(x_A, y_A, z_A) = (ax_B + by_B + cz_B + C) - (ax_A + by_A + cz_A + C)$
$= a(x_B - x_A) + b(y_B - y_A) + c(z_B - z_A) = \mathbf{F} \cdot \overrightarrow{BA}$

16.4 GREEN'S THEOREM IN THE PLANE

1. $M = -y = -a \sin t$, $N = x = a \cos t$, $dx = -a \sin t\, dt$, $dy = a \cos t\, dt \Rightarrow \frac{\partial M}{\partial x} = 0$, $\frac{\partial M}{\partial y} = -1$, $\frac{\partial N}{\partial x} = 1$, and $\frac{\partial N}{\partial y} = 0$;

   Equation (3): $\oint_C M\, dy - N\, dx = \int_0^{2\pi} [(-a \sin t)(a \cos t) - (a \cos t)(-a \sin t)]\, dt = \int_0^{2\pi} 0\, dt = 0$;
   $\iint_R \left(\frac{\partial M}{\partial x} + \frac{\partial N}{\partial y}\right) dx\, dy = \iint_R 0\, dx\, dy = 0$, Flux

   Equation (4): $\oint_C M\, dx + N\, dy = \int_0^{2\pi} [(-a \sin t)(-a \sin t) - (a \cos t)(a \cos t)]\, dt = \int_0^{2\pi} a^2\, dt = 2\pi a^2$;
   $\iint_R \left(\frac{\partial N}{\partial x} - \frac{\partial M}{\partial y}\right) dx\, dy = \int_{-a}^{a} \int_{-c}^{\sqrt{a^2 - x^2}} 2\, dy\, dx = \int_{-a}^{a} 4\sqrt{a^2 - x^2}\, dx = 4\left[\frac{x}{2}\sqrt{a^2 - x^2} + \frac{a^2}{2} \sin^{-1} \frac{x}{a}\right]_{-a}^{a}$
   $= 2a^2\left(\frac{\pi}{2} + \frac{\pi}{2}\right) = 2a^2\pi$, Circulation

3. $M = 2x = 2a \cos t$, $N = -3y = -3a \sin t$, $dx = -a \sin t\, dt$, $dy = a \cos t\, dt \Rightarrow \frac{\partial M}{\partial x} = 2$, $\frac{\partial M}{\partial y} = 0$, $\frac{\partial N}{\partial x} = 0$, and $\frac{\partial N}{\partial y} = -3$;

   Equation (3): $\oint_C M\, dy - N\, dx = \int_0^{2\pi} [(2a \cos t)(a \cos t) + (3a \sin t)(-a \sin t)]\, dt$
   $= \int_0^{2\pi} (2a^2 \cos^2 t - 3a^2 \sin^2 t)\, dt = 2a^2\left[\frac{t}{2} + \frac{\sin 2t}{4}\right]_0^{2\pi} - 3a^2\left[\frac{t}{2} - \frac{\sin 2t}{4}\right]_0^{2\pi} = 2\pi a^2 - 3\pi a^2 = -\pi a^2$;
   $\iint_R \left(\frac{\partial M}{\partial x} + \frac{\partial N}{\partial y}\right) = \iint_R -1\, dx\, dy = \int_0^{2\pi} \int_0^{a} -r\, dr\, d\theta = \int_0^{2\pi} -\frac{a^2}{2}\, d\theta = -\pi a^2$, Flux

   Equation (4): $\oint_C M\, dx + N\, dy = \int_0^{2\pi} [(2a \cos t)(-a \sin t) + (-3a \sin t)(a \cos t)]\, dt$
   $= \int_0^{2\pi} (-2a^2 \sin t \cos t - 3a^2 \sin t \cos t)\, dt = -5a^2\left[\frac{1}{2} \sin^2 t\right]_0^{2\pi} = 0$; $\iint_R 0\, dx\, dy = 0$, Circulation

5. $M = x - y$, $N = y - x \Rightarrow \frac{\partial M}{\partial x} = 1$, $\frac{\partial M}{\partial y} = -1$, $\frac{\partial N}{\partial x} = -1$, $\frac{\partial N}{\partial y} = 1 \Rightarrow$ Flux $= \iint_R 2\, dx\, dy = \int_0^1 \int_0^1 2\, dx\, dy = 2$;
   Circ $= \iint_R [-1 - (-1)]\, dx\, dy = 0$

7. $M = y^2 - x^2$, $N = x^2 + y^2 \Rightarrow \frac{\partial M}{\partial x} = -2x$, $\frac{\partial M}{\partial y} = 2y$, $\frac{\partial N}{\partial x} = 2x$, $\frac{\partial N}{\partial y} = 2y \Rightarrow$ Flux $= \iint_R (-2x + 2y)\, dx\, dy$
   $= \int_0^3 \int_0^x (-2x + 2y)\, dy\, dx = \int_0^3 (-2x^2 + x^2)\, dx = \left[-\frac{1}{3} x^3\right]_0^3 = -9$; Circ $= \iint_R (2x - 2y)\, dx\, dy$
   $= \int_0^3 \int_0^x (2x - 2y)\, dy\, dx = \int_0^3 x^2\, dx = 9$

9. $M = xy + y^2$, $N = x - y \Rightarrow \frac{\partial M}{\partial x} = y$, $\frac{\partial M}{\partial y} = x + 2y$, $\frac{\partial N}{\partial x} = 1$, $\frac{\partial N}{\partial y} = -1 \Rightarrow$ Flux $= \iint_R (y + (-1))\, dy\, dx$
   $= \int_0^1 \int_{x^2}^{\sqrt{x}} (y - 1)\, dy\, dx = \int_0^1 \left(\frac{1}{2}x - \sqrt{x} - \frac{1}{2}x^4 + x^2\right) dx = -\frac{11}{60}$; Circ $= \iint_R (1 - (x + 2y))\, dy\, dx$
   $= \int_0^1 \int_{x^2}^{\sqrt{x}} (1 - x - 2y)\, dy\, dx = \int_0^1 \left(\sqrt{x} - x^{3/2} - x - x^2 + x^3 + x^4\right) dx = -\frac{7}{60}$

11. $M = x^3y^2$, $N = \frac{1}{2}x^4y \Rightarrow \frac{\partial M}{\partial x} = 3x^2y^2$, $\frac{\partial M}{\partial y} = 2x^3y$, $\frac{\partial N}{\partial x} = 2x^3y$, $\frac{\partial N}{\partial y} = \frac{1}{2}x^4 \Rightarrow$ Flux $= \iint_R \left(3x^2y^2 + \frac{1}{2}x^4\right) dy\, dx$
   $= \int_0^2 \int_{x^2 - x}^{x} \left(3x^2y^2 + \frac{1}{2}x^4\right) dy\, dx = \int_0^2 \left(3x^5 - \frac{7}{2}x^6 + 3x^7 - x^8\right) dx = \frac{64}{9}$; Circ $= \iint_R (2x^3y - 2x^3y)\, dy\, dx = 0$

13. $M = x + e^x \sin y$, $N = x + e^x \cos y \Rightarrow \frac{\partial M}{\partial x} = 1 + e^x \sin y$, $\frac{\partial M}{\partial y} = e^x \cos y$, $\frac{\partial N}{\partial x} = 1 + e^x \cos y$, $\frac{\partial N}{\partial y} = -e^x \sin y$

$\Rightarrow \text{Flux} = \iint_R dx\,dy = \int_{-\pi/4}^{\pi/4} \int_0^{\sqrt{\cos 2\theta}} r\,dr\,d\theta = \int_{-\pi/4}^{\pi/4} \left(\frac{1}{2}\cos 2\theta\right) d\theta = \left[\frac{1}{4}\sin 2\theta\right]_{-\pi/4}^{\pi/4} = \frac{1}{2}$;

$\text{Circ} = \iint_R (1 + e^x \cos y - e^x \cos y)\,dx\,dy = \iint_R dx\,dy = \int_{-\pi/4}^{\pi/4} \int_0^{\sqrt{\cos 2\theta}} r\,dr\,d\theta = \int_{-\pi/4}^{\pi/4} \left(\frac{1}{2}\cos 2\theta\right) d\theta = \frac{1}{2}$

15. $M = xy$, $N = y^2 \Rightarrow \frac{\partial M}{\partial x} = y$, $\frac{\partial M}{\partial y} = x$, $\frac{\partial N}{\partial x} = 0$, $\frac{\partial N}{\partial y} = 2y \Rightarrow \text{Flux} = \iint_R (y + 2y)\,dy\,dx = \int_0^1 \int_{x^2}^{x} 3y\,dy\,dx$

$= \int_0^1 \left(\frac{3x^2}{2} - \frac{3x^4}{2}\right) dx = \frac{1}{5}$; $\text{Circ} = \iint_R -x\,dy\,dx = \int_0^1 \int_{x^2}^{x} -x\,dy\,dx = \int_0^1 (-x^2 + x^3)\,dx = -\frac{1}{12}$

17. $M = 3xy - \frac{x}{1+y^2}$, $N = e^x + \tan^{-1} y \Rightarrow \frac{\partial M}{\partial x} = 3y - \frac{1}{1+y^2}$, $\frac{\partial N}{\partial y} = \frac{1}{1+y^2}$

$\Rightarrow \text{Flux} = \iint_R \left(3y - \frac{1}{1+y^2} + \frac{1}{1+y^2}\right) dx\,dy = \iint_R 3y\,dx\,dy = \int_0^{2\pi} \int_0^{a(1+\cos\theta)} (3r\sin\theta)\,r\,dr\,d\theta$

$= \int_0^{2\pi} a^3(1+\cos\theta)^3(\sin\theta)\,d\theta = \left[-\frac{a^3}{4}(1+\cos\theta)^4\right]_0^{2\pi} = -4a^3 - (-4a^3) = 0$

19. $M = 2xy^3$, $N = 4x^2y^2 \Rightarrow \frac{\partial M}{\partial y} = 6xy^2$, $\frac{\partial N}{\partial x} = 8xy^2 \Rightarrow \text{work} = \oint_C 2xy^3\,dx + 4x^2y^2\,dy = \iint_R (8xy^2 - 6xy^2)\,dx\,dy$

$= \int_0^1 \int_0^{x^3} 2xy^2\,dy\,dx = \int_0^1 \frac{2}{3}x^{10}\,dx = \frac{2}{33}$

21. $M = y^2$, $N = x^2 \Rightarrow \frac{\partial M}{\partial y} = 2y$, $\frac{\partial N}{\partial x} = 2x \Rightarrow \oint_C y^2\,dx + x^2\,dy = \iint_R (2x - 2y)\,dy\,dx$

$= \int_0^1 \int_0^{1-x} (2x - 2y)\,dy\,dx = \int_0^1 (-3x^2 + 4x - 1)\,dx = \left[-x^3 + 2x^2 - x\right]_0^1 = -1 + 2 - 1 = 0$

23. $M = 6y + x$, $N = y + 2x \Rightarrow \frac{\partial M}{\partial y} = 6$, $\frac{\partial N}{\partial x} = 2 \Rightarrow \oint_C (6y + x)\,dx + (y + 2x)\,dy = \iint_R (2 - 6)\,dy\,dx$

$= -4(\text{Area of the circle}) = -16\pi$

25. $M = x = a\cos t$, $N = y = a\sin t \Rightarrow dx = -a\sin t\,dt$, $dy = a\cos t\,dt \Rightarrow \text{Area} = \frac{1}{2}\oint_C x\,dy - y\,dx$

$= \frac{1}{2}\int_0^{2\pi} (a^2\cos^2 t + a^2\sin^2 t)\,dt = \frac{1}{2}\int_0^{2\pi} a^2\,dt = \pi a^2$

27. $M = x = \cos^3 t$, $N = y = \sin^3 t \Rightarrow dx = -3\cos^2 t \sin t\,dt$, $dy = 3\sin^2 t \cos t\,dt \Rightarrow \text{Area} = \frac{1}{2}\oint_C x\,dy - y\,dx$

$= \frac{1}{2}\int_0^{2\pi} (3\sin^2 t\cos^2 t)(\cos^2 t + \sin^2 t)\,dt = \frac{1}{2}\int_0^{2\pi} (3\sin^2 t\cos^2 t)\,dt = \frac{3}{8}\int_0^{2\pi} \sin^2 2t\,dt = \frac{3}{16}\int_0^{4\pi} \sin^2 u\,du$

$= \frac{3}{16}\left[\frac{u}{2} - \frac{\sin 2u}{4}\right]_0^{4\pi} = \frac{3}{8}\pi$

29. (a) $M = f(x)$, $N = g(y) \Rightarrow \frac{\partial M}{\partial y} = 0$, $\frac{\partial N}{\partial x} = 0 \Rightarrow \oint_C f(x)\,dx + g(y)\,dy = \iint_R \left(\frac{\partial N}{\partial x} - \frac{\partial M}{\partial y}\right) dx\,dy = \iint_R 0\,dx\,dy = 0$

(b) $M = ky$, $N = hx \Rightarrow \frac{\partial M}{\partial y} = k$, $\frac{\partial N}{\partial x} = h \Rightarrow \oint_C ky\,dx + hx\,dy = \iint_R \left(\frac{\partial N}{\partial x} - \frac{\partial M}{\partial y}\right) dx\,dy$

$= \iint_R (h - k)\,dx\,dy = (h - k)(\text{Area of the region})$

31. The integral is 0 for any simple closed plane curve C. The reasoning: By the tangential form of Green's Theorem, with $M = 4x^3y$ and $N = x^4$, $\oint_C 4x^3y\,dx + x^4\,dy = \iint_R \left[\frac{\partial}{\partial x}(x^4) - \frac{\partial}{\partial y}(4x^3y)\right] dx\,dy$
$= \iint_R \underbrace{(4x^3 - 4x^3)}_{0}\, dx\,dy = 0.$

33. Let $M = x$ and $N = 0 \Rightarrow \frac{\partial M}{\partial x} = 1$ and $\frac{\partial N}{\partial y} = 0 \Rightarrow \oint_C M\,dy - N\,dx = \iint_R \left(\frac{\partial M}{\partial x} + \frac{\partial N}{\partial y}\right) dx\,dy \Rightarrow \oint_C x\,dy$
$= \iint_R (1+0)\,dx\,dy \Rightarrow$ Area of $R = \iint_R dx\,dy = \oint_C x\,dy$; similarly, $M = y$ and $N = 0 \Rightarrow \frac{\partial M}{\partial y} = 1$ and
$\frac{\partial N}{\partial x} = 0 \Rightarrow \oint_C M\,dx + N\,dy = \iint_R \left(\frac{\partial N}{\partial x} + \frac{\partial M}{\partial y}\right) dy\,dx \Rightarrow \oint_C y\,dx = \iint_R (0-1)\,dy\,dx \Rightarrow -\oint_C y\,dx$
$= \iint_R dx\,dy =$ Area of R

35. Let $\delta(x,y) = 1 \Rightarrow \overline{x} = \frac{M_y}{M} = \frac{\iint_R x\,\delta(x,y)\,dA}{\iint_R \delta(x,y)\,dA} = \frac{\iint_R x\,dA}{\iint_R dA} = \frac{\iint_R x\,dA}{A} \Rightarrow A\overline{x} = \iint_R x\,dA = \iint_R (x+0)\,dx\,dy$
$= \oint_C \frac{x^2}{2}\,dy$, $A\overline{x} = \iint_R x\,dA = \iint_R (0+x)\,dx\,dy = -\oint_C xy\,dx$, and $A\overline{x} = \iint_R x\,dA = \iint_R \left(\frac{2}{3}x + \frac{1}{3}x\right) dx\,dy$
$= \oint_C \frac{1}{3}x^2\,dy - \frac{1}{3}xy\,dx \Rightarrow \frac{1}{2}\oint_C x^2\,dy = -\oint_C xy\,dx = \frac{1}{3}\oint_C x^2\,dy - xy\,dx = A\overline{x}$

37. $M = \frac{\partial f}{\partial y}$, $N = -\frac{\partial f}{\partial x} \Rightarrow \frac{\partial M}{\partial y} = \frac{\partial^2 f}{\partial y^2}$, $\frac{\partial N}{\partial x} = -\frac{\partial^2 f}{\partial x^2} \Rightarrow \oint_C \frac{\partial f}{\partial y}\,dx - \frac{\partial f}{\partial x}\,dy = \iint_R \left(-\frac{\partial^2 f}{\partial x^2} - \frac{\partial^2 f}{\partial y^2}\right) dx\,dy = 0$ for such curves C

39. (a) $\nabla f = \left(\frac{2x}{x^2+y^2}\right)\mathbf{i} + \left(\frac{2y}{x^2+y^2}\right)\mathbf{j} \Rightarrow M = \frac{2x}{x^2+y^2}$, $N = \frac{2y}{x^2+y^2}$; since M, N are discontinuous at (0, 0), we compute $\int_C \nabla f \cdot \mathbf{n}\,ds$ directly since Green's Theorem does not apply. Let $x = a\cos t$, $y = a\sin t \Rightarrow dx = -a\sin t\,dt$, $dy = a\cos t\,dt$, $M = \frac{2}{a}\cos t$, $N = \frac{2}{a}\sin t$, $0 \le t \le 2\pi$, so $\int_C \nabla f \cdot \mathbf{n}\,ds = \int_C M\,dy - N\,dx$
$= \int_0^{2\pi} \left[\left(\frac{2}{a}\cos t\right)(a\cos t) - \left(\frac{2}{a}\sin t\right)(-a\sin t)\right] dt = \int_0^{2\pi} 2(\cos^2 t + \sin^2 t)\,dt = 4\pi$. Note that this holds for any $a > 0$, so $\int_C \nabla f \cdot \mathbf{n}\,ds = 4\pi$ for any circle C centered at (0, 0) traversed counterclockwise and $\int_C \nabla f \cdot \mathbf{n}\,ds = -4\pi$ if C is traversed clockwise.

(b) If K does not enclose the point (0, 0) we may apply Green's Theorem: $\int_C \nabla f \cdot \mathbf{n}\,ds = \int_C M\,dy - N\,dx$
$= \iint_R \left(\frac{\partial M}{\partial x} + \frac{\partial N}{\partial y}\right) dx\,dy = \iint_R \left(\frac{2(y^2-x^2)}{(x^2+y^2)^2} + \frac{2(x^2-y^2)}{(x^2+y^2)^2}\right) dx\,dy = \iint_R 0\,dx\,dy = 0$. If K does enclose the point (0, 0) we proceed as follows:
Choose a small enough so that the circle C centered at (0, 0) of radius a lies entirely within K. Green's Theorem applies to the region R that lies between K and C. Thus, as before, $0 = \iint_R \left(\frac{\partial M}{\partial x} + \frac{\partial N}{\partial y}\right) dx\,dy$
$= \int_K M\,dy - N\,dx + \int_C M\,dy - N\,dx$ where K is traversed counterclockwise and C is traversed clockwise.
Hence by part (a) $0 = \left[\int_K M\,dy - N\,dx\right] - 4\pi \Rightarrow 4\pi = \int_K M\,dy - N\,dx = \int_K \nabla f \cdot \mathbf{n}\,ds$. We have shown:
$$\int_K \nabla f \cdot \mathbf{n}\,ds = \begin{cases} 0 & \text{if } (0,0) \text{ lies inside K} \\ 4\pi & \text{if } (0,0) \text{ lies outside K} \end{cases}$$

41. $\int_{g_1(y)}^{g_2(y)} \frac{\partial N}{\partial x}\,dx\,dy = N(g_2(y), y) - N(g_1(y), y) \Rightarrow \int_c^d \int_{g_1(y)}^{g_2(y)} \left(\frac{\partial N}{\partial x}\,dx\right) dy = \int_c^d [N(g_2(y), y) - N(g_1(y), y)]\,dy$
$= \int_c^d N(g_2(y), y)\,dy - \int_c^d N(g_1(y), y)\,dy = \int_c^d N(g_2(y), y)\,dy + \int_d^c N(g_1(y), y)\,dy = \int_{C_2} N\,dy + \int_{C_1} N\,dy$
$= \oint_C dy \Rightarrow \oint_C N\,dy = \iint\limits_R \frac{\partial N}{\partial x}\,dx\,dy$

16.5 SURFACES AND AREA

1. In cylindrical coordinates, let $x = r\cos\theta$, $y = r\sin\theta$, $z = \left(\sqrt{x^2+y^2}\right)^2 = r^2$. Then $\mathbf{r}(r,\theta) = (r\cos\theta)\mathbf{i} + (r\sin\theta)\mathbf{j} + r^2\mathbf{k}$, $0 \le r \le 2$, $0 \le \theta \le 2\pi$.

3. In cylindrical coordinates, let $x = r\cos\theta$, $y = r\sin\theta$, $z = \frac{\sqrt{x^2+y^2}}{2} \Rightarrow z = \frac{r}{2}$. Then $\mathbf{r}(r,\theta) = (r\cos\theta)\mathbf{i} + (r\sin\theta)\mathbf{j} + \left(\frac{r}{2}\right)\mathbf{k}$. For $0 \le z \le 3$, $0 \le \frac{r}{2} \le 3 \Rightarrow 0 \le r \le 6$; to get only the first octant, let $0 \le \theta \le \frac{\pi}{2}$.

5. In cylindrical coordinates, let $x = r\cos\theta$, $y = r\sin\theta$; since $x^2+y^2 = r^2 \Rightarrow z^2 = 9 - (x^2+y^2) = 9 - r^2$ $\Rightarrow z = \sqrt{9-r^2}$, $z \ge 0$. Then $\mathbf{r}(r,\theta) = (r\cos\theta)\mathbf{i} + (r\sin\theta)\mathbf{j} + \sqrt{9-r^2}\,\mathbf{k}$. Let $0 \le \theta \le 2\pi$. For the domain of r: $z = \sqrt{x^2+y^2}$ and $x^2+y^2+z^2 = 9 \Rightarrow x^2+y^2+\left(\sqrt{x^2+y^2}\right)^2 = 9 \Rightarrow 2(x^2+y^2) = 9 \Rightarrow 2r^2 = 9$ $\Rightarrow r = \frac{3}{\sqrt{2}} \Rightarrow 0 \le r \le \frac{3}{\sqrt{2}}$.

7. In spherical coordinates, $x = \rho\sin\phi\cos\theta$, $y = \rho\sin\phi\sin\theta$, $\rho = \sqrt{x^2+y^2+z^2} \Rightarrow \rho^2 = 3 \Rightarrow \rho = \sqrt{3}$ $\Rightarrow z = \sqrt{3}\cos\phi$ for the sphere; $z = \frac{\sqrt{3}}{2} = \sqrt{3}\cos\phi \Rightarrow \cos\phi = \frac{1}{2} \Rightarrow \phi = \frac{\pi}{3}$; $z = -\frac{\sqrt{3}}{2} \Rightarrow -\frac{\sqrt{3}}{2} = \sqrt{3}\cos\phi$ $\Rightarrow \cos\phi = -\frac{1}{2} \Rightarrow \phi = \frac{2\pi}{3}$. Then $\mathbf{r}(\phi,\theta) = \left(\sqrt{3}\sin\phi\cos\theta\right)\mathbf{i} + \left(\sqrt{3}\sin\phi\sin\theta\right)\mathbf{j} + \left(\sqrt{3}\cos\phi\right)\mathbf{k}$, $\frac{\pi}{3} \le \phi \le \frac{2\pi}{3}$ and $0 \le \theta \le 2\pi$.

9. Since $z = 4 - y^2$, we can let $\mathbf{r}$ be a function of x and y $\Rightarrow \mathbf{r}(x,y) = x\mathbf{i} + y\mathbf{j} + (4-y^2)\mathbf{k}$. Then $z = 0$ $\Rightarrow 0 = 4 - y^2 \Rightarrow y = \pm 2$. Thus, let $-2 \le y \le 2$ and $0 \le x \le 2$.

11. When $x = 0$, let $y^2 + z^2 = 9$ be the circular section in the yz-plane. Use polar coordinates in the yz-plane $\Rightarrow y = 3\cos\theta$ and $z = 3\sin\theta$. Thus let $x = u$ and $\theta = v \Rightarrow \mathbf{r}(u,v) = u\mathbf{i} + (3\cos v)\mathbf{j} + (3\sin v)\mathbf{k}$ where $0 \le u \le 3$, and $0 \le v \le 2\pi$.

13. (a) $x + y + z = 1 \Rightarrow z = 1 - x - y$. In cylindrical coordinates, let $x = r\cos\theta$ and $y = r\sin\theta$ $\Rightarrow z = 1 - r\cos\theta - r\sin\theta \Rightarrow \mathbf{r}(r,\theta) = (r\cos\theta)\mathbf{i} + (r\sin\theta)\mathbf{j} + (1 - r\cos\theta - r\sin\theta)\mathbf{k}$, $0 \le \theta \le 2\pi$ and $0 \le r \le 3$.

    (b) In a fashion similar to cylindrical coordinates, but working in the yz-plane instead of the xy-plane, let $y = u\cos v$, $z = u\sin v$ where $u = \sqrt{y^2+z^2}$ and v is the angle formed by (x, y, z), $(x, 0, 0)$, and $(x, y, 0)$ with $(x, 0, 0)$ as vertex. Since $x + y + z = 1 \Rightarrow x = 1 - y - z \Rightarrow x = 1 - u\cos v - u\sin v$, then $\mathbf{r}$ is a function of u and v $\Rightarrow \mathbf{r}(u,v) = (1 - u\cos v - u\sin v)\mathbf{i} + (u\cos v)\mathbf{j} + (u\sin v)\mathbf{k}$, $0 \le u \le 3$ and $0 \le v \le 2\pi$.

15. Let $x = w\cos v$ and $z = w\sin v$. Then $(x-2)^2 + z^2 = 4 \Rightarrow x^2 - 4x + z^2 = 0 \Rightarrow w^2\cos^2 v - 4w\cos v + w^2\sin^2 v$ $= 0 \Rightarrow w^2 - 4w\cos v = 0 \Rightarrow w = 0$ or $w - 4\cos v = 0 \Rightarrow w = 0$ or $w = 4\cos v$. Now $w = 0 \Rightarrow x = 0$ and $y = 0$, which is a line not a cylinder. Therefore, let $w = 4\cos v \Rightarrow x = (4\cos v)(\cos v) = 4\cos^2 v$ and $z = 4\cos v\sin v$. Finally, let $y = u$. Then $\mathbf{r}(u,v) = (4\cos^2 v)\,\mathbf{i} + u\mathbf{j} + (4\cos v\sin v)\mathbf{k}$, $-\frac{\pi}{2} \le v \le \frac{\pi}{2}$ and $0 \le u \le 3$.

17. Let $x = r\cos\theta$ and $y = r\sin\theta$. Then $\mathbf{r}(r,\theta) = (r\cos\theta)\mathbf{i} + (r\sin\theta)\mathbf{j} + \left(\frac{2 - r\sin\theta}{2}\right)\mathbf{k}$, $0 \le r \le 1$ and $0 \le \theta \le 2\pi$
$\Rightarrow \mathbf{r}_r = (\cos\theta)\mathbf{i} + (\sin\theta)\mathbf{j} - \left(\frac{\sin\theta}{2}\right)\mathbf{k}$ and $\mathbf{r}_\theta = (-r\sin\theta)\mathbf{i} + (r\cos\theta)\mathbf{j} - \left(\frac{r\cos\theta}{2}\right)\mathbf{k}$

$$\Rightarrow \mathbf{r}_r \times \mathbf{r}_\theta = \begin{vmatrix} \mathbf{i} & \mathbf{j} & \mathbf{k} \\ \cos\theta & \sin\theta & -\frac{\sin\theta}{2} \\ -r\sin\theta & r\cos\theta & -\frac{r\cos\theta}{2} \end{vmatrix}$$

$$= \left(\frac{-r\sin\theta\cos\theta}{2} + \frac{(\sin\theta)(r\cos\theta)}{2}\right)\mathbf{i} + \left(\frac{r\sin^2\theta}{2} + \frac{r\cos^2\theta}{2}\right)\mathbf{j} + (r\cos^2\theta + r\sin^2\theta)\,\mathbf{k} = \frac{r}{2}\mathbf{j} + r\mathbf{k}$$

$$\Rightarrow |\mathbf{r}_r \times \mathbf{r}_\theta| = \sqrt{\frac{r^2}{4} + r^2} = \frac{\sqrt{5}\,r}{2} \Rightarrow A = \int_0^{2\pi}\int_0^1 \frac{\sqrt{5}\,r}{2}\,dr\,d\theta = \int_0^{2\pi}\left[\frac{\sqrt{5}\,r^2}{4}\right]_0^1 d\theta = \int_0^{2\pi} d\theta = \frac{\pi\sqrt{5}}{2}$$

19. Let $x = r\cos\theta$ and $y = r\sin\theta \Rightarrow z = 2\sqrt{x^2 + y^2} = 2r$, $1 \le r \le 3$ and $0 \le \theta \le 2\pi$. Then $\mathbf{r}(r,\theta) = (r\cos\theta)\mathbf{i} + (r\sin\theta)\mathbf{j} + 2r\mathbf{k} \Rightarrow \mathbf{r}_r = (\cos\theta)\mathbf{i} + (\sin\theta)\mathbf{j} + 2\mathbf{k}$ and $\mathbf{r}_\theta = (-r\sin\theta)\mathbf{i} + (r\cos\theta)\mathbf{j}$

$$\Rightarrow \mathbf{r}_r \times \mathbf{r}_\theta = \begin{vmatrix} \mathbf{i} & \mathbf{j} & \mathbf{k} \\ \cos\theta & \sin\theta & 2 \\ -r\sin\theta & r\cos\theta & 0 \end{vmatrix} = (-2r\cos\theta)\mathbf{i} - (2r\sin\theta)\mathbf{j} + (r\cos^2\theta + r\sin^2\theta)\,\mathbf{k}$$

$$= (-2r\cos\theta)\mathbf{i} - (2r\sin\theta)\mathbf{j} + r\mathbf{k} \Rightarrow |\mathbf{r}_r \times \mathbf{r}_\theta| = \sqrt{4r^2\cos^2\theta + 4r^2\sin^2\theta + r^2} = \sqrt{5r^2} = r\sqrt{5}$$

$$\Rightarrow A = \int_0^{2\pi}\int_1^3 r\sqrt{5}\,dr\,d\theta = \int_0^{2\pi}\left[\frac{r^2\sqrt{5}}{2}\right]_1^3 d\theta = \int_0^{2\pi} 4\sqrt{5}\,d\theta = 8\pi\sqrt{5}$$

21. Let $x = r\cos\theta$ and $y = r\sin\theta \Rightarrow r^2 = x^2 + y^2 = 1$, $1 \le z \le 4$ and $0 \le \theta \le 2\pi$. Then $\mathbf{r}(z,\theta) = (\cos\theta)\mathbf{i} + (\sin\theta)\mathbf{j} + z\mathbf{k} \Rightarrow \mathbf{r}_z = \mathbf{k}$ and $\mathbf{r}_\theta = (-\sin\theta)\mathbf{i} + (\cos\theta)\mathbf{j}$

$$\Rightarrow \mathbf{r}_\theta \times \mathbf{r}_z = \begin{vmatrix} \mathbf{i} & \mathbf{j} & \mathbf{k} \\ -\sin\theta & \cos\theta & 0 \\ 0 & 0 & 1 \end{vmatrix} = (\cos\theta)\mathbf{i} + (\sin\theta)\,\mathbf{j} \Rightarrow |\mathbf{r}_\theta \times \mathbf{r}_z| = \sqrt{\cos^2\theta + \sin^2\theta} = 1$$

$$\Rightarrow A = \int_0^{2\pi}\int_1^4 1\,dr\,d\theta = \int_0^{2\pi} 3\,d\theta = 6\pi$$

23. $z = 2 - x^2 - y^2$ and $z = \sqrt{x^2 + y^2} \Rightarrow z = 2 - z^2 \Rightarrow z^2 + z - 2 = 0 \Rightarrow z = -2$ or $z = 1$. Since $z = \sqrt{x^2 + y^2} \ge 0$, we get $z = 1$ where the cone intersects the paraboloid. When $x = 0$ and $y = 0$, $z = 2 \Rightarrow$ the vertex of the paraboloid is $(0, 0, 2)$. Therefore, z ranges from 1 to 2 on the "cap" $\Rightarrow$ r ranges from 1 (when $x^2 + y^2 = 1$) to 0 (when $x = 0$ and $y = 0$ at the vertex). Let $x = r\cos\theta$, $y = r\sin\theta$, and $z = 2 - r^2$. Then $\mathbf{r}(r,\theta) = (r\cos\theta)\mathbf{i} + (r\sin\theta)\mathbf{j} + (2 - r^2)\,\mathbf{k}$, $0 \le r \le 1$, $0 \le \theta \le 2\pi \Rightarrow \mathbf{r}_r = (\cos\theta)\mathbf{i} + (\sin\theta)\mathbf{j} - 2r\mathbf{k}$ and

$$\mathbf{r}_\theta = (-r\sin\theta)\mathbf{i} + (r\cos\theta)\mathbf{j} \Rightarrow \mathbf{r}_r \times \mathbf{r}_\theta = \begin{vmatrix} \mathbf{i} & \mathbf{j} & \mathbf{k} \\ \cos\theta & \sin\theta & -2r \\ -r\sin\theta & r\cos\theta & 0 \end{vmatrix}$$

$$= (2r^2\cos\theta)\,\mathbf{i} + (2r^2\sin\theta)\,\mathbf{j} + r\mathbf{k} \Rightarrow |\mathbf{r}_r \times \mathbf{r}_\theta| = \sqrt{4r^4\cos^2\theta + 4r^4\sin^2\theta + r^2} = r\sqrt{4r^2 + 1}$$

$$\Rightarrow A = \int_0^{2\pi}\int_0^1 r\sqrt{4r^2 + 1}\,dr\,d\theta = \int_0^{2\pi}\left[\frac{1}{12}\left(4r^2 + 1\right)^{3/2}\right]_0^1 d\theta = \int_0^{2\pi}\left(\frac{5\sqrt{5} - 1}{12}\right)d\theta = \frac{\pi}{6}\left(5\sqrt{5} - 1\right)$$

25. Let $x = \rho\sin\phi\cos\theta$, $y = \rho\sin\phi\sin\theta$, and $z = \rho\cos\phi \Rightarrow \rho = \sqrt{x^2 + y^2 + z^2} = \sqrt{2}$ on the sphere. Next, $x^2 + y^2 + z^2 = 2$ and $z = \sqrt{x^2 + y^2} \Rightarrow z^2 + z^2 = 2 \Rightarrow z^2 = 1 \Rightarrow z = 1$ since $z \ge 0 \Rightarrow \phi = \frac{\pi}{4}$. For the lower portion of the sphere cut by the cone, we get $\phi = \pi$. Then
$\mathbf{r}(\phi,\theta) = \left(\sqrt{2}\sin\phi\cos\theta\right)\mathbf{i} + \left(\sqrt{2}\sin\phi\sin\theta\right)\mathbf{j} + \left(\sqrt{2}\cos\phi\right)\mathbf{k}$, $\frac{\pi}{4} \le \phi \le \pi$, $0 \le \theta \le 2\pi$
$\Rightarrow \mathbf{r}_\phi = \left(\sqrt{2}\cos\phi\cos\theta\right)\mathbf{i} + \left(\sqrt{2}\cos\phi\sin\theta\right)\mathbf{j} - \left(\sqrt{2}\sin\phi\right)\mathbf{k}$ and $\mathbf{r}_\theta = \left(-\sqrt{2}\sin\phi\sin\theta\right)\mathbf{i} + \left(\sqrt{2}\sin\phi\cos\theta\right)\mathbf{j}$

$$\Rightarrow \mathbf{r}_\phi \times \mathbf{r}_\theta = \begin{vmatrix} \mathbf{i} & \mathbf{j} & \mathbf{k} \\ \sqrt{2}\cos\phi\cos\theta & \sqrt{2}\cos\phi\sin\theta & -\sqrt{2}\sin\phi \\ -\sqrt{2}\sin\phi\sin\theta & \sqrt{2}\sin\phi\cos\theta & 0 \end{vmatrix}$$

$$= (2\sin^2\phi\cos\theta)\,\mathbf{i} + (2\sin^2\phi\sin\theta)\,\mathbf{j} + (2\sin\phi\cos\phi)\mathbf{k}$$

$\Rightarrow |\mathbf{r}_\phi \times \mathbf{r}_\theta| = \sqrt{4\sin^4\phi\cos^2\theta + 4\sin^4\phi\sin^2\theta + 4\sin^2\phi\cos^2\phi} = \sqrt{4\sin^2\phi} = 2|\sin\phi| = 2\sin\phi$

$\Rightarrow A = \int_0^{2\pi}\int_{\pi/4}^{\pi} 2\sin\phi\, d\phi\, d\theta = \int_0^{2\pi}\left(2+\sqrt{2}\right)d\theta = \left(4+2\sqrt{2}\right)\pi$

27. The parametrization $\mathbf{r}(r,\theta) = (r\cos\theta)\mathbf{i} + (r\sin\theta)\mathbf{j} + r\mathbf{k}$ at $P_0 = \left(\sqrt{2},\sqrt{2},2\right) \Rightarrow \theta = \frac{\pi}{4}, r = 2,$

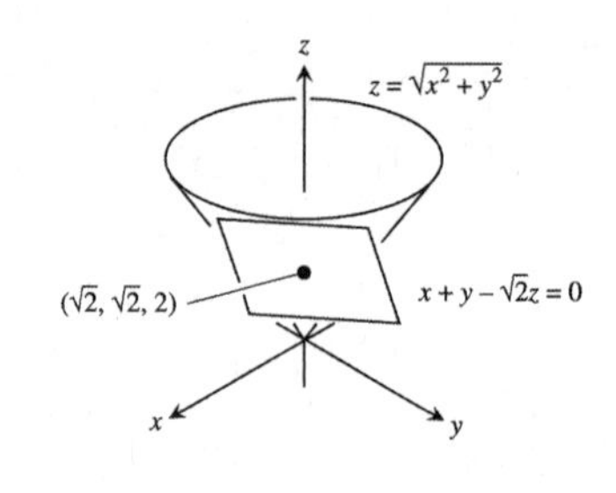

$\mathbf{r}_r = (\cos\theta)\mathbf{i} + (\sin\theta)\mathbf{j} + \mathbf{k} = \frac{\sqrt{2}}{2}\mathbf{i} + \frac{\sqrt{2}}{2}\mathbf{j} + \mathbf{k}$ and

$\mathbf{r}_\theta = (-r\sin\theta)\mathbf{i} + (r\cos\theta)\mathbf{j} = -\sqrt{2}\mathbf{i} + \sqrt{2}\mathbf{j}$

$\Rightarrow \mathbf{r}_r \times \mathbf{r}_\theta = \begin{vmatrix} \mathbf{i} & \mathbf{j} & \mathbf{k} \\ \sqrt{2}/2 & \sqrt{2}/2 & 1 \\ -\sqrt{2} & \sqrt{2} & 0 \end{vmatrix}$

$= -\sqrt{2}\mathbf{i} - \sqrt{2}\mathbf{j} + 2\mathbf{k} \Rightarrow$ the tangent plane is

$0 = \left(-\sqrt{2}\mathbf{i} - \sqrt{2}\mathbf{j} + 2\mathbf{k}\right)\cdot\left[\left(x-\sqrt{2}\right)\mathbf{i} + \left(y-\sqrt{2}\right)\mathbf{j} + (z-2)\mathbf{k}\right] \Rightarrow \sqrt{2}x + \sqrt{2}y - 2z = 0$, or $x + y - \sqrt{2}z = 0.$

The parametrization $\mathbf{r}(r,\theta) \Rightarrow x = r\cos\theta,\ y = r\sin\theta$ and $z = r \Rightarrow x^2+y^2 = r^2 = z^2 \Rightarrow$ the surface is $z = \sqrt{x^2+y^2}.$

29. The parametrization $\mathbf{r}(\theta,z) = (3\sin 2\theta)\mathbf{i} + (6\sin^2\theta)\mathbf{j} + z\mathbf{k}$ at $P_0 = \left(\frac{3\sqrt{3}}{2}, \frac{9}{2}, 0\right) \Rightarrow \theta = \frac{\pi}{3}$ and $z = 0$. Then

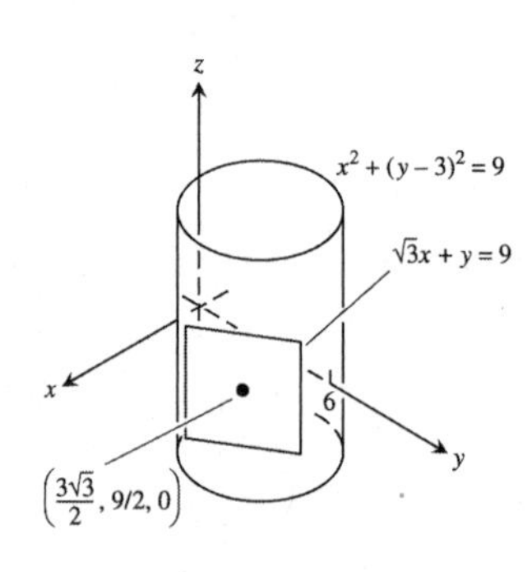

$\mathbf{r}_\theta = (6\cos 2\theta)\mathbf{i} + (12\sin\theta\cos\theta)\mathbf{j}$

$= -3\mathbf{i} + 3\sqrt{3}\mathbf{j}$ and $\mathbf{r}_z = \mathbf{k}$ at P_0

$\Rightarrow \mathbf{r}_\theta \times \mathbf{r}_z = \begin{vmatrix} \mathbf{i} & \mathbf{j} & \mathbf{k} \\ -3 & 3\sqrt{3} & 0 \\ 0 & 0 & 1 \end{vmatrix} = 3\sqrt{3}\mathbf{i} + 3\mathbf{j}$

$\Rightarrow$ the tangent plane is

$\left(3\sqrt{3}\mathbf{i} + 3\mathbf{j}\right)\cdot\left[\left(x - \frac{3\sqrt{3}}{2}\right)\mathbf{i} + \left(y - \frac{9}{2}\right)\mathbf{j} + (z-0)\mathbf{k}\right] = 0$

$\Rightarrow \sqrt{3}x + y = 9$. The parametrization $\Rightarrow x = 3\sin 2\theta$ and $y = 6\sin^2\theta \Rightarrow x^2 + y^2 = 9\sin^2 2\theta + \left(6\sin^2\theta\right)^2$

$= 9\left(4\sin^2\theta\cos^2\theta\right) + 36\sin^4\theta = 6\left(6\sin^2\theta\right) = 6y \Rightarrow x^2 + y^2 - 6y + 9 = 9 \Rightarrow x^2 + (y-3)^2 = 9$

31. (a) An arbitrary point on the circle C is $(x,z) = (R + r\cos u, r\sin u) \Rightarrow (x,y,z)$ is on the torus with $x = (R + r\cos u)\cos v,\ y = (R + r\cos u)\sin v$, and $z = r\sin u,\ 0 \le u \le 2\pi,\ 0 \le v \le 2\pi$

(b) $\mathbf{r}_u = (-r\sin u\cos v)\mathbf{i} - (r\sin u\sin v)\mathbf{j} + (r\cos u)\mathbf{k}$ and $\mathbf{r}_v = (-(R + r\cos u)\sin v)\mathbf{i} + ((R + r\cos u)\cos v)\mathbf{j}$

$\Rightarrow \mathbf{r}_u \times \mathbf{r}_v = \begin{vmatrix} \mathbf{i} & \mathbf{j} & \mathbf{k} \\ -r\sin u\cos v & -r\sin u\sin v & r\cos u \\ -(R + r\cos u)\sin v & (R + r\cos u)\cos v & 0 \end{vmatrix}$

$= -(R + r\cos u)(r\cos v\cos u)\mathbf{i} - (R + r\cos u)(r\sin v\cos u)\mathbf{j} + (-r\sin u)(R + r\cos u)\mathbf{k}$

$\Rightarrow |\mathbf{r}_u \times \mathbf{r}_v|^2 = (R + r\cos u)^2\left(r^2\cos^2 v\cos^2 u + r^2\sin^2 v\cos^2 u + r^2\sin^2 u\right) \Rightarrow |\mathbf{r}_u \times \mathbf{r}_v| = r(R + r\cos u)$

$\Rightarrow A = \int_0^{2\pi}\int_0^{2\pi}\left(rR + r^2\cos u\right)du\,dv = \int_0^{2\pi} 2\pi rR\,dv = 4\pi^2 rR$

33. (a) Let $w^2 + \frac{z^2}{c^2} = 1$ where $w = \cos\phi$ and $\frac{z}{c} = \sin\phi \Rightarrow \frac{x^2}{a^2} + \frac{y^2}{b^2} = \cos^2\phi \Rightarrow \frac{x}{a} = \cos\phi\cos\theta$ and $\frac{y}{b} = \cos\phi\sin\theta$

$\Rightarrow x = a\cos\theta\cos\phi,\ y = b\sin\theta\cos\phi$, and $z = c\sin\phi$

$\Rightarrow \mathbf{r}(\theta,\phi) = (a\cos\theta\cos\phi)\mathbf{i} + (b\sin\theta\cos\phi)\mathbf{j} + (c\sin\phi)\mathbf{k}$

(b) $\mathbf{r}_\theta = (-a\sin\theta\cos\phi)\mathbf{i} + (b\cos\theta\cos\phi)\mathbf{j}$ and $\mathbf{r}_\phi = (-a\cos\theta\sin\phi)\mathbf{i} - (b\sin\theta\sin\phi)\mathbf{j} + (c\cos\phi)\mathbf{k}$

$\Rightarrow \mathbf{r}_\theta \times \mathbf{r}_\phi = \begin{vmatrix} \mathbf{i} & \mathbf{j} & \mathbf{k} \\ -a\sin\theta\cos\phi & b\cos\theta\cos\phi & 0 \\ -a\cos\theta\sin\phi & -b\sin\theta\sin\phi & c\cos\phi \end{vmatrix}$

$= (\text{bc} \cos\theta \cos^2\phi)\,\mathbf{i} + (\text{ac} \sin\theta \cos^2\phi)\,\mathbf{j} + (\text{ab} \sin\phi \cos\phi)\mathbf{k}$

$\Rightarrow |\mathbf{r}_\theta \times \mathbf{r}_\phi|^2 = b^2c^2 \cos^2\theta \cos^4\phi + a^2c^2 \sin^2\theta \cos^4\phi + a^2b^2 \sin^2\phi \cos^2\phi$, and the result follows.

$A \Rightarrow \int_0^{2\pi}\int_0^{\pi} |\mathbf{r}_\theta \times \mathbf{r}_\phi|\, d\phi\, d\theta = \int_0^{2\pi}\int_0^{\pi} \left[a^2b^2 \sin^2\phi \cos^2\phi + b^2c^2 \cos^2\theta \cos^4\phi + a^2c^2 \sin^2\theta \cos^4\phi\right]^{1/2} d\phi\, d\theta$

35. $\mathbf{r}(\theta, u) = (5 \cosh u \cos\theta)\mathbf{i} + (5 \cosh u \sin\theta)\mathbf{j} + (5 \sinh u)\mathbf{k} \Rightarrow \mathbf{r}_\theta = (-5 \cosh u \sin\theta)\mathbf{i} + (5 \cosh u \cos\theta)\mathbf{j}$ and $\mathbf{r}_u = (5 \sinh u \cos\theta)\mathbf{i} + (5 \sinh u \sin\theta)\mathbf{j} + (5 \cosh u)\mathbf{k}$

$$\Rightarrow r_\theta \times \mathbf{r}_u = \begin{vmatrix} \mathbf{i} & \mathbf{j} & \mathbf{k} \\ -5\cosh u \sin\theta & 5 \cosh u \cos\theta & 0 \\ 5 \sinh u \cos\theta & 5 \sinh u \sin\theta & 5 \cosh u \end{vmatrix}$$

$= (25 \cosh^2 u \cos\theta)\,\mathbf{i} + (25 \cosh^2 u \sin\theta)\,\mathbf{j} - (25 \cosh u \sinh u)\mathbf{k}$. At the point $(x_0, y_0, 0)$, where $x_0^2 + y_0^2 = 25$ we have $5 \sinh u = 0 \Rightarrow u = 0$ and $x_0 = 25 \cos\theta$, $y_0 = 25 \sin\theta \Rightarrow$ the tangent plane is $5(x_0\mathbf{i} + y_0\mathbf{j}) \cdot [(x - x_0)\mathbf{i} + (y - y_0)\mathbf{j} + z\mathbf{k}] = 0 \Rightarrow x_0x - x_0^2 + y_0y - y_0^2 = 0 \Rightarrow x_0x + y_0y = 25$

37. $\mathbf{p} = \mathbf{k}$, $\nabla f = 2x\mathbf{i} + 2y\mathbf{j} - \mathbf{k} \Rightarrow |\nabla f| = \sqrt{(2x)^2 + (2y)^2 + (-1)^2} = \sqrt{4x^2 + 4y^2 + 1}$ and $|\nabla f \cdot \mathbf{p}| = 1$; $z = 2 \Rightarrow x^2 + y^2 = 2$; thus $S = \iint_R \frac{|\nabla f|}{|\nabla f \cdot \mathbf{p}|}\, dA = \iint_R \sqrt{4x^2 + 4y^2 + 1}\, dx\, dy$

$= \iint_R \sqrt{4r^2 \cos^2\theta + 4r^2 \sin^2\theta + 1}\, r\, dr\, d\theta = \int_0^{2\pi}\int_0^{\sqrt{2}} \sqrt{4r^2 + 1}\, r\, dr\, d\theta = \int_0^{2\pi} \left[\frac{1}{12}(4r^2 + 1)^{3/2}\right]_0^{\sqrt{2}} d\theta$

$= \int_0^{2\pi} \frac{13}{6}\, d\theta = \frac{13}{3}\pi$

39. $\mathbf{p} = \mathbf{k}$, $\nabla f = \mathbf{i} + 2\mathbf{j} + 2\mathbf{k} \Rightarrow |\nabla f| = 3$ and $|\nabla f \cdot \mathbf{p}| = 2$; $x = y^2$ and $x = 2 - y^2$ intersect at $(1, 1)$ and $(1, -1)$

$\Rightarrow S = \iint_R \frac{|\nabla f|}{|\nabla f \cdot \mathbf{p}|}\, dA = \iint_R \frac{3}{2}\, dx\, dy = \int_{-1}^{1}\int_{y^2}^{2-y^2} \frac{3}{2}\, dx\, dy = \int_{-1}^{1} (3 - 3y^2)\, dy = 4$

41. $\mathbf{p} = \mathbf{k}$, $\nabla f = 2x\mathbf{i} - 2\mathbf{j} - 2\mathbf{k} \Rightarrow |\nabla f| = \sqrt{(2x)^2 + (-2)^2 + (-2)^2} = \sqrt{4x^2 + 8} = 2\sqrt{x^2 + 2}$ and $|\nabla f \cdot \mathbf{p}| = 2$

$\Rightarrow S = \iint_R \frac{|\nabla f|}{|\nabla f \cdot \mathbf{p}|}\, dA = \iint_R \frac{2\sqrt{x^2+2}}{2}\, dx\, dy = \int_0^2\int_0^{3x} \sqrt{x^2 + 2}\, dy\, dx = \int_0^2 3x\sqrt{x^2 + 2}\, dx = \left[(x^2 + 2)^{3/2}\right]_0^2$

$= 6\sqrt{6} - 2\sqrt{2}$

43. $\mathbf{p} = \mathbf{k}$, $\nabla f = c\mathbf{i} - \mathbf{k} \Rightarrow |\nabla f| = \sqrt{c^2 + 1}$ and $|\nabla f \cdot \mathbf{p}| = 1 \Rightarrow S = \iint_R \frac{|\nabla f|}{|\nabla f \cdot \mathbf{p}|}\, dA = \iint_R \sqrt{c^2 + 1}\, dx\, dy$

$= \int_0^{2\pi}\int_0^1 \sqrt{c^2 + 1}\, r\, dr\, d\theta = \int_0^{2\pi} \frac{\sqrt{c^2+1}}{2}\, d\theta = \pi\sqrt{c^2 + 1}$

45. $\mathbf{p} = \mathbf{i}$, $\nabla f = \mathbf{i} + 2y\mathbf{j} + 2z\mathbf{k} \Rightarrow |\nabla f| = \sqrt{1^2 + (2y)^2 + (2z)^2} = \sqrt{1 + 4y^2 + 4z^2}$ and $|\nabla f \cdot \mathbf{p}| = 1$; $1 \le y^2 + z^2 \le 4$

$\Rightarrow S = \iint_R \frac{|\nabla f|}{|\nabla f \cdot \mathbf{p}|}\, dA = \iint_R \sqrt{1 + 4y^2 + 4z^2}\, dy\, dz = \int_0^{2\pi}\int_1^2 \sqrt{1 + 4r^2 \cos^2\theta + 4r^2 \sin^2\theta}\, r\, dr\, d\theta$

$= \int_0^{2\pi}\int_1^2 \sqrt{1 + 4r^2}\, r\, dr\, d\theta = \int_0^{2\pi} \left[\frac{1}{12}(1 + 4r^2)^{3/2}\right]_1^2 d\theta = \int_0^{2\pi} \frac{1}{12}\left(17\sqrt{17} - 5\sqrt{5}\right) d\theta = \frac{\pi}{6}\left(17\sqrt{17} - 5\sqrt{5}\right)$

47. $\mathbf{p} = \mathbf{k}$, $\nabla f = \left(2x - \frac{2}{x}\right)\mathbf{i} + \sqrt{15}\mathbf{j} - \mathbf{k} \Rightarrow |\nabla f| = \sqrt{\left(2x - \frac{2}{x}\right)^2 + \left(\sqrt{15}\right)^2 + (-1)^2} = \sqrt{4x^2 + 8 + \frac{4}{x^2}} = \sqrt{\left(2x + \frac{2}{x}\right)^2}$

$= 2x + \frac{2}{x}$, on $1 \le x \le 2$ and $|\nabla f \cdot \mathbf{p}| = 1 \Rightarrow S = \iint_R \frac{|\nabla f|}{|\nabla f \cdot \mathbf{p}|}\, dA = \iint_R (2x + 2x^{-1})\, dx\, dy$

$= \int_0^1\int_1^2 (2x + 2x^{-1})\, dx\, dy = \int_0^1 \left[x^2 + 2\ln x\right]_1^2 dy = \int_0^1 (3 + 2\ln 2)\, dy = 3 + 2\ln 2$

49. $f_x(x,y) = 2x,\ f_y(x,y) = 2y \Rightarrow \sqrt{f_x^2 + f_y^2 + 1} = \sqrt{4x^2 + 4y^2 + 1} \Rightarrow \text{Area} = \iint_R \sqrt{4x^2 + 4y^2 + 1}\, dx\, dy$
$= \int_0^{2\pi}\int_0^{\sqrt{3}} \sqrt{4r^2 + 1}\, r\, dr\, d\theta = \frac{\pi}{6}\left(13\sqrt{13} - 1\right)$

51. $f_x(x,y) = \frac{x}{\sqrt{x^2+y^2}},\ f_y(x,y) = \frac{y}{\sqrt{x^2+y^2}} \Rightarrow \sqrt{f_x^2 + f_y^2 + 1} = \sqrt{\frac{x^2}{x^2+y^2} + \frac{y^2}{x^2+y^2} + 1} = \sqrt{2} \Rightarrow \text{Area} = \iint_{R_{xy}} \sqrt{2}\, dx\, dy$
$= \sqrt{2}(\text{Area between the ellipse and the circle}) = \sqrt{2}(6\pi - \pi) = 5\pi\sqrt{2}$

53. $y = \frac{2}{3}z^{3/2} \Rightarrow f_x(x,z) = 0,\ f_z(x,z) = z^{1/2} \Rightarrow \sqrt{f_x^2 + f_z^2 + 1} = \sqrt{z+1};\ y = \frac{16}{3} \Rightarrow \frac{16}{3} = \frac{2}{3}z^{3/2} \Rightarrow z = 4$
$\Rightarrow \text{Area} = \int_0^4\int_0^1 \sqrt{z+1}\, dx\, dz = \int_0^4 \sqrt{z+1}\, dz = \frac{2}{3}\left(5\sqrt{5} - 1\right)$

55. $\mathbf{r}(x,y) = x\,\mathbf{i} + y\,\mathbf{j} + f(x,y)\,\mathbf{k} \Rightarrow \mathbf{r}_x(x,y) = \mathbf{i} + f_x(x,y)\,\mathbf{k},\ \mathbf{r}_y(x,y) = \mathbf{j} + f_y(x,y)\mathbf{k}$

$$\Rightarrow \mathbf{r}_x \times \mathbf{r}_y = \begin{vmatrix} \mathbf{i} & \mathbf{j} & \mathbf{k} \\ 1 & 0 & f_x(x,y) \\ 0 & 1 & f_y(x,y) \end{vmatrix} = -f_x(x,y)\,\mathbf{i} - f_y(x,y)\,\mathbf{j} + \mathbf{k}$$

$$\Rightarrow |\mathbf{r}_x \times \mathbf{r}_y| = \sqrt{(-f_x(x,y))^2 + (-f_y(x,y))^2 + 1^2} = \sqrt{f_x(x,y)^2 + f_y(x,y)^2 + 1}$$

$$\Rightarrow d\sigma = \sqrt{f_x(x,y)^2 + f_y(x,y)^2 + 1}\, dA$$

16.6 SURFACE INTEGRALS

1. Let the parametrization be $\mathbf{r}(x,z) = x\mathbf{i} + x^2\mathbf{j} + z\mathbf{k} \Rightarrow \mathbf{r}_x = \mathbf{i} + 2x\mathbf{j}$ and $\mathbf{r}_z = \mathbf{k} \Rightarrow \mathbf{r}_x \times \mathbf{r}_z = \begin{vmatrix} \mathbf{i} & \mathbf{j} & \mathbf{k} \\ 1 & 2x & 0 \\ 0 & 0 & 1 \end{vmatrix}$
$= 2x\mathbf{i} + \mathbf{j} \Rightarrow |\mathbf{r}_x \times \mathbf{r}_z| = \sqrt{4x^2 + 1} \Rightarrow \iint_S G(x,y,z)\, d\sigma = \int_0^3\int_0^2 x\sqrt{4x^2+1}\, dx\, dz = \int_0^3 \left[\frac{1}{12}(4x^2+1)^{3/2}\right]_0^2 dz$
$= \int_0^3 \frac{1}{12}\left(17\sqrt{17} - 1\right) dz = \frac{17\sqrt{17} - 1}{4}$

3. Let the parametrization be $\mathbf{r}(\phi,\theta) = (\sin\phi\cos\theta)\mathbf{i} + (\sin\phi\sin\theta)\mathbf{j} + (\cos\phi)\mathbf{k}$ (spherical coordinates with $\rho = 1$ on the sphere), $0 \le \phi \le \pi,\ 0 \le \theta \le 2\pi \Rightarrow \mathbf{r}_\phi = (\cos\phi\cos\theta)\mathbf{i} + (\cos\phi\sin\theta)\mathbf{j} - (\sin\phi)\mathbf{k}$ and
$\mathbf{r}_\theta = (-\sin\phi\sin\theta)\mathbf{i} + (\sin\phi\cos\theta)\mathbf{j} \Rightarrow \mathbf{r}_\phi \times \mathbf{r}_\theta = \begin{vmatrix} \mathbf{i} & \mathbf{j} & \mathbf{k} \\ \cos\phi\cos\theta & \cos\phi\sin\theta & -\sin\phi \\ -\sin\phi\sin\theta & \sin\phi\cos\theta & 0 \end{vmatrix}$
$= (\sin^2\phi\cos\theta)\,\mathbf{i} + (\sin^2\phi\sin\theta)\,\mathbf{j} + (\sin\phi\cos\phi)\mathbf{k} \Rightarrow |\mathbf{r}_\phi \times \mathbf{r}_\theta| = \sqrt{\sin^4\phi\cos^2\theta + \sin^4\phi\sin^2\theta + \sin^2\phi\cos^2\phi}$
$= \sin\phi;\ x = \sin\phi\cos\theta \Rightarrow G(x,y,z) = \cos^2\theta\sin^2\phi \Rightarrow \iint_S G(x,y,z)\, d\sigma = \int_0^{2\pi}\int_0^{\pi} (\cos^2\theta\sin^2\phi)(\sin\phi)\, d\phi\, d\theta$
$= \int_0^{2\pi}\int_0^{\pi} (\cos^2\theta)(1 - \cos^2\phi)(\sin\phi)\, d\phi\, d\theta;\ \begin{bmatrix} u = \cos\phi \\ du = -\sin\phi\, d\phi \end{bmatrix} \to \int_0^{2\pi}\int_1^{-1} (\cos^2\theta)(u^2 - 1)\, du\, d\theta$
$= \int_0^{2\pi} (\cos^2\theta)\left[\frac{u^3}{3} - u\right]_1^{-1} d\theta = \frac{4}{3}\int_0^{2\pi} \cos^2\theta\, d\theta = \frac{4}{3}\left[\frac{\theta}{2} + \frac{\sin 2\theta}{4}\right]_0^{2\pi} = \frac{4\pi}{3}$

5. Let the parametrization be $\mathbf{r}(x,y) = x\mathbf{i} + y\mathbf{j} + (4 - x - y)\mathbf{k} \Rightarrow \mathbf{r}_x = \mathbf{i} - \mathbf{k}$ and $\mathbf{r}_y = \mathbf{j} - \mathbf{k}$
$\Rightarrow \mathbf{r}_x \times \mathbf{r}_y = \begin{vmatrix} \mathbf{i} & \mathbf{j} & \mathbf{k} \\ 1 & 0 & -1 \\ 0 & 1 & -1 \end{vmatrix} = \mathbf{i} + \mathbf{j} + \mathbf{k} \Rightarrow |\mathbf{r}_x \times \mathbf{r}_y| = \sqrt{3} \Rightarrow \iint_S F(x,y,z)\, d\sigma = \int_0^1\int_0^1 (4 - x - y)\sqrt{3}\, dy\, dx$
$= \int_0^1 \sqrt{3}\left[4y - xy - \frac{y^2}{2}\right]_0^1 dx = \int_0^1 \sqrt{3}\left(\frac{7}{2} - x\right) dx = \sqrt{3}\left[\frac{7}{2}x - \frac{x^2}{2}\right]_0^1 = 3\sqrt{3}$

7. Let the parametrization be $\mathbf{r}(r,\theta) = (r\cos\theta)\mathbf{i} + (r\sin\theta)\mathbf{j} + (1-r^2)\mathbf{k}$, $0 \le r \le 1$ (since $0 \le z \le 1$) and $0 \le \theta \le 2\pi$

$\Rightarrow \mathbf{r}_r = (\cos\theta)\mathbf{i} + (\sin\theta)\mathbf{j} - 2r\mathbf{k}$ and $\mathbf{r}_\theta = (-r\sin\theta)\mathbf{i} + (r\cos\theta)\mathbf{j} \Rightarrow \mathbf{r}_r \times \mathbf{r}_\theta = \begin{vmatrix} \mathbf{i} & \mathbf{j} & \mathbf{k} \\ \cos\theta & \sin\theta & -2r \\ -r\sin\theta & r\cos\theta & 0 \end{vmatrix}$

$= (2r^2\cos\theta)\,\mathbf{i} + (2r^2\sin\theta)\,\mathbf{j} + r\mathbf{k} \Rightarrow |\mathbf{r}_r \times \mathbf{r}_\theta| = \sqrt{(2r^2\cos\theta)^2 + (2r^2\sin\theta) + r^2} = r\sqrt{1+4r^2}$; $z = 1 - r^2$ and

$x = r\cos\theta \Rightarrow H(x,y,z) = (r^2\cos^2\theta)\sqrt{1+4r^2} \Rightarrow \iint\limits_S H(x,y,z)\,d\sigma$

$= \int_0^{2\pi}\int_0^1 (r^2\cos^2\theta)\left(\sqrt{1+4r^2}\right)\left(r\sqrt{1+4r^2}\right) dr\,d\theta = \int_0^{2\pi}\int_0^1 r^3(1+4r^2)\cos^2\theta\,dr\,d\theta = \frac{11\pi}{12}$

9. The bottom face S of the cube is in the xy-plane $\Rightarrow z = 0 \Rightarrow G(x,y,0) = x + y$ and $f(x,y,z) = z = 0 \Rightarrow \mathbf{p} = \mathbf{k}$ and $\nabla f = \mathbf{k} \Rightarrow |\nabla f| = 1$ and $|\nabla f \cdot \mathbf{p}| = 1 \Rightarrow d\sigma = dx\,dy \Rightarrow \iint\limits_S G\,d\sigma = \iint\limits_R (x+y)\,dx\,dy$

$= \int_0^a\int_0^a (x+y)\,dx\,dy = \int_0^a \left(\frac{a^2}{2} + ay\right) dy = a^3$. Because of symmetry, we also get a^3 over the face of the cube in the xz-plane and a^3 over the face of the cube in the yz-plane. Next, on the top of the cube, $G(x,y,z)$ $= G(x,y,a) = x + y + a$ and $f(x,y,z) = z = a \Rightarrow \mathbf{p} = \mathbf{k}$ and $\nabla f = \mathbf{k} \Rightarrow |\nabla f| = 1$ and $|\nabla f \cdot \mathbf{p}| = 1 \Rightarrow d\sigma = dx\,dy$

$\iint\limits_S G\,d\sigma = \iint\limits_R (x+y+a)\,dx\,dy = \int_0^a\int_0^a (x+y+a)\,dx\,dy = \int_0^a\int_0^a (x+y)\,dx\,dy + \int_0^a\int_0^a a\,dx\,dy = 2a^3$.

Because of symmetry, the integral is also $2a^3$ over each of the other two faces. Therefore,

$\iint\limits_{\text{cube}} (x+y+z)\,d\sigma = 3(a^3 + 2a^3) = 9a^3$.

11. On the faces in the coordinate planes, $G(x,y,z) = 0 \Rightarrow$ the integral over these faces is 0.

On the face $x = a$, we have $f(x,y,z) = x = a$ and $G(x,y,z) = G(a,y,z) = ayz \Rightarrow \mathbf{p} = \mathbf{i}$ and $\nabla f = \mathbf{i} \Rightarrow |\nabla f| = 1$ and $|\nabla f \cdot \mathbf{p}| = 1 \Rightarrow d\sigma = dy\,dz \Rightarrow \iint\limits_S G\,d\sigma = \iint\limits_S ayz\,d\sigma = \int_0^c\int_0^b ayz\,dy\,dz = \frac{ab^2c^2}{4}$.

On the face $y = b$, we have $f(x,y,z) = y = b$ and $G(x,y,z) = G(x,b,z) = bxz \Rightarrow \mathbf{p} = \mathbf{j}$ and $\nabla f = \mathbf{j} \Rightarrow |\nabla f| = 1$ and $|\nabla f \cdot \mathbf{p}| = 1 \Rightarrow d\sigma = dx\,dz \Rightarrow \iint\limits_S G\,d\sigma = \iint\limits_S bxz\,d\sigma = \int_0^c\int_0^a bxz\,dx\,dz = \frac{a^2bc^2}{4}$.

On the face $z = c$, we have $f(x,y,z) = z = c$ and $G(x,y,z) = G(x,y,c) = cxy \Rightarrow \mathbf{p} = \mathbf{k}$ and $\nabla f = \mathbf{k} \Rightarrow |\nabla f| = 1$ and $|\nabla f \cdot \mathbf{p}| = 1 \Rightarrow d\sigma = dy\,dx \Rightarrow \iint\limits_S G\,d\sigma = \iint\limits_S cxy\,d\sigma = \int_0^b\int_0^a cxy\,dx\,dy = \frac{a^2b^2c}{4}$. Therefore,

$\iint\limits_S G(x,y,z)\,d\sigma = \frac{abc(ab+ac+bc)}{4}$.

13. $f(x,y,z) = 2x + 2y + z = 2 \Rightarrow \nabla f = 2\mathbf{i} + 2\mathbf{j} + \mathbf{k}$ and $G(x,y,z) = x + y + (2 - 2x - 2y) = 2 - x - y \Rightarrow \mathbf{p} = \mathbf{k}$, $|\nabla f| = 3$ and $|\nabla f \cdot \mathbf{p}| = 1 \Rightarrow d\sigma = 3\,dy\,dx$; $z = 0 \Rightarrow 2x + 2y = 2 \Rightarrow y = 1 - x \Rightarrow \iint\limits_S G\,d\sigma = \iint\limits_S (2 - x - y)\,d\sigma$

$= 3\int_0^1\int_0^{1-x} (2-x-y)\,dy\,dx = 3\int_0^1 \left[(2-x)(1-x) - \frac{1}{2}(1-x)^2\right] dx = 3\int_0^1 \left(\frac{3}{2} - 2x + \frac{x^2}{2}\right) dx = 2$

15. $f(x,y,z) = x + y^2 - z = 0 \Rightarrow \nabla f = \mathbf{i} + 2y\mathbf{j} - \mathbf{k} \Rightarrow |\nabla f| = \sqrt{4y^2+2} = \sqrt{2}\sqrt{2y^2+1}$ and $\mathbf{p} = \mathbf{k} \Rightarrow |\nabla f \cdot \mathbf{p}| = 1$

$\Rightarrow d\sigma = \frac{\sqrt{2}\sqrt{2y^2+1}}{1}\,dx\,dy \Rightarrow \iint\limits_S G\,d\sigma = \int_0^1\int_0^y (x + y^2 - x)\sqrt{2}\sqrt{2y^2+1}\,dx\,dy = \sqrt{2}\int_0^1\int_0^y y^2\sqrt{2y^2+1}\,dx\,dy$

$= \sqrt{2}\int_0^1 y^3\sqrt{2y^2+1}\,dy = \frac{6\sqrt{6}+\sqrt{2}}{30}$

17. $f(x,y,z) = 2x + y + z = 2 \Rightarrow \nabla f = 2\mathbf{i} + \mathbf{j} + \mathbf{k} \Rightarrow |\nabla f| = \sqrt{6}$ and $\mathbf{p} = \mathbf{k} \Rightarrow |\nabla f \cdot \mathbf{p}| = 1 \Rightarrow d\sigma = \frac{\sqrt{6}}{1}\,dy\,dx$

$\Rightarrow \iint\limits_S G\,d\sigma = \int_0^1\int_{1-x}^{2-2x} x\,y(2 - 2x - y)\sqrt{6}\,dy\,dx = \sqrt{6}\int_0^1\int_{1-x}^{2-2x} (2x\,y - 2x^2y - x\,y^2)\,dy\,dx$

$= \sqrt{6}\int_0^1 \left(\frac{2}{3}x - 2x^2 + 2x^3 - \frac{2}{3}x^4\right) dx = \frac{\sqrt{6}}{30}$

19. Let the parametrization be $\mathbf{r}(x, y) = x\mathbf{i} + y\mathbf{j} + (4 - y^2)\,\mathbf{k}, 0 \le x \le 1, -2 \le y \le 2; z = 0 \Rightarrow 0 = 4 - y^2$

$$\Rightarrow y = \pm 2; \mathbf{r}_x = \mathbf{i} \text{ and } \mathbf{r}_y = \mathbf{j} - 2y\mathbf{k} \Rightarrow \mathbf{r}_x \times \mathbf{r}_y = \begin{vmatrix} \mathbf{i} & \mathbf{j} & \mathbf{k} \\ 1 & 0 & 0 \\ 0 & 1 & -2y \end{vmatrix} = 2y\mathbf{j} + \mathbf{k} \Rightarrow \mathbf{F} \cdot \mathbf{n}\, d\sigma$$

$$= \mathbf{F} \cdot \frac{\mathbf{r}_x \times \mathbf{r}_y}{|\mathbf{r}_x \times \mathbf{r}_y|} |\mathbf{r}_x \times \mathbf{r}_y|\, dy\, dx = (2xy - 3z)\, dy\, dx = [2xy - 3(4 - y^2)]\, dy\, dx \Rightarrow \iint_S \mathbf{F} \cdot \mathbf{n}\, d\sigma$$

$$= \int_0^1 \int_{-2}^2 (2xy + 3y^2 - 12)\, dy\, dx = \int_0^1 \left[xy^2 + y^3 - 12y\right]_{-2}^{2} dx = \int_0^1 -32\, dx = -32$$

21. Let the parametrization be $\mathbf{r}(\phi, \theta) = (a\ \sin\phi \cos\theta)\mathbf{i} + (a \sin\phi \sin\theta)\mathbf{j} + (a \cos\phi)\mathbf{k}$ (spherical coordinates with $\rho = a, a \ge 0$, on the sphere), $0 \le \phi \le \frac{\pi}{2}$ (for the first octant) , $0 \le \theta \le \frac{\pi}{2}$ (for the first octant)

$\Rightarrow \mathbf{r}_\phi = (a \cos\phi \cos\theta)\mathbf{i} + (a \cos\phi \sin\theta)\mathbf{j} - (a \sin\phi)\mathbf{k}$ and $\mathbf{r}_\theta = (-a \sin\phi \sin\theta)\mathbf{i} + (a \sin\phi \cos\theta)\mathbf{j}$

$$\Rightarrow \mathbf{r}_\phi \times \mathbf{r}_\theta = \begin{vmatrix} \mathbf{i} & \mathbf{j} & \mathbf{k} \\ a\cos\phi\cos\theta & a\cos\phi\sin\theta & -a\sin\phi \\ -a\sin\phi\sin\theta & a\sin\phi\cos\theta & 0 \end{vmatrix}$$

$$= (a^2 \sin^2\phi \cos\theta)\,\mathbf{i} + (a^2 \sin^2\phi \sin\theta)\,\mathbf{j} + (a^2 \sin\phi \cos\phi)\,\mathbf{k} \Rightarrow \mathbf{F} \cdot \mathbf{n}\, d\sigma = \mathbf{F} \cdot \frac{\mathbf{r}_\phi \times \mathbf{r}_\theta}{|\mathbf{r}_\phi \times \mathbf{r}_\theta|} |\mathbf{r}_\phi \times \mathbf{r}_\theta|\, d\theta\, d\phi$$

$$= a^3 \cos^2\phi \sin\phi\, d\theta\, d\phi \text{ since } \mathbf{F} = z\mathbf{k} = (a\cos\phi)\mathbf{k} \Rightarrow \iint_S \mathbf{F} \cdot \mathbf{n}\, d\sigma = \int_0^{\pi/2} \int_0^{\pi/2} a^3 \cos^2\phi \sin\phi\, d\phi\, d\theta = \frac{\pi a^3}{6}$$

23. Let the parametrization be $\mathbf{r}(x, y) = x\mathbf{i} + y\mathbf{j} + (2a - x - y)\mathbf{k}, 0 \le x \le a, 0 \le y \le a \Rightarrow \mathbf{r}_x = \mathbf{i} - \mathbf{k}$ and $\mathbf{r}_y = \mathbf{j} - \mathbf{k}$

$$\Rightarrow \mathbf{r}_x \times \mathbf{r}_y = \begin{vmatrix} \mathbf{i} & \mathbf{j} & \mathbf{k} \\ 1 & 0 & -1 \\ 0 & 1 & -1 \end{vmatrix} = \mathbf{i} + \mathbf{j} + \mathbf{k} \Rightarrow \mathbf{F} \cdot \mathbf{n}\, d\sigma = \mathbf{F} \cdot \frac{\mathbf{r}_x \times \mathbf{r}_y}{|\mathbf{r}_x \times \mathbf{r}_y|} |\mathbf{r}_x \times \mathbf{r}_y|\, dy\, dx$$

$$= [2xy + 2y(2a - x - y) + 2x(2a - x - y)]\, dy\, dx \text{ since } \mathbf{F} = 2xy\mathbf{i} + 2yz\mathbf{j} + 2xz\mathbf{k}$$

$$= 2xy\mathbf{i} + 2y(2a - x - y)\mathbf{j} + 2x(2a - x - y)\mathbf{k} \Rightarrow \iint_S \mathbf{F} \cdot \mathbf{n}\, d\sigma$$

$$= \int_0^a \int_0^a [2xy + 2y(2a - x - y) + 2x(2a - x - y)]\, dy\, dx = \int_0^a \int_0^a (4ay - 2y^2 + 4ax - 2x^2 - 2xy)\, dy\, dx$$

$$= \int_0^a \left(\tfrac{4}{3} a^3 + 3a^2 x - 2ax^2\right) dx = \left(\tfrac{4}{3} + \tfrac{3}{2} - \tfrac{2}{3}\right) a^4 = \frac{13a^4}{6}$$

25. Let the parametrization be $\mathbf{r}(r, \theta) = (r\cos\theta)\mathbf{i} + (r\sin\theta)\mathbf{j} + r\mathbf{k}, 0 \le r \le 1$ (since $0 \le z \le 1$) and $0 \le \theta \le 2\pi$

$$\Rightarrow \mathbf{r}_r = (\cos\theta)\mathbf{i} + (\sin\theta)\mathbf{j} + \mathbf{k} \text{ and } \mathbf{r}_\theta = (-r\sin\theta)\mathbf{i} + (r\cos\theta)\mathbf{j} \Rightarrow \mathbf{r}_\theta \times \mathbf{r}_r = \begin{vmatrix} \mathbf{i} & \mathbf{j} & \mathbf{k} \\ -r\sin\theta & r\cos\theta & 0 \\ \cos\theta & \sin\theta & 1 \end{vmatrix}$$

$$= (r\cos\theta)\mathbf{i} + (r\sin\theta)\mathbf{j} - r\mathbf{k} \Rightarrow \mathbf{F} \cdot \mathbf{n}\, d\sigma = \mathbf{F} \cdot \frac{\mathbf{r}_\theta \times \mathbf{r}_r}{|\mathbf{r}_\theta \times \mathbf{r}_r|} |\mathbf{r}_\theta \times \mathbf{r}_r|\, d\theta\, dr = (r^3 \sin\theta \cos^2\theta + r^2)\, d\theta\, dr \text{ since}$$

$$\mathbf{F} = (r^2 \sin\theta \cos\theta)\,\mathbf{i} - r\mathbf{k} \Rightarrow \iint_S \mathbf{F} \cdot \mathbf{n}\, d\sigma = \int_0^{2\pi} \int_0^1 (r^3 \sin\theta \cos^2\theta + r^2)\, dr\, d\theta = \int_0^{2\pi} \left(\tfrac{1}{4} \sin\theta \cos^2\theta + \tfrac{1}{3}\right) d\theta$$

$$= \left[-\tfrac{1}{12} \cos^3\theta + \tfrac{\theta}{3}\right]_0^{2\pi} = \frac{2\pi}{3}$$

27. Let the parametrization be $\mathbf{r}(r, \theta) = (r\cos\theta)\mathbf{i} + (r\sin\theta)\mathbf{j} + r\mathbf{k}, 1 \le r \le 2$ (since $1 \le z \le 2$) and $0 \le \theta \le 2\pi$

$$\Rightarrow \mathbf{r}_r = (\cos\theta)\mathbf{i} + (\sin\theta)\mathbf{j} + \mathbf{k} \text{ and } \mathbf{r}_\theta = (-r\sin\theta)\mathbf{i} + (r\cos\theta)\mathbf{j} \Rightarrow \mathbf{r}_\theta \times \mathbf{r}_r = \begin{vmatrix} \mathbf{i} & \mathbf{j} & \mathbf{k} \\ -r\sin\theta & r\cos\theta & 0 \\ \cos\theta & \sin\theta & 1 \end{vmatrix}$$

$$= (r\cos\theta)\mathbf{i} + (r\sin\theta)\mathbf{j} - r\mathbf{k} \Rightarrow \mathbf{F} \cdot \mathbf{n}\, d\sigma = \mathbf{F} \cdot \frac{\mathbf{r}_\theta \times \mathbf{r}_r}{|\mathbf{r}_\theta \times \mathbf{r}_r|} |\mathbf{r}_\theta \times \mathbf{r}_r|\, d\theta\, dr = (-r^2 \cos^2\theta - r^2 \sin^2\theta - r^3)\, d\theta\, dr$$

$$= (-r^2 - r^3)\, d\theta\, dr \text{ since } \mathbf{F} = (-r\cos\theta)\mathbf{i} - (r\sin\theta)\mathbf{j} + r^2\mathbf{k} \Rightarrow \iint_S \mathbf{F} \cdot \mathbf{n}\, d\sigma = \int_0^{2\pi} \int_1^2 (-r^2 - r^3)\, dr\, d\theta = -\frac{73\pi}{6}$$

29. $g(x, y, z) = z, \mathbf{p} = \mathbf{k} \Rightarrow \nabla g = \mathbf{k} \Rightarrow |\nabla g| = 1$ and $|\nabla g \cdot \mathbf{p}| = 1 \Rightarrow \text{Flux} = \iint_S \mathbf{F} \cdot \mathbf{n}\, d\sigma = \iint_R (\mathbf{F} \cdot \mathbf{k})\, dA$

$$= \int_0^2 \int_0^3 3\, dy\, dx = 18$$

31. $\nabla g = 2x\mathbf{i} + 2y\mathbf{j} + 2z\mathbf{k} \Rightarrow |\nabla g| = \sqrt{4x^2+4y^2+4z^2} = 2a$; $\mathbf{n} = \frac{2x\mathbf{i}+2y\mathbf{j}+2z\mathbf{k}}{2\sqrt{x^2+y^2+z^2}} = \frac{x\mathbf{i}+y\mathbf{j}+z\mathbf{k}}{a} \Rightarrow \mathbf{F}\cdot\mathbf{n} = \frac{z^2}{a}$;
$|\nabla g\cdot\mathbf{k}| = 2z \Rightarrow d\sigma = \frac{2a}{2z}\,dA \Rightarrow \text{Flux} = \iint_R \left(\frac{z^2}{a}\right)\left(\frac{a}{z}\right) dA = \iint_R z\,dA = \iint_R \sqrt{a^2-(x^2+y^2)}\,dx\,dy$
$= \int_0^{\pi/2}\int_0^a \sqrt{a^2-r^2}\,r\,dr\,d\theta = \frac{\pi a^3}{6}$

33. From Exercise 31, $\mathbf{n} = \frac{x\mathbf{i}+y\mathbf{j}+z\mathbf{k}}{a}$ and $d\sigma = \frac{a}{z}\,dA \Rightarrow \mathbf{F}\cdot\mathbf{n} = \frac{xy}{a} - \frac{xy}{a} + \frac{z}{a} = \frac{z}{a} \Rightarrow \text{Flux} = \iint_R \left(\frac{z}{a}\right)\left(\frac{a}{z}\right) dA = \iint_R 1\,dA$
$= \frac{\pi a^2}{4}$

35. From Exercise 31, $\mathbf{n} = \frac{x\mathbf{i}+y\mathbf{j}+z\mathbf{k}}{a}$ and $d\sigma = \frac{a}{z}\,dA \Rightarrow \mathbf{F}\cdot\mathbf{n} = \frac{x^2}{a} + \frac{y^2}{a} + \frac{z^2}{a} = a \Rightarrow \text{Flux} = \iint_R a\left(\frac{a}{z}\right) dA = \iint_R \frac{a^2}{z}\,dA$
$= \iint_R \frac{a^2}{\sqrt{a^2-(x^2+y^2)}}\,dA = \int_0^{\pi/2}\int_0^a \frac{a^2}{\sqrt{a^2-r^2}}\,r\,dr\,d\theta = \int_0^{\pi/2} a^2\left[-\sqrt{a^2-r^2}\right]_0^a d\theta = \frac{\pi a^3}{2}$

37. $g(x,y,z) = y^2 + z = 4 \Rightarrow \nabla g = 2y\mathbf{j} + \mathbf{k} \Rightarrow |\nabla g| = \sqrt{4y^2+1} \Rightarrow \mathbf{n} = \frac{2y\mathbf{j}+\mathbf{k}}{\sqrt{4y^2+1}}$
$\Rightarrow \mathbf{F}\cdot\mathbf{n} = \frac{2xy-3z}{\sqrt{4y^2+1}}$; $\mathbf{p} = \mathbf{k} \Rightarrow |\nabla g\cdot\mathbf{p}| = 1 \Rightarrow d\sigma = \sqrt{4y^2+1}\,dA \Rightarrow \text{Flux} = \iint_R \left(\frac{2xy-3z}{\sqrt{4y^2+1}}\right)\sqrt{4y^2+1}\,dA$
$= \iint_R (2xy-3z)\,dA$; $z = 0$ and $z = 4-y^2 \Rightarrow y^2 = 4 \Rightarrow \text{Flux} = \iint_R [2xy - 3(4-y^2)]\,dA$
$= \int_0^1\int_{-2}^2 (2xy - 12 + 3y^2)\,dy\,dx = \int_0^1 \left[xy^2 - 12y + y^3\right]_{-2}^2 dx = \int_0^1 -32\,dx = -32$

39. $g(x,y,z) = y - e^x = 0 \Rightarrow \nabla g = -e^x\mathbf{i} + \mathbf{j} \Rightarrow |\nabla g| = \sqrt{e^{2x}+1} \Rightarrow \mathbf{n} = \frac{e^x\mathbf{i}-\mathbf{j}}{\sqrt{e^{2x}+1}} \Rightarrow \mathbf{F}\cdot\mathbf{n} = \frac{-2e^x-2y}{\sqrt{e^{2x}+1}}$; $\mathbf{p} = \mathbf{i}$
$\Rightarrow |\nabla g\cdot\mathbf{p}| = e^x \Rightarrow d\sigma = \frac{\sqrt{e^{2x}+1}}{e^x}\,dA \Rightarrow \text{Flux} = \iint_R \left(\frac{-2e^x-2y}{\sqrt{e^{2x}+1}}\right)\left(\frac{\sqrt{e^{2x}+1}}{e^x}\right) dA = \iint_R \frac{-2e^x-2e^x}{e^x}\,dA$
$= \iint_R -4\,dA = \int_0^1\int_1^2 -4\,dy\,dz = -4$

41. On the face $z = a$: $g(x,y,z) = z \Rightarrow \nabla g = \mathbf{k} \Rightarrow |\nabla g| = 1$; $\mathbf{n} = \mathbf{k} \Rightarrow \mathbf{F}\cdot\mathbf{n} = 2xz = 2ax$ since $z = a$;
$d\sigma = dx\,dy \Rightarrow \text{Flux} = \iint_R 2ax\,dx\,dy = \int_0^a\int_0^a 2ax\,dx\,dy = a^4$.
On the face $z = 0$: $g(x,y,z) = z \Rightarrow \nabla g = \mathbf{k} \Rightarrow |\nabla g| = 1$; $\mathbf{n} = -\mathbf{k} \Rightarrow \mathbf{F}\cdot\mathbf{n} = -2xz = 0$ since $z = 0$;
$d\sigma = dx\,dy \Rightarrow \text{Flux} = \iint_R 0\,dx\,dy = 0$.
On the face $x = a$: $g(x,y,z) = x \Rightarrow \nabla g = \mathbf{i} \Rightarrow |\nabla g| = 1$; $\mathbf{n} = \mathbf{i} \Rightarrow \mathbf{F}\cdot\mathbf{n} = 2xy = 2ay$ since $x = a$;
$d\sigma = dy\,dz \Rightarrow \text{Flux} = \int_0^a\int_0^a 2ay\,dy\,dz = a^4$.
On the face $x = 0$: $g(x,y,z) = x \Rightarrow \nabla g = \mathbf{i} \Rightarrow |\nabla g| = 1$; $\mathbf{n} = -\mathbf{i} \Rightarrow \mathbf{F}\cdot\mathbf{n} = -2xy = 0$ since $x = 0$
$\Rightarrow \text{Flux} = 0$.
On the face $y = a$: $g(x,y,z) = y \Rightarrow \nabla g = \mathbf{j} \Rightarrow |\nabla g| = 1$; $\mathbf{n} = \mathbf{j} \Rightarrow \mathbf{F}\cdot\mathbf{n} = 2yz = 2az$ since $y = a$;
$d\sigma = dz\,dx \Rightarrow \text{Flux} = \int_0^a\int_0^a 2az\,dz\,dx = a^4$.
On the face $y = 0$: $g(x,y,z) = y \Rightarrow \nabla g = \mathbf{j} \Rightarrow |\nabla g| = 1$; $\mathbf{n} = -\mathbf{j} \Rightarrow \mathbf{F}\cdot\mathbf{n} = -2yz = 0$ since $y = 0$
$\Rightarrow \text{Flux} = 0$. Therefore, Total Flux $= 3a^4$.

43. $\nabla f = 2x\mathbf{i} + 2y\mathbf{j} + 2z\mathbf{k} \Rightarrow |\nabla f| = \sqrt{4x^2+4y^2+4z^2} = 2a$; $\mathbf{p} = \mathbf{k} \Rightarrow |\nabla f\cdot\mathbf{p}| = 2z$ since $z \ge 0 \Rightarrow d\sigma = \frac{2a}{2z}dA$
$= \frac{a}{z}\,dA$; $M = \iint_S \delta\,d\sigma = \frac{\delta}{8}$ (surface area of sphere) $= \frac{\delta\pi a^2}{2}$; $M_{xy} = \iint_S z\delta\,d\sigma = \delta\iint_R z\left(\frac{a}{z}\right) dA = a\delta\iint_R dA$
$= a\delta\int_0^{\pi/2}\int_0^a r\,dr\,d\theta = \frac{\delta\pi a^3}{4} \Rightarrow \bar{z} = \frac{M_{xy}}{M} = \left(\frac{\delta\pi a^3}{4}\right)\left(\frac{2}{\delta\pi a^2}\right) = \frac{a}{2}$. Because of symmetry, $\bar{x} = \bar{y} = \frac{a}{2} \Rightarrow$ the centroid is $\left(\frac{a}{2}, \frac{a}{2}, \frac{a}{2}\right)$.

45. Because of symmetry, $\bar{x} = \bar{y} = 0$; $M = \iint_S \delta\, d\sigma = \delta \iint_S d\sigma = (\text{Area of S})\delta = 3\pi\sqrt{2}\,\delta$; $\nabla f = 2x\,\mathbf{i} + 2y\mathbf{j} - 2z\mathbf{k}$

$\Rightarrow |\nabla f| = \sqrt{4x^2+4y^2+4z^2} = 2\sqrt{x^2+y^2+z^2}$; $\mathbf{p} = \mathbf{k} \Rightarrow |\nabla f \cdot \mathbf{p}| = 2z \Rightarrow d\sigma = \frac{2\sqrt{x^2+y^2+z^2}}{2z}\, dA$

$= \frac{\sqrt{x^2+y^2+(x^2+y^2)}}{z}\, dA = \frac{\sqrt{2}\sqrt{x^2+y^2}}{z}\, dA \Rightarrow M_{xy} = \delta \iint_R z\left(\frac{\sqrt{2}\sqrt{x^2+y^2}}{z}\right) dA = \delta \iint_R \sqrt{2}\sqrt{x^2+y^2}\, dA$

$= \delta \int_0^{2\pi}\int_1^2 \sqrt{2}\, r^2\, dr\, d\theta = \frac{14\pi\sqrt{2}}{3}\delta \Rightarrow \bar{z} = \frac{\left(\frac{14\pi\sqrt{2}}{3}\delta\right)}{3\pi\sqrt{2}\,\delta} = \frac{14}{9} \Rightarrow (\bar{x},\bar{y},\bar{z}) = \left(0,0,\frac{14}{9}\right)$. Next, $I_z = \iint_S (x^2+y^2)\,\delta\, d\sigma$

$= \iint_R (x^2+y^2)\left(\frac{\sqrt{2}\sqrt{x^2+y^2}}{z}\right)\delta\, dA = \delta\sqrt{2}\iint_R (x^2+y^2)\, dA = \delta\sqrt{2}\int_0^{2\pi}\int_1^2 r^3\, dr\, d\theta = \frac{15\pi\sqrt{2}}{2}\delta \Rightarrow R_z = \sqrt{\frac{I_z}{M}} = \frac{\sqrt{10}}{2}$

47. (a) Let the diameter lie on the z-axis and let $f(x,y,z) = x^2+y^2+z^2 = a^2$, $z \ge 0$ be the upper hemisphere

$\Rightarrow \nabla f = 2x\mathbf{i} + 2y\mathbf{j} + 2z\mathbf{k} \Rightarrow |\nabla f| = \sqrt{4x^2+4y^2+4z^2} = 2a$, $a > 0$; $\mathbf{p} = \mathbf{k} \Rightarrow |\nabla f \cdot \mathbf{p}| = 2z$ since $z \ge 0$

$\Rightarrow d\sigma = \frac{a}{z}\, dA \Rightarrow I_z = \iint_S \delta(x^2+y^2)\left(\frac{a}{z}\right) d\sigma = a\delta \iint_R \frac{x^2+y^2}{\sqrt{a^2-(x^2+y^2)}}\, dA = a\delta \int_0^{2\pi}\int_0^a \frac{r^2}{\sqrt{a^2-r^2}}\, r\, dr\, d\theta$

$= a\delta \int_0^{2\pi}\left[-r^2\sqrt{a^2-r^2} - \frac{2}{3}(a^2-r^2)^{3/2}\right]_0^a d\theta = a\delta\int_0^{2\pi} \frac{2}{3}a^3\, d\theta = \frac{4\pi}{3}a^4\delta \Rightarrow$ the moment of inertia is $\frac{8\pi}{3}a^4\delta$ for the whole sphere

(b) $I_L = I_{c.m.} + mh^2$, where m is the mass of the body and h is the distance between the parallel lines; now, $I_{c.m.} = \frac{8\pi}{3}a^4\delta$ (from part a) and $\frac{m}{2} = \iint_S \delta\, d\sigma = \delta\iint_R \left(\frac{a}{z}\right) dA = a\delta \iint_R \frac{1}{\sqrt{a^2-(x^2+y^2)}}\, dy\, dx$

$= a\delta \int_0^{2\pi}\int_0^a \frac{1}{\sqrt{a^2-r^2}}\, r\, dr\, d\theta = a\delta\int_0^{2\pi}\left[-\sqrt{a^2-r^2}\right]_0^a d\theta = a\delta\int_0^{2\pi} a\, d\theta = 2\pi a^2\delta$ and $h = a$

$\Rightarrow I_L = \frac{8\pi}{3}a^4\delta + 4\pi a^2\delta a^2 = \frac{20\pi}{3}a^4\delta$

16.7 STOKES' THEOREM

1. $\text{curl}\,\mathbf{F} = \nabla \times \mathbf{F} = \begin{vmatrix} \mathbf{i} & \mathbf{j} & \mathbf{k} \\ \frac{\partial}{\partial x} & \frac{\partial}{\partial y} & \frac{\partial}{\partial z} \\ x^2 & 2x & z^2 \end{vmatrix} = 0\mathbf{i} + 0\mathbf{j} + (2-0)\mathbf{k} = 2\mathbf{k}$ and $\mathbf{n} = \mathbf{k} \Rightarrow \text{curl}\,\mathbf{F}\cdot\mathbf{n} = 2 \Rightarrow d\sigma = dx\, dy$

$\Rightarrow \oint_C \mathbf{F}\cdot d\mathbf{r} = \iint_R 2\, dA = 2(\text{Area of the ellipse}) = 4\pi$

3. $\text{curl}\,\mathbf{F} = \nabla \times \mathbf{F} = \begin{vmatrix} \mathbf{i} & \mathbf{j} & \mathbf{k} \\ \frac{\partial}{\partial x} & \frac{\partial}{\partial y} & \frac{\partial}{\partial z} \\ y & xz & x^2 \end{vmatrix} = -x\mathbf{i} - 2x\mathbf{j} + (z-1)\mathbf{k}$ and $\mathbf{n} = \frac{\mathbf{i}+\mathbf{j}+\mathbf{k}}{\sqrt{3}} \Rightarrow \text{curl}\,\mathbf{F}\cdot\mathbf{n}$

$= \frac{1}{\sqrt{3}}(-x-2x+z-1) \Rightarrow d\sigma = \frac{\sqrt{3}}{1}\, dA \Rightarrow \oint_C \mathbf{F}\cdot d\mathbf{r} = \iint_R \frac{1}{\sqrt{3}}(-3x+z-1)\sqrt{3}\, dA$

$= \int_0^1\int_0^{1-x}[-3x+(1-x-y)-1]\, dy\, dx = \int_0^1\int_0^{1-x}(-4x-y)\, dy\, dx = \int_0^1 -\left[4x(1-x)+\frac{1}{2}(1-x)^2\right] dx$

$= -\int_0^1\left(\frac{1}{2}+3x-\frac{7}{2}x^2\right) dx = -\frac{5}{6}$

5. $\text{curl}\,\mathbf{F} = \nabla \times \mathbf{F} = \begin{vmatrix} \mathbf{i} & \mathbf{j} & \mathbf{k} \\ \frac{\partial}{\partial x} & \frac{\partial}{\partial y} & \frac{\partial}{\partial z} \\ y^2+z^2 & x^2+y^2 & x^2+y^2 \end{vmatrix} = 2y\mathbf{i} + (2z-2x)\mathbf{j} + (2x-2y)\mathbf{k}$ and $\mathbf{n} = \mathbf{k}$

$\Rightarrow \text{curl}\,\mathbf{F}\cdot\mathbf{n} = 2x-2y \Rightarrow d\sigma = dx\, dy \Rightarrow \oint_C \mathbf{F}\cdot d\mathbf{r} = \int_{-1}^1\int_{-1}^1 (2x-2y)\, dx\, dy = \int_{-1}^1 \left[x^2-2xy\right]_{-1}^1 dy$

$= \int_{-1}^1 -4y\, dy = 0$

7. $x = 3\cos t$ and $y = 2\sin t \Rightarrow \mathbf{F} = (2\sin t)\mathbf{i} + (9\cos^2 t)\,\mathbf{j} + (9\cos^2 t + 16\sin^4 t)\sin e^{\sqrt{(6\sin t\cos t)(0)}}\mathbf{k}$ at the base of the shell; $\mathbf{r} = (3\cos t)\mathbf{i} + (2\sin t)\mathbf{j} \Rightarrow \mathbf{dr} = (-3\sin t)\mathbf{i} + (2\cos t)\mathbf{j} \Rightarrow \mathbf{F}\cdot\frac{d\mathbf{r}}{dt} = -6\sin^2 t + 18\cos^3 t$
$\Rightarrow \iint_S \nabla\times\mathbf{F}\cdot\mathbf{n}\,d\sigma = \int_0^{2\pi}(-6\sin^2 t + 18\cos^3 t)\,dt = \left[-3t + \frac{3}{2}\sin 2t + 6(\sin t)(\cos^2 t + 2)\right]_0^{2\pi} = -6\pi$

9. Flux of $\nabla\times\mathbf{F} = \iint_S \nabla\times\mathbf{F}\cdot\mathbf{n}\,d\sigma = \oint_C \mathbf{F}\cdot d\mathbf{r}$, so let C be parametrized by $\mathbf{r} = (a\cos t)\mathbf{i} + (a\sin t)\mathbf{j}$, $0 \le t \le 2\pi \Rightarrow \frac{d\mathbf{r}}{dt} = (-a\sin t)\mathbf{i} + (a\cos t)\mathbf{j} \Rightarrow \mathbf{F}\cdot\frac{d\mathbf{r}}{dt} = ay\sin t + ax\cos t = a^2\sin^2 t + a^2\cos^2 t = a^2$
$\Rightarrow$ Flux of $\nabla\times\mathbf{F} = \oint_C \mathbf{F}\cdot d\mathbf{r} = \int_0^{2\pi} a^2\,dt = 2\pi a^2$

11. Let S_1 and S_2 be oriented surfaces that span C and that induce the same positive direction on C. Then $\iint_{S_1} \nabla\times\mathbf{F}\cdot\mathbf{n}_1\,d\sigma_1 = \oint_C \mathbf{F}\cdot d\mathbf{r} = \iint_{S_2} \nabla\times\mathbf{F}\cdot\mathbf{n}_2\,d\sigma_2$

13. $\nabla\times\mathbf{F} = \begin{vmatrix} \mathbf{i} & \mathbf{j} & \mathbf{k} \\ \frac{\partial}{\partial x} & \frac{\partial}{\partial y} & \frac{\partial}{\partial z} \\ 2z & 3x & 5y \end{vmatrix} = 5\mathbf{i} + 2\mathbf{j} + 3\mathbf{k}$; $\mathbf{r}_r = (\cos\theta)\mathbf{i} + (\sin\theta)\mathbf{j} - 2r\mathbf{k}$ and $\mathbf{r}_\theta = (-r\sin\theta)\mathbf{i} + (r\cos\theta)\mathbf{j}$
$\Rightarrow \mathbf{r}_r\times\mathbf{r}_\theta = \begin{vmatrix} \mathbf{i} & \mathbf{j} & \mathbf{k} \\ \cos\theta & \sin\theta & -2r \\ -r\sin\theta & r\cos\theta & 0 \end{vmatrix} = (2r^2\cos\theta)\,\mathbf{i} + (2r^2\sin\theta)\,\mathbf{j} + r\mathbf{k}$; $\mathbf{n} = \frac{\mathbf{r}_r\times\mathbf{r}_\theta}{|\mathbf{r}_r\times\mathbf{r}_\theta|}$ and $d\sigma = |\mathbf{r}_r\times\mathbf{r}_\theta|\,dr\,d\theta$
$\Rightarrow \nabla\times\mathbf{F}\cdot\mathbf{n}\,d\sigma = (\nabla\times\mathbf{F})\cdot(\mathbf{r}_r\times\mathbf{r}_\theta)\,dr\,d\theta = (10r^2\cos\theta + 4r^2\sin\theta + 3r)\,dr\,d\theta \Rightarrow \iint_S \nabla\times\mathbf{F}\cdot\mathbf{n}\,d\sigma$
$= \int_0^{2\pi}\int_0^2 (10r^2\cos\theta + 4r^2\sin\theta + 3r)\,dr\,d\theta = \int_0^{2\pi}\left[\frac{10}{3}r^3\cos\theta + \frac{4}{3}r^3\sin\theta + \frac{3}{2}r^2\right]_0^2 d\theta$
$= \int_0^{2\pi}\left(\frac{80}{3}\cos\theta + \frac{32}{3}\sin\theta + 6\right)d\theta = 6(2\pi) = 12\pi$

15. $\nabla\times\mathbf{F} = \begin{vmatrix} \mathbf{i} & \mathbf{j} & \mathbf{k} \\ \frac{\partial}{\partial x} & \frac{\partial}{\partial y} & \frac{\partial}{\partial z} \\ x^2y & 2y^3z & 3z \end{vmatrix} = -2y^3\mathbf{i} + 0\mathbf{j} - x^2\mathbf{k}$; $\mathbf{r}_r\times\mathbf{r}_\theta = \begin{vmatrix} \mathbf{i} & \mathbf{j} & \mathbf{k} \\ \cos\theta & \sin\theta & 1 \\ -r\sin\theta & r\cos\theta & 0 \end{vmatrix}$
$= (-r\cos\theta)\mathbf{i} - (r\sin\theta)\mathbf{j} + r\mathbf{k}$ and $\nabla\times\mathbf{F}\cdot\mathbf{n}\,d\sigma = (\nabla\times\mathbf{F})\cdot(\mathbf{r}_r\times\mathbf{r}_\theta)\,dr\,d\theta$ (see Exercise 13 above)
$\Rightarrow \iint_S \nabla\times\mathbf{F}\cdot\mathbf{n}\,d\sigma = \iint_R (2ry^3\cos\theta - rx^2)\,dr\,d\theta = \int_0^{2\pi}\int_0^1 (2r^4\sin^3\theta\cos\theta - r^3\cos^2\theta)\,dr\,d\theta$
$= \int_0^{2\pi}\left(\frac{2}{5}\sin^3\theta\cos\theta - \frac{1}{4}\cos^2\theta\right)d\theta = \left[\frac{1}{10}\sin^4\theta - \frac{1}{4}\left(\frac{\theta}{2} + \frac{\sin 2\theta}{4}\right)\right]_0^{2\pi} = -\frac{\pi}{4}$

17. $\nabla\times\mathbf{F} = \begin{vmatrix} \mathbf{i} & \mathbf{j} & \mathbf{k} \\ \frac{\partial}{\partial x} & \frac{\partial}{\partial y} & \frac{\partial}{\partial z} \\ 3y & 5-2x & z^2-2 \end{vmatrix} = 0\mathbf{i} + 0\mathbf{j} - 5\mathbf{k}$; $\mathbf{r}_\phi\times\mathbf{r}_\theta = \begin{vmatrix} \mathbf{i} & \mathbf{j} & \mathbf{k} \\ \sqrt{3}\cos\phi\cos\theta & \sqrt{3}\cos\phi\sin\theta & -\sqrt{3}\sin\phi \\ -\sqrt{3}\sin\phi\sin\theta & \sqrt{3}\sin\phi\cos\theta & 0 \end{vmatrix}$
$= (3\sin^2\phi\cos\theta)\,\mathbf{i} + (3\sin^2\phi\sin\theta)\,\mathbf{j} + (3\sin\phi\cos\phi)\mathbf{k}$; $\nabla\times\mathbf{F}\cdot\mathbf{n}\,d\sigma = (\nabla\times\mathbf{F})\cdot(\mathbf{r}_\phi\times\mathbf{r}_\theta)\,d\phi\,d\theta$ (see Exercise 13 above) $\Rightarrow \iint_S \nabla\times\mathbf{F}\cdot\mathbf{n}\,d\sigma = \int_0^{2\pi}\int_0^{\pi/2} -15\cos\phi\sin\phi\,d\phi\,d\theta = \int_0^{2\pi}\left[\frac{15}{2}\cos^2\phi\right]_0^{\pi/2} d\theta = \int_0^{2\pi} -\frac{15}{2}\,d\theta = -15\pi$

19. (a) $\mathbf{F} = 2x\mathbf{i} + 2y\mathbf{j} + 2z\mathbf{k} \Rightarrow \text{curl }\mathbf{F} = \mathbf{0} \Rightarrow \oint_C \mathbf{F}\cdot d\mathbf{r} = \iint_S \nabla\times\mathbf{F}\cdot\mathbf{n}\,d\sigma = \iint_S 0\,d\sigma = 0$

(b) Let $f(x,y,z) = x^2y^2z^3 \Rightarrow \nabla\times\mathbf{F} = \nabla\times\nabla f = \mathbf{0} \Rightarrow \text{curl }\mathbf{F} = \mathbf{0} \Rightarrow \oint_C \mathbf{F}\cdot d\mathbf{r} = \iint_S \nabla\times\mathbf{F}\cdot\mathbf{n}\,d\sigma = \iint_S 0\,d\sigma = 0$

(c) $\mathbf{F} = \nabla\times(x\mathbf{i} + y\mathbf{j} + z\mathbf{k}) = \mathbf{0} \Rightarrow \nabla\times\mathbf{F} = \mathbf{0} \Rightarrow \oint_C \mathbf{F}\cdot d\mathbf{r} = \iint_S \nabla\times\mathbf{F}\cdot\mathbf{n}\,d\sigma = \iint_S 0\,d\sigma = 0$

(d) $\mathbf{F} = \nabla f \Rightarrow \nabla \times \mathbf{F} = \nabla \times \nabla f = \mathbf{0} \Rightarrow \oint_C \mathbf{F} \cdot d\mathbf{r} = \iint_S \nabla \times \mathbf{F} \cdot \mathbf{n}\, d\sigma = \iint_S 0\, d\sigma = 0$

21. Let $\mathbf{F} = 2y\mathbf{i} + 3z\mathbf{j} - x\mathbf{k} \Rightarrow \nabla \times \mathbf{F} = \begin{vmatrix} \mathbf{i} & \mathbf{j} & \mathbf{k} \\ \frac{\partial}{\partial x} & \frac{\partial}{\partial y} & \frac{\partial}{\partial z} \\ 2y & 3z & -x \end{vmatrix} = -3\mathbf{i} + \mathbf{j} - 2\mathbf{k}\,;\ \mathbf{n} = \frac{2\mathbf{i}+2\mathbf{j}+\mathbf{k}}{3}$

$\Rightarrow \nabla \times \mathbf{F} \cdot \mathbf{n} = -2 \Rightarrow \oint_C 2y\,dx + 3z\,dy - x\,dz = \oint_C \mathbf{F} \cdot d\mathbf{r} = \iint_S \nabla \times \mathbf{F} \cdot \mathbf{n}\, d\sigma = \iint_S -2\, d\sigma$

$= -2 \iint_S d\sigma$, where $\iint_S d\sigma$ is the area of the region enclosed by C on the plane S: $2x + 2y + z = 2$

23. Suppose $\mathbf{F} = M\mathbf{i} + N\mathbf{j} + P\mathbf{k}$ exists such that $\nabla \times \mathbf{F} = \left(\frac{\partial P}{\partial y} - \frac{\partial N}{\partial z}\right)\mathbf{i} + \left(\frac{\partial M}{\partial z} - \frac{\partial P}{\partial x}\right)\mathbf{j} + \left(\frac{\partial N}{\partial x} - \frac{\partial M}{\partial y}\right)\mathbf{k}$

$= x\mathbf{i} + y\mathbf{j} + z\mathbf{k}$. Then $\frac{\partial}{\partial x}\left(\frac{\partial P}{\partial y} - \frac{\partial N}{\partial z}\right) = \frac{\partial}{\partial x}(x) \Rightarrow \frac{\partial^2 P}{\partial x \partial y} - \frac{\partial^2 N}{\partial x \partial z} = 1$. Likewise, $\frac{\partial}{\partial y}\left(\frac{\partial M}{\partial z} - \frac{\partial P}{\partial x}\right) = \frac{\partial}{\partial y}(y)$

$\Rightarrow \frac{\partial^2 M}{\partial y \partial z} - \frac{\partial^2 P}{\partial y \partial x} = 1$ and $\frac{\partial}{\partial z}\left(\frac{\partial N}{\partial x} - \frac{\partial M}{\partial y}\right) = \frac{\partial}{\partial z}(z) \Rightarrow \frac{\partial^2 N}{\partial z \partial x} - \frac{\partial^2 M}{\partial z \partial y} = 1$. Summing the calculated equations

$\Rightarrow \left(\frac{\partial^2 P}{\partial x \partial y} - \frac{\partial^2 P}{\partial y \partial x}\right) + \left(\frac{\partial^2 N}{\partial z \partial x} - \frac{\partial^2 N}{\partial x \partial z}\right) + \left(\frac{\partial^2 M}{\partial y \partial z} - \frac{\partial^2 M}{\partial z \partial y}\right) = 3$ or $0 = 3$ (assuming the second mixed partials are equal). This result is a contradiction, so there is no field $\mathbf{F}$ such that curl $\mathbf{F} = x\mathbf{i} + y\mathbf{j} + z\mathbf{k}$.

25. $r = \sqrt{x^2 + y^2} \Rightarrow r^4 = \left(x^2 + y^2\right)^2 \Rightarrow \mathbf{F} = \nabla(r^4) = 4x(x^2 + y^2)\mathbf{i} + 4y(x^2 + y^2)\mathbf{j} = M\mathbf{i} + N\mathbf{j}$

$\Rightarrow \oint_C \nabla(r^4) \cdot \mathbf{n}\, ds = \oint_C \mathbf{F} \cdot \mathbf{n}\, ds = \oint_C M\,dy - N\,dx = \iint_R \left(\frac{\partial M}{\partial x} + \frac{\partial N}{\partial y}\right) dx\,dy$

$= \iint_R \left[4(x^2 + y^2) + 8x^2 + 4(x^2 + y^2) + 8y^2\right] dA = \iint_R 16(x^2 + y^2)\, dA = 16 \iint_R x^2\, dA + 16 \iint_R y^2\, dA$

$= 16I_y + 16I_x$.

16.8 THE DIVERGENCE THEOREM AND A UNIFIED THEORY

1. $\mathbf{F} = \frac{-y\mathbf{i} + x\mathbf{j}}{\sqrt{x^2 + y^2}} \Rightarrow \text{div}\,\mathbf{F} = \frac{xy - xy}{(x^2 + y^2)^{3/2}} = 0$

3. $\mathbf{F} = -\frac{GM(x\mathbf{i} + y\mathbf{j} + z\mathbf{k})}{(x^2 + y^2 + z^2)^{3/2}} \Rightarrow \text{div}\,\mathbf{F} = -GM\left[\frac{(x^2+y^2+z^2)^{3/2} - 3x^2(x^2+y^2+z^2)^{1/2}}{(x^2+y^2+z^2)^3}\right]$

$- GM\left[\frac{(x^2+y^2+z^2)^{3/2} - 3y^2(x^2+y^2+z^2)^{1/2}}{(x^2+y^2+z^2)^3}\right] - GM\left[\frac{(x^2+y^2+z^2)^{3/2} - 3z^2(x^2+y^2+z^2)^{1/2}}{(x^2+y^2+z^2)^3}\right]$

$= -GM\left[\frac{3(x^2+y^2+z^2)^2 - 3(x^2+y^2+z^2)(x^2+y^2+z^2)}{(x^2+y^2+z^2)^{7/2}}\right] = 0$

5. $\frac{\partial}{\partial x}(y - x) = -1,\ \frac{\partial}{\partial y}(z - y) = -1,\ \frac{\partial}{\partial z}(y - x) = 0 \Rightarrow \nabla \cdot \mathbf{F} = -2 \Rightarrow \text{Flux} = \int_{-1}^{1}\int_{-1}^{1}\int_{-1}^{1} -2\, dx\,dy\,dz = -2(2^3) = -16$

7. $\frac{\partial}{\partial x}(y) = 0,\ \frac{\partial}{\partial y}(xy) = x,\ \frac{\partial}{\partial z}(-z) = -1 \Rightarrow \nabla \cdot \mathbf{F} = x - 1;\ z = x^2 + y^2 \Rightarrow z = r^2$ in cylindrical coordinates

$\Rightarrow \text{Flux} = \iiint_D (x - 1)\, dz\,dy\,dx = \int_0^{2\pi}\int_0^2\int_0^{r^2} (r\cos\theta - 1)\, dz\, r\, dr\, d\theta = \int_0^{2\pi}\int_0^2 (r^3\cos\theta - r^2)\, r\, dr\, d\theta$

$= \int_0^{2\pi}\left[\frac{r^5}{5}\cos\theta - \frac{r^4}{4}\right]_0^2 d\theta = \int_0^{2\pi}\left(\frac{32}{5}\cos\theta - 4\right) d\theta = \left[\frac{32}{5}\sin\theta - 4\theta\right]_0^{2\pi} = -8\pi$

9. $\frac{\partial}{\partial x}(x^2) = 2x,\ \frac{\partial}{\partial y}(-2xy) = -2x,\ \frac{\partial}{\partial z}(3xz) = 3x \Rightarrow \text{Flux} = \iiint_D 3x\, dx\,dy\,dz$

$= \int_0^{\pi/2}\int_0^{\pi/2}\int_0^2 (3\rho\sin\phi\cos\theta)(\rho^2\sin\phi)\, d\rho\, d\phi\, d\theta = \int_0^{\pi/2}\int_0^{\pi/2} 12\sin^2\phi\cos\theta\, d\phi\, d\theta = \int_0^{\pi/2} 3\pi\cos\theta\, d\theta = 3\pi$

11. $\frac{\partial}{\partial x}(2xz) = 2z,\ \frac{\partial}{\partial y}(-xy) = -x,\ \frac{\partial}{\partial z}(-z^2) = -2z \Rightarrow \nabla \cdot \mathbf{F} = -x \Rightarrow \text{Flux} = \iiint_D -x\, dV$

$= \int_0^2 \int_0^{\sqrt{16-4x^2}} \int_0^{4-y} -x\, dz\, dy\, dx = \int_0^2 \int_0^{\sqrt{16-4x^2}} (xy - 4x)\, dy\, dx = \int_0^2 \left[\frac{1}{2}x(16-4x^2) - 4x\sqrt{16-4x^2}\right] dx$

$= \left[4x^2 - \frac{1}{2}x^4 + \frac{1}{3}(16-4x^2)^{3/2}\right]_0^2 = -\frac{40}{3}$

13. Let $\rho = \sqrt{x^2+y^2+z^2}$. Then $\frac{\partial \rho}{\partial x} = \frac{x}{\rho},\ \frac{\partial \rho}{\partial y} = \frac{y}{\rho},\ \frac{\partial \rho}{\partial z} = \frac{z}{\rho} \Rightarrow \frac{\partial}{\partial x}(\rho x) = \left(\frac{\partial \rho}{\partial x}\right)x + \rho = \frac{x^2}{\rho} + \rho,\ \frac{\partial}{\partial y}(\rho y) = \left(\frac{\partial \rho}{\partial y}\right)y + \rho$

$= \frac{y^2}{\rho} + \rho,\ \frac{\partial}{\partial z}(\rho z) = \left(\frac{\partial \rho}{\partial z}\right)z + \rho = \frac{z^2}{\rho} + \rho \Rightarrow \nabla \cdot \mathbf{F} = \frac{x^2+y^2+z^2}{\rho} + 3\rho = 4\rho$, since $\rho = \sqrt{x^2+y^2+z^2}$

$\Rightarrow \text{Flux} = \iiint_D 4\rho\, dV = \int_0^{2\pi}\int_0^{\pi}\int_1^{\sqrt{2}} (4\rho)(\rho^2 \sin\phi)\, d\rho\, d\phi\, d\theta = \int_0^{2\pi}\int_0^{\pi} 3\sin\phi\, d\phi\, d\theta = \int_0^{2\pi} 6\, d\theta = 12\pi$

15. $\frac{\partial}{\partial x}(5x^3 + 12xy^2) = 15x^2 + 12y^2,\ \frac{\partial}{\partial y}(y^3 + e^y \sin z) = 3y^2 + e^y \sin z,\ \frac{\partial}{\partial z}(5z^3 + e^y \cos z) = 15z^2 - e^y \sin z$

$\Rightarrow \nabla \cdot \mathbf{F} = 15x^2 + 15y^2 + 15z^2 = 15\rho^2 \Rightarrow \text{Flux} = \iiint_D 15\rho^2\, dV = \int_0^{2\pi}\int_0^{\pi}\int_1^{\sqrt{2}} (15\rho^2)(\rho^2 \sin\phi)\, d\rho\, d\phi\, d\theta$

$= \int_0^{2\pi}\int_0^{\pi}\left(12\sqrt{2} - 3\right)\sin\phi\, d\phi\, d\theta = \int_0^{2\pi}\left(24\sqrt{2} - 6\right) d\theta = \left(48\sqrt{2} - 12\right)\pi$

17. (a) $\mathbf{G} = M\mathbf{i} + N\mathbf{j} + P\mathbf{k} \Rightarrow \nabla \times \mathbf{G} = \text{curl}\,\mathbf{G} = \left(\frac{\partial P}{\partial y} - \frac{\partial N}{\partial z}\right)\mathbf{i} + \left(\frac{\partial M}{\partial z} - \frac{\partial P}{\partial x}\right)\mathbf{k} + \left(\frac{\partial N}{\partial x} - \frac{\partial M}{\partial y}\right)\mathbf{k} \Rightarrow \nabla \cdot \nabla \times \mathbf{G}$

$= \text{div}(\text{curl}\,\mathbf{G}) = \frac{\partial}{\partial x}\left(\frac{\partial P}{\partial y} - \frac{\partial N}{\partial z}\right) + \frac{\partial}{\partial y}\left(\frac{\partial M}{\partial z} - \frac{\partial P}{\partial x}\right) + \frac{\partial}{\partial z}\left(\frac{\partial N}{\partial x} - \frac{\partial M}{\partial y}\right)$

$= \frac{\partial^2 P}{\partial x \partial y} - \frac{\partial^2 N}{\partial x \partial z} + \frac{\partial^2 M}{\partial y \partial z} - \frac{\partial^2 P}{\partial y \partial x} + \frac{\partial^2 N}{\partial z \partial x} - \frac{\partial^2 M}{\partial z \partial y} = 0$ if all first and second partial derivatives are continuous

(b) By the Divergence Theorem, the outward flux of $\nabla \times \mathbf{G}$ across a closed surface is zero because

outward flux of $\nabla \times \mathbf{G} = \iint_S (\nabla \times \mathbf{G}) \cdot \mathbf{n}\, d\sigma$

$= \iiint_D \nabla \cdot \nabla \times \mathbf{G}\, dV$ [Divergence Theorem with $\mathbf{F} = \nabla \times \mathbf{G}$]

$= \iiint_D (0)\, dV = 0$ [by part (a)]

19. (a) $\text{div}(g\mathbf{F}) = \nabla \cdot g\mathbf{F} = \frac{\partial}{\partial x}(gM) + \frac{\partial}{\partial y}(gN) + \frac{\partial}{\partial z}(gP) = \left(g\frac{\partial M}{\partial x} + M\frac{\partial g}{\partial x}\right) + \left(g\frac{\partial N}{\partial y} + N\frac{\partial g}{\partial y}\right) + \left(g\frac{\partial P}{\partial z} + P\frac{\partial g}{\partial z}\right)$

$= \left(M\frac{\partial g}{\partial x} + N\frac{\partial g}{\partial y} + P\frac{\partial g}{\partial z}\right) + g\left(\frac{\partial M}{\partial x} + \frac{\partial N}{\partial y} + \frac{\partial P}{\partial z}\right) = g\nabla \cdot \mathbf{F} + \nabla g \cdot \mathbf{F}$

(b) $\nabla \times (g\mathbf{F}) = \left[\frac{\partial}{\partial y}(gP) - \frac{\partial}{\partial z}(gN)\right]\mathbf{i} + \left[\frac{\partial}{\partial z}(gM) - \frac{\partial}{\partial x}(gP)\right]\mathbf{j} + \left[\frac{\partial}{\partial x}(gN) - \frac{\partial}{\partial y}(gM)\right]\mathbf{k}$

$= \left(P\frac{\partial g}{\partial y} + g\frac{\partial P}{\partial y} - N\frac{\partial g}{\partial z} - g\frac{\partial N}{\partial z}\right)\mathbf{i} + \left(M\frac{\partial g}{\partial z} + g\frac{\partial M}{\partial z} - P\frac{\partial g}{\partial x} - g\frac{\partial P}{\partial x}\right)\mathbf{j} + \left(N\frac{\partial g}{\partial x} + g\frac{\partial N}{\partial x} - M\frac{\partial g}{\partial y} - g\frac{\partial M}{\partial y}\right)\mathbf{k}$

$= \left(P\frac{\partial g}{\partial y} - N\frac{\partial g}{\partial z}\right)\mathbf{i} + \left(g\frac{\partial P}{\partial y} - g\frac{\partial N}{\partial z}\right)\mathbf{i} + \left(M\frac{\partial g}{\partial z} - P\frac{\partial g}{\partial x}\right)\mathbf{j} + \left(g\frac{\partial M}{\partial z} - g\frac{\partial P}{\partial x}\right)\mathbf{j} + \left(N\frac{\partial g}{\partial x} - M\frac{\partial g}{\partial y}\right)\mathbf{k}$

$+ \left(g\frac{\partial N}{\partial x} - g\frac{\partial M}{\partial y}\right)\mathbf{k} = g\nabla \times \mathbf{F} + \nabla g \times \mathbf{F}$

21. The integral's value never exceeds the surface area of S. Since $|\mathbf{F}| \le 1$, we have $|\mathbf{F} \cdot \mathbf{n}| = |\mathbf{F}|\,|\mathbf{n}| \le (1)(1) = 1$ and

$\iiint_D \nabla \cdot \mathbf{F}\, d\sigma = \iint_S \mathbf{F} \cdot \mathbf{n}\, d\sigma$ [Divergence Theorem]

$\le \iint_S |\mathbf{F} \cdot \mathbf{n}|\, d\sigma$ [A property of integrals]

$\le \iint_S (1)\, d\sigma$ [$|\mathbf{F} \cdot \mathbf{n}| \le 1$]

= Area of S.

23. (a) $\frac{\partial}{\partial x}(x) = 1$, $\frac{\partial}{\partial y}(y) = 1$, $\frac{\partial}{\partial z}(z) = 1 \Rightarrow \nabla \cdot \mathbf{F} = 3 \Rightarrow \text{Flux} = \iiint_D 3\,dV = 3\iiint_D dV = 3(\text{Volume of the solid})$

(b) If **F** is orthogonal to **n** at every point of S, then $\mathbf{F}\cdot\mathbf{n} = 0$ everywhere $\Rightarrow \text{Flux} = \iint_S \mathbf{F}\cdot\mathbf{n}\,d\sigma = 0$. But the flux is 3 (Volume of the solid) $\neq 0$, so **F** is not orthogonal to **n** at every point.

25. $\iint_S \mathbf{F}\cdot\mathbf{n}\,d\sigma = \iiint_D \nabla\cdot\mathbf{F}\,dV = \iiint_D 3\,dV \Rightarrow \frac{1}{3}\iint_S \mathbf{F}\cdot\mathbf{n}\,d\sigma = \iiint_D dV = \text{Volume of D}$

27. (a) From the Divergence Theorem, $\iint_S \nabla f\cdot\mathbf{n}\,d\sigma = \iiint_D \nabla\cdot\nabla f\,dV = \iiint_D \nabla^2 f\,dV = \iiint_D 0\,dV = 0$

(b) From the Divergence Theorem, $\iint_S f\nabla f\cdot\mathbf{n}\,d\sigma = \iiint_D \nabla\cdot f\nabla f\,dV$. Now,

$f\nabla f = \left(f\frac{\partial f}{\partial x}\right)\mathbf{i} + \left(f\frac{\partial f}{\partial y}\right)\mathbf{j} + \left(f\frac{\partial f}{\partial z}\right)\mathbf{k} \Rightarrow \nabla\cdot f\nabla f = \left[f\frac{\partial^2 f}{\partial x^2} + \left(\frac{\partial f}{\partial x}\right)^2\right] + \left[f\frac{\partial^2 f}{\partial y^2} + \left(\frac{\partial f}{\partial y}\right)^2\right] + \left[f\frac{\partial^2 f}{\partial z^2} + \left(\frac{\partial f}{\partial z}\right)^2\right]$

$= f\nabla^2 f + |\nabla f|^2 = 0 + |\nabla f|^2$ since f is harmonic $\Rightarrow \iint_S f\nabla f\cdot\mathbf{n}\,d\sigma = \iiint_D |\nabla f|^2\,dV$, as claimed.

29. $\iint_S f\nabla g\cdot\mathbf{n}\,d\sigma = \iiint_D \nabla\cdot f\nabla g\,dV = \iiint_D \nabla\cdot\left(f\frac{\partial g}{\partial x}\mathbf{i} + f\frac{\partial g}{\partial y}\mathbf{j} + f\frac{\partial g}{\partial z}\mathbf{k}\right)dV$

$= \iiint_D \left(f\frac{\partial^2 g}{\partial x^2} + \frac{\partial f}{\partial x}\frac{\partial g}{\partial x} + f\frac{\partial^2 g}{\partial y^2} + \frac{\partial f}{\partial y}\frac{\partial g}{\partial y} + f\frac{\partial^2 g}{\partial z^2} + \frac{\partial f}{\partial z}\frac{\partial g}{\partial z}\right)dV$

$= \iiint_D \left[f\left(\frac{\partial^2 g}{\partial x^2} + \frac{\partial^2 g}{\partial y^2} + \frac{\partial^2 g}{\partial z^2}\right) + \left(\frac{\partial f}{\partial x}\frac{\partial g}{\partial x} + \frac{\partial f}{\partial y}\frac{\partial g}{\partial y} + \frac{\partial f}{\partial z}\frac{\partial g}{\partial z}\right)\right]dV = \iiint_D (f\nabla^2 g + \nabla f\cdot\nabla g)\,dV$

31. (a) The integral $\iiint_D p(t,x,y,z)\,dV$ represents the mass of the fluid at any time t. The equation says that the instantaneous rate of change of mass is flux of the fluid through the surface S enclosing the region D: the mass decreases if the flux is outward (so the fluid flows out of D), and increases if the flow is inward (interpreting **n** as the outward pointing unit normal to the surface).

(b) $\iiint_D \frac{\partial p}{\partial t}\,dV = \frac{d}{dt}\iiint_D p\,dV = -\iint_S p\mathbf{v}\cdot\mathbf{n}\,d\sigma = -\iiint_D \nabla\cdot p\mathbf{v}\,dV \Rightarrow \frac{\partial \rho}{\partial t} = -\nabla\cdot p\mathbf{v}$

Since the law is to hold for all regions D, $\nabla\cdot p\mathbf{v} + \frac{\partial p}{\partial t} = 0$, as claimed

CHAPTER 16 PRACTICE EXERCISES

1. Path 1: $\mathbf{r} = t\mathbf{i} + t\mathbf{j} + t\mathbf{k} \Rightarrow x = t, y = t, z = t, 0 \le t \le 1 \Rightarrow f(g(t),h(t),k(t)) = 3 - 3t^2$ and $\frac{dx}{dt} = 1$, $\frac{dy}{dt} = 1$, $\frac{dz}{dt} = 1 \Rightarrow \sqrt{\left(\frac{dx}{dt}\right)^2 + \left(\frac{dy}{dt}\right)^2 + \left(\frac{dz}{dt}\right)^2}\,dt = \sqrt{3}\,dt \Rightarrow \int_C f(x,y,z)\,ds = \int_0^1 \sqrt{3}(3 - 3t^2)\,dt = 2\sqrt{3}$

Path 2: $\mathbf{r}_1 = t\mathbf{i} + t\mathbf{j}, 0 \le t \le 1 \Rightarrow x = t, y = t, z = 0 \Rightarrow f(g(t),h(t),k(t)) = 2t - 3t^2 + 3$ and $\frac{dx}{dt} = 1$, $\frac{dy}{dt} = 1$, $\frac{dz}{dt} = 0 \Rightarrow \sqrt{\left(\frac{dx}{dt}\right)^2 + \left(\frac{dy}{dt}\right)^2 + \left(\frac{dz}{dt}\right)^2}\,dt = \sqrt{2}\,dt \Rightarrow \int_{C_1} f(x,y,z)\,ds = \int_0^1 \sqrt{2}(2t - 3t^2 + 3)\,dt = 3\sqrt{2}$;

$\mathbf{r}_2 = \mathbf{i} + \mathbf{j} + t\mathbf{k} \Rightarrow x = 1, y = 1, z = t \Rightarrow f(g(t),h(t),k(t)) = 2 - 2t$ and $\frac{dx}{dt} = 0$, $\frac{dy}{dt} = 0$, $\frac{dz}{dt} = 1$

$\Rightarrow \sqrt{\left(\frac{dx}{dt}\right)^2 + \left(\frac{dy}{dt}\right)^2 + \left(\frac{dz}{dt}\right)^2}\,dt = dt \Rightarrow \int_{C_2} f(x,y,z)\,ds = \int_0^1 (2 - 2t)\,dt = 1$

$\Rightarrow \int_C f(x,y,z)\,ds = \int_{C_1} f(x,y,z)\,ds + \int_{C_2} f(x,y,z) = 3\sqrt{2} + 1$

3. $\mathbf{r} = (a\cos t)\mathbf{j} + (a\sin t)\mathbf{k} \Rightarrow x = 0, y = a\cos t, z = a\sin t \Rightarrow f(g(t), h(t), k(t)) = \sqrt{a^2\sin^2 t} = a|\sin t|$ and $\frac{dx}{dt} = 0, \frac{dy}{dt} = -a\sin t, \frac{dz}{dt} = a\cos t \Rightarrow \sqrt{\left(\frac{dx}{dt}\right)^2 + \left(\frac{dy}{dt}\right)^2 + \left(\frac{dz}{dt}\right)^2}\, dt = a\, dt$

$\Rightarrow \int_C f(x, y, z)\, ds = \int_0^{2\pi} a^2|\sin t|\, dt = \int_0^{\pi} a^2\sin t\, dt + \int_{\pi}^{2\pi} -a^2\sin t\, dt = 4a^2$

5. $\frac{\partial P}{\partial y} = -\frac{1}{2}(x+y+z)^{-3/2} = \frac{\partial N}{\partial z}, \frac{\partial M}{\partial z} = -\frac{1}{2}(x+y+z)^{-3/2} = \frac{\partial P}{\partial x}, \frac{\partial N}{\partial x} = -\frac{1}{2}(x+y+z)^{-3/2} = \frac{\partial M}{\partial y}$

$\Rightarrow M\,dx + N\,dy + P\,dz$ is exact; $\frac{\partial f}{\partial x} = \frac{1}{\sqrt{x+y+z}} \Rightarrow f(x, y, z) = 2\sqrt{x+y+z} + g(y, z) \Rightarrow \frac{\partial f}{\partial y} = \frac{1}{\sqrt{x+y+z}} + \frac{\partial g}{\partial y}$

$= \frac{1}{\sqrt{x+y+z}} \Rightarrow \frac{\partial g}{\partial y} = 0 \Rightarrow g(y, z) = h(z) \Rightarrow f(x, y, z) = 2\sqrt{x+y+z} + h(z) \Rightarrow \frac{\partial f}{\partial z} = \frac{1}{\sqrt{x+y+z}} + h'(z)$

$= \frac{1}{\sqrt{x+y+z}} \Rightarrow h'(x) = 0 \Rightarrow h(z) = C \Rightarrow f(x, y, z) = 2\sqrt{x+y+z} + C \Rightarrow \int_{(-1,1,1)}^{(4,-3,0)} \frac{dx + dy + dz}{\sqrt{x+y+z}}$

$= f(4, -3, 0) - f(-1, 1, 1) = 2\sqrt{1} - 2\sqrt{1} = 0$

7. $\frac{\partial M}{\partial z} = -y\cos z \neq y\cos z = \frac{\partial P}{\partial x} \Rightarrow \mathbf{F}$ is not conservative; $\mathbf{r} = (2\cos t)\mathbf{i} + (2\sin t)\mathbf{j} - \mathbf{k}, 0 \le t \le 2\pi$

$\Rightarrow d\mathbf{r} = (-2\sin t)\mathbf{i} - (2\cos t)\mathbf{j} \Rightarrow \int_C \mathbf{F}\cdot d\mathbf{r} = \int_0^{2\pi} [-(-2\sin t)(\sin(-1))(-2\sin t) + (2\cos t)(\sin(-1))(-2\cos t)]\, dt$

$= 4\sin(1)\int_0^{2\pi} (\sin^2 t + \cos^2 t)dt = 8\pi\sin(1)$

9. Let $M = 8x\sin y$ and $N = -8y\cos x \Rightarrow \frac{\partial M}{\partial y} = 8x\cos y$ and $\frac{\partial N}{\partial x} = 8y\sin x \Rightarrow \int_C 8x\sin y\, dx - 8y\cos x\, dy$

$= \iint_R (8y\sin x - 8x\cos y)\, dy\, dx = \int_0^{\pi/2}\int_0^{\pi/2} (8y\sin x - 8x\cos y)\, dy\, dx = \int_0^{\pi/2} (\pi^2\sin x - 8x)\, dx = -\pi^2 + \pi^2 = 0$

11. Let $z = 1 - x - y \Rightarrow f_x(x, y) = -1$ and $f_y(x, y) = -1 \Rightarrow \sqrt{f_x^2 + f_y^2 + 1} = \sqrt{3} \Rightarrow$ Surface Area $= \iint_R \sqrt{3}\, dx\, dy$

$= \sqrt{3}$(Area of the circular region in the xy-plane) $= \pi\sqrt{3}$

13. $\nabla f = 2x\mathbf{i} + 2y\mathbf{j} + 2z\mathbf{k}, \mathbf{p} = \mathbf{k} \Rightarrow |\nabla f| = \sqrt{4x^2 + 4y^2 + 4z^2} = 2\sqrt{x^2 + y^2 + z^2} = 2$ and $|\nabla f \cdot \mathbf{p}| = |2z| = 2z$ since $z \ge 0 \Rightarrow$ Surface Area $= \iint_R \frac{2}{2z}\, dA = \iint_R \frac{1}{z}\, dA = \iint_R \frac{1}{\sqrt{1-x^2-y^2}}\, dx\, dy = \int_0^{2\pi}\int_0^{1/\sqrt{2}} \frac{1}{\sqrt{1-r^2}}\, r\, dr\, d\theta$

$= \int_0^{2\pi} \left[-\sqrt{1-r^2}\right]_0^{1/\sqrt{2}} d\theta = \int_0^{2\pi} \left(1 - \frac{1}{\sqrt{2}}\right) d\theta = 2\pi\left(1 - \frac{1}{\sqrt{2}}\right)$

15. $f(x, y, z) = \frac{x}{a} + \frac{y}{b} + \frac{z}{c} = 1 \Rightarrow \nabla f = \left(\frac{1}{a}\right)\mathbf{i} + \left(\frac{1}{b}\right)\mathbf{j} + \left(\frac{1}{c}\right)\mathbf{k} \Rightarrow |\nabla f| = \sqrt{\frac{1}{a^2} + \frac{1}{b^2} + \frac{1}{c^2}}$ and $\mathbf{p} = \mathbf{k} \Rightarrow |\nabla f \cdot \mathbf{p}| = \frac{1}{c}$ since $c > 0 \Rightarrow$ Surface Area $= \iint_R \frac{\sqrt{\frac{1}{a^2} + \frac{1}{b^2} + \frac{1}{c^2}}}{\left(\frac{1}{c}\right)}\, dA = c\sqrt{\frac{1}{a^2} + \frac{1}{b^2} + \frac{1}{c^2}} \iint_R dA = \frac{1}{2}abc\sqrt{\frac{1}{a^2} + \frac{1}{b^2} + \frac{1}{c^2}}$, since the area of the triangular region R is $\frac{1}{2}ab$. To check this result, let $\mathbf{v} = a\mathbf{i} + c\mathbf{k}$ and $\mathbf{w} = -a\mathbf{i} + b\mathbf{j}$; the area can be found by computing $\frac{1}{2}|\mathbf{v} \times \mathbf{w}|$.

17. $\nabla f = 2y\mathbf{j} + 2z\mathbf{k}, \mathbf{p} = \mathbf{k} \Rightarrow |\nabla f| = \sqrt{4y^2 + 4z^2} = 2\sqrt{y^2 + z^2} = 10$ and $|\nabla f \cdot \mathbf{p}| = 2z$ since $z \ge 0$

$\Rightarrow d\sigma = \frac{10}{2z}\, dx\, dy = \frac{5}{z}\, dx\, dy = \iint_S g(x, y, z)\, d\sigma = \iint_R (x^4y)(y^2 + z^2)\left(\frac{5}{z}\right) dx\, dy$

$= \iint_R (x^4y)(25)\left(\frac{5}{\sqrt{25-y^2}}\right) dx\, dy = \int_0^4\int_0^1 \frac{125y}{\sqrt{25-y^2}}\, x^4\, dx\, dy = \int_0^4 \frac{25y}{\sqrt{25-y^2}}\, dy = 50$

19. A possible parametrization is $\mathbf{r}(\phi, \theta) = (6\sin\phi\cos\theta)\mathbf{i} + (6\sin\phi\sin\theta)\mathbf{j} + (6\cos\phi)\mathbf{k}$ (spherical coordinates); now $\rho = 6$ and $z = -3 \Rightarrow -3 = 6\cos\phi \Rightarrow \cos\phi = -\frac{1}{2} \Rightarrow \phi = \frac{2\pi}{3}$ and $z = 3\sqrt{3} \Rightarrow 3\sqrt{3} = 6\cos\phi$

$\Rightarrow \cos\phi = \frac{\sqrt{3}}{2} \Rightarrow \phi = \frac{\pi}{6} \Rightarrow \frac{\pi}{6} \le \phi \le \frac{2\pi}{3}$; also $0 \le \theta \le 2\pi$

21. A possible parametrization is $\mathbf{r}(r, \theta) = (r \cos \theta)\mathbf{i} + (r \sin \theta)\mathbf{j} + (1 + r)\mathbf{k}$ (cylindrical coordinates); now $r = \sqrt{x^2 + y^2} \Rightarrow z = 1 + r$ and $1 \le z \le 3 \Rightarrow 1 \le 1 + r \le 3 \Rightarrow 0 \le r \le 2$; also $0 \le \theta \le 2\pi$

23. Let $x = u \cos v$ and $z = u \sin v$, where $u = \sqrt{x^2 + z^2}$ and v is the angle in the xz-plane with the x-axis $\Rightarrow \mathbf{r}(u, v) = (u \cos v)\mathbf{i} + 2u^2\mathbf{j} + (u \sin v)\mathbf{k}$ is a possible parametrization; $0 \le y \le 2 \Rightarrow 2u^2 \le 2 \Rightarrow u^2 \le 1$ $\Rightarrow 0 \le u \le 1$ since $u \ge 0$; also, for just the upper half of the paraboloid, $0 \le v \le \pi$

25. $\mathbf{r}_u = \mathbf{i} + \mathbf{j}, \mathbf{r}_v = \mathbf{i} - \mathbf{j} + \mathbf{k} \Rightarrow \mathbf{r}_u \times \mathbf{r}_v = \begin{vmatrix} \mathbf{i} & \mathbf{j} & \mathbf{k} \\ 1 & 1 & 0 \\ 1 & -1 & 1 \end{vmatrix} = \mathbf{i} - \mathbf{j} - 2\mathbf{k} \Rightarrow |\mathbf{r}_u \times \mathbf{r}_v| = \sqrt{6}$

$\Rightarrow$ Surface Area $= \iint\limits_{R_{uv}} |\mathbf{r}_u \times \mathbf{r}_v| \, du \, dv = \int_0^1 \int_0^1 \sqrt{6} \, du \, dv = \sqrt{6}$

27. $\mathbf{r}_r = (\cos \theta)\mathbf{i} + (\sin \theta)\mathbf{j}, \mathbf{r}_\theta = (-r \sin \theta)\mathbf{i} + (r \cos \theta)\mathbf{j} + \mathbf{k} \Rightarrow \mathbf{r}_r \times \mathbf{r}_\theta = \begin{vmatrix} \mathbf{i} & \mathbf{j} & \mathbf{k} \\ \cos \theta & \sin \theta & 0 \\ -r \sin \theta & r \cos \theta & 1 \end{vmatrix}$

$= (\sin \theta)\mathbf{i} - (\cos \theta)\mathbf{j} + r\mathbf{k} \Rightarrow |\mathbf{r}_r \times \mathbf{r}_\theta| = \sqrt{\sin^2 \theta + \cos^2 \theta + r^2} = \sqrt{1 + r^2} \Rightarrow$ Surface Area $= \iint\limits_{R_{r\theta}} |\mathbf{r}_r \times \mathbf{r}_\theta| \, dr \, d\theta$

$= \int_0^{2\pi} \int_0^1 \sqrt{1 + r^2} \, dr \, d\theta = \int_0^{2\pi} \left[\frac{r}{2} \sqrt{1 + r^2} + \frac{1}{2} \ln \left(r + \sqrt{1 + r^2} \right) \right]_0^1 d\theta = \int_0^{2\pi} \left[\frac{1}{2} \sqrt{2} + \frac{1}{2} \ln \left(1 + \sqrt{2} \right) \right] d\theta$
$= \pi \left[\sqrt{2} + \ln \left(1 + \sqrt{2} \right) \right]$

29. $\frac{\partial P}{\partial y} = 0 = \frac{\partial N}{\partial z}, \frac{\partial M}{\partial z} = 0 = \frac{\partial P}{\partial x}, \frac{\partial N}{\partial x} = 0 = \frac{\partial M}{\partial y} \Rightarrow$ Conservative

31. $\frac{\partial P}{\partial y} = 0 \ne ye^z = \frac{\partial N}{\partial z} \Rightarrow$ Not Conservative

33. $\frac{\partial f}{\partial x} = 2 \Rightarrow f(x, y, z) = 2x + g(y, z) \Rightarrow \frac{\partial f}{\partial y} = \frac{\partial g}{\partial y} = 2y + z \Rightarrow g(y, z) = y^2 + zy + h(z)$
$\Rightarrow f(x, y, z) = 2x + y^2 + zy + h(z) \Rightarrow \frac{\partial f}{\partial z} = y + h'(z) = y + 1 \Rightarrow h'(z) = 1 \Rightarrow h(z) = z + C$
$\Rightarrow f(x, y, z) = 2x + y^2 + zy + z + C$

35. Over Path 1: $\mathbf{r} = t\mathbf{i} + t\mathbf{j} + t\mathbf{k}, 0 \le t \le 1 \Rightarrow x = t, y = t, z = t$ and $d\mathbf{r} = (\mathbf{i} + \mathbf{j} + \mathbf{k})\,dt \Rightarrow \mathbf{F} = 2t^2\mathbf{i} + \mathbf{j} + t^2\mathbf{k}$
$\Rightarrow \mathbf{F} \cdot d\mathbf{r} = (3t^2 + 1)\,dt \Rightarrow \text{Work} = \int_0^1 (3t^2 + 1)\,dt = 2;$
Over Path 2: $\mathbf{r}_1 = t\mathbf{i} + t\mathbf{j}, 0 \le t \le 1 \Rightarrow x = t, y = t, z = 0$ and $d\mathbf{r}_1 = (\mathbf{i} + \mathbf{j})\,dt \Rightarrow \mathbf{F}_1 = 2t^2\mathbf{i} + \mathbf{j} + t^2\mathbf{k}$
$\Rightarrow \mathbf{F}_1 \cdot d\mathbf{r}_1 = (2t^2 + 1)\,dt \Rightarrow \text{Work}_1 = \int_0^1 (2t^2 + 1)\,dt = \frac{5}{3}; \mathbf{r}_2 = \mathbf{i} + \mathbf{j} + t\mathbf{k}, 0 \le t \le 1 \Rightarrow x = 1, y = 1, z = t$ and
$d\mathbf{r}_2 = \mathbf{k}\,dt \Rightarrow \mathbf{F}_2 = 2\mathbf{i} + \mathbf{j} + \mathbf{k} \Rightarrow \mathbf{F}_2 \cdot d\mathbf{r}_2 = dt \Rightarrow \text{Work}_2 = \int_0^1 dt = 1 \Rightarrow \text{Work} = \text{Work}_1 + \text{Work}_2 = \frac{5}{3} + 1 = \frac{8}{3}$

37. (a) $\mathbf{r} = (e^t \cos t)\,\mathbf{i} + (e^t \sin t)\,\mathbf{j} \Rightarrow x = e^t \cos t, y = e^t \sin t$ from $(1, 0)$ to $(e^{2\pi}, 0) \Rightarrow 0 \le t \le 2\pi$
$\Rightarrow \frac{d\mathbf{r}}{dt} = (e^t \cos t - e^t \sin t)\,\mathbf{i} + (e^t \sin t + e^t \cos t)\,\mathbf{j}$ and $\mathbf{F} = \frac{x\mathbf{i} + y\mathbf{j}}{(x^2 + y^2)^{3/2}} = \frac{(e^t \cos t)\,\mathbf{i} + (e^t \sin t)\,\mathbf{j}}{(e^{2t} \cos^2 t + e^{2t} \sin^2 t)^{3/2}}$
$= \left(\frac{\cos t}{e^{2t}}\right)\mathbf{i} + \left(\frac{\sin t}{e^{2t}}\right)\mathbf{j} \Rightarrow \mathbf{F} \cdot \frac{d\mathbf{r}}{dt} = \left(\frac{\cos^2 t}{e^t} - \frac{\sin t \cos t}{e^t} + \frac{\sin^2 t}{e^t} + \frac{\sin t \cos t}{e^t}\right) = e^{-t}$
$\Rightarrow \text{Work} = \int_0^{2\pi} e^{-t}\,dt = 1 - e^{-2\pi}$

(b) $\mathbf{F} = \frac{x\mathbf{i} + y\mathbf{j}}{(x^2 + y^2)^{3/2}} \Rightarrow \frac{\partial f}{\partial x} = \frac{x}{(x^2 + y^2)^{3/2}} \Rightarrow f(x, y, z) = -(x^2 + y^2)^{-1/2} + g(y, z) \Rightarrow \frac{\partial f}{\partial y} = \frac{y}{(x^2 + y^2)^{3/2}} + \frac{\partial g}{\partial y}$
$= \frac{y}{(x^2 + y^2)^{3/2}} \Rightarrow g(y, z) = C \Rightarrow f(x, y, z) = -(x^2 + y^2)^{-1/2}$ is a potential function for $\mathbf{F} \Rightarrow \int_C \mathbf{F} \cdot d\mathbf{r}$
$= f(e^{2\pi}, 0) - f(1, 0) = 1 - e^{-2\pi}$

39. $\nabla \times \mathbf{F} = \begin{vmatrix} \mathbf{i} & \mathbf{j} & \mathbf{k} \\ \frac{\partial}{\partial x} & \frac{\partial}{\partial y} & \frac{\partial}{\partial z} \\ y^2 & -y & 3z^2 \end{vmatrix} = -2y\mathbf{k}$; unit normal to the plane is $\mathbf{n} = \frac{2\mathbf{i}+6\mathbf{j}-3\mathbf{k}}{\sqrt{4+36+9}} = \frac{2}{7}\mathbf{i} + \frac{6}{7}\mathbf{j} - \frac{3}{7}\mathbf{k}$

$\Rightarrow \nabla \times \mathbf{F} \cdot \mathbf{n} = \frac{6}{7}y$; $\mathbf{p} = \mathbf{k}$ and $f(x, y, z) = 2x + 6y - 3z \Rightarrow |\nabla f \cdot \mathbf{p}| = 3 \Rightarrow d\sigma = \frac{|\nabla f|}{|\nabla f \cdot \mathbf{p}|}\, dA = \frac{7}{3}\, dA$

$\Rightarrow \oint_C \mathbf{F} \cdot d\mathbf{r} = \iint_R \frac{6}{7}y\, d\sigma = \iint_R \left(\frac{6}{7}y\right)\left(\frac{7}{3}\, dA\right) = \iint_R 2y\, dA = \int_0^{2\pi}\int_0^1 2r\sin\theta\, r\, dr\, d\theta = \int_0^{2\pi} \frac{2}{3}\sin\theta\, d\theta = 0$

41. (a) $\mathbf{r} = \sqrt{2}t\mathbf{i} + \sqrt{2}t\mathbf{j} + (4 - t^2)\mathbf{k},\ 0 \le t \le 1 \Rightarrow x = \sqrt{2}t,\ y = \sqrt{2}t,\ z = 4 - t^2 \Rightarrow \frac{dx}{dt} = \sqrt{2},\ \frac{dy}{dt} = \sqrt{2},\ \frac{dz}{dt} = -2t$

$\Rightarrow \sqrt{\left(\frac{dx}{dt}\right)^2 + \left(\frac{dy}{dt}\right)^2 + \left(\frac{dz}{dt}\right)^2}\, dt = \sqrt{4 + 4t^2}\, dt \Rightarrow M = \int_C \delta(x, y, z)\, ds = \int_0^1 3t\sqrt{4 + 4t^2}\, dt = \left[\frac{1}{4}(4 + 4t)^{3/2}\right]_0^1$

$= 4\sqrt{2} - 2$

(b) $M = \int_C \delta(x, y, z)\, ds = \int_0^1 \sqrt{4 + 4t^2}\, dt = \left[t\sqrt{1 + t^2} + \ln\left(t + \sqrt{1 + t^2}\right)\right]_0^1 = \sqrt{2} + \ln\left(1 + \sqrt{2}\right)$

43. $\mathbf{r} = t\mathbf{i} + \left(\frac{2\sqrt{2}}{3}t^{3/2}\right)\mathbf{j} + \left(\frac{t^2}{2}\right)\mathbf{k},\ 0 \le t \le 2 \Rightarrow x = t,\ y = \frac{2\sqrt{2}}{3}t^{3/2},\ z = \frac{t^2}{2} \Rightarrow \frac{dx}{dt} = 1,\ \frac{dy}{dt} = \sqrt{2}t^{1/2},\ \frac{dz}{dt} = t$

$\Rightarrow \sqrt{\left(\frac{dx}{dt}\right)^2 + \left(\frac{dy}{dt}\right)^2 + \left(\frac{dz}{dt}\right)^2}\, dt = \sqrt{1 + 2t + t^2}\, dt = \sqrt{(t + 1)^2}\, dt = |t + 1|\, dt = (t + 1)\, dt$ on the domain given.

Then $M = \int_C \delta\, ds = \int_0^2 \left(\frac{1}{t+1}\right)(t + 1)\, dt = \int_0^2 dt = 2$; $M_{yz} = \int_C x\delta\, ds = \int_0^2 t\left(\frac{1}{t+1}\right)(t + 1)\, dt = \int_0^2 t\, dt = 2$;

$M_{xz} = \int_C y\delta\, ds = \int_0^2 \left(\frac{2\sqrt{2}}{3}t^{3/2}\right)\left(\frac{1}{t+1}\right)(t + 1)\, dt = \int_0^2 \frac{2\sqrt{2}}{3}t^{3/2}\, dt = \frac{32}{15}$; $M_{xy} = \int_C z\delta\, ds$

$= \int_0^2 \left(\frac{t^2}{2}\right)\left(\frac{1}{t+1}\right)(t + 1)\, dt = \int_0^2 \frac{t^2}{2}\, dt = \frac{4}{3} \Rightarrow \overline{x} = \frac{M_{yz}}{M} = \frac{2}{2} = 1;\ \overline{y} = \frac{M_{xz}}{M} = \frac{\left(\frac{32}{15}\right)}{2} = \frac{16}{15};\ \overline{z} = \frac{M_{xy}}{M}$

$= \frac{\left(\frac{4}{3}\right)}{2} = \frac{2}{3}$; $I_x = \int_C (y^2 + z^2)\,\delta\, ds = \int_0^2 \left(\frac{8}{9}t^3 + \frac{t^4}{4}\right) dt = \frac{232}{45}$; $I_y = \int_C (x^2 + z^2)\,\delta\, ds = \int_0^2 \left(t^2 + \frac{t^4}{4}\right) dt = \frac{64}{15}$;

$I_z = \int_C (y^2 + x^2)\,\delta\, ds = \int_0^2 \left(t^2 + \frac{8}{9}t^3\right) dt = \frac{56}{9}$

45. $\mathbf{r}(t) = (e^t\cos t)\mathbf{i} + (e^t\sin t)\mathbf{j} + e^t\mathbf{k},\ 0 \le t \le \ln 2 \Rightarrow x = e^t\cos t,\ y = e^t\sin t,\ z = e^t \Rightarrow \frac{dx}{dt} = (e^t\cos t - e^t\sin t)$,

$\frac{dy}{dt} = (e^t\sin t + e^t\cos t),\ \frac{dz}{dt} = e^t \Rightarrow \sqrt{\left(\frac{dx}{dt}\right)^2 + \left(\frac{dy}{dt}\right)^2 + \left(\frac{dz}{dt}\right)^2}\, dt$

$= \sqrt{(e^t\cos t - e^t\sin t)^2 + (e^t\sin t + e^t\cos t)^2 + (e^t)^2}\, dt = \sqrt{3e^{2t}}\, dt = \sqrt{3}e^t\, dt$; $M = \int_C \delta\, ds = \int_0^{\ln 2} \sqrt{3}e^t\, dt$

$= \sqrt{3}$; $M_{xy} = \int_C z\delta\, ds = \int_0^{\ln 2} \left(\sqrt{3}e^t\right)(e^t)\, dt = \int_0^{\ln 2} \sqrt{3}e^{2t}\, dt = \frac{3\sqrt{3}}{2} \Rightarrow \overline{z} = \frac{M_{xy}}{M} = \frac{\left(\frac{3\sqrt{3}}{2}\right)}{\sqrt{3}} = \frac{3}{2}$;

$I_z = \int_C (x^2 + y^2)\,\delta\, ds = \int_0^{\ln 2} (e^{2t}\cos^2 t + e^{2t}\sin^2 t)\left(\sqrt{3}e^t\right) dt = \int_0^{\ln 2} \sqrt{3}e^{3t}\, dt = \frac{7\sqrt{3}}{3}$

47. Because of symmetry $\overline{x} = \overline{y} = 0$. Let $f(x, y, z) = x^2 + y^2 + z^2 = 25 \Rightarrow \nabla f = 2x\mathbf{i} + 2y\mathbf{j} + 2z\mathbf{k}$

$\Rightarrow |\nabla f| = \sqrt{4x^2 + 4y^2 + 4z^2} = 10$ and $\mathbf{p} = \mathbf{k} \Rightarrow |\nabla f \cdot \mathbf{p}| = 2z$, since $z \ge 0 \Rightarrow M = \iint_R \delta(x, y, z)\, d\sigma$

$= \iint_R z\left(\frac{10}{2z}\right) dA = \iint_R 5\, dA = 5(\text{Area of the circular region}) = 80\pi$; $M_{xy} = \iint_R z\delta\, d\sigma = \iint_R 5z\, dA$

$= \iint_R 5\sqrt{25 - x^2 - y^2}\, dx\, dy = \int_0^{2\pi}\int_0^4 \left(5\sqrt{25 - r^2}\right) r\, dr\, d\theta = \int_0^{2\pi} \frac{490}{3}\, d\theta = \frac{980}{3}\pi \Rightarrow \overline{z} = \frac{\left(\frac{980}{3}\pi\right)}{80\pi} = \frac{49}{12}$

$\Rightarrow (\overline{x}, \overline{y}, \overline{z}) = \left(0, 0, \frac{49}{12}\right)$; $I_z = \iint_R (x^2 + y^2)\,\delta\, d\sigma = \iint_R 5(x^2 + y^2)\, dx\, dy = \int_0^{2\pi}\int_0^4 5r^3\, dr\, d\theta = \int_0^{2\pi} 320\, d\theta = 640\pi$

49. $M = 2xy + x$ and $N = xy - y \Rightarrow \frac{\partial M}{\partial x} = 2y + 1,\ \frac{\partial M}{\partial y} = 2x,\ \frac{\partial N}{\partial x} = y,\ \frac{\partial N}{\partial y} = x - 1 \Rightarrow \text{Flux} = \iint_R \left(\frac{\partial M}{\partial x} + \frac{\partial N}{\partial y}\right) dx\,dy$
$= \iint_R (2y + 1 + x - 1)\,dy\,dx = \int_0^1 \int_0^1 (2y + x)\,dy\,dx = \frac{3}{2}$; $\text{Circ} = \iint_R \left(\frac{\partial N}{\partial x} - \frac{\partial M}{\partial y}\right) dx\,dy$
$= \iint_R (y - 2x)\,dy\,dx = \int_0^1 \int_0^1 (y - 2x)\,dy\,dx = -\frac{1}{2}$

51. $M = -\frac{\cos y}{x}$ and $N = \ln x \sin y \Rightarrow \frac{\partial M}{\partial y} = \frac{\sin y}{x}$ and $\frac{\partial N}{\partial x} = \frac{\sin y}{x} \Rightarrow \oint_C \ln x \sin y\,dy - \frac{\cos y}{x}\,dx$
$= \iint_R \left(\frac{\partial N}{\partial x} - \frac{\partial M}{\partial y}\right) dx\,dy = \iint_R \left(\frac{\sin y}{x} - \frac{\sin y}{x}\right) dx\,dy = 0$

53. $\frac{\partial}{\partial x}(2xy) = 2y,\ \frac{\partial}{\partial y}(2yz) = 2z,\ \frac{\partial}{\partial z}(2xz) = 2x \Rightarrow \nabla \cdot \mathbf{F} = 2y + 2z + 2x \Rightarrow \text{Flux} = \iiint_D (2x + 2y + 2z)\,dV$
$= \int_0^1 \int_0^1 \int_0^1 (2x + 2y + 2z)\,dx\,dy\,dz = \int_0^1 \int_0^1 (1 + 2y + 2z)\,dy\,dz = \int_0^1 (2 + 2z)\,dz = 3$

55. $\frac{\partial}{\partial x}(-2x) = -2,\ \frac{\partial}{\partial y}(-3y) = -3,\ \frac{\partial}{\partial z}(z) = 1 \Rightarrow \nabla \cdot \mathbf{F} = -4;\ x^2 + y^2 + z^2 = 2$ and $x^2 + y^2 = z \Rightarrow z = 1$
$\Rightarrow x^2 + y^2 = 1 \Rightarrow \text{Flux} = \iiint_D -4\,dV = -4 \int_0^{2\pi} \int_0^1 \int_{r^2}^{\sqrt{2-r^2}} dz\,r\,dr\,d\theta = -4 \int_0^{2\pi} \int_0^1 \left(r\sqrt{2 - r^2} - r^3\right) dr\,d\theta$
$= -4 \int_0^{2\pi} \left(-\frac{7}{12} + \frac{2}{3}\sqrt{2}\right) d\theta = \frac{2}{3}\pi\left(7 - 8\sqrt{2}\right)$

57. $\mathbf{F} = y\mathbf{i} + z\mathbf{j} + x\mathbf{k} \Rightarrow \nabla \cdot \mathbf{F} = 0 \Rightarrow \text{Flux} = \iint_S \mathbf{F} \cdot \mathbf{n}\,d\sigma = \iiint_D \nabla \cdot \mathbf{F}\,dV = 0$

59. $\mathbf{F} = xy^2\mathbf{i} + x^2y\mathbf{j} + y\mathbf{k} \Rightarrow \nabla \cdot \mathbf{F} = y^2 + x^2 + 0 \Rightarrow \text{Flux} = \iint_S \mathbf{F} \cdot \mathbf{n}\,d\sigma = \iiint_D \nabla \cdot \mathbf{F}\,dV$
$= \iiint_D (x^2 + y^2)\,dV = \int_0^{2\pi} \int_0^1 \int_{-1}^1 r^2\,dz\,r\,dr\,d\theta = \int_0^{2\pi} \int_0^1 2r^3\,dr\,d\theta = \int_0^{2\pi} \frac{1}{2}\,d\theta = \pi$

CHAPTER 16 ADDITIONAL AND ADVANCED EXERCISES

1. $dx = (-2 \sin t + 2 \sin 2t)\,dt$ and $dy = (2 \cos t - 2 \cos 2t)\,dt$; $\text{Area} = \frac{1}{2}\oint_C x\,dy - y\,dx$
$= \frac{1}{2}\int_0^{2\pi} [(2 \cos t - \cos 2t)(2 \cos t - 2 \cos 2t) - (2 \sin t - \sin 2t)(-2 \sin t + 2 \sin 2t)]\,dt$
$= \frac{1}{2}\int_0^{2\pi} [6 - (6 \cos t \cos 2t + 6 \sin t \sin 2t)]\,dt = \frac{1}{2}\int_0^{2\pi} (6 - 6 \cos t)\,dt = 6\pi$

3. $dx = \cos 2t\,dt$ and $dy = \cos t\,dt$; $\text{Area} = \frac{1}{2}\oint_C x\,dy - y\,dx = \frac{1}{2}\int_0^{\pi} \left(\frac{1}{2}\sin 2t \cos t - \sin t \cos 2t\right) dt$
$= \frac{1}{2}\int_0^{\pi} [\sin t \cos^2 t - (\sin t)(2\cos^2 t - 1)]\,dt = \frac{1}{2}\int_0^{\pi} (-\sin t \cos^2 t + \sin t)\,dt = \frac{1}{2}\left[\frac{1}{3}\cos^3 t - \cos t\right]_0^{\pi} = -\frac{1}{3} + 1 = \frac{2}{3}$

5. (a) $\mathbf{F}(x, y, z) = z\mathbf{i} + x\mathbf{j} + y\mathbf{k}$ is $\mathbf{0}$ only at the point $(0, 0, 0)$, and curl $\mathbf{F}(x, y, z) = \mathbf{i} + \mathbf{j} + \mathbf{k}$ is never $\mathbf{0}$.
(b) $\mathbf{F}(x, y, z) = z\mathbf{i} + y\mathbf{k}$ is $\mathbf{0}$ only on the line $x = t,\ y = 0,\ z = 0$ and curl $\mathbf{F}(x, y, z) = \mathbf{i} + \mathbf{j}$ is never $\mathbf{0}$.
(c) $\mathbf{F}(x, y, z) = z\mathbf{i}$ is $\mathbf{0}$ only when $z = 0$ (the xy-plane) and curl $\mathbf{F}(x, y, z) = \mathbf{j}$ is never $\mathbf{0}$.

7. Set up the coordinate system so that $(a, b, c) = (0, R, 0) \Rightarrow \delta(x, y, z) = \sqrt{x^2 + (y - R)^2 + z^2}$
$= \sqrt{x^2 + y^2 + z^2 - 2Ry + R^2} = \sqrt{2R^2 - 2Ry}$; let $f(x, y, z) = x^2 + y^2 + z^2 - R^2$ and $\mathbf{p} = \mathbf{i}$
$\Rightarrow \nabla f = 2x\mathbf{i} + 2y\mathbf{j} + 2z\mathbf{k} \Rightarrow |\nabla f| = 2\sqrt{x^2 + y^2 + z^2} = 2R \Rightarrow d\sigma = \frac{|\nabla f|}{|\nabla f \cdot \mathbf{i}|}\,dz\,dy = \frac{2R}{2x}\,dz\,dy$
$\Rightarrow \text{Mass} = \iint_S \delta(x, y, z)\,d\sigma = \iint_{R_{yz}} \sqrt{2R^2 - 2Ry}\left(\frac{R}{x}\right) dz\,dy = R \iint_{R_{yz}} \frac{\sqrt{2R^2 - 2Ry}}{\sqrt{R^2 - y^2 - z^2}}\,dz\,dy$

$$= 4R \int_{-R}^{R} \int_{0}^{\sqrt{R^2-y^2}} \frac{\sqrt{2R^2 - 2Ry}}{\sqrt{R^2 - y^2 - z^2}} \, dz \, dy = 4R \int_{-R}^{R} \sqrt{2R^2 - 2Ry} \sin^{-1}\left(\frac{z}{\sqrt{R^2 - y^2}}\right) \Bigg|_{0}^{\sqrt{R^2-y^2}} dy$$

$$= 2\pi R \int_{-R}^{R} \sqrt{2R^2 - 2Ry} \, dy = 2\pi R \left(\tfrac{-1}{3R}\right) \left(2R^2 - 2Ry\right)^{3/2} \Bigg|_{-R}^{R} = \frac{16\pi R^3}{3}$$

9. $M = x^2 + 4xy$ and $N = -6y \Rightarrow \frac{\partial M}{\partial x} = 2x + 4y$ and $\frac{\partial N}{\partial x} = -6 \Rightarrow$ Flux $= \int_0^b \int_0^a (2x + 4y - 6) \, dx \, dy$
$= \int_0^b (a^2 + 4ay - 6a) \, dy = a^2b + 2ab^2 - 6ab$. We want to minimize $f(a, b) = a^2b + 2ab^2 - 6ab = ab(a + 2b - 6)$. Thus, $f_a(a, b) = 2ab + 2b^2 - 6b = 0$ and $f_b(a, b) = a^2 + 4ab - 6a = 0 \Rightarrow b(2a + 2b - 6) = 0 \Rightarrow b = 0$ or $b = -a + 3$. Now $b = 0 \Rightarrow a^2 - 6a = 0 \Rightarrow a = 0$ or $a = 6 \Rightarrow (0, 0)$ and $(6, 0)$ are critical points. On the other hand, $b = -a + 3$ $\Rightarrow a^2 + 4a(-a + 3) - 6a = 0 \Rightarrow -3a^2 + 6a = 0 \Rightarrow a = 0$ or $a = 2 \Rightarrow (0, 3)$ and $(2, 1)$ are also critical points. The flux at $(0, 0) = 0$, the flux at $(6, 0) = 0$, the flux at $(0, 3) = 0$ and the flux at $(2, 1) = -4$. Therefore, the flux is minimized at $(2, 1)$ with value -4.

11. (a) Partition the string into small pieces. Let $\Delta_i s$ be the length of the ith piece. Let (x_i, y_i) be a point in the ith piece. The work done by gravity in moving the ith piece to the x-axis is approximately $W_i = (gx_iy_i\Delta_i s)y_i$ where $x_iy_i\Delta_i s$ is approximately the mass of the ith piece. The total work done by gravity in moving the string to the x-axis is $\sum_i W_i = \sum_i gx_iy_i^2\Delta_i s \Rightarrow$ Work $= \int_C gxy^2 \, ds$

(b) Work $= \int_C gxy^2 \, ds = \int_0^{\pi/2} g(2\cos t)(4\sin^2 t)\sqrt{4\sin^2 t + 4\cos^2 t} \, dt = 16g \int_0^{\pi/2} \cos t \sin^2 t \, dt$
$= \left[16g\left(\frac{\sin^3 t}{3}\right)\right]_0^{\pi/2} = \frac{16}{3} g$

(c) $\overline{x} = \frac{\int_C x(xy) \, ds}{\int_C xy \, ds}$ and $\overline{y} = \frac{\int_C y(xy) \, ds}{\int_C xy \, ds}$; the mass of the string is $\int_C xy \, ds$ and the weight of the string is $g\int_C xy \, ds$. Therefore, the work done in moving the point mass at $(\overline{x}, \overline{y})$ to the x-axis is
$W = \left(g\int_C xy \, ds\right)\overline{y} = g\int_C xy^2 \, ds = \frac{16}{3} g$.

13. (a) Partition the sphere $x^2 + y^2 + (z - 2)^2 = 1$ into small pieces. Let $\Delta_i \sigma$ be the surface area of the ith piece and let (x_i, y_i, z_i) be a point on the ith piece. The force due to pressure on the ith piece is approximately $w(4 - z_i)\Delta_i\sigma$. The total force on S is approximately $\sum_i w(4 - z_i)\Delta_i\sigma$. This gives the actual force to be $\iint_S w(4 - z) \, d\sigma$.

(b) The upward buoyant force is a result of the **k**-component of the force on the ball due to liquid pressure. The force on the ball at (x, y, z) is $w(4 - z)(-\mathbf{n}) = w(z - 4)\mathbf{n}$, where $\mathbf{n}$ is the outer unit normal at (x, y, z). Hence the **k**-component of this force is $w(z - 4)\mathbf{n} \cdot \mathbf{k} = w(z - 4)\mathbf{k} \cdot \mathbf{n}$. The (magnitude of the) buoyant force on the ball is obtained by adding up all these **k**-components to obtain $\iint_S w(z - 4)\mathbf{k} \cdot \mathbf{n} \, d\sigma$.

(c) The Divergence Theorem says $\iint_S w(z - 4)\mathbf{k} \cdot \mathbf{n} \, d\sigma = \iiint_D \operatorname{div}(w(z - 4)\mathbf{k}) \, dV = \iiint_D w \, dV$, where D is $x^2 + y^2 + (z - 2)^2 \le 1 \Rightarrow \iint_S w(z - 4)\mathbf{k} \cdot \mathbf{n} \, d\sigma = w\iiint_D 1 \, dV = \frac{4}{3}\pi w$, the weight of the fluid if it were to occupy the region D.

15. Assume that S is a surface to which Stokes's Theorem applies. Then $\oint_C \mathbf{E} \cdot d\mathbf{r} = \iint_S (\nabla \times \mathbf{E}) \cdot \mathbf{n} \, d\sigma$
$= \iint_S \left(-\frac{\partial \mathbf{B}}{\partial t}\right) \cdot \mathbf{n} \, d\sigma = -\frac{\partial}{\partial t} \iint_S \mathbf{B} \cdot \mathbf{n} \, d\sigma$. Thus the voltage around a loop equals the negative of the rate of change of magnetic flux through the loop.

17. $\oint_C f\nabla g \cdot d\mathbf{r} = \iint_S \nabla \times (f\nabla g) \cdot \mathbf{n}\, d\sigma$ (Stokes's Theorem)

$= \iint_S (f\nabla \times \nabla g + \nabla f \times \nabla g) \cdot \mathbf{n}\, d\sigma$ (Section 16.8, Exercise 19b)

$= \iint_S [(f)(\mathbf{0}) + \nabla f \times \nabla g] \cdot \mathbf{n}\, d\sigma$ (Section 16.7, Equation 8)

$= \iint_S (\nabla f \times \nabla g) \cdot \mathbf{n}\, d\sigma$

19. False; let $\mathbf{F} = y\mathbf{i} + x\mathbf{j} \neq \mathbf{0} \Rightarrow \nabla \cdot \mathbf{F} = \frac{\partial}{\partial x}(y) + \frac{\partial}{\partial y}(x) = 0$ and $\nabla \times \mathbf{F} = \begin{vmatrix} \mathbf{i} & \mathbf{j} & \mathbf{k} \\ \frac{\partial}{\partial x} & \frac{\partial}{\partial y} & \frac{\partial}{\partial z} \\ x & y & 0 \end{vmatrix} = 0\mathbf{i} + 0\mathbf{j} + 0\mathbf{k} = \mathbf{0}$

21. $\mathbf{r} = x\mathbf{i} + y\mathbf{j} + z\mathbf{k} \Rightarrow \nabla \cdot \mathbf{r} = 1 + 1 + 1 = 3 \Rightarrow \iiint_D \nabla \cdot \mathbf{r}\, dV = 3\iiint_D dV = 3V \Rightarrow V = \frac{1}{3}\iiint_D \nabla \cdot \mathbf{r}\, dV$

$= \frac{1}{3}\iint_S \mathbf{r} \cdot \mathbf{n}\, d\sigma$, by the Divergence Theorem